FIRST-ORDER LINEAR: $y' + p(x)y = q(x)$; $y(x) = \dfrac{1}{\sigma(x)}\left(\int \sigma(x)q(x)\,dx + C\right)$, $\sigma(x) = e^{\int p(x)\,dx}$

BERNOULLI EQUATION: $y' + p(x)y = q(x)y^n$; $v = y^{1-n}$, $v' + (1-n)p(x)v = (1-n)q(x)$

LEGENDRE EQUATION: $(1-x^2)y'' - 2xy' + \lambda y = 0$; polynomial solution if $\lambda = N(N+1)$ for $N = 0, 1, 2, \ldots$

BESSEL EQUATION OF ORDER ZERO: $xy'' + y' + xy = 0$; $y(x) = C_1 J_0(x) + C_2 Y_0(x)$

EULER'S METHOD: $y_{n+1} = y_n + f(x_n, y_n)h$

ERROR FUNCTION: $\operatorname{erf}(x) = \dfrac{2}{\sqrt{\pi}}\displaystyle\int_0^x e^{-t^2}\,dt$; $\operatorname{erf}(\infty) = 1$, $\operatorname{erf}(-x) = \operatorname{erf}(x)$

GAMMA FUNCTION:

$$\Gamma(p) = \int_0^\infty x^{p-1}e^{-x}\,dx \quad (p > 0); \quad \Gamma\left(\frac{1}{2}\right) = \sqrt{\pi}, \quad \Gamma(n) = (n-1)!, \quad \Gamma(p) = (p-1)\Gamma(p-1) \text{ for } p > 1$$

HEAVISIDE FUNCTION: $H(t-a) = \begin{cases} 0, & t < a \\ 1, & t > a. \end{cases}$ $\displaystyle\int_0^t H(\tau-a)f(\tau)\,d\tau = H(t-a)\int_a^t f(\tau)\,d\tau$

DELTA FUNCTION: $\displaystyle\int_0^\infty \delta(t-a)h(t)\,dt = h(a)$, $\displaystyle\int_0^t \delta(\tau-a)h(\tau)\,d\tau = H(t-a)h(a)$

SOME INTEGRALS

---

1. $\int x^a\,dx = x^{a+1}/(a+1) + C$ if $a \neq -1$
2. $\int dx/x = \ln|x| + C$
3. $\int e^{ax}\,dx = e^{ax}/a + C$ if $a \neq 0$
4. $\int xe^{ax}\,dx = (ax-1)e^{ax}/a^2 + C$ if $a \neq 0$
5. $\int x^2e^{ax}\,dx = (2-2ax + a^2x^2)e^{ax}/a^3$
6. $\int \sin ax\,dx = -(\cos ax)/a + C$ if $a \neq 0$
7. $\int \cos ax\,dx = (\sin ax)/a + C$ if $a \neq 0$
8. $\int \tan ax\,dx = -\dfrac{1}{a}\ln|\cos ax| + C$ if $a \neq 0$
9. $\int \cot ax\,dx = \dfrac{1}{a}\ln|\sin ax| + C$ if $a \neq 0$
10. $\int e^{ax}\sin bx\,dx = (a\sin bx - b\cos bx)e^{ax}/(a^2+b^2) + C$ if $a^2+b^2 \neq 0$
11. $\int e^{ax}\cos bx\,dx = (b\sin bx + a\cos bx)e^{ax}/(a^2+b^2) + C$ if $a^2+b^2 \neq 0$
12. $\int x\sin ax\,dx = (\sin ax - ax\cos ax)/a^2 + C$ if $a \neq 0$
13. $\int x\cos ax\,dx = (\cos ax + ax\sin ax)/a^2 + C$ if $a \neq 0$
14. $\int dx/(a^2+x^2) = \dfrac{1}{a}\arctan\dfrac{x}{a} + C$ if $a \neq 0$
15. $\int dx/\left(a^2-x^2\right) = \dfrac{1}{2a}\ln\left|\dfrac{a+x}{a-x}\right| + C$ if $a \neq 0$
16. $\int u\,dv = uv - \int v\,du$ (Integration by Parts)

Ordinary Differential Equations

Ordinary Differential Equations

Michael D. Greenberg
Department of Mechanical Engineering
University of Delaware
Newark, DE

Published by John Wiley & Sons, Inc., Hoboken, New Jersey.
Published simultaneously in Canada.

Library of Congress Cataloging-in-Publication Data:

Greenberg, Michael D., 1935–
Ordinary differential equations / Michael D. Greenberg.
p. cm.
Includes bibliographical references and index.
ISBN 978-1-118-23002-2 (hardback)
1. Differential equations—Textbooks. 2. Differential equations, Partial—Textbooks. I. Title.
QA372.G725 2012
515'.352—dc23 2011042287

Printed in the United States of America.

10 9 8 7 6 5 4 3 2 1

Contents

Preface

PURPOSE AND PREREQUISITES

This book is intended for students of science, engineering, and mathematics, as a textbook for a first course in differential equations, typically in the student's third or fourth semester. It is expected that the student has completed a differential and integral calculus sequence, but prior knowledge of linear algebra is not a prerequisite, and that material is provided here when it is needed.

TO THE INSTRUCTOR

The following points are to describe and explain some of the pedagogical decisions and approaches adopted in this text.

1. **Transition to higher-order equations.** The passage from first-order equations to equations of higher order is, we believe, often made more difficult for the student than necessary. Typically, the discussion of higher-order equations begins with the citing of an existence-and-uniqueness theorem and the introduction of linear independence and Wronskians, none of which is needed for the important case of equations with constant coefficients, which is indeed the first case to be studied. Consequently, second-order equations tend to look like a "new subject" to the student. Instead, we focus immediately on second-order equations with constant coefficients and derive their general solution in only a few pages (Theorem 2.2.1 of Section 2.2), using only results obtained in Chapter 1 for first-order equations. Proof of Theorem 2.2.1 is elementary, relying only on the factoring of the differential operator and the known solution of first-order equations with constant coefficients. The latter is not put forward as a solution method, but only to prove the theorem, and we are careful to caution the student that factorization cannot be expected to be useful for nonconstant-coefficient equations.

   The advantage of this approach is that the general solution of $y'' + p_1 y' + p_2 y = 0$ (in which the p_j's are constants) is obtained quickly and easily, without first introducing an existence-and-uniqueness theorem, linear independence, or Wronskians. The remainder of Sections 2.2 and 2.3 is devoted to familiarizing the student with the various solution forms: (a) the real exponentials and hyperbolic functions and (b) the complex exponentials and the circular functions. With that done, linear independence, Wronskians, existence, uniqueness, and general solution are introduced next, in Section 2.4, at which point the discussion can then be more readily grasped by the student, by virtue of the already completed discussion of the constant-coefficient case in Sections 2.2 and 2.3.

2. **Examples of additional pedagogical features.** Besides item 1 given above, pedagogical features include the following:
(a) Extensive margin notes provide a more informal dialogue with the student. These include additional explanation, emphasis, and cautions.
(b) To motivate the integrating factor method, a simple analogy is provided, regarding the solution of a quadratic equation by completing the square (page 15).
(c) Insight is provided for the special importance of the Riccati equation (Exercise 11 on page 112).
(d) The concept of linear independence is motivated by an analogy involving a painter's colors (page 119).
(e) It is noted how the solution form used in the method of reduction of order works even though it appears initially to be a "bad idea" (page 132).
(f) A relationship between human visual perception and the nonlinearity of the central nervous system is explored in Exercise 7 on page 154.
(g) In Chapter 3, attention to initial value and boundary value problems is balanced by devoting Sections 3.2–3.4 to initial value problems involving the harmonic oscillator, and Sections 3.5–3.6 to boundary value problems involving steady-state heat conduction and an introduction to the eigenvalue problem.
(h) Chapter 5 begins with a unique motivation for the idea of a "transform," and includes consideration of the "design" of the Laplace transform (Exercise 13 on page 346).

3. **Why not stop at second-order equations?** The material in Chapter 2, described above, emphasizes second-order equations but includes equations of higher order as well, mostly in Section 2.7. We suggest that including equations of order higher than two is valuable. First, the patterns are more readily grasped for $n > 2$. For instance, seeing that characteristic roots $r = -2, 5, 5, 5$ give $y(x) = C_1 e^{-2x} + (C_2 + C_3 x + C_4 x^2) e^{5x}$ permits the student to grasp the consequence of repeated roots better than does the case $r = 5, 5$, which gives $y(x) = (C_1 + C_2 x) e^{5x}$. Similarly for other concepts such as stability, which involves the location of the characteristic roots in the complex plane. Second, equations of fourth order are indeed encountered in the undergraduate science and engineering courses, and some of those applications are included in this text. These motivations notwithstanding, inclusion of the case $n > 2$ might be questionable if it were substantially more difficult. However, the only additional difficulty, for constant-coefficient and Cauchy–Euler equations, is that of solving nth-degree algebraic equations, and this computational difficulty is controlled through the choice of the text examples, although an occasional exercise may require the use of a computer algebra system to solve the characteristic equation.

4. **About the applications.** Besides applications to engineering systems, applications from the biological and life sciences are emphasized. Ecology and population dynamics are a natural opportunity for such coverage, and are es-

pecially attractive in that they involve both linear and nonlinear equations (in Sections 1.3 and 1.6, respectively) and the phase plane (in Chapter 7). These topics form one application thread that weaves through the chapters. Besides population dynamics, the diffusion of material and heat are important in biological and engineering systems and are covered in Section 3.5 for the steady case. Another thread consists of oscillators — mechanical and electrical, single and coupled.

5. **Applications separated as Chapter 3.** Chapter 1, on first-order equations, covers both theory and applications, but for higher-order equations we complete the theory first, in Chapter 2, and defer most of the applications to Chapter 3. Our reasoning is that the theory for higher-order equations is already long and if major applications were included, then the additional length and topic coverage would distract from the theoretical flow, and make it harder for the student to "see the forest for the trees." (Of course, there is nothing to prevent an instructor from jumping from Section 2.3, for instance, to Sections 3.2 and 3.3 before continuing in Chapter 2.)

   Section 3.5 on steady-state diffusion is an opportunity to include some material on diffusion even in the absence of coverage of the partial differential equation of unsteady diffusion, and Section 3.6 on column buckling is an opportunity to introduce the eigenvalue problem even without covering the more sophisticated Sturm–Liouville theory. The chapter is balanced between initial value problems and boundary value problems: Sections 3.2–3.4 on the harmonic oscillator involve initial value problems, and Sections 3.5–3.6 on diffusion and buckling involve boundary value problems.

6. **The method of assumed form.** There are numerous places, in the study of differential equations, where one assumes a solution of a certain form. Experience suggests that students may perceive such steps as unrelated and unmotivated "tricks." They become reluctant to "own" them, and their long-term memory of them is thereby reduced. To minimize this negative result, we've gathered these under a single umbrella and present them as special cases of what we call the *method of assumed form*; that is, it is indeed a *method*, just as the method of substitution is taught in the calculus as a method of integration. Then, we attempt to motivate the particular forms that are assumed in the different circumstances.

7. **Convenient stopping points.** Insofar as possible, chapters and sections are arranged with convenient stopping points. For instance, an abbreviated coverage of systems, in Chapter 4, might consist only of Sections 4.1 and 4.2, which cover solution by elimination and an important physical application to coupled oscillators. In Section 5.4 the application to periodic forcing functions is made into a subsection (Section 5.4.5), so it can be omitted if desired. Similarly, regarding Chapter 6 on series methods, we've placed the Legendre and Bessel equation section, Section 6.4, ahead of the final section on the method of Frobenius. That placement permits one to use Section 6.4

as a stopping point, without getting into the heavier and lengthier study of the method of Frobenius, yet still illustrating the idea of expansion about a singular point — for the Bessel equation.

8. **Additional Exercises.** In most sections, at the end of the exercises there are "Additional Exercises." These vary in nature but have in common that they are probably more challenging in one way or another. Solutions to them may be long or short, computer-oriented or not. The principal challenge may simply be in the length of the problem statement, and some additional guidance from the instructor may be desirable, bearing in mind the importance of helping the student to become more comfortable with problems that cannot be stated in a sentence or two. The Additional Exercises might also be useful as a source of supplemental material for lecture and class discussion.

CHOICE AND SEQUENCING OF MATERIAL

Some information regarding prerequisites, which may not be obvious, is as follows:

- Prerequisites for Section 4.2 on coupled oscillators are Sections 3.2, 3.3, and 3.4 on single oscillators.
- A suggested prerequisite for Chapter 7 on the phase plane is Section 1.3.4 on the phase line.
- Prerequisite for Sections 7.3–7.5 on the phase plane is Section 4.6 on the matrix eigenvalue problem.
- Prerequisite for Section 7.6 on the numerical solution of systems of differential equations is Section 1.9 on the numerical solution of single differential equations.

TO THE STUDENT

Overview. Typical mathematical "models" of physical and biological systems involve one or more "differential equations," which are equations containing one or more derivatives of the unknown function; thus, whereas in the calculus courses the answers are, typically, one or more numbers, in the subject of differential equations the answers are one or more *functions*. Thus, as you move from this course to courses on dynamics, heat and mass transfer, robotics, control theory, and neurophysiology, to name a few, expect to *routinely* encounter differential equations. Thus, both understanding the material covered in this course and retaining it will be important. Yet, experience indicates that it can slip away very quickly. The problem involves short-term versus long-term memory. If a telephone number is given verbally and you must walk across the room to make the call, a good and obvious strategy is to repeat the number to yourself until you've placed the call. Fine, but the number lives in memory only briefly and is probably then lost and gone forever. To get it into your long-term memory bin, a better strategy is to embed the number, if possible, into some logical framework, involving numerical relationships, birthdays, etc. Applying that idea to your study of this text and course, we suggest that you aim at deep understanding rather than memorization, insofar as possible, not only to maximize learning and recollection, but also to enjoy this material. You might consider explaining the concept you are studying to an "imaginary" colleague. Keep hearing your colleague reply "I still don't get it; can you explain it more simply and more briefly?"

Retaining the material. Just as it will be important to retain this material, to carry forward into your subsequent classes, it will also be important to bring into this class what you learned in the calculus, including the derivatives and integrals of simple and familiar functions, the helpful rules of differentiation and integration, and being on a first-name basis with the elementary functions and their graphs and most important properties.

Checking answers. We urge you to check your answers insofar as possible. This can be done in various ways. If the answer is numerical, does the value obtained look *reasonable*? If it is expressed as a formula, is it dimensionally correct — that is, does it have the correct units? Does your solution actually satisfy the differential equation? Does it behave correctly in any special cases or limiting cases that you can think of? Your instructor can help you with this.

Support materials. Be aware of the support materials in the text, the four pages of formulas given on the inside of the front and back covers, the four review-type appendices, the Answers to Exercises section at the end of the book, and the chapter reviews. This text is not tied to a specific Computer Algebra System (CAS), but your course may involve the use of one such system. It is tempting to ask why we need to study this subject at all if solutions can often be generated by computer software. To answer this question for yourself, you might try picking a few exercises in the text and working them out using software alone—without having studied or read that material. Really, we need both: (a) the analytical ability that is promoted by a text and (b) facility with a software as well.

The linear algebra used. Whereas it is expected that you have completed a calculus sequence, a prior course in linear algebra is not expected, and the linear algebra that will be needed will be introduced when it is needed. One exception is the topic of determinants. That material is relegated to an appendix since you have probably studied determinants before. With that expectation we've made it an appendix to keep the text as compact as possible.

Equation numbering. When referring to equation (7), for instance, in the text, we simply write "(7)," to avoid writing the word "equation" a great many times. In the Exercises we use a different system: The first numbered exercise in Exercise 5, say, is numbered as (5.1), the second as (5.2), and so on.

Use of the text for reference. In using this text for reference, after you've finished this course, be aware that the text sections are augmented by some additional material within the exercises, so if you do not find what you are looking for in the text, check the end-of-section exercises in the relevant sections.

EXERCISES

End-of-section exercises normally begin with routine drill-type problems and progress to ones that involve more thought or more analysis. In most sections there are also, at the end, "ADDITIONAL EXERCISES." These are expected to be more challenging in one way or another. Even a long problem statement can make a problem challenging, but we believe such problems are valuable and have not avoided them. Answers to problems for which the problem number is underlined are given in the ANSWERS TO EXERCISES, among the end papers. For drill-type exercises, answers are generally supplied for every other one, for instance, for parts (a), (c), (e), and so on. Answers are also supplied for other problems, but not on an every-other-problem pattern. For instance, a problem may be to show such and such, in which case the answer is simply "such and such" and need not be repeated in the answers section.

USE OF A COMPUTER ALGEBRA SYSTEM

As mentioned above in "To The Student," there exist a number of computer algebra systems, CAS's, and we urge you to use at least one of them to support your study of the material in this text. In fact, some of the exercises call for the use of such a system to solve a certain differential equation, or to plot the results, and so on. Matlab, Mathematica, and Maple tutorials are available in the supplemental materials.

ACKNOWLEGMENTS

I'm grateful to my colleagues Professors Tom Angell and George Hsiao at the University of Delaware for reading the final manuscript and providing valuable insights, and to these reviewers: Robert Carlson (University of Colorado, Colorado Springs), Andrew S. I. D. Lang (Oral Roberts University), Tom Power (Waterford Institute of Technology, Ireland), Steven Strogatz (Cornell University), and Linda Sundbye (Metropolitan State College of Denver). I also thank David Caldwell (University of Delaware), Anita Schwartz (University of Delaware), and Dr. David Weidner for their generous help with the LaTex manuscript and through various computer crises. I thank George Lobell for guiding me into the direction of writing on differential equations, and for his always-insightful advice along the way.

I'm grateful to my wife Yisraela for her love and encouragement — before, during, and after the writing of this book — and dedicate this book to her with all my love.

Above all, "From whence cometh my help? My help cometh from the Lord, who made heaven and earth." (Psalm 121)

Michael D. Greenberg
University of Delaware

Chapter 1

First-Order Differential Equations

1.1 MOTIVATION AND OVERVIEW

1.1.1 Introduction. Typically, phenomena in the natural sciences can be described, or "modeled," by equations involving derivatives of one or more unknown functions. Such equations are called **differential equations**.

To illustrate, consider the motion of a body of mass m that rests on an idealized frictionless table and is subjected to a force $F(t)$ where t is the time (Fig. 1). According to Newton's second law of motion, we have

$$m\frac{d^2x}{dt^2} = F(t), \tag{1}$$

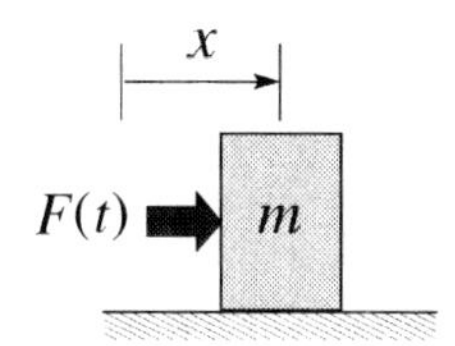

Figure 1. The motion of a mass on a frictionless table subjected to a force $F(t)$.

in which $x(t)$ is the mass's displacement. If we know the displacement history $x(t)$ and wish to determine the force $F(t)$ required to produce that displacement, the solution is simple: According to (1), merely differentiate the given $x(t)$ twice and multiply the result by m.

However, if we know the applied force $F(t)$ and wish to determine the displacement $x(t)$ that results, then we say that (1) is a "differential equation" governing the unknown function $x(t)$ because it involves derivatives of $x(t)$ with respect to t. Here, t is the independent variable and x is the dependent variable. The question is: *What function or functions $x(t)$, when differentiated twice with respect to t and then multiplied by m (which is a constant), give the prescribed function $F(t)$?*

To solve (1) for $x(t)$ we need to undo the differentiations; that is, we need to integrate (1) twice. To illustrate, suppose $F(t) = F_0$ is a constant, so

$$m\frac{d^2x}{dt^2} = F_0. \tag{2}$$

Integrating (2) once with respect to t gives

$$\int m\frac{d}{dt}\left(\frac{dx}{dt}\right) dt = \int F_0 \, dt$$

From the calculus, $\int \frac{du}{dt}\, dt = \int du = u$ plus an arbitrary constant.

or

$$m\frac{dx}{dt} + C_1 = F_0\,t + C_2 \tag{3}$$

in which C_1 and C_2 are the arbitrary constants of integration. Equivalently,

$$m\frac{dx}{dt} = F_0\,t + A, \tag{4}$$

in which the combined constant $A = C_2 - C_1$ is arbitrary. Integrating again gives $mx = F_0 t^2/2 + At + B$, so

$$x(t) = \frac{1}{m}\left(\frac{F_0}{2}t^2 + At + B\right). \tag{5}$$

It is a good habit to express the functional dependence explicitly, as we did in (5) when we wrote $x(t)$ instead of just x.

We say that a function is a **solution** *of a given differential equation, on an interval of the independent variable, if its substitution into the equation reduces that equation to an identity everywhere on that interval.* If so, we say that the function **satisfies** the differential equation on that interval. Accordingly, (5) is a solution of (2) on the interval $-\infty < t < \infty$ because if we substitute it into (2) we obtain $F_0 = F_0$, which is true for all t.

Actually, (5) is a whole "family" of solutions because A and B are arbitrary. Each choice of A and B in (5) gives one member of that family. That may sound confusing, for weren't we expecting to find "the" solution, not a whole collection of solutions? What's missing is that we haven't specified "starting conditions," for how can we expect to fully determine the ensuing motion $x(t)$ if we don't specify how it starts, namely, the displacement and velocity at the starting time $t = 0$? If we specify those values, say $x(0) = x_0$ and $x'(0) = x_0'$ where x_0 and x_0' are prescribed numbers, then the problem becomes

$$m\frac{d^2x}{dt^2} = F_0, \qquad (0 < t < \infty) \tag{6a}$$

$$x(0) = x_0, \quad \frac{dx}{dt}(0) = x_0', \tag{6b}$$

rather than consisting only of the differential equation (2). We seek a function or functions $x(t)$ that satisfy the differential equation $m\,d^2x/dt^2 = F_0$ on the interval $0 < t < \infty$ as well as the conditions $x(0) = x_0$ and $\dfrac{dx}{dt}(0) = x_0'$. We call (6b) **initial conditions**, and since the problem (6) includes one or more initial conditions we call it an **initial value problem** or **IVP**. Application of the initial conditions to the solution (5) gives

Initial value problem is often abbreviated as IVP.

$$x(0) = x_0 = \frac{1}{m}(0 + 0 + B) \qquad \text{[from (5)]}, \tag{7a}$$

$$\frac{dx}{dt}(0) = x_0' = \frac{1}{m}(0 + A) \qquad \text{[from (4)]}, \tag{7b}$$

so $A = mx'_0$ and $B = mx_0$, and we have the solution

$$x(t) = \frac{F_0}{2m}t^2 + x'_0 t + x_0 \tag{8}$$

of (6). Thus, from the differential equation (6a), which is a statement of Newton's second law, and the initial conditions (6b), we've been able to predict the displacement history $x(t)$ for all $t > 0$.

Whereas the differential equation (2), by itself, has the whole family of solutions given by (5), there is only one within that family that also satisfies the initial conditions (6b), the solution given by (8).

Unfortunately, most differential equations cannot be solved that readily, merely by undoing the derivatives by integration. For instance, suppose the mass is restrained by an ordinary coil spring that supplies a restoring force (i.e., in the direction opposite to the displacement) proportional to the displacement x, with constant of proportionality k (Fig. 2a). Then the total force on the mass when it is displaced to the right a distance x is $-kx + F(t)$, where the minus sign is because the kx force is in the negative x direction (Fig. 2b). Thus, now the differential equation governing the motion is

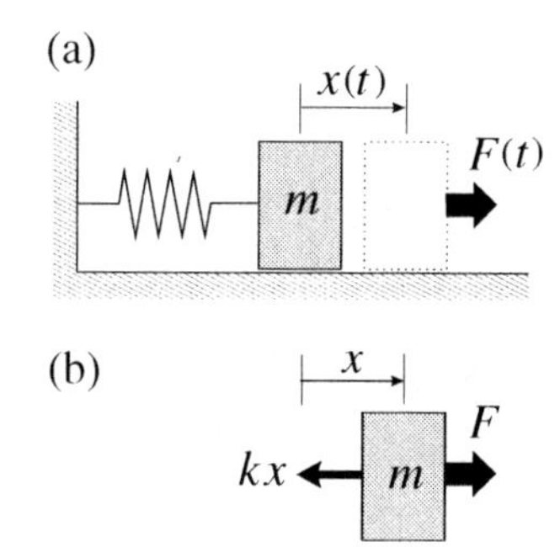

Figure 2. (a) The mass/spring system. (b) The forces on the mass. NOTE: The force kx exerted on the mass by the spring is proportional to the stretch in the spring, x, and the (empirically determined) constant of proportionality is k.

$$m\frac{d^2x}{dt^2} = -kx + F(t).$$

Finally, gathering all the unknown x terms on the left, as is customary, gives

$$m\frac{d^2x}{dt^2} + kx = F(t). \tag{9}$$

Let us try to solve (9) for $x(t)$ in the same way that we solved (2), by integrating twice with respect to t. One integration gives

$$m\frac{dx}{dt} + k\int x(t)\,dt = \int F(t)\,dt + A, \tag{10}$$

in which A is an arbitrary constant of integration. Since the function $F(t)$ is prescribed, the integral of $F(t)$ in (10) can be evaluated. However, *since the solution $x(t)$ is not yet known, the integral $\int x(t)\,dt$ cannot be evaluated, and we cannot proceed with our solution by repeated integration.*

Be sure to understand this point.

Thus, solving differential equations is, in general, not merely a matter of undoing the derivatives by integration. The theory and technique involved is considerable and will occupy us throughout this book. To develop that theory we will need to establish distinctions — definitions, some of which are given below.

1.1.2 Modeling. Besides *solving* the differential equations that arise in applications, we must *derive* them in the first place. Their derivation is called the *modeling* part of the analysis because it leads to the mathematical problem that is to be solved. To model the motion of the mass shown in Fig. 1, for instance, we defined the displacement variable x, identified the relevant logic as Newton's second law of motion, and arrived at the differential equation (1) that models the motion of the mass, subject to the approximations that the friction force exerted on the mass by the table and the force on it do to air resistance are negligible. The upshot is that mathematical models are not "off the shelf" items, they require thoughtful development.

1.1.3 The order of a differential equation. The **order** of a differential equation is the order of the highest derivative (of the unknown function or functions) in the equation. For instance, (9) is a second-order differential equation.

As additional examples,

$$\frac{dN}{dt} = r\left(1 - \frac{1}{K}N\right)N - \frac{N^2}{1+N^2} \qquad (0 < t < \infty) \tag{11}$$

for $N(t)$ and

$$EI\frac{d^4y}{dx^4} = -w(x) \qquad (0 < x < L) \tag{12}$$

for $y(x)$ are of first and fourth order, respectively.

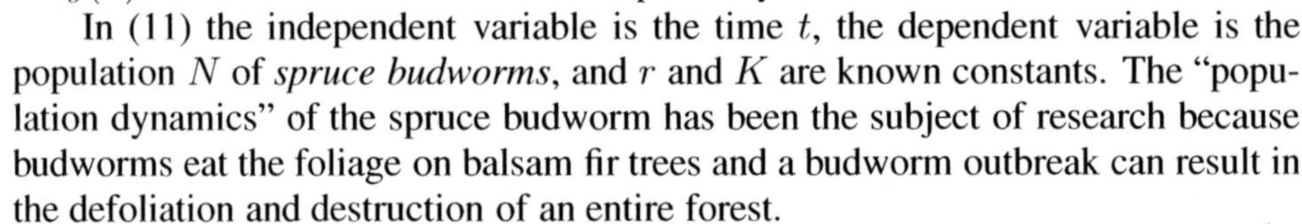

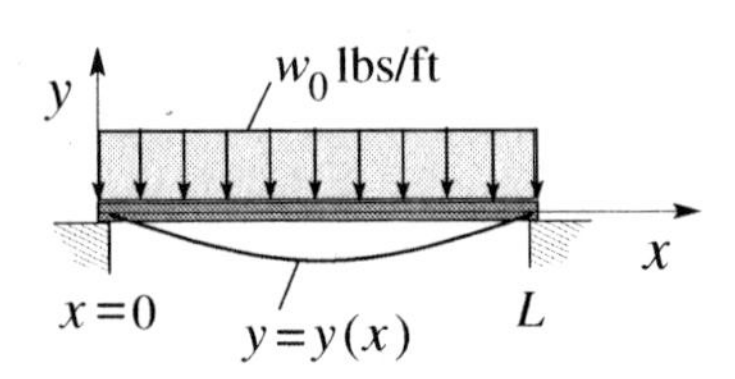

Figure 3. A beam subjected to a uniform load w lbs/ft; $y = y(x)$ is the deflection that results.

In (11) the independent variable is the time t, the dependent variable is the population N of *spruce budworms*, and r and K are known constants. The "population dynamics" of the spruce budworm has been the subject of research because budworms eat the foliage on balsam fir trees and a budworm outbreak can result in the defoliation and destruction of an entire forest.

Equation (12) governs the vertical deflection $y(x)$ of a beam of length L subjected to a prescribed load $w(x)$ lb/ft, and will be encountered in subsequent chapters. In Fig. 3 we've taken $w(x)$ to be a constant, w_0, so the total load is w_0L. Equation (12) is derived in a sophomore mechanical or civil engineering course on solid mechanics. In it, E and I are physical constants regarding the beam material and cross-sectional dimensions, respectively.

Equation (12) is similar to (2) in that it can be solved by repeated integration. To solve (2) we integrated (with respect to t) twice, and in doing so there arose two arbitrary "constants of integration." Similarly, to solve (12) we can integrate (with respect to x) four times, so there will be four arbitrary constants (Exercise 11).

These few examples hardly indicate the proliferation of differential equations that arise in applications — not just in engineering and physics, but in such diverse fields as biology, economics, psychology, chemistry, and agriculture. Since the applications are diverse, the independent and dependent variables differ from one application to another; for instance, in (2) the dependent variable is displacement and in (11) it is population. Often, though not necessarily, the independent variable will be a space coordinate x [as in (12)] or the time t [as in (1) and (11)]. As generic variables we will generally use x and y as the independent and dependent variables, respectively. With this notation, we can express our general nth-order differential equation for $y(x)$ as

We will often use x and y as generic independent and dependent variables, respectively.

$$\boxed{F\left(x, y, \frac{dy}{dx}, \ldots, \frac{d^ny}{dx^n}\right) = 0} \tag{13}$$

or, using the more compact prime notation for derivatives, $F(x, y, y', \ldots, y^{(n)}) = 0$, in which $y'(x)$ means dy/dx, $y''(x)$ means d^2y/dx^2, and so on. In (12), for instance, $F(x, y, y', ..., y'''')$ is $EIy'''' + w(x)$, and in (11), in which the variables are t and N instead of the generic variables x and y, we can identify $F(t, N, N')$ as $N' - r(1-N/K)N - N^2/(1+N^2)$.

1.1.4 Linear and nonlinear equations. In studying curves in the x, y plane, analytic geometry, one begins with straight lines, defined by equations of the form $ax + by = c$. And in studying surfaces in x, y, z space one begins with planes, defined by equations of the form $ax + by + cz = d$. Such equations are **linear** because the variables occur as a linear combination.

A **linear combination** of quantities $x_1, \ldots, x_n$ means a constant times x_1, plus a constant times $x_2, \ldots,$ plus a constant times x_n.

Likewise, to study differential equations it is best not to begin with the general case (13), but with *linear equations, ones in which the unknown function and its derivatives [namely, $y, y', ..., y^{(n)}$] occur as a linear combination,*

$$a_0(x)\frac{d^ny}{dx^n} + a_1(x)\frac{d^{n-1}y}{dx^{n-1}} + \cdots + a_n(x)y(x) = f(x), \tag{14}$$

in which the **coefficients** $a_0(x), \ldots, a_n(x)$ and the $f(x)$ on the right-hand side are prescribed functions of the independent variable x. An nth-order differential equation is **linear** if it is expressible in the form (14) and **nonlinear** if it is not. That is, (14) is a linear nth-order differential equation for $y(x)$ because it is in the form of a linear combination of y, y', $\ldots$, $y^{(n)}$ equaling some prescribed function of x.

To illustrate (14), (9) is a linear second-order equation [with $x(t)$ instead of $y(x)$] with $a_0(t) = m, a_1(t) = 0, a_2(t) = k$, and $f(t) = F(t)$, and (12) is a linear fourth-order equation with $a_0(x) = EI, a_1(x) = a_2(x) = a_3(x) = a_4(x) = 0$, and $f(x) = w(x)$. However, the first-order equation (11) is nonlinear; it cannot be put in the linear form $a_0(t)N' + a_1(t)N = f(t)$ because of the N^2 and $N^2/(1+N^2)$ terms, which we refer to as *nonlinear terms.*

Further, the linear differential equation (14) is **homogeneous** if $f(x)$ is zero and **nonhomogeneous** if $f(x)$ is not zero. For instance, (9) is nonhomogeneous because of the $F(t)$, and (12) is nonhomogeneous because of the $-w(x)$, but the linear second-order equation $y'' - e^x y' + 4y = 0$, for instance, is homogeneous because the right-hand side [after all of the y, y', and y'' terms are put on the left, as in (14)] is zero.

What physical or mathematical significance can we attach to the $f(x)$ term in (14)? In (1), for instance, $F(t)$ was an applied force that acted on the mass over the t interval of interest; in (12), $-w(x)$ was an applied force or load distribution that acted on the beam over the x interval of interest. Thus, it is common to call $f(x)$ in (14) a **forcing function** — even if it is not physically a force. For instance, in the linear differential equation governing the charge on a capacitor in an electrical circuit the forcing function will be seen in Section 1.3.5 to be an applied voltage, not an applied force, yet we will still call it a forcing function.

We call the right-hand side of (14) the "forcing function." Think of it as an "input," along with any initial conditions.

It is useful to think in terms of "inputs" and "outputs." If the linear equation (14) is augmented by initial conditions, for instance as (6a) was augmented by the initial conditions (6b), then both the forcing function $f(x)$ and the initial conditions are called **inputs**, and the response $y(x)$ to those inputs is the **output**. For instance, in the solution $x(t) = F_0 t^2/2m + x_0' t + x_0$ to the IVP (6), the term $F_0 t^2/2m$ is the response to the forcing function F_0 in (6a), and the term $x_0' t + x_0$ is the response to the initial conditions (6b). The idea is indicated schematically in Fig. 4.

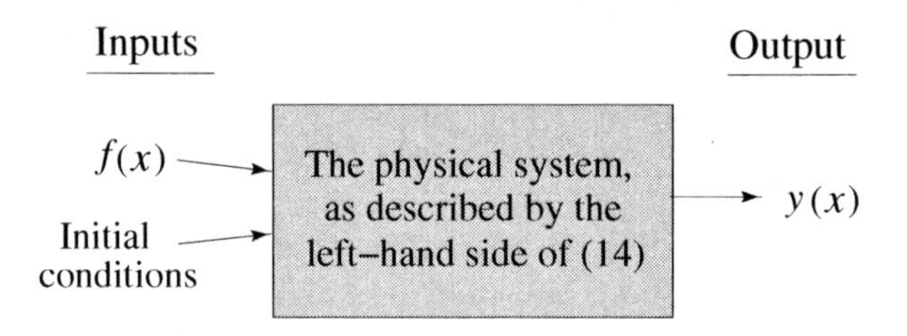

Figure 4. Schematic of the input/output nature of a linear initial value problem with differential equation (14).

1.1.5 Our plan. We will find that nonlinear differential equations are generally much more difficult than linear ones, and also that higher-order equations are more difficult than lower-order ones. Thus, *we will begin our study in Section 1.2 by considering differential equations that are both linear and of the lowest order — first order.*

To motivate our plan (which is typical, not unique to this text), think of one's early studies of algebra. Probably, it began with a single equation in one unknown, $ax = b$. From there, we proceeded in each of two different directions: higher-order algebraic equations in one unknown (quadratic, cubic, and so on), and also systems of linear equations in more than one unknown, such as the two equations $6x+y = 7$ and $2x - 8y = 5$ for x and y. The same is a good idea in differential equations. Following our study of first-order linear equations, in Chapter 1, we will proceed to higher-order linear equations in Chapters 2 and 3, and to **systems** of differential equations in Chapters 4 and 7.

We will develop three different approaches to solving and studying differential equations: **analytical**, **numerical**, and **qualitative**. Our derivation of the solution (8) of the problem (6) illustrates what we mean by *analytical*; that is, by carrying out a sequence of calculus-based steps we were able to end up with an expression for the unknown function. Most of our attention in this text is on analytical solution methods and the theory on which they are based.

Many differential equations, such as the budworm equation (11), are too difficult to solve analytically, but we can turn to a *numerical* method such as Euler's method. The idea, in numerical solution, is to give up on finding an expression for the solution $N(t)$ and to be content to numerically generate approximate values of $N(t)$ at a sequence of discrete t's, the spacing between them being called the *step size* of the calculation. To illustrate, let $r = K = 1$ in (11), let the initial condition be $N(0) = 3$, and let the step size be 0.2. The result of the Euler calculation is shown by the points in Fig. 5 along with the exact solution. Don't be concerned that the Euler-generated points are so inaccurate in this illustration, so far from the exact solution; one can increase the accuracy by reducing the step size.

Finally, by *qualitative* methods we mean methods that give information about solutions, without actually finding them analytically or numerically. One qualitative method that we will use is the "direction field," which we will use in Section 1.2.

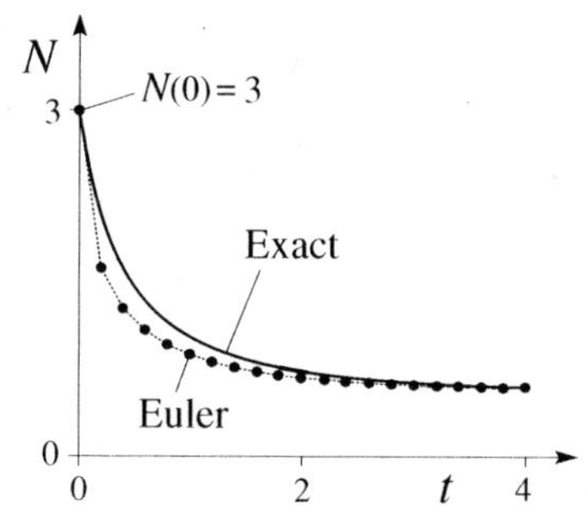

Figure 5. Solution of the budworm equation (11) for $r = K = 1$ and $N(0) = 3$. The dots are the approximate numerical solution (using Euler's method with a step size of 0.2) and the solid curve is the exact solution.

1.1.6 Direction field. If we can solve a given first-order equation $F(x, y, y') = 0$ for y', by algebra, we can express the equation in the form $y' = f(x, y)$, that is,

$$\frac{dy}{dx} = f(x, y), \tag{15}$$

which we take as our starting point.

To discuss the direction field of (15) we must first define the term "solution curve." A **solution curve** or **integral curve** of (15) is the graph of a solution $y(x)$ of that equation. Observe from (15) that at each point in the x, y plane at which $f(x, y)$ is defined, $f(x, y)$ gives the slope dy/dx of the solution curve through that point. For instance, for the differential equation

$$\frac{dy}{dx} = 4 - 3x - y \qquad (-\infty < x < \infty) \tag{16}$$

(16) is a linear first-order equation. Comparing it with (14) we see that n is 1, $a_0(x)$ is 1, $a_1(x)$ is 1, and $f(x)$ is $4 - 3x$.

the slope of the solution curve through the point $(2, 1)$ is given by $f(2, 1) = 4 - 3(2) - 1 = -3$.

In Fig. 6 we've plotted the **direction field** or **slope field** corresponding to (16), namely, a field of short line segments through a discrete set of points called a **grid**. Each line segment is called a *lineal element*, and the lineal element through any given grid point has the same slope as the solution curve through that point and is therefore a short tangent line to that solution curve. In computer graphics packages we can specify lines with or without arrowheads; we omitted arrowheads in Fig. 6.

In intuitive language, the direction field shows the overall "flow" of solution curves. Consider for instance the initial point $(0, -5)$ shown in Fig. 6 by the heavy dot; that is, consider the initial condition $y(0) = -5$ to be appended to (16). By following the direction field, we can sketch by hand the solution curve passing through that point. (Actually, we obtained that solution curve by computer, but we could just as well have sketched it by hand.) Four other solution curves are included as well.

You may wonder why we've shown the solution curve through $(0, -5)$ both to the right and to the left; if $(0, -5)$ is an "initial point," then shouldn't the solution through that point extend only to the right, over $0 < x < \infty$? If the independent variable is the time t, then the t interval of interest is usually to the right of the initial time. But in the present example the interval of interest of the independent variable x was stated in (16) to be $-\infty < x < \infty$. Hence, we extended the solution curve in Fig. 6 both to the right and to the left of the initial point.

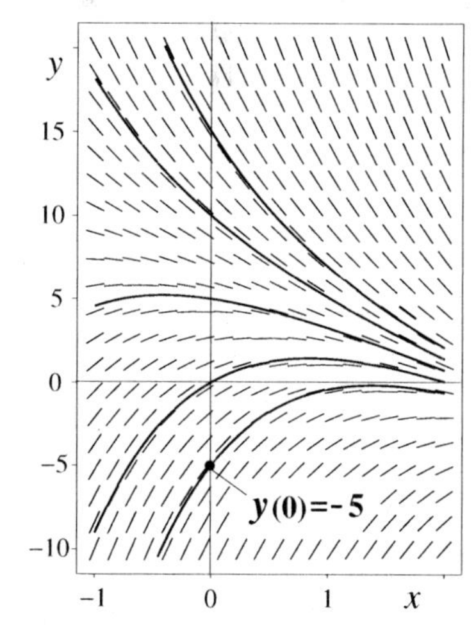

Figure 6. Direction field for $y' = 4 - 3x - y$, and representative solution curves.

Incidentally, (16) is linear [because it can be expressed in the form (14) as $dy/dx + y = 4-3x$], but (15) admits a direction field whether it is linear or nonlinear. In fact, direction fields are particularly valuable for nonlinear equations because those are more difficult, in general, and we may need all the help we can get to obtain information about their solutions.

1.1.7 Computer software. There are powerful computer software systems, such as *Maple*, *Mathematica*, and *MATLAB*, that can be used to implement much of the mathematics presented in this text — symbolically, numerically, and graphically. Though the reading is not tied to any particular software, it is anticipated that you will be using some such system as you go through this text. Thus, included among the exercises are some that call for the use of computer software, and the Student Solution Manual includes *Maple*, MATLAB, and *Mathematica* tutorials specifically for this text, chapter by chapter. Even if an exercise does not call for the use of software, and the answer is not given at the back of the book, you may be able to use computer software to solve the problem and check your work, and to plot your results if you wish.

Closure. We've introduced the idea of a differential equation and enough terminology to get us started. We defined the order of the equation as the order of the highest-order derivative in the equation, and we classified the equation as linear if it is expressible in the form (14), and nonlinear otherwise. We found that some differential equations, such as (2), can be solved merely by repeated integration, but in general that strategy does not work. However, whether or not the solution process proceeds by direct integrations, we can think of the arbitrary constants that will arise as "integration constants." The presence of these arbitrary constants makes it possible for the solution to satisfy initial conditions, such as the initial displacement and the initial velocity in (6b). Later, we will see that for differential equations of second order and higher it may be appropriate to specify conditions at *more than one point*. This case is illustrated in Exercise 11.

We've begun to classify different types of differential equations — for instance as linear or nonlinear, as homogeneous or nonhomogeneous, by order, and so on. Why do we do that? Because the most general differential equation is far too difficult for us to solve. Thus, we break the set of all possible differential equations into various categories and develop theory and solution strategies that are tailored to a given category. Historically, however, the early work on differential equations — by such great mathematicians as *Leonhard Euler* (1707 – 1783), *James Bernoulli* (1654–1705) and his brother *John* (1667–1748), *Joseph-Louis Lagrange* (1736–1813), *Alexis-Claude Clairaut* (1713–1765), and *Jean le Rond d'Alembert* (1717 – 1783) — generally involved attempts at solving specific equations rather than developing a general theory.

From the point of view of applications, we will find that in many cases diverse phenomena are modeled by the same differential equation. The remarkable conclusion is that *if one knows a lot about mechanical systems, for example, then one thereby knows a lot about electrical, biological, and social systems, for example, to whatever extent they are modeled by differential equations of the same form.* The significance of this fact can hardly be overstated as a justification for a careful study of the mathematical field of differential equations.

EXERCISES 1.1

NOTE: UNDERLINING OF AN EXERCISE NUMBER OR LETTER INDICATES THAT THAT EXERCISE IS INCLUDED AMONG THE ANSWERS TO THE SELECTED EXERCISES AT THE END OF THE TEXT.

1. Concepts of Order and Solution. State the order of each differential equation, and show whether or not the given functions are solutions of that equation.

(a) $y' = 3y$; $y_1(x) = e^{3x}$, $y_2(x) = 76e^{3x}$, $y_3(x) = e^{-3x}$
(b) $(y')^2 = 4y$; $y_1(x) = x^2$, $y_2(x) = 2x^2$, $y_3(x) = e^{-x}$
(c) $2yy' = 9\sin 2x$; $y_1(x) = \sin x$, $y_2(x) = 3\sin x$, $y_3(x) = e^x$
(d) $y'' - 9y = 0$; $y_1(x) = e^{3x} - e^x$, $y_2(x) = 3\sinh 3x$, $y_3(x) = 2e^{3x} - e^{-3x}$
(e) $(y')^2 - 4xy' + 4y = 0$; $y_1(x) = x^2 - x$, $y_2(x) = 2x - 1$
(f) $y'' + 9y = 0$; $y_1(x) = 4\sin 3x + 3\cos 3x$, $y_2(x) = 6\sin(3x + 2)$
(g) $y'' - y' - 2y = 6$; $y_1(x) = 5e^{2x} - 3$, $y_2(x) = -2e^{-x} - 3$
(h) $y''' - y'' = 6 - 6x$; $y_1(x) = 3e^x + x^3$
(i) $x^6 y''' = 6y^2$; $y_1(x) = x^3$, $y_2(x) = x^2$, $y_3(x) = 0$
(j) $y'' + y' = y^2 - 4$; $y_1(x) = x$, $y_2(x) = 1$, $y_3(x) = 2$
(k) $y' + 2xy = 1$; $y_1(x) = 4e^{-x^2}$, $y_2(x) = e^{-x^2}\left(\int_0^x e^{t^2}\,dt + A\right)$ for any value of A. HINT: For $y_2(x)$, recall the *fundamental theorem of the integral calculus*, that if $F(x) = \int_a^x f(t)\,dt$ and $f(t)$ is continuous on $a \le x \le b$, then $F'(x) = f(x)$ on $a \le x \le b$. [The reason we did not evaluate the integral in $y_2(x)$ is that it is too hard; it cannot be evaluated as a finite combination of elementary functions.]
(l) $y' - 4xy = x^2$; $y_1(x) = e^{2x^2}\int_1^x e^{-2t^2} t^2\,dt$
HINT: See the hint in part (k).

2. Including an Initial Condition; First-Order Equations. First, verify that the given function $y(x)$ is a solution of the given differential equation, for any value of A. Then, solve for A so that $y(x)$ satisfies the given initial condition.

(a) $y' + y = 1$; $y(x) = 1 + Ae^{-x}$; $y(0) = 3$
(b) $y' - y = x$; $y(x) = Ae^x - x - 1$; $y(2) = 5$
(c) $y' + 6y = 0$; $y(x) = Ae^{-6x}$; $y(4) = -1$
(d) $y' = 2xy^2$; $y(x) = -1/\left(x^2 + A\right)$; $y(0) = 5$
(e) $yy' = x$; $y(x) = \sqrt{x^2 + A}$; $y(1) = 10$

3. Second-Order Equations. First, verify that the given function is a solution of the given differential equation, for any constants A, B. Then, solve for A, B so that $y(x)$ satisfies the given initial conditions.

(a) $y'' + 4y = 8x^2$; $y(x) = 2x^2 - 1 + A\sin 2x + B\cos 2x$; $y(0) = 1$, $y'(0) = 0$
(b) $y'' - y = x^2$; $y(x) = -x^2 - 2 + A\sinh x + B\cosh x$; $y(0) = -2$, $y'(0) = 0$
(c) $y'' - 2y' + y = 0$; $y(x) = (A + Bx)e^x$; $y(0) = 0$, $y'(0) = 0$
(d) $y'' - y' = 0$; $y(x) = A + Be^x$; $y(0) = 1$, $y'(0) = 0$
(e) $y'' + 2y' = 4x$; $y(x) = A + Be^{-2x} + x^2 - x$; $y(0) = 0$, $y'(0) = 0$

4. Linear or Nonlinear? Classify each equation as linear or nonlinear:

(a) $y' + e^x y = 4$
(b) $yy' = x + y$
(c) $e^x y' = x - 2y$
(d) $y' - e^y = \sin x$
(e) $y'' + (\sin x)y = x^2$
(f) $y'' - y = e^x$
(g) $yy''' + 4y = 3x$
(h) $y''' = y$
(i) $\dfrac{y'' - y}{y' + y} = 4$
(j) $y''' + y^2 + 6y = x$
(k) $y'' = x^3 y'$
(l) $y''' + y''y' = 3x$
(m) $y'' - xy' = 3y + 4$
(n) $y''' = 4y$

5. Exponential Solutions. Each of the following is a homogeneous linear equation with constant coefficients [i.e., the coefficients $a_0(x), \ldots, a_n(x)$ in (14) are constants]. As we will see in Chapter 2, such equations necessarily admit solutions of exponential type, that is, of the form $y(x) = e^{rx}$ in which r is a constant. For the given equation, determine the value(s) of r such that $y(x) = e^{rx}$ is a solution. HINT: Put $y(x) = e^{rx}$ into the equation and determine any values of r such that the equation is satisfied, that is, reduced to an identity.

(a) $y' + 3y = 0$ (b) $2y' - y = 0$
(c) $y'' - 3y' + 2y = 0$ (d) $y'' - 2y' + y = 0$
(e) $y'' - 2y' - 3y = 0$ (f) $y'' + 5y' + 6y = 0$
(g) $y''' - y' = 0$ (h) $y''' - 2y'' - y' + 2y = 0$
(i) $y'''' - 6y'' + 5y = 0$ (j) $y'''' - 10y'' + 9y = 0$

6. Powers of x as Solutions. Unlike the equations in Exercise 5, the following equations admit solutions of the form $y(x) = x^r$, in which r is a constant. For the given equation determine the value(s) of r for which $y(x) = x^r$ is a solution.

(a) $xy' + y = 0$ (b) $xy' - y = 0$
(c) $xy'' + y' = 0$ (d) $xy'' - 4y' = 0$
(e) $x^2y'' + xy' - 9y = 0$ (f) $x^2y'' + xy' - y = 0$
(g) $x^2y'' + 3xy' - 2y = 0$ (h) $x^2y'' - 2y = 0$

7. Figure 6. Five representative solution curves are shown in Fig. 6. There is also one solution curve, not shown in the figure, that is a straight line. Find the equation of that straight-line solution. HINT: Seek a solution of (16) in the form $y(x) = mx + b$. Put that into (16) and see if you can find m and b such that the equation is satisfied. Does your result *look* correct — in terms of the direction field shown in the figure?

8. Straight-Line Solutions. First, read Exercise 7. For each given differential equation find any straight-line solutions, that is, of the form $y(x) = mx + b$. If there are none, state that.

(a) $y' + 2y = 2x - 1$ (b) $y' + 4y = 20$
(c) $y'' + y'^2 = 9$ (d) $y'' - 2y' + y = 0$
(e) $yy' + x = 0$ (f) $y' = y^2$
(g) $y' = y^2 - 4x^2 - 12x - 7$ (h) $yy' - y^2 = -x^2 + 3x - 2$
(i) $y' = y^2 - 4x^2 - 2$ (j) $y'' + y' + y = 3x$
(k) $y'' + y = x^2 + 7$ (l) $y'' - y' = 24x$
(m) A differential equation supplied by your instructor.

9. Grade This. Asked to solve the differential equation $\frac{dx}{dt} + x = 10t$, a student proposes this solution: By integrating with respect to t, obtain

$$x + xt = 5t^2 + A, \text{ so } x(t) = \frac{5t^2 + A}{1 + t}.$$

Is this correct? Explain.

10. No Solutions. (a) Show that the differential equation

$$\left|\frac{dy}{dx}\right| + |y| + 3 = 0 \tag{10.1}$$

has *no* solutions on any x interval. NOTE: This example shows that it is *possible* for a differential equation to have no solutions.
(b) Is (10.1) linear? Explain.

11. Deflection of a Loaded Beam; Boundary Conditions. Consider the beam shown in Fig. 3. Its deflection $y(x)$ is modeled by the fourth-order linear differential equation

$$EI\frac{d^4y}{dx^4} = -w_0. \tag{11.1}$$

(a) By repeated integration of (11.1), show that

$$y(x) = \frac{1}{EI}\Big(-\frac{w_0}{24}x^4 + \frac{A}{6}x^3 + \frac{B}{2}x^2 + Cx + D\Big). \tag{11.2}$$

(b) From Fig. 3 it is obvious that $y(0) = 0$ and $y(L) = 0$. Not so obvious (without some knowledge of Euler beam theory) is that $y''(0) = 0$ and $y''(L) = 0$ (because no moments are applied at the two ends). Use those four conditions to evaluate A, B, C, D in (11.2), and thus show that

$$y(x) = -\frac{w_0}{24EI}\left(x^4 - 2Lx^3 + L^3x\right). \tag{11.3}$$

NOTE: In this application the conditions are at two points, $x = 0$ and $x = L$, rather than one, so they are called **boundary conditions** rather than initial conditions, and the problem is a **boundary value problem** rather than an initial value problem.
(c) From (11.3), show that the largest deflection is $-5w_0L^4/384EI$.

1.2 LINEAR FIRST-ORDER EQUATIONS

We begin with the general *linear* first-order differential equation

$$a_0(x)\frac{dy}{dx} + a_1(x)y = f(x), \tag{1}$$

in which $a_0(x)$, $a_1(x)$, and $f(x)$ are prescribed. We assume $a_0(x)$ is nonzero on the x interval of interest, so we can divide (1) by $a_0(x)$ and obtain the simpler-looking version

$$\boxed{\frac{dy}{dx} + p(x)y = q(x),} \tag{2}$$

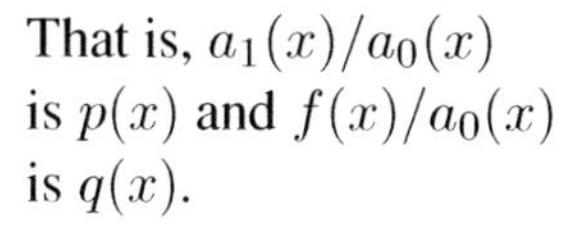
That is, $a_1(x)/a_0(x)$ is $p(x)$ and $f(x)/a_0(x)$ is $q(x)$.

which is the **standard form** of the linear first-order equation. It is assumed throughout this section that $p(x)$ and $q(x)$ are continuous on the x interval of interest. As noted in Section 1.1, we cannot solve (2) merely by integrating it because integration gives

$$y(x) + \int p(x)y(x)\,dx = \int q(x)\,dx + C, \tag{3}$$

and we don't yet know the $y(x)$ in the integrand of $\int p(x)y(x)\,dx$.

1.2.1 The simplest case. When stuck, it is good to simplify the problem temporarily, to get started. In this case we might do that by letting $p(x)$ or $q(x)$ be zero. If we let $p(x) = 0$, so the differential equation is simply

$$\frac{dy}{dx} = q(x), \tag{4}$$

then the $\int p(x)y(x)\,dx$ term causing the trouble in (3) drops out and we successfully obtain the solution by integrating (4) and obtaining

$$y(x) = \int q(x)\,dx + A, \tag{5}$$

in which the integration constant A is arbitrary. The integral in (5) does exist (i.e., converge) because we're assuming that $p(x)$ and $q(x)$ are continuous. Reversing our steps, differentiation of (5) shows that (5) does satisfy the original differential equation (4), because $\frac{dy}{dx} = \frac{d}{dx}\left(\int q(x)\,dx + A\right) = q(x)$.

We call (5) a **general solution** of (4) because it contains *all* solutions of (4). Put differently, (4) implies (5), and (5) implies (4), as we've seen. In fact, (5) is a whole "family" of solutions, a **one-parameter family** in which the parameter is the arbitrary constant A. Each choice of A gives a member of that family, called a **particular solution** of (4). For instance, if $q(x) = 6e^{2x}$, then the general solution is given by (5) as $y(x) = 3e^{2x} + A$, the graph of which is shown, for several values of A, in Fig. 1.

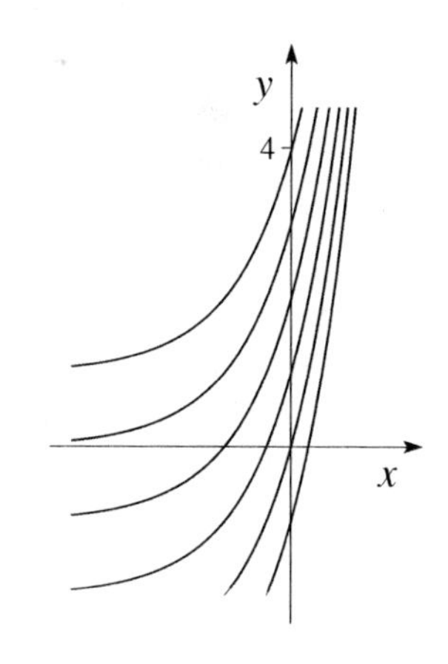

Figure 1. The solutions $y(x) = 3e^{2x} + A$ of the differential equation $\frac{dy}{dx} = 6e^{2x}$, for several values of A.

1.2.2 The homogeneous equation. Now consider the special case of (2) for which $q(x) = 0$ instead,

$$\boxed{\frac{dy}{dx} + p(x)y = 0.} \tag{6}$$

To solve (6), first divide both terms by y [which is permissible if $y(x) \neq 0$ on the x interval, which we tentatively assume], then integrate with respect to x:

$$\int \frac{1}{y}\frac{dy}{dx}\,dx + \int p(x)\,dx = 0, \tag{7a}$$

$$\int \frac{1}{y}\,dy + \int p(x)\,dx = 0, \tag{7b}$$

$$\ln|y| + \int p(x)\,dx = C, \tag{7c}$$

$$|y| = e^{-\int p(x)\,dx + C} = e^C e^{-\int p(x)\,dx}, \tag{7d}$$

and it follows from (7d) that

$$y(x) = \pm e^C e^{-\int p(x)\,dx}.$$

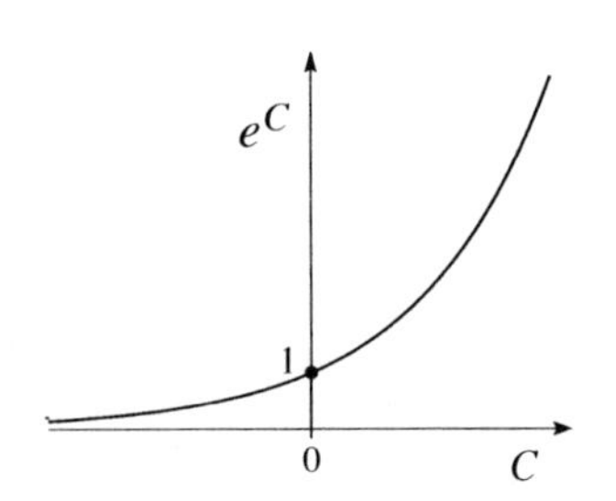

Figure 2. $0 < e^C < \infty$; e^C is not zero for any finite value of C.

The integration constant C is arbitrary so $-\infty < C < \infty$, and therefore $0 < e^C < \infty$ (Fig. 2). If we abbreviate $\pm e^C$ as A, then A is *any* number, positive or negative, but not zero because the exponential e^C is nonzero (Fig. 2). Thus, we can write $y(x)$ in the friendlier form

$$\boxed{y(x) = A\,e^{-\int p(x)\,dx}} \tag{8}$$

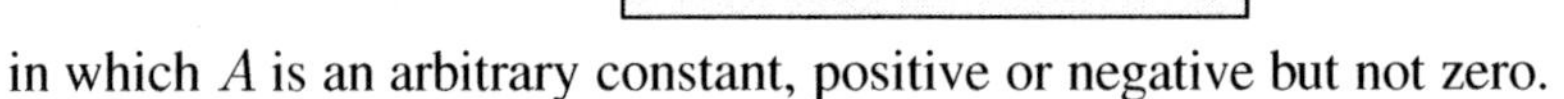

in which A is an arbitrary constant, positive or negative but not zero.

When we evaluate $\int p(x)\,dx$ in (8) we don't need to include an additive arbitrary integration constant; we already did in (7c).

Because we tentatively assumed that $y \neq 0$ in (7a), we must check the case $y = 0$ separately. In fact, we see that $y(x) = 0$ satisfies (6) because it reduces (6) to $0 + 0 = 0$. We can bring this additional solution under the umbrella of (8) if we now allow A to be zero. The upshot is that the general solution of (6) is given by (8) where A is an *arbitrary constant*: $-\infty < A < \infty$.

By (8) being the general solution of (6), we mean that it contains *all* solutions of (6). Each individual solution corresponds to a particular choice of the arbitrary constant A.

The preceding reasoning regarding the inclusion of the solution $y(x) = 0$ is similar to the reasoning involved in solving the algebraic equation $x^2 + 2x = 0$ for x. If we divide through by x, tentatively assuming that $x \neq 0$, then we obtain $x + 2 = 0$ and the root $x = -2$. Unless we then check the disallowed case $x = 0$, to see if it satisfies the equation $x^2 + 2x = 0$, we will have missed the root $x = 0$.

The key to our solution of (6) was dividing the equation by y because that step enabled us to end up [in (7b)] with one integral on y alone and one on x alone. The process of separating the x and y variables is called **separation of variables** and will be used again in Section 1.4 to solve certain *nonlinear* equations. Verification that (8) satisfies (6) is left for the exercises.

EXAMPLE 1. One to Remember Forever. If $p(x)$ is merely a constant in (6), then

$\int p(x)\,dx = px$, and (8) gives the general solution of

$$\boxed{\frac{dy}{dx} + py = 0.} \tag{9}$$

on $-\infty < x < \infty$, as

$$\boxed{y(x) = Ae^{-px},} \tag{10}$$

Roughly put, this example is as important in the study and application of differential equations as is the straight line in the study of curves.

with A an arbitrary constant. Recall that the graphs of the solutions, $y = Ae^{-px}$ in this case, are called the solution curves or integral curves. These are plotted in Fig. 3 for several representative values of A, along with the direction field. Notice, in the figure, how the solution curves follow the "flow" that is indicated by the direction field. ∎

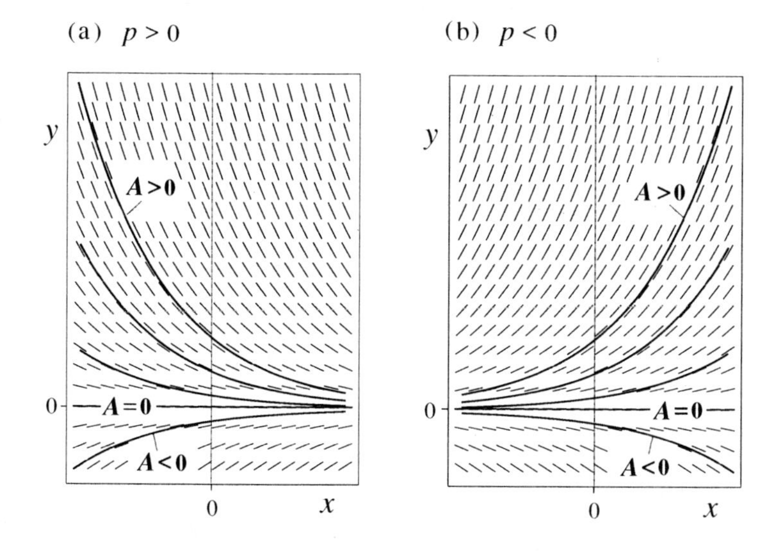

Figure 3. Representative solution curves $y(x) = Ae^{-px}$ for the equation $y' + py = 0$; direction field included.

EXAMPLE 2. Solve

$$y' + (\sin x)y = 0 \qquad (-\infty < x < \infty). \tag{11}$$

By comparing (11) with (6) we see that $p(x) = \sin x$. Then (8) gives

$$y(x) = Ae^{-\int \sin x\,dx} = Ae^{\cos x} \tag{12}$$

in which A is arbitrary.

Besides using the "off-the-shelf" formula (8), it is instructive to solve (11) by carrying out the separation of variables method that we used to derive (8) — as if stranded on a

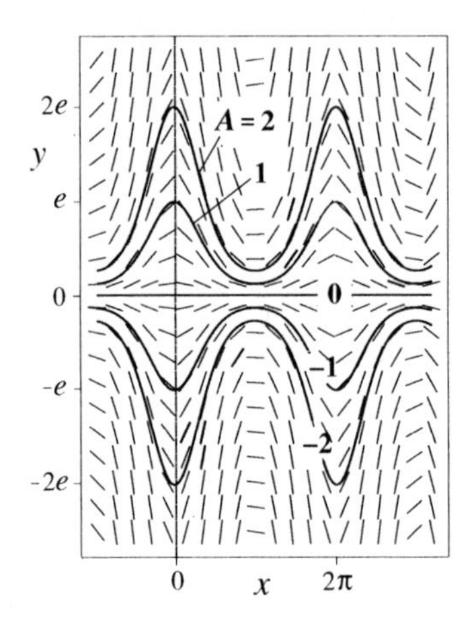

Figure 4. Representative solution curves $y(x) = Ae^{\cos x}$ for the equation $y' + (\sin x)y = 0$; direction field included.

desert island, our textbook having gone down with the ship:

$$\frac{dy}{y} + \sin x\, dx = 0, \tag{13a}$$

$$\int \frac{dy}{y} + \int \sin x\, dx = 0, \tag{13b}$$

$$\ln|y| - \cos x = C, \tag{13c}$$

$$|y| = e^{\cos x + C} = e^{\cos x} e^{C}, \tag{13d}$$

$$y(x) = \pm e^{C} e^{\cos x} = Ae^{\cos x}, \tag{13e}$$

which is the same result as we obtained in (12) by putting $p(x) = \sin x$ into (8). The solution curves are displayed for several values of A in Fig. 4. ∎

1.2.3 Solving the full equation by the integrating factor method.

We're now prepared to solve the full equation

$$\boxed{\frac{dy}{dx} + p(x)y = q(x),} \tag{14}$$

including both $p(x)$ and $q(x)$. This time our separation of variables technique fails because when we try to separate variables by re-expressing (14) as

$$\frac{1}{y}\, dy + p(x)\, dx = \frac{q(x)}{y}\, dx$$

the term on the right-hand side spoils the separation because $q(x)/y$ is a function not only of x but also of y. Instead of separation of variables, we will use an "integrating factor" method invented by the great mathematician *Leonhard Euler* (1707–1783).[1] We first motivate Euler's idea with an example:

(Pronounced "oiler," not "yuler.")

EXAMPLE 3. Motivating Euler's Integrating Factor Method. We wish to solve the equation

$$\frac{dy}{dx} + \frac{1}{x}y = 12x^2 \qquad (0 < x < \infty) \tag{15}$$

for $y(x)$. Notice that if we multiply (15) through by x and obtain

$$xy' + y = 12x^3, \tag{16}$$

[1]Euler is among the greatest and most productive mathematicians of all time. He contributed to virtually every branch of mathematics and to the application of mathematics to the science of mechanics. During the last 17 years of his life he was totally blind but produced several books and some 400 research papers. He knew by heart the entire *Aeneid* by Virgil, and he knew the first six powers of the first 100 prime numbers. If the latter does not seem impressive, note that the 100th prime number is 541 and its sixth power is 25,071,688,922,457,241.

then the left-hand side is the derivative of the product $xy(x)$ because $[xy(x)]' = y(x) + xy'(x)$. Thus, (16) can be expressed as

$$\frac{d}{dx}(xy) = 12x^3 \tag{17}$$

which can now be solved by integration:

$$\int d(xy) = \int 12x^3\,dx, \tag{18a}$$

$$xy = 12\frac{x^4}{4} + C, \tag{18b}$$

where C is arbitrary. Thus, we obtain the general solution

$$y(x) = 3x^3 + \frac{C}{x} \tag{19}$$

of (15). We can readily verify that substitution of (19) into (15) produces an identity (namely, $12x^2 = 12x^2$) on the interval $0 < x < \infty$ specified in (15). ▮

The integrating factor method is similar to the familiar method of solving a quadratic equation $ax^2 + bx + c = 0$ by completing the square: We add a suitable number to both sides so that the left-hand side becomes a "perfect square;" then the equation can be solved by the inverse operation — by taking square roots. *Analogously, in the integrating factor method we multiply both sides of (14) by a suitable function so that the left-hand side becomes a "perfect derivative;" then the equation can be solved by the inverse operation — by integration.* In Example 3 the integrating factor was x; when we multiplied (15) by x the left-hand side became the derivative $(xy)'$. Then $(xy)' = 12x^3$ could be solved [in (18)] by integration.

The integrating factor method is similar to the method of solving a quadratic equation by completing the square.

To apply Euler's method to the general equation (14), multiply (14) by a (not yet known) **integrating factor** $\sigma(x)$:

$$\underline{\sigma y' + \sigma p y} = \sigma q. \tag{20}$$

Our aim is to determine $\sigma(x)$ so the left-hand side of (20) is the derivative of σy, namely,

$$\frac{d}{dx}(\sigma y) \quad \text{or, written out,} \quad \underline{\sigma y' + \sigma' y}. \tag{21}$$

In (15), we "noticed" that $\sigma(x) = x$ works, but in general we cannot expect to find $\sigma(x)$ by inspection.

To match the underlined terms in (20) and (21), we need merely choose $\sigma(x)$ so that $\sigma p = \sigma'$:

$$\sigma' = \sigma p. \tag{22}$$

But the latter, rewritten as

$$\sigma' - p(x)\sigma = 0, \tag{23}$$

is of the same form as the equation $y' + p(x)y = 0$ that we solved in Section 1.2.2 [if we change $y(x)$ to $\sigma(x)$ and $p(x)$ to $-p(x)$], so its solution is given by (8) as

$$\sigma(x) = Ae^{\int p(x)\,dx}. \tag{24}$$

We don't need the most general integrating factor, we simply need *an* integrating factor, so we can choose $A = 1$ without loss. Then

This is an integrating factor for (14).

$$\boxed{\sigma(x) = e^{\int p(x)\,dx}.} \tag{25}$$

With $\sigma(x)$ so chosen, (20) becomes

$$(\sigma y)' = \sigma q \quad \text{or} \quad \frac{d(\sigma y)}{dx} = \sigma q,$$

which can be integrated to give

$$\int d(\sigma y) = \int \sigma(x)q(x)\,dx, \tag{26a}$$

$$\sigma y = \int \sigma(x)q(x)\,dx + C, \tag{26b}$$

CAUTION: (27) is *not* the same as $y(x) = \dfrac{1}{\sigma(x)}\displaystyle\int \sigma(x)q(x)\,dx + C$. That is, don't merely "tack on" an integration constant at the end of the analysis; carry it along from the point at which it arises.

so a general solution of $y' + p(x)y = q(x)$ is

$$\boxed{y(x) = \frac{1}{\sigma(x)}\left(\int \sigma(x)q(x)\,dx + C\right),} \tag{27}$$

with the integrating factor $\sigma(x)$ given by (25).

The abbreviation IVP.

EXAMPLE 4. Solution by Integrating Factor Method. Solve the initial value problem (IVP for brevity)

$$\frac{dy}{dx} + 3y = 9x \qquad (-\infty < x < \infty), \tag{28a}$$

$$y(2) = 1. \tag{28b}$$

To solve, we could simply use the solution formula (27), or we could carry out the steps of the integrating factor method that led to (27). To use (27) "off the shelf," first compare (28a) with $y' + p(x)y = q(x)$ to identify $p(x)$ and $q(x)$: $p(x) = 3$ and $q(x) = 9x$. Then, (25) gives

$$\sigma(x) = e^{\int p(x)\,dx} = e^{\int 3\,dx} = e^{3x}, \tag{29}$$

and (27) gives a general solution of (28a) as

$\int xe^{ax}\,dx = (ax-1)\dfrac{e^{ax}}{a^2}$.

$$y(x) = e^{-3x}\left(\int e^{3x}9x\,dx + C\right) = 3x - 1 + Ce^{-3x}. \tag{30}$$

Finally, apply the initial condition (28b) to (30) to determine C:

$$y(2) = 1 = 6 - 1 + Ce^{-6},$$

so $C = -4e^6$. Hence, the solution of the IVP (28) is

$$y(x) = 3x - 1 - 4e^{-3(x-2)}, \tag{31}$$

which is plotted as the solid curve in Fig. 5.

Alternatively, let us solve (28) using not the solution formula (27), but the **integrating factor method**. First, multiply (28a) through by $\sigma(x)$:

$$\underline{\sigma y' + 3\sigma y} = 9\sigma x. \tag{32}$$

We want to choose σ so the left side of (32) is a "perfect derivative" $(\sigma y)'$ or, written out,

$$\underline{\sigma y' + \sigma' y}. \tag{33}$$

For the underlined terms in (32) and (33) to be identical we need merely match the coefficients 3σ and σ' of y. Thus,

$$\sigma' = 3\sigma, \tag{34}$$

which gives $\sigma(x) = e^{3x}$. Then (32) is in the desired form $(\sigma y)' = 9x\sigma$, which can be integrated to give $\sigma y = \int 9x\sigma\, dx + C$, or,

$$y(x) = e^{-3x}\left(\int 9xe^{3x}\, dx + C\right) = 3x - 1 + Ce^{-3x}, \tag{35}$$

which is the same result as that given in (30).

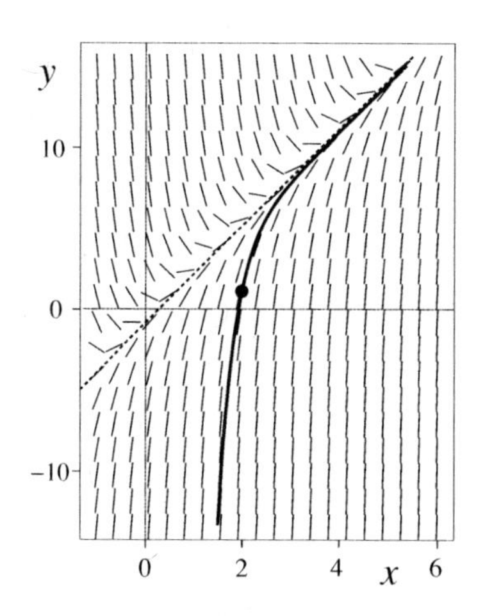

Figure 5. Graph of the solution (31) to the IVP (28), with the direction field. The heavy dot marks the initial point $y(2) = 1$.

COMMENT 1. Know and be comfortable with both approaches: memorizing and using (27) or, instead, using the integrating factor *method*.

COMMENT 2. We can see from (31) that the e^{-3x} term tends to zero as x increases, so every solution curve is asymptotic to the straight line $y = 3x - 1$. In fact, $y(x) = 3x - 1$ is itself a particular solution of (28a), corresponding to the choice $C = 0$ in (30), and is indicated in Fig. 5 by the dotted line. ∎

Can the integrating factor method fail? Perhaps for a given equation $y' + p(x)y = q(x)$ an integrating factor does not exist? No, $\sigma(x)$ is given by (25) and the only way that equation can fail to give $\sigma(x)$ is if the integral $\int p(x)\, dx$ does not exist. However, our assumption that $p(x)$ is continuous on the x interval of interest guarantees that the integral does exist.

1.2.4 Existence and uniqueness for the linear equation.

A fundamental question in the theory of differential equations is whether a given differential equation for $y(x)$ *has* a solution through a given initial point $y(a) = b$ in the x, y plane and, if so, on what x interval it is valid. That is the question of **existence**. If a solution does exist, then the next question is that of **uniqueness**: Is that solution unique? That is, is there only one solution or is there more than one?

For linear initial value problems we have the following result.

THEOREM 1.2.1 *Existence and Uniqueness for Linear Initial Value Problems*
The linear initial value problem

$$y' + p(x)y = q(x); \quad y(a) = b \tag{36}$$

has a solution

$$y(x) = \frac{1}{\sigma(x)} \left(\int_a^x \sigma(s)q(s)\,ds + b\sigma(a) \right), \tag{37}$$

where $\sigma(x) = e^{\int p(x)\,dx}$ is an integrating factor of the differential equation in (36). That solution exists and is unique *at least* on the broadest open x interval, containing the initial point $x = a$, on which $p(x)$ and $q(x)$ are continuous.

In (37), s is just a dummy integration variable. Partial check of (37): Setting $x = a$ in (37) gives $y(a) = \frac{1}{\sigma(a)}[0 + b\sigma(a)] = b$, so (37) does satisfy the initial condition $y(a) = b$.

Unlike (27), (37) includes a definite integral instead of an indefinite integral, and C has been chosen so that the initial condition $y(a) = b$ is satisfied. We leave the derivation of (37) to the exercises, and turn to applications of the theorem.

EXAMPLE 5. Existence on $-\infty < x < \infty$. Consider the IVP (28) again, in the light of Theorem 1.2.1: $p(x) = 3$ and $q(x) = 9x$ are continuous for all x, so Theorem 1.2.1 guarantees that there exists a unique solution of (28) on $-\infty < x < \infty$. That solution was given by (31) and was plotted as the solid curve in Fig. 5. ▮

EXAMPLE 6. The Possibilities of Existence on a Limited Interval, and of No Solution. Consider the IVP

$$x\frac{dy}{dx} + y = 12x^3, \tag{38a}$$

$$y(1) = b \tag{38b}$$

First, identify $p(x)$ and $q(x)$ by getting (38a) into the standard form $y' + p(x)y = q(x)$.

We've left b unspecified so we can consider several different b's. Here, $p(x) = 1/x$, $q(x) = 12x^2$, and $a = 1$. Although $q(x)$ is continuous for all x, $p(x) = 1/x$ is discontinuous at $x = 0$, so Theorem 1.2.1 guarantees the existence and uniqueness of a solution to the IVP (38) at *least* on $0 < x < \infty$, because that is the broadest open x interval, containing the initial point $x = 1$, on which both $p(x)$ and $q(x)$ are continuous.

In fact, the general solution of (38a) was found in Example 3 to be

$$y(x) = 3x^3 + \frac{C}{x}, \tag{39}$$

and for the representative initial conditions $y(1) = 0$, $y(1) = 3$, and $y(1) = 5$ we obtain $C = -3, 0$, and 2, respectively. These solutions are plotted in Fig. 6, and we see that we can *think* of the vertical line $x = 0$ as a barrier or wall; if the initial point $(1, b)$ is above the curve $y = 3x^3$ the solution "climbs the wall" to $+\infty$ as $x \to 0$ and if the initial point is below $y = 3x^3$ the solution approaches $-\infty$ as $x \to 0$, because of the C/x term in (39).

There is just one solution, corresponding to $y(1) = 3$, that manages to cross the barrier, for then we obtain $C = 0$; then $y(x) = 3x^3$ and the C/x term that "blows up" at $x = 0$ is not present. Thus, through the initial point $y(1) = 3$ the unique solution $y(x) = 3x^3$ exists for *all* x, on $-\infty < x < \infty$. The presence of this exceptional solution does not violate the theorem because of the words "at least" in the last sentence of the theorem.

Thus far we've considered initial conditions at $x = 1$. Since $p(x) = 1/x$ and $q(x) = 12x^2$ are both continuous at $x = 1$, the existence of unique solutions through those initial points was guaranteed, and the only question concerned their "intervals of existence." Now consider initial points at $x = 0$, at which $p(x) = 1/x$ is discontinuous. That is, consider initial points on the y axis. Since $p(x)$ is not continuous in any neighborhood of $x = 0$, Theorem 1.2.1 simply gives no information. In fact, through the initial point $y(0) = 0$ (the origin) there is the unique solution $y(x) = 3x^3$, which exists on $-\infty < x < \infty$, as noted above. But, through every other point on the y axis there is *no solution* because (39) gives $y(0) = b = 0 + C/0$, which cannot be satisfied by any value of C. ▮

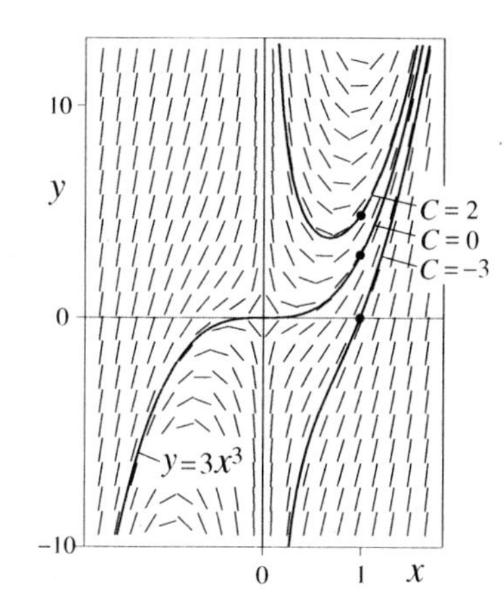

Figure 6. Representative solution curves $y(x) = 3x^3 + C/x$, with the direction field included.

Interval of existence.

The broadest interval on which a solution exists is called the **interval of existence** of that solution. For instance, in Example 6 consider the solution satisfying the initial condition $y(1) = 5$, its graph being the uppermost of the three shown in Fig. 6. Both $y(x) = 3x^3 + 2/x$ and $y' = 6x^2 - 2/x^2$ are undefined at $x = 0$, where they "blow up." Thus, the interval of existence of that solution is $0 < x < \infty$. In contrast, the initial condition $y(1) = 3$ gives $C = 0$ in (39), so the singular C/x term drops out and the solution $y(x) = 3x^3$ has, as its interval of existence, $-\infty < x < \infty$.

EXAMPLE 7. Occurrence of Nonuniqueness. The only case not illustrated in Examples 5 and 6 is that of nonuniqueness, so consider one more example,

$$x\frac{dy}{dx} = y, \tag{40a}$$

$$y(a) = b, \tag{40b}$$

so $p(x) = -1/x$ and $q(x) = 0$. Here, $p(x)$ is discontinuous at $x = 0$. The general solution of (40a) is found to be

$$y = Cx, \tag{41}$$

Use (8) and remember that $e^{\ln x} = x$.

and the initial condition (40b) gives $y(a) = Ca = b$. Now, if $a \neq 0$, the latter gives $C = b/a$ and we have the unique solution $y(x) = bx/a$ with interval of existence $-\infty < x < \infty$. [That interval happens to exceed the minimum interval of existence indicated by Theorem 1.2.1, which is $0 < x < \infty$ if $a > 0$ and $-\infty < x < 0$ if $a < 0$.]

However, consider the case $a = 0$ so the initial point lies on the y axis. If $b \neq 0$, then $(C)(0) = b$ has no solution for C and the IVP (40) has no solution. But if $b = 0$ (so the initial point is the origin), then $(C)(0) = 0$ is satisfied by *any* finite value of C, and (40) has the *nonunique* solution $y = Cx$ where C is an arbitrary finite value.

Summary: If the initial point is not on the y axis there is a unique solution, but if it is on the y axis [where $p(x) = -1/x$ is discontinuous] there are two cases: if it is not at the origin

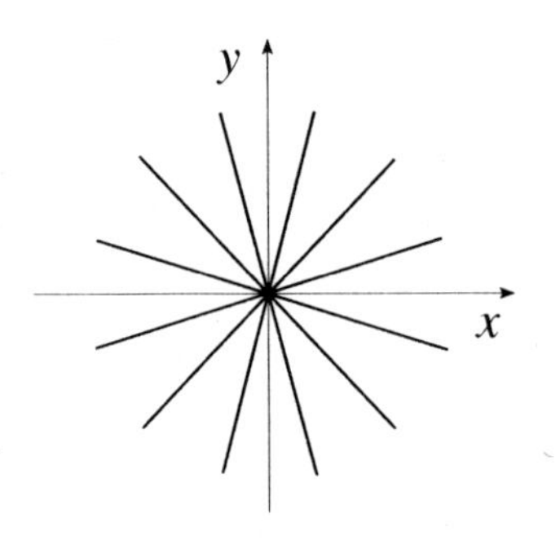

Figure 7. The solutions of (40).

there is no solution, and if it is at the origin there is a nonunique solution, namely, every line $y = Cx$ with finite slope C, as summarized in Fig 7. ▮

Closure. To study the general linear first-order equation $y' + p(x)y = q(x)$, we considered first the homogeneous case $y' + p(x)y = 0$, and used a separation of variables method to derive the general solution (8). For the nonhomogeneous case, separation of variables failed, but we were able to find a general solution by using an integrating factor. The result was the general solution (27), with the integrating factor $\sigma(x)$ given by (25).

Finally, we gave the fundamental existence and uniqueness theorem, Theorem 1.2.1, which states that a solution of the IVP (36) exists and is unique *at least* on the broadest open x interval, containing the initial point $x = a$, on which $p(x)$ and $q(x)$ are continuous; that solution is given by (37).

EXERCISES 1.2

CAUTION: The right-hand sides of equations (8) and (25) are similar, but have different signs in the exponents.

1. Verify, by direct substitution and with the help of chain differentiation, that
(a) (8) satisfies (i.e., is a solution of) $y' + p(x)y = 0$, for any value of A.
(b) (27) satisfies $y' + p(x)y = q(x)$, for any value of C.

2. Homogeneous Equations. Find the particular solution satisfying the initial condition $y(3) = 1$, and give its interval of existence.

(a) $y' = 6x^2y$ (b) $y' + 2(\sin x)y = 0$
(c) $y' - (\cos x)y = 0$ (d) $xy' - y = 0$
(e) $xy' + 3y = 0$ (f) $(\cos x)y' = (\sin x)y$
(g) $(\sin x)y' = (\cos x)y$ (h) $xy' + (1 + x)y = 0$
(i) $x^2y' - y = 0$ (j) $(2 + x)(6 - x)y' = 8y$
(k) $x(5 - x)y' = 5y$ (l) $(1 - x^2)y' - y = 0$
(m) $(2 + x)^2y' + 5y = 0$ (n) $(1 + x)y' - 2y = 0$
(o) $(1 + x)y' + 4y = 0$ (p) $(4 - x^2)y' - 2y = 0$

3. Nonhomogeneous Equations. Find the particular solution satisfying the initial condition $y(2) = 0$ and give its interval of existence.

(a) $y' - y = 3e^x$ (b) $y' + 4y = 8$
(c) $x^2y' + 3xy = 4$ (d) $xy' = 2y + 4x^3$
(e) $xy' + 2y = 10x^3$ (f) $y' - y = 8\sin x$
(g) $y' - 2x = -y - x$ (h) $2xe^xy' = 4 - 2e^xy$
(i) $xy' + y = \sin x + 2\cos x$ (j) $(9 - x^2)y' - 2xy = 10$
(k) $xy' = \sin x - y$ (l) $e^xy' + e^xy = 50$

4. The following equations are not linear, so the methods of this section seem not to apply. However, in these examples you will find that if you interchange the independent and dependent variables and consider $x(y)$ instead of $y(x)$, then the result will be a linear equation for $x(y)$. To do that, merely replace the dy/dx by $1/(dx/dy)$ and put the equation into the standard linear form. Solve it for $x(y)$, subject to the given initial condition. If you can, then solve for $y(x)$ from that result, and give its interval of existence.

(a) $y' = y/(4y - x);\ y(2) = 1$
(b) $y' = y^2/(4y^3 - 2xy);\ y(1) = -1$
(c) $(2y - x)y' = y;\ y(0) = 1$
(d) $(x + 2e^{-y})y' = 1;\ y(1) = 0$

5. Computer; Example 4. Obtain a computer plot of the direction field of (28a) and the solutions satisfying the initial conditions $y(1) = -10$, $y(3) = -10$, $y(1) = 20$, $y(3) = 20$, and $y(0) = -1$, within the rectangle $-2 \le x \le 6$ and $-10 \le y \le 20$.

6. Direction Fields. The following are direction fields of first-order linear differential equations. In each case sketch by hand, on a photocopy of the figure, the solution curve through each of the four initial points (that are denoted by heavy dots). To illustrate, we have shown the solution curve through the initial point $y(0) = 2$ in (a), and through $y(1) = 2$ in (b).

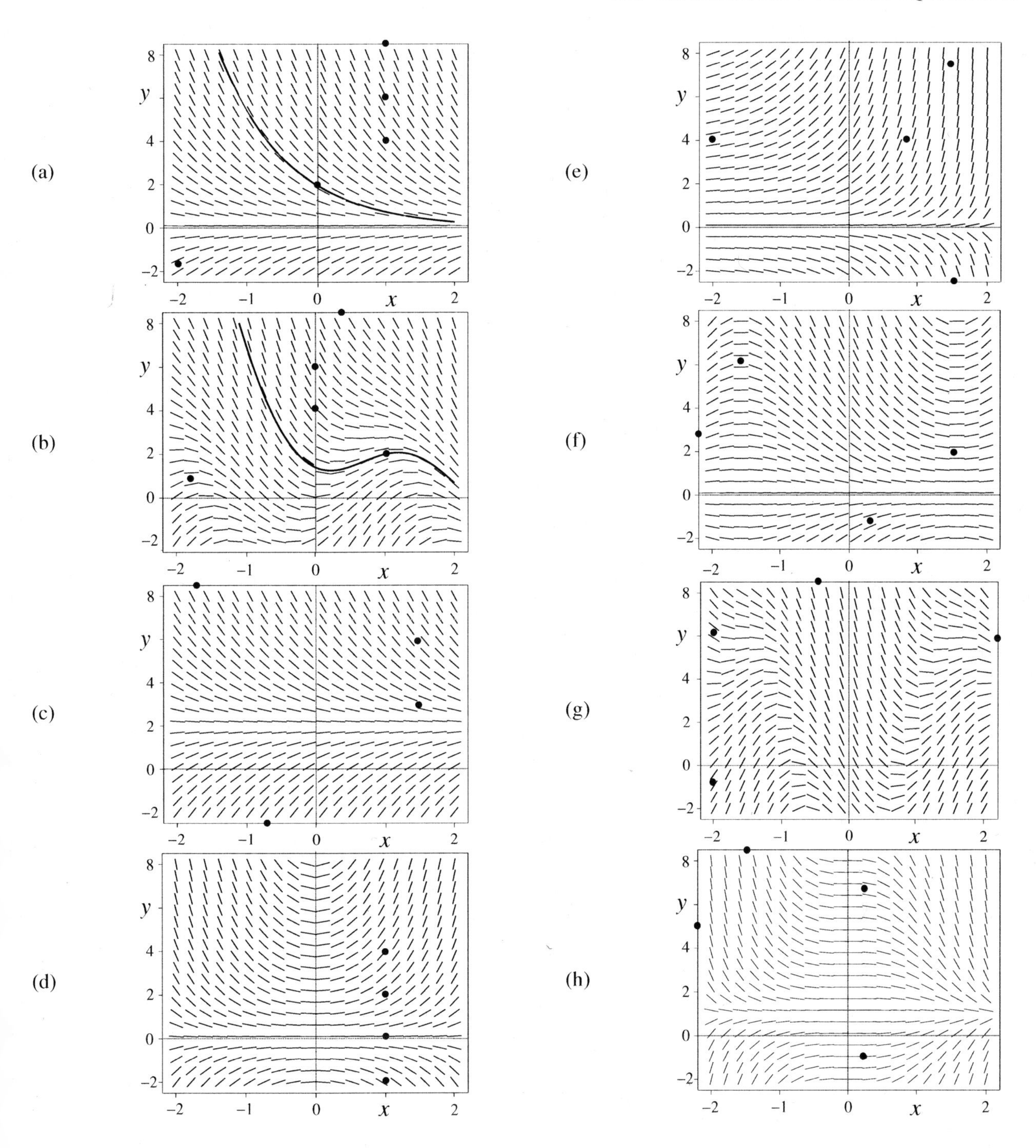
(a)
(b)
(c)
(d)
(e)
(f)
(g)
(h)
y
x

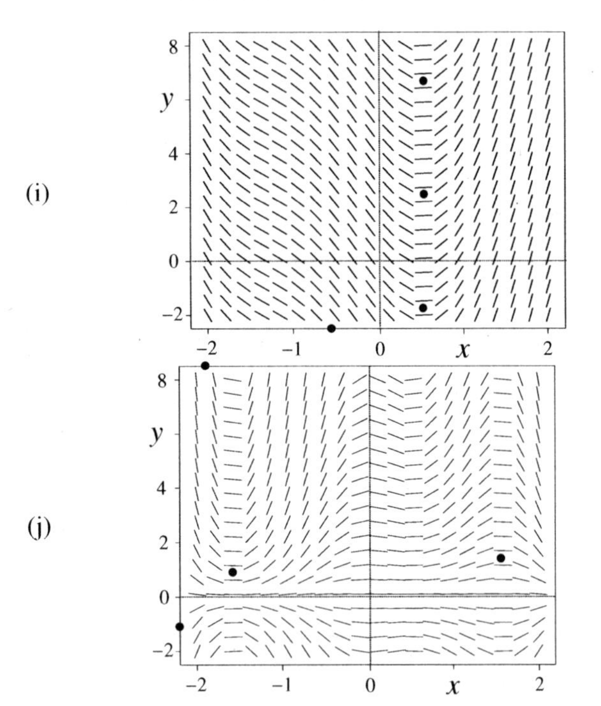

7. Matching. The differential equations whose direction fields are given in Exercise 6(a)–6(d) are these:

$$y' + y = 3\sin 2x, \tag{7.1}$$

$$y' + y = 2, \tag{7.2}$$

$$y' + y = 0, \tag{7.3}$$

$$y' = xy. \tag{7.4}$$

Match these four differential equations with the corresponding direction fields shown in 6(a)–6(d), and state your reasoning. HINT: Write the equation in the form $y' = f(x, y)$ and compare $f(x, y)$ with the directions shown in the figure. For example, (7.1) is $y' = 3\sin 2x - y$, so along the line $y = 0$, for instance, the slope $3\sin 2x$ should be oscillatory. Of the direction fields in 6(a)–(d), above, the only one with that property is (b), so we can match 6(b) with equation (7.1).

8. First, read Exercise 7. The differential equations whose direction fields are given in Exercise 6(e)–6(g) are these:

$$y' + y = -6\cos 2x \tag{8.1}$$

$$y' = e^x y, \tag{8.2}$$

$$y' + (\cos x)y = 0. \tag{8.3}$$

Match these differential equations with the corresponding direction fields shown in 6(e)–6(g), and state your reasoning.

9. First, read Exercise 7. The differential equations whose direction fields are given in Exercise 6(h)–6(j) are these:

$$y' + x^2(y-1) = 0, \tag{9.1}$$

$$y' + (1-2x)(\sin 2x)y = 0, \tag{9.2}$$

$$y' = 4(x - \cos 2x). \tag{9.3}$$

Match these differential equations with the corresponding direction fields shown in 6(h)–6(j), and state your reasoning.

10. Straight-Line Solutions. Straight-line solutions of $y' + p(x)y = q(x)$ are striking because of their simple form; for instance, in Example 4 there was one, and in Example 6 there were none. For the given differential equation, find all straight-line solutions, if any. HINT: You can find the general solution and then look within that family of solutions for any that are of the form $y = mx + b$, but it is more direct to seek solutions specifically in that form. This idea, of seeking solutions of a certain form, is prominent in the study of differential equations.

(a) $y' + 3xy = 6x^2 + 15x + 2$
(b) $y' + 3xy = 12x^2 + 15x + 2$
(c) $y' + e^x y = (1-3x)e^x - 3$
(d) $xy' + 2y = 15x - 4$
(e) $xy' = x^2 + y$
(f) $(x-1)y' - y = -3$
(g) $(x+3)y' = y + 1$
(h) $e^x y' + y = x + e^x - 2$

11. Form of General Solution. Observe that the form of the general solution (27) is $y(x) = F(x) + CG(x)$, in which the constant C is arbitrary. Show that $F(x)$ is a particular solution [i.e., of the full equation $y' + p(x)y = q(x)$] and that $G(x)$ is a homogeneous solution [i.e., of the "homogenized" version $y' + p(x)y = 0$]. HINT: Substitute $y(x) = F(x) + CG(x)$ into $y' + p(x)y = q(x)$ and use the fact that C is arbitrary.

12. Working Backwards. If possible, find an equation (or equations) $y' + p(x)y = q(x)$ that has the following functions among its solutions.

(a) $y_1(x) = 1,\ y_2(x) = x$
(b) $y_1(x) = e^x,\ y_2(x) = 5e^x$
(c) $y_1(x) = e^x,\ y_2(x) = e^{-x}$
(d) $y_1(x) = 0,\ y_2(x) = e^x,\ y_3(x) = 6e^x$
(e) $y_1(x) = 1,\ y_2(x) = x,\ y_3(x) = x^2$
(f) $y_1(x) = 1,\ y_2(x) = x,\ y_3(x) = 2x - 1$

13. Interval of Existence. (a) Make up any differential equation $y' + p(x)y = q(x)$ and initial condition that give a unique solution on $-1 < x < 1$ but not on any larger interval; give

that solution. Show your steps and reasoning.
(b) Make up another one.

14. Suppose an equation $y' + p(x)y = q(x)$ has solutions $y_1(x)$ and $y_2(x)$, the graphs of which cross at $x = a$. What can we infer, from that crossing, about the behavior of $p(x)$ and $q(x)$?

15. Change of Variables and the Bernoulli Equation. Sometimes it is possible to convert a nonlinear equation to a linear one (which is desirable because we know how to solve linear first-order equations). This idea will be developed in Section 1.8; but since you may not cover that section, we introduce the topic here as an exercise. The equation

$$\boxed{y' + p(x)y = q(x)y^n,} \tag{15.1}$$

in which n is a constant (not necessarily an integer), is called **Bernoulli's equation**, after the Swiss mathematician *James Bernoulli*. James (1654–1705), his brother John (1667–1748), and John's son Daniel (1700–1782) are the best known of the eight members of the Bernoulli family who were prominent mathematicians and scientists.

(a) Give the general solution of (15.1) for the special cases $n = 0$ and $n = 1$, in which case (15.1) is linear.
(b) If n is neither 0 nor 1, then (15.1) is *nonlinear* because of the y^n term. Nevertheless, show that by transforming the dependent variable from $y(x)$ to $v(x)$ according to

$$\boxed{v = y^{1-n}} \tag{15.2}$$

(for $n \neq 0, 1$), (15.1) can be converted to the equation

$$v' + (1-n)p(x)v = (1-n)q(x), \tag{15.3}$$

which is *linear* and which can be solved by the methods developed in this section. This method of solution was discovered by *Gottfried Wilhelm Leibniz* (1646–1716) in 1696.

16. Use the method suggested in Exercise 15(b) to solve each of the following. Give the interval of existence. HINT: To solve, identify n, $p(x)$, and $q(x)$, then use (15.3).

(a) $y' + y = -3e^x y^2; \quad y(0) = 1$
(b) $y' + 2y = -12e^{3x}y^{3/2} \quad (y > 0); \quad y(0) = 1$
(c) $(1 + x)y' + 2y = 2\sqrt{y} \quad (y > 0); \quad y(3) = 4$
(d) $xy' - 2y = 5x^3y^2; \quad y(1) = 4$
(e) $3y' + y = x/\sqrt{y} \quad (y > 0); \quad y(3) = 1$

17. (a)–(p) For the corresponding part of Exercise 2, what minimum interval of existence and uniqueness is predicted by Theorem 1.2.1 for the initial condition $y(0.7) = 2$?

18. (a)–(l) For the corresponding part of Exercise 3, what minimum interval of existence and uniqueness is predicted by Theorem 1.2.1 for the initial condition $y(-2) = 5$?

19. Proof of Existence Part of Theorem 1.2.1. To prove existence it suffices to put forward a solution, and (37) is indeed a solution. Thus, to prove the existence part of the theorem you need merely verify that (37) satisfies the differential equation and the boundary condition:

(a) Verify that (37) satisfies the differential equation in (36). HINT: Since $p(x)$ is continuous, (25) shows that $\sigma(x)$ is also continuous and nonzero. Also, $\frac{d}{dx}\int_a^x \sigma(s)q(s)\,ds = \sigma(x)p(x)$ because $\sigma(x)$ and $q(x)$ are continuous.
(b) Verify that (37) also satisfies the initial condition $y(a) = b$.
(c) We wrote (37) without derivation. Derive it. HINT: Instead of using indefinite integrals when you integrate $(\sigma y)' = \sigma q$, use definite integrals, from a to x.

20. Proof of Uniqueness Part of Theorem 1.2.1. To prove that a problem has a unique solution, the standard approach is to consider any two solutions and to show that their difference must be identically zero, so the two solutions must be identical and hence the solution must be unique. Accordingly, suppose $y_1(x)$ and $y_2(x)$ satisfy (36), in which $p(x)$ and $q(x)$ satisfy the continuity condition stated in the theorem. Then

$$y_1' + p(x)y_1 = q(x); \quad y_1(a) = b, \tag{20.1}$$

$$y_2' + p(x)y_2 = q(x); \quad y_2(a) = b. \tag{20.2}$$

Denote the difference $y_1(x) - y_2(x)$ as $u(x)$.

(a) By subtracting (20.2) from (20.1), show that $u(x)$ satisfies the "homogenized" problem

$$u' + p(x)u = 0; \quad u(a) = 0. \tag{20.3}$$

[We say (20.3) is homogeneous because both the forcing function on the right-hand side of the differential equation is zero and the initial condition is zero as well; there are no "inputs."]

(b) Solve (20.3) and show that its only solution is $u(x) = 0$. It follows that $y_1(x) - y_2(x) = 0$ so $y_1(x) = y_2(x)$. Hence, the solution of (36) is unique. HINT: Use an integrating factor $\sigma(x) = e^{\int_a^x p(s)\,ds}$.

21. Alternative Solution Method: Variation of Parameters. We derived a general solution of the linear first-order equation

$$y' + p(x)y = q(x) \tag{21.1}$$

by the integrating factor method. An alternative method of solution is as follows. First, recall that the *homogeneous* equation $y' + p(x)y = 0$ has the general solution

$$y(x) = Ae^{-\int p(x)\,dx}, \tag{21.2}$$

where A is an arbitrary constant. To solve the *nonhomogeneous* equation (21.1), seek $y(x)$ in the form

$$y(x) = A(x)e^{-\int p(x)\,dx}; \tag{21.3}$$

that is, let the constant A in the homogeneous solution (21.2) vary. (The motivation behind this step is not obvious, but we will see that it works.) Substitute (21.3) into (21.1) and show, after canceling two terms, that you obtain

$$A'(x) = e^{\int p(x)\,dx} q(x) \tag{21.4}$$

so

$$A(x) = \int e^{\int p(x)\,dx} q(x)\,dx + C \tag{21.5}$$

and

$$y(x) = e^{-\int p\,dx}\left(\int e^{\int p\,dx} q\,dx + C\right), \tag{21.6}$$

which agrees with the solution (27) obtained earlier by using an integrating factor. This method is called **variation of parameters** because the key is in letting the parameter A vary.

1.3 APPLICATIONS OF LINEAR FIRST-ORDER EQUATIONS

Having solved the first-order linear differential equation, we now give representative physical applications — to population dynamics, radioactive decay, mixing problems, and electrical circuits, with additional applications in the exercises.

1.3.1 Population dynamics; exponential model. We want to model the population dynamics of a certain species, such as bass in a lake or the malaria parasite introduced into the host's bloodstream. That is, we want to develop a mathematical problem that governs the variation of the population $N(t)$ of that species with the time t.

That is, the birth and death rates β and κ are *per capita*.

Let β be the birth rate (births per individual per unit time) and δ the death rate, with β and δ assumed to be known constants over the time of interest. Then, for any time interval Δt,

$$N(t + \Delta t) = N(t) + \beta N(t)\Delta t - \delta N(t)\Delta t, \tag{1}$$

or,

$$\frac{N(t + \Delta t) - N(t)}{\Delta t} = (\beta - \delta)N(t). \tag{2}$$

Equations (1) and (2) hold for *any* time interval Δt, so it is permissible to let $\Delta t \to 0$ in (2), which step gives the differential equation (3).

Equation (1) is simply bookkeeping: The number of individuals at time $t + \Delta t$ equals the number that we start with at time t plus the number that are born minus the number that die over the Δt time interval. If we let $\Delta t \to 0$ in (2), and denote the *net* birth/death rate $\beta - \delta$, called the **growth rate**, as κ, we obtain[1]

$$\boxed{N' = \kappa N,} \tag{3}$$

which is a linear first-order homogeneous differential equation for $N(t)$, homogeneous because it is $N' - \kappa N = 0$.

The latter is of the same form as $y' + py = 0$, studied in Section 1.2, with N in place of y, t in place of x, and $p = -\kappa$, so its general solution is

$$N(t) = Ae^{\kappa t}, \tag{4}$$

with A an arbitrary constant. Alternative to obtaining (4) by using the memorized solution formula, let us solve (3) by separation of variables, as review:

$\ln N$ rather than $\ln |N|$ in (5) because the population $N(t)$ cannot be negative, so there is no need for absolute values.

$$\int \frac{dN}{N} = \int \kappa \, dt, \quad \ln N = \kappa t + C, \quad N(t) = e^{\kappa t + C} = e^{C} e^{\kappa t} = Ae^{\kappa t}, \tag{5}$$

as in (4). If we have an initial condition

$$N(0) = N_0, \tag{6}$$

then $N(0) = N_0 = Ae^{(\kappa)(0)} = A$ so $A = N_0$, and (4) becomes

$$\boxed{N(t) = N_0 e^{\kappa t},} \tag{7}$$

which we've plotted in Fig. 1 for several values of κ.

COMMENT 1. Equation (3) is often called the **Malthus model** after the British economist *Thomas Malthus* (1766–1834), who observed that many biological populations change at a rate that is proportional to their population. It is also known as the **exponential model** because of the exponential form of its solution.

COMMENT 2. If the growth rate κ is negative, then (7) predicts an exponential decrease to zero as $t \to \infty$, which seems reasonable (although when N becomes small enough our approximation of N as a continuous and differentiable function of t comes into question). But if κ is positive, then (7) predicts exponential growth, with $N(t)$ tending to infinity as $t \to \infty$. Such sustained growth is not reasonable because if N becomes sufficiently large then other factors will no doubt come into play, such as insufficient food, factors that have not been accounted for in our

[1]Strictly speaking, $N(t)$ is integer-valued since one cannot have a population of 28.37, say. Its graph develops in a stepwise manner so $N(t)$ is a discontinuous function of t. Hence, it is not differentiable and the $N'(t)$ in (3) does not exist. However, if N is sufficiently large *so that the steps are sufficiently small compared to* N, then we can regard $N(t)$ as a continuous function of t.

Recall that $\kappa < 0$ if the death rate exceeds the birth rate, and $\kappa > 0$ if the birth rate exceeds the death rate.

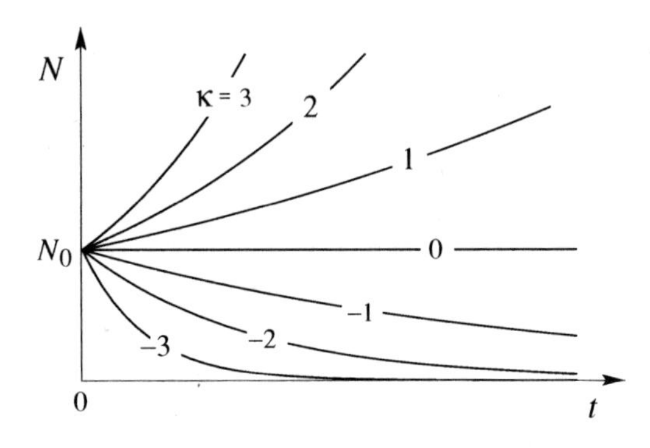

Figure 1. Exponential growth and decay given by (7); $N(t) = N_0 e^{\kappa t}$.

The logistic equation (8) is also called the **Verhulst equation** after the Belgian mathematician *P. F. Verhulst* (1804–1849).

simple model. Specifically, we expect κ not really to be a constant but to be a function of N, decreasing as N increases. As a first approximation of such behavior, suppose κ varies linearly with N: $\kappa = a - bN$, where a and b are positive constants. Then (3) is replaced by

$$\frac{dN}{dt} = (a - bN)N, \tag{8}$$

which is well known as the **logistic equation**. However, the logistic equation is *nonlinear* because of the N^2 term, and will be studied later, in Section 1.6.[1]

1.3.2 Radioactive decay; carbon dating. Another classical application of linear first-order equations involves radioactive decay and carbon dating.

Radioactive materials, such as carbon-14, plutonium-241, radium-226, and thorium-234, are observed to decay at a rate that is proportional to the amount of radioactive material present. Thus, the number of nuclei disintegrating per unit time will be proportional to the number of nuclei present, so

$$\frac{dN}{dt} = -kN, \tag{9}$$

in which $N(t)$ is the number of atoms of the radioactive element at time t, and the positive constant k is the *decay rate*, which we assume is known. However, it is inconvenient to work with N since one cannot count the number of atoms in a given batch of material. Thus, multiply both sides of (9) by the atomic mass (mass per atom). Since the atomic mass times $N(t)$ is the mass $m(t)$ of the radioactive material, (9) gives

$$\boxed{\frac{dm}{dt} = -km} \tag{10}$$

[1]Verhulst studied human population but did not have sufficient census data to test the accuracy of his model. Later researchers turned to species with much shorter life spans, such as *Drosophila melanogaster* (fruit fly), which could be accurately monitored in the laboratory over many generations, and they did obtain good agreement using Verhulst's logistic model.

for $m(t)$, which is more readily measured than $N(t)$. Solving (10) gives

$$m(t) = m_0 e^{-kt}, \tag{11}$$

where $m(0) = m_0$ is the initial amount of the radioactive mass (Fig. 2). This result agrees well with experiment.

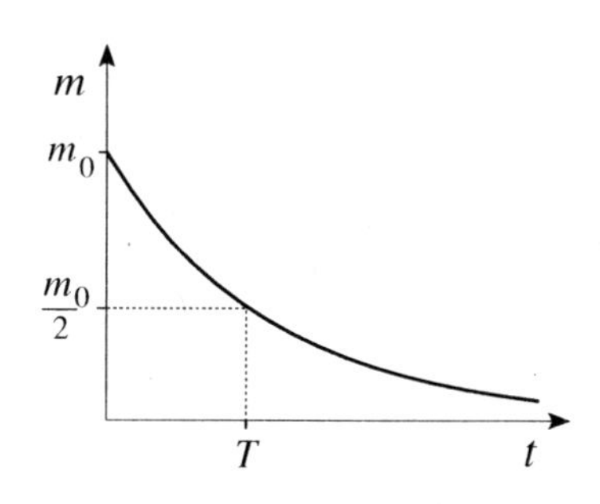

Figure 2. The exponential decay $m(t) = m_0 e^{-kt}$ and the half-life T.

The decay rate k determines the **half-life** T of the material, the time required for any initial amount of mass m_0 to be reduced by half, to $m_0/2$. It is more common and more convenient to work with T than k, so we will eliminate k from (11), in favor of T, as follows: When t is T, in (11), $m(t)$ is $m_0/2$, so $m_0/2 = m_0 e^{-kT}$, which gives $\boldsymbol{k = (\ln 2)/T}$. If we put the latter into (11), $m(t)$ can be re-expressed in terms of the half-life T as

$$m(t) = m_0 2^{-t/T}. \tag{12}$$

For instance, at $t = 0, T, 2T, 3T, 4T, \ldots$, (12) gives $m(t) = m_0, m_0/2, m_0/4, m_0/8$, and so on.

The steps leading from (11) to (12) involve the properties of the exponential and logarithmic functions, which are among the review formulas on the inside cover of this book. We leave those steps for the exercises.

Radioactivity has had an important archaeological application in connection with **dating**. The idea behind any dating technique is to identify a physical process that proceeds at a known rate. If we measure the state of the system now, and we know its state at the initial time, then from these two quantities together with the known rate of the process we can infer how much time has elapsed; the mathematics enables us to "travel back in time as easily as a wanderer walks up a frozen river."[1]

For instance, consider carbon dating, developed by the American chemist *Willard Libby* in the 1950's. Cosmic rays consisting of high-velocity nuclei penetrate the earth's lower atmosphere. Collisions of these nuclei with atmospheric gases produce free neutrons. These collide with nitrogen, changing it to carbon-14, which is radioactive and which decays to nitrogen-14 with a half-life of around 5,570 years. Thus, some of the carbon dioxide (CO_2) in the atmosphere contains this radioactive C-14. Plants take in both radioactive and nonradioactive CO_2, and humans and animals inhale both and eat the plants. Consequently, the plants and animals living today contain both the nonradioactive C-12 and, to a much lesser extent, its radioactive isotope C-14, in a ratio that is the same from one plant or animal to another. When a plant or animal dies its C-12 remains fixed but its C-14 decreases with time by radioactive decay. The resulting "shortage" of C-14 at any given time is a measure of how long ago the plant or animal died.

Radioactive carbon, C-14, is called radiocarbon because it decays radioactively.

For instance, suppose we wish to carbon date a given sample of wood, that is, to determine how long ago it died. To do so we make two assumptions: First, assume that the ratio of radioactive to nonradioactive carbon (C-14 to C-12) in living material at the time the tree died was the same as it is in living material today. Second, assume that the rate of radioactive decay of C-14 has been constant over that period of time. Subject to these assumptions (which cannot be verified because they are historical in nature), here is how the method works. Measure the mass of C-14 present in the sample now, which is the $m(t)$ on the left-hand side of (12), and assume that the initial mass of C-14 (when the tree died), m_0, is the same

The first assumption establishes the initial condition, the second establishes the differential equation.

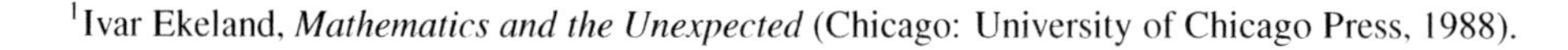

[1]Ivar Ekeland, *Mathematics and the Unexpected* (Chicago: University of Chicago Press, 1988).

as the mass of C-14 in a sample of the same weight that is alive today. Knowing $m(t)$, m_0, and the half-life T, solve (12) for t, which is the time that has elapsed since the tree died.

EXAMPLE 1. Carbon Dating a Sample of Petrified Wood.
Consider a petrified wood sample that we wish to date. Since C-14 emits approximately 15 beta particles (i.e., high-speed electrons) per minute per gram, we can determine how many grams of C-14 are contained in the sample by measuring the rate of beta particle emission. Suppose we find that the sample contains 0.002 grams of C-14, whereas if it were alive today it would, based upon its weight, contain around 0.0045 grams. Assuming it contained 0.0045 grams of C-14 when it died, then that mass of C-14 will have decayed, over the subsequent time t, to 0.002 grams. Then (12) gives

$$0.002 = (0.0045)\, 2^{-t/5570},$$

and, solving for t, we determine the sample to be around $t = 6,520$ years old. ∎

1.3.3 Mixing problems; a one-compartment model. Consider a mixing tank, as in a chemical plant, with a constant inflow of Q gallons per minute and an equal outflow (Fig. 3). The inflow is at a constant concentration c_i (pounds per gallon) of a particular solute such as salt, and the tank is stirred so the concentration $c(t)$ is uniform throughout the tank; t is the time. Hence, the outflow is at concentration $c(t)$. Let v be the liquid volume within the tank, in gallons; v is constant because the inflow and outflow rates are equal. We want to determine the solute concentration $c(t)$.

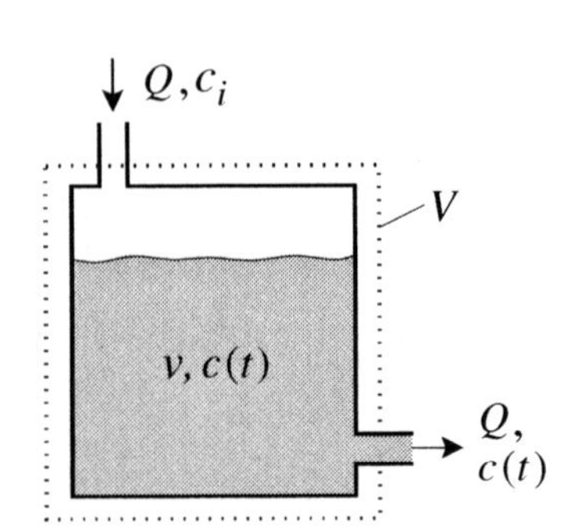

Figure 3. Mixing tank.

To derive a differential equation for $c(t)$, carry out a mass balance for the "control volume" V (dashed lines in the figure):

$$\text{Rate of increase of mass of solute within } V = \text{Rate in} - \text{Rate out}, \tag{13}$$

$$\frac{d}{dt}[c(t)v] = Qc_i - Qc(t) \tag{14}$$

The units of each term in (14) are lb/min. For instance, the first term on the right is $(Q\frac{\text{gal}}{\text{min}})(c_i\frac{\text{lb}}{\text{gal}})$, or $Qc_i\frac{\text{lb}}{\text{min}}$.

or, since v is constant,

$$\boxed{\frac{dc}{dt} + \frac{Q}{v}c = \frac{c_i Q}{v},} \tag{15}$$

which is a first-order linear differential equation for $c(t)$.

The tank in Fig. 3 could, literally, be a mixing tank in a chemical plant, but in some applications the figure may be only schematic. For instance, in biological applications it is common to represent the interacting parts of the biological system as one or more interconnected **compartments**, with inflows, outflows, and exchanges between compartments.[1] One compartment could be an organ such as the liver; another could be all the blood in the circulatory system. The system represented in Fig. 3 is an example of a one-compartment system; the compartment is the tank.

EXAMPLE 2. Mixing Tank; Approach to Steady-State Operation.
Let the initial concentration in the tank be $c(0) = 0$, so

$$\frac{dc}{dt} + \frac{Q}{v}c = \frac{c_i Q}{v}; \qquad c(0) = 0. \tag{16}$$

The integrating factor is $\sigma(t) = e^{\int p(t)\,dt} = e^{\int (Q/v)\,dt} = e^{Qt/v}$ and the general solution of the differential equation in (16) is

$$\begin{aligned} c(t) &= e^{-Qt/v}\left(\int e^{Qt/v}\frac{c_i Q}{v}\,dt + C\right) \\ &= c_i + Ce^{-Qt/v}. \end{aligned} \tag{17}$$

Finally, $c(0) = 0 = c_i + C$ gives $C = -c_i$, so

$$\boxed{c(t) = c_i\left(1 - e^{-Qt/v}\right).} \tag{18}$$

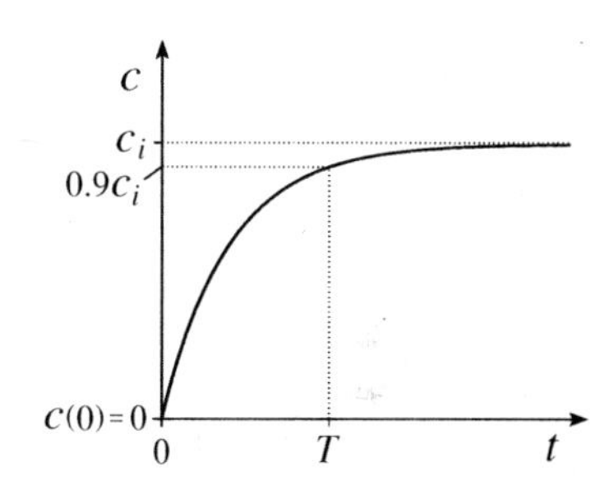

Figure 4. Exponential approach of $c(t)$ to its steady-state value c_i.

Since $e^{-Qt/v} \to 0$ as $t \to \infty$, it follows from (18) that $c(t) \to c_i$ as $t \to \infty$, as we might have expected since the inflow is maintained at that concentration. This asymptotic behavior is seen in Fig. 4.

The time T that it takes for $c(t)$ to reach $0.9c_i$, say, can be found from (18):

$$c(T) = 0.9c_i = c_i\left(1 - e^{-QT/v}\right),$$

which gives

$$T = (\ln 10)\frac{v}{Q}. \tag{19}$$

Thus far we've taken $c(0) = 0$, but now suppose it is not necessarily zero. Let $c(0) = c_0$, with $c_0 \geq 0$ because a negative concentration $c_0 < 0$ is impossible. Then (17) gives $c(0) = c_0 = c_i + C$ so $C = c_0 - c_i$, and in place of (18) we have

$$c(t) = \underbrace{(c_0 - c_i)e^{-Qt/v}}_{\text{transient}} + \underbrace{c_i.}_{\text{steady state}} \tag{20}$$

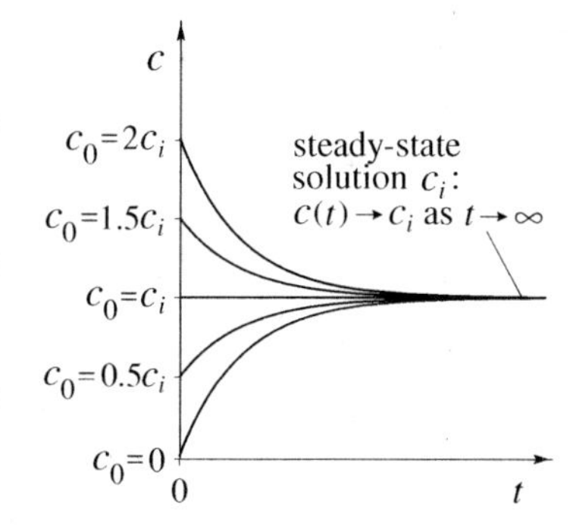

Figure 5. Varying the initial condition c_0, with Q fixed.

As $t \to \infty$, the exponential term in (20) tends to zero and $c(t) \to c_i$. Thus we call the $(c_0 - c_i)e^{-Qt/v}$ term the **transient** part of the solution, and we call the c_i term the **steady-state** solution. Graphs of $c(t)$ in Fig. 5 show the approach to steady state for several different initial conditions. The bottom curve corresponds to the one in Fig. 4. ∎

[1]For a discussion of compartmental analysis in biology see L. Edelstein-Keshet, *Mathematical Models in Biology* (New York: Random House, 1988) or John A. Jacquez, *Compartmental Analysis in Biology and Medicine*, 3rd ed (Ann Arbor, MI: BioMedware, 1996).

The $\equiv$ in (21) means *equal to by definition*. That is, let $(Q/v)(c_i - c)$ be called "$f(c)$."

1.3.4 The phase line, equilibrium points, and stability.

1.3.4 The phase line, equilibrium points, and stability. Another graphical idea is useful. If we write the differential equation in (16) as

$$\frac{dc}{dt} = \frac{Q}{v}(c_i - c) \equiv f(c), \tag{21}$$

we see that the equation is **autonomous**, which means that the right-hand side contains no explicit t dependence, for it is of the form $f(c)$ rather than $f(c,t)$.[1] It is informative to plot $f(c)$ versus c, as we've done in Fig. 6a. Since c is a function of t we can imagine each point on the c axis as moving, with time, along that axis, with its velocity equal to the value of $f(c)$ at that point [because $c' = f(c)$], to the right if $f(c) > 0$ and to the left if $f(c) < 0$. The point c_i is not moving because $f(c_i) = 0$, points to the left of c_i are moving rightward, and points to the left of c_i are moving leftward, as indicated by the two arrows in Fig. 6a.

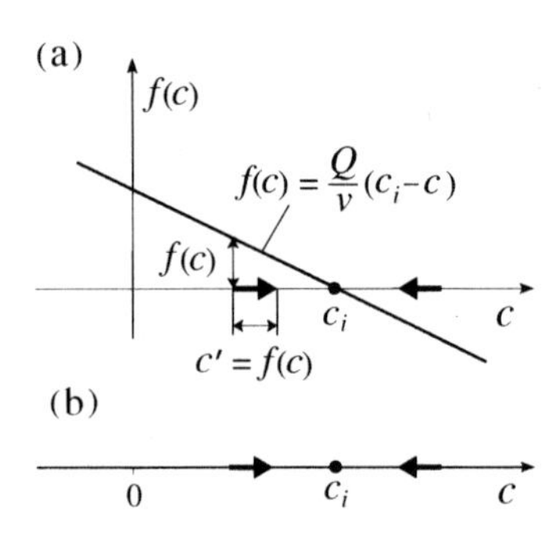

Figure 6. The "flow" along the phase line, implied by (21). The phase line is shown in (b).

Thus, we can think of the movement of points along the c axis as a one-dimensional "flow," and we call the line along which that flow takes place the **phase line**. From that point of view the steady-state solution $c = c_i$ in Fig. 5 corresponds to an **equilibrium point** of the flow along the phase line (Fig. 6b) because the flow velocity dc/dt is zero there.

We could show many arrows on the phase line, rather than just the two in Fig. 6b, and could even scale them according to their magnitude, but we will keep phase line displays simple and just show any equilibrium points (with heavy dots) and single arrows to indicate flow directions.

To see the connection between Fig. 5 and Fig. 6b, we've shown them together in Fig. 7, with the phase line arranged vertically at the left, and we've included arrows on the solution curves in the c, t plot — in the direction of increasing time. To see how the flow on the phase line is related to the flow in the c, t plane, imagine the c, t plot as resulting if (on our imaginary computer screen) we click on the phase line and drag it to the right, in time, as suggested by the two large arrows at the right. Conversely, if we drag that dotted line back to the left, then the c, t graphs get "squashed," and all we're left with is the flow along the phase line, shown at the left of the figure.

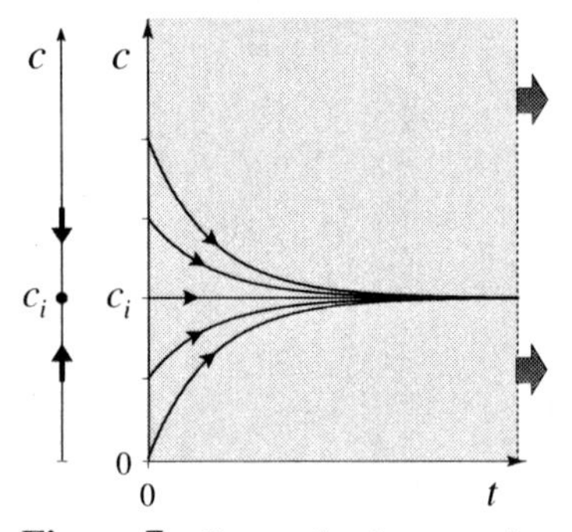

Figure 7. Connection between the phase line flow and the solution curves in the t, c plane.

Along with the concept of equilibrium comes the concept of **stability**.

For instance, the equilibrium of a marble on a hilltop is "unstable" and the equilibrium of a marble in a valley is "stable." To define the stability of an equilibrium point on the phase line, let c_{eq} be an equilibrium point on the phase line of $\dfrac{dc}{dt} = f(c)$; that is, $f(c_{eq}) = 0$. We say that c_{eq} is **stable** if points that start out close to it remain close to it, and **unstable** if it is not stable.[2]

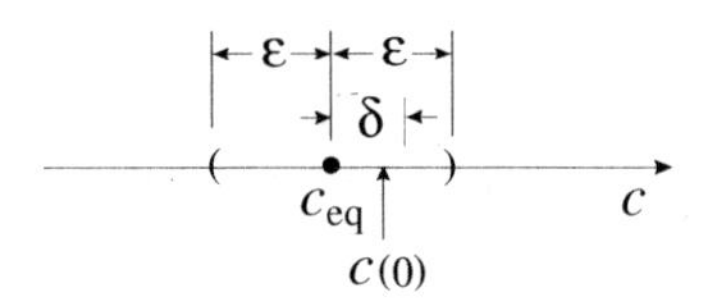

Figure 8. Stability of an equilibrium point c_{eq} on the phase line.

---

[1]If any of Q, c_i, v were functions of time, then $\dfrac{dc}{dt} = \dfrac{Q(t)}{v(t)}[c_i(t) - c] = f(c,t)$ would not be autonomous, it would be *nonautonomous*.

[2]Let us make that intuitively stated definition precise: an equilibrium point c_{eq} is **stable** if, for any $\epsilon > 0$ (i.e., no matter how small), there corresponds a $\delta > 0$ such that $c(t)$ remains closer to c_{eq} than ϵ for all $t > 0$ if $c(0)$ is closer to c_{eq} than δ (Fig. 8). That is, if $|c(0) - c_{eq}| < \delta$ then $|c(t) - c_{eq}| < \epsilon$ for all $t > 0$. If c_{eq} is not stable, it is **unstable**.

We can see that the equilibrium point c_i in Fig. 6b is stable because the flow approaches c_i from both sides, so if we start close to c_i then we remain close to it for all $t > 0$. Actually, the stability of an equilibrium point c_{eq} does not require $c(t)$ to *approach* c_{eq} as $t \to \infty$, but only to remain close to it. If it does approach c_{eq}, that is, if $c(t) \to c_{\text{eq}}$ as $t \to \infty$, then the equilibrium point is not only stable, it is **asymptotically stable**. In Fig. 7, c_i is a steady-state solution because that point is an asymptotically stable equilibrium point of the phase line.

Realize that we obtained the graph in Fig. 6a, and hence the phase line in Fig. 6b, merely by plotting the right-hand side of the differential equation $c' = (Q/v)(c_i - c)$ versus c; *we did not need to solve the differential equation.*

There is good news and bad news regarding the phase line: The bad news is that it contains less information than the traditional plots of c versus t. Of *course*: To get the phase line we merely plotted the right-hand side of the differential equation versus c, we did not solve it, so it makes sense that the detailed time history, contained in the c, t plot, is not available from the phase line. But the good news is that the phase line is readily obtained and contains key information. The key information in this example is the equilibrium point c_i and its stability. After all, from the c, t plot we see that after some time goes by, the line $c(t) = c_i$ is where all the solution curves "end up," for it is approached as $t \to \infty$. Furthermore, from the phase line, at the left in Fig. 7, we could even sketch the solution curves in the c, t plane, if only qualitatively, without actually solving the differential equation.

Thus, the phase line concept is more *qualitative* than quantitative. It is not so impressive for linear equations because linear equations can be solved analytically anyhow, but nonlinear equations are much more difficult in general, and in that case we will need to rely more heavily on other approaches — qualitative ones such as direction fields and the phase line, and quantitative ones involving numerical solution by computer. In any case, remember that the phase line method applies only if the differential equation is autonomous.

1.3.5 Electrical circuits.

Consider electrical circuits consisting of closed wire loops and a number of circuit elements such as resistors, inductors, capacitors, and voltage sources such as batteries.

An electric current is a flow of charges: The *current* through a given control surface, such as the cross section of a wire, is the charge per unit time crossing that surface. Each electron carries a negative charge of 1.6×10^{-19} *coulomb*, and each proton carries an equal positive charge. Current is measured in *amperes*, one ampere being a flow of one coulomb per second. A current is counted as positive in a given direction if it is the flow of positive charge in that direction. While, in general, currents can involve the flow of positive or negative charges, the flow, typically, is of negative charges, free electrons. Thus, when one speaks of a current of one ampere in a given direction in an electrical circuit, one really means the flow of one coulomb per second of negative charges (electrons) in the opposite direction.

Just as heat flows due to a temperature difference, from one point to another, an electric current flows due to a difference in electric potential, or *voltage*, measured

Resistor:

i

$e_1 \quad R \quad e_2$

$e_1 - e_2 = e = Ri$

Inductor:

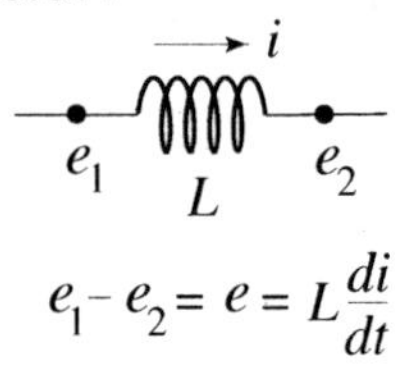

Capacitor:

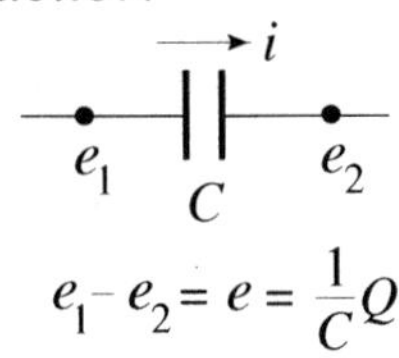

Figure 9. Circuit elements.

in *volts*. Thus, we will need to know the relationship between the voltage difference across a given circuit element and the corresponding current flow through it. The circuit elements considered here are resistors, inductors, and capacitors.

For a **resistor**, the voltage drop $e(t)$, where t is the time, is proportional to the current $i(t)$ through it:

$$\boxed{e(t) = Ri(t),} \tag{22}$$

where the constant of proportionality R is called the *resistance* and is measured in *ohms*; (22) is called **Ohm's law**. By a resistor we mean an electrical device, often made of carbon, that offers a specified resistance — such as 100 ohms, 500 ohms, and so on. The standard symbolic representation of a resistor is shown in Fig. 9.

For an **inductor**, the voltage drop is proportional to the time rate of change of current through it:

$$\boxed{e(t) = L\frac{di(t)}{dt},} \tag{23}$$

in which the constant of proportionality L is called the **inductance** and is measured in *henrys*. Physically, most inductors are coils of wire, hence the symbolic representation in Fig. 9.

For a **capacitor**, the voltage drop is proportional to the charge $Q(t)$ on the capacitor:

$$\boxed{e(t) = \frac{1}{C}Q(t),} \tag{24}$$

where C is called the **capacitance** and is measured in *farads*. Physically, a capacitor consists of two plates separated by a gap across which no current flows, and $Q(t)$ is the charge on one plate relative to the other. Though no current flows across the gap, there will be a current $i(t)$ that flows through the (closed) circuit that links the two plates and is equal to the time rate of change of charge on the capacitor:

$$i(t) = \frac{dQ(t)}{dt}. \tag{25}$$

Equations (22)–(24) give the behavior of the respective circuit elements, but we also need to know the physics of the circuit itself, which consists of two laws named after the German physicist *Gustav Robert Kirchhoff* (1824–1887):

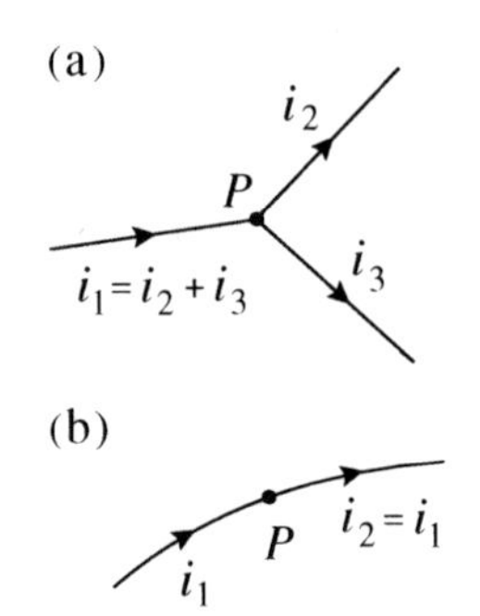

Figure 10. Kirchhoff's current law.

Kirchhoff's current law states that the sum of the currents approaching any point P of a circuit equals the sum of the currents leaving that point. The latter is a *conservation law*, namely, that electric charge is conserved; it is neither created nor destroyed at P. To illustrate, consider the portion of a circuit shown in Fig. 10a. Application of Kirchhoff's current law to point P, say, gives

$$i_1 = i_2 + i_3.$$

Kirchhoff's voltage law states that the algebraic sum of the voltage drops around any loop of a circuit is zero. Since voltage is an electric potential (i.e., electric potential energy), the statement that the potential drops around any loop of a circuit is zero is also a conservation law, this time the conservation of *energy*.

To apply these ideas, consider the circuit shown in Fig. 11, consisting of a single loop containing a resistor, an inductor, a capacitor, a voltage source (such as a battery or generator), and the necessary wiring. Take the current $i(t)$ to be positive clockwise; if it flows counterclockwise, its numerical value will be negative. In this case, Kirchhoff's current law simply says that the current i is a constant from point to point within the circuit and therefore varies only with time. That is, the current law states that at *every* point P in the circuit the currents i_1 and i_2 (Fig. 10b) are the same, namely, $i(t)$. Next, Kirchhoff's voltage law gives

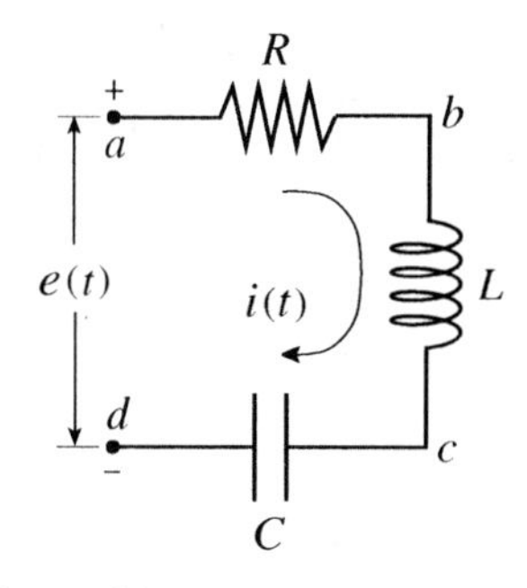

Figure 11. RLC circuit.

$$(e_a - e_d) + (e_b - e_a) + (e_c - e_b) + (e_d - e_c) = 0, \tag{26}$$

which, canceling terms, is simply an algebraic identity. If we use (22) – (24), (26) gives

$$e(t) - Ri - L\frac{di}{dt} - \frac{1}{C}Q(t) = 0. \tag{27}$$

If we differentiate (27) with respect to t and use (25) to eliminate $Q(t)$ in favor of $i(t)$, we obtain

$$\boxed{L\frac{d^2i}{dt^2} + R\frac{di}{dt} + \frac{1}{C}i = \frac{de(t)}{dt},} \tag{28}$$

If the voltage source is a battery, then $e(t)$ is a constant. More generally, a generator can be a time-varying voltage source.

which is a linear second-order differential equation for $i(t)$, in which $e(t)$ is known — prescribed. Alternatively, we could use $Q(t)$ instead of $i(t)$ as our dependent variable. In that case we again use (25) in (27), but this time to eliminate the $i(t)$'s in favor of $Q(t)$, and we obtain the differential equation

$$\boxed{L\frac{d^2Q}{dt^2} + R\frac{dQ}{dt} + \frac{1}{C}Q = e(t)} \tag{29}$$

for $Q(t)$. Either way, we have a linear second-order differential equation.

In this chapter our interest is in *first*-order equations, but we do obtain first-order equations in the following two special cases.

EXAMPLE 3. *RC* Circuit.
If $L = 0$ (i.e., if we remove the inductor from the circuit in Fig. 11, as shown in Fig. 12a, then (28) reduces to the linear first-order equation

$$R\frac{di}{dt} + \frac{1}{C}i = \frac{de(t)}{dt} \tag{30}$$

for $i(t)$. ▮

EXAMPLE 4. *RL* Circuit.
If, instead of removing the inductor from the circuit shown in Fig. 11, we remove the capacitor (Fig. 12b), then (27) gives the first-order equation[1]

$$\boxed{L\frac{di}{dt} + Ri = e(t)} \tag{31}$$

for $i(t)$. ▮

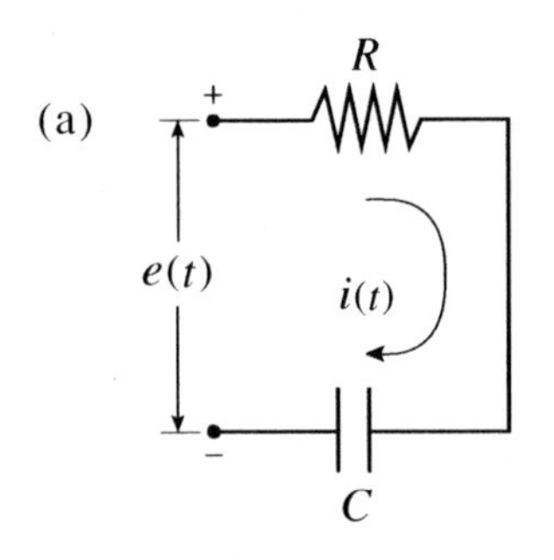

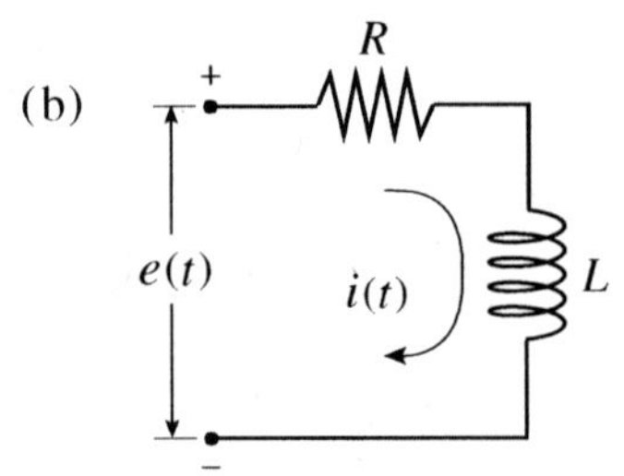

Figure 12. Two special cases of the RLC circuit shown in Fig. 11; RC and RL circuits.

Although the RC and RL circuits are different, their governing equations are of the same form — first-order linear equations with constant coefficients on the left-hand side. Thus, it will suffice to consider just one of the two circuits in Fig. 12, for instance the RL circuit in Fig. 12a, modeled by (31). Hence, we've highlighted (31). Dividing by L to put the equation into the standard form $i' + p(t)i = q(t)$, and appending an initial condition, consider the IVP

$$i' + \frac{R}{L}i = \frac{1}{L}e(t); \quad i(0) = i_0. \tag{32}$$

Identifying $p(t)$ as R/L and $q(t)$ as $e(t)/L$, the results in Section 1.2 give a general solution of the differential equation as

$$i(t) = e^{-Rt/L}\Big(\int e^{Rt/L}\,\frac{e(t)}{L}\,dt + A\Big), \tag{33}$$

in which A can be found by applying the initial condition $i(0) = i_0$. In Examples 5–7 we will specify several typical $e(t)$'s and complete the solution.

EXAMPLE 5. *RL* Circuit with Constant Applied Voltage.
Suppose the applied voltage is a constant, $e(t) = \text{constant} = E_0$. Then (33), together with the initial condition, gives

$$i(t) = \underbrace{\Big(i_0 - \frac{E_0}{R}\Big)e^{-Rt/L}}_{\text{transient}} + \underbrace{\frac{E_0}{R}}_{\text{steady state}}, \tag{34}$$

and representative solution curves are plotted in Fig. 13.

[1]It would be natural to expect that removing the capacitor is equivalent to setting $C = 0$, yet in that case the capacitor term in (27) becomes infinite rather than zero. Rather, to remove the capacitor, move its plates together until they touch. The capacitance C is *inversely* proportional to the gap dimension, so as the gap diminishes to zero $C \to \infty$ and the capacitor term in (27) does indeed drop out because in that limit the $1/C$ factor becomes zero.

Does Fig. 13 look familiar? It should, for the IVP (32) [with $e(t)$ = constant = E_0] and its solution (34) are identical to the IVP (16) and its solution (20), respectively, with the correspondences

$$c(t) \leftrightarrow i(t), \quad c_0 \leftrightarrow i_0, \quad c_1 \leftrightarrow \frac{E}{R}, \quad \frac{Q}{v} \leftrightarrow \frac{R}{L}. \tag{35}$$

As $t \to \infty$, the exponential term in (34) tends to zero and $i(t)$ tends to the steady-state value E_0/R. Outside of name changes, the only difference between Figs. 5 and 13 is that in Fig. 13 we've added the case $i_0 = -0.5E/R$ because whereas the concentration c_0 cannot be negative, the current i_0 *can* be negative. ∎

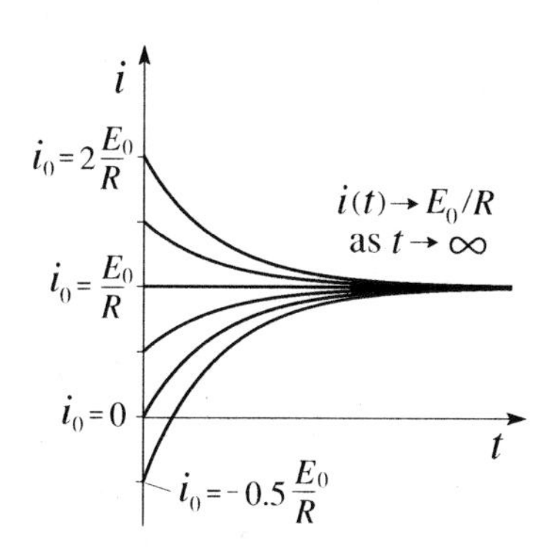

Figure 13. Response $i(t)$ for the case $e(t) = \text{constant} = E_0$, for six different i_0's. It is convenient to express the i_0's as multiples of the steady-state value E_0/R, which is a natural reference value.

The correspondence just noted between the mixing tank and the RL circuit is important, for if two different systems are modeled by the same IVP, to within name changes, then their solutions are identical to within those name changes. Such systems are called **analogs** of each other.

EXAMPLE 6. *RL* Circuit with Sinusoidal Applied Voltage.
Now let the applied voltage be oscillatory instead, for instance $e(t) = E_0 \sin \omega t$, with amplitude E_0 and frequency ω. With this expression for $e(t)$ we can evaluate the integral in (33) and then apply the initial condition to evaluate A. Doing so, we obtain

$$i(t) = \underbrace{\left(i_0 + \frac{E_0 \omega L}{R^2 + (\omega L)^2}\right) e^{-Rt/L}}_{\text{transient}} + \underbrace{\frac{E_0 R}{R^2 + (\omega L)^2}\left(\sin \omega t - \frac{\omega L}{R} \cos \omega t\right)}_{\text{steady state}}. \tag{36}$$

By "steady state" we don't necessarily mean that the dependent variable is constant; in this example it is a "steady oscillation."

Once again we have a transient response, transient in that it tends to zero as $t \to \infty$ because of the negative exponential function, plus a steady-state response, namely, that which is left after the transients have died out. Note *that steady state does not necessarily mean constant.* In Example 5 the applied voltage was constant and the steady state was, likewise, a constant or "steady" current, but in this example the applied voltage is oscillatory and the steady state is oscillatory as well. The response is plotted, for representative values of various parameters, in Fig. 14; the transient and steady-state parts are shown

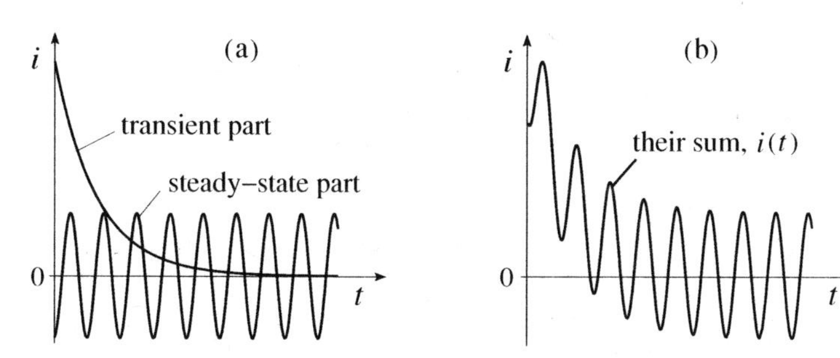

Figure 14. The response (36), using the representative values $E_0 = 1$, $R = 1$, $L = 4$, $\omega = 2$, and $i_0 = 0.3$.

in (a) and the total response $i(t)$, their sum, is shown in (b). For the choice of parameters used to generate Fig. 14, we see that the transient part of $i(t)$ has practically died out after five or six cycles. ▮

Closure. We studied representative applications of linear first-order equations, relying on solution techniques and formulas that were derived in Section 1.2, but two new ideas did arise.

First, in some problems the solution $x(t)$, say, approaches a steady state that is a constant or a steady oscillation as $t \to \infty$. In that case it is convenient to express

$$x(t) = x_{\text{tr}}(t) + x_{\text{ss}}(t) \tag{37}$$

where the transient part $x_{\text{tr}}(t)$ tends to zero as $t \to \infty$ and the steady-state part $x_{\text{ss}}(t)$ is a constant or a steady oscillation. The steady-state part is of particular interest because that is, after all, what we end up with after the transient part has died out.

Second, we introduced the idea of a one-dimensional phase line for autonomous systems, namely, systems of the form

$$\frac{dx}{dt} = f(x). \tag{38}$$

The idea is to plot the derivative x' versus x; that is, $f(x)$ versus x. Where $f(x) > 0$ the flow on the x axis phase line is rightward, where $f(x) < 0$ it is leftward, and where $f(x) = 0$ there is an equilibrium point. This qualitative flow diagram helps us to sketch the solution curves, without solving the differential equation.

EXERCISES 1.3

1. If a population governed by the exponential model has 4,500 members after five years and 6,230 after 10 years, what is its growth rate? Its initial population?

2. If a population governed by the exponential model has 500 members after two years and 460 after five years, what is its growth rate? Its initial population?

3. The world population is increasing at approximately 1.3% per year. If its growth rate remains constant, how many years will it take for its population to double? To triple?

4. If a population governed by the exponential model doubles after m days, after how many days will it have tripled?

5. A certain population is initially 1,000, grows to 1,200 after 10 years, and to 1,400 after another five years. Do you think it might be well described by the exponential model (3)? Explain.

6. ***E. Coli* Cultures.** The bacterium *Escherichia coli*, which inhabits the human intestine, multiplies by cell division. Since it is capable of rapid growth and can be grown in the laboratory it is a useful subject for experiments on population dynamics. It can be grown in culture and the population can be estimated indirectly, by measuring the turbidity of the culture through its scattering of incident light. The population $N(t)$ of a colony of *E. coli* cells can be modeled by the Malthus equation $N' = kN$. Suppose a colony of *E. coli* is grown in a culture having a growth rate $k = 0.2$ per hour. (From

$N' = kN$ we can see that the units of k are 1/time.) At the end of 5 hours the culture conditions are modified (e.g. by increasing the nutrient concentration in the medium) so that the new growth rate is $k = 0.5$ per hour. If the initial population is $N(0) = 500$, evaluate N at $t = 20$.

7. Allowing for Migration. Thus far we've used $N' = kN$ to model the population dynamics of a single species. Implicit in that equation is that the system is *closed*; that is, its borders are closed to influx or efflux of that species due, for instance, to migration. How would you modify that equation to account for a known migration rate $r(t)$ (individuals per unit time, counted as positive if it is immigration and negative if it is emigration)?

8. Radioactive Decay. We claimed that if we put $k = (\ln 2)/T$ into (11) we obtain (12). Fill in the missing steps.

9. (a) A seashell contains 90% as much C-14 as a living shell of the same size. How old is it? NOTE: The half-life of C-14 is $T = 5,570$ years.
(b) How many years did it take for its C-14 content to diminish from its initial value to 99% of that value?

10. Suppose 10 grams of some radioactive substance reduces to 8 grams in 60 years.
(a) How many more years until 2 grams are left?
(b) What is its half-life?

11. If 20% of a radioactive substance disappears in 70 days, what is its half-life?

12. Suppose an element X decays radioactively to an element Y with a half-life T_{xy}, that Y in turn decays to an element Z with a half-life T_{yz}, and that Z is not radioactive.

(a) Let $x(t), y(t), z(t)$ denote the masses of X, Y, Z, respectively, in a given sample. Write down a set of three differential equations for $x(t), y(t), z(t)$. NOTE: Recall that the rate constant k in (10) is expressible in terms of the half-life T as $k = (\ln 2)/T$.
(b) By adding the three differential equations, show that $x(t) + y(t) + z(t)$ is a constant, and explain why that result makes sense.
(c) Let the initial conditions be $x(0) = 100$ g, $y(0) = 50$ g, $z(0) = 20$ g, and let $T_{xy} = 50$ yr, and $T_{yz} = 200$ yr. Solve the three IVPs for $x(t), y(t), z(t)$.

13. A radioactive substance having a mass m_1 at time t_1 decays to a mass m_2 at time t_2. Use that information to solve for its half-life T in terms of m_1, m_2, t_1, t_2.

14. Mixing Tank. In Example 2, let $v = 500$ gal. For $c(t)$ to diminish to 98% of c_i in one hour, what flow rate Q (gal/min) is required?

15. For the mixing tank in Fig. 3, let the initial concentration in the tank be $c(0) = 0$. Beginning at time T the inflow concentration is changed from c_i to zero.

(a) Solve for $c(t)$, both for $t < T$ and for $t > T$. HINT: Break the problem into two parts: for $t < T$ solve $c' + (Q/v)c = c_iQ/v$; $c(0) = 0$, and for $t > T$ solve $c' + (Q/v)c = 0$ subject to an initial condition that $c(T)$ that is the final value (i.e., at $t = T$) from the first solution (i.e., on $t < T$).
(b) Give a labeled hand sketch of the graph of $c(t)$,

16. For the mixing tank shown in Fig. 3, let $c(0) = 0$. Beginning at time T the flow rate Q is increased to $10Q$. Solve for $c(t)$, both for $t < T$ and for $t > T$. Is $c(t)$ continuous at $t = T$? How about $c'(t)$? Explain. HINT: See the hint in Exercise 15. The idea in this problem is the same.

17. **Runoff Into Your Pond.** Your garden pond is 300 ft^2, with an average depth of 3 ft. It rains hard for one hour, during which time the pond receives runoff from your neighbor at a rate $Q = 20$ ft^3/hr, with a concentration of a weed killer, Di-Bolic, equal to 0.01 lb/ft^3. The pond volume remains constant because there is an overflow pipe. Considering the concentration of Di-Bolic in the pond to be spatially uniform (hence, a function only of the time t), calculate its value $c(t)$ at the end of the hour if $c(0) = 0$.

18. Inflow and Outflow Rates Unequal. If, for the mixing tank shown in Fig. 3, the inflow is $Q_i = 5$ gal/min and the outflow is $Q_o = 12$ gal/min, then the liquid volume v in the tank is *not constant, so (15) does not apply*. But in place of (15) you can use (14), which becomes

$$[v(t)c(t)]' = Q_i c_i - Q_o c(t), \tag{18.1}$$

because it holds even if v is a function of t.

(a) Let the liquid volume $v(t)$ be $1,000$ gal at $t = 0$, so $v(t) = 1,000 + (5 - 12)t = 1,000 - 7t$, let $c_i = 2$ lb/gal, and let $c(0) = 0$. Write down the IVP for $c(t)$ and solve for $c(t)$.
(b) What is $c(t)$ at the instant when the last bit of liquid is draining from the tank?
(c) You should have found, in (b), that $c(t)$ tends to the incoming concentration $c_i = 2$ as the last bit of liquid is draining from the tank. Is that result is a coincidence? Explain.

19. A tank initially contains 100 gal of fresh water. Brine containing 0.5 lb/gal of salt flows in at the rate of 8 gal/min and brine at concentration $c(t)$ flows out at the rate of 5 gal/min.

(a) Solve for the concentration $c(t)$ in the tank. HINT: The volume v is not a constant, so equation (15) does not apply; use (18.1) instead, in Exercise 18.
(b) How long will it take for there to be 40 lb of salt in the tank?

If a difficult algebraic equation arises, use computer software to solve it.

20. For the mixing tank shown in Fig. 3, let the initial concentration in the tank be $c(0) = c_0$. Beginning at time T the inflow is shut off, while the outflow rate Q is maintained. Solve for $c(t)$ for $t < T$ and for $t > T$. HINT: See the hint for Exercise 15. The idea here is the same. Also, for $t > T$ the liquid volume v is not a constant, so use (14) instead of (15).

21. Mixing Tanks in Series. Consider two tanks in series, as shown below, with an inflow of Q gal/min of solution containing c_i lb/gal of solute and an equal outflow rate Q. Let the liquid volume in each be v gal, and let $c_1(0) = c_2(0) = 0$.

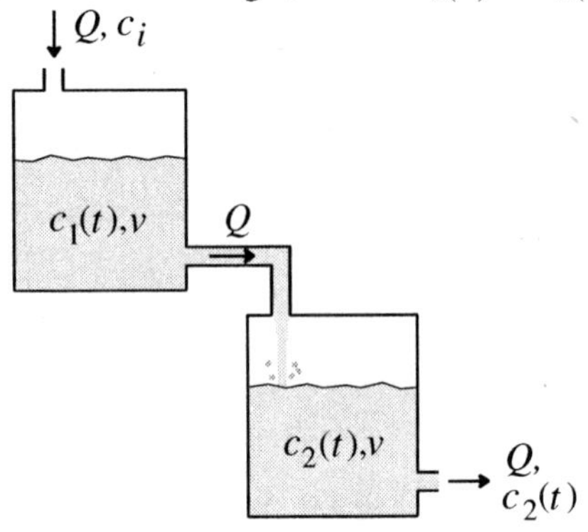

(a) Use a mass balance for each tank to derive the IVPs

$$c_1'(t) + \frac{Q}{v}c_1(t) = \frac{Q}{v}c_i; \qquad c_1(0) = 0, \tag{21.1a}$$

$$c_2'(t) + \frac{Q}{v}c_2(t) - \frac{Q}{v}c_1(t) = 0; \quad c_2(0) = 0. \tag{21.1b}$$

(b) Solve (21.1a,b) and show that

$$c_2(t) = c_i\left[1 - \left(1 + \frac{Qt}{v}\right)e^{-Qt/v}\right]. \tag{21.2}$$

NOTE: Actually, (21.1) is a *system* of coupled differential equations for c_1 and c_2, and systems are not studied until Chapter 4. However, although both c_1 and c_2 are present in (21.1b), only c_1 is present in (21.1a), so you can solve (21.1a) for c_1. Then, put that result into (21.1b) and solve the latter for c_2.

(c) From (21.2), show that $c_2(t) \to c_i$ as $t \to \infty$; that is, show that the steady-state outflow concentration equals the inflow concentration c_i. Further, show that $c_2(t) \to c_i$ and $c_1(t) \to c_i$ as $t \to \infty$, directly from (21.1). HINT: By the definition of steady state, set $c_1'(t)$ and $c_2'(t)$ to zero in the differential equations, and solve the resulting algebraic equations for the steady-state concentrations.

22. Computer. This exercise is to provide experience with some of the differential equation software.

(a) Use computer software to obtain an analytic solution of the IVP

$$100i' + 500i = 25(1+t)/(2+t); \quad i(0) = 0. \tag{22.1}$$

(You will find that the solution is messy, and involves a nonelementary function denoted as Ei and called the *exponential integral function*.) Also, use computer software to generate the graph of $i(t)$ on $0 \le t \le 25$. Finally, obtain a computer tabulation of the solution values at $t = 0, 5, 10$.

(b) Same as (a), for the IVP

$$(2+t)i' + i = -50\sin t/(4+t); \quad i(0) = 2. \tag{22.2}$$

(For the analytical solution you will again run into nonelementary functions, this time the *cosine integral* and the *sine integral* functions Ci and Si, respectively.)

23. Phase Line. Develop the phase line, as we did in Fig. 6, identify any equilibrium points, and state whether each is stable or unstable. Then, use that phase line to develop a hand sketch of the solutions corresponding to a handful of representative initial conditions, as we did in Fig. 7 (without the shaded area and large dragging arrows, of course).

(a) $x' = x^2 - x$ (b) $x' = x + x^2$ (c) $x' = x^3 - x$
(d) $x' = (x-1)^2$ (e) $x' = \sin x$
(f) an equation supplied by your instructor

24. Light Extinction; Lambert's Law. Consider window glass subjected to light rays normal to its surface, and let x be a coordinate normal to that surface with $x = 0$ at the incident face. It is found that the light intensity I in the glass is not constant, but decreases with the penetration distance x, as light is "absorbed" by the glass. According to *Lambert's law*, the fractional loss in intensity between x and $x + dx$, $-dI/I$ (with the minus sign included because dI is negative), is proportional to dx: $-dI/I = k\,dx$, where k is a positive constant. Thus, $I(x)$ satisfies the differential equation

$$\boxed{\frac{dI}{dx} = -kI.} \tag{24.1}$$

(a) If 80% of the light penetrates a 1-inch thick slab of this glass, how thin must the glass be to let 95% penetrate?
(b) If 50% of the light penetrates five inches, how far does 25% penetrate? How far does only 1% penetrate?

25. Modeling Mothballs and "Mothcylinders." (a) A spherical mothball evaporates with time. (For a mothball that is not a defect; it is *supposed to* evaporate.) Does it completely disappear in finite time or only as $t \to \infty$? NOTE: You will need to model the evaporation process in some simple and reasonable way so as to derive a differential equation for the radius $r(t)$. Solve it for $r(t)$, assuming an initial radius $r(0) = r_0$.

(b) Suppose that instead of being spherical the mothball is a circular cylinder of radius $r(t)$, with an initial radius $r(0) = r_0$ and an initial length L that is much larger than r_0. Again, model this problem so as to obtain a differential equation for $r(t)$. Solve for $r(t)$ in terms of r_0. Does this type mothball (really, "mothcylinder") evaporate in finite time? HINT: Use the fact that the initial length is much larger than the initial radius to help you to obtain an *approximate* differential equation for $r(t)$. Indicate the approximations that you adopt in obtaining your differential equation. For instance, do you need to take into account evaporation at the two ends, or only along the lateral surface?

26. Compound Interest. If a sum of money S earns interest at a rate k per unit time, compounded continuously, then in time dt we have $dS/S = kdt$, so $S(t)$ satisfies

$$\frac{dS}{dt} = kS. \tag{26.1}$$

Thus, if $S(0) = S_0$, then

$$S(t) = S_0 e^{kt}. \tag{26.2}$$

If, instead, interest is compounded yearly, then after t years

$$S(t) = S_0(1+k)^t. \tag{26.3}$$

Finally, if it is compounded n times per year, then

$$S(t) = S_0\Big(1 + \frac{k}{n}\Big)^{nt}. \tag{26.4}$$

(a) Show that if we let $n \to \infty$ in (26.4), then we do obtain the continuous compounding result (26.2). HINT: Recall, from the calculus, that $\lim_{m\to\infty}\left(1+\frac{1}{m}\right)^m = e$.

(b) Let $k = 0.05$ (i.e., 5% interest) and compare $S(t)/S_0$ after 1 year [i.e., at $t = 1$] if interest is compounded yearly, monthly, weekly, daily, and continuously.

27. Mass Sliding Down a Lubricated Plane. A block of mass m slides down a plane that is at an angle α with respect to the horizontal, under the action of gravity and friction, air resistance being negligible. Applying Newton's second law to the motion in the tangential and normal directions gives

$$mx'' = -f + mg\sin\alpha, \tag{27.1a}$$

$$0 = N - mg\cos\alpha, \tag{27.1b}$$

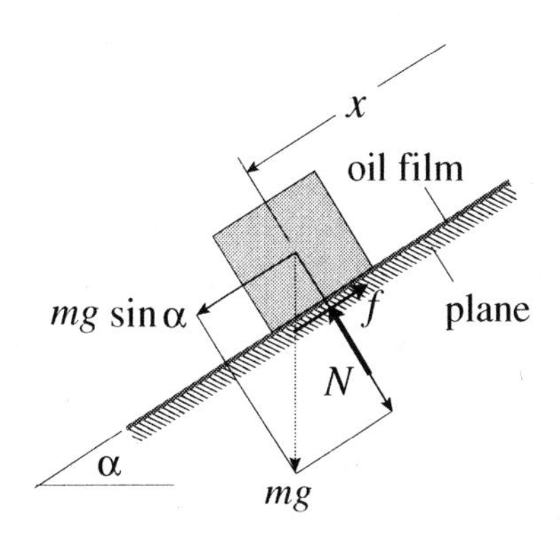

respectively. The left-hand side of (27.1b) is zero because the acceleration normal to the plane is zero. If the plane and block are lubricated then the friction force is, to a good approximation, proportional to the block's velocity $x'(t)$, so $f \approx cx'$ where c is a constant, and (27.1a) gives the second-order differential equation

$$mx'' + cx' = mg\sin\alpha \tag{27.2}$$

for the motion $x(t)$. Assume that $x(0) = 0$ and $x'(0) = 0$.

(a) It's true that (27.2) is a second-order equation whereas this chapter is about first-order equations, but you can integrate it once with respect to t to obtain a first-order equation. Do that, then solve that equation for $x(t)$ and show that

$$x(t) = \frac{m^2 g\sin\alpha}{c^2}\Big(\frac{ct}{m} + e^{-ct/m} - 1\Big). \tag{27.3}$$

(b) The constant of proportionality c in $f = cx'$ is $\mu A/h$ where A is the area of the bottom of the block, h is the normal distance between the bottom of the block and the plane (i.e., the oil film thickness), and μ is the viscosity of the oil. Suppose, in an experiment, $m = 0.5$ slugs, $A = 0.6$ ft^2, $\alpha = 30°$, $h = 0.003$ ft, and that when $t = 5$ sec we measure x to be 114 ft. Also, $g = 32.2$ ft/sec^2. Use that data in (27.3) to solve for the viscosity of the oil, μ, by computer if necessary. NOTE: That is, think of this as an experiment aimed at the determination of the viscosity of a given lubricating oil.

28. Sliding With Dry Friction. In the preceding exercise the block/plane interface was lubricated and the friction force f in (27.1a) was of the form $f = cx'$, proportional to the velocity x'. Suppose instead that the interface is dry (not lubricated). Then (27.1a,b) still hold, but in that case (if α is large enough for slipping to be initiated in the first place) the friction force f is proportional to the normal force N: $f = \mu N$ where the constant of proportionality μ is the coefficient of sliding friction. With $f = \mu N$ in (27.1a), and N given by (27.1b), solve for $x(t)$ subject to the conditions $x(0) = 0$ and $x'(0) = 0$, and show that

$$x(t) = \frac{g}{2}(\sin\alpha - \mu\cos\alpha)t^2. \tag{28.1}$$

29. The R, L Circuit in Fig. 12b. The R, L circuit in Fig. 12b was modeled by (32) and its solution was given by (33). Let $R = 2$ ohms, $L = 10$ henrys, and $i_0 = 0$ amperes.
(a) Let $e(t) = 5\sin t$ volts for $0 < t < 6\pi$ and 0 for $t > 6\pi$ seconds. Solve for $i(t)$, both for $0 < t < 6\pi$ and for $t > 6\pi$, either by hand or by computer. HINT: First, for $0 < t < 6\pi$ solve $i' + 0.2i = (0.1)(5)\sin t$ with $i(0) = 0$, and call the solution $i_1(t)$. Then, for $6\pi < t < \infty$ solve $i' + 0.2i = 0$ with initial condition $i(6\pi) = i_1(6\pi)$.
(b) Obtain a computer plot of the solution obtained in part (a). HINT: How can we plot the two parts of the solution together? Denote the solutions on $0 < t < 6\pi$ and $6\pi < t < \infty$ as $i_1(t)$ and $i_2(t)$, respectively. Then a single expression valid on $0 < t < \infty$ is

$$i(t) = i_1(t) + H(t - 6\pi)[i_2(t) - i_1(t)] \tag{29.1}$$

in which $H(t)$ is the **Heaviside function** which is defined as 0 for $t < 0$ and 1 for $t > 0$. Thus, $H(t - 6\pi)$ is 0 for $t < 6\pi$ and 1 for $t > 6\pi$, so the right-hand side of (29.1) is $i_1(t)$ for $t < 6\pi$ and $i_2(t)$ for $t > 6\pi$. In *Maple*, for instance, $H(t)$ is entered as Heaviside(t).
(c) Now let $e(t) = 5t$ volts for $0 < t < 10$ and 10 for $t > 10$. Solve for $i(t)$, both for $0 < t < 10$ and for $t > 10$, by hand or by computer.
(d) Obtain a computer plot of the solution obtained in part (c).

30. Newton's Law of Cooling. Newton's law of cooling states that a body that is hotter than its environment will cool at a rate proportional to the temperature difference between the body and its environment, so that the temperature $u(t)$ of the body is modeled by the differential equation

$$\boxed{\frac{du}{dt} = k(U - u),} \tag{30.1}$$

in which U is the temperature of the environment (assumed here to be a constant), t is the time, and k is a positive constant of proportionality. NOTE: Parts (b), (c), (d), below, are *independent problems* that apply the results of part (a) to different situations. [Note also that if $u > U$ then (30.1) does indeed model the *cooling* of the body by Newton's law of cooling. But, (30.1) holds for "Newton heating" as well, that is, if $U > u$. In that case the right-hand side of (30.1) is positive, so $u(t)$ is an increasing function of t, as the body is being heated by the environment.]

(a) Derive the general solution of (30.1),

$$u(t) = U + Ae^{-kt}. \tag{30.2}$$

(b) A cup of coffee in a room that is at 70° F is initially at 200° F. After 10 minutes it has cooled to 180° F. How long will it take to cool to 100° F? What will its temperature be three hours after it was poured?

(c) Yoshiko takes a cup of tea, initially at 200°F, outdoors at noon. By 12:06 pm it has cooled to 188° and by 12:12 pm it has cooled to 177°. By what time will it have cooled to 130°, assuming that the ambient temperature remains constant over that time period?

(d) An interesting application of (30.1) and its solution (30.2) is in connection with estimating the time of death in a homicide. For instance, suppose a body is discovered at a time T after death and its temperature is measured to be 90°F. We wish to determine T. Suppose the ambient temperature is $U = 70$°F and assume that the temperature of the body at the time of death was $u_0 = 98.6$°F. If we put this information into (30.2) we can solve for T, provided that we know k, but we don't. Proceeding indirectly, we can infer the value of k by taking one more temperature reading. Thus, suppose we wait an hour and again measure the temperature of the body, and find that $u(T+1) = 87$° F. [$u(T+1)$ is u at time $T+1$, not u *times* $T+1$.] Use this information to solve for T (in hours).

ADDITIONAL EXERCISES

31. Newton Heating and Cooling of a House. First, read the introduction to Exercise 30. Let $u(t)$ in (30.1) be the temperature inside a house that is subjected to a time-varying outside temperature $U(t) = 70 - 15\cos(\pi t/12)$ degrees Fahrenheit, where t is in hours and $t = 0$ corresponds to midnight. [If unclear about our choice of $U(t)$, sketch its graph and see that it is a reasonable choice of a daily temperature fluctuation, from a low of 55° at midnight to a high of 85° at noon.] Suppose that neither heating nor cooling are being used inside the house. Then (30.1) applies, with $U(t)$ as given above. Let the initial condition be $u(0) = 50$° although, looking ahead, this value will not affect the steady-state temperature fluctuation, which is our chief interest. Then we have the IVP

$$\begin{aligned}\frac{du}{dt} &= k\,[U(t) - u(t)] \\ &= k\left[70 - 15\cos\frac{\pi t}{12} - u(t)\right]; \quad u(0) = 50.\end{aligned} \tag{31.1}$$

To understand the physical significance of k, consider the extreme cases in which $k \to \infty$ and $k \to 0$. As $k \to \infty$ it follows from (31.1) that $U(t) - u(t) \to 0$. Since in that case $u(t) = U(t)$ we see that $k = \infty$ corresponds to there being no insulation at all in the walls of the house. At the other extreme, in which $k \to 0$, it follows from (31.1) that $du/dt = 0$ so $u(t) = \text{constant} = u(0)$, which indicates that the house is "infinitely" insulated so there is no heat exchange at all with the outside. Here, let $k = 0.05$.

(a) Solve (31.1) by computer (with $k = 0.05$). You should find that

$$u(t) = -19.45e^{-0.05t} + 70 \\ -0.5275\cos 0.2618t - 2.764\sin 0.2618t. \tag{31.2}$$

As $t \to \infty$ the exponential term tends to zero and leaves a steady oscillation which we call the *steady-state solution*,

$$u_s(t) = 70 - (0.5275\cos 0.2618t + 2.764\sin 0.2618t). \tag{31.3}$$

(b) Obtain computer plots of the outside temperature $U(t)$ and the inside temperature $u(t)$, over a long enough time for the transient part of the response to die out.

(c) **Time Lag.** Verify that (31.3) can be re-expressed in the form

$$u_s(t) = 70 - 2.814\cos[0.2618(t - 5.280)]. \tag{31.4}$$

Our purpose in converting (31.3) to the form (31.4) is that in the latter form we can more readily compare it with the outside temperature $U(t) = 70 - 15\cos 0.2618t$. Doing so, we can see two effects. First, the presence of insulation causes a reduction in the amplitude of the temperature variation, from 15° outside to the more comfortable value of around 2.8° inside. Second, we see that there is a **time lag** of 5.28 hours in the response (see the figure, below), so although the outside

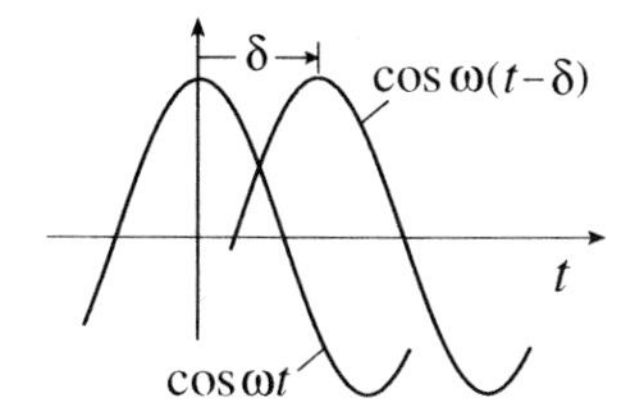

temperature peaks at noon, the temperature inside does not peak until after 5pm. Do these two results agree with your computer plots obtained in part (a)? HINT: To verify (31.4) use the trigonometric identity $\cos(A - B) = \cos A\cos B + \sin A\sin B$.

(d) Now consider the more general case in which $U(t) = U_0 + a\cos(\pi t/12)$, for any values of the average outdoor temperature U_0, the amplitude a, the constant k, and the initial temperature $u(0)$. Solve the differential equation

$$\frac{du}{dt} = k\left[U_0 + a\cos\frac{\pi t}{12} - u(t)\right] \tag{31.5}$$

and discuss the effect on the amplitude and time lag of the steady-state response in the limits as $k \to \infty$ (no insulation) and as $k \to 0$ (perfect insulation).

32. Drug Delivery in Pharmacology. Suppose we take a dose of a certain drug, either orally or intravenously. As the blood circulates, the drug will disperse and its concentration will tend to become uniform throughout the circulatory system. That will probably happen so quickly (particularly if the dose is administered intravenously), compared to the time T between doses (such as 24 hours), that we can idealize the situation and regard the concentration $C(t)$ (the mass of the drug per unit blood volume) as rising instantaneously to C_1 (which is the dosage divided by the total blood volume), and then diminishing relatively slowly as the drug passes through the the walls of the circulatory system into the muscles and organs of the body. Clinical studies show that $C(t)$ will diminish with time, approximately, according to the differential equation

$$\boxed{\frac{dC}{dt} = -kC,} \tag{32.1}$$

in which k is a positive experimentally known constant. The solution to (32.1), subject to the initial condition $C(0) = C_1$ is

$$C(t) = C_1 e^{-kt}. \tag{32.2}$$

At time T the concentration has fallen to $C_1 e^{-kT}$, so when we administer another dose the concentration jumps up "instantaneously" from that value by an additional C_1, to a new peak given by $C_2 = C_1 e^{-kT} + C_1$, as shown in the figure (in which all the vertical rises are of the same magnitude, C_1). The phar-

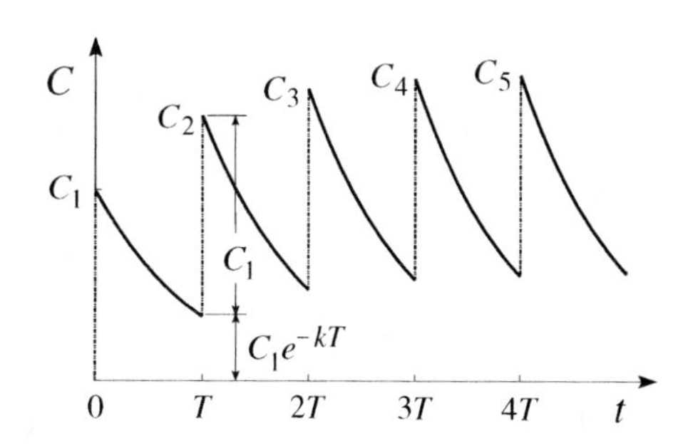

macology problem that we want to solve is to determine the correct dosage (which is the concentration C_1 times the patient's total blood volume v) and the time interval T between doses. The constraints are that the drug is helpful if its concentration is above some known value $C_{\min}$ and harmful if it exceeds some known value $C_{\max}$.

(a) Show that the successive peaks are

$$C_2 = C_1 + C_1 e^{-kT},$$
$$C_3 = C_1 + C_2 e^{-kT} = C_1\left(1 + e^{-kT} + e^{-2kT}\right),$$

and so on, so

$$C_n = C_1\left(1 + e^{-kT} + e^{-2kT} + \cdots + e^{-nkT}\right). \tag{32.3}$$

(b) We can see that $C_1, C_2, \ldots$ is an increasing sequence because C_2 is C_1 plus the positive quantity $C_1 e^{-kT}$, C_3 is C_2 plus the positive quantity $C_2 e^{-kT}$, and so on. If that increasing sequence diverges to infinity then $C_{\max}$ will be exceeded, so we are concerned with whether or not it converges and, if it does, to what value. To see if it does, recall that the geometric series $1 + x + x^2 + \cdots$ converges to $1/(1-x)$ if $|x| < 1$, and hence show that

$$\lim_{n\to\infty} C_n = C_1/\left(1 - e^{-kT}\right) \equiv C_\infty. \tag{32.4}$$

That is, as $t \to \infty$ the sequence of peaks C_n converges, and $C(t)$ tends to a steady-state oscillation, with peak values C_∞ given by (32.4).

(c) Following such a peak, the concentration diminishes to a minimum value that occurs immediately before the next dose. Show that that minimum value is $C_\infty e^{-kT}$.

(d) Now that we understand the time history of $C(t)$, we can design the drug protocol: Set $C_\infty = C_{\max}$ and $C_\infty e^{-kT} = C_{\min}$. Thus, solve for the time T between doses and the dosage D, say, in terms of the known values $C_{\max}$, $C_{\min}$, the blood volume v, and the empirical constant k.

(e) If we forget to take a pill, we're tempted to take two the next time, but the instructions tell us not to do that, but to return to taking one pill every T hours. Explain the reasoning behind those instructions.

NOTE: Observe how important the figure was in facilitating this analysis. It is difficult to imagine successfully analyzing this problem without it. More generally, be aware of the importance of supporting your work with suitable sketches.

33. Belt Friction. We know from experience that if a belt (or rope) is wrapped around a cylinder such as a tree trunk, then a large force on one end of the belt can be supported, without the belt slipping, by a relatively small force on the other end of the belt, thanks to the friction between the cylinder and the belt. For instance, consider a flexible belt hanging over a fixed horizontal circular cylinder, as shown in the figure.

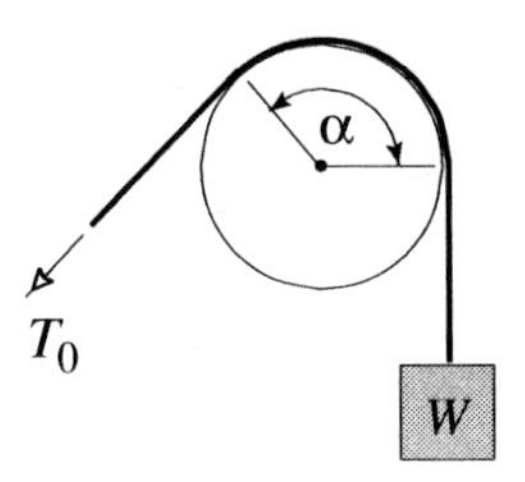

A weight force W is applied at one end and the problem is to find the tension T_0 at the other end, as a function of the wrap angle α, that will keep the belt from slipping. Consider a typical infinitesimal element of the belt and the forces that act upon it, as shown below. The force exerted on the element by the cylinder can be broken into a radial component

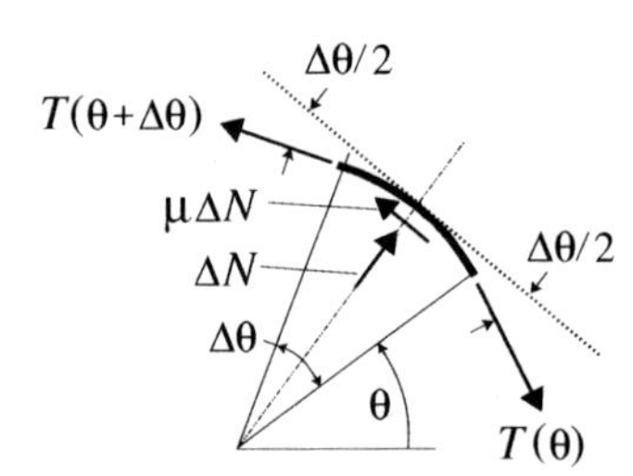

or normal force ΔN, say, and a tangential component due to the friction between the cylinder and the belt. It is known from physics that the friction force that can be sustained, without the belt slipping, is proportional to the normal force. The constant of proportionality is the coefficient of static friction μ, which is an experimentally known constant that depends upon the two materials (the belt and the cylinder). Thus, the tangential friction force is $\mu\Delta N$, as shown in the figure. Assume that the belt is sufficiently light (compared to W) for us to neglect its weight. Thus, we have not included a weight force on the belt element in the figure. For static equilibrium the net force on the element must be zero, so the net tangential force and the net radial force must each be zero:

$$\Sigma F_{\text{tang}} = T(\theta+\Delta\theta)\cos\frac{\Delta\theta}{2} - T(\theta)\cos\frac{\Delta\theta}{2} + \mu\Delta N = 0, \tag{33.1a}$$

$$\Sigma F_{\text{radial}} = \Delta N - T(\theta+\Delta\theta)\sin\frac{\Delta\theta}{2} - T(\theta)\sin\frac{\Delta\theta}{2} = 0. \tag{33.1b}$$

(a) We're not interested in the normal force distribution, so eliminate ΔN between (33.1a) and (33.1b) by algebra. In the resulting equation let $\Delta\theta \to 0$ and thus show that

$$\boxed{\frac{dT}{d\theta} = -\mu T,} \tag{33.2}$$

which is a differential equation for the tension T in the belt as a function of angular position θ.

(b) From (33.2) show that the tension T_0 in the first figure, needed to support the weight without slipping, is

$$T_0 = We^{-\mu\alpha}, \tag{33.3}$$

which is our final result.

COMMENTS: (i) Thus, the force T_0 that is needed decreases *exponentially* with the wrapping angle α. For instance, suppose that $\mu = 0.4$, corresponding to leather on metal. If $\alpha = \pi$ then $T_0 = We^{-(0.4)\pi} = 0.285W$, and if $\alpha = 5\pi$ (so the belt is wrapped around the cylinder two and one half times), then $T_0 = We^{-(0.4)5\pi} = 0.0019W$. For instance, if $W = 1000$ lbs, then the force T_0 needed to support it is only around 2 lb.

(ii) Our derivation of (33.2) is typical of the method used in engineering courses and textbooks on mechanics: Isolate a typical infinitesimal element of the system, show the forces acting on it, write down the governing physics (Newton's second law of motion in this case), and then let the infinitesimal increment ($\Delta\theta$ in this case) tend to zero.

(iii) Instead of setting the radial and tangential force components equal to zero, as we did in (33.1), we could have set the horizontal and vertical force components equal to zero, but the former was more convenient.

(iv) The foregoing derivation of (33.2) is typical of the derivation of the differential equations governing the variety of phenomena encountered in undergraduate engineering curricula: Isolate a typical arbitrarily small element; indicate the forces, fluxes, etc.; write down the governing physical principle(s); and take the limit as the spatial or temporal increment tends to zero. In this case the physical law was Newton's second law of motion which, for static equilibrium, amounts to the sum of the forces being zero.

34. Differential equations of the form $y' + py = 0$, in which p is a constant, have arisen, in this section, in modeling a wide range of applications. List as many as you can find, in both the text and the exercises. For instance, one would be (33.2) in Exercise 33, for the tension in a belt. Refer to texts on application areas, such as bioengineering, if you wish.

1.4 NONLINEAR FIRST-ORDER EQUATIONS THAT ARE SEPARABLE

Having thus far studied only the linear equation

$$\frac{dy}{dx} + p(x)y = q(x), \tag{1}$$

we now consider the general equation

$$F(x, y, y') = 0,$$

and assume that we can solve it by algebra for y' and thus express it in the **standard form**

$$\boxed{\frac{dy}{dx} = f(x, y).} \tag{2}$$

Standard form.

Of course, (2) includes the linear equation (1) as a special case, but we've already studied that case, so here we focus on the case where (2) is *nonlinear*.

For the linear equation (1), we were successful in deriving an explicit formula for its general solution, in terms of $p(x)$ and $q(x)$. If an initial condition was prescribed, then the solution thus obtained was unique, and a minimum interval of existence could be determined for it in advance, by examining $p(x)$ and $q(x)$. For the nonlinear equation (2) we are not so fortunate. It is not possible to obtain an explicit solution formula in terms of $f(x, y)$, and the issues of existence, uniqueness, and interval of existence are more subtle.

Therefore, we consider only some special cases of (2), for which solution methods are available. The most prominent is the case where (2) is **separable**, which means that $f(x, y)$ can be factored as a function of x times a function of y:

$$\boxed{\frac{dy}{dx} = X(x)Y(y).} \tag{3}$$

For instance, $y' = xe^{x+2y}$ is separable because it can be written as $y' = (xe^x)(e^{2y})$, but $y' = 3x + y$ is not, because $3x + y$ cannot be written as a function of x times a function of y.

Actually, the linear homogeneous equation $y' + p(x)y = 0$ that we studied in Section 1.2.2 was separable because it can be expressed as $y' = X(x)Y(y)$ with $X(x) = -p(x)$ and $Y(y) = y$. In that case we solved by *separation of variables*: We divided both sides by y, multiplied both sides by dx, and integrated:

Or $X(x)=p(x)$ and $Y(y)=-y$, of course.

$$\int \frac{dy}{y} = -\int p(x)\,dx. \tag{4}$$

We can use that same **separation of variables** method to solve (3), whether it is linear or not: Divide both sides by $Y(y)$ [tentatively assuming that $Y(y) \neq 0$ because division by zero is not permissible], multiply both sides by dx, and integrate:

Thanks to the separable form of (3), there are no x's in the y integral and no y's in the x integral.

$$\boxed{\int \frac{dy}{Y(y)} = \int X(x)\,dx.} \tag{5}$$

Then evaluate the integrals in (5), if we can, including the usual additive arbitrary constant of integration.

EXAMPLE 1. Solution by Separation of Variables. Solve

$$\frac{dy}{dx} = 2(x-1)e^{-y}. \tag{6}$$

First, identify (6) as separable, with $X(x) = 2(x-1)$ and $Y(y) = e^{-y}$. In this case $Y(y) \neq 0$ for all y so we can divide both sides of (6) by e^{-y} (or, equivalently, multiply by e^y), multiply by dx, and integrate. Thus,

$$\int e^y\,dy = 2\int (x-1)\,dx, \tag{7}$$

so

$$e^y = x^2 - 2x + C, \tag{8}$$

in which C is an arbitrary constant. Solving (8) for y gives

$$y(x) = \ln(x^2 - 2x + C). \tag{9}$$

The latter is plotted in Fig. 1 for the representative values $C = 1, 3, 5, 7$, together with the direction field.

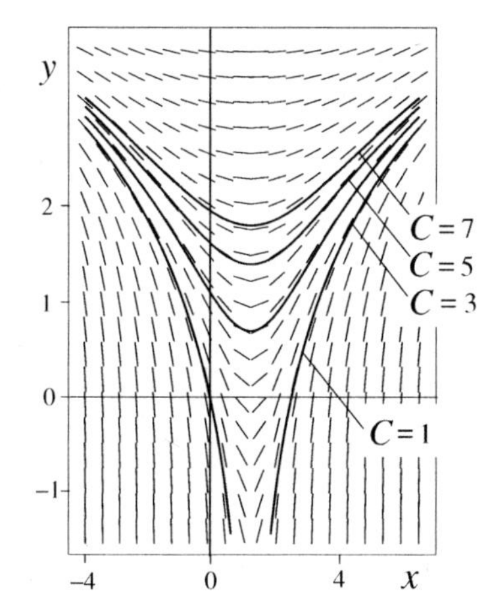

Figure 1. Representative members of the family of solutions (9), and the direction field.

Now consider applying initial conditions, say $y(4) = 5$ and $y(0) = 0$, in turn.

$\boldsymbol{y(4) = 5}$**:** Applying this condition to (9) gives $5 = \ln(16 - 8 + C)$ so $C = e^5 - 8$. Hence,

$$\begin{aligned} y(x) &= \ln(x^2 - 2x + e^5 - 8) \\ &= \ln\left[(x-1)^2 + e^5 - 9\right], \end{aligned} \tag{10}$$

the graph of which is plotted in Fig. 2.

What is its interval of existence? It's true that the logarithm function tends to $-\infty$ as its argument tends to zero (Fig. 3), but the argument $(x-1)^2 + e^5 - 9$ is positive for all x, its smallest value being $e^5 - 9$. Thus, the right-hand side of (10) is defined for all x, and the interval of existence of (10) is $-\infty < x < \infty$.

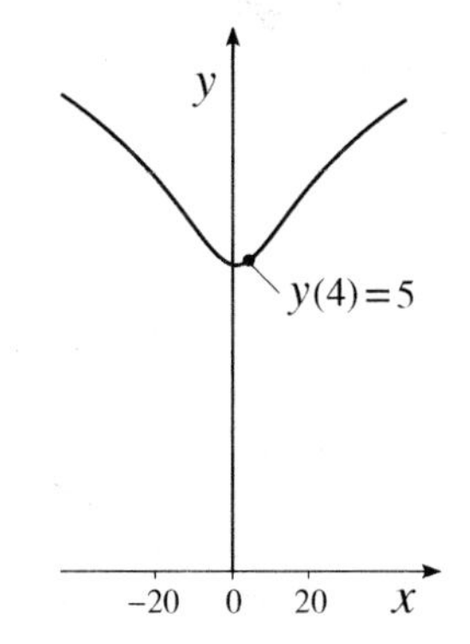

Figure 2. Particular solution (10), satisfying $y(4) = 5$, with domain of existence $-\infty < x < \infty$.

$\boldsymbol{y(0) = 0}$**:** In this case, (9) gives $y(0) = 0 = \ln C$, so $C = 1$ and

$$y(x) = \ln(x^2 - 2x + 1) = \ln(x-1)^2 = 2\ln|x-1|, \tag{11}$$

the graph of which is given in Fig. 1, for $C = 1$, and is displayed by itself in Fig. 4. This time the logarithm does "blow up," namely, at $x = 1$. That is, $2\ln|x-1| \to -\infty$ as $x \to 1$, so we can think of the graph as consisting of two branches, one to the left of $x = 1$ (labeled L in Fig. 4) and one to the right of $x = 1$ (labeled R). The solution through $(0, 0)$ can be extended arbitrarily far to the left, along L, but it cannot be extended to the right up to or beyond $x = 1$ because the solution (11) tends to $-\infty$ as $x \to 1$ from the left, and becomes undefined at $x = 1$. Thus, the solution through the initial point $(0, 0)$ consists only of the left-hand branch L, and its domain of existence is $-\infty < x < 1$. The right-hand branch R is to be discarded, as we've suggested in Fig. 4 by using a dotted line for its graph.

COMMENT 1. To solve for C we applied the initial condition $y(4) = 5$ to the solution (9), but it would have been slightly simpler to apply the initial condition to (8).

COMMENT 2. A potential error is to omit the constant C in (8) and then include it in (9), writing $y(x) = \ln(x^2 - 2x) + C$ instead of (9). That is incorrect, and the two are not equivalent; C is an integration constant so it must be inserted immediately upon doing the integrations, in (8). ∎

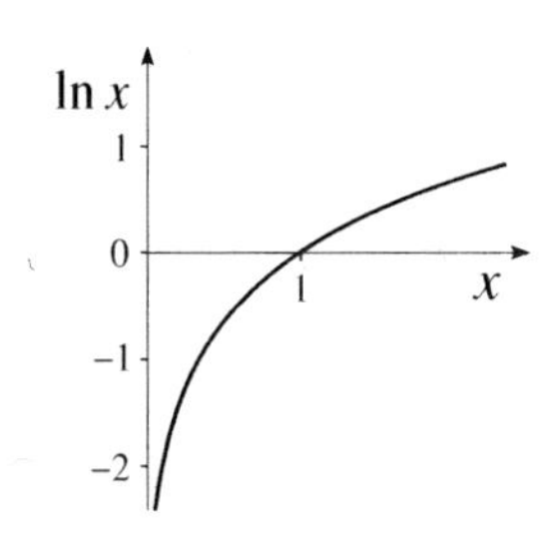

Figure 3. Recall: $\ln x$ tends to $-\infty$ as $x \to 0$ and to $+\infty$ as $x \to \infty$.

EXAMPLE 2. Solve

$$\frac{dy}{dx} = -y^2. \tag{12}$$

It might appear that the right-hand side of (12) is not of the form $X(x)Y(y)$ because we see no x's, but it is; we can take $X(x) = -1$ and $Y(y) = y^2$. Now proceed. If $y \neq 0$, we

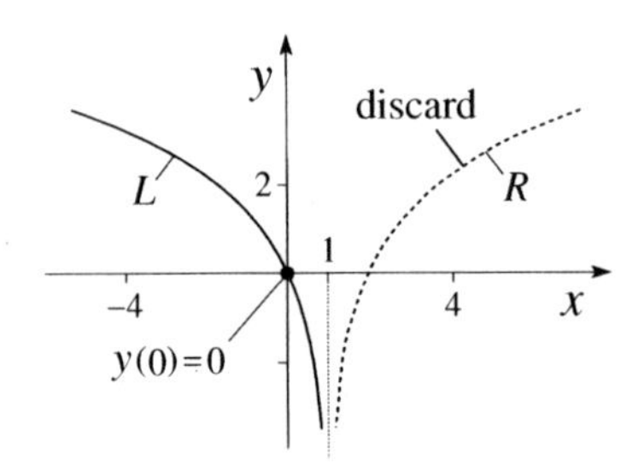

Figure 4. Particular solution (11) corresponding to the initial condition $y(0)=0$. These curves are also in Fig. 1, for $C=1$.

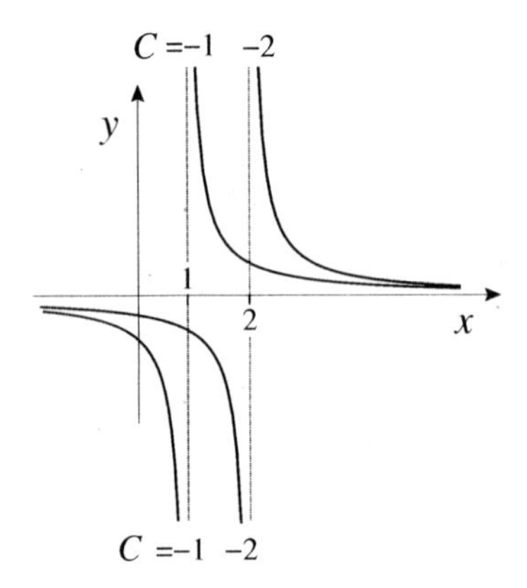

Figure 5. The graph of (13) for representative values of C; note the infinite jump discontinuities.

can divide (12) by y^2, separate the variables, and obtain

$$\int \frac{dy}{y^2} = -\int dx.$$

Evaluating the integrals and solving for y gives

$$y(x) = \frac{1}{x+C}, \tag{13}$$

with C arbitrary ($-\infty < C < \infty$). The solution (13) has an infinite jump discontinuity at $x=-C$ and is plotted in Fig. 5 for representative values of C.

Having tentatively assumed that $y \neq 0$ when we divided (12) by y^2, we must consider that case separately. In fact, we see that

$$y(x) = 0 \tag{14}$$

satisfies (12) because it reduces (12) to the identity $0=0$. The solution (14) is not contained in (13) by any finite choice of C, so it is an additional solution, in addition to (13). In summary, the solutions of (12) consist of the set of functions (13), for all values of C in $-\infty < C < \infty$, together with the additional solution $y(x)=0$.

Now consider appending representative initial conditions to (12): $y(0)=1$, $y(2)=-3$, and $y(1)=0$, in turn.

$\boldsymbol{y(0)=1}$**:** Applying this condition to (13) gives $C=1$, so $y(x)=1/(x+1)$, which is displayed in Fig. 6a. We can see that the solution "blows up" at $x=-1$; $y \to +\infty$ as $x \to -1$ from the right. Surely (12) is not satisfied by $y(x)=1/(x+1)$ *at* $x=-1$ because neither the y' on the left nor the y on the right of (12) is defined at $x=-1$. Thus, keep the right-hand branch in Fig. 6a, discard the left-hand branch, and conclude that the interval of existence of the solution $y(x)=1/(x+1)$ through the initial point $y(0)=1$ is $-1 < x < \infty$.

$\boldsymbol{y(2)=-3}$**:** Applying this condition to (13) gives $C=-7/3$, so $y(x)=1/(x-7/3)$, which is plotted in Fig. 6b. This time keep the left-hand branch, discard the right-hand branch, and conclude that the interval of existence of the solution $y(x)=1/(x-7/3)$ is $-\infty < x < 7/3$.

$\mathbf{y(1)=0}$: Applying this condition to (13) gives $y(1)=0=1/(1+C)$, but the latter cannot be solved for C. However, the additional solution $y(x)=0$, given by (14), satisfies this initial condition and the graph of that solution is shown in Fig. 6c. This solution exists on $-\infty < x < \infty$.

COMMENT. The factorization $X(x)=-1$ and $Y(y)=y^2$ is unique only to within an inconsequential scale factor. For instance, we could have taken $X(x)=1$ and $Y(y)=-y^2$, or $X(x)=378$ and $Y(y)=-y^2/378$, and so on. ∎

In summary, the separation of variables process is this: *Identify the factors* $X(x)$ *and* $Y(y)$ *(if the equation is separable), divide both sides by* $Y(y)$ *under the tentative assumption that* $Y(y) \neq 0$*, multiply both sides by* dx*, and integrate.*

The case where $Y(y) = 0$ has one or more roots for y must be treated separately. If y_0 is any root of $Y(y) = 0$, then $y(x) = y_0$ is a solution of $y' = X(x)Y(y)$ because setting $y(x) = y_0$ reduces the differential equation to the identity $d(y_0)/dx = X(x)Y(y_0)$, namely, $0 = X(x)(0)$. Thus, first assume that $Y(y) \neq 0$ and obtain the solution family obtained from (5). Then solve $Y(y) = 0$. If there are any (real) roots $y_1, \ldots, y_k$, then besides the solution family (5) there are the additional constant solutions $y(x) = y_1, \ldots, y(x) = y_k$. In Example 1 there were no such additional solutions, and in Example 2 there was one, namely, $y(x) = 0$.

There is no analog of these "additional solutions" for the *linear* differential equation $y' + p(x)y = q(x)$. In that case we were able to obtain the *general solution* with confidence that it contained *all* solutions. For nonlinear equations, however, being certain that we have the set of all solutions is a more subtle matter. *To avoid calling a solution a "general solution" without proving that it does contain all solutions, we will not use the term general solution for nonlinear differential equations.* Thus, in Example 2, for instance, we did not call (13) a general solution; we said that we found the "family of solutions" (13) plus the "additional solution" (14).

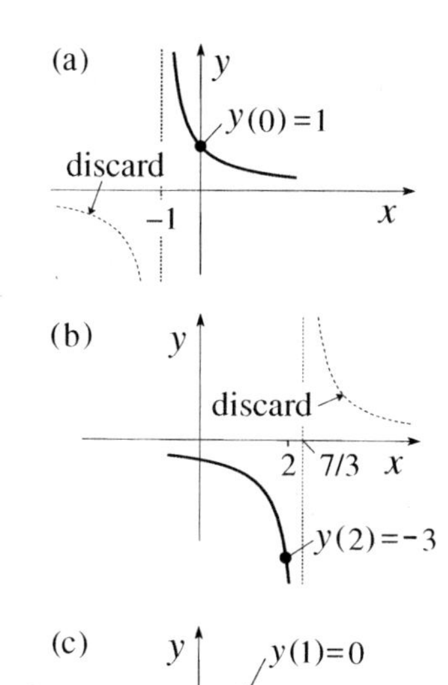

Figure 6. Particular solutions of (12) for $y(0) = 1$, $y(2) = -3$, and $y(1) = 0$.

EXAMPLE 3. Implicit Solution. Solve the IVP

$$\frac{dy}{dx} = \frac{(\sin x - 3x^2)(y-3)}{y-2}; \qquad y(0) = 5. \tag{15}$$

Separating variables and integrating gives

$$\int \frac{y-2}{y-3}\, dy = \int (\sin x - 3x^2)\, dx, \tag{16}$$

$$y-3 + \ln|y-3| = -\cos x - x^3 + C. \tag{17}$$

To integrate, let $y - 3 = z$ and get $\int (1 + \frac{1}{z})\, dz$.

Unfortunately, we cannot solve (17) for y. Nevertheless, we can apply the initial condition to (17) to evaluate C: $2 + \ln 2 = -1 + C$ gives $C = 3 + \ln 2$, so the solution of (15) is given by

$$y-3 + \ln|y-3| = -\cos x - x^3 + 3 + \ln 2. \tag{18}$$

Since we are not able to solve the latter for y, we accept it as it is, a **relation** on x and y that defines $y(x)$ only implicitly rather than explicitly. Thus, we say that the solution (18) is in **implicit form** rather than explicit form.

In spite of its implicit form, (18) can be used to obtain a computer plot of y versus x. The result is shown in Fig. 7.

COMMENT. In this example, $Y(y) = (y-3)/(y-2) = 0$ has the root $y = 3$ so, in addition to the family of solutions given implicitly by (18), $y(x) = 3$ is a solution as well. That additional solution turns out not to be relevant in that it does not satisfy the initial condition $y(0) = 5$, but if the initial condition were $y(0) = 3$ instead, then the solution would be that additional solution $y(x) = 3$. By the way, the solution shown in Fig. 7 is *not* identically 3 beyond $x \approx 2$; it is simply extremely close to 3. For instance, $y(2.5) \approx 3.0000147$. ∎

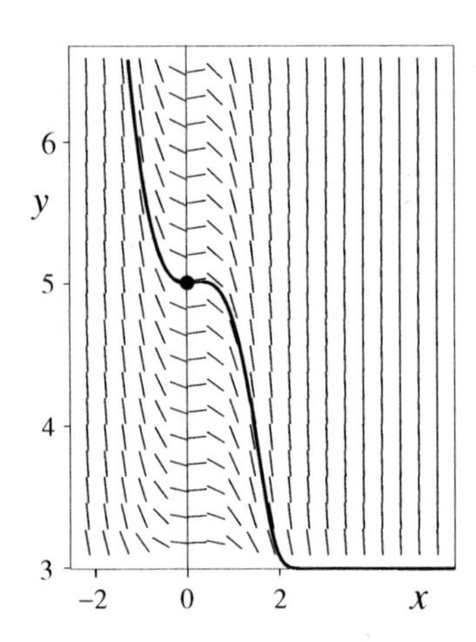

Figure 7. The solution (18) of the IVP (15). The dot shows the initial condition $y(0) = 5$.

Closure. We began our study of nonlinear equations in this section by considering the important special case in which the equation is separable, namely, of the form $y' = X(x)Y(y)$. Assuming first that $Y(y) \neq 0$, separation of variables gives

$$\int \frac{dy}{Y(y)} = \int X(x)\,dx, \tag{19}$$

but the latter *may* give $y(x)$ only in implicit rather than explicit form, as occurred in Example 3. In contrast, the general linear equation $y' + p(x)y = q(x)$ can always be solved in explicit form, and its explicit solution was given in Section 1.2.

If $Y(y) = 0$ has (real) roots $y_1, \dots, y_k$, then besides the family of solutions obtained from (19) there are additional constant solutions $y(x) = y_1, \dots, y(x) = y_k$. When we are done we must go back and recover them, and include them as "additional solutions."

If that point is unclear, it may help to consider a simple algebraic analogy, such as the equation $x^2 - 5x = 2x$ or, $x(x-5) = 2x$. If we cancel x's we obtain $x - 5 = 2$ and hence $x = 7$. However, canceling the two x's (i.e., dividing both sides of the equation by x) is permissible only if $x \neq 0$, so after obtaining $x = 7$ we must return to the original equation to check the case $x = 0$. Indeed, $x = 0$ *is* a solution, so we must augment the solution set to $x = 7, 0$.

In this section we've emphasized the separation of variables solution method for equations $y' = f(x, y)$ that are separable. In the next section we continue to consider the general case $y' = f(x, y)$, and give a fundamental existence/uniqueness theorem.

EXERCISES 1.4

1. Solution by Separation of Variables. Solve the given IVP by separation of variables. Sketch the graph of the solution(s) or, if you prefer, use computer software to obtain both the graph of the solution(s) and also the direction field; indicate initial conditions by heavy dots. Determine the interval(s) of existence of the solution(s). If more than one initial condition is given, consider each, in turn.

(a) $y' - 3x^2 e^{-y} = 0;\quad y(0) = 0$
(b) $xyy' = 2;\quad y(1) = 2$
(c) $y' + 4y^3 = 0;\quad y(-1) = -1$
(d) $y' - 4y^3 = 0;\quad y(1) = 1$
(e) $y' = y^2 + 1;\quad y(0) = 1$
(f) $y' = (\sin x)y;\quad y(1) = 0,\ y(1) = 1$
(g) $y' = (y+1)^2;\quad y(0) = -3,\ y(0) = -1,\ y(0) = 3$
(h) $y' = 4y^2\quad y(1) = -1,\ y(1) = 0,\ y(1) = 1$
(i) $y' + e^{y-x} = 0;\quad y(0) = 0$
(j) $y' = e^{y-x};\quad y(0) = 0$
(k) $2xy' = y;\quad y(3) = -1$
(l) $y' = e^{x-y};\quad y(-1) = 0$
(m) $[\tan^2(y-4) + 1]y' = 1;\quad y(3) = 4$
(n) $2(1+y)y' = 1;\quad y(3) = -2$
(o) $y = \ln y';\quad y(-2) = 0$

2. Implicit Solutions. The problems in Exercise 1 led to explicit solutions. The following lead to solutions in implicit form — although you *may* be able to solve for y and thus convert your solution to explicit form. Solve the IVP and determine its interval of existence, for each initial condition that is given. NOTE: Implicit solutions are more challenging regarding determination of the interval of existence, and plotting and examining their graphs may be particularly helpful.

(a) $(\cos y)y' + e^x = 0;\ y(0) = 0$ and $y(0) = 2\pi$.
(b) $(1 + e^y)y' = 1;\ y(0) = 0$. Give asymptotic expressions for $y(x)$ as $x \to \pm\infty$.
(c) $x(1 + y)y' = -y;\ y(1) = 2$
(d) $(1 + e^y)y' = 1 + \sin x;\ y(0) = 0$
(e) $(1 + e^y)y' = e^{x+y};\ y(2) = 0$. Show that $y(x) \sim 1.874$ as $x \to -\infty$ and that $y(x) \sim e^x$ as $x \to +\infty$.
(f) $(2 - \cos y)y' = 2 - \cos x;\ y(0) = 0$.
(g) $(3y^4 - 1)y' = 0.2;\ y(0) = 0$
(h) $(30y^4 - 1)y' = x;\ y(0) = 0$
(i) $(2y + 0.06y^5)y' + 8x = 0;\ y(0) = -1$
(j) $x(1 - y)y' = (x - 1)y;\ y(1) = 0.1$

3. By separation of variables, solve the IVP

$$y' = y(y - 2) \tag{3.1}$$

subject to the initial conditions $y(0) = -1, 0, 1, 2, 2.4$, in turn. Sketch the graphs of the solutions, or use computer software to obtain both the graphs of the solutions and also the direction field; indicate initial conditions by heavy dots. Determine the interval of existence of each solution. HINT: To integrate $\int dy/[y(y - 2)]$, subject to the condition that $y \neq 0$ and $y \neq 2$ [which cases correspond to additional solutions], use partial fractions and obtain

$$y(x) = 2/(1 - Ae^{2x}). \tag{3.2}$$

4. Consider the equation

$$xy' = y \ln y \quad (y > 0). \tag{4.1}$$

(a) Derive the solution family $y(x) = e^{Cx}$ of (4.1), in which the constant C is arbitrary.
(b) Sketch (or plot) the solutions on $-\infty < x < \infty$ for representative values of C.
(c) Show that *one and only one* of those solutions satisfies $y(a) = b$ for any $a \neq 0$ and for any $b > 0$, that *none* of those solutions satisfies $y(0) = b$ if $b \neq 1$, and that *infinitely many* solutions in the family $y(x) = e^{Cx}$ satisfy the initial condition $y(0) = 1$.

5. Find the solutions, if any, of $x(1 + y)y' + y = 0$ subject to the initial conditions $y(5) = 0.4$, and $y(5) = -0.2$, in turn. What are their intervals of existence?

6. Relative Rates of Growth. It is important to understand relative orders of magnitude. For instance, if two populations $N_1(t)$ and $N_2(t)$ are given by $N_1(t) = 100t^{50}$ and $N_2(t) = 100e^{0.0001t}$, it is evident that both tend to infinity as $t \to \infty$, but which one grows more quickly? [Of course, whether the independent variable is temporal (t), spatial (x), or whatever, doesn't matter in this discussion.] Let us compare three common types of growth (as $t \to \infty$):

logarithmic growth,	$\ln t$;
algebraic growth,	$t^\alpha \ (\alpha > 0)$
exponential growth,	$e^{\beta t} \ (\beta > 0)$

(a) Show that *algebraic growth dominates logarithmic growth*, as $t \to \infty$. Namely, show that $t^\alpha / \ln t \to \infty$ as $t \to \infty$, for any $\alpha > 0$ no matter how small. Since t^α dominates $\ln t$ for *any* α, even $\alpha = 10^{-12}$, say, logarithmic growth is extremely weak in comparison with algebraic growth.
(b) Show that *exponential growth dominates algebraic growth*, as $t \to \infty$, no matter howsmall β is and no matter how large α is. Thus, algebraic growth is extremely weak in comparison with exponential growth. For instance $N_2(t)$, given above, dominates $N_1(t)$ as $t \to \infty$; i.e., $N_2(t)/N_1(t) \to \infty$ as $t \to \infty$.

7. Relative Rates of Decay. Analogous to Exercise 6, compare these types of decay (as $t \to \infty$):

algebraic decay,	$t^{-\alpha} \ (\alpha > 0)$;
exponential decay,	$e^{-\beta t} \ (\beta > 0)$.

Show that *exponential decay dominates algebraic decay*, as $t \to \infty$, namely, that $e^{-\beta t}/t^{-\alpha} \to 0$ as $t \to \infty$, no matter how small β is and no matter how large α is.

8. Algebraic, Exponential, and Explosive Growth. We saw in Section 1.3 that the population model

$$\frac{dN}{dt} = \kappa N \qquad (\kappa > 0) \tag{8.1}$$

gives the exponential growth $N(t) = Ae^{\kappa t}$, so $N \to \infty$ as $t \to \infty$ (if $A > 0$). More generally, consider the model

$$\boxed{\frac{dN}{dt} = \kappa N^p, \qquad (\kappa > 0)} \tag{8.2}$$

in which p is a positive constant. Our purpose in this exercise is to examine how the rate of growth of $N(t)$ varies with the exponent p in (8.2).

(a) Solve (8.2) and show that if $0 < p < 1$, then the solution exhibits **algebraic growth** [i.e., $N(t) \sim at^b$ as $t \to \infty$, where a and b are positive constants that depend upon p].
(b) Show that as $p \to 0$ the exponent b tends to unity, and as $p \to 1$ the exponent b tends to infinity. (Of course, when $p = 1$ we have **exponential growth**, as mentioned above, so we can

think, crudely, of exponential growth as a limiting case of algebraic growth, in the limit as the exponent b becomes infinite. Thus, exponential growth is powerful indeed.)
(c) If p is increased beyond 1 then we expect the growth to be even more spectacular. Show that if $p > 1$ then the solution exhibits **explosive growth**, explosive in the sense that not only does $N \to \infty$, but it does so in *finite* time, namely as $t \to T$ where

$$T = \frac{1}{\kappa(p-1)N_0^{p-1}} \tag{8.3}$$

and N_0 denotes the initial value $N(0)$. Observe that not only does the growth become explosive when p is increased beyond 1, but that the time T until "blowup" decreases as p increases and tends to 0 as $p \to \infty$.

9. Exponential Decay Versus Explosive Growth. We know that $N' + N = 0$ gives exponential decay and [Exercise 8(c)] that $N' = N^2$ gives explosive growth, as $t \to \infty$. If we combine both forms and write

$$N' + N = N^2; \quad N(0) = N_0, \tag{9.1}$$

which one wins? That is, does $N(t)$ exhibit exponential decay or explosive growth as $t \to \infty$, or perhaps a different behavior altogether? [Think of N as population, so N_0 and $N(t)$ are nonnegative.]

1.5 EXISTENCE AND UNIQUENESS

1.5.1 An existence and uniqueness theorem. Recall from Theorem 1.2.1 that if $p(x)$ and $q(x)$ are continuous at a, then the *linear* equation $y' + p(x)y = q(x)$ admits a solution through an initial point $y(a) = b$ that exists at least on the broadest open x interval containing $x = a$, on which p and q are continuous, and is unique. What can be said about existence and uniqueness for the initial value problem

$$y' = f(x, y); \quad y(a) = b$$

if the latter is *nonlinear*, that is, if $f(x, y)$ is not a function of x times y plus a function of x? We have the following theorem.[1]

Theorem 1.5.1 applies whether (1) is nonlinear *or* linear, but we already have the stronger Theorem 1.2.1 for the linear case. Thus, our interest in Theorem 1.5.1 is for the nonlinear case.

THEOREM 1.5.1 *Existence and Uniqueness for Initial Value Problems*
If f and $\partial f/\partial y$ are continuous functions of x and y in an open disk D about the initial point (a, b) (Fig. 1), then the initial value problem

$$\boxed{\frac{dy}{dx} = f(x, y); \quad y(a) = b} \tag{1}$$

has a unique solution at *least* on the open x interval $x_1 < x < x_2$, where x_1 and x_2 denote the x locations of the points at which the solution curve intersects the

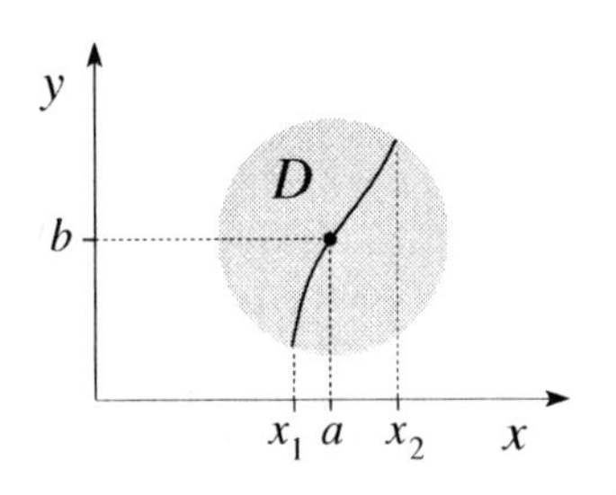

Figure 1. The disk D in Theorem 1.5.1.

[1] For further discussion of this fundamental theorem and its proof, at about the same level as this text, see Section 2.11 of W. E. Boyce and R. C. DiPrima, *Elementary Differential Equations and Boundary Value Problems*, 6th ed. (NY: John Wiley, 1997). We also recommend J. Polking, A. Boggess, and D. Arnold, "Differential Equations with Boundary Value Problems," 2nd ed. [Upper Saddle River, NJ: Pearson, 2005]. See Theorems 7.6 and 7.16 and the related discussion.

circular boundary of D.

Whereas Theorem 1.2.1 gave a minimum interval of existence and uniqueness for the linear equation [namely, the largest open x interval containing $x = a$ on which both $p(x)$ and $q(x)$ are continuous], Theorem 1.5.1 ensures existence and uniqueness for the general equation $y' = f(x, y)$ on an interval $x_1 < x < x_2$, but it does not tell us what x_1 and x_2 are. In fact, the interval $x_1 < x < x_2$ can be arbitrarily small if the solution curve through (a, b) happens to be steep. Thus, Theorem 1.5.1 is a *local* result in the sense that it tells us that under the stipulated conditions there does exist a unique solution in *some* neighborhood of the initial point $x = a$, but it does not tell us the size of that neighborhood.

Of course, Theorem 1.5.1 applies also if the differential equation in (1) happens to be linear, but there is little point in using it for linear equations because, as we've mentioned, Theorem 1.2.1 for linear equations is much more informative.

1.5.2 Illustrating the theorem.

We will illustrate Theorem 1.5.1 with two examples before discussing the significance of existence and uniqueness in a physical application in the next subsection.

EXAMPLE 1. Consider the IVP

$$4y\frac{dy}{dx} = -x; \quad y(3) = 1. \tag{2}$$

To identify $f(x, y)$, first put the differential equation into standard form by dividing both sides by $4y$.

To apply Theorem 1.5.1 to (2), observe that both $f(x, y) = -x/(4y)$ and $\partial f/\partial y = x/(4y^2)$ are continuous everywhere in the plane except on the line $y = 0$ (the x axis). Since the initial point $y(3) = 1$ is not on that line, it follows from the theorem that the IVP (2) does have a solution, a unique solution, passing through the largest disk D centered at (3,1), throughout which both f and $\partial f/\partial y$ are continuous, namely the shaded disk of unit radius shown in Fig. 2. The size of that disk is limited by the presence of the line $y = 0$ along which the theorem's continuity conditions are not met.

In fact, the differential equation in (2) is separable and gives the solution in implicit form as the ellipse $x^2 + 4y^2 = C$. The initial condition $y(3) = 1$ then gives $C = 13$, so

$$y(x) = \pm\frac{1}{2}\sqrt{13 - x^2}. \tag{3}$$

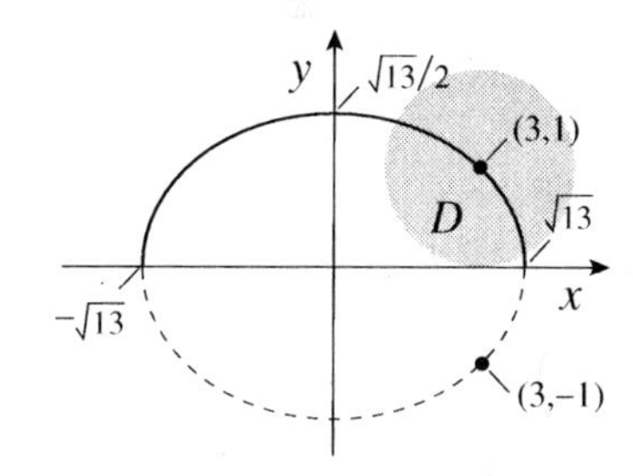

Figure 2. The solution curves through $(3, 1)$ and $(3, -1)$, shown as solid and dotted, respectively. The largest possible disk D at (3,1) is shown as shaded. Its radius is 1.

Choose the plus sign in (3) because it gives the upper half of the ellipse, which passes through the initial point (3,1), whereas the minus sign gives the lower half of the ellipse, which does not pass through the initial point. The graph of the solution is the solid curve in Fig. 2.

Theorem 1.5.1 assures us of the existence of a unique solution in some interval about $x = 3$, but it does not give a minimum size of that interval. However, in this example we did not need a prediction of the interval of existence because we were able to *solve* (2); we can now simply examine its solution, the graph of which is the upper half of the ellipse in Fig. 2. We can see from the figure that the solution exists on $-\sqrt{13} < x < \sqrt{13}$; the endpoints $x = \pm\sqrt{13}$ are not included because the slope y' is undefined (infinite) at those points.

COMMENT 1. Similarly, through $(3, -1)$ there is the unique solution

$$y = -\frac{1}{2}\sqrt{13-x^2}$$

on $-\sqrt{13} < x < \sqrt{13}$, corresponding to the lower branch of the ellipse (the dotted curve in Fig. 2).

COMMENT 2. If the initial point is *on* the x axis then the theorem does not guarantee the existence or uniqueness of a solution. It simply gives *no information* because then the continuity conditions on f and $\partial f/\partial y$ are not satisfied in any open disk D centered at the initial point. In fact, if the initial point is on the x axis then there is *no solution* through that point because the slope of the ellipse passing through that point is infinite there; that is, y' is undefined, so the differential equation in (2) cannot be satisfied there or in any x interval containing that point.[1] ∎

In Example 1, Theorem 1.5.1 assured us that there exists a unique solution through the given initial point, although it did not guarantee that that solution would exist at *least* on "such and such" an x interval. However, we were able to solve (2), and hence to determine the interval of existence simply by examining the solution. In other cases the IVP may be too difficult for us to solve, and our interest is in determining some guaranteed interval of existence, even in the absence of having the solution in hand. In the next example, we will use Theorem 1.5.1 to see what we can do about determining an interval of existence.

EXAMPLE 2. Consider the IVP

$$y' = y^2; \quad y(0) = 1. \tag{4}$$

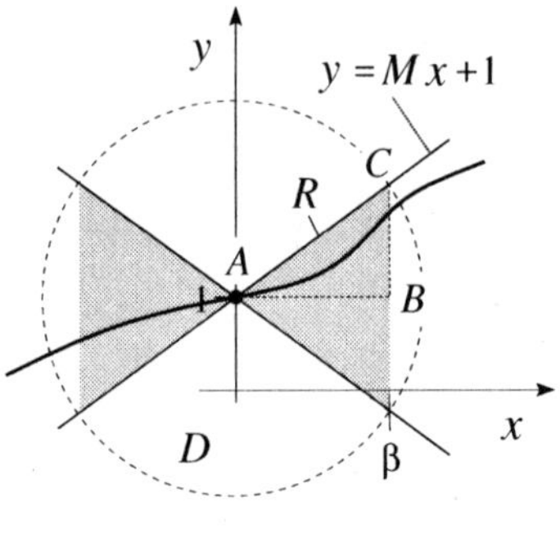

Here, $f(x, y) = y^2$ and $\partial f/\partial y = 2y$ are continuous everywhere in the x, y plane, so Theorem 1.5.1 assures us that there is a unique solution of the IVP (4) — in *some* interval about the initial point.

What we can learn (without peeking at the solution, as we did in Example 1) about the interval of existence and uniqueness of that solution? [Actually, (4) can be solved readily by separation of variables, but let us see what we can determine even in the absence of having the solution in hand to examine.]

Since the continuity conditions are satisfied throughout the plane, we can make the disk D any size we like. Begin by drawing the disk D, of radius R, about the initial point (0,1), as in Fig. 3. Everywhere in D, $|y'| = |y^2| < (R+1)^2$ because the maximum y is at the top of the disk, where $y = R + 1$. Thus, the absolute magnitude of the slope of

Figure 3. Interval of existence and uniqueness in Example 2.

[1] In Section 1.1.6 we stated that an **integral curve** is simply the graph of a solution. Actually, it can be the union of such graphs. For instance, the entire ellipse in Fig. 2 is called an integral curve of $4yy' = -x$, even though it is not the graph of a single solution curve but, rather, the union of the upper and lower solution curves.

the solution curve through (0,1) is less than $(R+1)^2$, which we will denote as M, so the solution curve must fall within the shaded "bow tie" region. After all, for the solution curve to break out of the bow tie its slope would have to exceed M at the point of break out, and that cannot happen because $|y'| < M$ everywhere in D.

Hence, the interval of existence and uniqueness is *at least* $-\beta < x < \beta$. To determine β, write the Pythagorean theorem for the right triangle ABC: $AB^2 + BC^2 = R^2$ or, $\beta^2 + (M\beta)^2 = R^2$, which can be solved for β as

$$\beta = \frac{R}{\sqrt{1+M^2}} = \frac{R}{\sqrt{1+(1+R)^4}}. \tag{5}$$

For instance, $R = 1$ gives $\beta = 0.2425$. Since we can choose R as large or small as we like, we might as well choose R so as to maximize the right hand side of (5). To do that, set $d\beta/dR = 0$ and obtain $R^4 + 2R^3 - 2R - 2 = 0$ which (using computer software) gives $R = 1.1069$; putting that into (5) then gives $\beta = 0.2031$. Thus, we have shown that the interval of existence and uniqueness is *at least* $-0.2031 < x < 0.2031$.

In fact, (4) is readily solved, its solution being

$$y(x) = \frac{1}{1-x}, \tag{6}$$

the graph of which is given in Fig. 4. Thus, the actual interval of existence and uniqueness is $-\infty < x < 1$, so the interval $-0.2031 < x < 0.2031$ is correct, but falls well short of capturing the full interval of existence. ∎

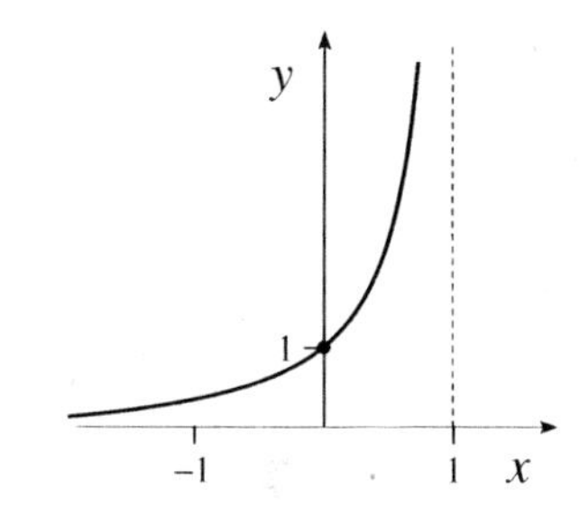

Figure 4. The exact solution (6) of the IVP (4).

1.5.3 Application to free fall; physical significance of nonuniqueness. It is important to give a physical application as well, so the impression is not left that the questions of existence and uniqueness are only of theoretical interest. Such an application can be found even in the simple problem of a body of mass m that is dropped from rest at time $t = 0$. Let the mass's downward displacement from the point of release be $x(t)$ (Fig. 5). Neglecting air resistance, Newton's second law gives $mx'' = mg$, so we have the IVP

$$x'' = g, \qquad 0 \le t < \infty, \tag{7a}$$

$$x(0) = 0, \quad x'(0) = 0. \tag{7b}$$

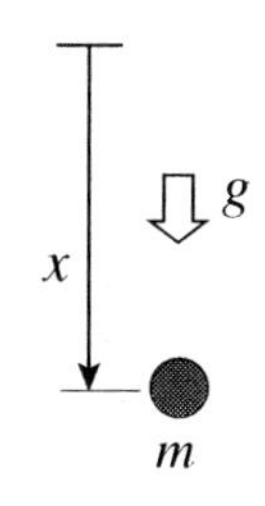

Figure 5. Free fall under the influence of gravity, neglecting air resistance.

We can integrate (7a) twice with respect to t and use the initial conditions in (7b) to evaluate the two constants of integration. Doing so gives the solution

$$x(t) = \frac{1}{2}gt^2 \tag{8}$$

that is probably familiar from a first course in physics. The graph of (8) is the parabola shown in Fig. 6.

However, it will be instructive to work not with Newton's second law but with an "energy equation." To derive an energy equation, multiply Newton's law $mx'' = mg$ not by dt but by dx:[2]

[2]Multiplying the terms in Newton's law by dx will lead to an energy equation because dx is distance, force times distance is work (mg in Newton's law is the force), and work is manifested as energy.

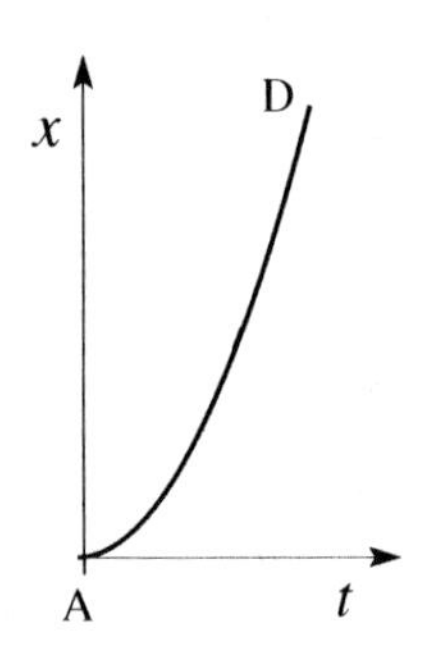

Figure 6. The solution $x(t) = gt^2/2$ of (6).

We say that (9e) is a "first integral" of (7a). In place of the second-order equation (7a) we now have the first-order equation (10a). Our starting point is now the IVP (10), not (7).

$$mx''\,dx = mg\,dx, \tag{9a}$$

$$m\frac{dx'}{dt}\,dx = mg\,dx, \tag{9b}$$

$$mdx'\,\frac{dx}{dt} = mg\,dx, \tag{9c}$$

$$\int mx'\,dx' = \int mg\,dx, \tag{9d}$$

$$\frac{1}{2}mx'^2 = mgx + A \tag{9e}$$

or $\frac{1}{2}mx'^2 + (-mgx) = A$. The latter is a statement of conservation of energy: The kinetic energy $mx'^2/2$ plus the (gravitational) potential energy $-mgx$ is a constant. Putting $t = 0$ in (9e) gives $0 = 0 + A$ so $A = 0$, and it follows from (9e) that $x' = \sqrt{2gx}$. The latter is a first-order differential equation, so append the single initial condition $x(0) = 0$. Then we have the IVP

$$\frac{dx}{dt} = \sqrt{2g}\,x^{1/2}, \qquad 0 \le t < \infty, \tag{10a}$$

$$x(0) = 0. \tag{10b}$$

Our interest here is in considering the IVP (10) in the light of Theorem 1.5.1. Solve (10a) by separation of variables. If $x \neq 0$ we can divide both sides by $x^{1/2}$, multiply by dt, integrate, and obtain

$$x(t) = \frac{1}{4}\left(\sqrt{2g}\,t + C\right)^2. \tag{11}$$

Then the initial condition $x(0) = 0$ gives $C = 0$ so

$$x(t) = \frac{1}{2}gt^2, \tag{12}$$

which is the same as (8). However, recall from Section 1.4 that a separable equation $y' = X(x)Y(y)$ can have solutions $y(x) = constant$ coming from any roots of $Y(y) = 0$, in addition to the family of solutions obtained by separation of variables. In the present case (where the variables are t, x instead of x, y) the root $x = 0$ of $x^{1/2} = 0$ gives the solution $x = 0$ [i.e., $x(t) = 0$] of (10a), and that solution is additional since it is not contained in (11) by any choice of the constant C. That solution also satisfies the initial condition (10b), so besides the solution (12) of (10) (which corresponds to AD in Fig. 7) we also have the solution $x(t) = 0$ (which corresponds to the positive t axis in Fig. 7). Thus, the solution of (10) is nonunique.

With that result in mind, turn to Theorem 1.5.1 to see what it can tell us about the existence and uniqueness of solutions of the IVP (10). Realizing that x and y in the theorem correspond here to t and x, respectively, observe that $\partial f/\partial x = \partial(\sqrt{2g}\,x^{1/2})/\partial x = \sqrt{2g}/(2\sqrt{x}\,)$ is not continuous in any neighborhood of the initial point $(0, 0)$ in the t, x plane because it "blows up" to infinity at $x = 0$, that

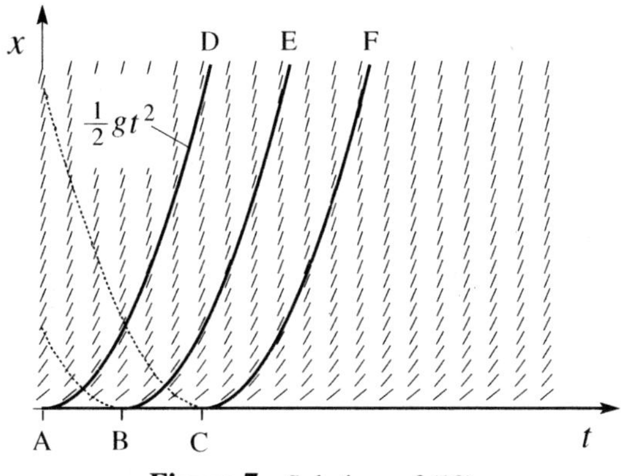

Figure 7. Solutions of (10).

is, all along the t axis. Thus, the conditions of the theorem are not met, so the theorem simply gives no information for the IVP (10).[1]

COMMENT 1. The additional solution $x(t) = 0$ of (10) amounts to the mass levitating: it does not fall! While the latter is a legitimate solution of (10) and does not violate the conservation of energy expressed by (10a), it *does* violate Newton's second law (7a) because if we put $x(t) = 0$ into (7a) we obtain the contradiction $0 = g$. Thus, the levitation solution can be discarded, finally, as nonphysical.

This is a general situation: *energy formulations may lead to solutions that are unacceptable.* Still, there is a nagging question: If we derived (10a) from Newton's law, in equations (9a) through (9e), then how did this nonphysical solution get its foot in the door? It entered in (9a), for if $x(t) = \text{constant}$ then dx is zero, so when we multiplied both sides of $mx'' = mg$ by dx, the resulting equation $mx''dx = mgdx$ does not imply that $mx'' = g$, because the dx's are zero.

COMMENT 2. Actually, (10) admits not only the two solutions $x(t) = gt^2/2$ (the curve AD in Fig. 7) and $x(t) = 0$, but an infinite number of other solutions as well. For instance, the segment AB of the t axis (in Fig. 7) followed by the half-parabola BE is also a solution curve, as is AC followed by CF, and so on. That is, the energy equation (10a) and initial condition (10b) are both satisfied if the mass levitates for a while, and *then* falls.

Closure. Theorem 1.5.1 gives sufficient conditions for the existence of a unique solution to the IVP (1). It is less informative than the corresponding Theorem 1.2.1 for the linear case: Theorem 1.2.1 gave a formula for the solution and a minimium x interval on which that solution exists and is unique. Theorem 1.5.1 assures existence but does not give the solution, and whereas it assures existence and uniqueness on "some" x interval, it does not indicate how broad that interval will be.

[1]The lack of satisfaction of the conditions of the theorem does *not* imply that (10) does not have a unique solution, because the theorem says "if," not "if and only if." That is, its conditions are sufficient, not necessary and sufficient.

EXERCISES 1.5

1. Application of the Theorems. First, solve the given IVP. If there is no solution state that. Is the solution unique? If possible, give the interval of existence of each solution. Then, show that your findings are consistent with the existence and uniqueness theorem; if the equation is nonlinear use Theorem 1.5.1, and if it is linear use the more informative Theorem 1.2.1. HINT: In difficult cases a computer plot of the direction field may help.

(a) $y' = 2xy;\quad y(0) = 2$ (b) $y' = x + y;\quad y(0) = 3$
(c) $yy' = x;\quad y(0) = 2$ (d) $yy' = x;\quad y(0) = 0$
(e) $yy' = x;\quad y(0) = -2$ (f) $xy' + y = 0;\quad y(1) = 1$
(g) $xy' + y = 0;\quad y(0) = 5$ (h) $xy' + y = 0;\quad y(2) = 0$

(i) $xy' + 2y = 0;\quad y(-1) = -4$
(j) $xy' - 2y = 0;\quad y(0) = 0$
(k) $2yy' = 1;\quad y(3) = -1$
(l) $x^2y' + y^2 = 0;\quad y(-2) = -1$
(m) $y' = \tan y;\quad y(0) = -3$
(n) $y' = 6y^{1/3};\quad y(0) = 0\ (x \geq 0)$
(o) $y' = 6y^{1/3};\quad y(0) = 1$

2. Estimating a Minimum Interval of Existence and Uniqueness. For the nonlinear IVP (4) we used the "bow tie" idea to estimate the interval of existence and uniqueness; we showed that it is *at least* $-0.2031 < x < 0.2031$. For the problems in this exercise, follow that same idea to obtain a formula for $\beta(R)$ analogous to that in (5).

(a) For the IVP $y' = 1 + y^2$ with $y(0) = 0$, obtain $\beta = R/\sqrt{R^4 + 2R^2 + 2}$ and show that the maximum β is 0.45509, so that existence and uniqueness is assured *at least* on $-0.45509 < x < 0.45509$. Further, solve the IVP for $y(x)$ and show that the actual interval of existence and uniqueness is $-\pi/2 < x < \pi/2$.
(b) For the IVP $yy' = x$ with $y(0) = -2$, obtain $\beta = R(2-R)/\sqrt{2R^2 - 2R + 4}$. You need not maximize the latter, but show that it is at least 0.5, so that existence and uniqueness is assured *at least* on $-0.5 < x < 0.5$. Further, solve the IVP for $y(x)$ and show that the actual interval of existence and uniqueness is $-\infty < x < \infty$.
(c) For the IVP $yy' = x$ with $y(2) = 3$, obtain $\beta = R(3-R)/\sqrt{2R^2 - 2R + 13}$. You need not maximize the latter, but show that it is at least 0.5, so that existence and uniqueness is assured *at least* on $1.5 < x < 2.5$. Further, solve the IVP for $y(x)$ and show that the actual interval of existence and uniqueness is $-\infty < x < \infty$.
(d) Given the IVP $y' = y^2/x^2$ with $y(1) = 0.5$, we simply want to be assured that a unique solution exists on $|x - 1| < 0.1$. Show that that is the case. Further, solve the IVP for $y(x)$ and determine its actual interval of existence and uniqueness.

3. Envelopes. In this exercise we introduce the geometric concept of the "envelope" of a one-parameter family of curves in a plane, and in subsequent exercises we will show what envelopes have to do with first-order nonlinear differential equations. Consider a one-parameter family of curves

$$g(x, y, c) = 0, \tag{3.1}$$

in which c is the parameter. For instance, $x^2 + y^2 - c^2 = 0$ is the family of concentric circles centered at the origin, each one corresponding to a different value of the parameter c. Such a family of curves may, but need not, have an envelope, such as the curve Γ in the left-hand figure. (A curve Γ is an **envelope**

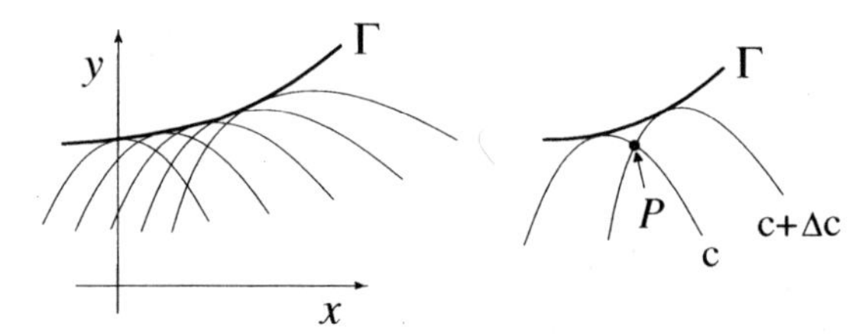

of a family of curves if every member of the family is tangent to Γ and if Γ is tangent, at each of its points, to some member of the family.) If we are given $g(x, y, c)$, how can we *find* any such envelopes? The coordinates x, y of point P (in the right-hand figure) must satisfy both $g(x, y, c) = 0$ and $g(x, y, c + \Delta c) = 0$ or, equivalently,

$$g(x, y, c) = 0 \tag{3.2}$$

and

$$\frac{g(x, y, c + \Delta c) - g(x, y, c)}{\Delta c} = 0. \tag{3.3}$$

Equation (3.3) is valid for Δc arbitrarily small, so it must hold in the limit as $\Delta c \to 0$, in which limit P approaches Γ. Thus (3.2) and (3.3) become

$$\boxed{g(x, y, c) = 0, \qquad \frac{\partial g}{\partial c}(x, y, c) = 0.} \tag{3.4a,b}$$

Eliminating c between (3.4a) and (3.4b) gives the desired equation of Γ, if the family does indeed have an envelope. To illustrate, consider the family of circles $(x-c)^2+y^2=9$. Equations (3.4a) and (3.4b) give $g=(x-c)^2+y^2-9=0$ and $\partial g/\partial c=-2(x-c)=0$, and eliminating c between these gives the two straight line envelopes $y=+3$ and $y=-3$ which, from a sketch of the family of circles, is seen to be correct. On the other hand, consider the family of parallel lines $y=x+c$, which has no envelope. (To see that, sketch the lines for several different c's.) In this case (3.4a) and (3.4b) give $g=y-x-c=0$ and $\partial g/\partial c=-1=0$. These cannot be satisfied (because $-1=0$ cannot be satisifed), so the family $y=x+c$ has no envelopes.

The problem: In each case use (3.4a) and (3.4b) to determine all envelopes, if any, of the given family of curves, and illustrate with a labeled sketch (or computer plot).

(a) $y=cx+1/c$ (b) $(x-c)^2+y^2=c^2/2$
(c) $y=(x-c)^3$ (d) $y=(x-c)^2+x$

4. Envelope Solutions of Differential Equations. Let the differential equation

$$y'=f(x,y) \tag{4.1}$$

have a one-parameter family of solutions

$$g(x,y,c)=0, \tag{4.2}$$

where c is a constant of integration, and suppose that the family of solution curves (4.2) has an envelope Γ (as in the figure in Exercise 3). At each point on Γ the values of x,y and the slope y' are such that (4.1) is satisfied, so Γ *itself is a solution curve*. That solution is not contained in (4.2) because Γ is not itself a member of the family (4.2).

The problem: Show that the levitation solution $x(t)=0$ in Section 1.5.3 is such an envelope solution.

NOTE: The concept of the envelope of a family of curves is of interest not only in connection with solutions of differential equations but also in optics and acoustics. The next exercise illustrates its relevance in optics.

ADDITIONAL EXERCISES

5. Application of Envelopes to Caustics in a Coffee Cup. In this exercise we apply the geometric idea of envelopes (Exercise 3) — not to differential equations but, for fun, to the reflected light pattern in a coffee cup. In the morning, when the sun is low, the sunlight striking the inside lip of a coffee cup is reflected by the inside of the lip. Continuing its slightly downward trajectory it strikes the coffee surface and is reflected to our eye, revealing a bright geometric pattern called a **caustic**. The latter is the envelope of the light rays reflected off the lip, such as those labeled 1, 2, and 3 in the following figure:

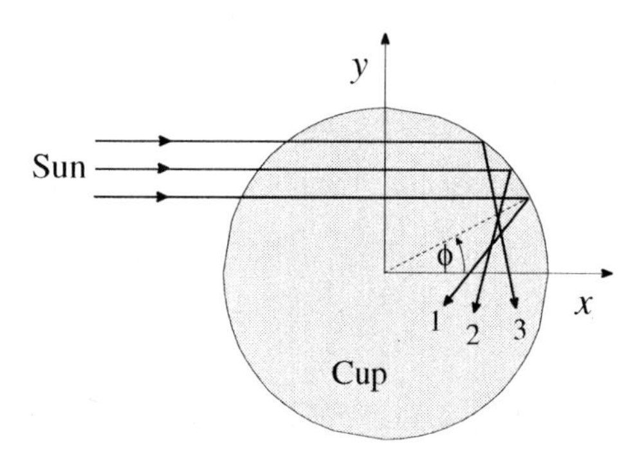

(a) Let the inner radius of the cup be 1 inch and assume that the angle of reflection off the lip equals the angle of incidence. Show that the equation of the typical reflected ray is

$$y=\big(\tan 2\phi\big)x+(\sin\phi-\tan 2\phi\cos\phi). \tag{5.1}$$

(b) Using computer graphics, plot enough of the lines defined by (5.1) to reveal the shape of the caustic, as we have in the second figure. It suffices to plot them on $0\le y\le 1$

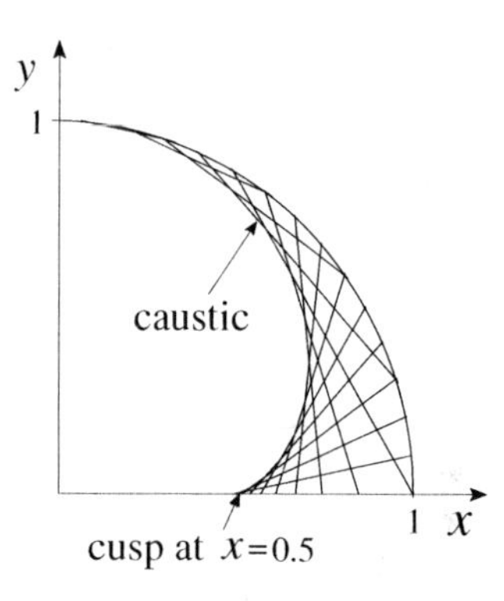

rather than on $-1\le y\le 1$ because surely the caustic will be symmetric about the x axis.

(c) Show that the caustic has a cusp at $x=0.5$, that is, that the slope of the caustic tends to zero as $y\to 0$.

NOTE: The envelope of light rays amounts to their mutual reinforcement to the extent that the caustic becomes visible.

6. Iterative Solution; Picard's Method. Suppose we try to solve the IVP

$$y'=f(x,y);\quad y(a)=b \tag{6.1}$$

by integrating the differential equation from the initial point a to an arbitrary point x, obtaining

$$y(x)-y(a)=\int_a^x f(s,y(s))\,ds$$

in which s is a dummy integration variable, or, imposing the initial condition $y(a) = b$,

$$y(x) = b + \int_a^x f(s, y(s))\,ds. \tag{6.2}$$

Unfortunately, the latter is not the solution of (6.1) because the unknown y appears inside the integral. Rather, it is an **integral equation** for y because the unknown y appears under the integral sign. Thus, all we've accomplished is the conversion of the differential equation IVP (6.1) to an integral equation. (Integral equations will not be studied in this text except insofar as they occur here and in the chapter on the Laplace transform.) Consider the solution of (6.2) by "iteration." That is, since the complication in (6.2) is the presence of y inside the integral, let us approximate that y (inside the integral) and integrate. Let $y_0(x) = b$ be a first approximation of the desired solution $y(x)$; the latter probably doesn't satisfy the differential equation in (6.1) but at least it satisfies the initial condition $y(a) = b$. Because of this approximation on the right, the $y(x)$ on the left will likewise not (in general) satisfy (6.1) exactly, but only approximately. Hopefully, it will be an improvement over the initial approximation $y_0(x) = b$. If we denote that new approximation as $y_1(x)$ then

$$y_1(x) = b + \int_a^x f(s, y_0(s))\,ds. \tag{6.3}$$

We can repeat the process and use the function $y_1(x)$ computed from (6.3) as, hopefully, a better approximation of the y inside the integrand, obtaining

$$y_2(x) = b + \int_a^x f(s, y_1(s))\,ds,$$

and so on. That is, beginning with $y_0(x) = b$ we can use the iterative formula

$$\boxed{y_{n+1}(x) = b + \int_a^x f(s, y_n(s))\,ds} \tag{6.4}$$

with $n = 0, 1, 2, \ldots$ in turn, to generate a sequence of iterates $y_0(x), y_1(x), y_2(x)$, and so on. Hopefully, if (6.1) has a unique solution then the $y_n(x)$ sequence thus generated will converge to that solution. This iterative method is due to the French mathematician *Emile Picard* (1856–1941) and is known as **Picard's method**. If we assume that $f(x, y)$ satisfies the conditions of Theorem 1.5.1 then it can be shown that that sequence does converge to the exact solution $y(x)$,

$$\lim_{n\to\infty} y_n(x) = y(x), \tag{6.5}$$

on some open interval containing the initial point a; in fact, Picard iteration is a traditional method of proof of the existence part of Theorem 1.5.1. See, for instance, Section 2.11 of W. Boyce and R. DiPrima, *Elementary Differential Equations and Boundary Value Problems*, 6th ed. (NY: John Wiley, 1997). Our purpose here is not to attempt that proof but only to explore the idea of iterative solution and to provide guidance through some exploratory examples.

The problem: For the example

$$y' = -y; \quad y(0) = 1, \tag{6.6}$$

beginning with $y_0(x) = y(0) = 1$, use (6.4) to generate the first several iterates:

$$y_1(x) = 1 - x, \tag{6.7a}$$
$$y_2(x) = 1 - x + \tfrac{1}{2}x^2, \tag{6.7b}$$
$$y_3(x) = 1 - x + \tfrac{1}{2}x^2 - \tfrac{1}{6}x^3, \tag{6.7c}$$
$$y_4(x) = 1 - x + \tfrac{1}{2}x^2 - \tfrac{1}{6}x^3 + \tfrac{1}{24}x^4. \tag{6.7d}$$

NOTE: Since the Taylor expansion of the exact solution is

$$y(x) = e^{-x} = 1 - x + \frac{1}{2!}x^2 - \frac{1}{3!}x^3 + \frac{1}{4!}x^4 - \cdots \quad (-\infty < x < \infty) \tag{6.8}$$

it appears that the Picard sequence is indeed converging to the exact solution.

7. Another Example of Picard's Method. Consider the IVP

$$y' = 2xy^2; \quad y(0) = 1. \tag{7.1}$$

(a) Derive, by separation of variables, the solution $y(x) = 1/(1 - x^2)$, which exists on $-1 < x < 1$.
(b) Now use the Picard method given above, beginning with $y_0(x) = 1$, to generate $y_n(x)$ for $n = 1, 2$, and 3. You can do this by hand, or using computer software.
(c) Plot those iterates, together with the exact solution, on the interval of existence $-1 < x < 1$.

8. One More. Consider the IVP

$$y' = e^x y; \quad y(0) = 1. \tag{8.1}$$

(a) Derive the solution $y(x) = \exp(e^x - 1)$, which exists on $-\infty < x < \infty$.

(b) Use the Picard method given above, beginning with $y_0(x) = 1$, to generate $y_n(x)$ for $n = 1, 2$, and 3.

(c) Plot those iterates, together with the exact solution, on $-3 < x < 3$, for instance.

1.6 APPLICATIONS OF NONLINEAR FIRST-ORDER EQUATIONS

In this section we consider the logistic model of population dynamics as a representative application of nonlinear first-order differential equations, and we give a variety of other applications in the exercises. Use of the phase line, from Section 1.3, will continue to be prominent, and we will introduce one new idea: linearized stability analysis.

1.6.1 The logistic model of population dynamics. In Section 1.3.1 we studied the simple exponential population model

$$\frac{dN}{dt} = \kappa N; \qquad N(0) = N_0, \tag{1}$$

with solution

$$N(t) = N_0 e^{\kappa t}. \tag{2}$$

We noted that the exponential model is not necessarily realistic for long time intervals if the net birth/death rate κ is positive, because in that case (2) indicates unbounded growth. As a more realistic model we suggested the logistic equation

$$\boxed{\frac{dN}{dt} = (a - bN)N,} \tag{3}$$

The well-known **logistic equation**.

which we wrote down in Section 1.3.1 but did not solve. In (3), a and b are known positive constants and $N(t)$ is the population, such as the number of bass in a lake. [Alternatively, we could take $N(t)$ to be the total mass, the *biomass*, of bass in the lake, or some other measure of the population.]

We can solve (3) by separation of variables:

$$\int \frac{dN}{(a-bN)N} = \int dt \qquad (\text{if } N \neq 0 \text{ and } N \neq a/b),$$

$$\int \left(-\frac{1}{a}\frac{1}{N-a/b} + \frac{1}{a}\frac{1}{N} \right) dN = t + C \qquad (-\infty < C < \infty),$$

We've expanded the $1/[(a-bN)N]$ in partial fractions.

$$-\ln|N - a/b| + \ln|N| = at + aC,$$

$$\left| \frac{N}{N-a/b} \right| = e^{at+aC} = e^{aC}e^{at},$$

$$\frac{N}{N-a/b} = \pm e^{aC}e^{at} \equiv Ae^{at} \tag{4}$$

or, solving (4) for N,

$$N(t) = \frac{aAe^{at}}{1+bAe^{at}}. \tag{5}$$

Since $-\infty < C < \infty$, the constant A can be any value other than zero (because $A = \pm e^{aC}$, and e^{aC} is not zero for any finite value of C).

Besides the solution family (5), $(a-bN)N = 0$ gives the additional solutions $N(t) = 0$ and $N(t) = a/b$. The former can be included in (5) if we allow the arbitrary-but-nonzero constant A to be zero, because setting $A = 0$ in (5) gives $N(t) = 0$. But $N(t) = a/b$ cannot be obtained from (5) by any finite choice of A so it is an additional solution of (3), in addition to (5).

If we apply an initial condition $N(0) = N_0$ to (5), we can solve for A and obtain $A = N_0/(a-bN_0)$. Then, after some algebra, (5) becomes

To solve for A it is simpler to apply $N(0) = N_0$ to (4) than to (5); try it both ways and see.

$$N(t) = \frac{a}{b}\frac{N_0}{N_0 + (\frac{a}{b} - N_0)e^{-at}}. \tag{6}$$

At this point we could use (6) to plot N versus t, for representative values of a, b, and N_0. Instead, put the solution (6) aside and return to the differential equation (3), to see what we can learn using a more qualitative appfoach. We see that (3) is autonomous, of the form

Autonomous differential equations and the phase line were discussed in Section 1.3.4.

$$\frac{dN}{dt} = (a-bN)N \equiv f(N), \tag{7}$$

so consider the phase line. Accordingly, we've plotted $N' = f(N) = (a-bN)N$ versus N in Fig. 1, from which we find equilibrium points at $N = 0$ and at $N = a/b$. (Of *course* $N = 0$ is an equilibrium point, because if we begin with no fish, we will never have any fish.)

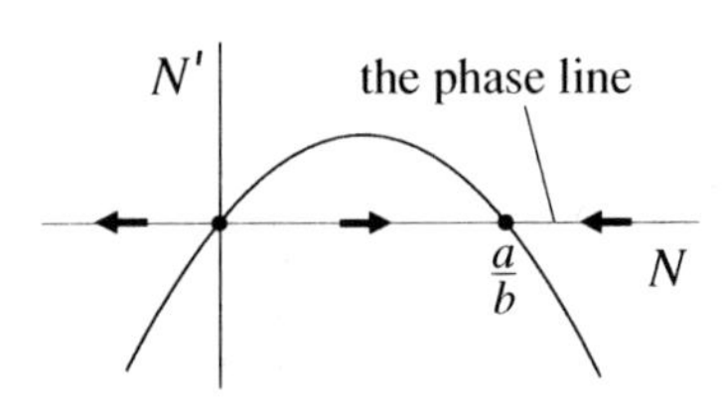

Figure 1. $N' = (a-bN)N$ and the resulting N-axis phase line. The two heavy dots denote equilibrium points.

If we take the phase line in Fig. 1 and place it vertically at the left of the N axis in a Cartesian t, N plane, as in Fig. 2, we can infer the qualititative shape of the solution curves in the t, N plane directly from the phase line flow. For instance, the dot at $N = a/b$ on the phase line indicates an equilibrium point there, so the solution curve springing from $N(0) = a/b$ in the t, N plane is simply a horizontal line. The downward flow on the phase line above a/b and the upward flow below a/b imply that the solution curves in the t, N plane approach the equilibrium solution $N(t) = a/b$ from above and below, respectively. And the dot at $N = 0$ gives the constant equilibrium solution $N(t) = 0$.

To sketch the solution curves in Fig. 2 it would help to find the inflection points, if any. Like the phase line, that information can be obtained directly from the differential equation (3). Inflection points are points at which N'' vanishes and changes sign, so differentiate (3) and set $N'' = 0$:

Following the second equality, replace each N' by $(a-bN)N$ and simplify.

$$\begin{aligned} N'' &= \frac{d}{dt}[(a-bN)N] \\ &= -bN'N + (a-bN)N' \\ &= (a-2bN)(a-bN)N = 0, \end{aligned} \tag{8}$$

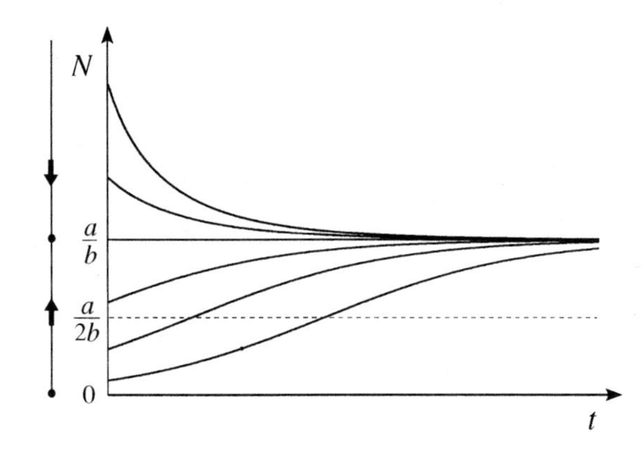

Figure 2. Using the phase line at the left to sketch solution curves for (3). Solution curves below the t axis are omitted because $N \geq 0$..

which gives the three roots $N = a/2b$, a/b, and 0. Of these, we can discard the last two, which are simply the horizontal-line equilibrium solutions $N(t) = a/b$ and $N(t) = 0$. Consider the first root, $a/2b$. The $a - 2bN$ factor in (8) changes sign at $a/2b$ but the $(a - bN)N$ part does not (since it is positive for $0 < N < a/b$). Thus, N'' both vanishes and changes sign at $a/2b$, so $N = a/2b$ is an inflection point. That is, all along the horizontal line $N = a/2b$ (dotted in Fig. 2) the solution curves have inflection points. That information enables us to complete our sketch of the solution curves in Fig. 2.[1]

Merely from the phase line, at the left in Fig. 2, we can see that the equilibrium points $N = 0$ and $N = a/b$ are unstable and stable, respectively, as can also be seen from the solution curves in the t, N plane. $N = a/b$ is an important quantity, the **environmental carrying capacity**, the population that can be supported by the environment.

The logistic equation (3) is often written, instead, as $N' = r(1 - \frac{N}{K})N$, in which r is called the intrinsic growth rate and K is the environmental carrying capacity.

1.6.2 Stability of equilibrium points and linearized stability analysis.

Let us review and extend the definitions of equilibrium points and stability given in Section 1.3.4. Recall that N_{eq} is an **equilibrium point** of $N'(t) = f(N)$ if $f(N_{\text{eq}}) = 0$. The equilibrium point is **stable** if $N(t)$ can be kept arbitrarily close to N_{eq} for all $t \geq 0$ by taking it to be sufficiently close initially (at $t = 0$);[2] otherwise it is **unstable**. For the logistic model (3), we can see from Fig. 2 that the equilibrium points $N_{\text{eq}} = 0$ and $N_{\text{eq}} = a/b$ are unstable and stable, respectively.[3]

[1] We didn't really sketch Fig. 2 by hand; we plotted computer generated solutions, but we *could* have sketched it from the information that we've discussed.

[2] That is, corresponding to each number $\epsilon > 0$ (i.e., no matter how small) there exists a number $\delta > 0$ such that $|N(t) - N_{\text{eq}}| < \epsilon$ for all $t \geq 0$ if $|N(0) - N_{\text{eq}}| < \delta$.

[3] The former is unstable because we *cannot* keep $|N(t) - N_{\text{eq}}| = |N(t) - 0| = N(t) < \epsilon$ for all $t \geq 0$, where ϵ is arbitrarily small, no matter how close $N(0)$ is to $N_{\text{eq}} = 0$, for no matter how close the initial point is to the t axis, in Fig. 2, the solution curve moves upward, tending to the asymptote a/b as $t \to \infty$. And the equilibrium point $N_{\text{eq}} = a/b$ is stable because we *can* keep $|N(t) - N_{\text{eq}}| = |N(t) - a/b| < \epsilon$ for all $t \geq 0$, where ϵ is arbitrarily small, simply by starting out closer to a/b than ϵ.

Further, we classify a stable equilibrium point as **asymptotically stable** if $N(t)$ not only remains arbitrarily close to N_{eq} for all $t \geq 0$ but if it actually *tends to* N_{eq} as $t \to \infty$, that is, if $N(t) \to N_{\text{eq}}$ as $t \to \infty$. From Fig. 2 it seems evident that not only is $N_{\text{eq}} = a/b$ stable, but that it is asymptotically stable.

Proceeding one step further, we introduce the idea of "linearized stability analysis." The idea is simple. Suppose N_{eq} is an equilibrium point of an autonomous equation

$$\boxed{\frac{dN}{dt} = f(N),} \tag{9}$$

and that we wish to examine its stability. Since the stability concept used here is a "local" one, why not simplify the function $f(N)$ in (9) by approximating it in the neighborhood of the point N_{eq}? Specifically, expand $f(N)$ in a Taylor series about N_{eq} and cut off after the linear (i.e., the first-degree) term:

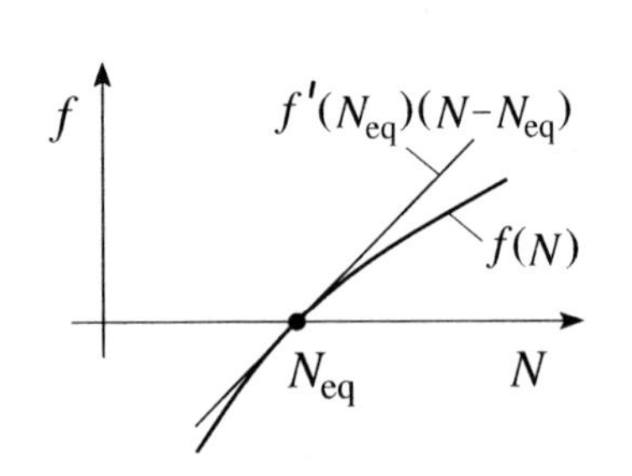

Figure 3. Local approximation of $f(N)$ in neighborhood of the equilibrium point N_{eq}.

$$\begin{aligned} f(N) &= f(N_{\text{eq}}) + f'(N_{\text{eq}})(N - N_{\text{eq}}) + \frac{1}{2!}f''(N_{\text{eq}})(N - N_{\text{eq}})^2 + \cdots \\ &\approx f'(N_{\text{eq}})(N - N_{\text{eq}}), \end{aligned} \tag{10}$$

in which we've also dropped the leading term $f(N_{\text{eq}})$ because $f(N_{\text{eq}}) = 0$ by the definition of equilibrium point; see Fig. 3.[1]

The approximation (10) reduces (9) to the simple *linear* equation

$$\boxed{\frac{dN}{dt} = f'(N_{\text{eq}})(N - N_{\text{eq}}),} \tag{11}$$

which is called the **linearized** version of (9). If we define the **deviation** from the equilibrium point as $\delta(t) \equiv N(t) - N_{\text{eq}}$, then $\delta'(t) = N'(t)$ and (11) becomes

$\delta(t) \equiv N(t) - N_{\text{eq}}$.

$$\frac{d\delta}{dt} = f'(N_{\text{eq}})\delta, \tag{12}$$

with solution

$$\delta(t) = \delta_0 e^{f'(N_{\text{eq}})t}. \tag{13}$$

Everything hinges on the sign of the number $f'(N_{\text{eq}})$ in the exponent: If

$$\boxed{f'(N_{\text{eq}}) < 0,} \tag{14a}$$

then (13) shows that $\delta(t) \to 0$ as $t \to \infty$ so the equilibrium point N_{eq} is evidently *asymptotically stable*, and if

$$\boxed{f'(N_{\text{eq}}) > 0,} \tag{14b}$$

[1] We've assumed that $f'(N_{\text{eq}}) \neq 0$ so that the approximation (10) does capture the first *nonvanishing* term of the series.

then $\delta(t)$ grows, instead, so the equilibrium point N_{eq} is evidently *unstable*. If $f'(N_{\text{eq}}) = 0$, the criterion gives no information. In that case we must go back to (10) and proceed farther into the Taylor series, to keep the first *non*vanishing term. This point is pursued in the exercises.

We said "evidently," above, because although it is reasonable to expect the *original* differential equation $dN/dt = f(N)$ to have the same behavior, near N_{eq}, as its linearized version, we have not proved that it does. However, the expected result is true, and we state it as a theorem:

THEOREM 1.6.1 *Stability Criterion for $dN/dt = f(N)$*
Let N_{eq} be an equilibrium point of $dN/dt = f(N)$, where $f(N)$ is differentiable at N_{eq}.
(a) If $f'(N_{\text{eq}}) < 0$, then N_{eq} is *asymptotically stable*.
(b) If $f'(N_{\text{eq}}) > 0$, then N_{eq} is *unstable*.

The criterion simply echos what we have already seen from our phase line pictures such as Fig. 1, namely, that if the slope f' at N_{eq} is negative then the flow is toward N_{eq} and the latter is stable, and if the slope there is positive then the flow is away from N_{eq} and the latter is unstable.

Realize that we can find the equilibrium points [by solving $f(N) = 0$ for N] and can then determine their stability [by determining the sign of f' at each equilibrium point] *without ever solving the differential equation (8) — even without plotting $f'(N)$ versus N and obtaining the phase line*! Indeed, (9) might be too difficult to solve, or its solution might be obtainable but intractably messy, or we might not be *interested* in the solution, but only in the equilibrium points and their stability.

EXAMPLE 1. Application of the Stability Criterion (14). To illustrate, we will apply the linearization procedure to the differential equation

$$\frac{dx}{dt} = \frac{1-x^2}{1+x^2}. \tag{15}$$

This example is just a made-up differential equation, not a population problem, so in place of $N(t)$ we revert to our generic $x(t)$ notation.

If we set $f(x) = (1-x^2)/(1+x^2) = 0$ to find the equilibrium points we obtain $x_{\text{eq}} = \pm 1$. To determine the stability of $x_{\text{eq}} = +1$, expand f in a Taylor series about that point and linearize,

$$\begin{aligned} \frac{dx}{dt} = \frac{1-x^2}{1+x^2} &= -(x-1) + \frac{1}{2}(x-1)^2 - \frac{1}{4}(x-1)^4 + \cdots \\ &\approx -(x-1) \end{aligned} \tag{16}$$

Recall that the Taylor series of f about $x = 1$ is $f(x) = f(1) + \frac{f'(1)}{1!}(x-1) + \frac{f''(1)}{2!}(x-1)^2 + \cdots$.

so, with $\delta = x - x_{\text{eq}} = x - 1$, (16) gives

$$\delta' = -\delta, \tag{17}$$

$$\delta(t) = \delta_0 e^{-t} \tag{18}$$

and because of the *negative* exponential in (18), $x_{\text{eq}} = +1$ is stable. Actually, we didn't need to carry out the solution of (17), or even the Taylor expansion in (16); we could simply have evaluated $f'(1)$ and examined its sign. Since $f'(1) = -1 < 0$, we could have concluded from Theorem 1.6.1 that $x_{\text{eq}} = +1$ is asymptotically stable.

For $x_{\text{eq}} = -1$ we will take the shortcut: We find that $f'(-1) = +1 > 0$, so $x_{\text{eq}} = -1$ is unstable. ∎

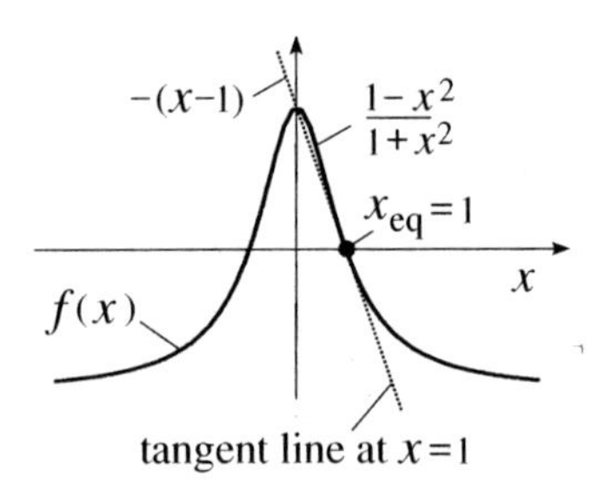

Figure 4. The linearization (16) is a tangent-line approximation of the function $f(x) = (1-x^2)/(1+x^2)$ at $x_{\text{eq}} = 1$.

To understand the linearization idea, think geometrically. Specifically, observe that the linearization of $f(x)$ about x_{eq} amounts to replacing the nonlinear function $f(x)$ in $x' = f(x)$ by its tangent-line approximation at that point. For instance, the approximation of $(1-x^2)/(1+x^2)$ by $-(x-1)$ in (16) amounts to the tangent-line approximation shown in Fig. 4. Just as the tangent line faithfully approximates $f(x)$ in the neighborhood of x_{eq}, so does the flow corresponding to the linearized differential equation faithfully approximate the flow corresponding to the original nonlinear differential equation in the neighborhood of x_{eq}.

Closure. In this section we used already-developed solution techniques and qualitative phase line methods to study representative problems involving separable first-order differential equations. The only new mathematical idea was that of linearization about an equilibrium point, which we used in Section 1.6.2 to determine the stability of equilibrium solutions of nonlinear autonomous differential equations.

EXERCISES 1.6

1. Incorporating Harvesting. Let $N(t)$ denote the fish population in a commercial fish pond. If we harvest fish at a rate h fish per unit time, we must modify the logistic equation (3) as

$$\boxed{N' = (a - bN)N - h.} \tag{1.1}$$

To maximize profits, we want to make h as large as possible, but if we make it too large then we will drive the fish population to zero and be out of business. Thus, our interest is not so much in solving (1.1) and obtaining traditional plots of $N(t)$ versus t, for instance, but in determining the maximum sustainable harvesting rate h. *The problem*: Determine that rate. NOTE: The phase line contains all the information that is needed. Sketch the phase line [i.e., the graph of $N' = (a - bN)N - h$ versus N] for $h = 0$, and again for $h > 0$, and see the effect of h on the flow along the phase line. This exercise illustrates the simplicity and value of the phase line — not to replace standard solution methods, but to complement them.

2. Incorporating a Threshold Population. Field studies indicate that if the population of a certain fish in a lake falls below a critical level, say P, then it will decline to zero (i.e., to extinction). Thus, to successfully stock the lake with that species one must supply enough fish so that the initial population is more than P fish. To incorporate this behavior, it is proposed that we modify the logistic equation (3) to the form

$$N' = -\kappa(P - N)(Q - N)N, \tag{2.1}$$

in which κ, P, and Q are positive constants and $Q > P$. Does the form of (2.1) seem reasonable? HINT: Consider the phase line.

3. Gompertz Growth Model. Let $W(t)$ be the weight of an organism as a function of the time t. One model of the growth of the organism is given by the **Gompertz** equation

$$W' = rW \ln \frac{K}{W} \tag{3.1}$$

or, equivalently, $W' = r(\ln K - \ln W)W$, in which r and K are positive constants.

(a) Solve (3.1) by separation of variables and show that

$$W(t) = Ke^{Ae^{-rt}}, \tag{3.2}$$

in which A is an arbitrary constant. HINT: To evaluate the integral that arises, the substitution $u = \ln W$ will help.
(b) Verify, by substitution, that (3.2) satisfies (3.1).
(c) Show that if the initial condition is $W(0) = W_0 > 0$, then (3.2) gives

$$W(t) = K\left(\frac{W_0}{K}\right)^{(e^{-rt})}, \tag{3.3}$$

that is, W_0/K to the e^{-rt} power, times K.
(d) Proceeding qualitatively instead, we can use the phase line because (3.1) is autonomous. Give labeled sketches analogous to Figs. 1 and 2 [which were for the logistic equation (3)]; consider only $W > 0$. Include any equilibrium points, sketch representative solution curves (as in Fig. 2), and show that the solution curves have inflection points at $W = 0.3679K$ (analogous to the line $N = a/2b$ in Fig. 2). Finally, classify each equilibrium point as stable or unstable.

4. Qualitative Analysis. Consider the autonomous equation $x' = f(x)$ on $0 \leq t < \infty$ and for $-\infty < x < \infty$, where $f(x)$ is given below. Determine the equilibrium points, if any, sketch the graph of x' versus x and the phase line, and classify each equilibrium point as stable or unstable.

(a) $f(x) = e^x - 10$ (b) $f(x) = (x-2)^3$
(c) $f(x) = x^4 - 5x^2 + 4$ (d) $f(x) = x^3 + 8$
(e) $f(x) = x^4 - 1$ (f) $f(x) = 3x - \sin x$
(g) $f(x)$ supplied by your instructor

5. (a)–(g) Applying Theorem 1.6.1. For the $f(x)$ given in the corresponding part of Exercise 4, use Theorem 1.6.1 to determine the stability or instability of each equilibrium point. If the theorem gives no information, state that.

6. Speed of Approach to Equilibrium. For both equations

$$x' = -x \quad \text{and} \quad x' = -x^3, \tag{6.1a,b}$$

$x = 0$ is a stable equilibrium point. How does the *speed* of approach to the equilibrium point $x = 0$ compare, for (6.1a) and (6.1b)? Explain.

7. One-Compartment Biological Systems. Consider a one-compartment biological system, such as a single cell, or an organ such as a kidney, and consider a particular chemical with concentration $c(t)$ within it. If the difference between the concentration $c(t)$ inside the compartment and the concentration c_0 outside of it is sufficiently small, the transport of the chemical across the boundary of the compartment can be modeled as being proportional to the difference $c(t) - c_0$:

$$c'(t) = -k(c - c_0), \tag{7.1}$$

in which k is an empirically determined positive constant of proportionality. For larger concentration differences, a better model is probably the **Michaelis–Menten** equation

$$\boxed{c'(t) = -\frac{a(c - c_0)}{b + (c - c_0)},} \tag{7.2}$$

which contains two empirically determined positive parameters a and b. The right-hand side of (7.2) is designed so that for small concentration differences $b + (c - c_0) \sim b$ and (7.2) reduces to

$$c'(t) \sim -\frac{a}{b}(c - c_0), \tag{7.3}$$

which is of the form of (7.1), but for large concentration differences $b + (c - c_0) \sim (c - c_0)$ and (7.1) reduces to

$$c'(t) \sim -a. \tag{7.4}$$

That is, the membrane cannot accomodate an arbitrarily large flow rate (any more than one can consume pizza at an arbitrarily large rate); the flow rate levels off as the concentration difference approaches infinity. NOTE: The right-hand side of the Michaelis–Menten equation is not derived, it is *designed* to be simple and to exhibit the two limiting behaviors indicated in (7.3) and (7.4).

(a) Determine any equilibrium points of (7.2) and their stability.
(b) If the initial concentration is a prescribed value $c(0)$, solve (7.2) and obtain the following solution, in implicit form,

$$b \ln\left|\frac{c(t) - c_0}{c(0) - c_0}\right| + c(t) = at + c(0). \tag{7.5}$$

(c) With $a = b = 10$ and $c_0 = 2$, say, obtain a computer plot of $c(t)$ versus t for each of the three initial conditions $c(0) = 1$, $c(0) = 2$, and $c(0) = 3$.

8. Orthogonal Families of Plane Curves. In a variety of applications, one is interested in two coplanar families of curves that intersect each other at right angles. Such families of curves are said to be **orthogonal**. For instance, the families of all concentric circles centered at the origin of an x, y plane and of all straight lines through the origin are orthogonal. (These are the constant-r and constant-θ curves of a polar coordinate system.) Consider two representative curves, one from each of the two families, defined by $y_1(x)$ and $y_2(x)$, and suppose they cross at P, as in the figure. With the help of a labeled

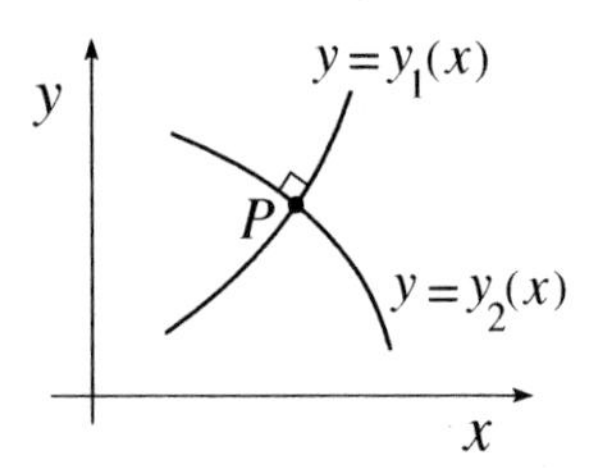

sketch, show that their slopes at P are negative inverses of each other:

$$y_1'(x) = -\frac{1}{y_2'(x)}. \tag{8.1}$$

9. Exercise 8, Continued. Suppose one family is comprised of the solutions of a given differential equation

$$y' = f(x, y), \tag{9.1}$$

and that we want to find the corresponding family of orthogonal curves. According to (8.1), to do so we must solve the differential equation

$$y' = -\frac{1}{f(x, y)}. \tag{9.2}$$

To illustrate, suppose we are given the differential equation

$$y' = y. \tag{9.3}$$

The family of solutions of (9.3) is the set of exponentials $y = Ce^x$, where C is an arbitrary constant. Then, to find the corresponding orthogonal curves form the negative inverse of the slope $y' = y$ given in (9.3) and solve

$$y' = -1/y. \tag{9.4}$$

That step gives the family of curves $y = \sqrt{2}\sqrt{D - x}$, in which D is an arbitrary constant. Representative members of the two orthogonal families are shown in the figure. Do the same for

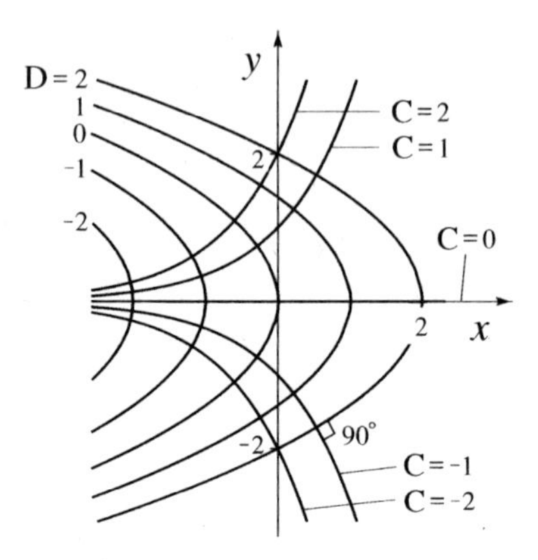

each given differential equation: find the two families of curves and give a hand sketch or computer plot of representative members of each family.

(a) $y' = -4y$ (b) $y' = 2y/x$
(c) $y' = -y/x$ (d) $y' = -x/y$

10. Exercises 8 and 9, Continued. Instead of being given the differential equation of one of the families, as we were in Exercise 9, in this exercise we give the family itself and ask you to find a second family, that is orthogonal to the one that is given. HINT: Work backwards and find a differential equation (9.1) for which the given family is the solution. Then proceed as in Exercise 9.

(a) $y = Ce^{2x}$ (b) $y = 1/(x+C)$ (c) $y = Cx^3$

11. The Draining of a Tank; Torricelli's Law. A tank, of uniform cross sectional area A, has a leak at the bottom due to a hole of cross-sectional area a, so the liquid depth x will diminish with time. We wish to predict the time T it will take the tank to empty if the initial liquid depth is x_0. According to

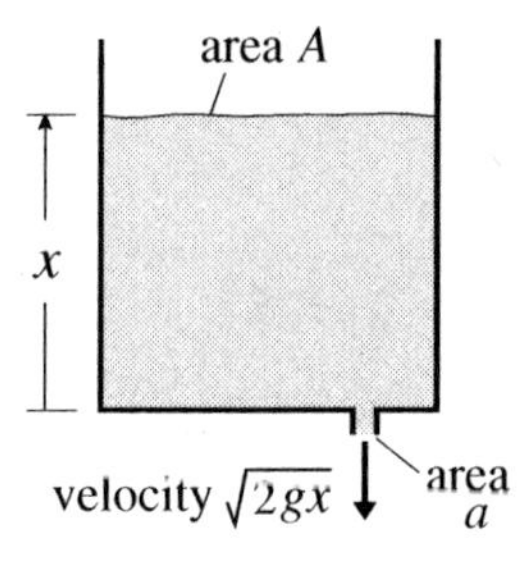

Torricelli's law, the efflux velocity from the hole is $\sqrt{2gx}$, which is the same as the velocity that would result from free fall, from rest, through the vertical distance x. (Actually, it will be $\beta\sqrt{2gx}$ for some positive constant $\beta < 1$, due to frictional losses, but we will neglect such effects and take $\beta = 1$.)

(a) Derive the IVP

$$x' = -\frac{a}{A}\sqrt{2gx}; \quad x(0) = x_0. \tag{11.1}$$

(b) Solve (11.1) and show that the tank empties not as $t \to \infty$, but in the finite time

$$T = \frac{A}{a}\sqrt{\frac{2x_0}{g}}. \tag{11.2}$$

12. A Conical Tank. First, read the introduction to Exercise 11, through Equation (11.1). Suppose the tank is not of uniform cross section but is conical, as shown below and, as in Exercise 11, suppose it has a hole of area a at the bottom.

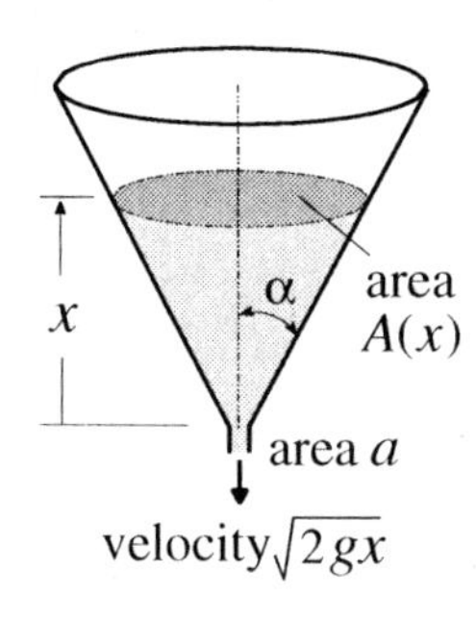

(a) Recalling that the volume of a cone of base radius r and altitude h is $\pi r^2 h/3$, show that the IVP for the depth $x(t)$ is

$$x' = -\left(\frac{a\sqrt{2g}}{\pi \tan^2 \alpha}\right)x^{-3/2}; \quad x(0) = x_0. \tag{12.1}$$

(b) Solve (12.1) and show that the draining time is

$$T = \frac{2\pi \tan^2 \alpha}{5a\sqrt{2g}} x_0^{5/2}. \tag{12.2}$$

13. Streamline Pattern. Let

$$\frac{dx}{dt} = (1-y)x, \tag{13.1a}$$

$$\frac{dy}{dt} = (x-1)y \tag{13.1b}$$

be the x and y velocity components of a certain fluid flow in the first quadrant of the x, y plane. We wish to find the streamlines, that is, the paths of the fluid particles, and these are given by the integral curves of

$$\frac{dy}{dx} = \frac{(x-1)y}{(1-y)x}, \tag{13.2}$$

obtained by dividing (13.1b) by (13.1a).

(a) Derive the implicit solution

$$xy\, e^{-(x+y)} = C \tag{13.3}$$

of (13.2), and verify by differentiating (13.3), that it does satisfy (13.2).

(b) To see the streamline pattern, obtain computer-generated streamlines through the points (1,0), (1,0.1), (1,0.3), (1,0.5), (1,0.5), (1,0.9), and (0,1), and add flow direction arrows. Plot on the square $0 \le x \le 3$, $0 \le y \le 3$.

ADDITIONAL EXERCISES

14. Liquid Level Feedback Control. Liquid flows into a tank of horizontal cross sectional area A ft^2 at a constant rate Q ft^3/sec and leaves at the rate $q = \alpha\sqrt{x}$ ft^3/sec, where $x(t)$ is the liquid depth, t is the time, and α is an empirically known constant. Torricelli's law gives the exit velocity as $\sqrt{2gx}$,

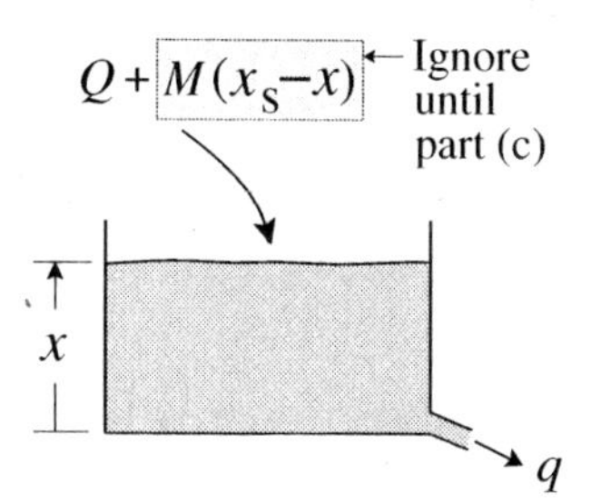

and when we multiply that by the exit area we get a flow rate of the form $\kappa\sqrt{x}$, in which κ is a known positive constant. Equating the rate of increase of liquid volume in the tank to the rate in minus the rate out gives the differential equation

$$Ax' = Q - \kappa\sqrt{x} \tag{14.1}$$

for $x(t)$, with an equilibrium or steady-state x_s found from $0 = Q - \kappa\sqrt{x_s}$ as

$$x_s = (Q/\kappa)^2. \tag{14.2}$$

The goal, in the operation of this "chemical plant," is to maintain $x(t)$ at its equilibrium value x_s, and to return it quickly to that value following any "disturbance" of x from its desired value x_s.

(a) From the phase line, show that x_s is a stable equilibrium point of (14.1).

(b) **Linearization.** If x is close to x_s, then it seems justified to linearize the nonlinear equation (14.1) about x_s.

Thus, expand the nonlinear $\sqrt{x}$ term in (14.1) in a Taylor series, about x_s, and cut it off after the first-order term, which amounts to using the tangent-line approximation illustrated in the next figure. Show that (14.1) is thus simplified to the

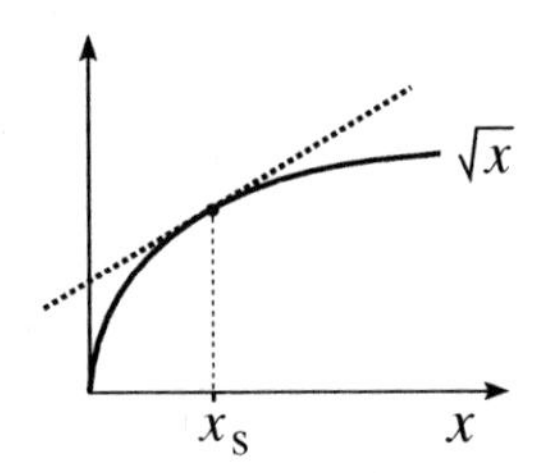

linearized equation

$$Ax' = -\beta(x - x_s) \quad (\beta = \frac{\kappa}{2\sqrt{x_s}}), \tag{14.3}$$

and, taking $x(0) = x_0 \neq x_s$, derive the solution of (14.3) as

$$x(t) = x_s + (x_0 - x_s)e^{-\beta t/A}. \tag{14.4}$$

(c) **Feedback Control.** Although (14.4) shows that $x(t) \to x_s$ as $t \to \infty$, that approach may be too slow for successful plant operation if β/A is small. To speed up the return to x_s suppose, using suitable equipment, that we continuously monitor x, compare its measured value with x_s to determine the instantaneous error

$$e(t) = x_s - x(t), \tag{14.5}$$

and augment the inflow Q by an amount proportional to that error, $M(x_s - x)$, as indicated in the first figure. Accordingly, re-write (14.3) as

$$Ax' + \beta(x - x_s) = M(x_s - x), \tag{14.6}$$

show that $e(t)$ satisfies a linear differential equation with initial condition $e(0) = x_s - x_0$, with solution

$$e(t) = e(0)e^{-(\beta+M)t/A}. \tag{14.7}$$

COMMENT 1. The upshot is that β/A is enhanced to $(\beta + M)/A$, so disturbances from equilibrium die out exponentially faster with the "feedback" than without it.

COMMENT 2. This is an example of a **feedback control** system, because the error is fed back to the input as is indicated schematically in the next figure. Since the feedback $Me(t)$ is proportional to $e(t)$ it is an example of **proportional control**,

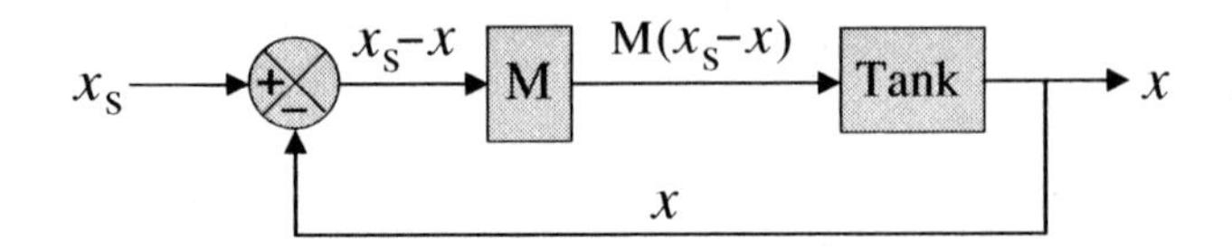

the amplification M being the **gain**. Also used are **derivative control** [proportional to $e'(t)$], **integral control** [proportional to $\int_0^t e(t)\,dt$], and combinations of the three. The human body is a whole hierarchy of control systems that control body temperature, heart rate, respiration, and so on. Control theory is normally taught, at the undergraduate and graduate levels, in engineering and bioengineering departments.

15. Free Fall and Terminal Velocity; Drag Proportional to Velocity Squared. The equation of motion of a body of mass m falling vertically in a fluid (such as air or water) follows from Newton's second law as

$$m\frac{dv}{dt} = mg - B - D, \tag{15.1}$$

in which $v(t)$ is the velocity (so dv/dt is the acceleration), B is the "buoyant force," and D is the "drag force" exerted on the body by the fluid. The buoyant force, by Archimedes' principle, is constant and equal to the weight of the fluid displaced by the body. The drag force is more complicated. For definiteness, suppose the body is spherical. It is shown, in a course on fluid mechanics, that if the *Reynolds number* parameter $\text{Re} = \rho v d/\mu$ is *small*, in which $d = 2r$ is the diameter of the sphere, and ρ and μ are the mass density and viscosity of the fluid, respectively, then the drag force D in (15.1) is proportional to the velocity v. In that case (15.1) is linear. Here, we consider instead the case of *large* Reynolds number, in the range

$$10^3 < \text{Re} < 10^5. \tag{15.2}$$

If (15.2) is satisfied then the drag force D on the sphere is approximately

$$D \approx 0.23\pi r^2 \rho v^2. \tag{15.3}$$

This time the quadratic dependence of D on v expressed by (15.3) results in the differential equation (15.1) for v being *nonlinear*. Take $g = 32.2$ ft/sec^2, and let the mass density of the body be 5 slugs/ft^3 (typical of stone). Let the fluid be water, with $\rho = 1.94$ slugs/ft^3 and $\mu = 2.36 \times 10^{-5}$ slugs/ft sec (at 60° F). With these values (15.1) becomes

$$6.67\pi r^3 \frac{dv}{dt} = 215\pi r^3 - 83.3\pi r^3 - (0.23)(1.94)\pi r^2 v^2,$$

or

$$\frac{dv}{dt} = 19.7 - \frac{0.067}{r}v^2. \tag{15.4}$$

(a) Using a phase line approach to (15.4), show that there is a steady-state value

$$v_s = 17.1\sqrt{r} \text{ ft/sec}, \tag{15.5}$$

called the *terminal velocity*, and that the latter is stable. [Note from (15.5) that large spheres, of a given mass density, fall faster than smaller ones.]

(b) Also, solve (15.4) with $v(0) = 0$ and show that

$$v(t) = 17.1\sqrt{r} \tanh(1.15t/\sqrt{r}). \tag{15.6}$$

Sketch the graph of $v(t)$, labeling any key values.

(c) Find the time, in terms of r, that it takes for the body to attain 90% of its terminal velocity.

(d) Remember that the approximate formula (15.3) for the drag force is accurate only for large Reynolds numbers in the interval defined by (15.2). In terms of the size of the sphere, (15.2) implies that our analysis is valid only if the diameter d of the sphere falls within certain limits. Show that with v given by (15.5), with $d = 2r$, and with the values of ρ and μ given above, the inequality (15.2) requires that 0.0050 ft $< r < 0.108$ ft or

$$0.06 \text{ in} < r < 1.3 \text{ in}. \tag{15.6}$$

For a stone of radius 0.06 inches (0.005 ft), (15.5) gives the terminal velocity as 1.21 ft/sec, and for a stone of radius 1.3 inches (0.108 ft) the terminal velocity is 5.62 ft/sec.

16. Curve of Pursuit. The following is a classical problem of pursuit — for instance, of one ship by another. Denote the pursued ship by point B and the pursuing ship by point A in the figure. Suppose B is at $(x, y) = (1, 0)$ at time $t = 0$ and moves

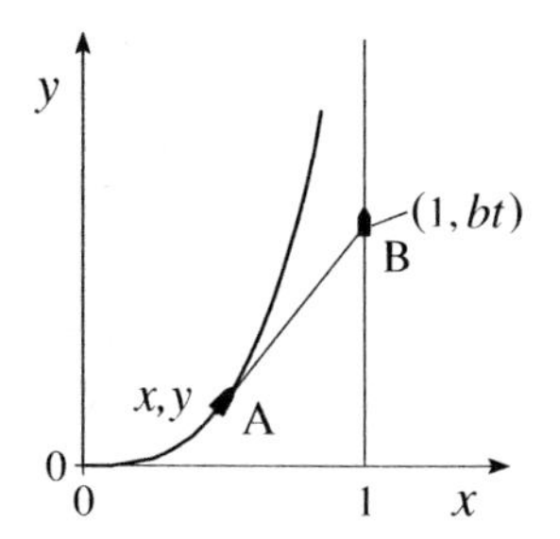

in the positive y direction with a constant speed b, and that A begins at the origin at $t = 0$, steers a course that is always directed at B, and moves with a constant speed a that is greater than b.

(a) Show, from the figure, that

$$\frac{dy}{dx} = \frac{bt - y}{1 - x}. \tag{16.1}$$

(b) Think about the variables used in (16.1): It is natural to think of x and y as functions of the time t, but in (16.1) the dy/dx implies that we are instead regarding x and y as independent and dependent variables, respectively. Fine, but then the t in (16.1) is not welcome. To eliminate it, differentiate (16.1) with respect to x. That step gives rise to a dt/dx term, and to obtain an expression for dt/dx use the formula

$$\frac{ds}{dt} = a = \frac{ds}{dx}\frac{dx}{dt} \tag{16.2}$$

in which s is the arclength along the curve of pursuit, from the origin to A. Show, from (16.2), that $dt/dx = \sqrt{1 + y'^2}/a$, and that (16.1) becomes

$$(1 - x)y'' = \frac{b}{a}\sqrt{1 + y'^2}. \tag{16.3}$$

(c) The latter is a *second*-order equation, and we haven't studied second-order equations yet, but the substitution $u(x) = y'(x)$ reduces it to a first-order equation for $u(x)$. Do that, solve for $u(x)$ by separation of variables, and obtain the solution (in implicit form)

$$u + \sqrt{u^2 + 1} = C(1 - x)^{-r}, \tag{16.4}$$

where $r = b/a < 1$ and C is an arbitrary constant. Applying the initial condition $u(0) = y'(0) = 0$, evaluate C.

(d) Solve (16.4) by algebra for u, replace u by $y'(x)$, and integrate again to show that

$$y(x) = -\frac{1}{2}\frac{(1 - x)^{1-r}}{1 - r} + \frac{1}{2}\frac{(1 - x)^{1+r}}{1 + r} + D, \tag{16.5}$$

where D is an arbitrary constant. Finally, apply the initial condition $y(0) = 0$ to solve for D, and thus show that the curve of pursuit is given by

$$y(x) = \frac{1}{2}\left[\frac{(1 - x)^{1+r}}{1 + r} - \frac{(1 - x)^{1-r}}{1 - r}\right] + \frac{r}{1 - r^2}. \tag{16.6}$$

(e) Determine the location of B when it is caught by A, and show that capture occurs at the time $T = a/(a^2 - b^2)$. Sketch the curve of pursuit up to the time of capture, with suitable labeling and with key features clearly rendered.

(f) For A to catch B, is it really an optimal strategy for it to always steer so as to be aiming at B? What would be the *optimal* pursuit path, optimal in the sense of overtaking B in minimal time, and what would be the time and place when that occurs? NOTE: This is the strategy used by a (good) baseball player in catching a "fly ball."

17. Projectile Dynamics and Escape Velocity. Consider a classical problem in Newtonian mechanics, the motion of a projectile subject to a gravitational force field. **Newton's law of gravitation** states that the force of attraction F exerted by one point mass M on another point mass m is

$$F = G\frac{Mm}{d^2}, \tag{17.1}$$

where d is the distance between them and $G = 6.67 \times 10^{-8}\,\text{cm}^3/\text{gm sec}^2$ is called the universal gravitational constant; (17.1) is called an **inverse-square law** because the force F varies as the inverse square of the distance d. (By M and m being point masses, we mean that their sizes are negligible compared to the distance between them; even an elephant could be a "point mass.")

Consider the linear motion of a projectile of mass m launched from the surface of the earth, as sketched in the figure, where M and R are the mass and radius of the earth respectively, and where any air resistance is neglected. From Newton's second law of motion and his law of gravitation (17.1), it follows that the equation of motion of the projectile is

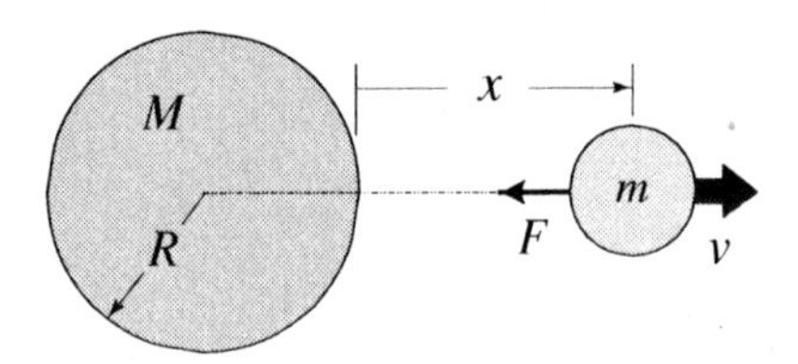

$$m\frac{d^2x}{dt^2} = -G\frac{Mm}{(x+R)^2}, \tag{17.2}$$

the minus sign because the force is in the negative x direction, and $x + R$ being the distance between mass centers. (It is not at all obvious, but can be proved, and indeed *was* proved by Newton and published in his *Principia Mathematica*, that the force of attraction of a spherical homogeneous mass M at any point outside that mass is the same as if the entire mass M were compressed to a point, at the center of that sphere.)
Here is the problem:

(a) Now, (17.2) is of second order and we have not yet studied equations of second order, but we can reduce it to a first-order equation as follows: If we integrate both sides with respect to t, the left side is simple; it merely gives $m\,dx/dt$, but the right-hand integral cannot be evaluated because the $x(t)$ in the integrand is not yet known. If we integrate both sides with respect to x instead, then the right-hand integral can be evaluated, but what can we do with the integral on the left, $\int mx''\,dx$? Proceeding as in Equation (9) of Section 1.5, show that the result of integrating (17.2) with respect to x is

$$\frac{1}{2}mx'^2 = \frac{GMm}{x+R} + C, \tag{17.3}$$

in which C is the arbitrary constant of integration. NOTE: Physically, if we express the latter in the form

$$\underbrace{\frac{1}{2}mx'^2}_{\text{KE}} + \underbrace{\left(-\frac{GMm}{x+R}\right)}_{\text{PE}} = C \tag{17.4}$$

we can understand it as a statement of **conservation of energy**, for it says the kinetic energy plus the gravitational potential energy is a constant over the course of the motion. If we solve (17.3) for $x'(t)$ by taking square roots of both sides, the resulting differential equation for $x(t)$ will be separable, but the x integration is difficult and leads to a messy solution in implicit form. Thus, let us forego the solution for $x(t)$ and see what we can learn, in the remainder of this exercise, directly from (17.3).

(b) Denote the velocity $x'(t)$ as $v(t)$ and let the launch velocity be $v(0) = V$. Applying that initial condition, solve (17.3) for C, and show that

$$v = \sqrt{V^2 - \frac{2GM}{R}\frac{x}{x+R}}. \tag{17.5}$$

It is possible to eliminate the universal gravitational constant G in favor of the more familiar constant g, the gravitational acceleration g at the earth's surface, by noting that when $x = 0$ the right-hand side of (17.2) must be the weight force $-mg$. Thus, show that $G = R^2g/M$ so (17.5) becomes

$$v = \sqrt{V^2 - 2gR\frac{x}{x+R}}. \tag{17.6}$$

(c) Show, from (17.6), that if V is less than a certain critical velocity, the **escape velocity** V_e, then the projectile reaches a maximum distance $x_{\max}$ from the earth and then returns to the earth, but if $V > V_e$, then the projectile escapes and does not return. Show that

$$x_{\max} = \frac{V^2 R}{2gR - V^2} \quad \text{and} \quad V_e = \sqrt{2gR}. \tag{17.7}$$

(d) Sketch the graph of v versus x for two representative launch velocities V, one smaller than V_e and one greater than V_e, and label any key values.

(e) Calculate V_e in km/sec and in miles/hr, using $R = 6378$ km = 3960 mi, and $g = 9.81$ m/sec$^2 = 32.2$ ft/sec^2.

HISTORICAL NOTE: Newton inferred (17.1) from **Kepler's laws** of planetary motion, which were, in turn, inferred empirically by Kepler from the voluminous measurements recorded by the Danish astronomer *Tycho Brahe* (1546–1601). Usually, in applications, one knows the force exerted on a mass and determines the motion by twice integrating Newton's second law of motion. In deriving (17.1), however, Newton worked "backwards:" The motion of the planets was described by Kepler's laws, and Newton used those laws to infer the force needed to cause that motion. It turned out to be an inverse-square force directed toward the sun. Newton then proposed the bold generalization that (17.1) holds not only between each planet and the sun, but between *any* two bodies in the universe; hence the name **universal law of gravitation**. Imagine how the idea of a force *acting at a distance*, rather than through physical contact, must have been incredible when first proposed. In fact, such great scientists and mathematicians as Huygens, Leibniz, and John Bernoulli called Newton's idea of gravitation absurd and revolting! But, Newton stood upon the results of his mathematics, in inferring the concept of gravitation, even in the face of such distinguished opposition. Remarkably, **Coulomb's law** subsequently stated an inverse-square type of attraction or repulsion between two electric charges. But, although the forms of the two laws are identical, the magnitudes of the forces are staggeringly different. Specifically, the electrical force of repulsion between two electrons is, independent of the distance of separation, 4.17×10^{42} times stronger than their gravitational attraction due to their mass!

1.7 EXACT EQUATIONS AND EQUATIONS THAT CAN BE MADE EXACT

Thus far we've developed solution techniques for first-order differential equations that are linear or separable. In this section we consider another important case, equations that are "exact." The method that we develop will be a version of the integrating factor method used in Section 1.2 to solve the linear equation $y' + p(x)y = q(x)$.

1.7.1 Exact differential equations. To motivate the idea of exact equations, consider

$$\frac{dy}{dx} = \frac{\sin y}{2y - x\cos y} \tag{1}$$

or, in differential form,

$$\sin y\, dx + (x\cos y - 2y)\, dy = 0. \tag{2}$$

The left-hand side is the differential of $F(x, y) = x\sin y - y^2$ because, by the chain rule,

$$dF = \frac{\partial F}{\partial x}dx + \frac{\partial F}{\partial y}dy = \sin y\, dx + (x\cos y - 2y)\, dy, \tag{3}$$

so (2) is simply $dF = 0$, which is readily integrated and gives $F =$ constant. Thus,

$$F(x, y) = x \sin y - y^2 = C, \tag{4}$$

with C an arbitrary constant. Equation (4) is the solution to (1), in implicit form.

Here we are regarding x as independent variable and y as dependent variable.

To generalize the method outlined above, consider the differential equation

$$\frac{dy}{dx} = -\frac{M(x, y)}{N(x, y)}, \tag{5}$$

in which the minus sign is included merely so that when we re-express (5) in the differential form

$$\boxed{M(x, y)dx + N(x, y)dy = 0} \tag{6}$$

we end up with a plus sign in (6). For (1), for instance, we see by comparing (2) and (6) that $M = \sin y$ and $N = x \cos y - 2y$.

Here we change our viewpoint temporarily and regard both x and y as independent variables.

Before proceeding, notice that in equation (5) y is regarded as a function of x, as is implied by the presence of the derivative dy/dx; x is the independent variable and y is the dependent variable. But upon re-expressing (5) in the form (6) we've changed our viewpoint, and now consider x and y as having the same status: now, both are independent variables.

We've seen that integration of (6) is simple if $Mdx + Ndy$ happens to be the differential of some function $F(x, y)$, for if there is a function $F(x, y)$ such that

$$dF(x, y) = M(x, y)dx + N(x, y)dy, \tag{7}$$

then (6) is simply

$$dF(x, y) = 0, \tag{8}$$

which gives the solution

$$F(x, y) = C \tag{9}$$

of (6), with C an arbitrary constant.

Given $M(x, y)$ and $N(x, y)$, which we can identify when we write the given differential equation in the differential form (6), suppose there does exist an $F(x, y)$ such that $Mdx + Ndy = dF$. If so, we call $Mdx + Ndy$ an **exact differential**, and we call (6) an **exact differential equation**.

Two questions arise: *Given a first-order differential equation, expressed in the differential format (6), does such an $F(x, y)$ exist* and, *if so, how do we find it?* The first is answered by the following theorem.

THEOREM 1.7.1 *Test for Exactness*
Let $M(x, y)$, $N(x, y)$, $\partial M/\partial y$, and $\partial N/\partial x$ be continuous within a rectangle R in the x, y plane. Then $Mdx + Ndy$ is an exact differential in R if and only if

$$\boxed{\frac{\partial M}{\partial y} = \frac{\partial N}{\partial x}} \tag{10}$$

everywhere in R.

Partial Proof: Suppose $M dx + N dy$ is exact, so there is a function F such that $dF = M dx + N dy$. Then, by the chain rule,

$$M = \frac{\partial F}{\partial x} \tag{11a}$$

and

$$N = \frac{\partial F}{\partial y}. \tag{11b}$$

Differentiating (11a) partially with respect to y, and (11b) partially with respect to x, gives

$$M_y = F_{xy}, \tag{12a}$$

and

$$N_x = F_{yx}. \tag{12b}$$

In (12a) and below, we use subscripts for partial derivatives, for compactness. For example, $M_y = \frac{\partial M}{\partial y}$ and $F_{xy} = (F_x)_y = \frac{\partial}{\partial y}\left(\frac{\partial F}{\partial x}\right)$.

Since M, N, M_y, and N_x have been assumed continuous in R, it follows from (11) and (12) that F_x, F_y, F_{xy}, and F_{yx} are too, so $F_{xy} = F_{yx}$.[1] Then it follows from (12) that $M_y = N_x$, which is equation (10). Because of the "if and only if" wording in the theorem, we must also prove the reverse, that the truth of (10) implies the existence of F, but we will omit that part.[2] ■

Assuming that the conditions of the theorem are met, so we are assured that such an F exists, how do we *find* it? From (11a) and (11b). We will illustrate the procedure by revisiting our introductory example.

EXAMPLE 1. Solving an Exact Equation. Consider equation (1) again, in differential form,

$$\sin y \, dx + (x \cos y - 2y) dy = 0. \tag{13}$$

Compare (13) with (6) and identify $M = \sin y$ and $N = x \cos y - 2y$. Clearly, M, N, M_y, and N_x are continuous in the whole plane, so turn to the exactness condition (10):

---

[1] Recall that the partial derivative notation F_{xy} means $(F_x)_y$: differentiate first with respect to x and then with respect to y. It is shown in the calculus that a sufficient condition for $F_{xy} = F_{yx}$ is that F_x, F_y, F_{xy}, and F_{yx} are all continuous within the region in question. This is typically true in applications.

[2] See, for example, William E. Boyce and Richard C. DiPrima, *Elementary Differential Equations and Boundary Value Problems*, 6th ed. (NY: Wiley, 1997), page 85. In applications, of course, the *existence* of F follows when we actually *find* F, as in our Example 1.

$M_y = \cos y$, and $N_x = \cos y$, so (10) is satisfied, and it follows from Theorem 1.7.1 that there does exist an $F(x, y)$ such that the left-hand side of (13) is dF. *To find F use (11)*:

$$\frac{\partial F}{\partial x} = \sin y, \tag{14a}$$

$$\frac{\partial F}{\partial y} = x \cos y - 2y. \tag{14b}$$

The "partial integration" notation $\int (\ ) \partial x$ is not standard. We use it here to remind us that any y's in the integrand are to be treated as constants.

Integrating (14a) partially, with respect to x, gives

$$F(x, y) = \int \sin y \, \partial x = x \sin y + A(y), \tag{15}$$

The $\sin y$ integrand was treated as a constant in the integration because we performed a "partial integration" on x, holding y fixed [just as y was held fixed in computing $\partial F/\partial x$ in (14a)]. The constant of integration A must therefore be allowed to depend on y since y was held fixed and was therefore constant. As a check, taking a partial derivative of (15) with respect to x does recover (14a). Next, (14b) will determine $A(y)$: Putting (15) into (14b) gives $\frac{\partial}{\partial y}[x \sin y + A(y)] = x \cos y - 2y$, or,

$$x \cos y + A'(y) = x \cos y - 2y, \tag{16}$$

in which $A'(y)$ denotes dA/dy. Canceling terms gives $A'(y) = -2y$, so

$$A(y) = -\int 2y \, dy = -y^2 + B. \tag{17}$$

[The integration in (17) was not a "partial integration;" it was an ordinary integration on y because $A'(y)$ in $A'(y) = -2y$ was an ordinary derivative.] Putting (17) into (15) gives

$$F(x, y) = x \sin y - y^2 + B = \text{constant}. \tag{18}$$

Finally, absorb B into the constant, and call the resulting constant C. Thus, we have the solution

$$x \sin y - y^2 = C \tag{19}$$

of (13), in implicit form.

How can the method *fail*? Be sure to understand this point.

COMMENT 1. It would be natural to wonder how this method can *fail*. After all, even if $M_y \neq N_x$ can't we integrate (11) to find F? The clue, in this example, is in (16). For suppose (16) were $2x \cos y + A'(y) = x \cos y - 2y$, for instance, instead. Then the $x \cos y$ terms would not cancel, as they did in (16), and we would have $A'(y) = -x \cos y - 2y$, which is impossible because it expresses a relationship between x and y, whereas x and y are independent variables. Put differently, $A'(y)$ is a function of y only, so it cannot depend on x. Thus, the cancelation of the $x \cos y$ terms in (16) was crucial and was not an accident, but was a consequence of M and N satisfying the exactness condition (10).

COMMENT 2. We integrated (14a) and then (14b), but the order doesn't matter. ∎

1.7.2 Making an equation exact; integrating factors.

If M and N fail to satisfy (10), so the equation

$$M(x,y)dx + N(x,y)dy = 0 \tag{20}$$

is *not* exact, we can try to find a function $\sigma(x,y)$ so that if we multiply (20) by that function, then the new equation,

$$\sigma(x,y)M(x,y)dx + \sigma(x,y)N(x,y)dy = 0, \tag{21}$$

is exact.

Here, σM is our new "M" and σN is our new "N." We are seeking a function $\sigma(x,y)$ so that the exactness condition

$$\boxed{\frac{\partial}{\partial y}(\sigma M) = \frac{\partial}{\partial x}(\sigma N)} \tag{22}$$

is satisfied for (21). If we can find a function $\sigma(x,y)$ satisfying (22), we call it an **integrating factor** of (20) because then the left-hand side of (21) is dF, the differential of some function $F(x,y)$. Then (21) is simply $dF = 0$, which gives the solution of (20) as $F(x,y) =$ constant.

How can we find σ? It is any (nontrivial) solution of (22), that is, of

By trivial solution we mean $\sigma(x,y) = 0$.

$$\sigma_y M + \sigma M_y = \sigma_x N + \sigma N_x. \tag{23}$$

in which subscripts denote partial derivatives. Since (23) contains partial derivatives of $\sigma(x,y)$ it is not an ordinary differential equation but a *partial differential equation* for σ. Partial differential equations are beyond the scope of this text, so we have made dubious headway: To solve the original ordinary differential equation on $y(x)$ we now need to solve the partial differential equation (23) for $\sigma(x,y)$!

However, perhaps an integrating factor σ can be found that is a function of x alone, $\sigma(x)$. In that case $\sigma_y = 0$ and (23) reduces to the *ordinary* differential equation

$$\sigma M_y = \frac{d\sigma}{dx}N + \sigma N_x$$

or

$$\frac{d\sigma}{dx} = \left(\frac{M_y - N_x}{N}\right)\sigma. \tag{24}$$

This idea succeeds if and only if the $(M_y - N_x)/N$ in (24) is a function of x only, for if it contained any y dependence, then (24) would amount to a contradiction: a function of x equaling a function of x *and* y, where x and y are independent variables. Thus, if

$$\boxed{\frac{M_y - N_x}{N} = \text{function of } x \text{ alone,}} \tag{25}$$

For $\sigma(x)$.

If $M_y = N_x$, then (26) gives $\sigma(x) = 1$. After all, if $M_y = N_x$ then the equation was exact in the first place.

then (24) is separable and gives

$$\boxed{\sigma(x) = e^{\int \frac{M_y - N_x}{N}\, dx}.} \tag{26}$$

Actually, the general solution of (24) for $\sigma(x)$ is an arbitrary constant times the right-hand side of (26), but the constant can be taken to be 1 without loss since all we need is *an* integrating factor.

If $(M_y - N_x)/N$ is *not* a function of x alone, then an integrating factor $\sigma(x)$ does not exist, but perhaps we can find σ as a function of y alone, $\sigma(y)$. In that case, (23) reduces to

$$\frac{d\sigma}{dy} M + \sigma M_y = \sigma N_x$$

or

$$\frac{d\sigma}{dy} = -\left(\frac{M_y - N_x}{M}\right)\sigma.$$

This time, if

For $\sigma(y)$.

$$\boxed{\frac{M_y - N_x}{M} = \text{function of } y \text{ alone},} \tag{27}$$

then

$$\boxed{\sigma(y) = e^{-\int \frac{M_y - N_x}{M}\, dy}.} \tag{28}$$

EXAMPLE 2. Finding and Using an Integrating Factor. Consider the equation $y' = 2xe^y/(e^y - 4)$, or

$$2xe^y\, dx + (4 - e^y)\, dy = 0. \tag{29}$$

Then $M(x,y) = 2xe^y$ and $N(x,y) = 4 - e^y$, so (10) is not satisfied and (29) is not exact. If we seek an integrating factor that is a function of x alone, we find that

$$\frac{M_y - N_x}{N} = \frac{2xe^y - 0}{4 - e^y} \neq \text{function of } x \text{ alone}, \tag{30}$$

so $\sigma(x)$ is not possible. Seeking instead an integrating factor that is a function of y alone,

$$\frac{M_y - N_x}{M} = \frac{2xe^y - 0}{2xe^y} = 1 = \text{function of } y \text{ alone}, \tag{31}$$

so $\sigma(y)$ *is* possible, and

$$\sigma(y) = e^{-\int \frac{M_y - N_x}{M}\, dy} = e^{-\int 1\, dy} = e^{-y}. \tag{32}$$

Thus, multiply (29) through by $\sigma(y) = e^{-y}$ and obtain

$$2x\, dx + \left(4e^{-y} - 1\right) dy = 0, \tag{33}$$

which *is* exact. Then, (33) gives

$$\frac{\partial F}{\partial x} = 2x \tag{34a}$$

and

$$\frac{\partial F}{\partial y} = 4e^{-y} - 1, \tag{34b}$$

Actually, the steps (34)–(37) are overkill, for if we have the form $f(x)\,dx + g(y)\,dy = 0$, as we do in (33), we can simply integrate. Doing so gives $x^2 - 4e^{-y} - y = C$, which is the same result as (37).

and (34a) gives

$$F(x,y) = \int 2x\,\partial x = x^2 + A(y). \tag{35}$$

Next, put the right-hand side of (35) into the left-hand side of (34b):

$$\frac{\partial}{\partial y}\left[x^2 + A(y)\right] = 4e^{-y} - 1$$

or

$$A'(y) = 4e^{-y} - 1. \tag{36}$$

Thus,

$$A(y) = -4e^{-y} - y + B$$

and

$$F(x,y) = x^2 + A(y) = x^2 - 4e^{-y} - y + B = \text{constant}$$

or

$$x^2 = y + 4e^{-y} + C, \tag{37}$$

with C an arbitrary constant; (37) is the desired solution of (29), in implicit form.

COMMENT. Suppose we impose an initial condition $y(2) = 0$. Then (37) becomes $4 = 0 + 4 + C$, so $C = 0$. Thus, in implicit form, we have the particular solution

$$x^2 = y + 4e^{-y}, \tag{38}$$

which is plotted in Fig.1. The curve consists of two branches, an upper branch AB and a lower branch AC. The initial point is on the lower branch so discard the upper branch AB and keep the lower branch AC. Point A can be determined from the fact that the slope is infinite there. By setting the denominator in the differential equation $y' = 2xe^y/(e^y - 4)$ equal to zero we obtain $e^y = 4$, so A is at $y = \ln 4$ and, as follows from (38), at $x = \sqrt{\ln 4 + 1}$. Thus, the interval of existence of the solution satisfying the initial condition $y(2) = 0$ is $\sqrt{\ln 4 + 1} < x < \infty$. ∎

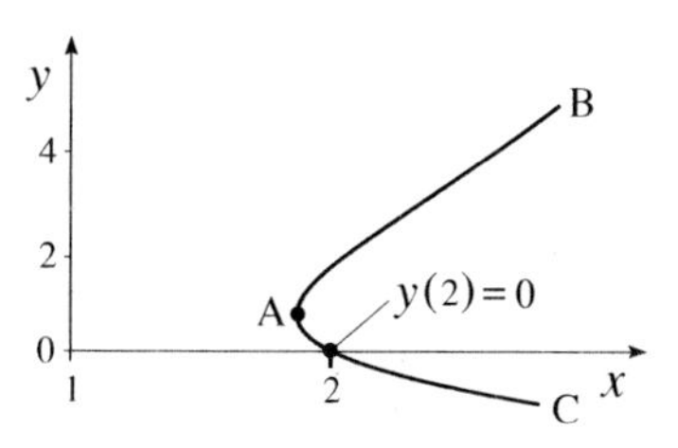

Figure 1. Graph of the relation (38).

EXAMPLE 3. Application to General Linear First-Order Equation. We've already solved the general *linear* first-order equation

$$\frac{dy}{dx} + p(x)y = q(x) \tag{39}$$

in Section 1.2, but let us see if we can solve it again, using the ideas in this section. First, put (39) into the form $Mdx + Ndy = 0$ by writing it as

$$[p(x)y - q(x)]\,dx + dy = 0. \tag{40}$$

Thus, $M = p(x)y - q(x)$ and $N = 1$, so $M_y = p(x)$ and $N_x = 0$. Hence $M_y \neq N_x$, so (40) is not exact [unless $p(x) = 0$]. Since

$$\frac{M_y - N_x}{N} = \frac{p(x) - 0}{1} = p(x) = \text{ function of } x \text{ alone,}$$

$$\frac{M_y - N_x}{M} = \frac{p(x) - 0}{p(x)y - q(x)} \neq \text{ function of } y \text{ alone,}$$

we can find an integrating factor that is a function of x alone, but not one that is a function of y alone. We leave it for the exercises to show that the integrating factor is

$$\sigma(x) = e^{\int p(x)\,dx},$$

and that the final solution (this time obtainable in explicit form) is

$$y(x) = e^{-\int p\,dx} \left(\int e^{\int p\,dx} q\,dx + C \right), \tag{41}$$

as we found in Section 1.2. ∎

Closure. Summary of the method of exact differentials:

To check your solution, a differential of $F(x, y) = C$ should give back the original equation $Mdx + Ndy = 0$.

1. Express the equation in the differential form $M(x, y)dx + N(x, y)dy = 0$. If M, N, M_y, and N_x are continuous in the x, y region of interest, check the exactness condition (10). If it is satisfied, the equation is exact, and its solution is $F(x, y) = C$, with F found from (11a) and (11b).

2. If the equation is not exact, see if $(M_y - N_x)/N$ is a function of x alone. If it is, an integrating factor $\sigma(x)$ can be found from (26). Multiply the given equation $Mdx + Ndy = 0$ through by that $\sigma(x)$ so the new equation is exact, then proceed as outlined in step 1.

3. If $(M_y - N_x)/N$ is not a function of x alone, see if $(M_y - N_x)/M$ is a function of y alone. If it is, an integrating factor $\sigma(y)$ can be found from (28). Multiply $Mdx + Ndy = 0$ through by that $\sigma(y)$ so the new equation is exact, then proceed as outlined in step 1.

4. If $M_y \neq N_x$, $(M_y - N_x)/N$ is not a function of x alone, and $(M_y - N_x)/M$ is not a function of y alone, then perhaps an integrating factor σ can be found that is a function of both x and y. Some examples of this type are included in the exercises.

Thus far, we've studied three types of first-order equation: the linear equation $y' + p(x)y = q(x)$ (Section 1.2), separable equations $y' = X(x)Y(y)$ (Section 1.4), and equations that are exact or can be made exact by the methods of this secction. Are these cases mutually exclusive? No. For instance, a subset of linear equations is also separable, namely, if $q(x)$ is zero or if $p(x)$ is a constant times $q(x)$. Further, *every* separable equation is exact if we write it in the form

$$X(x)dx - \frac{1}{Y(y)}dy = 0,$$

and every linear equation can be made exact (as we did in Example 3). These results are indicated schematically in Fig. 2.

We see from Fig. 2 that in principle it would suffice to study only equations that are exact [or can be made exact by $\sigma(x)$ or $\sigma(y)$] since that set *includes* linear and separable equations. However, it is important and traditional to study these cases separately — the linear equation because its theory is so complete and because it is so prominent in applications, and separable equations because the separation-of-variables solution method is so simple and, like the linear equation, it is so important in applications.

In fact, given a first-order differential equation, we suggest first checking to see if it is separable. If it is, solve by separation of variables. If not, see if it is linear or exact, whichever of these methods you prefer. If it is none of these, see if you can make it exact.

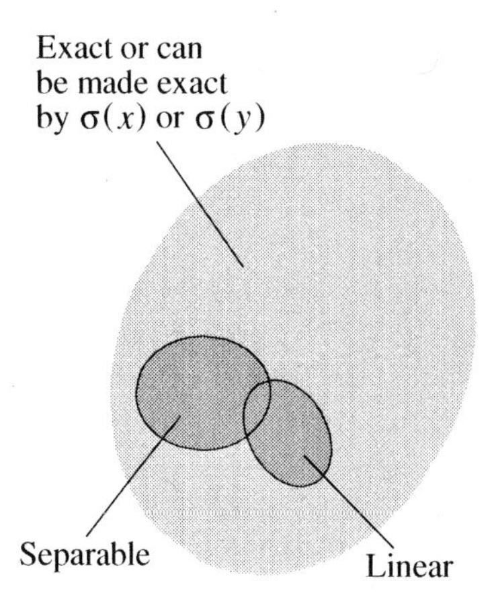

Figure 2. Schematic of the sets of first-order equations that are exact (or can be made exact), separable, or linear.

EXERCISES 1.7

1. Exact Equations. Show that the equation is exact, and obtain its solution. You may leave the answer in implicit form. If an initial condition is specified, also obtain a particular solution satisfying that condition.

(a) $3dx - dy = 0$
(b) $xdx - 4ydy = 0; \quad y(0) = -1$
(c) $4\cos 2x\,dx - e^{-5y}dy = 0$
(d) $(y^2e^x + 1)dx + 2ye^x dy = 0; \quad y(0) = -3$
(e) $(e^x + y)dx + (x - \sin y)dy = 0$
(f) $(x - 2y)dx + (y - 2x)dy = 0$
(g) $(\sin y + y\cos x)dx + (\sin x + x\cos y)dy = 0;$ $y(\pi/2) = 2\pi$
(h) $e^y dx + (xe^y - 1)dy = 0$
(i) $2xydx + [(y+1)e^y + x^2]dy = 0$
(j) $(ye^{xy} + 1)dx + xe^{xy}dy = 0; \quad y(2) = 0$
(k) $2xy\ln y\,dx + [x^2(\ln y + 1) + 2y]dy = 0 \quad (y > 0)$

2. Describe a way to make up exact equations, such as those in Exercise 1, and give an example to illustrate your procedure.

3. Determine whatever conditions, if any, must be satisfied by the constants $a, b, \ldots, f, A, B, \ldots, F$ for the equation to be exact.

(a) $(ax + by + c)dx + (Ax + By + C)dy = 0$
(b) $(ax^2 + by^2 + cxy + dx + ey + f)dx + (Ax^2 + By^2 + Cxy + Dx + Ey + F)dy = 0$

4. An Integrating Factor Needed. Find an integrating factor $\sigma(x)$ or $\sigma(y)$, if possible, and use it to solve the given differential equation. If neither is possible, state that.

(a) $3ydx + dy = 0$
(b) $ydx + x\ln x\,dy = 0 \quad (x > 0)$
(c) $y\ln y\,dx + (x + y)dy = 0 \quad (y > 0)$
(d) $dx + (x - e^{-y})dy = 0$
(e) $dx + xdy = 0$
(f) $(ye^{-x} + 1)dx + (xe^{-x})dy = 0$
(g) $xy\,dx + \sin x\cos y\,dy = 0$
(h) $\sin x\,dx + y\cos x\,dy = 0$
(i) $(3x - 2y)dx - xdy = 0$
(j) $2xy\,dx + (y^2 - x^2)dy = 0$
(k) $(y\ln y + 2xy^2)dx + (x + x^2y)dy = 0$

5. First-Order Linear Equation. Use the integrating factor

$\sigma(x) = e^{\int p(x)\,dx}$ of (40) to derive the general solution (41). That is, fill in the missing steps between (40) and (41).

6. Cases Requiring $\sigma(x,y)$. Show that the following equations are not exact, nor do they admit an integrating factor that is a function of x alone or of y alone. If possible, find an integrating factor in the form $\sigma(x,y) = x^a y^b$, where a and b are suitably chosen constants. You need not solve the equation, just find σ; if such a σ cannot be found, state that.

(a) $ydx + (x - x^2y)dy = 0$
(b) $(x + y^2)dx + (x - y)dy = 0$
(c) $(3xy + 2y^2)dx + (3x^2 + 4xy)dy = 0$

7. Nonuniqueness of σ. Of course, if σ is an integrating factor of a given equation then so is any nonzero constant times σ. But, integrating factors may be nonunique beyond an arbitrary scale factor. To illustrate, show that the equation

$$2ydx + 3xdy = 0 \tag{7.1}$$

has integrating factors $\sigma(x) = x^{-1/3}$ and also $\sigma(x,y) = 1/xy$. You need not derive these; just verify them.

8. Integrating Factors for Separable Equations. Show that

$$P(x)Q(y)dx + R(x)S(y)dy = 0 \tag{8.1}$$

has an integrating factor $\sigma(x,y) = 1/[R(x)Q(y)]$ and, after multiplying (8.1) by that integrating factor, that the solution can be found from

$$\int \frac{P(x)}{R(x)}\,dx + \int \frac{S(y)}{Q(y)}\,dy = 0. \tag{8.2}$$

NOTE: Actually, (8.1) is a *separable* equation, and using the integrating factor $\sigma(x,y) = 1/[R(x)Q(y)]$ simply amounts to separating the variables. Having thus shown the connection between separable equations and the method of integrating factors, we suggest that if an equation is separable it is simplest to just separate the variables and integrate, rather than to invoke the integrating factor method. For instance, in Exercise 4 the equations in parts (a), (b), (e), (g), and (h) could have been solved more readily by separation of variables than by the integrating mactor method. Note also that a special case of (8.1) is $P(x)dx + S(y)dy = 0$, which is exact.

9. Solve, using the methods of this section. HINT: First re-express the equation in differential form.

(a) $\dfrac{dy}{dx} = \dfrac{x-y}{x+y}$ (b) $\dfrac{dr}{d\theta} = -\dfrac{r^2\cos\theta}{2r\sin\theta + 1}$

(c) $t\dfrac{dv}{dt} = 2te^v + 1$ (d) $(x\cos y + x^2)\dfrac{dy}{dx} = \sin y$

10. If $Mdx + Ndy = 0$ and $Pdx + Qdy = 0$ are exact, does it follow that $(M+P)dx + (N+Q)dy = 0$ is exact? Explain.

11. We solved (1) by using the fact that (2) is exact. Alternatively, observe that although (1) is neither separable nor first-order linear, it *is* first-order linear if we change our viewpoint and regard x as a function of y. Use that idea to solve for $x(y)$ and verify that your solution agrees with (4).

12. Grade This. Asked to solve

$$(3x - 2y)dx - xdy = 0, \tag{12.1}$$

a student writes this: "If we can find an $F(x,y)$ such that (12.1) is $dF = F_x dx + F_y dy = 0$, then the general solution of (12.1) is $F(x,y) = C$. Integrating $F_x = 3x - 2y$ gives $F(x,y) = 3x^2/2 - 2xy + A(y)$, and then $F_y = 0 - 2x + A'(y) = -x$ gives $A'(y) = x$ and $A(y) = xy + B$. Then, $F(x,y) = 3x^2/2 - 2xy + xy + B =$ constant gives the general solution as $3x^2/2 - xy = C$." Grade that response, based on 10 points, and explain your grade.

13. Thermodynamics; the Entropy of an Ideal Gas. Consider an *ideal gas*, namely, a gas for which

$$pv = RT, \tag{13.1}$$

in which p is the pressure, v is the specific volume (i.e., the volume per mole), T is the absolute temperature, and R is the universal gas constant. The first law of thermodynamics for one mole of an ideal gas can be expressed in differential form as

$$\begin{aligned} dq &= pdv + c_v\,dT \\ &= RT\frac{dv}{v} + c_v(T)dT \end{aligned} \tag{13.2}$$

in which dq is the heat input and the known function $c_v(T)$ is the specific heat at constant volume. (Here, the independent variables are v and T rather than the generic x and y.) Show that the right-hand side of (13.2) is not an exact differential. Show that an integrating factor that is a function of v does not exist, but that an integrating factor that is a function of T does exist, $\sigma(T) = 1/T$, so that

$$\frac{dq}{T} = R\frac{dv}{v} + \frac{c_v(T)}{T}dT \tag{13.3}$$

is an exact differential, which we will call $ds(v, T)$; s is the *entropy* and we have just shown that it can be defined by the integral

$$s(v, T) = \int \frac{dq}{T}. \tag{13.4}$$

The latter formula is fundamental in the study of thermodynamics.

1.8 SOLUTION BY SUBSTITUTION

The integral $I = \int \dfrac{x^2\,dx}{(x^3+5)^2}$ may look difficult, but with the substitution $u = x^3 + 5$ we obtain $I = \frac{1}{3}\int u^{-2}du$, which readily gives $I = -1/3u + C$. Finally, replace u by x^3+5 and obtain $I = -1/[3(x^3+5)] + C$. The same idea of *substitution* can be used to solve differential equations.

1.8.1 Bernoulli's equation.

The differential equation

$$\frac{dy}{dx} + p(x)y = q(x)y^n, \tag{1}$$

in which n is a constant (not necessarily an integer), is called **Bernoulli's equation** after the Swiss mathematician **James Bernoulli**.[1] If n is 0 or 1, then (1) is linear and readily solved, so our interest is in the case where n is neither 0 nor 1.

Following Leibniz, change the dependent variable from y to v by the substitution

$$v(x) = y(x)^{1-n}, \tag{2}$$

keeping x as the independent variable. To substitute (2) into (1), we will need y and dy/dx in terms of v and dv/dx:

It is always important to be clear as to which variables are the independent and dependent variables.

$$y(x) = v(x)^{1/(1-n)} \quad \text{[from (2)]} \tag{3a}$$

and

$$\frac{dy}{dx} = \frac{1}{1-n} v^{\left(\frac{1}{1-n}-1\right)} \frac{dv}{dx} \quad \text{[by chain differentiation of (3a)]} \tag{3b}$$

[1]James (1654–1705), his brother John (1667–1748), and John's son Daniel (1700–1782) are the best known of the eight members of the Bernoulli family who were mathematicians and scientists. James proposed equation (1) as a challenge to the mathematicians of his day in 1695 and solved it himself in 1696. Other solutions were put forward by his brother John and by *Gottfried Leibniz* (1646–1716), and it is Leibniz's substitution method that we will discuss. The Bernoulli equation (1) is not related to the Bernoulli (energy) equation that one studies in a course in fluid mechanics.

so (1) becomes

$$\frac{1}{1-n} v^{n/(1-n)} \frac{dv}{dx} + p(x)\, v^{1/(1-n)} = q(x)\, v^{n/(1-n)}. \tag{4}$$

Finally, multiplying (4) by $(1-n)v^{-n/(1-n)}$ gives

$$\boxed{\frac{dv}{dx} + (1-n)p(x)\, v = (1-n)q(x).} \tag{5}$$

The upshot is that the substitution (2) works — in the sense that it reduces the nonlinear equation (1) to a simpler one, the *linear* equation (5). We can solve (5) for $v(x)$, then return from $v(x)$ to $y(x)$ by (3a).

First, put (6) in the form (1) by multiplying through by $1/x$.

EXAMPLE 1. Solve

$$x\frac{dy}{dx} - y = -2xy^2. \tag{6}$$

The latter is a Bernoulli equation with $p(x) = -1/x$, $q(x) = -2$, and $n = 2$. Then, (2) gives $v = 1/y$. Assuming that $y \neq 0$, for $v = 1/y$ to be meaningful, (5) is the linear equation

$$\frac{dv}{dx} + \frac{1}{x}v = 2, \tag{7}$$

with solution

$$v(x) = x + \frac{C}{x}. \tag{8}$$

But $v = 1/y$, so (8) gives $1/y = x + C/x$, and hence

$$y(x) = \frac{x}{x^2 + C} \tag{9}$$

is the solution of (6).

COMMENT. Since we assumed that $y \neq 0$, we must check $y = 0$ separately. In fact, $y(x) = 0$ does satisfy (6), and it cannot be obtained from (9) by any (finite) choice of C. Thus, besides the one-parameter family of solutions (9) we have the additional solution $y(x) = 0$. ∎

The sequence of steps in solving a differential equation by the method of substitution is as follows: *Find a substitution, if possible, that converts the given differential equation to one we can solve. Make the substitution, obtain the new differential equation, solve it, and return to the original variables.*

1.8.2 Homogeneous equations. An equation $y' = f(x, y)$ is **homogeneous** if $f(x, y)$ can be expressed as a function of the ratio y/x alone, in which case we can express the equation as

$$\boxed{\frac{dy}{dx} = F\left(\frac{y}{x}\right).} \tag{10}$$

For instance,

$$\frac{dy}{dx} = \frac{2x + 2y}{3x + y} \tag{11}$$

is homogeneous because if we divide the numerator and denominator on the right-hand side by x we can express (11) as

$$\frac{dy}{dx} = \frac{2 + 2\frac{y}{x}}{3 + \frac{y}{x}} = F\left(\frac{y}{x}\right). \tag{12}$$

However, the equation

$$\frac{dy}{dx} = \frac{x + y + 2x^2}{x + 4y} = \frac{1 + \frac{y}{x} + 2x}{1 + 4\frac{y}{x}} \tag{13}$$

is *not* homogeneous because the right-hand side is a function of y/x and x, not of y/x alone.

If the equation is homogeneous, of the form (10), it seems natural to let y/x be a single variable, say v, so $v = y/x$. That is, $v(x) = y(x)/x$, or

$$\boxed{y(x) = xv(x).} \tag{14}$$

CAUTION: Earlier, we defined a linear equation $y' + p(x)y = q(x)$ to be homogeneous if the forcing function $q(x)$ is zero. That was a different use of the word homogeneous and is not relevant in the present discussion.

To put (14) into (10) we need dy/dx, so differentiate (14):

$$\frac{dy}{dx} = v(x) + x\frac{dv}{dx}. \tag{15}$$

Using (14) and (15), equation (10) becomes

$$v + x\frac{dv}{dx} = F(v)$$

or

$$\boxed{x\frac{dv}{dx} = F(v) - v,} \tag{16}$$

which is simple because it is separable. Thus,

$$\boxed{\int \frac{dv}{F(v) - v} = \int \frac{dx}{x}.} \tag{17}$$

As discussed in Section 1.4, if v_0 is any root of $F(v) - v = 0$ then $v(x) = v_0$ is an additional solution of (16), in addition to the solutions found from (17). This point will come up, below, in COMMENT 2.

Evaluate the integrals in (17) and, in the result, replace v by y/x.

EXAMPLE 2. Solve the homogeneous equation (11),

$$\frac{dy}{dx} = \frac{2x+2y}{3x+y} = \frac{2+2(y/x)}{3+(y/x)} = F\left(\frac{y}{x}\right). \tag{18}$$

In this case $F(v) = \dfrac{2+2v}{3+v}$, so (16) is

$$\begin{aligned} x\frac{dv}{dx} &= \frac{2+2v}{3+v} - v \\ &= -\frac{v^2+v-2}{3+v}. \end{aligned} \tag{19}$$

If $v^2 + v - 2 \neq 0$, separation of variables (and partial fractions) gives

$$-\int \frac{(3+v)\,dv}{v^2+v-2} = \int \frac{dx}{x}, \tag{20a}$$

$$\frac{1}{3}\int \frac{dv}{v+2} - \frac{4}{3}\int \frac{dv}{v-1} = \int \frac{dx}{x}, \tag{20b}$$

$$\frac{1}{3}\ln|v+2| - \frac{4}{3}\ln|v-1| = \ln|x| + A, \tag{20c}$$

$$\ln\left|\frac{(v+2)^{1/3}}{x(v-1)^{4/3}}\right| = A, \tag{20d}$$

$$\left|\frac{(v+2)^{1/3}}{x(v-1)^{4/3}}\right| = e^A, \tag{20e}$$

$$\frac{(v+2)^{1/3}}{x(v-1)^{4/3}} = \pm e^A \equiv B, \tag{20f}$$

$$(v+2) = B^3x^3(v-1)^4. \tag{20g}$$

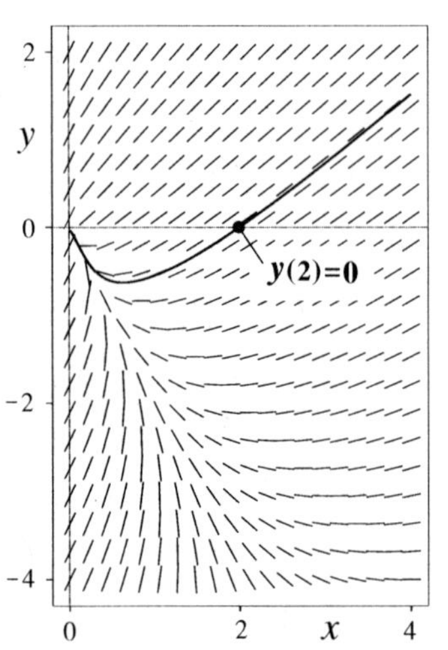

Figure 1. Solution of (18) with the initial condition $y(2) = 0$, together with the direction field.

Putting $v = y/x$, renaming B^3 as C and simplifying gives the solution in implicit form as

$$y + 2x = C(y-x)^4. \tag{21}$$

If we have an initial condition $y(2) = 0$, for instance, (21) gives $C = 1/4$. The corresponding solution, and the direction field, are plotted in Fig. 1.

COMMENT 1. Is C arbitrary in (21)? We need to track A, B, C in (20). In (20c), A is arbitrary ($-\infty < A < \infty$). Consequently, $B = \pm e^A$ is arbitrary but nonzero, because $e^A \neq 0$. Finally, since B is arbitrary but nonzero and $C = B^3$, then C is arbitrary but nonzero in (21). Conclusion: C in (21) is arbitrary but nonzero.

COMMENT 2. Besides the family of solutions given by (21), we must see if there are any additional solutions from the roots of $v^2 + v - 2 = 0$, which we assumed was nonzero when we proceeded from (19) to (20a). The roots are $v = 1$ and $v = -2$, which, recalling that $v = y/x$, correspond to straight-line solutions $y = x$ and $y = -2x$ of (18). Of these, $y = -2x$ is contained within (21) if we allow $C = 0$, but $y = x$ cannot be obtained from (21) by any finite choice of C. We conclude that the solutions of (18) are those defined implicitly by (21) (with $-\infty < C < \infty$), plus the line $y = x$. It is tempting to not fuss with such details as whether C in (21) is arbitrary, or arbitrary but nonzero, and whether there are any additional solutions from the roots of $v^2 + v - 2 = 0$, but if we did not fuss with those details we would have missed the solution $y = x$. That loss would be fatal if an initial condition were prescribed on that line.

A number of solutions of (18) are plotted in Fig. 2. ∎

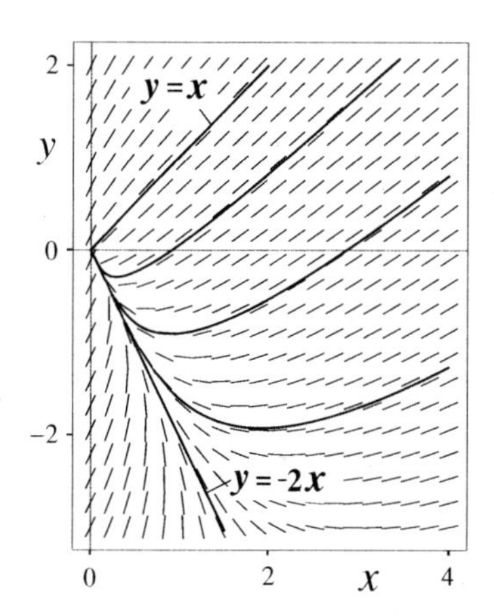

Figure 2. Representative solutions of (18), together with the direction field.

Closure. In earlier sections we attacked the differential equation $y' = f(x, y)$ by considering only special cases that are tractable: equations that are linear, separable, or exact (or can be made exact by an integrating factor that is a function of x or y). Those cases by no means cover all possible equations $y' = f(x, y)$, but they do cover a great many equations that arise in applications. Similarly, for the method of substitution one tries to develop substitutions that work for various types of equations. In this section we considered only two: Bernoulli equations and homogeneous equations. Additional types are included in the exercises.

EXERCISES 1.8

1. Bernoulli Equations. Show that the equation is of Bernoulli type; that is, identify $p(x)$, $q(x)$, and n. Then, solve it accordingly. If the equation happens to also be separable you can solve it by separation of variables as well, if you wish, to check your results.

(a) $xy' - 2y = x^3y^2$ (b) $y' + 2y = -6e^{3x}y^{3/2}$
(c) $(1+x)y' + 2y = 2\sqrt{y}$ (d) $xy' = y - xy^2$
(e) $\sqrt{y}\,(3y' + y) = x$ (f) $y' = y^2$
(g) $y' = x\sqrt{y}$ (h) $xy' - y = -12x^3y^2$

2. Inventing Leibniz's Substitution. In case you regard a substitution such as Leibniz's substitution (2) as a "miracle," let us illustrate how that this (and other substitutions) might be developed in a reasonable and systematic way. To begin, we observe that the difficulty with (1) is the y^n term. Thus, try letting y^n be a new variable, for surely the substitution $v = y^n$ will simplify the right-hand side of (1). But it is possible that while simplifying the right-hand side it might complicate the left-hand side.

(a) Try it. Let $v = y^n$ and show that it does not work.

(b) Not discouraged, try $v = y^r$ instead, where this time the exponent r is not prescribed in advance. Make that substitution in (1) and choose r, if possible, so that the equation for $v(x)$ is simple, such as linear or separable. Show that these steps lead to the choice $r = 1 - n$ and hence to Leibniz's substitution (2).

3. Homogeneous. (a) Solve $(2x - y)y' = x - 2y$ and find a particular solution for each of the initial conditions $y(2) = 0$, $y(2) = 4 - 2\sqrt{3}$, $y(2) = 1$, $y(2) = 7$, $y(2) = 4 + 2\sqrt{3}$, $y(2) = 8$ and determine its interval of existence. Obtain computer-generated graphs of those solutions.
(b) Solve $y' = (4x^2 + 3y^2)/2xy$ with the initial condition $y(1) = 2$, and determine the interval of existence.
(c) Solve $y' = (xy + 2y^2)/x^2$ $(x > 0)$, find a particular solution satisfying $y(1) = 2$, and determine its interval of existence. NOTE: The differential equation is both homogeneous and a Bernoulli equation. Solve it both ways and show that your results are the same.
(d) Solve $y' = e^{y/x} + y/x$ $(x > 0)$ with the initial condition $y(1) = 0$, and determine the solution's interval of existence.
(e) Solve $y' = \tan(y/x) + y/x$ $(x > 0)$ with the initial condition $y(1) = \pi/6$, and determine the solution's interval of existence.

4. Almost Homogeneous. The equation

$$\boxed{y' = \frac{ax + by + c}{dx + ey + f},} \tag{4.1}$$

in which $a, b, \ldots, f$ are constants, is "almost homogeneous" in the sense that it would be homogeneous if c and f were not present.

(a) Change variables from $x, y(x)$ to $X, Y(X)$ according to the "translation"

$$x = X + h, \quad y = Y + k, \tag{4.2}$$

and choose the constants h and k so as to knock out the c and the f. Show that the result is

$$\frac{dY}{dX} = \frac{aX + bY}{dX + eY}, \tag{4.3}$$

with $h = (bf - ce)/(ae - bd)$ and $k = (cd - af)/(ae - bd)$, provided that $ae - bd \neq 0$.
(b) Use the idea in part (a) to solve $y' = (y+1)/(2x-y-3)$. (This problem is continued in Exercise 10.)
(c) Similarly, solve $y' = (x-y-4)/(x+y-4)$.
(d) Similarly, solve $y' = (y+x+2)/(y-x)$.
(e) As stated in part (a), the change of variables (4.2) fails if $ae - bd = 0$. Devise a substitution that *will* work in that case, and use it to solve $y' = (2x+4y+1)/(4x+8y-2)$.

5. The equation $y' = x^3 e^{y/x} + y/x$ is not homogeneous because of the x^3. Show that the change of variables $v = y/x$ from $y(x)$ to $v(x)$ works nevertheless, and use it to solve for $y(x)$.

6. To solve equations of the form

$$\boxed{y' = f(ax + by)} \tag{6.1}$$

it seems reasonable to try the substitution $v = ax + by$; that is,

$$v(x) = ax + by(x). \tag{6.2}$$

(a) Show that (6.2) simplifies (6.1) to the separable equation

$$v' = bf(v) + a. \tag{6.3}$$

Use this idea to solve the following equations.

(b) $y' = (2x+y)^2 - 2$
(c) $(x-y)y' = 4$
(d) $y' = (x + y + 2)/(x + y)$

7. Riccati's Equation. The equation

$$\boxed{y' = p(x)y^2 + q(x)y + r(x)} \tag{7.1}$$

is called **Riccati's equation** after the Italian mathematician *Jacopo Francesco Riccati* (1676 – 1754). The latter is made difficult by the y^2 term, which makes the equation nonlinear [unless $p(x)$ is zero]. The key to being able to solve a Riccati equation is to find any one solution, hopefully by inspection. For suppose $y = \eta(x)$ is any particular solution of (7.1). Then, show that by changing the dependent variable from y to v according to

$$y = \eta(x) + \frac{1}{v} \tag{7.2}$$

the Riccati equation (7.1) is converted to

$$v' + \left[2p(x)\eta(x) + q(x)\right]v = -p(x), \tag{7.3}$$

which is *linear*. This solution method was discovered by *Leonhard Euler* (1707 – 1783) in 1760. Note that if $r(x) = 0$ in (7.1), then an obvious choice for a particular solution is simply $y = \eta(x) = 0$, although in that case (7.1) is also a Bernoulli equation.

8. Read Exercise 7 and use the method described there to solve the following Riccati equations. HINT: In part (b) try $\eta(x)$ in the form ax and determine an a that works.

(a) $y' = 4y + y^2$
(b) $y' = y^2 - 2xy + 1 + x^2$
(c) $y' = e^{-x}y^2 - y$
(d) $y' = x^2y^2 - y$

9. Smorgasbord. You don't need to solve these. Just identify any solution method that will work and give any substitutions that will be needed. You may find ideas in the preceding exercises.

(a) $xy' = \sqrt{xy} + y$
(b) $y' = (y+x)^3$
(c) $(2x-y)y' = y$
(d) $y' = x(y-x)^2 + 1$
(e) $(xy)' = y^{3.2}$
(f) $y' = x^3y^2 - y$
(g) $y^2y' = (x+y)^2$
(h) $yy' = x - y$
(i) $y' - y^2 = 3ye^x$
(j) $y' = 4 + (y-5x)^4$

10. Continuation of Exercise 4(b). For the differential equation in Exercise 4(b), use computer graphics to plot representative solution curves, being sure to include the two straight line solutions.

ADDITIONAL EXERCISES

11. Swimming in a Current. A river is a miles wide and has a current W mi/hr. Two swimmers, Maifeng and Yuan, are to race, from point A on one bank to point B on the opposite bank (see the figure). They are equally fast, able to swim

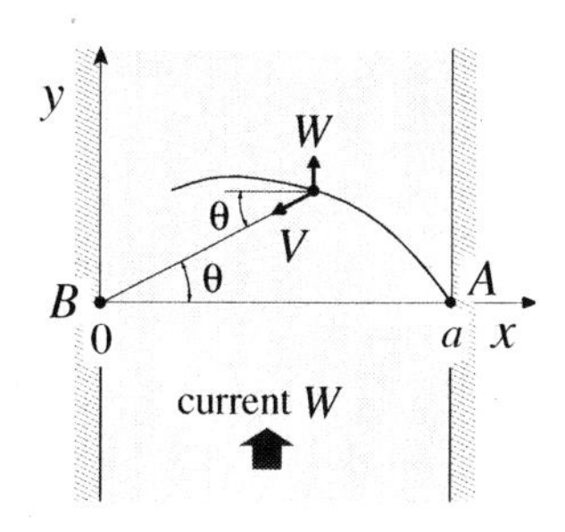

with a constant speed V relative to the water, so the outcome will be determined by strategy. Yuan elects to swim so as to always aim at the destination B, as indicated in the figure, and Maifeng elects to aim upstream so as to swim straight across from A to B. Assume that they swim faster than the current, so $V > W$.

(a) Compute and compare the crossing time T for each of them. The calculation for Yuan is the harder so we'll get you started. Yuan's x and y velocity components are

$$x' = -V\cos\theta = -V\frac{x}{\sqrt{x^2+y^2}}, \tag{11.1a}$$

$$y' = W - V\sin\theta = W - V\frac{y}{\sqrt{x^2+y^2}}, \tag{11.1b}$$

and dividing (11.1b) by (11.1a) gives

$$\frac{dy}{dx} = \frac{y - r\sqrt{x^2+y^2}}{x} \tag{11.2}$$

with $r = W/V$. Solve (11.2) and show that the path traversed by Yuan is

$$y(x) = \frac{a}{2}\left[\left(\frac{x}{a}\right)^{1-r} - \left(\frac{x}{a}\right)^{1+r}\right]. \tag{11.3}$$

Put (11.3) into (11.1a) and get a differential equation for $x(t)$ alone. Solve that equation for $x(t)$, and compute Yuan's crossing time. Then compute Maifeng's time — which should be a much simpler calculation. You should obtain the two times

$$T = \frac{a}{V}\frac{1}{\sqrt{1-r^2}} \quad \text{and} \quad \frac{a}{V}\frac{1}{1-r^2}, \tag{11.4}$$

but we leave it for you to determine which one corresponds to Yuan and which one to Maifeng. As a partial check on (11.4) show that both results are correct in the limit as $r \to 0$.

(b) Show that Maifeng wins if $W < V$ $(r < 1)$, as assumed, but that if $W \geq V$ $(r \geq 1)$ then neither swimmer can reach point B.

NOTE: This problem is based on one given in Ralph Palmer Agnew's dated-but-still-excellent text *Differential Equations* (NY: McGraw Hill, 1942).

1.9 NUMERICAL SOLUTION BY EULER'S METHOD

The preceding sections have been devoted mostly to analytical solution methods: for instance, solving the linear equation $y' + p(x)y = q(x)$ by using an integrating factor, the use of separation of variables, exactness, and so on. Our use of direction fields and phase line analysis in those sections has been more qualitative. Here, we complement those analytical and qualitative approaches with a brief introduction to *quantitative* methods. By a quantitative method we mean one that uses a numerical algorithm to generate the solution numerically, approximately and only at discrete points, the calculations normally being carried out on a computer rather than by hand.

1.9.1 Euler's method.

We consider the IVP

$$\boxed{y' = f(x, y); \quad y(a) = b} \tag{1}$$

Our chief interest is the case in which $y' = f(x, y)$ is nonlinear, because we have an exact solution for the linear equation $y' + p(x)y = q(x)$.

for $y(x)$.

To motivate the simplest numerical solution method, Euler's method, consider a specific example,

$$y' = y + 2x - x^2; \quad y(0) = 1 \quad (0 \le x < \infty) \tag{2}$$

which has the exact solution (Exercise 1)

$$y(x) = x^2 + e^x. \tag{3}$$

Of course, we don't *need* to solve (2) numerically because we know its solution, given by (3), but our aim here is only to illustrate the method. In fact, it is good to begin with a problem for which we do know the exact solution because then we can compare our numerical results with that solution.

In Fig. 1 we display the direction field defined by $f(x, y) = y + 2x - x^2$, and the exact solution $y(x)$ given by (3). In graphical terms, Euler's method amounts to using the direction field as a "road map" to develop an approximate solution to (2) in a step-by-step manner. Beginning at the initial point P, namely $(0, 1)$, we strike out in the direction dictated by the lineal element at that point. As seen from the figure, the farther we move along that line, from the starting point P to a stopping point Q, the more we can expect our path to deviate from the exact solution. Thus, the idea is not to move very far. Stopping at $x = 0.5$, for the sake of illustration, we then revise our direction according to the slope of the lineal element there, at Q. Moving in that new direction until $x = 1$, we revise our direction again at R, and so on, in x increments of 0.5.

Thus, our strategy is this: First, **discretize** the problem (1) by seeking the solution $y(x)$ not everywhere on the x interval, but only at discrete points $x_0 = a$, $x_1 = x_0 + h$, $x_2 = x_1 + h$, and so on, where h is our chosen **step size** for the calculation; in Fig. 1, $h = 0.5$. That is, rather than seek the function $y(x)$ we seek only the discrete approximate values $y_1, y_2, \ldots$ at $x_1, x_2, \ldots$, respectively. According to

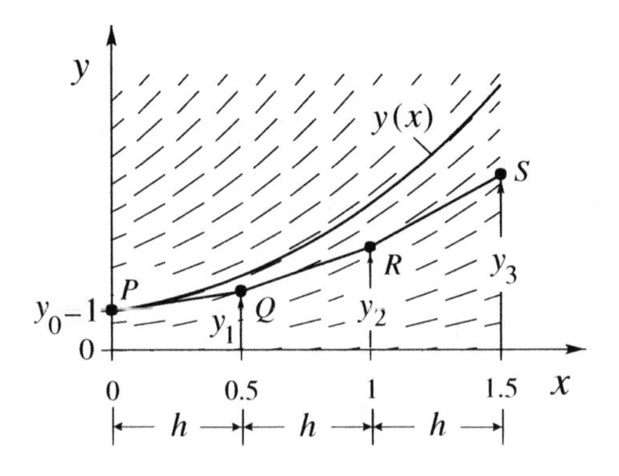

Figure 1. Direction field motivation of Euler's method, for the initial value problem (2); "going with the flow."

the algorithm known as **Euler's method**, we compute y_{n+1} as the preceding value y_n plus the slope $f(x_n, y_n)$ at (x_n, y_n) times the step size h:

From Fig. 1, we see that the values y_1, y_2, ... generated by the Euler algorithm (4) do not, in general, fall on the solution curve and are only approximate.

$$\boxed{y_{n+1} = y_n + f(x_n, y_n)h, \qquad (n = 0, 1, 2, \ldots)} \tag{4}$$

in which f is the function on the right-hand side of the given differential equation (1), $x_0 = a$, $y_0 = b$, h is the chosen step size, and $x_n = x_0 + nh$.

Euler's method is also known as the **tangent-line method** because each straight-line segment of the approximate solution, emanating from (x_n, y_n), is tangent to the exact solution curve through that point.

We will illustrate the calculation, using the IVP (2).

EXAMPLE 1. Application of Euler's Method to (2). In (2), $f(x, y) = y + 2x - x^2$, $x_0 = 0$, and $y_0 = 1$. With $h = 0.5$, Euler's method (4) gives

$$y_1 = y_0 + \left(y_0 + 2x_0 - x_0^2\right) h = 1 + (1 + 0 - 0)(0.5) = 1.5,$$
$$y_2 = y_1 + \left(y_1 + 2x_1 - x_1^2\right) h = 1.5 + (1.5 + 1 - 0.25)(0.5) = 2.625,$$
$$y_3 = y_2 + \left(y_2 + 2x_2 - x_2^2\right) h = 2.625 + (2.625 + 2 - 1)(0.5) = 4.4375,$$

and so on. ▮

We can see from Example 1 that Euler's method is simple and readily implemented, even by hand calculation.

Evidently, the greater the step size the less accurate the results, in general. For instance, we can see that the first point Q in Fig. 1 deviates more and more from the exact solution as the step size is increased — that is, as the segment PQ is extended. Conversely, we expect the approximate solution to approach the exact solution curve as $h \to 0$. We are encouraged in this expectation by the results shown in Table 1 for the IVP (2), obtained by Euler's method with step sizes $h = 0.5, 0.1$, and 0.02; we've included the exact solution given by (3) in the final column, for comparison. To keep the tabulation short, many intermediate y_n and $y(x)$ values were omitted for the cases $h = 0.1$ and $h = 0.02$. For instance, for $h = 0.1$ we omitted y_1 through y_4, y_6 through y_9, and y_{11} through y_{14}.

Table 1. Comparison of numerical solution of (2) using Euler's method, with the exact solution (3).

x	$h = 0.5$	$h = 0.1$	$h = 0.02$	Exact $y(x)$
0	$y_0 = 1$	$y_0 = 1$	$y_0 = 1$	$y(0) = 1$
0.5	$y_1 = 1.5$	$y_5 = 1.7995$	$y_{25} = 1.8778$	$y(0.5) = 1.8987$
1.0	$y_2 = 2.625$	$y_{10} = 3.4344$	$y_{50} = 3.6578$	$y(1.0) = 3.7183$
1.5	$y_3 = 4.4375$	$y_{15} = 6.1095$	$y_{75} = 6.5975$	$y(1.5) = 6.7317$

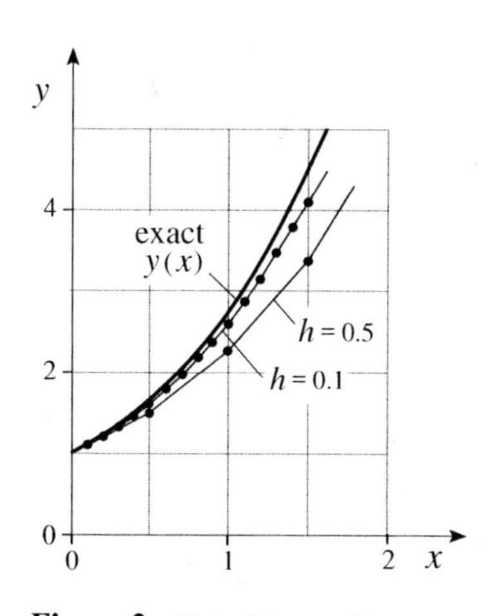

Figure 2. Plot of the results reported in Table 1. Note that the approximate solution approaches the exact solution as h is decreased.

If we scan across each of the bottom three rows of the tabulation, we see that the approximate solution values do appear to be converging to the exact solution as $h \to 0$ (though we cannot be certain of that, no matter how much we reduce h). Besides tabulating the results, above, we have also plotted them in Fig. 2 for $h = 0.5$ and 0.1.

1.9.2 Convergence of Euler's method. Two questions follow: Does the method really give convergence to the exact solution as $h \to 0$ and, if so, how fast? By the method being **convergent** we mean that *for any fixed x value in the x interval of interest the sequence of y values, obtained from (4) using smaller and smaller step size h, tends to the exact solution $y(x)$ of (1), at that point x, as $h \to 0$.*

In Table 1, this amounts to moving from left to right across a given row. Understand this definition.

There are two sources of error in the numerical solution. One is the tangent-line approximation upon which the method is based, and the other is an accumulation of numerical **roundoff errors** that result from the machine carrying only a finite number of significant figures, after which it rounds (or chops) off. Typically, roundoff errors are negligible, and they will be ignored in this discussion.

The single-step error. Although we are interested in the accumulation of error after many steps have been carried out, to reach a given x, it is logical to begin by investigating the error incurred *in a single step*, from x_n to x_{n+1}.

To distinguish between the exact and the approximate solutions, we will denote the exact solution at any computation point x_n as $y(x_n)$, and the approximate numerical solution there by a subscript notation, as y_n.

Notation: We denote the exact solution at x_n as $y(x_n)$ and the numerical solution there as y_n.

If we start out at x_n with the correct value, so $y(x_n) = y_n$, what is the error $y(x_{n+1}) - y_{n+1}$ after that one step? Write expressions for $y(x_{n+1})$ and y_{n+1}, and then subtract them. First, the Taylor series expansion of y about x_n gives

$$\begin{aligned} y(x_{n+1}) &= y(x_n) + y'(x_n)(x_{n+1} - x_n) + \frac{y''(x_n)}{2!}(x_{n+1} - x_n)^2 + \cdots \\ &= y(x_n) + f(x_n, y(x_n))h + \frac{y''(x_n)}{2!}h^2 + \cdots \\ &= y_n + f(x_n, y_n)h + \frac{y''(x_n)}{2!}h^2 + \cdots . \end{aligned} \tag{5}$$

In the final equality in (5) we use the fact that we begin our step, at x_n, with y_n equal to the exact solution there, $y(x_n)$.

Next, the Euler algorithm gives

$$y_{n+1} = y_n + f(x_n, y_n)h, \tag{6}$$

and if we subtract (6) from (5), we obtain the **single-step error** (SSE), as

$$\text{SSE} = y(x_{n+1}) - y_{n+1} = \frac{y''(x_n)}{2!}h^2 + \frac{y'''(x_n)}{3!}h^3 + \cdots \sim \frac{y''(x_n)}{2}h^2 \tag{7}$$

The single-step error, denoted here as SSE, is the error at x_{n+1} if $y(x_n) = y_n$.

as $h \to 0$. We are interested in how fast SSE $\to 0$ as $h \to 0$ and (7) tells us that it tends to zero proportional to h^2 so we write, more simply than (7),

$$\boxed{\text{SSE} = O(h^2).} \tag{8}$$

According to (8), the SSE for Euler's method is of order h^2.

In words, (8) means that the single-step error SSE tends to zero proportional to h^2. That is, SSE $\sim Ch^2$ for some nonzero constant C, as $h \to 0$. The notation used in (8) is the standard **big oh notation** used to express *order of magnitude*. Read (8) as follows: SSE is of order h^2 as $h \to 0$.

The error and convergence. Of ultimate interest, however, is not the error incurred in a single step, but the error that accumulates over *all* the steps (from x_0 to any given point x_N). That error is the difference between the exact solution and the computed solution at x_N, and we will simply call it the error E:

$$E \equiv y(x_N) - y_N. \tag{9}$$

Let us illustrate, using Table 1. For $h = 0.02$, for instance, the error at $x = 1$ is $y(1) - y_{50} = 3.7183 - 3.6578 = 0.0605$, and at $x = 1.5$ it is $y(1.5) - y_{50} = 6.7317 - 6.5975 = 0.1342$.

We can estimate E, at least insofar as its order of magnitude, as the single-step error SSE times the number of steps N. Since SSE $= O(h^2)$, that idea gives

$$\begin{aligned} E = O(h^2) \cdot N = O(h^2)\frac{Nh}{h} &= O(h^2)\frac{x_N - x_0}{h} \\ &= O(h)(x_N - x_0) = O(h). \end{aligned} \tag{10}$$

In the last equality in (10) we absorbed the $x_N - x_0$ factor into the $O(h)$ because $x_N - x_0$ is simply a constant, and the big oh notation is insensitive to (nonzero) constant scale factors. Thus,

Regarding the last step in (10), the big oh notation is insensitive to nonzero scale factors. For instance, both $\sin x = x - \frac{x^3}{3!} + \cdots = O(x)$ and $\sin 8x = 8x - \frac{8^3x^3}{3!} + \cdots = O(x)$, as $x \to 0$. We do not distinguish between $O(8x)$ and $O(x)$.

$$\boxed{E = O(h),} \tag{11}$$

which tells us how fast the numerical solution converges to the exact solution, at any fixed x location, as $h \to 0$: $E \sim Ch$, as $h \to 0$, for some constant C. It tends to zero proportional to h.

Our steps in (10) were formal, not rigorous, but the result (11) is correct and indicates that the Euler method (4) is convergent because $E = O(h)$ does tend to zero as $h \to 0$. More generally, if, for a given method, $E = O(h^p)$ as $h \to 0$, then the method is convergent if $p > 0$ (because $h^p \to 0$ as $h \to 0$ if $p > 0$), and the method is said to be **of order p**. With $p = 1$ in (11), we see that **Euler's method is a first-order method**.

The order of the method and the order of the differential equation are distinct and unrelated.

How do we know how small to choose h in a given application? As a rule of thumb, repeat the calculation, reducing h until the results settle down to the desired accuracy. The foregoing is the same idea normally used in summing an infinite series, adding more terms until successive partial sums settle down to the desired number of significant figures.

1.9.3 Higher-order methods. Though convergent and readily implemented, Euler's method may be too inaccurate because it is only a first-order method. That is, since the error at any given x point is proportional to h to the first power, we must make h extremely small if the error is to be extremely small, and if h is extremely small then the number of computational steps is extremely large. Consequently, it is common to favor higher-order methods.

If a method is of order p, then $E = O(h^p)$ as $h \to 0$ or, equivalently,

$$\boxed{E(h) \sim Ch^p} \tag{12}$$

as $h \to 0$.

There are higher-order methods available, such as the **Runge–Kutta methods** of orders 2, 3, and 4, which are abbreviated as rk2, rk3, and rk4, respectively. As their names indicate, they are of orders $p = 2, 3$, and 4, respectively. We will not give those methods here, but merely highlight the dramatic increase in accuracy afforded by such higher-order methods, in Table 2. There, we tabulate the values of the numerical solution of the IVP (2), obtained using the first-order Euler method and the second- and fourth-order Runge–Kutta methods, at $x = 1$, for step sizes of $h = 0.1$ and 0.01. We've tabulated the first eleven digits and have indicated by bold fonts as many digits as agree with those in the exact solution. We see from

The accuracy increases if for a given method we decrease h, or if for a given h we turn to higher-order methods.

Table 2. Increasing the order of the method.

	Euler	rk2	rk4
$h = 0.1$	**3.**4343682140	**3.7**059185568	**3.71827**63403
$h = 0.01$	**3.**6877656911	**3.718**1513781	**3.7182818**278
Exact $y(1)$	**3.7182818285**	**3.7182818285**	**3.7182818285**

the table that the accuracy of fourth-order methods is quite impressive. In fact, rk4 is one of the most widely used methods.

Closure. The Euler method is given by (4). It is readily implemented, either using a hand-held calculator or running it on a computer. The method is convergent but is only of first order, so it is generally not very accurate unless the step size h is made extremely small. Thus, for "serious computation," higher-order methods such as rk4 are normally used and are available within CAS software. Of course, in a given application we may not even have a very accurate differential equation model in the first place and/or may not know the physical parameters very accurately, in which case there may be little point in demanding extreme accuracy from the numerical algorithm. By the way, when studying the calculus, one is not in a position to appreciate the many important applications of Taylor series. They are invaluable in developing numerical solution methods for differential equations, as we begin to see here in Section 1.9.2.

EXERCISES 1.9

1. Filling in a Gap. Derive the particular solution (3) of the IVP (2).

2. By Hand. Use the Euler method to compute, by hand, y_1, y_2, and y_3 for the specified IVP using $h = 0.2$.

(a) $y' = -y;\quad y(0) = 1$
(b) $y' = 2xy;\quad y(0) = 4$
(c) $y' = 1 + 2xy^2;\quad y(-1) = 2$
(d) $y' = 2xe^y;\quad y(1) = -1$
(e) $y' = x^2 - y^2;\quad y(3) = 0$
(f) $y' = x \sin y;\quad y(0) = 1$
(g) $y' = 5y - 2\sqrt{y};\quad y(1) = 3$
(h) $y' = \sqrt{x+y};\quad y(0) = 3$

3. By Computer. Use computer software to solve the given IVP for $y(x)$ by Euler's method, with a step size of 0.001. Print your Euler-computed values y_{100}, y_{200}, and y_{300} (i.e., at $x = 1, 2, 3$) as well as the exact solution (which is given) at those same points, for comparison.

(a) $y' = -y,\ y(0) = 1;\ y(x) = e^{-x}$
(b) $y' = 0.1(x+1)e^y,\ y(0) = 0;\ y(x) = \ln \dfrac{20}{20 - 2x - x^2}$
(c) $y' = x^2 - y,\ y(0) = 2;\ y(x) = x^2 - 2x + 2$
(d) $y' = 0.01xy^2,\ y(0) = 5;\ y(x) = 200/(40 - x^2)$
(e) $y' = x^2/y^2,\ y(0) = 4;\ y(x) = (x^3 + 64)^{1/3}$
(f) $y' = \dfrac{y}{x-5} - 2y^2,\ y(0) = 1;\ y(x) = \dfrac{x-5}{x^2 - 10x - 5}$
(g) $y' = -y^2,\ y(0) = -0.2;\ y(x) = 1/(x-5)$

4. (a)–(g) **Convergence.** For the corresponding part of Exercise 3, use computer software to solve the given IVP for $y(x)$ at $x = 3$. Obtain computed values of $y(3)$ using step sizes of $0.1, 0.01, 0.001$, and 0.0001, as well as the exact value of $y(3)$. Do your results appear to be consistent with Euler's method being a first-order method? Explain.

5. Negative Steps? Thus far we've taken the step h to be positive, and therefore we've developed solutions to the right of the initial point. Is Euler's method valid if we use a negative step, $h < 0$, and develop a solution to the left? Explain.

6. Variable Step Size? In this section we've taken the step size h to be a constant from one step to the next. Is there any reason why we could not vary h from one step to the next? Why might we *want* to use a variable step size? Explain.

7. (a) What is meant by discretizing an IVP?
(b) What is meant by a method being convergent?

8. Verifying Convergence for a Simple Example. For the simple IVP

$$y' = Ay; \qquad y(0) = y_0, \tag{8.1}$$

in which A is a constant, we can actually show that its Euler solution converges to the exact solution $y(x) = y_0 e^{Ax}$ as $h \to 0$. For (8.1), Euler's method gives

$$y_{n+1} = y_n + hAy_n = (1 + Ah)y_n; \qquad y_0 \text{ given} \tag{8.2}$$

(a) Show that (8.2) gives the solution for y_n as

$$y_n = (1 + Ah)^n y_0. \tag{8.3}$$

(b) To see if the latter converges to the exact solution, consider a *fixed* point x. To arrive at x after n steps we must choose h to be x/n so (8.3) becomes

$$y_n = \left(1 + \frac{Ax}{n}\right)^n y_0. \tag{8.4}$$

With x fixed, h tending to zero corresponds to n tending to infinity. Show that the limit of (8.4) as $n \to \infty$ is indeed the exact solution $y_0 e^{Ax}$. HINT: Recall from the calculus that

$$\lim_{h \to 0} \left(1 + \frac{1}{h}\right)^h = e.$$

9. Empirical Determination of the Order of Euler's Method. Suppose we do not know Euler's method is a first-order method, and want to determine its order empirically. Make up a simple "test equation" such as $y' = -y$, with initial condition $y(0) = 1$, so the known exact solution is $y(x) = e^{-x}$. Next, suppose we solve that IVP by Euler's method, for various h's, each time computing the solution at $x = 1$, at which point the known exact solution is $y(1) = 0.3678794412$. For $h = 0.1$, 0.01, 0.001, and 0.0001 the results at $x = 1$ are: $y|_{h=0.1} = 0.3486784401$, $y|_{h=0.01} = 0.3660323413$, $y|_{h=0.001} = 0.3676954248$, and $y|_{h=0.0001} = 0.3678610464$.

(a) From the computed values of y at $x = 1$ for $h = 0.1$ and $h = 0.01$, and the exact value of $y(1)$, use (12) to solve for p. That is, write $0.3678794412 - 0.3486784401 = C(0.1)^p$ and $0.3678794412 - 0.3660323413 = C(0.01)^p$, and solve those two equations for p. Show that $p = 1.01683$.
(b) Repeat the empirical evaluation of p, this time using the results for $h = 0.01$ and $h = 0.001$, and show that $p = 1.00163$.
(c) Finally, repeat the evaluation of p again, this time using

the results for $h = 0.001$ and $h = 0.0001$, and show that $p = 1.00016$.
NOTE: The results for p in parts (a), (b), (c) are not exact, but only approach the exact value $p = 1$ as $h \to 0$ because (12) is only an asymptotic result, valid as $h \to 0$.

10. Formula for Empirical Determination of the Order of Any Given Method. Exercise 9 pertained specifically to Euler's method. More generally, suppose we are using any numerical solution algorithm and want to determine its order empirically, for instance as a check against possible programming errors that would no doubt reduce its order. To do so, consider a simple test equation such as $y' = -y$ with initial condition $y(0) = 1$ and exact solution $y(x) = e^{-x}$. Suppose we use our algorithm to solve that problem, for two different h's, h_1 and h_2, obtaining the values $y|_{h_1}$ and $y|_{h_2}$, respectively, at $x = 1$. Then the errors are $E(h_1) = y(1) - y|_{h_1}$ and $E(h_2) = y(1) - y|_{h_2}$, respectively. From (12), show that

$$p \approx \frac{\ln\left[\dfrac{E(h_1)}{E(h_2)}\right]}{\ln\left[\dfrac{h_1}{h_2}\right]}. \tag{10.1}$$

11. Use of (10.1), Above, for Some Other Methods. First, read Exercise 10. A few well known higher-order methods are as follows: the **improved Euler method** (Heun's method), the fourth-order Runge–Kutta method **rk4**, and the Fehlberg fourth-fifth order Runge–Kutta method **rkf4-5**. Suppose we use each of these methods, which are available in CAS systems, to solve the test problem given in Exercise 10.

(a) If, at $x = 1$, the improved Euler method gives $y|_{h=0.001} = 0.367879502531$ and $y|_{h=0.0001} = 0.3678794417846$, use (10.1) to evaluate the order of the method. You will need to know the exact solution at $x = 1$, which is $y(1) = e^{-1} = 0.367879441171$.
(b) If the rk4 method gives $y|_{h=0.1} = 0.367879774412$ and $y|_{h=0.05} = 0.367879461148$, use (10.1) to evaluate the order of the method. (Of course, your answer should be close to 4.)
(c) If the rkf4-5 method gives $y(1)|_{h=0.1} = 0.367879437559$ and $y|_{h=0.05} = 0.367879441063$, use (10.1) to evaluate the order of the method.

ADDITIONAL EXERCISES

12. Convergence Theorem and Error Bound for Euler's Method. More informative than the formula $E = O(h)$ is the following:

THEOREM 1.9.1. Let y_n be the approximate solution of

$$y' = f(x, y); \quad y(a) = b \tag{12.1}$$

by Euler's method (4), and let $y(x)$ denote the exact solution. If $y'' = f_x + f_y y' = f_x + f_y f$ is continuous on the interval I of interest, $a \le x \le X$, and there are constants M and N such that

$$|f_y| \le M \text{ and } |f_x + f_y f| \le N \tag{12.2}$$

on I, then the error $e_n = y(x_n) - y_n$ at any *fixed* point $x_n = a + nh$ in I is bounded as follows:

$$|e_n| \le \frac{N}{2M}\left(e^{(x_n - a)M} - 1\right)h. \tag{12.3}$$

Before continuing, you may be wondering how the "x_n" in Theorem 1.9.1 can be a "fixed point," since it appears to increase with n. The idea is that once we choose a point x_n we keep it fixed by decreasing h and increasing n so that the nh in $x_n = a + nh$ remains constant. To continue, (12.3) is of the form $|e_n| \le Ch$ in which C is a constant, so $e_n \to 0$ at the fixed point x_n, as $h \to 0$, and it follows that Euler's method is indeed convergent.
Here is the problem: To illustrate (12.3), consider the Euler solution of the IVP (2), on $0 \le x \le 1.5$, for which results were given in Table 1. Use (12.3) to obtain a bound on the error at the fixed point $x_n = 1.5$, for each of the three h's used (namely, 0.5, 0.1, and 0.02), and verify that the actual errors (determined from Table 1) are indeed consistent with those bounds. You may use the fact that the solution is known to be $y(x) = x^2 + e^x$. HINT: Show that you can take $M = 1$ and $N = 2 + e^{1.5}$. Thus, show that (12.3) gives $|e_n| \le 11.284h$ at $x_n = 1.5$. NOTE: For a derivation of (12.3) see S. D. Conte, *Elementary Numerical Analysis*, 3rd ed (Auckland: McGraw-Hill International Book Company, 1980), Chapter 8. NOTE: This calculation does not prove (12.3), it only illustrates its use.

13. Extrapolation. Suppose we use Euler's method to find an approximation y_N to the exact solution $y(x_N)$ at some point x_N. The error $E = y(x_N) - y_N$ there satisfies

$$y(x_N) - y_N \sim Ch \tag{13.1}$$

as $h \to 0$. If we don't actually let $h \to 0$ but merely choose a small h and compute y_N, then the *asymptotic* formula (13.1) becomes the *approximate* formula

$$y(x_N) - y_N \approx Ch. \tag{13.2}$$

In (13.2) there are two unknowns, C and $y(x_N)$. We're not particularly interested in C, but are after the exact solution $y(x_N)$. For definiteness, consider the illustrative IVP (2):

(a) Suppose we wish to find the solution $y(x)$ at $x = 0.5$. If we run the Euler method with $h = 0.1$ through $N = 5$, and with $h = 0.02$ through $N = 25$, we obtain the results $y_5 = 1.7995$ and $y_{25} = 1.8778$, respectively (Table 1). Thus, (13.2) gives

$$y(0.5) - 1.7995 \approx C(0.1), \tag{13.3a}$$

$$y(0.5) - 1.8778 \approx C(0.02), \tag{13.3b}$$

which are two (approximate) equations in the two unknowns $y(0.5)$ and C. Solve (13.3a,b) for $y(0.5)$, and show that the result is much more accurate than the value 1.8778 obtained using $h = 0.02$.

NOTE: The latter is called an **extrapolation method** because if we know how the error dies out as $h \to 0$ [namely, according to (13.1) for Euler's method], and we run the method for two small but different h's, then we can "extrapolate" those two results to solve for $y(x_N)$. Though the method gives an improved result it does not yield the *exact* solution $y(x_N)$, except by coincidence, because we've used (13.2) and (13.4) which are approximate, not exact.

(b) Use this same procedure to obtain improved estimates of $y(1.0)$ and $y(1.5)$.

14. Possible Existence of a Critical Step Size *h*. First, read Exercise 8. Suppose $A = -50$, for instance, and $y_0 = 1$. Then the exact solution of (8.1) is $y(x) = e^{-50x}$, which is unity at $x = 0$ and tends rapidly to zero as x increases. Yet, we see from (8.3) that the Euler approximate solution y_n dies out as n increases *only* if $|1 - 50h| < 1$, that is, only if $-1 < 1 - 50h < 1$; otherwise it *grows* with n.

(a) Show from the two foregoing inequalities that we must have $0 < h < 0.04$ if we expect meaningful results.

(b) As a numerical experiment, use the computer to run the Euler solution of (8.1), with $A = -50$ and $y_0 = 1$, on $0 \le x \le 1$, using various values of h, and report your results. The upshot in this example is that even though Euler's method gives convergence to the exact solution at any fixed x as $h \to 0$, we must reduce h below some threshold before that convergence begins to be manifested, before it "kicks in."

(c) Show, from computer results, that if $h > 0.04$ then instead of y_n dying out as n increases, as it should, it exhibits a rapidly growing oscillation. This behavior in the output, a growing oscillation, with a sign change at each successive calculation point — even if the step size is changed — should always suggest to us that the results are not meaningful.

CHAPTER 1 REVIEW

This chapter was devoted to first-order differential equations. We began with the homogeneous linear equation

$$\frac{dy}{dx} + p(x)y = 0 \tag{1}$$

and obtained its general solution

$$y(x) = Ae^{-\int p(x)\,dx} \tag{2}$$

by separation of variables, then considered the general linear equation

$$\frac{dy}{dx} + p(x)y = q(x). \tag{3}$$

Using an integrating factor we obtained a general solution, in explicit form, as

$$y(x) = \frac{1}{\sigma(x)}\Big(\int \sigma(x)q(x)\,dx + C\Big), \tag{4}$$

in which

$$\sigma(x) = e^{\int p(x)\, dx} \tag{5}$$

is an integrating factor of (3) and C is an arbitrary constant.

If an initial condition $y(a) = b$ is appended to (3), we can evaluate the constant C in (4) and obtain

$$y(x) = \frac{1}{\sigma(x)} \left(\int_a^x \sigma(s) q(s)\, ds + b\sigma(a) \right). \tag{6}$$

The Existence and Uniqueness Theorem 1.2.1 assured us that the solution (6) exists and is unique at *least* on the broadest x interval, containing the initial point a, on which $p(x)$ and $q(x)$ are continuous.

The equation

$$\frac{dy}{dx} = f(x, y), \tag{7}$$

which in general is nonlinear, is much more difficult and we were able to solve it only in special cases. First, if $f(x, y)$ is of the separable form $X(x)Y(y)$, we can separate variables and obtain the solution from

$$\int \frac{dy}{Y(y)} = \int X(x)\, dx. \tag{8}$$

If we can do the integrations and solve the resulting equation for $y(x)$, then we have the solution in explicit form; otherwise (8) gives it in implicit form.

We can divide $y' = X(x)Y(y)$ by $Y(y)$ to obtain (8) only if $Y(y) \neq 0$, so the case where $Y(y) = 0$ must be treated separately. If $y = y_0$ is any root of $Y(y) = 0$, then $y(x) = y_0$ is a solution of $y' = X(x)Y(y)$, because it reduces the latter to the identity $0 = 0$, so we must include such constant solutions in addition to the one-parameter family of solutions given by (8).

After separable equations, we considered exact equations. Expressed in the differential form

$$M(x, y)dx + N(x, y)dy = 0, \tag{9}$$

(9) is exact if there exists a function $F(x, y)$ such that $Mdx + Ndy = dF$, in which case integration of $dF = 0$ gives the one-parameter family of solutions, in implicit form, as $F(x, y) = C$. If (9) is not exact, we may nevertheless be able to make it exact using an integrating factor that is a function only of x or only of y.

Finally, we studied the use of substitutions in converting a given differential equation to one that we can solve, such as one that is linear or separable. For instance, the nonlinear Bernoulli equation

$$\frac{dy}{dx} + p(x)y = q(x)y^n \tag{10}$$

can be converted to a linear equation for $v(x)$ by the substitution $v = y^{1-n}$, and a homogeneous equation

$$\frac{dy}{dx} = F\left(\frac{y}{x}\right) \tag{11}$$

can be converted to a separable equation for $v(x)$ by letting $v = y/x$.

The Existence and Uniqueness Theorem 1.5.1 for the IVP

$$\frac{dy}{dx} = f(x,y); \qquad y(a) = b \tag{12}$$

was less informative than the corresponding Theorem 1.2.1 for the linear case, for whereas Theorem 1.2.1 gave an explicit solution and predicted a minimum x interval on which that solution exists and is unique, Theorem 1.5.1 gave sufficient conditions ensuring existence but did not give the solution. Also, whereas Theorem 1.5.1 assured existence and uniqueness on "some" x interval, it did not indicate how broad that interval will be.

We closed our discussion of first-order equations with an introduction to numerical solution methods in Section 1.9, covering only the Euler algorithm for the approximate solution of the IVP $y' = f(x,y)$ with initial condition $y(a) = b$. The method generates the approximating sequence $y_1, y_2, \ldots$ according to the algorithm $y_{n+1} = y_n + f(x_n, y_n)h$ for $n = 0, 1, 2, \ldots,$ where $y_0 = b$ and h is the chosen step size. The latter is a first-order method, which means that at a given computational x point, the error E is asymptotic to some constant times h to the first power as $h \to 0$. We mentioned higher-order methods such as the fourth-order Runge–Kutta method, but did not study them in this brief introduction.

Chapter 2

HIGHER-ORDER LINEAR EQUATIONS

2.1 LINEAR DIFFERENTIAL EQUATIONS OF SECOND ORDER

2.1.1 Introduction. In Chapter 1 we studied first-order equations. For the linear equation $y' + p(x)y = q(x)$ we were able to obtain a general solution, but for nonlinear first-order equations we were able to solve only in special cases such as equations that are separable or exact.

As we mentioned in Section 1.1.5, two reasonable alternative plans present themselves at this point: to turn our attention next to *systems* of first-order equations, or to single equations of *higher order*. We will do both, but we begin with the latter case, a single equation of higher order. We will defer nonlinear higher-order equations and nonlinear systems of first-order equations to the final chapter, Chapter 7, so linear higher-order quations, mostly of second order, will occupy us in all of Chapters 2 through 6.

Recall from the definitions in Section 1.1 that the general **second-order linear** differential equation is an equation of the form

$$a_0(x)\frac{d^2y}{dx^2} + a_1(x)\frac{dy}{dx} + a_2(x)y = F(x), \tag{1}$$

in which the **coefficients** $a_0(x)$, $a_1(x)$, $a_2(x)$, and the **forcing function** $F(x)$ are given. If $F(x) = 0$, then (1) is **homogeneous**. Throughout this chapter we require the coefficients and forcing function to be continuous functions on the x interval I. Further, if $a_0(x) \neq 0$ on I, then we can divide the terms in (1) by $a_0(x)$, to normalize the leading coefficient, and obtain the **standard form**

In (2), we assume that $p_1(x)$, $p_2(x)$, and $f(x)$ are continuous on I.

$$\frac{d^2y}{dx^2} + p_1(x)\frac{dy}{dx} + p_2(x)y = f(x). \tag{2}$$

For instance, if $y'' - 3e^x y = \sin x$, then $p_1(x) = 0$, $p_2(x) = -3e^x$, and the forcing function is $f(x) = \sin x$.

2.1.2 Operator notation and linear differential operators. We will devote the first seven sections of this chapter to the case in which (2) is homogeneous. That is, we first consider

$$Ly = 0, \tag{3}$$

The simplest differential operator is the derivative operator $L = d/dx$ with which you are familiar from the calculus.

in which L denotes

$$L = \frac{d^2}{dx^2} + p_1(x)\frac{d}{dx} + p_2(x) \tag{4}$$

and is called a second-order **differential operator**. When L "operates" on a function y, it produces a new function Ly, namely, $y'' + p_1(x)y' + p_2(x)y$. For instance, if $Ly = y'' + 2xy' - e^x y$ and $y(x) = x^2$, then $Ly = (x^2)'' + 2x(x^2)' - e^x(x^2)$ is the function $2 + 4x^2 - x^2e^x$. The L notation not only enables us to express our differential equation in the compact form (3), it also brings the concept of the differential operator into view.

Just as we know from the calculus that the derivative d/dx admits the property

$$\frac{d}{dx}(\alpha u + \beta v) = \alpha\frac{du}{dx} + \beta\frac{dv}{dx} \tag{5}$$

for any differentiable functions $u(x)$, $v(x)$ and any constants α, β, it is also true that L given by (4) admits the same property,

$$L(\alpha u + \beta v) = \alpha Lu + \beta Lv, \tag{6}$$

for any functions $u(x), v(x)$ that are twice differentiable (so L can operate on them in the first place), and for any constants α, β. The property (6) is called the **linearity property**. Hence, L is called a **linear differential operator** and (3) is called a **linear differential equation**.

The linearity property (6) will be central in all of Chapter 2. It holds for more than two functions as well; see Exercise 10.

To verify that (4) does indeed satisfy (6), just work out the left-hand side:

$$\begin{aligned} L(\alpha u + \beta v) &= \left(\frac{d^2}{dx^2} + p_1(x)\frac{d}{dx} + p_2(x)\right)(\alpha u + \beta v) \\ &= (\alpha u + \beta v)'' + p_1(\alpha u + \beta v)' + p_2(\alpha u + \beta v) \\ &= \alpha u'' + \beta v'' + p_1\alpha u' + p_1\beta v' + p_2\alpha u + p_2\beta v \\ &= \alpha\left(u'' + p_1u' + p_2u\right) + \beta\left(v'' + p_1v' + p_2v\right) \\ &= \alpha Lu + \beta Lv, \end{aligned} \tag{7}$$

as claimed.

If a differential operator is not linear [that is, if it does not satisfy the linearity property (6)], then it is **nonlinear**.

A differential operator is one or the other, linear or nonlinear.

EXAMPLE 1. The Pendulum Equation; a Nonlinear Operator. To better understand the distinction between linear and nonlinear, let us give an example of a nonlinear operator. Using Newton's second law, the differential equation governing the motion $\theta(t)$ of the pendulum shown in Fig. 1, in the absence of friction at the pivot and air resistance, is found to be

$$\theta'' + (g/l)\sin\theta = 0. \tag{8}$$

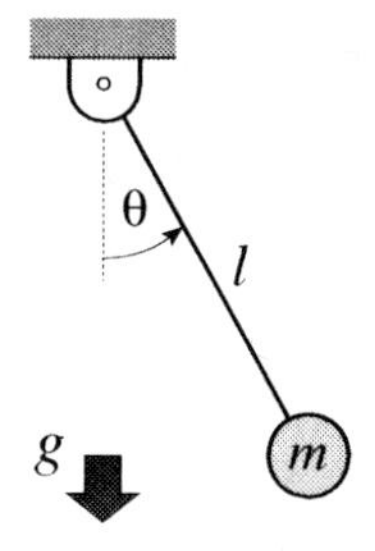

Figure 1. The pendulum modeled by equation (8). The mass m does not occur in (8) because it multiplies both terms and hence cancels out.

Evidently, (8) is nonlinear because of the $\sin\theta$ term; that is, (8) is not of the form $\theta'' + p_1(t)\theta' + p_2(t)\theta = 0$. In case the nonlinearity of (8) is not obvious, let us show that the operator L in $L\theta = \theta'' + (g/l)\sin\theta$ does not satisfy the linearity property (6). To do that, we will show that the difference $L(\alpha u + \beta v) - (\alpha Lu + \beta Lv)$ is not necessarily zero, as it must be for a linear operator. With $g/l = \kappa$, for brevity,

$$\begin{aligned} &L(\alpha u + \beta v) - \alpha Lu - \beta Lv \\ &= (\alpha u + \beta v)'' + \kappa \sin(\alpha u + \beta v) - \alpha(u'' + \kappa \sin u) - \beta(v'' + \kappa \sin v) \\ &= \kappa\big[\sin(\alpha u + \beta v) - \alpha \sin u - \beta \sin v\big]. \end{aligned} \tag{9}$$

The right-hand side of (9) is not necessarily zero — that is, not for all possible choices of $u(x)$, $v(x)$, α, and β. For instance, if $\beta = 0$ and $\alpha \neq 0$, then the right-hand side is $\kappa[\sin(\alpha u) - \alpha \sin u]$, which is not identically zero . After all, one sine is of frequency α and the other is of frequency 1, and α need not equal 1. Thus, equation (8) and the operator L are nonlinear.

Nonlinear differential equations are generally much more difficult than linear ones, and we will not study the pendulum equation (8) until Chapter7. ∎

2.1.3 Superposition principle.

The linearity property of L will be crucial in this chapter in helping us to build up general solutions of (3), for there follows from (6) a superposition principle regarding the solutions of (3).

THEOREM 2.1.1 *Superposition Principle for $Ly = 0$*
If $y_1(x)$ and $y_2(x)$ are solutions of a *linear homogeneous* equation $Ly = 0$ on an x interval I, then

$$\boxed{y(x) = C_1 y_1(x) + C_2 y_2(x)} \tag{10}$$

is also a solution on I, for any constants C_1 and C_2.

$C_1y_1(x) + C_2y_2(x)$ is called a **linear combination** of $y_1(x)$ and $y_2(x)$.

Proof: By y_1 and y_2 being solutions of $Ly = 0$ on I, we mean that $Ly_1 = 0$ and $Ly_2 = 0$ on I. Then, the linearity property (6) gives

$$\begin{aligned} L\left(C_1 y_1 + C_2 y_2\right) &= C_1 L y_1 + C_2 L y_2 \\ &= (C_1)(0) + \cdots + (C_k)(0) = 0, \end{aligned} \tag{11}$$

so $C_1 y_1 + C_2 y_2$ is also a solution on I. ∎

The next two examples are to emphasize that the theorem applies only to linear homogeneous equations, *not* to ones that are nonlinear or nonhomogeneous.

The superposition principle in Theorem 2.1.1 is only for equations that are linear and homogeneous.

EXAMPLE 2. A Nonlinear Equation. The functions $y_1(x) = 1$ and $y_2(x) = x^2$ are solutions of the equation $x^3 y'' - yy' = 0$, which is *nonlinear* because of the yy' term. But their linear combination $4y_1(x) + 3y_2(x) = 4 + 3x^2$, for instance, is not, because substituting it into the differential equation gives $x^3(6) - (4+3x^2)(6x) = -12x^3 - 24x$, rather than zero. ∎

EXAMPLE 3. A Nonhomogeneous Equation. The functions $y_1(x) = 4e^{3x} - 2$ and $y_2(x) = e^{3x} - 2$ are solutions of the linear *nonhomogeneous* equation $y'' - 9y = 18$, but their linear combination $5y_1(x) - 2y_2(x) = 18e^{3x} - 6$, for instance, is not because substituting it into the differential equation gives $18(9)e^{3x} - 9(18e^{3x} - 6) = 54$, rather than 18. ∎

Closure. We indicated our plan to limit this chapter to linear equations, primarily of second order. We discussed the linearity property (6) and an important consequence, namely, that if $y_1(x)$ and $y_2(x)$ are solutions of a linear homogeneous equation $Ly = 0$, then so is any linear combination of them. We will not consider nonhomogeneous equations until Section 2.8.

EXERCISES 2.1

1. Show that the linearity property (7) is equivalent to the two properties

$$L(u + v) = Lu + Lv \quad \text{and} \quad L(\alpha u) = \alpha L u. \tag{1.1a,b}$$

That is, show that the truth of (7) implies the truth of (1.1a) and (1.1b), and vice versa.

2. For $y'' - y' - 6y = 0$ verify by substitution that e^{-2x} and e^{3x} are solutions, for all x. Then verify, again by substitution, that any linear combination $C_1 e^{-2x} + C_2 e^{3x}$ is as well.

3. For $y'' - 3y' + 2y = 0$ verify by substitution that e^x and e^{2x} are solutions, for all x. Then verify, again by substitution, that any linear combination $C_1 e^x + C_2 e^{2x}$ is as well.

4. (a) For $y'' - 9y = 0$ verify by substitution that e^{3x} and e^{-3x} are solutions, for all x. Then verify, again by substitution, that any linear combination $C_1 e^{3x} + C_2 e^{-3x}$ is as well.
(b) Is any linear combination $C_1 e^{3x} + C_2 e^{-3x} + C_3 \sinh 3x$ a solution as well?

5. For $y'' + 16y = 0$ verify by substitution that $\cos 4x$ and $\sin 4x$ are solutions, for all x. Verify, again by substitution, that any linear combination $C_1 \cos 4x + C_2 \sin 4x$ is as well.

6. For $xy'' + 2y' = 0$ verify by substitution that 1 and $1/x$ are solutions for $x > 0$. Then verify by substitution that any linear combination $C_1 + C_2/x$ is as well.

7. For $x^2 y'' + 4xy' + 2y = 0$ verify by substitution that $1/x$ and $1/x^2$ are solutions for $x > 0$. Then verify by substitution that any linear combination $C_1/x + C_2/x^2$ is as well.

8. If $y_1(x)$ and $y_2(x)$ are solutions of a linear differential equation $Ly = 0$, is the following a solution as well? Explain.

(a) $(y_1(x))^2$ (b) $(y_1(x)+y_2(x))^2$
(c) $y_1(x)y_2(x)$ (d) $3y_1(x)-y_2(x)$

9. Violation of Theorem 2.1.1? Given is a differential equation, solutions y_1 and y_2 of it, and a certain linear combination of them. Verify by substitution that y_1 and y_2 are indeed solutions, and show that the given linear combination of them is *not*. Does that result violate Theorem 2.1.1? Explain.

(a) $y''+y=2$; $y_1(x)=5\sin x+2$, $y_2(x)=\cos x+2$; $2y_1(x)-5y_2(x)$
(b) $y''+\sin y=0$; $y_1(x)=\pi$, $y_2(x)=3\pi$; $\sqrt{2}y_1(x)+y_2(x)$
(c) $y''+y'^2=0$ on $0<x<\infty$; $y_1(x)=\ln x$, $y_2(x)=1$; $3y_1(x)+2y_2(x)$
(d) $y''-y'^2=0$ on $0<x<\infty$; $y_1(x)-\ln x$, $y_2(x)=-\ln(x+1)$; $y_1(x)+y_2(x)$
(e) $y''+y'=2e^x$; $y_1(x)=1+e^x$, $y_2(x)=1+2\cosh x$; $y_1(x)-y_2(x)$
(f) $(1-x^4)y''+2y^2=2$; $y_1(x)=1$, $y_2(x)=x^2$; $3y_1(x)+0y_2(x)$
(g) $y''=4$; $y_1(x)=3+2x^2$, $y_2(x)=x+2x^2$; $y_1(x)-y_2(x)$

10. Linearity property for more than two functions. The linearity property (6) holds for any two (twice-differentiable) functions $u(x)$ and $v(x)$ and scalars α and β. It *follows from (6)* that

$$\boxed{L(\alpha_1 u_1+\cdots+\alpha_n u_n)=\alpha_1 Lu_1+\cdots+\alpha_n Lu_n,} \tag{10.1}$$

for any n (twice-differentiable) functions $u_1(x),\ldots,u_n(x)$ and scalars $\alpha_1,\ldots,\alpha_n$, for $n>2$ as well. Prove that claim for $n=3$ and for $n=4$.

2.2 CONSTANT-COEFFICIENT EQUATIONS

2.2.1 Constant coefficients. We begin our study of second-order linear equations by considering the simplest case, namely, ones in which the coefficients (on the left-hand side) are constants. Examples of such equations are

$$L\frac{d^2Q}{dt^2}+R\frac{dQ}{dt}+\frac{1}{C}Q=e(t) \tag{1}$$

for the charge $Q(t)$ on the capacitor in the electrical circuit in Fig. 1, and

$$m\frac{d^2x}{dt^2}+c\frac{dx}{dt}+kx=f(t), \tag{2}$$

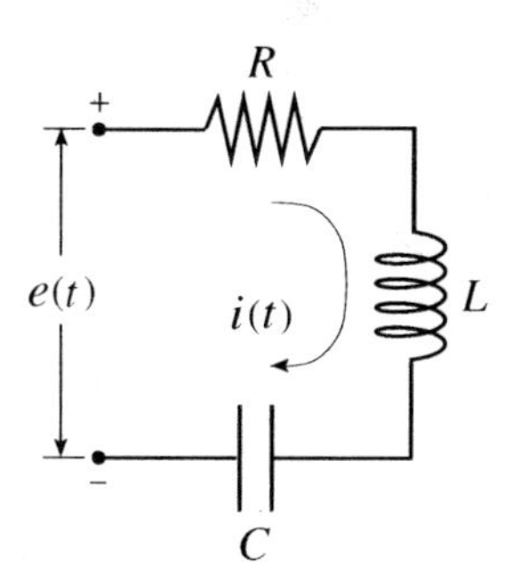

Figure 1. RLC circuit. R,L,C are the resistance, inductance, and capacitance, respectively.

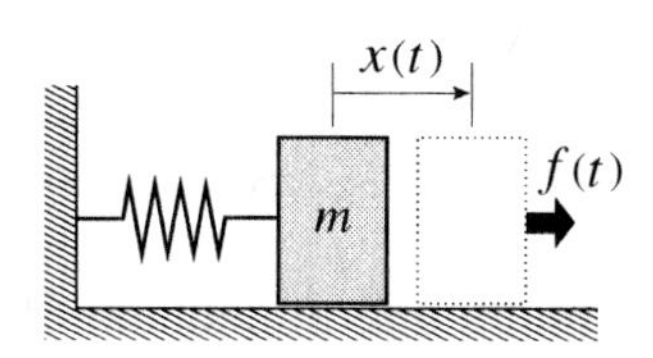

Figure 2. Mechanical oscillator.

for the displacement $x(t)$ of the mechanical oscillator in Fig. 2. The first was derived in Section 1.3.5, and the second will be derived in Chapter 3. In (1), L, R, and C are constants, and $e(t)$ is the applied voltage; in (2), the mass m, damping coefficient c, and spring stiffness k are constants, and $f(t)$ is the applied force. In (1) and (2) the independent variable is the time t rather than our generic x.

For the time being, we are considering homogeneous equations, so the $e(t)$ in (1) and the $f(t)$ in (2) will be zero. By the way, you may be thinking that homogeneous equations are simple and uninteresting because the solution is necessarily zero, that is, "nothing happens." Not so. For instance, consider the physical system in Fig. 2. Let $f(t)=0$ so there is no force applied to the mass. With $f(t)=0$, we can nevertheless set the mass in motion by giving it nonzero initial conditions. For instance, if we move the mass to the right, hold it still, and then let go, the mass will of course go into a vibratory motion even though $f(t)=0$. These points will become clear as we proceed.

2.2.2 Seeking a general solution. Thus, we begin this chapter with the linear, second-order, homogeneous equation

$$\boxed{y'' + p_1 y' + p_2 y = 0,} \tag{3}$$

in which p_1 and p_2 are real constants. Here, we've reverted to our generic independent variable x and dependent variable y.

How can we solve (3)? The solution methods that we developed for first-order equations, such as separation of variables and the method of exact equations, cannot be applied. But, since the *first*-order constant-coefficient equation $y' + py = 0$ has the exponential solution $y(x) = Ce^{-px}$, we wonder if (3) might also admit exponential solutions. Accordingly, let us seek a solution (or solutions) $y(x)$ of (3) in the form

$$y(x) = e^{rx}, \tag{4}$$

in which r a constant that is not yet known. If e^{rx} is to be a solution of (3), it must reduce that equation to an identity, so put (4) into (3) and see if one or more r's can be found so the result is an identity. That step gives $r^2 e^{rx} + p_1 r e^{rx} + p_2 e^{rx} = 0$ or

$$\left(r^2 + p_1 r + p_2\right) e^{rx} = 0. \tag{5}$$

The e^{rx} factor is nonzero, so (5) is satisfied if and only if

$$r^2 + p_1 r + p_2 = 0. \tag{6}$$

Some authors use "auxiliary equation" and "auxiliary polynomial" instead.

This equation and its left-hand side are called the **characteristic equation** and **characteristic polynomial**, respectively, for the differential equation (3). Equation (6) has two roots, say r_1 and r_2, given by the quadratic formula as

$$r = \frac{-p_1 \pm \sqrt{p_1^2 - 4p_2}}{2}. \tag{7}$$

To review: Having sought solutions in the form e^{rx}, we found two: $e^{r_1 x}$ and $e^{r_2 x}$. From those two, we used Theorem 2.1.1 to form the set of solutions given by (8).

Suppose, first, that those roots are real and distinct. Since (1) is linear and homogeneous, it follows from Theorem 2.1.1, that if $e^{r_1 x}$ and $e^{r_2 x}$ are solutions of (3), then so is every linear combination of them,

$$y(x) = C_1 e^{r_1 x} + C_2 e^{r_2 x}, \tag{8}$$

in which C_1 and C_2 are arbitrary constants.

Remember from Chapter 1 that $y(x) = Ce^{-px}$ is a *general* solution of $y' + py = 0$ (when p is a constant) in the sense that it contains *all* solutions of that equation. Likewise, we desire a general solution of (3). Is (8) a general solution of (3)? Unfortunately, our solution procedure does not permit us to make such a bold claim because we sought solutions only in exponential form, and it is possible that there are solutions of (3) that are *not* of exponential form and, hence, are not represented within (8).

It turns out that, without needing any theory beyond what we developed in Chapter 1, we can derive a general solution of (3) by a different method, one that is not based on any specific assumed solution form. The result obtained is that (8) is indeed a general solution of (3) if the roots r_1, r_2 are distinct. If they are not distinct [that is, if $p_1^2 = 4p_2$, so the two roots in (7) coalesce into one] then the general solution is a modified version of (8). These results are stated in the following theorem, and the derivation of that general solution is given in the proof that follows.

THEOREM 2.2.1 *General Solution of* $y'' + p_1 y' + p_2 y = 0$
The linear homogeneous constant-coefficient equation

$$\boxed{y'' + p_1 y' + p_2 y = 0,} \tag{9}$$

in which p_1 and p_2 are real constants, has the characteristic equation

$$r^2 + p_1 r + p_2 = 0. \tag{10}$$

The result (11a) holds, as a general solution of (9), even if the roots are complex, but we will have to show how to evaluate the complex exponentials in that case, and will do so in the next section.

Distinct roots: If (10) has distinct roots r_1, r_2, then a general solution of (9), on $-\infty < x < \infty$ or any subinterval thereof, is

$$\boxed{y(x) = C_1 e^{r_1 x} + C_2 e^{r_2 x},} \tag{11a}$$

in which C_1, C_2 are arbitrary constants.
Repeated roots.: If (10) has repeated roots $r_1 = r_2 \equiv r$, then a general solution of (9), on $-\infty < x < \infty$ or any subinterval thereof, is

$$\boxed{y(x) = (C_1 + C_2 x) e^{rx},} \tag{11b}$$

in which C_1, C_2 are arbitrary constants.

Proof: We will put aside steps (4)–(8) and pursue a different approach. If we denote $\dfrac{d}{dx}$ as D, then the operator $L = \dfrac{d^2}{dx^2} + p_1 \dfrac{d}{dx} + p_2$ in (9) can be expressed as $L = D^2 + p_1 D + p_2$, and (9) can be expressed as

D is $\dfrac{d}{dx}$, D^2 is $\dfrac{d^2}{dx^2}$, and so on.

$$\left(D^2 + p_1 D + p_2\right) y = 0. \tag{12}$$

This form encourages us to think of the differential operator L more algebraically, as a "polynomial in D," even though D is a derivative operator, not a number. We can factor polynomials, so we wonder if that "operator polynomial" can be *factored* as well. To show that it can, observe that for any constants a and b,

$$\begin{aligned}(D-a)(D-b)\,y &= (D-a)(y'-by)\\ &= (y''-by')-a(y'-by)\\ &= y''-(a+b)y'+aby\\ &= \big(D^2-(a+b)D+ab\big)\,y.\end{aligned} \tag{13}$$

On the other hand, if we formally multiply the factors $D-a$ and $D-b$ in the operator on the left-hand side of (13), as though the D's were numbers, we obtain

$$(D-a)(D-b) = D^2-(a+b)D+ab, \tag{14}$$

which is the same as the operator on the right-hand side of (13). Thus, *for operators with constant coefficients* the operator polynomial *can* be factored as if the D's were constants. (The same is true for higher-order linear constant-coefficient operators as well.)

Thus, if the characteristic polynomial $r^2+p_1r+p_2$ can be factored as $(r-r_1)(r-r_2)$, then the operator $L = D^2+p_1D+p_2$ in (12) can be factored as $(D-r_1)(D-r_2)$, and (12) can be expressed, equivalently, as

$$(D-r_1)\underbrace{(D-r_2)y}_{u} = 0. \tag{15}$$

If we define $(D-r_2)y$ as u, as indicated in (15), we can reduce (15) to the two *first-order* equations $(D-r_1)u = 0$ and $(D-r_2)y = u$; that is,

$$u' - r_1u = 0, \tag{16a}$$

$$y' - r_2y = u. \tag{16b}$$

These are simple first-order linear equations, and we can solve them one at a time. A general solution of (16a) is

$$u(x) = Ae^{r_1x} \tag{17}$$

in which A is an arbitrary constant. Then, (16b) becomes

$$y' - r_2y = Ae^{r_1x}. \tag{18}$$

$\int e^{ax}\,dx = e^{ax}/a$ if $a \neq 0$, but if $a=0$ the integral equals x. In (19), $a=r_1-r_2$ is nonzero if $r_1 \neq r_2$, and zero if $r_1 = r_2$. We must consider the two cases in evaluating the integral.

A general solution of (18) is, from equation (27) in Section 1.2,

$$y(x) = e^{r_2x}\Big(\int e^{-r_2x}Ae^{r_1x}\,dx + B\Big). \tag{19}$$

We must consider two cases: First, if $r_2 \neq r_1$, then (19) gives

$$y(x) = e^{r_2x}\left(A\frac{e^{(r_1-r_2)x}}{r_1-r_2} + B\right) = \frac{A}{r_1-r_2}e^{r_1x} + Be^{r_2x}, \tag{20}$$

in which A and B are arbitrary constants. If we rename $A/(r_1 - r_2)$ as C_1 and B as C_2, (20) becomes

$$y(x) = C_1 e^{r_1 x} + C_2 e^{r_2 x}, \tag{21}$$

which is identical to (8), that we found by seeking y in the form e^{rx}.

We will generally prefer subscripted C's for the arbitrary constants in a general solution, as in (21).

Now we can see that the result (21), which is the same as (8), is indeed a **general solution** of (3). After all, (15) is just a re-statement of (9), and (16) a re-statement of (15); (17) gives all possible u's that satisfy (16a), and then (19) gives all possible y's that satisfy (16b). Thus, (19) gives all possible y's that satisfy (9) and is therefore a general solution of (9).

Similarly, for the case of repeated roots where $r_1 = r_2 \equiv r$, (19) gives

$$y(x) = (C_1 + C_2 x)e^{rx}, \tag{22}$$

in which C_1 and C_2 are arbitrary constants. ■

Observe that (16) was a "system" of two linear differential equations for the two unknowns $u(x)$ and $y(x)$ in the same sense that $3x + 5y = 7$ and $2x - y = 12$, for instance, are two linear algebraic equations for x and y. As there are special theory and solution methods for systems of linear algebraic equations, there are also special theory and solution methods for systems of linear differential equations, and they will be the subject of Chapter 4. In this case (16) was simple enough for us to solve without needing that theory, but we mention this as a "heads up" as we look ahead to studying systems in Chapter 4.

Solution method, using Theorem 2.2.1. To solve equations of the form (3), in which the coefficients p_1 and p_2 are constants, *seek $y(x)$ in the form e^{rx}, put that into the differential equation and obtain its characteristic equation. Solve for the roots r_1 and r_2: if they are distinct, the superposition principle successfully gives a general solution as the arbitrary linear combination of solutions in (11a); if they are repeated, a general solution is given by (11b).*

Solution method for a constant-coefficient equation $y'' + p_1 y' + p_2 y = 0$.

Looking ahead. We look ahead to equations not covered by Theorem 2.2.1: equations with nonconstant coefficients, and equations of order higher than two. The more substantial theory that will be needed for such equations will be undertaken in Section 2.4. Meanwhile, Theorem 2.2.1 will enable us to begin solving second-order equations, in this section and the next, and to become familiar with the various solution forms that arise.

EXAMPLE 1. General Solution; Distinct Roots. Obtain a general solution of

$$y'' + 2y' - 3y = 0. \tag{23}$$

Seeking solutions in the form e^{rx}, put $y(x) = e^{rx}$ into (24), factor out an e^{rx}, and obtain

$$(r^2 + 2r - 3)e^{rx} = 0. \tag{24}$$

Since $e^{rx} \neq 0$, cancel it in (24) and obtain the characteristic equation

$$r^2 + 2r - 3 = 0, \tag{25}$$

with roots $r_1 = 1$ and $r_2 = -3$. Since these are distinct, a general solution of (23) is

$$y(x) = C_1 e^x + C_2 e^{-3x}. \quad \blacksquare \tag{26}$$

EXAMPLE 2. General Solution; Repeated Roots. Obtain a general solution of

$$y'' + 6y' + 9y = 0. \tag{27}$$

Seeking solutions in the form e^{rx} leads to the characteristic equation

$$r^2 + 6r + 9 = 0, \tag{28}$$

with the repeated roots $r_1 = r_2 = -3$. Thus, by (11b), a general solution of (27) is

$$y(x) = (C_1 + C_2 x)e^{-3x}. \tag{29}$$

We will return to the case of repeated roots in Section 2.5, using a method of **reduction of order**. ∎

EXAMPLE 3. CAUTION. Let us try to solve the equation

$$y'' + 2xy' + y = 0 \tag{30}$$

by assuming a solution form $y(x) = e^{rx}$, in which r is a constant. To put the latter into (30) we compute the derivatives $y'(x) = re^{rx}$ and $y''(x) = r^2 e^{rx}$, which formulas are correct only if r is a constant, as we assumed. Putting those derivatives into (30) and canceling the common (nonzero) factor e^{rx} leads to

$$r^2 + 2xr + 1 = 0, \tag{31}$$

which we hope to solve for the constant r. However, the quadratic formula gives the roots $r = -x \pm \sqrt{x^2 - 1}$, which are *not constants*, as we assumed, they are functions of x. That is, we assumed that $y(x) = e^{rx}$, in which r is a constant, and found that no such constant values of r exist. Thus, solutions of (31), of the form $y(x) = e^{rx}$, do not exist, and the method failed in the case of equation (30). ∎

Be sure to understand this example. To claim that the solution of (30) is $y(x) = C_1 e^{(-x+\sqrt{x^2-1})x} + C_2 e^{(-x-\sqrt{x^2-1})x}$ would not just be wrong, it would be "very" wrong.

Thus, keep in mind that the method of assumed exponential fails, in general, for nonconstant-coefficient equations! Could we fall back on the idea of factoring the operator, as we did in our proof of Theorem 2.2.1? In principle yes, but, in general, the factorization steps prove to be prohibitively difficult; see Exercise 11.

Since exponential functions will continue to be prominent, remember from the calculus that the exponential function is related to the hyperbolic cosine and hyperbolic sine functions by the definitions

$$\cosh x = \frac{e^x + e^{-x}}{2}, \quad \sinh x = \frac{e^x - e^{-x}}{2}. \tag{32a,b}$$

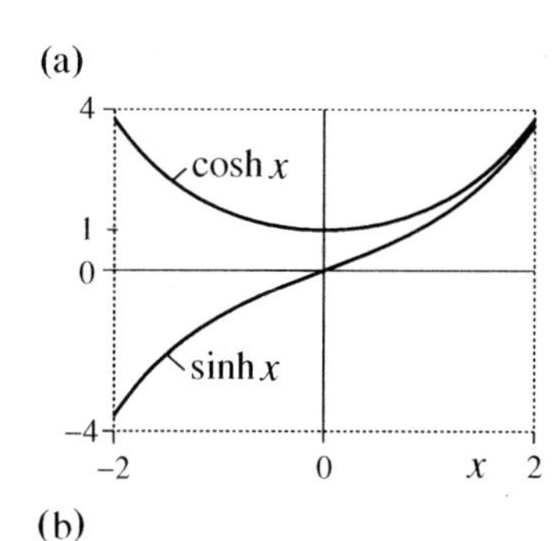

Or, by adding and subtracting these,

$$e^x = \cosh x + \sinh x, \quad e^{-x} = \cosh x - \sinh x. \tag{33a,b}$$

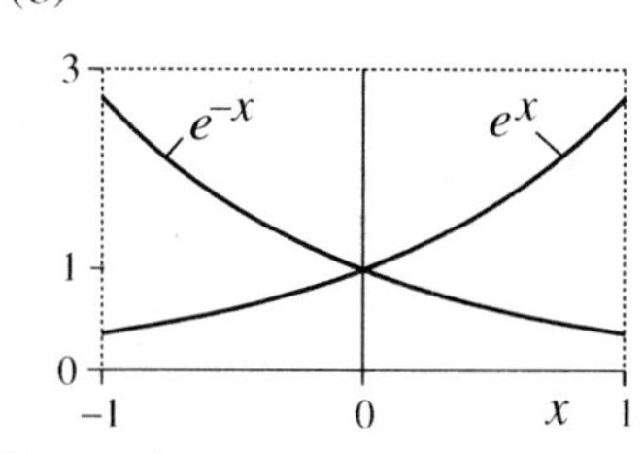

Figure 3. Graphs of $\cosh x$, $\sinh x$, e^x, and e^{-x}.

Keep in mind the graphs of $\cosh x$, $\sinh x$, e^x, and e^{-x}, given in Fig. 3.

EXAMPLE 4. Obtain a general solution of

$$y'' - 9y = 0. \tag{34}$$

Seeking $y(x) = e^{rx}$, we obtain the characteristic equation

$$r^2 - 9 = 0.$$

Thus $r = +3$ and $r = -3$, and a general solution of (34) is

$$y(x) = C_1 e^{3x} + C_2 e^{-3x}. \tag{35}$$

Alternatively, we can express the general solution in terms of the hyperbolic cosine and sine functions since

$$\begin{aligned} y(x) &= C_1 e^{3x} + C_2 e^{-3x} \\ &= C_1(\cosh 3x + \sinh 3x) + C_2(\cosh 3x - \sinh 3x) \\ &= (C_1 + C_2)\cosh 3x + (C_1 - C_2)\sinh 3x \\ &= C_3 \cosh 3x + C_4 \sinh 3x, \end{aligned} \tag{36}$$

To verify that $\cosh 3x$ and $\sinh 3x$ satisfy (34), you will need to recall the derivatives $\frac{d}{dx}\cosh ax = a \sinh ax$ and $\frac{d}{dx}\sinh ax = a\cosh ax$.

in which $C_3 = C_1 + C_2$ and $C_4 = C_1 - C_2$ are arbitrary constants. That is, an arbitrary linear combination of e^{3x} and e^{-3x} is equivalent to an arbitrary linear combination of $\cosh 3x$ and $\sinh 3x$, so we can use whichever general solution form we prefer, (35) or (36).

More generally, *an arbitrary linear combination of* e^{ax} *and* e^{-ax} *is equivalent to an arbitrary linear combination of* $\cosh ax$ *and* $\sinh ax$. ∎

2.2.3 Initial value problem. Suppose that besides the differential equation we also have initial conditions at some point $x = a$. Since the general solution, given by (22), contains two arbitrary constants, it seems appropriate to specify two initial conditions to determine them — that is, not only $y(a)$ as in Chapter 1, but also $y'(a)$.

In fact, the following theorem follows easily:

THEOREM 2.2.2 *Existence/Uniqueness for the Initial Value Problem*
The initial value problem (IVP)

$$y'' + p_1 y' + p_2 y = 0; \tag{37a}$$
$$y(a) = b_1, \ y'(a) = b_2, \tag{37b}$$

For instance, if $a = 2$, $y(2) = 5$, and $y'(2) = -3$, then b_1 is 5 and b_2 is -3.

in which p_1 and p_2 are constants, has a unique solution on $-\infty < x < \infty$, given by (11a) if $r_1 \neq r_2$ and by (11b) if $r_1 = r_2 = r$, with C_1, C_2 determined by the initial conditions (37b).

Proof: All solutions of (37a) are contained in its general solution (11a,b), so we need only show that C_1, C_2 are uniquely determined by the initial conditions. First suppose $r_1 \neq r_2$, and apply the conditions to (11a),

$$y(a) = y_0 = C_1 e^{r_1 a} + C_2 e^{r_2 a} \tag{38a}$$
$$y'(a) = y_0' = r_1 C_1 e^{r_1 a} + r_2 C_2 e^{r_2 a}, \tag{38b}$$

Determinants and systems of linear algebraic equations are reviewed in Appendices B and C.

which is a set of two linear algebraic equations for C_1, C_2. We know *a set of n linear algebraic equations in n unknowns has a unique solution if and only if the determinant of the coefficients is nonzero.* The determinant of the coefficients in (38) is

$$\det \mathbf{A} = \begin{vmatrix} e^{r_1 a} & e^{r_2 a} \\ r_1 e^{r_1 a} & r_2 e^{r_2 a} \end{vmatrix} = (r_2 - r_1) e^{(r_1 + r_2)a}, \tag{39}$$

which is indeed nonzero because $r_2 \neq r_1$ by assumption, and the exponential function is nonzero. Thus, (39) has a unique solution for C_1 and C_2. Proof for the case where $r_1 = r_2$ is left for the exercises. ■

EXAMPLE 5. Solve the IVP

$$y'' - 5y' + 6y = 0; \qquad y(0) = 4, \ y'(0) = -1. \tag{40}$$

Seek $y(x) = e^{rx}$. The characteristic equation is $r^2 - 5r + 6 = 0$, with roots $r = 2, 3$, so a general solution of the differential equation is

$$y(x) = C_1 e^{2x} + C_2 e^{3x}. \tag{41}$$

The initial conditions give

$$y(0) = 4 = C_1 + C_2, \tag{42a}$$
$$y'(0) = -1 = 2C_1 + 3C_2, \tag{42b}$$

with the unique solution $C_1 = 13$ and $C_2 = -9$. Thus,

$$y(x) = 13e^{2x} - 9e^{3x} \tag{43}$$

is the unique solution of (40), on $-\infty < x < \infty$. ▮

Closure. This section and the next are particularly important because constant-coefficient second-order equations are prominent in applications and also because they provide a transition from first-order equations to equations of higher order.

We derived the general solution of the linear second-order constant-coefficient equation $y'' + p_1 y' + p_2 y = 0$. If the characteristic equation $r^2 + p_1 r + p_2 = 0$ has distinct roots r_1, r_2, then the general solution is $y(x) = C_1 e^{r_1 x} + C_2 e^{r_2 x}$; and if there are repeated roots $r_1 = r_2 \equiv r$, then the general solution is $y(x) = (C_1 + C_2 x)e^{rx}$. If the differential equation is augmented by initial conditions $y(a) = b_1, y'(a) = b_2$ at $x = a$, then those conditions serve to uniquely determine C_1 and C_2 and a unique solution of the IVP for $y(x)$. The initial point a is often 0, but it need not be. We also noted that if the characteristic roots are of the form $r_1 = b$, $r_2 = -b$, with $b \neq 0$ so that they are distinct, then the general solution can be expressed either as a linear combination of e^{bx} e^{-bx} or, equivalently, as a linear combination of $\cosh bx$ and $\sinh bx$.

EXERCISES 2.2

1. General Solution. Obtain a general solution of the given differential equation.

(a) $y'' + 5y' = 0$	(b) $y'' - y' = 0$
(c) $y'' + 2y' + y = 0$	(d) $y'' - 2y' + y = 0$
(e) $y'' - 3y' + 2y = 0$	(f) $y'' - 4y' - 5y = 0$
(g) $y'' + 8y' + 16y = 0$	(h) $4y'' - 4y' + y = 0$
(i) $y'' + 6y' + 8y = 0$	(j) $y'' + y' - 12y = 0$
(k) $y'' + 7y' + 12y = 0$	(l) $y'' - y' - 20y = 0$
(m) $2y'' - 5y' - 3y = 0$	(n) $3y'' + 5y' + 2y = 0$
(o) $y'' + 3y' + y = 0$	(p) $y'' + 8y' + 3y = 0$
(q) $y'' + 5y' + y = 0$	(r) $y'' - 6y' - 2y = 0$

2. (a)–(r) **Initial Value Problem.** Find the solution of the differential equation given in the corresponding part of Exercise 1, subject to the initial conditions $y(0) = 3$, $y'(0) = -1$.

3. (a)–(r) **Initial Value Problem.** Find the solution of the differential equation given in the corresponding part of Exercise 1, subject to the initial conditions $y(5) = 0$, $y'(5) = 4$.

4. Archeology: Inferring the Equation. Here we give the roots of the characteristic equation and ask you to infer from them the original differential equation. NOTE: Of course this is "backwards," but the exercises are to support understanding of the material in this section, not always to practice steps that are likely to be needed in applications.

(a) $2, 6$	(b) $1, 2/3$	(c) $0, 5$	(d) $3, -4$
(e) $1/5, -2/5$	(f) $0, 5/4$	(g) $-2, -2$	(h) $-1, -2$
(i) $20, 20$	(j) $0, 0$	(k) $0.1, 0.2$	(l) $0.5, -0.5$

5. Inferring the Equation. If C_1 and C_2 are arbitrary constants, then the given $y(x)$ is a general solution of what constant-coefficient equation $y'' + p_1 y' + p_2 y = 0$? That is, determine p_1 and p_2.

(a) $y(x) = C_1 + C_2 e^{-x}$	(b) $y(x) = C_1 e^x + C_2 e^{2x}$
(c) $y(x) = (C_1 + C_2 x)e^{7x}$	(d) $y(x) = C_1 + C_2 x$
(e) $y(x) = C_1 e^x + C_2 \sinh x$	(f) $y(x) = C_1 e^{3x} + C_2 \cosh 3x$

(g) $y(x) = C_1 \cosh x + C_2 \sinh x$
(h) $y(x) = C_1 \cosh 5x + C_2 \sinh 5x$

6. Show that if a differential equation

$$y'' + p_1(x)y' + p_2(x)y = 0 \tag{6.1}$$

has exponential solutions $y_1(x) = e^{r_1 x}$ and $y_2(x) = e^{r_2 x}$, with $r_1 \neq r_2$, then it *must* be of constant-coefficient type; that is, $p_1(x)$ and $p_2(x)$ must be constants. HINT: If you use the fact that the given functions $y_1(x)$ and $y_2(x)$ are solutions of (6.1), you should be able to solve for $p_1(x)$ and $p_2(x)$, and show that they are constants.

7. Show that if a differential equation

$$y'' + p_1(x)y' + p_2(x)y = 0 \tag{7.1}$$

has solutions $y_1(x) = e^{rx}$ and $y_2(x) = xe^{rx}$, then it *must* be of constant-coefficient type; that is, $p_1(x)$ and $p_2(x)$ must be constants.

8. A Subtle Point in Example 4. Be careful about the reasoning in Example 4. Specifically, does it follow from the arbitrariness of C_1 and C_2 in (34) that $C_3 = C_1 + C_2$ and $C_4 = C_1 - C_2$ are arbitrary as well, as we stated?

9. Factoring the Operator. To be sure the steps in our proof of Theorem 2.2.1 are understood, we ask you to use that method, based on factoring the operator, to derive a general solution of each of the following equations. That is, don't merely use the results given in Theorem 2.2.1, go through the solution steps that we used to derive the results stated in that theorem.

(a) $y'' + y' - 6y = 0$
(b) $y'' + 3y' + 2y = 0$
(c) $y'' - y - 2y = 0$
(d) $y'' + 5y' + 4y = 0$
(e) $y'' + 6y' + 5y = 0$
(f) $y'' - 3y' + 2y = 0$

10. Does the Exponential Solution Form Necessarily Fail for Nonconstant-Coefficient Equations? Seek solutions in the form $y(x) = e^{rx}$, and report on your results.

(a) $y'' + (x-1)y' - xy = 0$
(b) $y'' + (1-x)y' - xy = 0$
(c) $y'' + (3-x^2)y' - 3x^2 y = 0$
(d) $y'' + (e^x - 1)y' - e^x y = 0$

ADDITIONAL EXERCISES

11. Can We Solve Nonconstant-Coefficient Equations by Factoring the Operator? We saw, in equations (12)–(19), that if we can factor the differential operator into successive first-order operators, then the differential equation can be expressed, equivalently, as a system of linear first-order differential equations, which can be solved successively, using the solution formulas given in Section 1.2. The key to the success of this method was in being able to factor the operator. For instance, we can factor $D^2 - D - 6$ as $(D+2)(D-3)$. But, can we use this method to solve linear *nonconstant*-coefficient equations as well? Let us see.

(a) To illustrate, consider the nonconstant-coefficient equation

$$y'' - x^2 y = (D^2 - x^2)y = 0. \tag{11.1}$$

It might appear that we can factor the operator $D^2 - x^2$ in (11.1) as $(D+x)(D-x)$ [or as $(D-x)(D+x)$]. However, show that

$$(D+x)(D-x)y = y'' - (x^2+1)y, \tag{11.2}$$

so $(D+x)(D-x)$ is actually $D^2 - (x^2+1)$, *not* $D^2 - x^2$ as we might have thought. NOTE: The difficulty, in this nonconstant-coefficient case, is that not only do the $D+x$ and $D-x$ factors operate on the y to their right, the D in the $D+x$ also operates on the $-x$ in the $D-x$ factor to its right, thus leading to the $-y$ term in $y'' - (x^2+1)y$ and causing the result to differ from $y'' - x^2 y$.

(b) Evidently, successfully factoring a nonconstant-coefficient operator is more difficult. The question is this: Given a nonconstant-coefficient operator, *can* it be factored and, if so, *how*? That is, given $p_1(x)$ and $p_2(x)$ we seek functions $a(x)$ and $b(x)$ so that

$$y'' + p_1(x)y' + p_2(x)y = [D - a(x)][D - b(x)]y. \tag{11.3}$$

Write out the right-hand side of (11.3) and compare coefficients of y'', y', and y on both sides. Doing so, show that (11.3) holds if $a(x)$ and $b(x)$ satisfy the equations

$$a + b = -p_1, \tag{11.4a}$$

$$ab - b' = p_2. \tag{11.4b}$$

By eliminating b between these equations, show that a must satisfy the first-order differential equation

$$\boxed{a' = a^2 + (p_1)a + (p_2 - p_1').} \tag{11.5}$$

That's the end of the exercise.
CONCLUSION: If we can solve (11.5) for $a(x)$, then (11.4a) gives $b(x)$ and we can solve the nonconstant-coefficient equation $y'' + p_1(x)y' + p_2(x)y = 0$ by factorization, namely, by re-expressing $[D - a(x)][D - b(x)]y = 0$ as two first-order linear equations. Unfortunately, (11.5) is difficult in general,

because it is *nonlinear* due to the a^2 term. In fact, it is a **Riccati equation**, namely, of the form

$$y' = p(x)y^2 + q(x)y + r(x), \tag{11.6}$$

and was discussed in the exercises for Section 1.8. Thus, we can make the following interesting qualitative observation: *The class of linear second-order equations with nonconstant coefficients, $y'' + p_1(x)y' + p_2(x)y = 0$, is equivalent in difficulty to the class of nonlinear first-order Riccati equations,* because if we can solve for $a(x)$ and $b(x)$, then we can proceed without further difficulty to the solution (except possibly if the integrals that arise when we solve the first-order linear equations are difficult). Further, the fact that a Riccati equation stands between every equation $y''+p_1(x)y'+p_2(x)y = 0$ and its solution reveals the fundamental importance of the Riccati equation in the theory of differential equations.

12. A Taylor Series Idea. To illustrate an interesting Taylor series solution method, consider for the IVP

$$y'' - 3y' + 2y = 0; \qquad y(0) = y_0,\ y'(0) = y_0'. \tag{12.1}$$

(We've taken it to be of constant-coefficient type so we can solve it first by using expnential solutions, and then again using a Taylor series idea.) First, use the method of assumed exponential form and show that the solution is

$$y(x) = (2y_0 - y_0')e^x + (y_0' - y_0)e^{2x}. \tag{12.2}$$

Next, try a different strategy. Namely, seek a solution $y(x)$ in the form of a *Taylor series* about the initial point $x = 0$:

$$y(x) = y(0) + y'(0)x + \frac{y''(0)}{2!}x^2 + \cdots. \tag{12.3}$$

We must evaluate the coefficients $y(0)$, $y'(0)$, ... in (12.3). The first two coefficients are known from the initial conditions given in (12.1). Next, $y''(0)$ can be found in terms of $y(0)$ and $y'(0)$ *from the differential equation itself,* and similarly for $y'''(0)$, $y''''(0)$ and so on. That is, repeated differentiation of (12.1) gives

$$\begin{aligned} y'' &= -2y + 3y', \\ y''' &= -2y' + 3y'' = -2y' + 3(-2y + 3y') = -6y + 7y', \end{aligned}$$

and so on. Thus,

$$\begin{aligned} y''(0) &= -2y(0) + 3y'(0), \\ y'''(0) &= -6y(0) + 7y'(0), \end{aligned}$$

and so on. Proceeding in this manner, show that if we put these expressions for $y''(0)$, $y'''(0)$, and so on into (12.3), we obtain

$$\begin{aligned} y(x) &= y(0)\left(1 - x^2 - x^3 - \tfrac{7}{12}x^4 + \cdots\right) \\ &\quad + y'(0)\left(x + \tfrac{3}{2}x^2 + \tfrac{7}{6}x^3 + \tfrac{5}{8}x^4 + \cdots\right). \end{aligned} \tag{12.4}$$

Finally, verify that the series solution (12.4) agrees with the closed form solution (12.2), through terms of fourth order. NOTE: This exercise should help us to appreciate the initial conditions $y(a)$, $y'(a)$ as being "perfect" for leading to a unique solution, for if we think in terms of Taylor series then we can see that those initial conditions give the first two coefficients in the Taylor series about the initial point $x = a$, and then the differential equation and its derivatives give the subsequent coefficients.

2.3 COMPLEX ROOTS

2.3.1 Complex exponential function. In Section 2.2 we sought solutions of $y'' + p_1y' + p_2y = 0$ in the form e^{rx} and determined r from the characteristic equation. Realize that even though we restricted p_1, p_2 to be real, the roots of the characteristic equation can be complex. For instance, $y'' - 4y' + 13y = 0$ gives $r^2 - 4r + 13 = 0$. This characteristic equation has real coefficients, yet its roots are the complex conjugates $r = 2+3i$ and $r = 2-3i$, in which i denotes $\sqrt{-1}$.

Complex numbers and the complex plane are reviewed in Appendix D. Incidentally, electrical engineers generally use j rather than i for $\sqrt{-1}$.

All results in Section 2.2 hold, even if the characteristic equation has complex roots, but a computational question arises if the r's are complex: How do we evaluate the e^{rx} in the solution if r is complex?

We need to define the **complex exponential function** e^z, in which $z = x + iy$ is complex. By "defining" a complex valued function we mean giving a formula for computing it in the standard Cartesian form $a + ib$. To motivate a definition, proceed as follows:

$$\begin{aligned} e^z &= e^{x+iy} = e^x e^{iy} \\ &= e^x \left[1 + iy + \frac{1}{2!}(iy)^2 + \frac{1}{3!}(iy)^3 + \frac{1}{4!}(iy)^4 + \cdots\right] \\ &= e^x \left[\left(1 - \frac{1}{2!}y^2 + \frac{1}{4!}y^4 - \cdots\right) + i\left(y - \frac{1}{3!}y^3 + \frac{1}{5!}y^5 - \cdots\right)\right] \\ &= e^x(\cos y + i \sin y). \end{aligned} \tag{1}$$

In the second equality in (1) we used the familiar formula $e^{c+d} = e^c e^d$; in the third we used the Taylor series $e^u = 1 + \frac{1}{1!}u + \frac{1}{2!}u^2 + \cdots$ with $u = iy$; in the fourth we used the formulas $i^2 = -1$, $i^3 = -i$, and so on, and we grouped real and imaginary terms; and in the fifth we recognized the two series as the Taylor series of $\cos y$ and $\sin y$.

The result $(e^x \cos y) + i(e^x \sin y)$ in (1) is in the desired Cartesian form. The steps in (1) were not all rigorous.[1] They were only to *motivate* a definition of the function e^{x+iy}, and we will now take (1) as our *definition*,

$$\boxed{e^{x+iy} \equiv e^x \left(\cos y + i \sin y\right).} \tag{2}$$

As a special case, let $x = 0$. Then (2) becomes

$$e^{iy} = \cos y + i \sin y, \tag{3a}$$

known as **Euler's formula**, after *Leonhard Euler* (1707–1783) whose many contributions to mathematics include the systematic development of the theory of linear constant-coefficient differential equations. We will refer to (3a) and the more general formula (2) collectively as "Euler's formula."

To illustrate, let us evaluate e^{2-3i} and $e^{\pi i}$: (2) gives $e^{2-3i} = e^2 \left(\cos 3 - i \sin 3\right) = 7.39(-0.990 - 0.141i) = -7.32 - 1.04i$, and (3a) gives $e^{\pi i} = \cos \pi + i \sin \pi = -1 + 0i = -1$.

Since (3a) holds for all y, it must hold also with the y's changed to $-y$'s: $e^{-iy} = \cos(-y) + i \sin(-y)$, and since $\cos(-y) = \cos y$ and $\sin(-y) = -\sin y$ (because the graphs of $\cos y$ and $\sin y$ are symmetric and antisymmetric, respectively, about the origin), it follows that

$$e^{-iy} = \cos y - i \sin y, \tag{3b}$$

which is a "companion formula" for (3a).

If not convinced that $\cos(-x) = \cos x$, write the Taylor series of $\cos x$ and then change al the x's in that equation to $-x$'s. Similarly for $\sin(-x) = -\sin x$.

[1]For instance, the formula $e^{c+d} = e^c e^d$ is for the case where c and d are real numbers, whereas the exponent $x + iy$ in (1) is a single complex number.

Conversely, we can express $\cos y$ and $\sin y$ as linear combinations of the complex exponentials e^{iy} and e^{-iy} because, by adding and subtracting (3a) and (3b), we obtain $\cos y = \left(e^{iy} + e^{-iy}\right)/2$ and $\sin y = \left(e^{iy} - e^{-iy}\right)/(2i)$, respectively. We will frame these important formulas for emphasis:

$$\boxed{\begin{aligned} e^{iy} &= \cos y + i \sin y, \\ e^{-iy} &= \cos y - i \sin y \end{aligned}} \tag{4a,b}$$

and

$$\boxed{\begin{aligned} \cos y &= \frac{e^{iy} + e^{-iy}}{2}, \\ \sin y &= \frac{e^{iy} - e^{-iy}}{2i}. \end{aligned}} \tag{5a,b}$$

Note, in (5a,b), that there is an i in one denominator but not in the other. That is not a misprint. Note also that there are *no* i's in the analogous formulas for cosh and sinh (framed in Section 2.2).

We've obtained all four of these formulas from the single formula (3a). There is nothing special about the name of the variable. For instance, $e^{it} = \cos t + i \sin t$, $e^{i\theta} = \cos\theta + i\sin\theta$, and so on.

How do we differentiate and integrate e^{rx} if r is a complex number? We know that

$$\frac{d}{dx} e^{rx} = r e^{rx} \quad \text{and} \quad \int e^{rx}\, dx = \frac{1}{r} e^{rx} + \text{constant} \tag{6}$$

if r is a real constant (nonzero in the latter because of the $1/r$), but do these hold if $r = a + ib$ is complex? They do. Verification follows readily from (2) and is left for the exercises.

One more property of the exponential function e^z, which we will need, is that

$$\boxed{e^z \neq 0} \tag{7}$$

for all z, because

$$\begin{aligned} |e^z| &= \left|e^{x+iy}\right| = |e^x (\cos y + i \sin y)| \\ &= |e^x|\, |\cos y + i \sin y| = e^x\, |\cos y + i \sin y| \\ &= e^x \sqrt{\cos^2 y + \sin^2 y} = e^x(1) = e^x. \end{aligned} \tag{8}$$

The fourth equality in (8) follows because e^x is positive for all x. Equation (8) shows that $|e^z| > 0$ for all z. Thus $e^z \neq 0$ for all z, as claimed, and we say that e^z *has no zeros*.

e^z has no zeros. That is, $e^z = 0$ has no roots, real or complex.

2.3.2 Complex characterisitc roots. Now that we know how to evaluate complex exponentials, we can use the results found in Section 2.2 for the case in which the characteristic roots are complex.

EXAMPLE 1. Complex Roots. Find a general solution of

$$y'' + 9y = 0. \tag{9}$$

Seek $y(x)$ in the exponential form $y(x) = e^{rx}$ and obtain the characteristic equation

$$r^2 + 9 = 0,$$

with roots $r = \pm 3i$. Thus, a general solution of (9) is

$$y(x) = C_1 e^{i3x} + C_2 e^{-i3x}, \tag{10}$$

CAUTION: Do not forget the x's and, incorrectly, write $y(x) = C_1 e^{i3} + C_2 e^{-i3}$ in place of (10).

in which C_1 and C_2 are arbitrary constants. The right-hand side of (10) is a perfectly acceptable general solution of (8), but we prefer to use Euler's formula to re-express our solutions in terms of real valued functions:

$$\begin{aligned} y(x) &= C_1 e^{i3x} + C_2 e^{-i3x} && \text{(11a)}\\ &= C_1(\cos 3x + i \sin 3x) + C_2(\cos 3x - i \sin 3x) && \text{(11b)}\\ &= (C_1 + C_2)\cos 3x + i\,(C_1 - C_2)\sin 3x && \text{(11c)}\\ &= C_3 \cos 3x + C_4 \sin 3x, && \text{(11d)} \end{aligned}$$

The steps in (11), where we express the complex exponentials in terms of cosine and sine, are analogous to the steps in equation (36) of Section 2.2, where we expressed the real exponentials in terms of the hyperbolic cosine and sine.

in which $C_3 = C_1 + C_2$ and $C_4 = i\,(C_1 - C_2)$ are arbitrary constants; each of them can have any numerical value, real or complex. Our final result is (11d).

COMMENT. If we have initial conditions $y(0) = 3$ and $y'(0) = -1$, say, then (11d) gives

$$\begin{aligned} y(0) &= \ \ 3 = \ \ C_3 \cos 0 + C_4 \sin 0 \ \ = C_3, && \text{(12a)}\\ y'(0) &= -1 = -3C_3 \sin 0 + 3C_4 \cos 0 = 3C_4, && \text{(12b)} \end{aligned}$$

which (since $\cos 0 = 1$ and $\sin 0 = 0$) give $C_3 = 3$ and $C_4 = -1/3$. The result,

$$y(x) = 3\cos 3x - \frac{1}{3}\sin 3x, \tag{13}$$

is plotted in Fig. 1. ∎

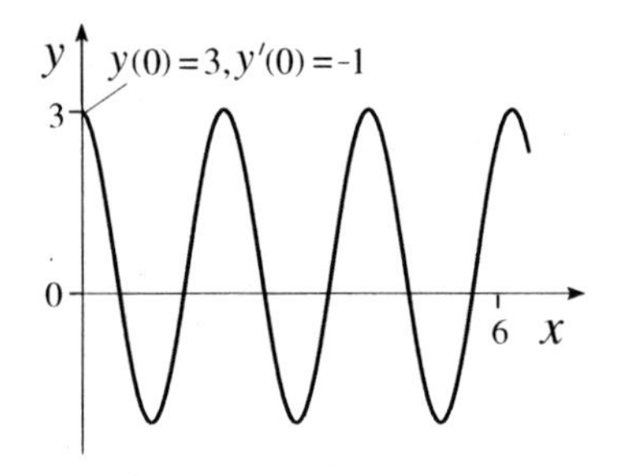

Figure 1. The solution (13) of (9), with the initial conditions $y(0) = 3$ and $y'(0) = -1$.

EXAMPLE 2. Solve the IVP

$$\begin{aligned} y'' + 2y' + 65y &= 0, && \text{(14a)}\\ y(0) = 0, \quad y'(0) &= 2. && \text{(14b)} \end{aligned}$$

Seeking $y(x) = e^{rx}$ gives $r^2 + 2r + 65 = 0$ and $r = -1 \pm 8i$. Thus, a general solution of (14a) is

$$y(x) = C_1 e^{(-1+8i)x} + C_2 e^{(-1-8i)x}, \tag{15}$$

which we normally prefer to re-express in terms of cosine and sine as follows:

$$\begin{aligned} y(x) &= C_1 e^{-x} e^{i8x} + C_2 e^{-x} e^{-i8x}\\ &= e^{-x}\left(C_1 e^{i8x} + C_2 e^{-i8x}\right)\\ &= e^{-x}\left[(C_1 + C_2)\cos 8x + i(C_1 - C_2)\sin 8x\right]\\ &= e^{-x}\left(C_3 \cos 8x + C_4 \sin 8x\right). && \text{(16)} \end{aligned}$$

Now apply the initial conditions (14b) to the latter:

$$y(0) = 0 = C_3, \tag{17a}$$
$$y'(0) = 2 = -C_3 + 8C_4, \tag{17b}$$

so $C_3 = 0$ and $C_4 = 1/4$. [Of course, we must differentiate (16) first, before applying $y'(0) = 2$.] Thus,

$$y(x) = \frac{1}{4}e^{-x}\sin 8x, \tag{18}$$

which is plotted in Fig. 2. ▮

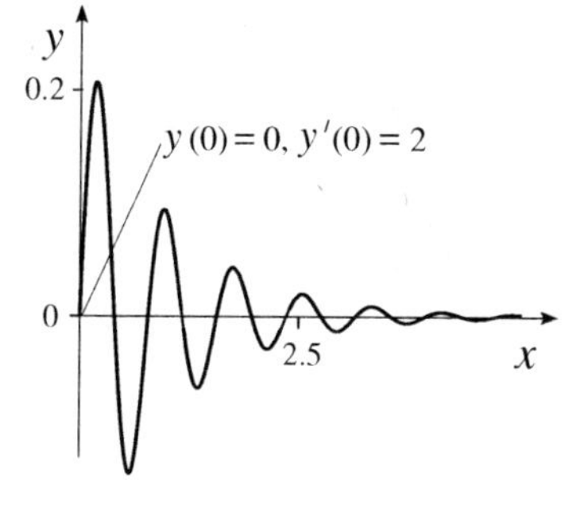

Figure 2. The solution (18) of the IVP (14).

Closure. This section is a completion of Section 2.2 for the case of complex characterisitc roots, and has been separated only to keep that section from growing too long.

If a characteristic root r is complex we must know how to evaluate the corresponding complex exponential solution e^{rx}. We defined the general complex exponential function e^{x+iy} by Euler's formula $e^{x+iy} \equiv e^x(\cos y + i \sin y)$. That was all we needed; no new theory was required.

EXERCISES 2.3

1. Evaluate (i.e., in the standard Cartesian form $a + ib$)

(a) $e^{2+\pi i}$ (b) e^{1-3i} (c) $e^{\pi i/4}$
(d) $e^{-\pi i/3}$ (e) $e^{(1+i)^2}$ (f) $e^{(1-i)^3}$
(g) $(e^{-\pi i/5})^2$ (h) $e^{(1+i)^2}e^{3i}$ (i) $ie^{(1-i)^3}$
(j) $(e^{-\pi i/7})^{14}$ (k) $e^{1+2i}e^{4-5i}$ (l) $e^{(1-i)^4}$

2. Evaluate

(a) $|e^{3-2i}|$ (b) $|e^{1-25i}|$ (c) $|e^{27.36i}|$
(d) $|e^{(2-3i)^2}|$ (e) $|3e^{(1+i)^2}|$ (f) $|5ie^{\pi i/8}|$

3. Find all roots, if any.

(a) $e^{2z} = 0$ (b) $e^{z^2} = 0$ (c) $e^{z+3} = 0$
(d) $e^{z^2-3z+2} = 0$ (e) $e^{iz} = 0$ (f) $e^{-(z+1)^2} = 0$

4. Show that, in general, $|e^z| \neq e^{|z|}$.

5. Show that $e^z = 1$ if and only if $z = 2n\pi i$, where n is any integer.

6. Can one plot the graph of e^{ix} versus x? Explain.

7. Deriving (6). Let r be a complex number.
(a) Derive the derivative formula in (6). HINT: Let $r = a + ib$ and show that $e^{rx} = e^{ax}\cos bx + ie^{ax}\sin bx$, then differentiate the latter with respect to x and show that the result can be identified as re^{rx}.
(b) Derive the integral formula in (6). HINT: The same hint as in part (a), but integrate instead of differentiate, using computer software to do the integration if you wish. Show that the result of the integration can be identified as e^{rx}/r.

8. Complex Roots; General Solution. Derive a general solution of the given equation, and express your answer in terms of real valued functions.

(a) $y'' - 8y' + 25y = 0$ (b) $y'' + 2y' + 2y = 0$
(c) $y'' + 2y' + 3y = 0$ (d) $y'' - y' + y = 0$
(e) $y'' + 2y' + 5y = 0$ (f) $y'' - y' + 7y = 0$
(g) $2y'' + y' + y = 0$ (h) $y'' + y' + y = 0$
(i) $5y'' + 4y' + y = 0$ (j) $4y'' + y = 0$
(k) $y'' - 2y' + 2y = 0$ (l) $y'' + y' + 10y = 0$

9. Exercise 8 as an IVP. Solve the IVP consisting of the differential equation from the corresponding part of Exercise 8, subject to the specified initial conditions, and leave your answer in terms of real valued functions.

(a) $y(0) = 3,\ y'(0) = 6$ (b) $y(\pi) = 0,\ y'(\pi) = 10$
(c) $y(0) = 1,\ y'(0) = 1$ (d) $y(0) = 2,\ y'(0) = 1$
(e) $y(5) = 0,\ y'(5) = 2$ (f) $y(0) = 0,\ y'(0) = 9$
(g) $y(0) = 10,\ y'(0) = 0$ (h) $y(0) = 10,\ y'(0) = 1$

(i) $y(2\pi) = 2,\ y'(2\pi) = 0$ (j) $y(0) = 4,\ y'(0) = 0$
(k) $y(0) = y'(0) = 0$ (l) $y(0) = 1,\ y'(0) = 0$

10. Give conditions (one or more inequalities or equalities) on the coefficients p_1, p_2 in $y'' + p_1 y' + p_2 y = 0$ for the characteristic roots to be as follows. That is, give the most general such conditions, not merely an example.

(a) complex (b) repeated
(c) both positive and distinct (d) both negative and distinct
(e) one positive, one negative (f) one zero, one not zero

11. Working Backwards. If the characteristic roots of a given second-order, linear, homogeneous, constant-coefficient differential equation are as given, write down a general solution for $y(x)$, in terms of real valued functions. Further, infer from the given roots the original differential equation. NOTE: In applications we normally don't work "backwards" to infer the differential equation from the characteristic roots, but this exercise is included to promote understanding of the material in this section.

(a) $3 \pm 4i$ (b) $-1 \pm 2i$ (c) $3 \pm 5i$
(d) $\pm 6i$ (e) $-2 \pm i$ (f) $3 \pm 4i$

12. Inferring the Differential Equation. If the C_j's are arbitrary constants, then the given $y(x)$ is a general solution of what second-order linear homogeneous differential equation, with real constant coefficients? If such an equation does not exist, state that and explain your reasoning.

(a) $y(x) = e^{3x}(C_1 \cos 2x + C_2 \sin 2x)$
(b) $y(x) = e^{x}(C_1 \cos x + C_2 \sin x)$
(c) $y(x) = e^{-2x}(C_1 \cos 5x + C_2 \sin 5x)$
(d) $y(x) = C_1 e^{i5x} + C_2 e^{-i5x}$
(e) $y(x) = (C_1 e^{i3x} + C_2 e^{-i3x})e^{4x}$
(f) $y(x) = C_1 e^{ix} + C_2 e^{i2x}$
(g) $y(x) = e^{-5x}(C_1 e^{ix} + C_2 e^{-ix})$
(h) $y(x) = e^{x}(C_1 e^{i7x} + C_2 e^{-i7x})$
(i) $y(x) = C_1 e^{i5x} + C_2 e^{i6x}$
(j) $y(x) = e^{x}(C_1 e^{ix} + C_2 e^{-i3x})$
(k) $y(x) = (C_1 + C_2 x)e^{ix}$
(l) $y(x) = (C_1 + C_2 x)(e^{ix} + e^{-ix})$

2.4 LINEAR INDEPENDENCE; EXISTENCE, UNIQUENESS, GENERAL SOLUTION

Having solved the constant-coefficient equation $y'' + p_1 y' + p_2 y = 0$, we turn to the more general equation

We now allow p_1 and p_2 to be functions of x. Of course, they *may* be constants but, in this section, that is just a special (simple) case.

$$\boxed{y'' + p_1(x)y' + p_2(x)y = 0.} \tag{1}$$

in which the coefficients p_1 and p_2 are now permitted to be functions of x.

The general solution, obtained in Section 2.2, does not apply now because our derivation depended on the assumption of constant coefficients. We will discuss solution strategies for (1) later, but before we can do that we must establish the necessary theory.

A key concept will be that of "linear dependence and independence" of sets of functions, which we introduce next. When that is completed, we can return to our consideration of equation (1).

2.4.1 Linear dependence and linear independence. Asked how many different colors he had, a painter replied "Five: red, blue, yellow, green, and purple." One could say that his count was inflated because green can be made by mixing blue and yellow, and purple by mixing red and blue. That is, there is some redundancy in his set of five colors; they are related. As this idea is important in painting it is also important in mathematics and involves a concept of the linear dependence or linear independence of a given set of functions.

DEFINITION 2.4.1 *Linear Dependence and Linear Independence*
A set of k functions $\{u_1(x), \dots, u_k(x)\}$ defined on an x interval I is **linearly dependent** on I if there exist constants $a_1, \dots, a_k$, not all zero, such that

$$a_1u_1(x) + a_2u_2(x) + \cdots + a_ku_k(x) = 0 \tag{2}$$

for all x in I. If the set is not linearly dependent on I, then it is **linearly independent** on I.

We will use **LD** and **LI** to denote linearly dependent and linearly independent, respectively.

In this book we adopt the convenient abbreviations LD and LI for linearly dependent and linearly independent, respectively, but these abbreviations are not standard.

The definition holds even if $k = 1$, but, typically, our interest is in sets of two or more functions, $k \geq 2$. In that case, the significance of linear dependence is that *at least one of the u_j's can be expressed as a linear combination of the others*, for if the a_j's are not all zero then at least one of them must be nonzero. If, for instance, $a_2 \neq 0$, we can divide (2) by a_2 (because it is nonzero) and solve for $u_2(x)$:

Just as green, in our paint analogy, can be made as a combination of red, blue, and yellow.

$$u_2(x) = \frac{a_1}{a_2}u_1(x) - \frac{a_3}{a_2}u_3(x) - \cdots - \frac{a_k}{a_2}u_k(x). \tag{3}$$

The latter gives u_2 as a linear combination of the other u_j's.

We will need to be able to test sets of functions to determine if they are LD or LI.

EXAMPLE 1. By Inspection. Let $k = 4$, and consider the set $\{1, x, e^x, 3x-2\}$. This set is LD (on any interval) because, as seen by inspection, the fourth is a linear combination of the first two: $3x-2 = 3(x) - 2(1)$. Thus, we can write

$$2(1) - 3(x) + 0e^x + 1(3x-2) = 0. \tag{4}$$

The latter is of the form (1) with the a_j's *not all* zero, for $a_1 = 2$, $a_2 = -3$, $a_3 = 0$, and $a_4 = 1$ so, we repeat, the set is LD. In fact, we can use (4) to express the first, second, and fourth functions as linear combinations of the others:

$$1 = \frac{3}{2}(x) - \frac{1}{2}(3x-2) + 0(e^x),$$
$$x = \frac{2}{3}(1) + \frac{1}{3}(3x-2) + 0(e^x),$$
$$3x-2 = -2(1) + 3(x) + 0(e^x).$$

We happen not to be able to solve for the third, e^x, as a linear combination of the others because its coefficient in (4) is zero, but the set is LD nonetheless because for the linear dependence of a set of two or more functions we need only be able to express *at least* one of the functions as a linear combination of the others. ▮

In Example 1 we saw, merely by inspection, that the set was LD, but inspection is generally not feasible, so a systematic test is needed. We will give such tests below, but we first mention two particularly simple cases:

TEST: Any set containing the 0 function is LD.

If one of the functions in the set is 0, the set is LD, for suppose in the set $\{u_1, \ldots, u_k\}$ the function $u_1(x) = 0$. Then we can express $u_1 = 0u_2 \cdots + 0u_k$, which can be written as $1u_1 + 0u_2 + \cdots + 0u_k = 0$, in which not all the coefficients are zero. For instance, $\{x, 0, \sin x, e^{2x}\}$ is LD because of the 0.

TEST: If there are only two functions, the set is LD if and only if either one of them is a constant multiple of the other.

Also, *if there are only two functions, then the set is LD if and only if at least one of them is a constant multiple of the other.* Proof is left for the exercises. For instance, $\{2e^x, -6e^x\}$ is LD because $-6e^x$ is -3 times $2e^x$, and $\{x, \sin x\}$ is LI because x is *not* a constant times $\sin x$, nor is $\sin x$ a constant times x.

There is a convenient test, in terms of the **Wronskian determinant**, that can be applied to any given set of functions $\{u_1, \ldots, u_k\}$ that are differentiable through order $k-1$. The Wronskian determinant, or "Wronskian," of the set is the $k \times k$ determinant

$$W[u_1, \ldots, u_k](x) = \begin{vmatrix} u_1(x) & u_2(x) & \cdots & u_k(x) \\ u_1'(x) & u_2'(x) & \cdots & u_k'(x) \\ \vdots & \vdots & & \vdots \\ u_1^{(k-1)}(x) & u_2^{(k-1)}(x) & \cdots & u_k^{(k-1)}(x) \end{vmatrix}. \tag{5}$$

Named after the Polish mathematician *Josef Wronski* (1778–1853), the Wronskian W is a function of x; hence the "(x)" argument on the left side of (5).

EXAMPLE 2. Wronskian. The Wronskian of the set $\{x, x^2, x^3\}$ is

$$W[x, x^2, x^3](x) = \begin{vmatrix} x & x^2 & x^3 \\ 1 & 2x & 3x^2 \\ 0 & 2 & 6x \end{vmatrix} = 2x^3. \quad ▮ \tag{6}$$

Here is the test:

If $W \neq 0$ anywhere in I, the set is LI. If not, Theorem 2.4.1 gives no information.

THEOREM 2.4.1 *Wronskian Sufficient Condition for LI*
Let $u_1(x), \ldots, u_k(x)$ be $(k-1)$-times differentiable on an open interval I. If their

Wronskian $W[u_1, \ldots, u_k](x)$ is nonzero anywhere in I, then the set is LI on I.

Proof: Consider the equation

$$a_1 u_1(x) + a_2 u_2(x) + \cdots + a_k u_k(x) = 0 \tag{7}$$

on I. Differentiate it $k-1$ times, then set $x = \xi$ in each equation, where ξ is a point at which the Wronskian is nonzero, obtaining the system of k linear algebraic equations

$$\begin{aligned} a_1 u_1(\xi) + a_2 u_2(\xi) + \cdots + a_k u_k(\xi) &= 0, \\ a_1 u_1'(\xi) + a_2 u_2'(\xi) + \cdots + a_k u_k'(\xi) &= 0, \\ &\vdots \\ a_1 u_1^{(k-1)}(\xi) + a_2 u_2^{(k-1)}(\xi) + \cdots + a_k u_k^{(k-1)}(\xi) &= 0 \end{aligned} \tag{8}$$

The unknowns in (8) are $a_1, \ldots, a_k$.

for $a_1, \ldots, a_k$. The determinant of the coefficients in (8) is $W[u_1 \ldots, u_k](\xi)$, which is nonzero by assumption, so (8) has only the trivial solution (Appendix C). Thus, (7) holds only if $a_1 = \cdots = a_k = 0$, so $\{u_1, \ldots, u_k\}$ is LI on I. ∎

EXAMPLE 3. Application of Theorem 2.4.1. Is the set $\{x, x^2, x^3\}$ on $-\infty < x < \infty$ LD or LI? The three functions are indeed twice differentiable on the interval I, and we found in Example 2 that their Wronskian is $2x^3$. The latter is nonzero everywhere in I except at $x = 0$, which is even more than we need since Theorem 2.4.1 asks only that W be zero *anywhere* in I, that is, at least at one point in I. Thus, it follows from Theorem 2.4.1 that $\{x, x^2, x^3\}$ is LI on I. ∎

2.4.2 Existence, uniqueness, and general solution. Before we can develop the idea of a general solution of (1), we lay the foundation with the following theorem for the existence and uniqueness of a solution to the initial value problem — that is, the differential equation (1), together with initial conditions:

It's true that we've already solved (9) in Section 2.2, but that was only for the simpler case in which p_1 and p_2 are constants. The case of non-constant coefficients is much more difficult, in general, so Theorem 2.4.2 provides welcome assurance, in advance, of the existence and uniqueness of a solution.

THEOREM 2.4.2 *Existence/Uniqueness for the IVP*
For the IVP

$$y'' + p_1(x)y' + p_2(x)y = 0; \tag{9a}$$
$$y(a) = b_1, \; y'(a) = b_2, \tag{9b}$$

a solution exists and is unique, at *least* on the broadest open x interval containing the initial point a, on which the p_j's are continuous.

It is instructive to compare Theorem 2.4.2 with the result obtained in Section 2.1 for the *first*-order IVP $y' + p(x)y = 0$ with initial condition $y(a) = b$. There, we showed that the solution exists and is unique on the broadest open x interval containing a, on which $p(x)$ is continuous, which is analogous to Theorem 2.4.2. The difference, however, is that we were able to determine that solution, namely, $y(x) = b\exp\left(-\int_a^x p(s)\,ds\right)$, whereas (9) is too difficult and we cannot give an analogous solution of (9) in terms of $p_1(x)$, $p_2(x)$, b_1, and b_2.

Now that we know, from Theorem 2.4.2, that solutions of (1) exist, we can turn to the form of a general solution of (1).

Equation (1) has *two* LI solutions: no more, no less.

Use them, and the principle of superposition, to construct a general solution.

THEOREM 2.4.3 *General Solution of (1)*
(a) *Number of LI Solutions*: Let the $p_j(x)$ coefficients in (1) be continuous on an open interval I. Then, (1) has exactly two LI solutions on I: no more than two, and no less than two.
(b) *Form of General Solution*: If $y_1(x)$ and $y_2(x)$ are LI solutions of (1) on I, and C_1, C_2 are arbitrary constants, then

$$\boxed{y(x) = C_1 y_1(x) + C_2 y_2(x)} \tag{10}$$

is a **general solution** of (1) on I; that is, every solution of (1) is expressible in the form (10) by suitable choice of C_1, C_2.
(c) *Unique Representation by (10)*: Not only can any given solution $y(x)$ of (1) be represented in the form (10), that representation is unique; that is, C_1, C_2 are uniquely determined for the given $y(x)$.

The term **fundamental set** is also used.

Such a set of any two LI solutions of (1), $\{y_1, y_2\}$, is called a **basis** of solutions for (1) because every solution of (1) *can be* expressed as a linear combination of them, and *uniquely* so, that is, with the coefficients C_1, C_2 uniquely determined.

EXAMPLE 4. IVP and General Solution. Solve the IVP

$$y'' - 7y' + 10y = 0; \tag{11a}$$

$$y(0) = 3, \ y'(0) = -2. \tag{11b}$$

In (1) we allow for the coefficients p_1 and p_2 to be nonconstant, but in this introductory example we take them to be constants.

Since $p_1 = -7$ and $p_2 = 10$ are continuous for all x, Theorem 2.4.2 assures us that a solution of (11) exists, and is unique, on $-\infty < x < \infty$, which interval we denote as I. To find it, we will develop a general solution, and then apply the initial conditions.

If we seek solutions of (11a) in the exponential form $y(x) = e^{rx}$, we obtain the characteristic equation $r^2 - 7r + 10 = 0$, with solutions $r = 2, 5$. We know from Section 2.2 that $y(x) = C_1e^{2x} + C_2e^{5x}$ is therefore a general solution of (11a), but the results in Section 2.2 were for the special case of constant-coefficient equations. Here, we wish to illustrate the general theory, given above, so let us rely instead on the theorems of this section.

Thus, here is where we stand: We have found solutions $y_1(x) = e^{2x}$ and $y_2(x) = e^{5x}$ and can use the superposition principle from Section 2.1 to write

$$y(x) = C_1 e^{2x} + C_2 e^{5x}. \tag{12}$$

To test y_1 and y_2 for linear independence, evaluate their Wronskian,

$$W[y_1, y_2](x) = W[e^{2x}, e^{5x}] = \begin{vmatrix} e^{2x} & e^{5x} \\ 2e^{2x} & 5e^{5x} \end{vmatrix} = (5-2)e^{7x} = 3e^{7x}. \tag{13}$$

The latter is nonzero at $x = 5$, for instance (and indeed for all x). Thus, by Theorem 2.4.1, the solutions y_1 and y_2 are LI on I. Then, by part (b) of Theorem 2.4.3, (12) is a general solution of (11a).

First obtain a general solution, then apply the initial conditions to it.

Next, apply the initial conditions (11b) to that general solution:

$$\begin{aligned} y(0) = 3 = {} & C_1 y_1(0) + C_2 y_2(0) = C_1(1) + C_2(1), \\ y'(0) = -2 = {} & C_1 y_1'(0) + C_2 y_2'(0) = C_1(2) + C_2(5), \end{aligned} \tag{14}$$

which give the unique solution $C_1 = -8/3, C_2 = 17/3$ for the C_j's. Thus, the unique solution of (11) is $y(x) = -(8/3)e^{2x} + (17/3)e^{5x}$.

COMMENT. The basis $\{e^{2x}, e^{3x}\}$ for (11a) is by no means unique, because an infinite number of bases can be formed by linear combinations of e^{2x} and e^{3x}, such as $\{e^{2x}, 3e^{2x} + 17e^{5x}\}$ and $\{e^{2x} - e^{5x}, e^{5x}\}$. The latter four functions are of the form (12) and are therefore solutions of (11a), and the Wronskian of each of the two pairs is nonzero, so each of the pairs is LI. Of course, not *every* pair of linear combinations of e^{2x} and e^{3x} is a basis. For instance, $\{e^{2x} + 4e^{3x},\ 2e^{2x} + 8e^{3x}\}$ is not, because the latter two functions are not LI. ∎

EXAMPLE 5. This Time, Nonconstant Coefficients. Consider the IVP

$$x^2 y'' - xy' - 3y = 0; \tag{15a}$$

$$y(1) = 3,\ y'(1) = 1. \tag{15b}$$

To apply Theorem 2.4.2, first put (15a) in the form (1) by multiplying through by $1/x^2$. We obtain $y'' - (1/x)y' - (3/x^2)y = 0$ and can now identify the coefficients in (1) as $p_1(x) = -1/x$ and $p_2(x) = 3/x^2$. The largest interval (containing the initial point $x = 1$) on which they are continuous is $0 < x < \infty$, so Theorem 2.4.2 guarantees that a unique solution of (15) exists at *least* on that interval, which we will call I.

We will learn how to solve (15a) in Section 2.5, but for the present let it suffice to give, without deriving them, two solutions on I: $y_1(x) = 1/x$ and $y_2(x) = x^3$. To verify their linear independence, we can use Theorem 2.4.1: $W[y_1, y_2](x) = 2x$, which is nonzero on I, so $y_1(x) = 1/x$ and $y_2(x) = x^3$ are LI on I. Thus, a general solution of (15a) on I is, by Theorem 2.4.3,

$$y(x) = C_1 \frac{1}{x} + C_2 x^3. \tag{16}$$

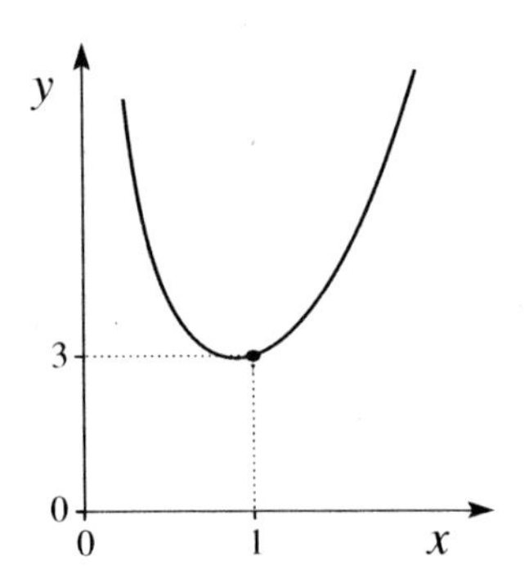

Figure 1. Solution of (15).

Next, application of the initial conditions to (16) gives

$$\begin{aligned} y(1) = 3 = {} & C_1 y_1(1) + C_2 y_2(1) = C_1(1) + C_2(1), \\ y'(1) = 1 = {} & C_1 y_1'(1) + C_2 y_2'(1) = C_1(-1) + C_2(3). \end{aligned} \tag{17}$$

The latter gives $C_1 = 2$ and $C_2 = 1$, so the unique solution of (15) on I is $y(x) = 2/x + x^3$. The interval of existence/uniqueness, $0 < x < \infty$, cannot be extended any further to the left because the $2/x$ is undefined at the left endpoint $x = 0$, where it "blows up" to infinity (Fig. 1). ▮

2.4.3 Abel's formula and Wronskian test for linear independence.

There is a formula for the Wronskian of *any* two solutions of (1), be they LI or LD, in terms only of the coefficient $p_1(x)$ in the differential equation.

Equation (18) is known as **Abel's formula**, after the Norwegian mathematician *Niels Henrik Abel* (1802–1829). It is striking that y_1 and y_2 do not take part in the calculation on the right-hand side of (18).

THEOREM 2.4.4 *Abel's Formula and Wronskian Test for LI*
Let the $p_j(x)$ coefficients in (1) be continuous on an open interval I, and let $y_1(x)$ and $y_2(x)$ be any two solutions of (1) on I.
(a) **Abel's Formula**: Then, their Wronskian is

$$\boxed{W[y_1, y_2](x) = Ce^{-\int p_1(x)\, dx},} \tag{18}$$

in which the constant C depends upon the two solutions $y_1(x)$ and $y_2(x)$ that are used. For brevity, now denote $W[y_1, y_2](x)$ as W:
(b) W is either zero everywhere in I, or else nonzero everywhere in I.
(c) **Wronskian Test**: If $W = 0$ in I, then $\{y_1, y_2\}$ is LD on I, and if $W \neq 0$ in I then $\{y_1, y_2\}$ is LI on I.

The result:
LI if and only if $W \neq 0$.

Proof: Note first that (18) holds for every pair of solutions $y_1(x)$ and $y_2(x)$ of (1), be they LI or LD. It expresses a specific relatedness between every such solution pair, by virture of their being "fellow solutions" of the same differential equation (1).

Derivation of Abel's formula in part (a) is outlined in the exercises, so we turn to part (b). Since p_1 is continuous, the integral $\int p_1(x)\, dx$ exists and is finite. Thus, the exponential function in (18) is nonzero, so W *is zero for all x in I if $C = 0$, and is nonzero for all x in I if $C \neq 0$.* For part (c), the truth of the second half (that is, the sufficiency of $W[y_1, y_2](x) \neq 0$ on I for $\{y_1, y_2\}$ to be LI on I) is simply Theorem 2.4.1. Thus, it remains to prove the first half, that if $W[y_1, y_2](x) = 0$ on I then $\{y_1, y_2\}$ is LD on I.

To do so, let x_0 be any point in I. Then $W[y_1, y_2](x_0) = 0$, so the equations

$$\begin{aligned} C_1 y_1(x_0) + C_2 y_2(x_0) = {} & 0, \\ C_1 y_1'(x_0) + C_2 y_2'(x_0) = {} & 0 \end{aligned} \tag{19}$$

have nontrivial solutions for C_1, C_2, since the determinant of the coefficients is $W[y_1, y_2](x_0)$, which is zero because $W = 0$ on I by assumption. Let C_1, C_2 be such a nontrivial pair and, using those C_j's, define $u(x) = C_1y_1(x) + C_2y_2(x)$. Then u satisfies both (1) and, by virtue of (19), also the initial conditions $u(x_0) = 0$, $u'(x_0) = 0$. Surely $u(x) = 0$ satisfies (1) and those conditions and, by the uniqueness part of Theorem 2.4.2, it is the *only* such solution of (1). Thus, we have $u(x) = C_1y_1(x) + C_2y_2(x) = 0$ on I with the C_j's not both zero, so y_1 and y_2 must be LD on I. ■

To put Theorem 2.4.4 in perspective, recall that Theorem 2.4.1 gave the condition of the nonvanishing of the Wronskian of a function set as a *sufficient* condition for the linear independence of the set. However, Theorem 2.4.4 tells us that *if the functions y_1 and y_2 are not just "any old" functions, but are fellow solutions of (1), then the nonvanishing of their Wronskian is not only sufficient for their linear independence, it is both necessary and sufficient.*

Thus, to test two solutions y_1 and y_2 of (1) use the stronger of the two theorems, Theorem 2.4.4.

In fact, it is possible for the Wronskian of two functions that are *not* fellow solutions of (1) to be identically zero on an interval I, and yet for them to be LI on I. Such an example is given in Exercise 7.

EXAMPLE 6. Abel's Formula. To illustrate Abel's formula, consider solutions $y_1(x) = e^x \cos 3x$ and $y_2(x) = e^x \sin 3x$ of

$$y'' - 2y' + 10y = 0 \tag{20}$$

on $-\infty < x < \infty$, which does have continuous coefficients $p_1 = -2$ and $p_2 = 10$. Then

$$W[y_1, y_2](x) = \begin{vmatrix} e^x \cos 3x & e^x \sin 3x \\ e^x(\cos 3x - 3\sin 3x) & e^x(\sin 3x + 3\cos 3x) \end{vmatrix} = 3e^{2x}. \tag{21}$$

The latter is indeed nonzero on I [which is consistent with part (b)], so y_1, y_2 are LI on I by part (c) of Theorem 2.4.4. In contrast, Theorem 2.4.1 gives no information in this case.

Further, let us verify (18) for this example:

$$W[y_1, y_2](x) = Ce^{-\int(-2)dx} = Ce^{2x}, \tag{22}$$

which does agree with (21), with $C = 3$. ▮

2.4.4 Building a solution method on these results. Based upon the foregoing results, the recommended solution method is as follows.

SOLUTION METHOD

To solve a given equation

$$y'' + p_1(x)y' + p_2(x)y = 0 \tag{23}$$

on an interval I:

1. First, generate solutions $y_1(x)$ and $y_2(x)$. If p_1 and p_2 are not both constants, that step may be difficult, because the exponential solution form e^{rx} will, in general, not work (see Exercise 14), and factoring the operator is probably too difficult to implement. One may need to give up on obtaining closed form analytical solutions and to seek numerical or infinite series solutions; these methods are covered in Section 4.11 and Chapter 6, respectively.

Using Theorems 2.4.3 and 2.4.4(c).

2. If solutions $y_1(x)$ and $y_2(x)$ are in hand, evaluate their Wronskian determinant $W[y_1, y_2](x)$. If the latter is nonzero on I, then y_1 and y_2 are LI on I [Theorem 2.4.4(c)] and form a basis of solutions of (23). Then, by Theorem 2.4.3, a general solution of (23) is $y(x) = C_1y_1(x) + C_2y_2(x)$.

Closure. The focus of this section is Theorem 2.4.2 on the existence and uniqueness of solutions of the IVP, and Theorem 2.4.3 on the form of a general solution of (1), both of which assume that the p_j coefficients are continuous on the interval of interest. In terms of solution technique, the two itemized steps given above summarize our results.

It is true that for a set of only two functions the Wronskian test is "overkill," because we need merely see if one is a constant multiple of the other. If so, the set is LD; if not, the set is LI. But Abel's formula and the Wronskian test are fundamental and will be helpful in developing the theory of equations of order higher than two, in Section 2.7.

Regarding Theorem 2.4.4(c), how can we remember if $W = 0$ corresponds to linear dependence or independence? Here is a suggestion: If we take $y_1(x) = y_2(x)$, then surely $y_1(x)$ and $y_2(x)$ are LD and surely their Wronskian is zero. Those cases go together, with $W = 0$ indicating LD and $W \neq 0$ indicating LI.

EXERCISES 2.4

1. Determine whether the set is LD or LI on $-\infty < x < \infty$. If you find that it is LD, give a linear relationship among them, of the form of (2), in which the a_j's are not all zero.

(a) $\{e^x,\ 0\}$ (b) $\{e^x,\ 1\}$
(c) $\{1,\ 5\}$ (d) $\{\sin x\cos x,\ \sin 2x\}$
(e) $\{e^x,\ e^{2x},\ e^{3x}\}$ (f) $\{x,\ x^2,\ x+3x^2\}$
(g) $\{3,\ \cos 2x,\ \cos^2 x\}$ (h) $\{e^x,\ \sinh x,\ \cosh x\}$
(i) $\{x,\ 3x,\ e^x,\ e^{2x}\}$ (j) $\{e^x,\ e^{-x},\ \sinh 2x\}$
(k) $\{x^4,\ x^3,\ x^2\}$ (l) $\{1,\ -x,\ x^2,\ -x^3\}$
(m) $\{x^2,\ -x^2,\ x^4,\ x^5\}$ (n) $\{0,\ x^2,\ x^4,\ x^6\}$
(o) $\{3,\ 1+x,\ 2+3x\}$ (p) $\{1,\ e^x,\ e^{-x}\}$

2. Sets of Exponentials. Use Theorem 2.4.1 to show that the following sets are LI on every interval. The r's are constants.

(a) $\{e^{r_1x}, e^{r_2x}\}$ with $r_1 \neq r_2$ (b) $\{e^{rx}, xe^{rx}\}$
(c) $\{e^{r_1x}, e^{r_2x}, e^{r_3x}\}$, where r_1, r_2, r_3 are distinct
(d) $\{e^{r_1x}, \dots, e^{r_kx}\}$, where $r_1, \dots, r_k$ are distinct. HINT: The determinant properties D6 and D7 in Appendix B will be useful.

3. Only Two Functions. Prove the "Test" given above equation (5), namely, that if there are only two functions, then the set is LD if and only if either one of them is a constant multiple of the other.

4. Caution. Consider a set $\{u_1(x), \dots, u_k(x)\}$ on an open interval I. Show that even if every pair of functions in the set (i.e., two at a time) is LI on I, it does *not* follow that the set is LI on I. HINT: To show that an assertion is true, a single example does not suffice; one must show that it is true for all cases. However, to show that an assertion is *not* true in general, it suffices to give a single example for which it fails.

5. Caution. It is tempting to think that if a function set is LI on $-\infty < x < \infty$, then it is LI on every subinterval. Show that the latter is *false*. HINT: Consider an example, such as the set $\{x^2, x|x|\}$.

6. Grade This. In an exam a student claims that "the set $\{e^x, 3\cos x\}$ is LD on $-\infty < x < \infty$ because at $x = 0$ they equal 1 and 3, respectively, which are constant multiples of one another." Evaluate that response.

7. About Theorem 2.4.1. We noted that the conditions in Theorem 2.4.1 are only *sufficient* for linear independence. Thus, if $W = 0$ everywhere on I, it does *not* follow from the theorem that the set is LD. This exercise is to prove that point by constructing a set of functions, the Wronskian of which is zero on some interval I, but which is LI on I.

(a) Let I be $-\infty < x < \infty$, let $u(x)$ be 0 for $x \le 0$ and x^4 for $x \ge 0$, and let $v(x)$ be x^4 for $x \le 0$ and 0 for $x \ge 0$. Show that u and v are LI on I even though $W[u, v](x) = 0$ on I.
(b) It follows then from Theorem 2.4.4(c) that u and v must not be solutions of an equation of the form (1), with coefficients $p_1(x)$ and $p_2(x)$ that are continuous on I. In fact, determine $p_1(x)$ and $p_2(x)$ so that (1) has solutions $u(x)$ and $v(x)$, and show that, indeed, they are not continuous on I.

8. Linear Independence and Matching Coefficients. A useful result of linear independence is this: If $\{u_1, \dots, u_k\}$ is LI on I, and

$$a_1u_1(x) + \cdots + a_ku_k(x) = b_1u_1(x) + \cdots + b_ku_k(x) \quad (8.1)$$

on I, then it must be true that $a_j = b_j$ for $j = 1, \dots, k$. That is, surely (8.1) holds if each a_j equals each b_j, but the point here is that there is no other way for (8.1) to be true on I: the b_j's *must* equal the a_j's. To prove the claim, express (8.1) as

$$(a_1 - b_1)u_1(x) + \cdots + (a_k - b_k)u_k(x) = 0. \quad (8.2)$$

Since the u_j's are LI by assumption, it follows from (8.2) that each $a_j - b_j = 0$. Thus, each $a_j = b_j$.
Here is the problem: If possible, satisfy the given equation, on $0 < x < \infty$, by suitable choice of A, B, and C. If it cannot be satisfied by any choice of those constants then say so and give your reasoning.

(a) $A + Bx + Cx^2 = (2 + 3x)^2$
(b) $A + Bx + C\cos x = 3 + 5\cos x - 4x$
(c) $Ax^2 + Bx^3 + Cx^4 = (x - 2x^2)^2$
(d) $(A + Be^x)^2 = 4 + 3x + 12e^x + 9e^{2x}$
(e) $(A + Be^x)^2 = 1 + 9e^x + e^{2x}$
(f) $A + B\cos x + C\sin x = \cos x - 15$
(g) $(A+B)e^x + (A-B)e^{-x} + C = 1 + 4\cosh x - \sinh x$
(h) $Ae^x + B\sinh x + Cx = 3e^x + 5e^{-x} - 6x$
(i) $A + (A + B)x + (A + B + C)x^2 = (2 + x)^2$
(j) $A + B + Ax + Cx^3 = 1 - x - x^3$
(k) $Ae^{3x} + Be^{-3x} = 4(\cosh 3x - 1)$
(l) $Ax + Bx^2 + Cx^3 = C + x + x^2$

9. Basis for (1). Let the p_j's in (1) be continuous on I, and suppose a is in I. Let $y_1(x)$ and $y_2(x)$ denote the solutions of (1) corresponding to the first and second set of initial conditions, respectively. Is $\{y_1, y_2\}$ a basis for (1) on I? Explain.

(a) $y(a) = 1,\ y'(a) = 0$ and $y(a) = 1,\ y'(a) = 1$
(b) $y(a) = 0,\ y'(a) = 1$ and $y(a) = 1,\ y'(a) = -1$
(c) $y(a) = 0,\ y'(a) = 0$ and $y(a) = 1,\ y'(a) = 1$
(d) $y(a) = 1,\ y'(a) = 2$ and $y(a) = 4,\ y'(a) = 8$
(e) $y(a) = 1,\ y'(a) = 0$ and $y(a) = -1,\ y'(a) = 0$
(f) $y(a) = 4,\ y'(a) = -5$ and $y(a) = 0,\ y'(a) = 2$

10. Basis. Given are a differential equation, functions y_1 and y_2, and an interval I. Determine whether or not $\{y_1, y_2\}$ is a basis (fundamental set) for the given equation, on I.

(a) $y'' - 2y' + y = 0,\ -\infty < x < \infty,\ y_1(x) = 3e^x,$ $y_2(x) = xe^x - 4e^x$
(b) $y'' - 4y = 0,\ -\infty < x < \infty,\ y_1(x) = e^{2x},$ $y_2(x) = \sinh 2x$
(c) $y'' + 9y = 0,\ -\infty < x < \infty,\ y_1(x) = \cos 3x,$ $y_2(x) = 0$
(d) $y'' + 4y = 0,\ 0 < x < 1,\ y_1(x) = \cos 2x,$ $y_2(x) = \cos 2x + \sin 2x$
(e) $x^2y'' - 2xy' - 10y = 0,\ 0 < x < \infty,$ $y_1(x) = 1/x^2,\ y_2(x) = x^4$
(f) $x^2y'' - 6y = 0,\ 0 < x < \infty,\ y_1(x) = 3/x^2,\ y_2(x) = 2x^3$
(g) $xy'' + y' = 0,\ 0 < x < \infty,\ y_1(x) = 1,\ y_2(x) = 1 + \ln x$
(h) $(x-1)y'' - xy' + y = 0,\ 2 < x < 5,$ $y_1(x) = x,\ y_2(x) = e^x$
(i) $(\cos x)y'' + (\sin x)y' = 0,\ -1 < x < 1,$ $y_1(x) = 1,\ y_2(x) = \sin x$
(j) $(x^2 - 2x)y'' + (2 - x^2)y' + (2x - 2)y = 0,\ 2 < x < \infty,$ $y_1(x) = x^2,\ y_2(x) = e^x$
(k) $(x^2 - 2x)y'' + (2 - x^2)y' + (2x - 2)y = 0,\ 2 < x < 3,$ $y_1(x) = x^2 - e^x,\ y_2(x) = x^2 + e^x$

11. Abel's Formula. Let $y_1(x)$ and $y_2(x)$ be solutions of (1) on an open interval I, and let $p_1(x)$ and $p_2(x)$ be continuous on I. Derive **Abel's formula** (18) for $W[y_1, y_2]$. HINT: Since y_1, y_2 are solutions, write

$$y_1'' + p_1y_1' + p_2y_1 = 0, \text{ and } y_2'' + p_1y_2' + p_2y_2 = 0. \quad (11.2)$$

Multiply the first by y_2, the second by y_1, subtract, and show that the resulting equation is

$$W' + p_1W = 0. \quad (11.3)$$

12. Verification of Abel's Formula (18). Given are a second-order equation $Ly = 0$ and two solutions of it. Verify Abel's formula (18) by evaluating both sides of the equation and showing that a constant C can be chosen to make the equation an identity. Let the interval be $0 < x < 1$.

(a) $y'' + 4y = 0$, $y_1(x) = \sin 2x$, $y_2(x) = \cos 2x$
(b) $y'' - y = 0$, $y_1(x) = e^x$, $y_2(x) = 3e^x$
(c) $y'' - 2y' + 26y = 0$, $y_1(x) = e^x \sin 5x$, $y_2(x) = e^x \cos 5x$
(d) $y'' - y' = 0$, $y_1(x) = 2 + e^x$, $y_2(x) = 4 + 2e^x$
(e) $xy'' + y' = 0$, $y_1(x) = 1$, $y_2(x) = \ln x$
(f) $x^2y'' - 2xy' - 10y = 0$, $y_1(x) = x^{-2}$, $y_2(x) = x^5$
(g) $x^2y'' - 5xy' + 9y = 0$, $y_1(x) = x^3$, $y_2(x) = x^3 \ln x$
(h) $y'' + (\tan x)y' = 0$, $y_1(x) = 1$, $y_2(x) = \sin x$

13. (a) For a given constant-coefficient equation $y'' + p_1y' + p_2y = 0$, suppose that one of the two functions $\cosh x$, $\sinh x$ is a solution. Show whether or not the other is necessarily a solution as well. HINT: The linear independence of e^x and e^{-x} is relevant.
(b) Now suppose one of the functions $\cos x$, $\sin x$ is a solution. Show whether or not the other is necessarily a solution as well.

14. Below (23) we stated that if p_1 and p_2 are not both constants, then the solution form $y(x) = e^{rx}$ will, in general, fail.
(a) To show that it does not *necessarily* fail, make up a differential equation of the form (23) that has e^x and x as solutions and show that it has nonconstant coefficients.
(b) Next, for that equation, seek solutions in the form e^{rx} and show that you do indeed find the solution e^x (that has been "planted").

2.5 REDUCTION OF ORDER

2.5.1 Deriving the formula.

We continue to study the equation

$$Ly = y'' + p_1(x)y' + p_2(x)y = 0. \tag{1}$$

Both the theory and strategy for finding a general solution were covered in Section 2.4, but if the coefficients are not both constants then *finding* two LI solutions may be very difficult. In this section we show that if we can at least find *one* solution, then a method of reduction of order can be used to find a second LI solution.

To illustrate, suppose we try to solve

$$y'' - 2y' + y = 0 \tag{2}$$

by seeking solutions in the form $y(x) = e^{rx}$. Putting the latter into (2) gives $r^2 - 2r + 1 = 0$, with repeated roots $r = 1, 1$. Thus, we have found

$$y(x) = C_1e^x + C_2(?), \tag{3}$$

the question mark indicating that we've come up short, because we're missing a second LI solution. Evidently, the missing solution is not of the assumed exponential form, or else we would have found it when we sought solutions in that form. (In this example we happen to know from Theorem 2.2.1 that the missing solution is xe^x, but for the sake of this illustration let us make believe we do not know that.)

To continue, suppose we know one solution $y_1(x)$ of (1). Then $y(x) = C_1y_1(x)$ is a solution for any constant C_1. According to the method of **reduction of order**, we can achieve the desired reduction in order by seeking a second solution of (1) in the form

$$\boxed{y(x) = C(x)y_1(x),} \tag{4}$$

in which $C(x)$ is to be determined. That is, let the constant C_1 vary; we will call it $C(x)$.

Next, substitute (4) into (1) and see if we can determine $C(x)$. To do that, differentiate (4) twice, obtaining $y' = C'y_1 + Cy_1'$ and $y'' = C''y_1 + 2C'y_1' + Cy_1''$. Putting these into (1) gives this differential equation for $C(x)$:

$$y_1 C'' + (2y_1' + p_1 y_1)C' + \underbrace{(y_1'' + p_1 y_1' + p_2 y_1)}_{Ly_1,\ \text{which is } 0} C = 0. \tag{5}$$

The latter looks just as difficult as (1)! However, *the underbraced terms in (5) cancel to zero because y_1 is a solution of (1) by assumption*, so (5) becomes

$$C'' + \left(\frac{2y_1'}{y_1} + p_1\right)C' = 0. \tag{6}$$

The point, and the reason for the method's name, is that the latter is really only a *first*-order equation for C'. That is, if we let $v(x) \equiv C'(x)$, then (6) reduces to the first-order equation

$$v' + \left(\frac{2y_1'}{y_1} + p_1\right)v = 0. \tag{7}$$

We know from Section 1.2 that a general solution of $y' + p(x)y = 0$ is $y(x) = A\exp\left(-\int p_1(x)\,dx\right)$, so a general solution of (7) is

$$\begin{aligned} v(x) &= Ae^{-\int\left(\frac{2y_1'}{y_1} + p_1\right)dx} = Ae^{-\int 2\frac{dy_1}{y_1}}e^{-\int p_1\,dx} \\ &= Ae^{-2\ln|y_1|}e^{-\int p_1\,dx} = A\frac{1}{y_1^2}e^{-\int p_1\,dx}. \end{aligned} \tag{8}$$

Since $v = C'$, (8) gives C', which, in turn, can be integrated to give $C(x)$, so

$$C(x) = A\int \frac{e^{-\int p_1\,dx}}{y_1^2}\,dx. \tag{9}$$

We can set $A = 1$ without loss because the final solution $y_2(x)$ can be scaled arbitrarily anyhow. Finally, if we put (9) into (4) and call the resulting second solution $y_2(x)$, we have

$$\boxed{y_2(x) = y_1(x)\int \frac{e^{-\int p_1(x)\,dx}}{[y_1(x)]^2}\,dx.} \tag{10}$$

If we wish to solve the algebraic equation $x^3 - 2x^2 - 19x + 20 = 0$, and know that one root is $x = 1$, then we can divide $x - 1$ into $x^3 - 2x^2 - 19x + 20$, obtaining $x^2 - x - 20$. Having reduced the degree of the equation by one, to the quadratic equation $x^2 - x - 20 = 0$, we can readily solve for the other two roots. Can we do something similar for differential equations? Knowing one solution, can we use that solution to reduce the order of the differential equation by one? Yes, as we demonstrate in this section.

The fourth equality in (8) involves these steps: $e^{-2\ln|y_1|} = e^{\ln|y_1|^{-2}} = |y_1|^{-2} = \frac{1}{y_1^2}$.

CAUTION: If the given equation is of the form $a(x)y'' + b(x)y' + c(x)y = 0$ then $p_1(x)$ is not $b(x)$, it is $b(x)/a(x)$.

EXAMPLE 1. Using (10) to Find a Second Solution of (2). Recall that if we seek solutions of (2) in the form $y(x) = e^{rx}$, we obtain the repeated roots $r = 1, 1$ and hence only the one LI solution $y_1(x) = e^x$. Let us use (10) to find the missing LI solution of (2). Since $p_1(x) = -2$ and $y_1(x) = e^x$, (10) gives

$$y_2(x) = e^x \int \frac{e^{2x}}{e^{2x}}\,dx = e^x(x + A). \tag{11}$$

If we set the arbitrary constant A equal to zero in (11), we obtain

$$y_2(x) = xe^x \tag{12}$$

as a second solution. The Wronskian of $y_1 = e^x$ and $y_2 = xe^x$ is

$$W[y_1, y_2](x) = \begin{vmatrix} e^x & xe^x \\ e^x & (x+1)e^x \end{vmatrix} = e^{2x} \neq 0, \tag{13}$$

so y_1 and y_2 are LI, and a general solution of (2) is $y(x) = C_1 e^{2x} + C_2 x e^{2x}$. There was nothing lost in setting $A = 0$ in (11) because the additive Ae^x term merely corresponds to the already-known solution $y_1(x) = e^x$. ∎

EXAMPLE 2. A More Difficult Equation. Of course, the integration in (10) may be difficult, as we now illustrate. The equation

$$x^2 y'' + 2x^3 y' + (2x^2 - 2)y = 0 \qquad (0 < x < \infty) \tag{14}$$

has a solution $y(x) = 1/x$. Use (10) to find a second LI solution.

To identify $p_1(x)$, multiply (14) through by $1/x^2$. Thus, $p_1(x) = 2x$, and $y_1(x) = 1/x$, so (10) gives

$$y_2(x) = \frac{1}{x} \int \frac{e^{-\int 2x\,dx}}{(1/x)^2}\,dx = \frac{1}{x} \int x^2 e^{-x^2}\,dx. \tag{15}$$

Unfortunately, the integral in (15) is nonelementary; that is, it cannot be evaluated in closed form in terms of elementary functions. To evaluate it in *series* form, use the Taylor series of the e^{-x^2} about $x = 0$. We know, from the calculus, that

$$e^{-t} = 1 - \frac{1}{1!}t + \frac{1}{2!}t^2 - \frac{1}{3!}t^3 + \frac{1}{4!}t^4 - \cdots, \tag{16}$$

so we can let $t = x^2$ in (16) and write

$$\begin{aligned} e^{-x^2} &= 1 - \frac{1}{1!}(x^2) + \frac{1}{2!}(x^2)^2 - \frac{1}{3!}(x^2)^3 + \frac{1}{4!}(x^2)^4 - \cdots \\ &= 1 - \frac{1}{1!}x^2 + \frac{1}{2!}x^4 - \frac{1}{3!}x^6 + \frac{1}{4!}x^8 - \cdots. \end{aligned} \tag{17}$$

Since (16) holds for all t, and $t = x^2$, (17) holds for all x. If we put (17) into (15) and formally integrate term by term we obtain[1]

$$y_2(x) = \frac{1}{x} \int \left(x^2 - \frac{1}{1!}x^4 + \frac{1}{2!}x^6 - \frac{1}{3!}x^8 + \frac{1}{4!}x^{10} - \cdots\right) dx$$

[1]Is it corrrect to integrate the infinite series, within the integral following the first equal sign in (18), term by term, as we did? Yes, it is the Taylor series of $x^2 e^{-x^2}$, and it is known from the calculus that a Taylor series can be integrated term by term and that the resulting series will converge in the same interval as the original series, which, in this case, is $-\infty < x < \infty$.

$$= \frac{1}{x}\left(\int x^2 dx - \int \frac{x^4}{1!}dx + \int \frac{x^6}{2!}dx - \int \frac{x^8}{3!}dx + \int \frac{x^{10}}{4!}dx - \cdots\right)$$

$$= \frac{1}{3}x^2 - \frac{1}{5}x^4 + \frac{1}{14}x^6 - \frac{1}{54}x^8 + \frac{1}{264}x^{10} - \cdots \tag{18}$$

The second equality in (18) follows from term-by-term integration of the series.

or

$$y_2(x) = \sum_{n=0}^{\infty} \frac{(-1)^n}{n!(2n+3)} x^{2n} \tag{19}$$

in which $(-1)^0 \equiv 1$ and $0! \equiv 1$.

Graphs of $y_1(x)$ and $y_2(x)$ are given in Fig. 1.

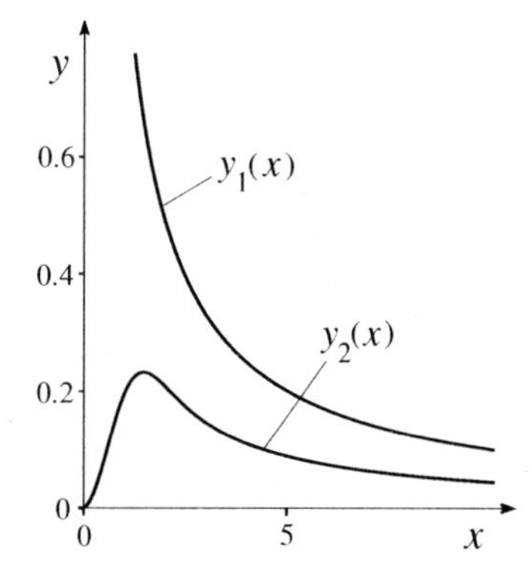

Figure 1. The solutions $y_1(x)$ and $y_2(x)$ in Example 2.

COMMENT. Actually, the integral in (15) can be expressed in terms of the **error function**; that point is left for the exercises. ∎

2.5.2 The method rather than the formula.

In Examples 1 and 2 we relied on the solution formula (10), but we encourage you to use the reduction-of-order *method* rather than (10).

EXAMPLE 3. Using the Method Rather than (10). Suppose we know that one solution of

$$x^2y'' - 5xy' + 9y = 0 \quad (0 < x < \infty) \tag{20}$$

is $y_1(x) = x^3$. To find a second LI solution, by reduction of order, seek $y(x)$ in the form $C(x)x^3$. Then,

$$\begin{aligned} y &= Cx^3, \\ y' &= C'x^3 + 3Cx^2, \\ y'' &= C''x^3 + 6C'x^2 + 6Cx, \end{aligned} \tag{21}$$

and putting these into (20) gives

$$x^5C'' + x^4C' + \underbrace{(6x^3 - 15x^3 + 9x^3)\,C}_{Ly_1 = 0} = 0 \tag{22}$$

or, dropping the underbraced terms (which cancel), $x^5C'' + x^4C' = 0$. Let $C' = v$ and obtain the first-order equation $xv' + v = 0$. The latter is separable and can be solved by separating variables, but it is simpler to notice that it can be written as $(xv)' = 0$. Then, $xv = A$, $v = C' = A/x$, and $C(x) = A\ln x + B$, so $y(x) = (A\ln x + B)x^3$. We can set $B = 0$ because the Bx^3 term reproduces the already-known solution $y_1(x) = x^3$, and without loss we can set $A = 1$. Thus, $y_2(x) = x^3 \ln x$. The Wronskian of y_1 and y_2 is

$$W[y_1, y_2](x) = \begin{vmatrix} y_1 & y_2 \\ y_1' & y_2' \end{vmatrix} = \begin{vmatrix} x^3 & x^3 \ln x \\ 3x^2 & x^2(1 + 3\ln x) \end{vmatrix} = x^5, \tag{23}$$

which is nonzero on the interval $(0 < x < \infty)$, so the solutions y_1 and y_2 are LI on I and a general solution of (20) is $y(x) = C_1x^3 + C_2x^3 \ln x$. ∎

2.5.3 About the method of reduction of order. It is interesting that the method of reduction of order looks, at first, like a "bad idea." After all, recall the solution form $y(x) = e^{rx}$ that worked so well for constant-coefficient equations. The reason it worked so well is that out of the huge set of all possible (twice-differentiable) functions, we looked in the relatively tiny subset consisting only of exponential functions. Having looked in the "right place," the remaining steps were few and simple: Substitution gave the characteristic equation for r, which was readily solved by the quadratic formula.

Think of it as the eye-glass principle:

If we lose our glasses, it is easier to find them if we know they are somewhere on our desk than if we know only that they are somewhere in the universe.

In contrast, observe that the solution form $y(x) = C(x)y_1(x)$ does not narrow the search at all, because *any* given function $y(x)$ can be expressed in that form simply by choosing $C(x)$ to be $y(x)/y_1(x)$. The reason the method works so well is that, rather than narrowing the search, the undifferentiated C terms in (5) cancel, thus giving the reduction of order to a first-order equation for C' — in fact, a first-order equation that we can solve because it is separable. The cancelation of the C terms in (6) was not accidental, because suppose C were a constant. Then the C'' and C' in (5) would be zero, so the remaining terms must be CLy_1, which is zero because y_1 is a solution of (1) by assumption.

Closure. If we know one solution of an equation of the form (1), say $y_1(x)$ or equivalently $C_1y_1(x)$, but are missing a second LI solution, we can find it by the method of reduction of order: If we let the constant C_1 vary, and seek $y(x)$ in the form $C(x)y_1(x)$, then the second-order equation for $C(x)$ is, inevitably, missing a C term, so its order can be reduced by one to a first-order equation, indeed, one that is separable and hence readily solved (though its solution involves an integration that *may* be difficult, as it was in Example 2).

We carried out those steps for the general equation (1) and obtained the formula (10) for the missing solution $y_2(x)$, but we suggest using the method rather than the formula.

EXERCISES 2.5

1. Verify the claim, below (12), that the solutions $y_1(x) = e^x$ and $y_2(x) = xe^x$ of (2) are LI (on $-\infty < x < \infty$).

2. Knowing One Solution, Find a Second LI Solution. Find a second LI solution by using (10). Then re-work the problem by the method of reduction of order. That is, seek $y(x) = C(x)y_1(x)$ and go through the steps of solving for $C(x)$. For parts (h)–(m), let the interval be $0 < x < \infty$.

(a) $y'' - 6y' + 9y = 0, \quad y_1(x) = e^{3x}$
(b) $y'' + 4y' + 4y = 0, \quad y_1(x) = e^{-2x}$
(c) $y'' - 3y' + 2y = 0, \quad y_1(x) = e^{2x}$
(d) $y'' - 9y = 0, \quad y_1(x) = e^{3x}$
(e) $y'' + 2\sqrt{2}y' + 2y = 0, \quad y_1(x) = e^{-\sqrt{2}x}$
(f) $y'' - 10y' + 25y = 0, \quad y_1(x) = e^{5x}$
(g) $y'' - 2\sqrt{5}y' + 5y = 0, \quad y_1(x) = e^{\sqrt{5}x}$
(h) $x^2y'' + xy' - 4y = 0, \; y_1(x) = x^2$
(i) $x^2y'' + 3xy' + y = 0, \; y_1(x) = 1/x$
(j) $x^2y'' - xy' + y = 0, \; y_1(x) = x$
(k) $x^2y'' - 3xy' + 4y = 0, \; y_1(x) = x^2$
(l) $x^2y'' + 5xy' + 4y = 0, \; y_1(x) = x^{-2}$
(m) $x^2y'' - 4xy' + 4y = 0, \; y_1(x) = x$

3. Use (10) to find a second LI solution of $x^2y''+xy'+y=0$, given that one solution is $y_1(x)=\cos(\ln x)$. HINT: To evaluate the integral, try the substitution $t=\ln x$.

4. Finding a Second LI Solution. In each case one solution is given. Find a second LI solution.

(a) $(x-1)y''-xy'+y=0,\ 1<x<\infty,\ y_1(x)=e^x$
(b) $xy''-(x+1)y'+y=0,\ 0<x<\infty,\ y_1(x)=1+x$
(c) $xy''-y'+4x^3y=0,\ 0<x<\infty,\ y_1(x)=\sin(x^2)$
(d) $xy''-y'+(1-x)y=0,\ 0<x<\infty,\ y_1(x)=e^x$
(e) $(1-x^2)y''-2xy'+2y=0,\ -1<x<1,\ y_1(x)=x$
NOTE: The differential equation in (e) is a **Legendre equation**, which will be the subject of Section 6.4.2.

5. One solution of

$$(1-3x+2x^2)y''+2xy'-2y=0 \tag{5.1}$$

is $y_1(x)=x$.
(a) Use (10) to find a second LI solution and hence a general solution.
(b) Computer software gives a general solution of (5.1) as

$$y(x)=C_1\frac{1}{x-1}+C_2\frac{(2x-1)^2}{x-1}. \tag{5.2}$$

Show that your solution in part (a) is consistent with (5.2).

ADDITIONAL EXERCISES

6. Example 2 and the Error Function. Knowing one solution of (13), $y_1(x)=1/x$, we used the reduction-of-order formula (10) to derive the second LI solution

$$y_2(x)=\frac{1}{x}\int x^2e^{-x^2}dx. \tag{6.1}$$

(a) Explain why, in place of the indefinite integral in (6.1), we can use the definite integral form

$$y_2(x)=\frac{1}{x}\int_0^x t^2e^{-t^2}dt. \tag{6.2}$$

(b) The function defined by the integral formula

$$\boxed{\text{erf}(x)=\frac{2}{\sqrt{\pi}}\int_0^x e^{-t^2}dt} \tag{6.3}$$

is known as the **error function**, its name deriving from its application to error analysis in statistics and probablility. The error function is one of numerous functions that arise, in applications, in the form of integrals that cannot be evaluated in closed form in terms of the familiar elementary functions. Thus, we define them as new functions, called "special functions," and proceed to derive their properties and to develop formulas for their calculation. Thus, they can be added to our list of known functions, like sines and exponentials, even though they may be less familiar to us.
The problem: Using a change of variables and integration by parts, show that the solution $y_2(x)$ in (6.2) can be expressed in terms of the error function as

$$y_2(x)=-\frac{1}{2}e^{-x^2}+\frac{\sqrt{\pi}}{4x}\text{erf}(x). \tag{6.4}$$

The graph of erf(x) is shown below. The scale factor $2/\sqrt{\pi}$ is included in the definition (6.3) to "normalize" the function so that erf(∞) $=1$, as seen in the figure.

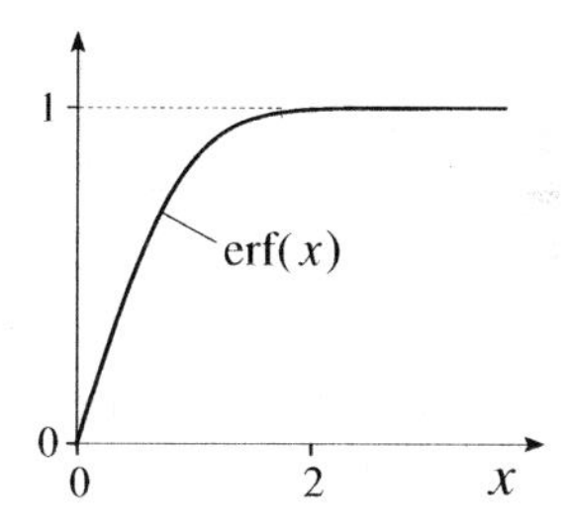

(c) A power series representation of erf(x) can be obtained from (6.3) by expanding the e^{-x^2} in the integrand in a Maclaurin series and integrating term by term. Show that those steps give the first several terms as

$$\text{erf}(x)=\frac{2}{\sqrt{\pi}}\left(x-\frac{1}{3}x^3+\frac{1}{10}x^5-\frac{1}{42}x^7+\cdots\right). \tag{6.5}$$

(d) In fact, show that the whole series can be expressed as

$$\text{erf}(x)=\frac{2}{\sqrt{\pi}}\sum_{n=0}^{\infty}\frac{(-1)^n}{n!(2n+1)}x^{2n+1}, \tag{6.6}$$

with $0!\equiv 1$ and $(-1)^0\equiv 1$, and use the ratio test to show that the series converges for $-\infty<x<\infty$.
(e) Various properties of erf(x) can be deduced directly from its integral definition in (6.3). For instance, show from (6.3) that

$$\text{erf}(-x)=-\text{erf}(x) \tag{6.7}$$

so that the graph of the error function (shown above for $0\le x<\infty$) is antisymmetric about $x=0$.
(f) **Computer.** Besides the series representation of the error

function, given by (6.6), accurate approximate formulas are also available. For instance, the following approximation due to C. Hastings, Jr.,

$$\text{erf}(x) = 1 - (a_1 t + a_2 t^2 + a_3 t^3)e^{-x^2}, \tag{6.8}$$

where $t = 1/(1 + px)$, $p = 0.47047$, $a_1 = 0.3480242$, $a_2 = -0.0958798$, and $a_3 = 0.7478556$, is accurate to within $\pm 2.5 \times 10^{-5}$ for $0 \leq x < \infty$. As a partial check of Hastings' claim that the discrepancy is within $\pm 2.5 \times 10^{-5}$, compute erf(1) two ways. First, use his approximation (6.8); second, use computer software to evaluate the integral in (6.3), with $x = 1$. NOTE: For a compilation of such approximations see Y. L. Luke's *Mathematical Functions and their Approximation* (NY: Academic, 1975). Such formulas are built into computer software for the evaluation of these special functions.

7. More About Integral Representations. The error function was defined by the integral in (6.3), above. In case the idea of an integral definition of a function is new to you, it might help to point out that even elementary functions can be introduced in that manner. For example, one can *define* the logarithm $\ln x$ by the integral

$$\ln x = \int_1^x \frac{dt}{t}, \qquad (x > 0) \tag{7.1}$$

from which formula the values of $\ln x$ can be derived by numerical integration, and its properties derived as well.

(a) To illustrate, use (7.1) to derive the property $\ln x^a = a \ln x$ of the logarithm. HINT: Make a change of variables from t to τ, according to $\tau = t^a$.

(b) Likewise, use (7.1) to derive the property $\ln(xy) = \ln x + \ln y$. HINT: Express the integral from 1 to xy as the integral from 1 to x plus the integral from x to xy, and in the latter integral make a suitable change of variables.

2.6 CAUCHY–EULER EQUATIONS

A differential equation is said to be "elementary" if it can be solved in closed form in terms of elementary functions. We've seen in Sections 2.2 and 2.3 that constant-coefficient equations of second order are elementary since they can be solved in closed form in terms of (real and complex) exponentials and powers of x.

In this section we study the **Cauchy–Euler equation** of second-order,

$$x^2 \frac{d^2y}{dx^2} + a_1 x \frac{dy}{dx} + a_2 y = 0, \tag{1}$$

The Cauchy–Euler pattern is this: the order of the derivative is the same as the power of x multiplying it. Cauchy–Euler equations can be of any order; this one is of second order.

which also turns out to be elementary, even though it has nonconstant coefficients. In the notation of Section 2.4, we see, after dividing by x^2, that $p_1(x) = a_1/x$ and $p_2(x) = a_2/x^2$. These are undefined at $x = 0$ so (except in some of the exercises) we will consider (1) on the interval $0 < x < \infty$, or a subinterval thereof.

Cauchy–Euler equations occur particularly in applications involving *circular, cylindrical, or spherical geometries.* For instance, consider the temperature distribution within a hollow metal cylinder such as a steam pipe, a cross section of which is the annular region $a < r < b$ in Fig. 1. If its temperature is held at constant values T_1 at the inner edge $r = a$ and T_2 at the outer edge $r = b$, then the steady-state temperature distribution $T(r)$ in the annulus can be modeled by the differential equation

$$\frac{d^2T}{dr^2} + \frac{1}{r}\frac{dT}{dr} = 0 \quad (a < r < b), \tag{2}$$

together with the "boundary conditions" $T(a) = T_1$ and $T(b) = T_2$. Equation (2) may not look like a Cauchy–Euler equation, but if we multiply through by r^2 we see that it is indeed of the form (1), with $a_1 = 1$ and $a_2 = 0$. This and other applications of Cauchy–Euler equations will be studied in Chapter 3.

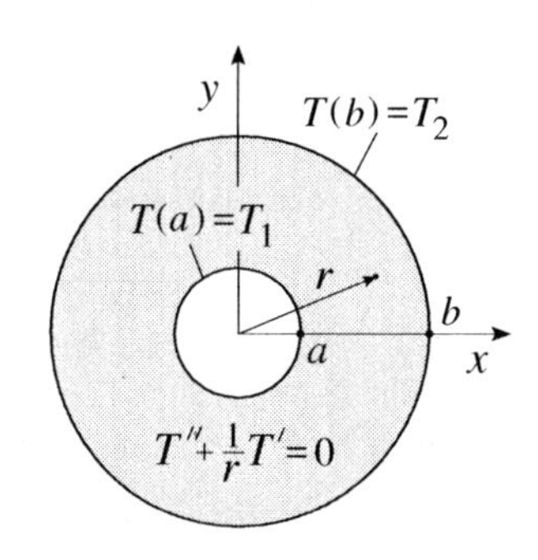

Figure 1. Steady-state temperature distribution $T(r)$ in a hollow metal cylinder.

2.6.1 General solution. Recall that to solve the second-order constant-coefficient equation, in Section 2.2, we began by noting that the first-order version $y' + p_1 y = 0$ has the exponential solution $y(x) = Ce^{-p_1 x}$. That inspired us to seek solutions of constant-coefficient equations of second order in that form as well. We will do the same here: Begin with the *first*-order Cauchy–Euler equation

$$x\frac{dy}{dx} + a_1 y = 0, \tag{3}$$

solve by separation of variables, and obtain

$$y(x) = Cx^{-a_1}. \tag{4}$$

The solution is a power of x, not an exponential, so this time we are inspired to seek solutions of (1) in the form

$$\boxed{y(x) = x^r,} \tag{5}$$

in which r is a constant. Put (5) into (1), and see if we can choose r to get it to work. That step gives

$$x^2 r(r-1)x^{r-2} + xra_1 x^{r-1} + a_2 x^r = [(r(r-1) + ra_1 + a_2]x^r = 0 \tag{6}$$

or, canceling x^r,

$$\boxed{r^2 + (a_1 - 1)r + a_2 = 0,} \tag{7}$$

which can be solved for r by the quadratic formula.

Thus, the form (5) is perfect for Cauchy–Euler equations just as e^{rx} is perfect for constant-coefficient equations. The reason is this: If $y = x^r$, then the third term in (1) is an x^r term; likewise the second term because the derivative knocks the exponent of x down by one, but then multiplying by x builds it back up by one. And likewise the first term, because the double derivative knocks the exponent down by two, but multiplying by x^2 builds it back up by two. Thus, all three terms in (6) are of the same type, x^r, and by suitable choice of r they can be made to cancel to zero.

Equation (7) is analogous to the characteristic equation obtained when we seek solutions of a constant-coefficient equation in the exponential form e^{rx}. In this text we call it the ***r*-equation** corresponding to (1), to distinguish it from the "characteristic equation" that we obtain in the constant-coefficient equation case. If (7) gives

We will call (7) the "r-equation" corresponding to the Cauchy–Euler equation (1).

distinct roots r_1 and r_2, then we have found solutions x^{r_1} and x^{r_2} and, thanks to the linearity of (1) and the superposition principle expressed in Theorem 2.1.1, we can say that an arbitrary linear combination of them,

$$y(x) = C_1 x^{r_1} + C_2 x^{r_2}, \tag{8}$$

is a solution as well. These two solutions are verified as LI on the interval $(0 < x < \infty)$ by the Wronskian test in Theorem 2.4.4(c) so, by Theorem 2.4.3, (8) is a general solution of (1).

2.6.2 Repeated roots and reduction of order.

However, if the discriminant $(a_1 - 1)^2 - 4a_2$ in

$$r = \left[1 - a_1 \pm \sqrt{(a_1 - 1)^2 - 4r_2}\right] / 2 \tag{9}$$

is zero, then we have a repeated root

$$r_1 = r_2 \equiv r = (1 - a_1)/2 \tag{10}$$

and have found only the one LI solution x^r, with r given by (10). That is,

$$y(x) = C_1 x^r + C_2(?), \tag{11}$$

with the question mark used again to remind us that we are missing a second LI solution. Why did we miss it? Evidently, it is not of the assumed form (5), or we would have found it.

To find a second LI solution, use reduction of order. That is, having found one solution of (1), x^r, seek another in the form

$$\boxed{y(x) = C(x)x^r.} \tag{12}$$

Next, put (12) into (1). With $y' = C'x^r + Crx^{r-1}$ and $y'' = C''x^r + 2C'rx^{r-1} + Cr(r-1)x^{r-2}$, substitution in (1) gives

$$C''x^{r+2} + (2r + a_1)C'x^{r+1} + \underbrace{\left[r^2 + (a_1 - 1)r + a_2\right]}_{0 \text{ by } (7)} Cx^r = 0. \tag{13}$$

The underbraced terms cancel by virtue of (7). Dropping that term and canceling x^r gives $x^2 C'' + (2r + a_1)xC' = 0$. Finally, the $2r + a_1$ is 1, by (10), so

$$x^2 C'' + xC' = 0, \tag{14}$$

or, letting $C'(x) \equiv p(x)$,

$$x^2 \frac{dp}{dx} + xp = 0. \tag{15}$$

The latter can be solved by separation of variables, but it is simpler to cancel x's and write $xp' + p = (xp)' = 0$, which gives $xp = \text{constant} = A$. Thus, $p = \dfrac{dC}{dx} = \dfrac{A}{x}$, and integration gives $C(x) = A \ln x + B$. Putting the latter into (12) and renaming B as C_1 and A as C_2 according to our usual notation, we finally have

$$y(x) = (C_1 + C_2 \ln x)\, x^r. \tag{16}$$

Of course, the $C_1 x^r$ part we already knew; what we have found, now, is the second solution $(\ln x)x^r$. Further, $W[x^r, x^r \ln x](x) = x^{2r-1} \neq 0$ on I, so the solutions x^r and $x^r \ln x$ are LI on I. Thus, (16) is a general solution of (1).

The $(\ln x)x^r$ term is not of the form x^r. Hence, the assumed form (5) did not lead us to it.

In summary, a general solution of (1) is this:

$$y(x) = \begin{cases} C_1 x^{r_1} + C_2 x^{r_2} & \text{if } r_1 \neq r_2, \\ (C_1 + C_2 \ln x)x^r & \text{if } r_1 = r_2 \equiv r. \end{cases} \tag{17}$$

Thus, if (7) gives repeated roots, one can seek the missing solution as $C(x)x^r$ and go through the reduction of order steps, as above. Of course, we've done that once and for all, so we can simply use the result given in (17).

EXAMPLE 1. Distinct Roots. To solve the Cauchy–Euler equation

$$x^2 y'' - 2xy' - 10y = 0, \tag{18}$$

seek $y = x^r$. That form gives the r-equation $r^2 - 3r - 10 = 0$, with distinct roots $r = -2$ and 5, so, according to (17), a general solution of (18) is

Be sure you get that same r-equation, $r^2 - 3r - 10 = 0$, *not* $r^2 - 2r - 10 = 0$.

$$y(x) = \frac{C_1}{x^2} + C_2 x^5. \tag{19}$$

COMMENT 1. Be prepared for solutions of (1) to be undefined at $x = 0$ because, re-expressing (1) in the standard form $y'' + p_1(x)y' + p_2(x)y = 0$, the coefficients $p_1(x) = a_1/x$ and $p_2(x) = a_2/x^2$ "blow up" as $x \to 0$. We can think of the unboundedness of the coefficients in (18) as a "virus" at $x = 0$, that may be passed on in some form to the solutions of the equation. In (19), the C_1/x^2 term is indeed unbounded as $x \to 0$.

However, the solution of a Cauchy–Euler equation is not *necessarily* undefined at $x = 0$, because if both r_1 and r_2 turn out to be distinct nonnegative integers, then the x^{r_1} and x^{r_2} in the general solution, and their derivatives, are well defined at $x = 0$, and the interval of existence of every solution is $-\infty < x < \infty$ rather than only $0 < x < \infty$.

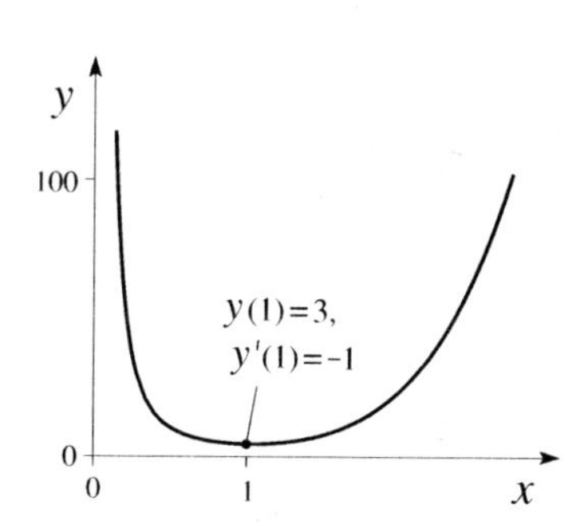

Figure 2. The solution of (18), with initial conditions $y(1) = 3$ and $y'(1) = -1$. The interval of existence is $0 < x < \infty$.

COMMENT 2. To illustrate the inclusion of initial conditions, suppose $y(1) = 3$ and $y'(1) = -1$. Application of these conditions to the general solution (19) gives $C_1 = 16/7$ and $C_2 = 5/7$, so $y(x) = \dfrac{16}{7}\dfrac{1}{x^2} + \dfrac{5}{7}x^5$, the graph of which is given in Fig. 2. The solution is unbounded as $x \to 0$ due to the x^{-2} term, so the interval of existence of the solution is $0 < x < \infty$. However, if we were to change the initial conditions to $y(1) = 2$ and

$y'(1) = 10$, we would obtain $C_1 = 0$ and $C_2 = 5$, and the interval of existence of the solution $y(x) = 5x^5$ would then be $-\infty < x < \infty$. ▮

EXAMPLE 2. Repeated Roots. Solve

$$x^2 y'' - 5xy' + 9y = 0. \tag{20}$$

Setting $y(x) = x^r$ gives the r-equation $r^2 - 6r + 9 = 0$ with repeated roots $r = 3, 3$. Then (17) gives the general solution of (20) as

$$y(x) = (C_1 + C_2 \ln x)x^3. \tag{21}$$

COMMENT. If we subject (21) to the initial conditions $y(2){=}0$ and $y'(2){=}4$, for instance, we find that $C_1 = -\ln 2$ and $C_2 = 1$, so $y(x) = (-\ln 2 + \ln x)x^3$. Unlike the solution (19) in Example 1, the latter does not grow unboundedly as $x \to 0$, in fact, it tends to zero as $x \to 0$ (Exercise 2). Nevertheless, the interval of existence is still $0 < x < \infty$. The solution cannot be extended leftward to $x{=}0$, or beyond, because the functions $y(x)$, $y'(x)$, and $y''(x)$ in (21) contain $\ln x$ terms, and $\ln x$ is undefined at $x = 0$. ▮

2.6.3 Complex roots. There is one last point to discuss. Realize that the roots given by (9) can be distinct real numbers (if the discriminant is positive), repeated real numbers (if the discriminant is zero), or complex conjugates (if the discriminant is negative). The question arises as to how to evaluate the x^{r_1} and x^{r_2} in (8) if the r's are complex.

Let us write those complex roots as $r = \alpha \pm i\beta$. Just as we were challenged by the complex exponential function that arose in Section 2.3, now we are challenged by the $x^{\alpha \pm i\beta}$ terms in (8). We will define

In (22) we are formally using the identities $x^{a+b} = x^a x^b$, $a = e^{\ln a}$, and $\ln a^b = b \ln a$ that hold if a and b are real numbers. The last step in (22) follows from Euler's formula.

$$\begin{aligned} x^{\alpha \pm i\beta} &= x^\alpha x^{\pm i\beta} = x^\alpha e^{\ln (x^{\pm i\beta})} = x^\alpha e^{\pm i\beta \ln x} \\ &= x^\alpha [\cos (\beta \ln x) \pm i \sin (\beta \ln x)]. \end{aligned} \tag{22}$$

With the help of (22), we can now write the general solution as

$$\begin{aligned} y(x) &= Ax^{\alpha+i\beta} + Bx^{\alpha-i\beta} = x^\alpha \left(Ax^{i\beta} + Bx^{-i\beta} \right) \\ &= x^\alpha \{ A[\cos (\beta \ln x) + i \sin (\beta \ln x)] \\ &\qquad + B[\cos (\beta \ln x) - i \sin (\beta \ln x)] \} \\ &= x^\alpha \left[(A{+}B) \cos (\beta \ln x) + i(A{-}B) \sin (\beta \ln x) \right]. \end{aligned} \tag{23}$$

Renaming $A{+}B$ as C_1 and $i(A{-}B)$ as C_2, we can express the general solution of (1), for the case of complex roots, as

If the r-equation gives $r = \alpha \pm i\beta$, then the general solution of (1) is given by (24).

$$\boxed{y(x) = x^\alpha \Big[C_1 \cos (\beta \ln x) + C_2 \sin (\beta \ln x) \Big],} \tag{24}$$

in which C_1 and C_2 are arbitrary constants. (See Exercise 3.)

EXAMPLE 3. Complex Roots. To solve the Cauchy–Euler equation

$$x^2y'' - 3xy' + 20y = 0, \tag{25}$$

seek $y = x^r$. That step gives $r^2 - 4r + 20 = 0$, so $r = 2 \pm 4i$. Thus, $\alpha = 2$ and $\beta = 4$ in (24), so a general solution of (25) is

$$y(x) = x^2\left[C_1 \cos(4\ln x) + C_2 \sin(4\ln x)\right]. \quad \blacksquare \tag{26}$$

Although the graphs of the trigonometric and logarithmic functions are, individually, familiar, the graphs of $\cos(\beta \ln x)$ and $\sin(\beta \ln x)$ in (24) are probably not. To prepare for the graph of $\cos(\beta \ln x)$, for instance, rewrite $\cos(\beta \ln x)$ as $\cos\left[\left(\beta \frac{\ln x}{x}\right) x\right] = \cos\omega x$, because the form $\cos\omega x$ is more familiar. Here, the frequency ω is not a constant, it varies with x:

$$\omega(x) = \beta\frac{\ln x}{x}. \tag{27}$$

As $x \to 0$, both the $\ln x$ and $1/x$ factors become infinite, so $\omega(x)$ does too. And as $x \to \infty$, $\ln x/x \to \infty/\infty$ is indeterminate, but l'Hôpital's rule shows that $\omega \to 0$ as $x \to \infty$. Thus, the $\cos(\beta \ln x)$ term represents an oscillation between $+1$ and -1, but with a frequency that varies with x, becoming infinite as $x \to 0$ and zero as $x \to \infty$. With $\beta = 10$, say, a computer plot of $\cos(\beta \ln x)$ is given in Fig. 3a.

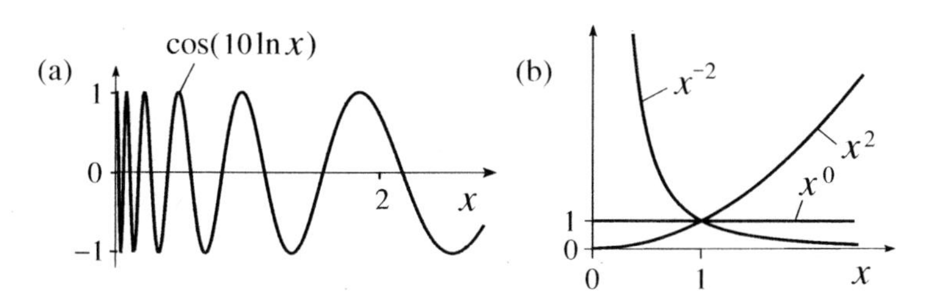

Figure 3. Getting ready to plot $x^\alpha \cos(\beta \ln x)$.

Since the $\cos(10 \ln x)$ in Fig. 3a is scaled by x^α in (24), we have also plotted x^α for representative values of α in Fig. 3b. Finally, the products of x^α and $\cos(10 \ln x)$, for those α's, are shown in Fig. 4.

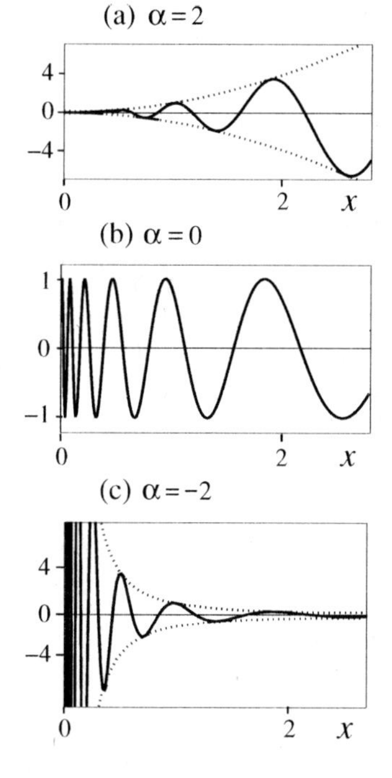

Figure 4. Graphs of the function $x^\alpha \cos(\beta \ln x)$ for representative values of α and β: $\alpha = 2$, 0, -2, and $\beta = 10$.

Closure. Besides constant-coefficient equations, Cauchy–Euler equations admit general solutions in terms of elementary functions: powers of x, sines, cosines, and logarithms. Throughout this section we considered the x interval to be $0 < x < \infty$. We did not permit it to include or straddle the origin because the coefficients

$p_1(x) = a_1/x$ and $p_2(x) = a_2/x^2$ are undefined there, so the solution $y(x)$ may be undefined there as well. We derived the general solution (17) and, for the case of complex roots $r = \alpha \pm i\beta$, we used Euler's formula to re-express the solution $C_1 x^{\alpha+i\beta} + C_2 x^{\alpha-i\beta}$ in terms of real valued functions; the result is given in (24).

It turns out that we can think of Cauchy–Euler equations as constant-coefficient equations in disguise, since every Cauchy–Euler equation can be converted to a constant-coefficient equation by the change of variables $x = e^t$. This point is covered in the exercises.

EXERCISES 2.6

1. (a) We successfully sought solutions of (1) in the form (5). Instead, try $y(x) = e^{rx}$, and show that it doesn't work. Explain your reasoning.
(b) Similarly, show that $y(x) = x^r$ does not work for the constant-coefficient equation $y'' + p_1 y' + p_2 y = 0$. Explain.

2. In the Comment below (21), we stated that the limit of $x^3 \ln x$, as $x \to 0$, is 0. Use l'Hôpital's rule to verify that result.

3. Recall that A and B in (23) were arbitrary constants. Then, we renamed $A+B = C_1$ and $i(A-B) = C_2$ and said that C_1 and C_2 are also arbitrary constants. Show that the arbitrariness of A, B does imply the arbitrariness of C_1, C_2, as claimed. HINT: If the equations were $A+B = C_1$ and $2A+2B = C_2$, for instance, the arbitrariness of A and B would *not* imply the arbitrariness of C_1 and C_2. Rather, C_2 would necessarily be twice C_1. The issue is whether any desired values of C_1 and C_2 can be obtained by suitable choices of A and B.

4. Solve These. Derive a general solution for the given Cauchy–Euler equation on $0 < x < \infty$. Then find a particular solution satisfying the given initial condition(s) and state its interval of existence, which will be $0 < x < \infty$ at the least, and perhaps $-\infty < x < \infty$.

REAL ROOTS:
(a) $x^2y'' - 6y = 0;\ y(1) = 1, y'(1) = 3$
(b) $x^2y'' - 4xy' + 6y = 0;\ y(2) = 4, y'(2) = 8$
(c) $x^2y'' - 3xy' + 4y = 0;\ y(1) = 3, y'(1) = 7$
(d) $x^2y'' - 9xy' + 25y = 0;\ y(2) = 32, y'(2) = 0$
(e) $x^2y'' - 2xy' + 2y = 0;\ y(4) = 20, y'(4) = 9$
(f) $x^2y'' + xy' - 4y = 0;\ y(2) = 4, y'(2) = 8$
(g) $y'' + (3/x)y' + (1/x^2)y = 0;\ y(3) = 6, y'(3) = -2$
(h) $xy'' - 4y' + (4/x)y = 0;\ y(3) = 18, y'(3) = 6$
(i) $x^2y'' + 7xy' + 9y = 0;\ y(1) = 0, y'(1) = 1$
(j) $x^2y'' + xy' = 0;\ y(2) = 3 - 2\ln 2, y'(2) = -1$
(k) $x^2y'' - 6y = 0;\ y(2) = 1, y'(2) = -1$
(l) $y'' + (1/x)y' = 0;\ y(2) = 3, y'(2) = 3/(2\ln 2)$

COMPLEX ROOTS:
(m) $x^2y'' + xy' + y = 0;\ y(1) = 3, y'(1) = 2$
(n) $x^2y'' - xy' + 10y = 0;\ y(1) = 9, y'(1) = 6$
(o) $4x^2y'' + 5y = 0;\ y(1) = 6, y'(1) = 4$
(p) $x^2y'' + xy' + 4y = 0;\ y(1) = 3, y'(1) = 2$
(q) $x^2y'' - 5xy' + 10y = 0;\ y(1) = 1, y'(1) = 4$

5. Change of Variables. Make the change of variables

$$\boxed{x = e^t} \tag{5.1}$$

in (1), and show that the new differential equation is

$$\boxed{\frac{d^2Y}{dt^2} + (a_1 - 1)\frac{dY}{dt} + a_2 Y = 0,} \tag{5.2}$$

where $y(x(t))$ is being called $Y(t)$. HINT: You will need to express the d/dx and d^2/dx^2 in (1) in terms of t. Use chain differentiation to show that $d/dx = e^{-t}d/dt$ and then use the fact that d^2/dx^2 is $(d/dx)(d/dx)$, that is, a "double action" of d/dx.
NOTE 1: Thus, we see that every Cauchy–Euler equation $x^2y'' + a_1xy' + a_2y = 0$ is, in a sense, a "constant-coefficient equation in disguise," since the simple change of variables (5.1) converts the former to the latter.
NOTE 2: How can we motivate (5.1)? Look at it this way: What change of variables will change the generic solution x^r of Cauchy–Euler equations to the generic solution e^{rt} of constant-coefficient equations? $x = e^t$, for it gives $x^r = (e^t)^r = e^{rt}$.

6. Using the Change of Variables Given in Exercise 5. Derive a general solution of the given equation using the results of Exercise 5. That is, obtain a general solution of the constant-coefficient equation (5.2), in terms of t, and then substitute

$t = \ln x$ (or, equivalently, $e^t = x$) in your result to recover $y(x)$.

(a) $x^2y'' - 6y = 0$
(b) $x^2y'' - 4xy' + 6y = 0$
(c) $x^2y'' - 3xy' + 4y = 0$
(d) $x^2y'' - 9xy' + 25y = 0$
(e) $x^2y'' - 2xy' + 2y = 0$
(f) $x^2y'' + xy' - 4y = 0$
(g) $y'' + (3/x)y' + (1/x^2)y = 0$
(h) $xy'' - 4y' + (4/x)y = 0$
(i) $x^2y'' + xy' + y = 0$
(j) $x^2y'' - xy' + 10y = 0$
(k) $4x^2y'' + 5y = 0$
(l) $x^2y'' + xy' + 4y = 0$

7. Working Backward. As in the sections on constant-coefficient equations, we include here a problem in which we ask you to work backward, from the roots of the r-equation back to the corresponding Cauchy–Euler differential equation. Not that this problem is apt to arise in applications, but the exercise is to promote understanding of the material in this section by looking at it from different points of view. *Here is the problem*:

(a) If the r-equation has distinct roots r_1, r_2, what is the original Cauchy–Euler equation? That is, determine a_1 and a_2, in (1), in terms of r_1 and r_2.
(b) If the r-equation has repeated roots $r_1 = r_2 \equiv r$, then what is the original Cauchy–Euler equation? That is, determine a_1 and a_2, in (1), in terms of r.

8. What if $x < 0$? Thus far we've considered the x-interval to be $x > 0$, but if initial conditions are given at a point to the left of the origin then we must consider the interval $-\infty < x < 0$ instead. If the roots of the r-equation are distinct integers, we can proceed as before, obtaining a general solution and then applying the two boundary conditions. But if any roots are repeated, then we obtain $\ln x$ functions, and $\ln x$ is undefined (at least, as a real-valued function) if $x < 0$. Similarly, if any roots are nonintegers. For instance, if $r = 1/3$ we obtain $x^{1/3}$, which is undefined (as a real-valued function) if $x < 0$. In such cases, one can proceed by changing the independent variable from x to z according to $x = -z$, so the new differential equation, obtained by the substitution $x = -z$, is on $0 < z < \infty$, with which case we're familiar. To obtain the new differential equation, note that under the change of variables $x = -z$ chain differentiation shows that dy/dx becomes $-dY/dz$ and d^2y/dx^2 becomes d^2Y/dz^2, if we use "$Y(z)$" to denote $y(-z)$.
The problem: Use the foregoing idea to solve the following for $y(x)$, and evaluate $y(-10)$.

(a) $x^2y'' + xy' = 0;\ \ y(-1) = 0,\ y'(-1) = 5$
(b) $2x^2y'' + xy' = 0;\ \ y(-4) = 9,\ y'(-4) = 1$
(c) $2x^2y'' - xy' + y = 0;\ \ y(-9) = 9,\ y'(-9) = 0$
(d) $x^2y'' - xy' + y = 0;\ \ y(-1) = 5,\ y'(-1) = 4$
(e) $9x^2y + 9xy' - y = 0;\ \ y(-8) = 120,\ y'(-8) = -3$

9. Electric Potential. Various phenomena, such as the distribution of steady-state temperature or the electric potential (i.e., the voltage) within a region, are governed by a so-called **Laplace equation**. If that equation is written in cylindrical coordinates r, θ, z and there is no variation with θ or z, or if it is written in spherical coordinates ρ, θ, ϕ and there is no variation with θ or ϕ, then the governing equation is of Cauchy–Euler type. Consider, for instance, the problem of electric potential V.

(a) **Cylindrical geometry.** Specifically, consider the electric potential $V(r)$ within an annular region such as a metal pipe or the axon of a nerve cell, of inner radius a and outer radius b; V satisfies the differential equation

$$\frac{1}{r}\frac{d}{dr}\left(r\frac{dV}{dr}\right) = \frac{d^2V}{dr^2} + \frac{1}{r}\frac{dV}{dr} = 0. \quad (a < r < b) \qquad (9.1)$$

If the inner and outer surfaces are held at potentials V_1 and V_2, respectively, solve for the distribution of potential within the annulus; see the figure. That is, solve (9.1) for $V(r)$, subject to the boundary conditions $V(a) = V_1$ and $V(b) = V_2$.

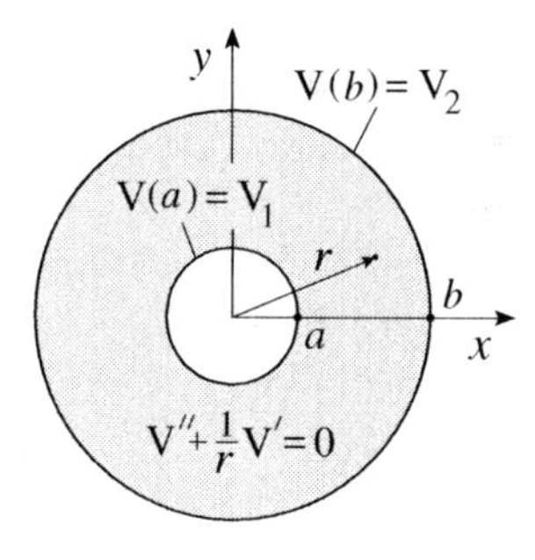

NOTE: Up to this point we've considered only IVPs — in which conditions are specified at a single point, as in Exercise 4. This problem is different because conditions are prescribed at two points (a and b), not one. Thus, it is a *boundary value problem* (BVP), namely, a differential equation with conditions prescribed at each of two points, which are normally the endpoints of the interval under consideration. *Boundary value problems were not encountered in Chapter 1 because for first-order equations only one condition is appropriate*; it is only for equations of second order and higher that BVPs can occur, because only for such equations are two or more conditions appropriate. We will study BVPs in Chapter 3 but, meanwhile, do not be intimidated: merely obtain a general solution of (9.1) and then apply the two boundary conditions.

(b) **Spherical geometry.** The problem in part (a) is for an annular region. Another important case is that where the region is a hollow sphere, of inner radius a and outer radius b, with the potential held at V_1 and V_2 on those surfaces, respectively. In that case the BVP governing the potential $V(\rho)$ within the hollow sphere is

$$\frac{1}{\rho^2}\frac{d}{d\rho}\left(\rho^2\frac{dV}{d\rho}\right)=\frac{d^2V}{d\rho^2}+\frac{2}{\rho}\frac{dV}{d\rho}=0, \quad (a<\rho<b) \quad (9.2)$$

with $V(a)=V_1$, and $V(b)=V_2$. Solve for the potential distribution $V(\rho)$ in the sphere, that is, in $a<\rho<b$.
NOTE: The radial variables in parts (a) and (b) are not the same. In (a), r is the perpendicular distance from a given point in the pipe to the central axis of the pipe, namely, the r variable in *cylindrical coordinates*. In (b), ρ is the distance from the origin (i.e., the center of the sphere) to the point, and is the radial variable in *spherical coordinates*.

ADDITIONAL EXERCISES

10. Factorization of Cauchy–Euler Equations. (a) Try to factor the Cauchy–Euler equation (1) as

$$\left(x\frac{d}{dx}-\alpha\right)\left(x\frac{d}{dx}-\beta\right)y=0, \quad (10.1)$$

and show that α can be computed as either root of $\alpha^2+(a_1-1)\alpha+a_2=0$, with β then computed from $\alpha\beta=a_2$.
NOTE: We are not suggesting that this solution method is better than the one given in the text. The exercise is simply to follow up on the factorization idea that was used for constant-coefficient equations in Section 2.2. It's interesting that, either way, the same quadratic equation must be solved.
(b) As an example, use the results in part (a) to solve

$$x^2y''-2xy'-10y=0.$$

(c) Same as (b), for $x^2y''-3xy'+4y=0$.

2.7 THE GENERAL THEORY FOR HIGHER-ORDER EQUATIONS

In this chapter, we've shown how to obtain general solutions for constant-coefficient and Cauchy–Euler equations of second-order, and we've given Theorems 2.4.2–2.4.4 for the more general equation $y''+p_1(x)y'+p_2(x)y=0$, in which the p_j coefficients can be any functions of x that are continuous on the interval. In this section we extend those results to equations of any order n, with $n\geq 2$. There is nothing fundamentally new, so we will move quickly, just indicating the extensions.

Following the same pattern as in Section 2.4, we first give an existence/uniqueness theorem for the IVP

$y^{(n)}$, for instance, means d^ny/dx^n.

$$\begin{gathered} y^{(n)}+p_1(x)y^{(n-1)}+\cdots+p_{n-1}(x)y'+p_n(x)y=0, \\ y(a)=b_1,\ y'(a)=b_2,\ \ldots,\ y^{(n-1)}(a)=b_n, \end{gathered} \quad (1)$$

and then establish the form of a general solution of the nth-order equation

$$y^{(n)}+p_1(x)y^{(n-1)}+\cdots+p_{n-1}(x)y'+p_n(x)y=0. \quad (2)$$

The general solution will involve the linear independence of a set of n solutions of (2), so we will give a test for linear independence, analogous to the Wronskian test in Theorem 2.4.4(c), which was for the case $n = 2$.

2.7.1 Theorems for nth-order linear equations. First, a fundamental existence/uniqueness theorem for the IVP (1):

THEOREM 2.7.1 *Existence and Uniqueness for the IVP (1)*
A solution of the IVP (1) exists and is unique, at *least* on the broadest open x interval containing the initial point a, on which all the p_j's are continuous.

Next, the general solution of (2):

THEOREM 2.7.2 *Linear, Homogeneous Equation of Order n*
Consider the linear differential equation (2) on an open interval I.
(a) *Number of LI Solutions*: If all the p_j coefficients are continuous on I, then (2) has n LI solutions on I, no more and no less.
(b) *Form of General Solution*: If $y_1(x), \ldots, y_n(x)$ are any n LI solutions of (2), and $C_1, \ldots, C_n$ are arbitrary constants, then

$$\boxed{y(x) = C_1 y_1(x) + \cdots + C_n y_n(x)} \tag{3}$$

is a **general solution** of (2) on I.

The number n of solutions in (3) is the same as the order of equation (2).

Such a set of any n LI solutions of (2), $\{y_1, \ldots, y_n\}$, is called a **basis** of solutions for (2) because every solution of (2) *can be* expressed as a linear combination of them, *uniquely*, that is, with the coefficients $C_1, \ldots, C_n$ uniquely determined.

To use part (b) of Theorem 2.7.2 we must be able to test a set of n solutions to verify that they are LI. For that, we have the following extension of Theorem 2.4.4:

THEOREM 2.7.3 *A Necessary and Sufficient Wronskian Test for LI/LD*
Let all the p_j coefficients in (2) be continuous on an open interval I, and let $y_1(x), \ldots, y_n(x)$ be solutions of (2) on I. Denote $W[y_1, \ldots, y_n](x)$ as W. Then:
(a) W is either zero everywhere in I, or else nonzero everywhere in I.
(b) **Wronskian Test**: If $W = 0$ in I then $\{y_1, \ldots, y_n\}$ is LD on I, and if $W \neq 0$ in I then $\{y_1, \ldots, y_n\}$ is LI on I.

The result: LI if and only if $W \neq 0$.

We will now apply these results to constant-coefficient equations and to Cauchy–Euler equations.

2.7.2 Constant-coefficient equations. If the coefficients $p_1(x), \ldots, p_n(x)$ in (2) are *constants*, then we can solve (2) in the same way that we did for the case where $n = 2$: Seeking solutions in the exponential form

$$y(x) = e^{rx}, \tag{4}$$

obtain, by substituting (4) into (1), the **characteristic equation**

$$r^n + p_1 r^{n-1} + \cdots + p_{n-1} r + p_n = 0, \tag{5}$$

which is now an nth-degree polynomial equation for r.

We have the familiar quadratic formula for solving quadratic equations, and analogous formulas for cubic and quartic equations, but not for polynomial equations of degree five or higher. Thus, there is the technical problem of solving (5), but that "detail" is not our focus here, and we will consider examples in which (5) can be solved by hand. At worst, we can use a computer algebra system to find the characteristic roots. Thus, consider the roots of (5) as known.

Suppose, first, that (5) gives n distinct roots $r_1, \ldots, r_n$, so we have found n solutions $e^{r_1 x}, \ldots, e^{r_n x}$ of (1). To test these for linear independence, calculate their Wronskian and use Theorem 2.7.3(b):

Proving the linear independence of $e^{r_1 x}, \ldots, e^{r_n x}$.

$$W[e^{r_1 x}, e^{r_2 x}, \ldots, e^{r_n x}](x) = \begin{vmatrix} e^{r_1 x} & e^{r_2 x} & \cdots & e^{r_n x} \\ r_1 e^{r_1 x} & r_2 e^{r_2 x} & \cdots & r_n e^{r_n x} \\ \vdots & \vdots & & \vdots \\ r_1^{n-1} e^{r_1 x} & r_2^{n-1} e^{r_2 x} & \cdots & r_n^{n-1} e^{r_n x} \end{vmatrix}$$

$$= e^{r_1 a} \cdots e^{r_n a} \begin{vmatrix} 1 & 1 & \cdots & 1 \\ r_1 & r_2 & \cdots & r_n \\ \vdots & \vdots & & \vdots \\ r_1^{n-1} & r_2^{n-1} & \cdots & r_n^{n-1} \end{vmatrix}. \tag{6}$$

In the second equality we used determinant property D6 in Appendix B to pull out the common factors $e^{r_1 x}$ from the first column, $e^{r_2 x}$ from the second, and so on. The last determinant in (6) is of *Vandermonde* type and, by property D7 in Appendix B, it is nonzero because $r_1, \ldots, r_n$ are distinct by assumption. Since that determinant is nonzero and the factor $e^{(r_1 + \cdots + r_n)x}$ is nonzero as well, it follows that the Wronskian is nonzero. Then, by Theorem 2.7.3(b), *the set of n exponential solutions is LI* and, by Theorem 2.7.2, a general solution of (2) is

In (7) the r_j's are distinct.

$$y(x) = C_1 e^{r_1 x} + \cdots + C_n e^{r_n x}. \tag{7}$$

Next, what if (5) has repeated roots? Recall that if, for second-order equations, there is a repeated root $r_1 = r_2 \equiv r$, then two LI solutions are e^{rx} and xe^{rx}. That is, the general solution is $(C_1 + C_2 x)e^{rx}$, a two-term polynomial times e^{rx}. The same pattern holds for higher-order equations, as we now explain.

First, we say that a characteristic root r_j is of **multiplicity** k if the exponent of $r-r_j$ in the factored polynomial is k; in other words, that root occurs k times. For example, if the characteristic polynomial can be factored as $(r-1)(r-5)^3(r+4)^2$, then the roots 1, 5, and -4 are of multiplicity 1, 3, and 2, respectively. We state that if r is a characteristic root of multiplicity k, then it contributes k LI solutions: e^{rx}, xe^{rx}, and so on, up to $x^{k-1}e^{rx}$, namely,

$$\boxed{\Big(\underbrace{C_1 + C_2x + \cdots + C_kx^{k-1}}_{k\text{-term polynomial}}\Big)e^{rx}.} \tag{8}$$

A "string" of solutions corresponding to a root r of multiplicity k.

SUMMARY: *To form a general solution of (2), include a $C_je^{r_jx}$ term for each nonrepeated characteristic root r_j (i.e., for each root of multiplicity one), and for each root of multiplicity k include a string of solutions of the form (8).*

EXAMPLE 1. General Solution. To make up an example, we don't even need to write down the differential equation because everything hinges on the characteristic roots. So, suppose that for a certain sixth-order differential equation (linear, homogeneous, with constant coefficients) its characteristic roots are found to be

$$r = 1, 4, 4, 4, -5, -5, \tag{9}$$

That is, $r=1$ is of multiplicity 1, $r=4$ is of multiplicity 3, and $r=-5$ is of multiplicity 2. Then, a general solution of the differential equation is

$$y(x) = C_1e^x + \left(C_2+C_3x+C_4x^2\right)e^{4x} + (C_5+C_6x)e^{-5x}. \quad ∎ \tag{10}$$

EXAMPLE 2. General Solution. Solve

$$y'''' + 8y'' + 16y = 0. \tag{11}$$

Its characteristic equation $r^4+8r^2+16=0$ can be solved by noting that it can be factored as $(r^2+4)^2=0$, or by considering it as a quadratic equation in r^2 and using the quadratic formula to solve for r^2, and then for r. Either way, we obtain the factored form $(r-2i)^2(r+2i)^2=0$ and hence the repeated roots $r=2i,\ 2i,\ -2i,\ -2i$. Thus,

Each of the roots $+2i$ and $-2i$ is of multiplicity two.

$$\begin{aligned} y(x) &= (C_1+C_2x)\,e^{i2x} + (C_3+C_4x)\,e^{-i2x} \\ &= \left(C_1e^{i2x}+C_3e^{-i2x}\right) + x\left(C_2e^{i2x}+C_4e^{-i2x}\right) \\ &= C_5\cos 2x + C_6\sin 2x + x(C_7\cos 2x + C_8\sin 2x) \end{aligned} \tag{12}$$

is a general solution of (11). The first line of (12) was perfectly acceptable as a general solution, but we preferred to re-express it in terms of real valued functions. ∎

EXAMPLE 3. IVP. Solve the IVP

$$y''' - 2y'' + y' - 2y = 0; \quad y(0)=0,\ y'(0)=0,\ y''(0)=10. \tag{13}$$

The characteristic equation is $r^3 - 2r^2 + r - 2 = 0$, and the roots are $r = 2, i, -i$, so a general solution of the differential equation is

$$y(x) = C_1 e^{2x} + C_2 e^{ix} + C_3 e^{-ix} = C_1 e^{2x} + C_4 \cos x + C_5 \sin x. \tag{14}$$

Then, $y' = 2C_1 e^{2x} - C_4 \sin x + C_5 \cos x$ and $y'' = 4C_1 e^{2x} - C_4 \cos x - C_5 \sin x$, so the initial conditions give the algebraic equations $y(0) = 0 = C_1 + C_4$, $y'(0) = 0 = 2C_1 + C_5$, and $y''(0) = 10 = 4C_1 - C_4$ for C_1, C_4, C_5. These give $C_1 = 2$, $C_4 = -2$, and $C_5 = -4$, so the solution of (9) is $y(x) = 2e^{2x} - 2\cos x - 4\sin x$. ▮

The constant coefficients in the differential equation are continuous for all x, so, by Theorem 2.7.3, the solution (11) exists and is unique on $-\infty < x < \infty$.

2.7.3 Cauchy–Euler equations. The nth-order Cauchy–Euler equation is this:

$$x^n y^{(n)} + a_1 x^{n-1} y^{(n-1)} + \cdots + a_{n-1} y' + a_n y = 0, \tag{15}$$

in which the a_j's are constants. Consider (15) on the interval $0 < x < \infty$.

Putting (15) into the form (2), the coefficients are $p_1(x) = a_1/x, \ldots, p_n(x) = a_n/x^n$, which are continuous on the interval, as assumed in Theorem 2.7.2.

As in Section 2.5, seek solutions in the form

$$y(x) = x^r, \tag{16}$$

put the latter into (15), and obtain an nth-degree polynomial equation for r, which we call the r-equation corresponding to (15). If the latter has n distinct roots $r_1, \ldots, r_n$, then we have obtained the n solutions $x^{r_1}, \ldots, x^{r_n}$ which, we state without proof, are LI (Exercise 8) on I. Thus, if there are n distinct r's, then we have the general solution

In (17) the r_j's are distinct.

$$y(x) = C_1 x^{r_1} + \cdots + C_n x^{r_n}. \tag{17}$$

What if there are repeated roots of the r-equation? If r_j is a root of multiplicity k, then it contributes to the general solution a k-term polynomial, not in powers of x but in powers of $\ln x$, times x^{r_j},

In (18) the root r of the r-equation is of multiplicity k.

$$\Big[\underbrace{C_1 + C_2(\ln x) + \cdots + C_k(\ln x)^{k-1}}_{k\text{-term polynomial in } \ln x}\Big] x^r. \tag{18}$$

SUMMARY: *To form a general solution of (15), include a $C_j x^{r_j}$ term for each non-repeated characteristic root r_j of the r-equation, and for each root of multiplicity k include a "string" of terms of the form (18).*

EXAMPLE 4. Cauchy–Euler. Solve

$$x^4 y'''' - 5x^3 y''' + 10x^2 y'' + 10xy' - 64y = 0. \tag{19}$$

Its r-equation is $r(r-1)(r-2)(r-3) - 5r(r-1)(r-2) + 10r(r-1) + 10r - 64 = 0$, or $r^4 - 11r^3 + 36r^2 - 16r - 64 = 0$, with roots $r = -1, 4, 4, 4$. Thus, a general solution of (19) is

$$y(x) = C_1 \frac{1}{x} + \left[C_2 + C_3(\ln x) + C_4(\ln x)^2\right] x^4. \quad \blacksquare \tag{20}$$

Actually, we worked backward to make up Example 4, choosing the four roots and working backward to the corresponding Cauchy–Euler equation (19).

Closure. The solutions and theorems for higher-order constant-coefficient and Cauchy–Euler equations are a natural extension of those for the second-order versions of those equations, that were studied in Sections 2.2–2.6. Some applications of fourth-order equations, in particular, are given in the exercises and subsequent chapters.

EXERCISES 2.7

CONSTANT-COEFFICIENT EQUATIONS, EXERCISES 1–7

1. Find a general solution. HINT: Determine the characteristic roots and, from them, write down a general solution. In (b), (c), (e), and (g), one root is 1. Note that $\sqrt{i} = \pm(1+i)/\sqrt{2}$ and $\sqrt{-i} = \pm(1-i)/\sqrt{2}$

(a) $y''' - 2y' = 0$
(b) $y''' - 3y'' + 3y' - y = 0$
(c) $y''' - 5y'' + 8y' - 4y = 0$
(d) $y''' - 9y' = 0$
(e) $y''' - y'' + 9y' - 9y = 0$
(f) $y''' + y'' + 3y' = 0$
(g) $y''' - 7y'' + 11y' - 5y = 0$
(h) $y'''' + y = 0$
(i) $y'''' - 5y'' + 4y = 0$
(j) $y'''' - 3y'' + 2y = 0$
(k) $y'''' - y = 0$
(l) $y'''' + 4y'' + 3y = 0$

2. Using Reduction of Order. This exercise is to illustrate the use of reduction of order for equations of higher order. For each given differential equation, one characteristic root is $r = 1$. Use that root and the method of reduction of order to find a general solution. That is, seek $y(x) = C(x)e^x$. NOTE: You should obtain a reduction of order equal to the multiplicity of that root, $r = 1$. For instance, if $r = 1, 1, 5$ (the multiplicity in that case being 2), then the reduction will be from a third-order equation to a first-order equation; that is, letting $u'' = v$ will give a first-order equation for v.

(a) $y''' - 6y'' + 11y' - 6y = 0$
(b) $y''' - 5y'' + 7y' - 3y = 0$
(c) $y''' - 3y'' + 3y' - y = 0$
(d) $y'''' - 6y''' + 12y'' - 10y' + 3y = 0$

3. Inferring the Equation. Given the roots of the characteristic equation, infer from them the original constant-coefficient differential equation.

(a) $1, -1, 2$
(b) $1, 2, -2$
(c) $3, -4, 0$
(d) $0, 2, 3$
(e) $-1, -2, -3$
(f) $1, 2, 3, 4$
(g) $1, -1, 4, -4$
(h) $0, 1, -1, 2, 4$
(i) $0, 3 \pm 5i$
(j) $1 \pm i, \pm 3i$
(k) $-2, -2 \pm i$
(l) $\pm 4i, 3 \pm 4i$

4. Inferring the Equation. If C_1, C_2, and C_3 are arbitrary constants, then the given $y(x)$ is a general solution of what constant-coefficient equation $y''' + p_1 y'' + p_2 y' + p_3 y = 0$? That is, determine p_1, p_2, p_3.

(a) $y(x) = C_1 + C_2 e^{-x} + C_3 e^{2x}$
(b) $y(x) = C_1 e^x + (C_2 + C_3 x)e^{2x}$
(c) $y(x) = C_1 + C_2 \cosh x + C_3 \sinh x$
(d) $y(x) = C_1 e^{2x} + C_2 \cosh x + C_3 \sinh x$
(e) $y(x) = (C_1 + C_2 x + C_3 x^2)e^x$
(f) $y(x) = (C_1 + C_2 x)e^{-x} + C_3 e^x$
(g) $y(x) = C_1 e^x + C_2 \cos 3x + C_3 \sin 3x$
(h) $y(x) = e^{4x}(C_1 \sin 3x + C_2 \cos 3x) + C_3 e^{4x}$
(i) $y(x) = e^x(C_1 \sin x + C_2 \cos x + C_3)$
(j) $y(x) = C_1 - e^x(C_2 \sin 7x + C_3 \cos 7x)$

5. Repeated Roots; General Solution. Let the characteristic roots be as given. Write down a general solution for the differential equation, in terms of real valued functions.

(a) $1,\ 1,\ 1$
(b) $2,\ -1,\ -1$
(c) $0,\ 0,\ -4$
(d) $0,\ 0,\ 0,\ 5$
(e) $\pm 3i,\ \pm 3i$
(f) $4,\ 4,\ 3 \pm i$
(g) $\pm i,\ +i$
(h) $2,\ 2,\ 2,\ \pm 5i$
(i) $1,\ 1,\ 2,\ 2,\ 2$
(j) $2 \pm 3i,\ 2 \pm 3i$

6. Can the characteristic roots $r = i, 2i, 3i, 4i$ be obtained for a fourth-order, linear, homogeneous differential equation with real constant coefficients? Explain.

7. Basis for (2). Let a be a point inside an interval on which all the p_j's in (2) are continuous. Suggest any n sets of initial conditions at a, that can be used, with equation (2), to generate n LI solutions $y_1, \ldots, y_n$ so that $C_1 y_1(x) + \cdots + C_n y_n(x)$ is a general solution of (2). Give your reasoning.

CAUCHY–EULER EQUATIONS, EXERCISES 8–11

8. LI of the Powers of x. We stated, above (17), that $\{x^{r_1}, \ldots, x^{r_n}\}$ is LI on $0 < x < \infty$ if the r_j's are distinct. Prove that claim for the case $n = 3$. Extension to the general case should be evident, but in this exercise consider only $n = 3$.

9. Find a General Solution. Find a general solution for the given third-order Cauchy–Euler equation. HINT: Determine the roots of the r equation and, from them, write down a general solution.
(a) $x^3y''' - x^2y'' - 3xy' = 0$
(b) $x^3y''' - 3x^2y'' + 6xy' - 6y = 0$
(c) $x^3y''' + x^2y'' - 2xy' + 2y = 0$
(d) $x^3y''' - 3x^2y'' + 4xy' - 4y = 0$
(e) $x^3y''' - 12x^2y'' + 61xy' - 125y = 0$
(f) $x^3y''' + 6x^2y'' + 7xy' + y = 0$
(g) $x^3y''' - 2x^2y'' + 13xy' - 13y = 0$

10. Inferring the Equation. Given the roots of the r-equation, infer from them the original Cauchy–Euler equation.

(a) 1, 2, 3 (b) 1, 1, 2 (c) 2, i, $-i$
(d) 0, $1+3i$, $1-3i$ (e) 0, 1, -1 (f) 0, 0, 5

11. Inferring the Equation. If C_1, C_2, and C_3 are arbitrary constants, then the given $y(x)$ is a general solution of what Cauchy–Euler equation $x^3y''' + ax^2y'' + bxy' + cy = 0$? That is, determine a, b, c. HINT for parts (g)–(j): Roots $r = a \pm ib$ contribute solutions $y(x) = x^{a \pm ib} = x^a x^{\pm ib} = x^a e^{\pm ib \ln x}$, or, $x^a \cos(b \ln x)$ and $x^b \sin(b \ln x)$, to the general solution.

(a) $y(x) = C_1x + C_2x^2 + C_3x^4$
(b) $y(x) = C_1 + C_2x^3 + C_3x^{-3}$
(c) $y(x) = C_1 + (C_2 + C_3 \ln x)x^2$
(d) $y(x) = C_1x + (C_2 + C_3 \ln x)x^4$
(e) $y(x) = [C_1 + C_2 \ln x + C_3(\ln x)^2]x$
(f) $y(x) = C_1 + C_2 \ln x + C_3(\ln x)^2$
(g) $y(x) = C_1 + C_2 \cos(\ln x) + C_3 \sin(\ln x)$
(h) $y(x) = C_1x + x^3[C_2 \cos(2 \ln x) + C_3 \sin(2 \ln x)]$
(i) $y(x) = C_1x^2 + x[C_2 \cos(3 \ln x) + C_3 \sin(3 \ln x)]$
(j) $y(x) = C_1x^{-1} + x^5[C_2 \cos(3 \ln x) + C_3 \sin(3 \ln x)]$

12. Abel's Formula for Solutions of (2). It turns out that **Abel's formula**, given by (18) in Section 2.4 for the case $n = 2$, carries over directly to equation (2) of this section, as

$$W[y_1, \ldots, y_n](x) = Ce^{-\int p_1(x)\,dx}, \tag{12.1}$$

in which C is a constant that depends on the n solutions $y_1, \ldots, y_n$ that are used. Here, we ask you to prove it, not for arbitrary n but only for the "next" n, $n = 3$. HINT: You could use the same method as outlined in Exercise 11 of Section 2.4, but it will be simpler to write the Wronskian and differentiate it with the help of property D9 in Appendix B. Then use property D5 and the equation $y_j''' + p_1y_j'' + p_2y_j' + p_3y_j = 0$ to replace the y_j''''s that occur by $-p_1y_j'' - p_2y_j' - p_3y_j$.

13. Verification of Abel's Formula for $n = 3$. Given are a third-order equation $Ly = 0$ and three solutions of it. Verify Abel's formula (12.1) by evaluating both sides of the equation (the left-hand side being the Wronskian determinant), and showing that a constant C can be chosen to make the equation an identity. Let the interval be $0 < x < 1$.

(a) $y''' + 4y' = 0$, $y_1(x) = 1$, $y_2(x) = \sin 2x$, $y_3(x) = \cos 2x$
(b) $y''' - y' = 0$, $y_1(x) = 1$, $y_2(x) = e^x$, $y_3(x) = e^{-x}$
(c) $y''' - 2y'' - y' + 2y = 0$, $y_1(x) = e^x$, $y_2(x) = e^{-x}$, $y_3(x) = e^{2x}$
(d) $x^3y''' - x^2y'' - 3xy' = 0$, $y_1(x) = 1$, $y_2(x) = x^4$, $y_3(x) = \ln x$
(e) $y'''' = 0$, $y_1(x) = 1$, $y_2(x) = x$, $y_3(x) = x^2$, $y_4(x) = x^3$

14. Stokes' Law of Resistance for a Sphere. In the exercises for Section 1.3 we considered the free fall and terminal velocity of a shperical particle in a fluid. Under certain conditions of "slow flow," the drag force on the sphere is modeled by Stokes' law of resistance, which is derived in a course in fluid mechanics. The key differential equation in the analysis is

$$\rho^4 f'''' - 4\rho^2 f'' + 8\rho f' - 8f = 0, \quad (a < \rho < \infty) \tag{14.1}$$

in which ρ is a radial coordinate, $f(\rho)$ is related to the "stream function," and a is the radius of the sphere.

(a) Derive the general solution of (14.1) as

$$f(\rho) = A\rho + B\rho^2 + C\rho^4 + D/\rho. \tag{14.2}$$

(b) The boundary conditions are $f(a) = 0$, $f'(a) = 0$, and $f(\rho) \sim U\rho^2/2$ as $\rho \to \infty$. Thus, this is a boundary value problem, not an initial value problem, since there are conditions at both ends of the ρ interval, which is $a < \rho < \infty$. The one "at infinity" is called an *asymptotic* boundary condition. Here, U is the speed of the sphere through the fluid.

Problem: Show that application of the boundary conditions to (14.2) gives

$$f(\rho) = \frac{U}{2}\left(\rho^2 - \frac{3a}{2}\rho + \frac{a^3}{2}\frac{1}{\rho}\right). \tag{14.3}$$

15. Solution by Factoring the Operator. For second-order equations with constant coefficients we used operator factorization to derive the general solution in Section 2.2. The same approach can be used to solve higher-order equations. This exercise is to illustrate that alternative solution method for a representative example. Namely, consider $y''' - 4y'' + 5y' - 2y = 0$, with the characteristic roots $r = 1, 1, 2$ (which you need not derive). Thus, we know in advance that a general solution is $y(x) = C_1 e^{2x} + (C_2 + C_3 x)e^{x}$. *Problem*: Derive that solution by factorization. HINT: Using operator notation, write

$$(D-1)\underbrace{(D-1)(D-2)y}_{u} = 0, \tag{15.1}$$

$$(D-1)\underbrace{(D-2)y}_{v} = u, \tag{15.2}$$

$$(D-2)y = v. \tag{15.3}$$

Then solve (15.1)–(15.3) for u, v, y, in turn. That is, solve $u' - u = 0$, $v' - v = u$, and, finally, $y' - 2y = v$.

2.8 NONHOMOGENEOUS EQUATIONS

In Sections 2.1–2.7 we studied only linear homogeneous equations $Ly = 0$, and we now turn to *nonhomogeneous* linear equations

$$\boxed{Ly = f(x).} \tag{1}$$

That is, we now allow for a forcing function $f(x)$.

We did the same in studying $y' + p(x)y = q(x)$ in Section 1.2, solving the homogeneous equation first, and then the nonhomogeneous equation.

2.8.1 General solution. Like many of the concepts developed in this chapter, the theory that follows will rest on the *linearity* of (1), namely, the property

$$L(\alpha u + \beta v) = \alpha Lu + \beta Lv \tag{2}$$

of L, for any functions $u(x), v(x)$ and any constants α, β.

Of course, if L is an nth-order operator, then u and v must be n-times differentiable.

We want to determine how to construct a general solution of nonhomogeneous equations. Suppose $y_h(x)$ is a general solution of the homogeneous version of (1), so $Ly_h = 0$. And suppose $y_p(x)$ is any particular solution of (1), so $Ly_p = f(x)$. That is, $y_p(x)$ is any function which, when put into the left-hand side of (1), gives $f(x)$. We call $y_h(x)$ and $y_p(x)$ **homogeneous** and **particular solutions** of (1), respectively. If we know $y_h(x)$ and $y_p(x)$ we can use them to construct a general solution of (1) as follows:

THEOREM 2.8.1 *General Solution of* $Ly = f(x)$
If $y_h(x)$ is a general solution of the linear homogeneous equation $Ly = 0$ and $y_p(x)$

is a particular solution of the nonhomogeneous equation $Ly = f(x)$, on an interval I, then their sum

$$y(x) = y_h(x) + y_p(x) \tag{3}$$

is a **general solution** of $Ly = f(x)$ on I.

Proof: The right-hand member of (3) satisfies $Ly = f(x)$ because of the linearity of (1):

$$\begin{aligned} L(y_h + y_p) &= Ly_h + Ly_p \\ &= 0 + f(x) = f(x). \end{aligned} \tag{4}$$

Thus, (3) is a solution of (1). To show that it is a *general* solution, let $y(x)$ be *any* solution of (1). Again using the linearity of L,

$$L(y - y_p) = Ly - Ly_p = f(x) - f(x) = 0. \tag{5}$$

That is, $L(y - y_p) = 0$, the general solution of which is y_h. Thus, $y(x) - y_p(x) = y_h(x)$, so *every* solution $y(x)$ of (1) is of the form $y(x) = y_h(x) + y_p(x)$, Hence, the latter is a general solution of (1). ■

The ingredients in the general solution (3) are $y_h(x)$ and $y_p(x)$. We know how to find a general solution $y_h(x)$ of the homogeneous equation, at least for constant-coefficient and Cauchy–Euler equations, so it remains to learn how to find particular solutions, $y_p(x)$. We will do that in Sections 2.9 and 2.10.

EXAMPLE 1. Application of Theorem 2.8.1. Consider the linear nonhomogeneous equation

$$y'' + 4y = 8x \tag{6}$$

on $-\infty < x < \infty$. First, solve the associated homogeneous equation $y'' + 4y = 0$, obtaining

$$y_h(x) = C_1 \cos 2x + C_2 \sin 2x. \tag{7}$$

Next, we need a particular solution $y_p(x)$ so we can add it to $y_h(x)$, to form a general solution of (6). Since we've not yet shown how to find particular solutions, we've made (6) simple enough so a particular solution can be found by inspection. What function, put into $y'' + 4y$ will give $8x$? The function

$$y_p(x) = 2x \tag{8}$$

works because it reduces (6) to an identity, $0 + 8x = 8x$. Then, Theorem 2.8.1 tells us that

$$y(x) = y_h(x) + y_p(x) = C_1 \cos 2x + C_2 \sin 2x + 2x \tag{9}$$

is a general solution of (6); *every* solution of (6) is given by (9), for some choice of C_1 and C_2.

Notice that $y_p(x) = 2x$ is not the only particular solution of (6), for if any homogeneous solution is added to it we still a particular solution. For instance, $y_p(x) = 2x + 37\cos 2x$ is also a paricular solution because $L(2x + 37\cos 2x) = L(2x) + 37L(\cos 2x) = 8x + (37)(0) = 8x$. However, the simplest and most natural particular solution is $y_p(x) = 2x$, given in (8).

If we impose initial conditions $y(0) = 5$ and $y'(0) = -4$, say, we can evaluate C_1 and C_2. We obtain $C_1 = 5$ and $C_2 = -3$, so

$$y(x) = 5\cos 2x - 3\sin 2x + 2x, \tag{10}$$

which is plotted in Fig. 1. ∎

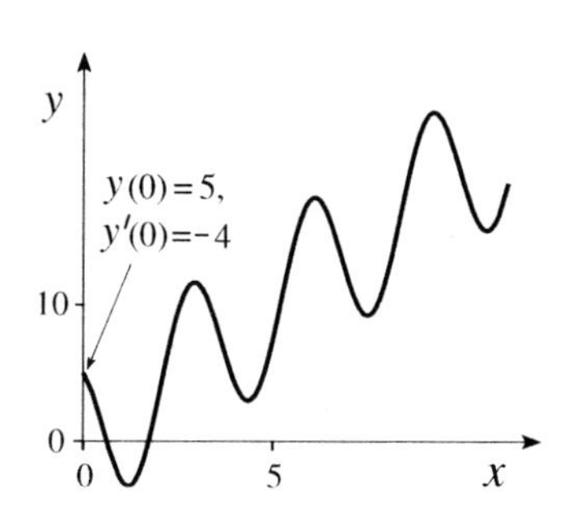

Figure 1. Graph of the solution (10).

2.8.2 The scaling and superposition of forcing functions. The linearity of the differential equation (1) results in the following two important properties:

THEOREM 2.8.2 *Scaling and Superposition of Forcing Functions*
Let L be a linear differential operator.
(a) If $Ly = f(x)$ has a particular solution $y_p(x)$ on an interval I, then $Ly = \alpha f(x)$ has a particular solution $\alpha y_p(x)$ on I.

Scaling.

(b) If $y_{p1}(x), \ldots, y_{pk}(x)$ are particular solutions of $Ly = f_1(x), \ldots, Ly = f_k(x)$ on I, respectively, then a particular solution of

Superposition.

$$Ly = f_1(x) + \cdots + f_k(x) \tag{11}$$

on I is

$$y_p(x) = y_{p1}(x) + \cdots + y_{pk}(x). \tag{12}$$

Theorem 2.8.2 will be used when we develop the method of undetermined coefficients in the next section.

Proof: Both results follow from the linearity of L. To prove (a), the linearity of L gives $L(\alpha y_p(x)) = \alpha L y_p(x) = \alpha f(x)$, and to prove (b), the linearity of L gives $L(y_{p1} + \cdots + y_{pk}) = Ly_{p1} + \cdots + Ly_{pk} = f_1(x) + \cdots + f_k(x)$. ∎

In words, part (a) says that if, for a *linear* system, we scale the forcing function by α, then the response to it is scaled by α. For instance, if we double the forcing function, then the response to it is doubled. Part (b) says that if two or more forcing functions are applied to the system, then the response to their combined application is the sum of the responses to their individual application.

To give a visual representation of part (b), think of the diving board shown in Fig. 2, in which the applied forces F_1 and F_2 are the weights of two divers standing at those points. Without writing the differential equation governing the deflection $y(x)$, we simply state that it is linear, and that F_1 and F_2 appear as forcing functions $f_1(x)$ and $f_2(x)$ on the right-hand side of that equation, as in (11). In this case, (12) means that the deflection $y(x)$ due to F_1 and F_2 acting together is simply the sum of the deflections $y_1(x)$ and $y_2(x)$ due to F_1 and F_2 acting individually, as

Divers on a diving board.

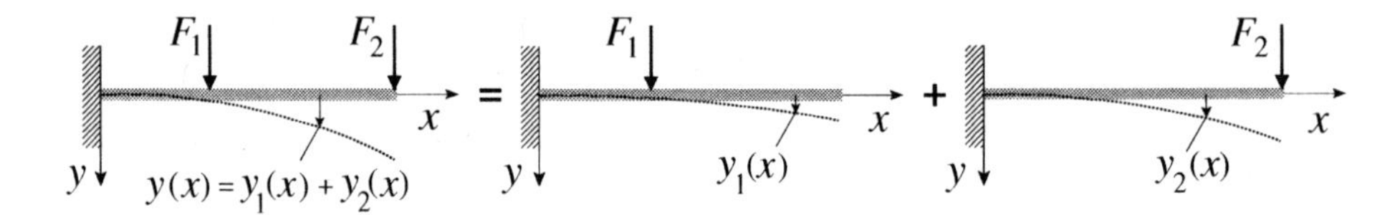

Figure 2. Property (b) in Theorem 2.8.2: $y(x) = y_1(x) + y_2(x)$.

indicated in the sketch. This result is probably consistent with one's intuition.

Here are two more illustrations of Theorem 2.8.2.

EXAMPLE 2. Part (a) of Theorem 2.8.2. Consider

$$y'' - y' = x \tag{13}$$

on $-\infty < x < \infty$. Suppose we obtain the particular solution $y_p(x) = -x^2/2 - x$, which can be verified by substitution. Then a particular solution of

$$y'' - y' = 8x \tag{14}$$

is $y_p(x) = 8(-x^2/2 - x) = -4x^2 - 8x$, which can be verified by substitution. ▮

EXAMPLE 3. Part (b) of Theorem 2.7.2. Consider

$$y'' + 2y = 12e^x - 3\sin x + 10 \tag{15}$$

on $-\infty < x < \infty$. Particular solutions corresponding to $f_1(x) = 12e^x$, $f_2(x) = -3\sin x$, and $f_3(x) = 10$ are $y_{p1}(x) = 4e^x$, $y_{p2}(x) = -3\sin x$, and $y_{p3}(x) = 5$. Then, according to part (b) of the theorem, a particular solution of (15) is $y_p(x) = y_{p1}(x) + y_{p2}(x) + y_{p3}(x) = 4e^x - 3\sin x + 5$. ▮

Closure. The chief result of this section is that for a linear differential equation (1) a general solution can be constructed as a general solution of the homogeneous equation plus any particular solution of (1). If two or more forcing functions are applied, then the response to their combined action is the sum of their individual responses. In this section, L can be of any order $n > 1$ but, since most of our interest in this chapter is on second-order equations, our three examples were for equations of second order. Methods for determining $y_p(x)$ are given in Sections 2.9 and 2.10.

EXERCISES 2.8

1. General Solution. A particular solution $y_p(x)$ of the differential equation is given. Verify that it is a particular solution, as claimed. Then obtain a general solution of the differential equation. Let the interval under consideration be $0 < x < \infty$ in all cases.

(a) $y' - y = -x^2, \quad y_p(x) = x^2 + 2x + 2$
(b) $y' + 3y = 4e^x, \quad y_p(x) = e^x$
(c) $y' + 2y = 8x^3, \quad y_p(x) = 4x^3 - 6x^2 + 6x - 3$
(d) $y'' + y' = 4x + 4, \quad y_p(x) = 2x^2$
(e) $y'' - y = 3e^{2x} + 8e^{3x}, \quad y_p(x) = e^{2x} + e^{3x}$
(f) $y'' - 2y' = e^x, \quad y_p(x) = -e^x$
(g) $y'' - 2y' + y = 6\sin x, \quad y_p(x) = 3\cos x$
(h) $y'' + 2y' + y = 6\cos x, \quad y_p(x) = 3\sin x$
(i) $y'' - 3y' + 2y = 2x^2 - 6x, \quad y_p(x) = x^2 - 1$
(j) $y'' + 4y' + 3y = (6 + 8x)e^x, \quad y_p(x) = xe^x$
(k) $y''' - y'' = -6x, \quad y_p(x) = x^3 + 3x^2$
(l) $y''' + 4y' = 48x, \quad y_p(x) = 6x^2$
(m) $y'''' - y = x^2, \quad y_p(x) = -x^2$
(n) $x^2y'' - 2xy' - 10y = -24x, \ (x > 0) \ y_p(x) = 2x$
(o) $x^2y'' - 5xy' + 9y = 9x^6, \ (x > 0) \ y_p(x) = x^6$
(p) $x^2y'' - 6y = -16x^2, \ (x > 0) \ y_p(x) = 4x^2$
(q) $x^2y'' - 4xy' + 6y = 20x, \ (x > 0) \ y_p(x) = 10x$

2. The following equations are simple enough so that we don't need to know how to find particular solutions. After all, (a)–(e) can be solved just by repeated integration, and (f)–(h) by using the general solution formula given in Section 1.2. Doing that, obtain a general solution, and verify that it is of the form $y_h(x) + y_p(x)$. That is, identify $y_h(x)$ and $y_p(x)$.

(a) $y'' = 8e^{2x} - 12$
(b) $y'' = 4\sin 2x$
(c) $y''' = 24$
(d) $y''' = 96x^2$
(e) $y'''' = 48x$
(f) $xy' + y = 12x^3$
(g) $xy' - y = 12x^2$
(h) $y' - y = 24e^{2x}$

3. Find a particular solution by inspection or trial and error.

(a) $y'' + 2y = 9$
(b) $y'' + 3y = 12x$
(c) $y'' + y' = 10$
(d) $y'' + 3y' + 2y = 40$
(e) $y'' + y' = 3e^x$
(f) $y''' + 2y = 6x^2$
(g) $y''' + y'' + 2y = 7e^x$
(h) $y''' - 4y' + 5y = 100e^{3x}$

4. Using a Trial Form for $y_p(x)$. Find a particular solution by assuming a solution in the suggested form and finding constants $A, B, \ldots$ that will work. NOTE: How do we make up such forms for $y_p(x)$? That will be explained in the next section. For now, accept them as given.

(a) $y'' + y = 4x^2, \ y_p(x) = Ax^2 + Bx + C$
(b) $y'' + y = e^{2x}, \ y_p(x) = Ae^{2x}$
(c) $y'' + 3y' + y = 6e^{2x}, \ y_p(x) = Ae^{2x}$
(d) $y'' - 2y' = 4\sin x, \ y_p(x) = A\sin x + B\cos x$
(e) $y'' - 2y' + y = 3\cos x, \ y_p(x) = A\cos x + B\sin x$
(f) $y'' + 3y = 2x^3, \ y_p(x) = Ax^3 + Bx^2 + Cx + D$
(g) $y'' + y' + y = 5\cos x, \ y_p(x) = A\cos x + B\sin x$
(h) $y'' + 2y = 3xe^x, \ y_p(x) = Axe^x + Be^x$
(i) $y'' - y' = 8 - 8x, \ y_p(x) = Ax^2 + Bx + C$
(j) $y'' - 9y = 10\sin x + 45, \ y_p(x) = A\sin x + B$
(k) $y'' + 2y' = 4e^{2x} + 8, \ y_p(x) = Ae^{2x} + Bx + C$

5. Show that it follows from Theorem 2.8.2 that if $Ly = f_1(x), \ldots, Ly = f_k(x)$ have particular solutions $y_{p1}(x), \ldots, y_{pk}(x)$, respectively, then a particular solution of

$$\boxed{Ly = \alpha_1 f_1(x) + \cdots + \alpha_k f_k(x)} \tag{5.1}$$

is

$$\boxed{y_p(x) = \alpha_1 y_{p1}(x) + \cdots + \alpha_k y_{pk}(x).} \tag{5.2}$$

6. Using Theorem 2.8.2. For this problem use Theorem 2.8.2 or, equivalently, (5.2) given above.

(a) Suppose particular solutions of $y'' + 2y = f(x)$ for $f(x) = 1, e^x, e^{3x}$ are $y_p(x) = 1/2,\ e^x/3,\ e^{3x}/11$, respectively. Then what is a particular solution of $y'' + 2y = 50 - 12e^x - 22e^{3x}$?
(b) Suppose particular solutions of $y'' - y = f(x)$ for $f(x) = 1, x, x^2$ are $-1,\ -x,\ -x^2 - 2$, respectively. Then what is a particular solution of $y'' - y = (x-2)^2$?
(c) Suppose particular solutions of $y'' - 4y' + 3y = f(x)$ for $f(x) = x^2, e^{2x}, e^{-2x}$ are $x^2/3 + 8x/9 + 26/27,\ -e^{2x},\ e^{-2x}/15$, respectively. Then what is a particular solution of $y'' - 4y' + 3y = 27x^2 + 30\sinh 2x$?
(d) Suppose particular solutions of $y'' - 2y' + y = f(x)$ for $f(x) = e^x, e^{-x}$ are $x^2e^x/2,\ e^{-x}/4$, respectively. Then what is a particular solution of $y'' - 2y' + y = 16\cosh x$?
(e) Suppose particular solutions of $y'' + y' = f(x)$ for $f(x) = 1, \sin x$ are $x,\ -(\sin x + \cos x)/2$, respectively. Then what is a particular solution of $y'' + y' = 3 - 4\sin x$?
(f) Suppose particular solutions of $y'''' - y = f(x)$ for

$f(x) = \sin x, \cos 2x$ are $(x \cos x - \sin x)/4, (\cos 2x)/15$, respectively. Then what is a particular solution of $y'''' - y = 12 \sin x + 60 \cos 2x$?

(g) Suppose particular solutions of $y'''' + 2y'' = f(x)$ for $f(x) = 1, x, x^2$ are $x^2/4$, $x^3/12$, $x^4/24 - x^2/4$, respectively. Then what is a particular solution of $y''''+2y'' = 24(2x-1)^2$?

ADDITIONAL EXERCISES

7. Human Visual Perception: Linear or Nonlinear? The human central nervous system is a complex network of around 10^{12} neurons, with many difficult responsibilities, such as sensory perception, memory, muscular coordination, and so on. Of these, consider visual perception, to which fully a third of that system is devoted. Heuristically, think of the system as being modeled by a set of differential equations that can be summarized as "$Ly = f$," in which f is the sum [as on the right-hand side of (11)] of the many bits of light information falling upon the retinas and y, in some sense, is our resulting visual perception — what we *see*. Thus, think in terms of a complex input/output system, in which the bits of incoming light energy are the input, and our visual perception is the output, as indicated schematically in the figure. Until around 1912 the *Molecularists* represented the accepted school of thought in psychology and their view was that one's perception is equal to the sum of its parts. That is, our perception is the sum of the perceptions induced by the individual bits of incoming light energy falling upon our retinas. They did not think

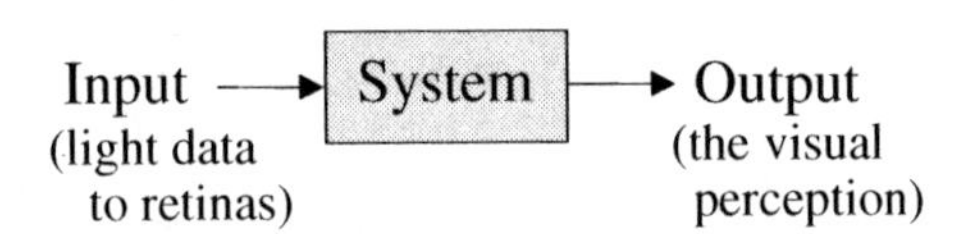

in terms of "linear and nonlinear" in those days, but a fundamental question that comes to mind now, as we study differential equations, is whether the system is linear or nonlinear. We don't have a literal differential equation or set of differential equations to examine, but we can look at some pairs of inputs and outputs and see if the system *behaves* as a linear system or not. Here are three such bits of evidence:

(i) In 1912 *Max Wertheimer* conducted his classic *psi experiment*: He projected two parallel lines of light on the screen, sequentially, and found that when they were projected close enough, in space and time, they were perceived not as a sequence of two lines, but as a single *moving* line. Besides giving birth to the moving picture industry, this result showed clearly that perception is not merely the sum of its parts; rather, "the whole is greater than the sum of its parts," which became the now-familiar hallmark of Wertheimer's *Gestalt* school of psychology that proceeded to supplant the molecularists.

(ii) Here is another experiment: Hold your hands in front of your face, one around one foot and the other around two feet from your eyes. The retinal image of the closer is twice the size of the retinal image of the farther. Do they *look* that way to you, or do they look approximately the same size?

(iii) Finally, consider the basic building block of the whole central nervous system — the single neuron, or nerve cell, of which the rods and cones are specialized examples. If a voltage is applied to the input end of a neuron the neuron will not "fire" — that is, send a series of voltage pulses called *action potentials* down its axon to the output end — unless the input voltage is equal to or greater than a critical threshold value. (Once the voltage pulses reach the terminal end of the neuron, they serve as inputs for neighboring neurons, and so on.) If we consider the magnitude of the action potential as the output, the input/output data looks something like this:

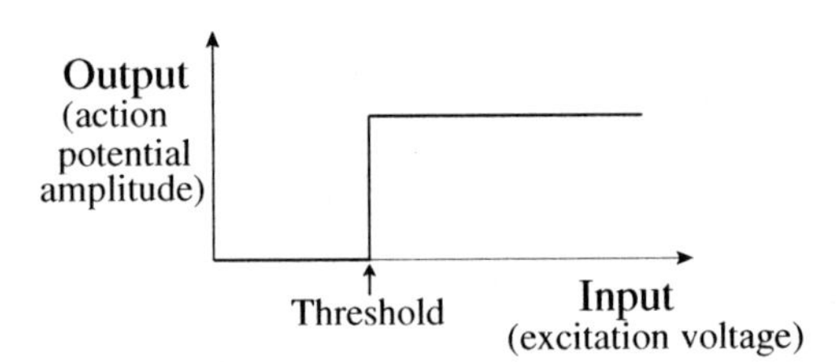

Here is the question:

(a) Does the evidence in (i) and (ii) indicate that the perceptual system (which is a part of the central nervous system) is linear or nonlinear? Explain.

(b) Does the second figure indicate that the individual neuron is a linear or nonlinear device? Does your answer to this part of the problem fit together with your answer to part (a)? Discuss.

2.9 PARTICULAR SOLUTION BY UNDETERMINED COEFFICIENTS

2.9.1 Undetermined coefficients. In Section 2.8, we learned that a general solution of a linear nonhomogeneous differential equation

$$Ly = f(x) \tag{1}$$

is given by $y(x) = y_h(x) + y_p(x)$, in which $y_h(x)$ is a general solution of the associated homogeneous equation $Ly = 0$ and $y_p(x)$ is a particular solution of (1). The simplest method for finding a particular solution is inspection, so it is good to try that first. If inspection doesn't work, and the equation is of constant-coefficient type, we suggest using the **method of undetermined coefficients**.

The interval of interest will be $-\infty < x < \infty$, unless stated otherwise.

The method is designed for linear *constant-coefficient* equations.

EXAMPLE 1. Find a general solution of

$$y'' - 4y = 3x^2. \tag{2}$$

First, obtain a general solution of the corresponding homogeneous equation $Ly = y'' - 4y = 0$, which is

$$y_h(x) = C_1e^{2x} + C_2e^{-2x}. \tag{3}$$

Next, to find a particular solution $y_p(x)$ by the method of undetermined coefficients, begin by listing the functions generated by repeated differentiation of the forcing function, beginning with the function itself:

$$f(x) = 3x^2 \to \{3x^2,\ 6x,\ 6,\ 0,\ 0,\ \ldots\}. \tag{4}$$

Observe that the listed functions are linear combinations of the three LI functions $\{x^2, x, 1\}$. Let us call those three functions the **LI family generated by** $3x^2$. Tentatively, seek a particular solution of (2) as a linear combination of those functions:

$$y_p(x) = Ax^2 + Bx + C, \tag{5}$$

The last term is really C times 1.

in which A, B, C are the "undetermined coefficients." They are by no means arbitrary; they simply have not yet been determined.

Next, check that none of the terms in (5) are solutions of the homogeneous equation, that is, they are not contained in (3). They are not, because surely x^2 cannot be expressed as a linear combination of e^{2x} and e^{-2x}, nor can x, nor can 1. [If that isn't obvious, you can proceed differently: Just substitute x^2 into the left-hand side of (2) and see that it does not give 0; thus, it is not a homogeneous solution. Similarly for x and 1.] Since there is no such redundancy, between any of the terms in (5) and those in the homogeneous solution, accept (5) and proceed.

For (5) to be a solution of (2), substitute it into (2) and try to choose A, B, C to make the result,

$$2A - 4(Ax^2 + Bx + C) = 3x^2, \tag{6}$$

an identity. We can do that by matching the coefficients of the LI terms x^2, x, and 1 on the left- and right-hand sides (Exercise 8 in Section 2.4). Thus, set

$$\begin{aligned} x^2: \quad & -4A = 3, \\ x: \quad & -4B = 0, \\ 1: \quad & 2A - 4C = 0, \end{aligned} \tag{7}$$

so $A = -3/4$, $B = 0$, $C = -3/8$. Hence, (5) becomes

It is readily verified, by substitution, that (8) satisfies (2).

$$y_p(x) = -\frac{3}{4}x^2 - \frac{3}{8}, \tag{8}$$

so a general solution of (2) is

$$\begin{aligned} y(x) &= y_h(x) + y_p(x) \\ &= C_1 e^{2x} + C_2 e^{-2x} - \frac{3}{4}x^2 - \frac{3}{8}. \end{aligned} \tag{9}$$

COMMENT. The guarantee that the method of undetermined coefficients works amounts to a guarantee that the system (7) of linear algebraic equations obtained for the unknown coefficients is *solvable* for those coefficients, that is, that the system is "consistent." For suppose we had not followed the undetermined coefficient procedure correctly — for instance by seeking $y_p(x)$ in the form $Ax^2 + Bx$ rather than in the correct form given by (5). Then, in place of (7) we would have obtained the system

$$\begin{aligned} x^2: \quad & -4A = 3, \\ x: \quad & -4B = 0, \\ 1: \quad & 2A = 0, \end{aligned} \tag{10}$$

which is inconsistent because A cannot equal $-3/4$ *and* 0. That is, $y_p(x) = Ax^2 + Bx$ would not have worked. ▮

For the method of undetermined coefficients to apply, the LI family generated by the forcing function $f(x)$ must contain only a *finite* number of members. For instance, if $f(x) = 1/x$, then repeated differentiation generates the list x^{-1}, $-x^{-2}$, $2x^{-3}$, and so on, without end. A linear combination of the LI functions $x^{-1}, x^{-2}, x^{-3}, \ldots$ would be an infinite series, with an infinite number of undetermined coefficients. To avoid that situation, *we restrict the method of undetermined coefficients to forcing functions that generate only a finite family of LI functions.*

EXAMPLE 2. Find a general solution of

$$y'' + y = 4xe^x. \tag{11}$$

First, the homogeneous solution is

$$y_h(x) = C_1 \sin x + C_2 \cos x. \tag{12}$$

Next, write the functions generated by repeated differentiation of the forcing function $4xe^x$:

$$f(x) = 4xe^x \rightarrow \{4xe^x,\ 4e^x + 4xe^x,\ 8e^x + 4xe^x,\ \dots\}. \tag{13}$$

Every function in the list can be expressed as a linear combination of the two LI functions e^x and xe^x, so we can take the LI family generated by $f(x) = 4xe^x$ to be those two functions, e^x and xe^x. Then, by the the method of undetermined coefficients, tentatively seek $y_p(x)$ in the form

$4xe^x$ generates the LI family e^x and xe^x.

$$y_p(x) = Ae^x + Bxe^x. \tag{14}$$

Next, check (14) for any redundancy with the terms in the homogeneous solution (12). There is none because neither the e^x nor the xe^x in (14) can be expressed as a linear combination of the $\sin x$ and $\cos x$ in (12). That is, neither e^x nor xe^x is a solution of the homogeneous equation, as is readily verified.

Next, put (14) into (11), obtaining

$$Ae^x + B(2e^x + xe^x) + Ae^x + Bxe^x = 4xe^x. \tag{15}$$

As in Example 1, make (15) an identity by matching the coefficients of the LI terms e^x and xe^x on the left- and right-hand sides. That is,

$$\begin{aligned} e^x:&\quad 2A + 2B = 0,\\ xe^x:&\quad 2B = 4, \end{aligned} \tag{16}$$

so $B = 2$ and $A = -2$. Thus,

$$y_p(x) = -2e^x + 2xe^x, \tag{17}$$

and a general solution of (11) is

$$\begin{aligned} y(x) &= y_h(x) + y_p(x)\\ &= C_1 \sin x + C_2 \cos x + 2(x-1)e^x. \quad ∎ \end{aligned} \tag{18}$$

EXAMPLE 3. Redundancy. Solve

$$y'' + y = 12 \sin x. \tag{19}$$

The homogeneous solution is

$$y_h(x) = C_1 \sin x + C_2 \cos x. \tag{20}$$

Next,

$$f(x) = 12 \sin x \rightarrow \{12 \sin x,\ 12 \cos x,\ -12 \sin x, \dots\}. \tag{21}$$

Every function in the list is a linear combination of the two LI functions $\sin x$ and $\cos x$, so tentatively take

$$y_p(x) = A \sin x + B \cos x. \tag{22}$$

Tentative form.

Unlike Examples 1 and 2, this time we *do* find redundancy between terms in (22) and terms in the homogeneous solution (20). Such a solution form is pointless because, if substituted into the left side of (19), it will inevitably produce zero, not the nonzero forcing function. In any case, the general solution of the homogeneous equation is already accounted for in (20).

According to the method of undetermined coefficients, we must therefore revise the tentative form (22) by multiplying the right-hand side by the lowest power of x needed so that none of the terms is a homogeneous solution. Let us begin by multiplying the right-hand side of (22) by x to the first power, so

Revised final form.

$$y_p(x) = x(A \sin x + B \cos x). \tag{23}$$

Now there is no redundancy because neither the $x \sin x$ nor the $x \cos x$ in (23) can be formed as a linear combination of the homogeneous solutions $\sin x$ and $\cos x$ in (20). In other words, neither $x \sin x$ nor $x \cos x$ is a homogeneous solution.

Thus, accept the revised form (23) and put it into (19). That step gives

$$2A \cos x - 2B \sin x = 12 \sin x, \tag{24}$$

and the linear independence of $\sin x$ and $\cos x$ implies that

$$\begin{aligned} \sin x: &\quad -2B = 12, \\ \cos x: &\quad 2A = 0. \end{aligned} \tag{25}$$

Thus, $A = 0$ and $B = -6$, so (23) gives $y_p(x) = -6x \cos x$, and we have

$$\begin{aligned} y(x) &= y_h(x) + y_p(x) \\ &= C_1 \sin x + C_2 \cos x - 6x \cos x \quad ▮ \end{aligned} \tag{26}$$

EXAMPLE 4. Redundancy Again. Find a general soluton of

$$y'''' - y'' = 24x^2. \tag{27}$$

If you skipped Section 2.7 and therefore don't know how to obtain (28), accept it as the general solution of the homogeneous equation and proceed.

The latter is of fourth-order, but the method is the same. First, the homogeneous solution is

$$y_h(x) = C_1 + C_2 x + C_3 e^x + C_4 e^{-x}. \tag{28}$$

The forcing function generates the sequence

$$24x^2 \to \{24x^2, 48x, 48, 0, 0, \dots\} \tag{29}$$

containing the LI family $\{x^2, x, 1\}$, so tentatively seek

Tentative form.

$$y_p(x) = Ax^2 + Bx + C. \tag{30}$$

Comparing (30) with (28), we see that the Bx and C terms are redundant with the $C_2 x$ and C_1 terms in (28), respectively. Thus, revise (30) by multiplying the whole expression, not just the redundant terms, by x:

$$y_p(x) = x(Ax^2 + Bx + C). \tag{31}$$ Revised form.

However, some redundancy remains, because the Cx term is redundant with the C_2x in (28). Thus, multiply by x again:

$$y_p(x) = x^2(Ax^2 + Bx + C). \tag{32}$$ Final form.

Now all redundancy is gone, so accept (32), substitute it in $y'''' - y'' = 24x^2$, and obtain

$$24A - (12Ax^2 + 6Bx + 2C) = 24x^2. \tag{33}$$

Matching coefficients,

$$\begin{aligned} x^2: &\quad -12A = 24, \\ x: &\quad -6B = 0, \\ 1: &\quad 24A - 2C = 0, \end{aligned} \tag{34}$$

so $A = -2$, $B = 0$, $C = -24$. Thus, $y_p(x) = -2x^4 - 24x^2$, so a general solution of (27) is

$$y(x) = C_1 + C_2x + C_3e^x + C_4e^{-x} - 2x^4 - 24x^2. \quad \blacksquare$$

In summary, here are the steps:

THE METHOD OF UNDETERMINED COEFFICIENTS FOR A LINEAR CONSTANT-COEFFICIENT EQUATION $Ly = f(x)$

1. Obtain a general solution $y_h(x)$ of the homogeneous equation $Ly = 0$.
2. Suppose, first, that the forcing function f is a single term. Verify that repeated differentiation of f produces only a finite number of LI functions.
3. Identify the LI family generated by f.
4. Seek $y_p(x)$, tentatively, as a linear combination of those LI functions.
5. See if there is redundancy between any of the terms in $y_p(x)$ and those in $y_h(x)$. If there is, multiply the entire tentative particular solution by the lowest positive integer power of x necessary to remove all such redundancy, as we did in altering the initial form (30) to the final form (32).
6. Substitute the final version of the form assumed for $y_p(x)$ into the left-hand side of $Ly = f$, and match coefficients of LI functions on both sides, as we did in (33) and (34).
7. Solve the resulting system of linear algebraic equations for the undetermined coefficients. That step completes the determination of $y_p(x)$.

8. If $f(x)$ is not one term but a sum of terms, $f_1(x) + \cdots + f_k(x)$, carry out steps 2–7 for each term, obtaining $y_{p1}(x), \ldots, y_{pk}(x)$. Then, by Theorem 2.8.2(b), $y_p(x) = y_{p1}(x) + \cdots + y_{pk}(x)$.

9. Finally, a general solution of $Ly = f(x)$ is $y(x) = y_h(x) + y_p(x)$.

2.9.2 A special case; the complex exponential method. If the forcing function happens to be a sine or a cosine, it is possible to simplify the method of undetermined coefficients by replacing the forcing function by a single complex exponential.

For instance, suppose we seek a particular solution $y_p(x)$ of a differential equation

$$Ly = F\cos\omega x, \tag{35}$$

in which L is a linear differential operator with constant coefficients, and F and ω are constants. By the method of undetermined coefficients, we can obtain $y_p(x)$ in the form $A\cos\omega x + B\sin\omega x$ [if $\cos\omega x$ and $\sin\omega x$ are not homogeneous solutions of (35)]. However, it is simpler to change the $\cos\omega x$ to $e^{i\omega x}$ and to solve the modified problem

$$Lv = Fe^{i\omega x} \tag{36}$$

for a particular solution $v_p(x)$ and then to obtain $y_p(x)$ as the real part of $v_p(x)$,

$$y_p(x) = \text{Re}\, v_p(x). \tag{37}$$

The reason (36) is simpler than (35) is that by the method of undetermined coefficients a particular solution $v_p(x)$ of (36) can be found in the form of a *single* term $Ae^{i\omega x}$, whereas a *two*-term form $A\cos\omega x + B\sin\omega x$ is needed for equation (35).

To verify our claim that (37) is a particular solution of (35), break all quantities in (36) into real and imaginary parts and write

$$L(\text{Re}\, v + i\,\text{Im}\, v) = F\cos\omega x + iF\sin\omega x,$$

Linearity again.

or, *because L is linear*,

$$L(\text{Re}\, v) + iL(\text{Im}\, v) = F\cos\omega x + iF\sin\omega x. \tag{38}$$

Then, equating real and imaginary parts on the left- and right-hand sides gives

$$L(\text{Re}\, v) = F\cos\omega x \tag{39a}$$

and

$$L(\text{Im}\, v) = F\sin\omega x. \tag{39b}$$

By comparing (39a) with (35) we see that $\text{Re}\, v$ is a particular solution of (35), so (37) follows, as we claimed.

If the right-hand side of (36) were $F \sin \omega x$ instead of $F \cos \omega x$, then we see from (39b) that in place of (37) we would find $y_p(x)$ from $y_p(x) = \text{Im}\, v(x)$.

EXAMPLE 5. An *RLC* Circuit. If the applied voltage is $E(t) = E_0 \sin \omega t$, then the current $i(t)$ in the electrical circuit shown in Fig. 1 is modeled by the differential equation

$$Li'' + Ri' + \frac{1}{C} i = \frac{dE(t)}{dt} = \omega E_0 \cos \omega t, \tag{40}$$

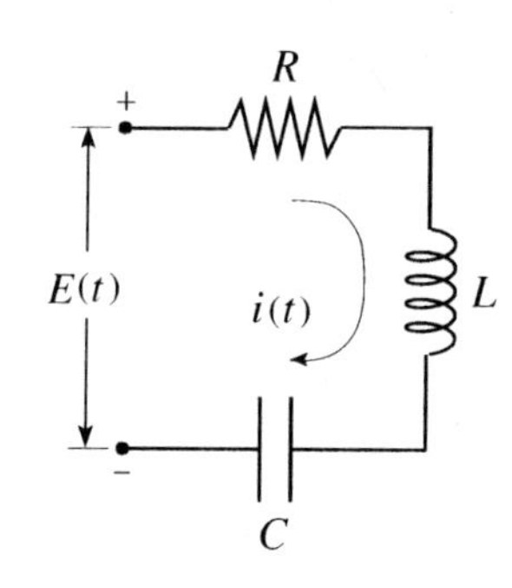

Figure 1. *RLC* circuit.

in which L, R, C, E_0, ω are constants, and the inductance L is not to be confused with the differential operator symbol L used in (35)–(39). In this case the independent and dependent variables are t and i in place of the generic x and y that we've been using.

According to the **complex exponential method**, to find a particular solution $i_p(t)$ of (40), consider instead

$$Lv'' + Rv' + \frac{1}{C} v = \omega E_0 e^{i\omega t} \tag{41}$$

and seek

$$v_p(t) = Ae^{i\omega t}. \tag{42}$$

Putting (42) into (41) gives

$$\left(- L\omega^2 + iR\omega + \frac{1}{C} \right) Ae^{i\omega t} = \omega E_0 e^{i\omega t}. \tag{43}$$

Matching coefficients of $e^{i\omega t}$ gives $A = \omega E_0 / \left(- L\omega^2 + iR\omega + \frac{1}{C} \right)$, so (42) becomes

$$v_p(t) = \frac{\omega E_0 C}{(1 - LC\omega^2) + iRC\omega} e^{i\omega t}. \tag{44}$$

Finally, $i_p(t)$ is the real part of $v_p(t)$:

$$\begin{aligned} i_p(t) &= \text{Re}\, v_p(t) \\ &= \omega E_0 C\, \text{Re} \left(\frac{\cos \omega t + i \sin \omega t}{(1 - LC\omega^2) + iRC\omega} \frac{(1 - LC\omega^2) - iRC\omega}{(1 - LC\omega^2) - iRC\omega} \right) \\ &= \frac{\omega E_0 C}{(1 - LC\omega^2)^2 + R^2C^2\omega^2} \left[(1 - LC\omega^2) \cos \omega t + RC\omega \sin \omega t \right]. \end{aligned} \tag{45}$$

Recall (Appendix D) that

$$\frac{1}{a+ib} = \frac{1}{a+ib} \frac{a-ib}{a-ib} = \frac{a}{a^2+b^2} - i\frac{b}{a^2+b^2}.$$

COMMENT. In this example our objective was only to find a particular solution, so we did not bother to write down a general solution $i_h(t)$ of the homogeneous equation. Therefore, we could not check (42) to see if it was redundant with $i_h(t)$. However, evidently there was no such redundancy because (42) *worked*, which is the "bottom line" in any case. ▌

Closure. The method of undetermined coefficients is used to find particular solutions of constant-coefficient equations. It requires that repeated differentiation of the forcing function generates only a finite number of LI functions. The key is in choosing the appropriate form to assume, and the steps are itemized above. A special case of the method occurs if the forcing function is a cosine (or sine). In that case it is simpler to change the cosine (sine) forcing function to a complex exponential, and to find a particular solution of the original equation by taking the real (imaginary) part of the particular solution of the modified equation.

A final comment about the method of undetermined coefficients: If the steps are not followed correctly, and the adopted solution form is incorrect, it will not lead to an "incorrect solution;" it simply won't work, as was illustrated in the Comment at the end of Example 1. Thus, if you obtain a system of algebraic equations (for the unknown coefficients) that is unsolvable, check your assumed solution form.

EXERCISES 2.9

1. In Example 1 the method of undetermined coefficients led us to the assumed form (5). Try, instead, the form given below, see if it works, or not, and comment on your results.

(a) $Ax^2 + Bx$ (b) $Ae^x + Be^{2x} + Ce^{3x}$
(c) $Ax^3 + Bx^2 + Cx + D$ (d) $Ax^2 + Bx + Ce^x$

2. Use the method of undetermined coefficients to find a particular solution of the given equation.

(a) $y' - 3y = 27x^3$ (b) $y' - y = 5e^x$
(c) $y' - 2y = 6x^2e^{2x}$ (d) $y' - 3y = 9x + 10\sin x$
(e) $y'' - 3y' + 2y = 8x^2 + 4x$ (f) $y'' + y' = 60\cos^2 x$
(g) $y'' - 9y = 18 + 216xe^{3x}$ (h) $y'' - y' = 3 - 4\sin x$
(i) $y'' - 2y' + y = 100e^x$ (j) $y'' + 2y' + y = 28 - 12x$
(k) $y'' - 2y' = 48(x - x^2)$ (l) $y'' - y' = 24x^3$
(m) $y''' + y'' = 24x^2$ (n) $y'''' - y'' = 40\cos x$
(o) $y'''' - 2y''' = 240x^2$ (p) $y'''' - y = 12e^x$
(q) $y'' + 4y' + 3y = 6 + 130\cos 2x$
(r) $y'' - 4y' + 3y = (8 - 4x)e^x$
(s) $y'' - 2y' + 10y = 60 + 26\sin 2x$
(t) $y''' - 2y'' - y' + 2y = 48e^{-2x}$

3. Complex Exponential Method. Use the complex exponential method to find a particular solution of the given equation.

(a) $y' + y = 4\cos x$ (b) $y' - 2y = 2\sin 5x$
(c) $y' + 3y = \cos x + 13\cos 2x$ (d) $y' - y = 10(\sin x + \sin 3x)$
(e) $y'' - 2y' + y = 100\cos x$ (f) $y'' + 3y' + 2y = -20\sin 2x$
(g) $y'' + 4y' + 3y = 60\sin 3x$ (h) $y'' + y' = 14\cos x$
(i) $y''' - 2y' + y = 5\cos 2x$ (j) $y'''' + y' + y = \sin 2x$

4. Example 5. (a) Show that the homogeneous solution of (40) tends to zero as $t \to \infty$, so (45) is actually the steady-state solution, a steady-state oscillation. NOTE: R, L, C are positive constants.
(b) Does that steady-state oscillation depend on the initial conditions $i(0)$ and $i'(0)$? Explain.

5. Particular Solutions for Cauchy–Euler Equations. The method of undetermined coefficients is for constant-coefficient equations, so we cannot expect it to work for Cauchy–Euler equations. It's true that the change of variables $x = e^t$ will convert any given Cauchy–Euler equation to an equation with constant coefficients, a particular solution of which can then be found by the method of undetermined coefficients (see Exercise 5 in Section 2.6). However, a simple idea to keep in mind is that if the Cauchy–Euler equation has a forcing function of the form ax^b, where a and b are constants, then a particular solution can readily be found in the form $y_p(x) = Ax^b$, unless x^b happens to be a homogeneous solution, in which case we can fall back on converting the equation to a constant-coefficient equation. Thus, obtain particular solutions of the following equations.

(a) $xy' + 4y = 24x^3$ (b) $xy' - 2y = 24x^6 + 12$
(c) $x^2y'' - 2xy' - 10y = 20x^3$ (d) $x^2y'' - 5xy' + 9y = 48x^2$
(e) $x^2y'' - 2xy' - 10y = 4x^5$ (f) $x^2y'' - xy' + y = 10x$

ADDITIONAL EXERCISES

6. Deflection of a Uniformly Loaded Circular Plate. Suppose a horizontal circular plate of radius R is subjected to a uniform constant load p lbs/in^2, which results in a downward deflection $w(r)$ that is a function of the radial variable r. In a course on structural mechanics one finds that the differential equation for $w(r)$ is the fourth-order equation

$$r^4 w'''' + 2r^3 w''' - r^2 w'' + rw' = \frac{p}{D} r^4. \tag{6.1}$$

The constant D is the "stiffness" of the plate, which depends on the plate's thickness and material, and is considered here as known.

(a) Obtain a general solution of (6.1). HINT: Read Exercise 5, which will tell you how to find a particular solution.
(b) If the plate is clamped at its outer edge so that its slope there is zero, then we have the boundary conditions $w(R) = 0$ and $w'(R) = 0$. But those are only two boundary conditions, whereas the equation is of fourth order. We state that the missing boundary conditions are these: $w'(0) = 0$, and $w''(0)$ must be bounded (i.e., finite). Applying these four conditions to your general solution, show that the deflection is

$$w(r) = \frac{p}{64D}\left(R^2 - r^2\right)^2, \tag{6.2}$$

and sketch its graph. Determine the maximum deflection, and the r at which it occurs. NOTE: It is interesting that the maximum deflection is found to be proportional to R^4. Thus, if we double the radius, we increase the maximum deflection by a factor of 16!

7. Sunken Treasure. A treasure chest containing closed-form solutions to many difficult differential equations sinks in water 300 ft deep. There is a current, with velocity $u(y)$ that varies linearly from 0 at the bottom to 15 ft/sec at the surface, as indicated in the sketch. If the fluid drag force on the chest is proportional to the velocity of the chest relative to the water, with constant of proportionality α, its motion $x(t), y(t)$ is modeled by the IVP

$$my'' = -mg - \alpha y'; \quad y(0) = 0,\ y'(0) = 0, \tag{7.1a}$$

$$mx'' = \alpha[u(y) - x']; \quad x(0) = 0,\ x'(0) = 0. \tag{7.1b}$$

If $m = 1$ slug, $\alpha = 1$ lb-sec/ft, and $g = 32$ ft/sec^2, determine the location of the chest on the sea floor, that is, its x location.

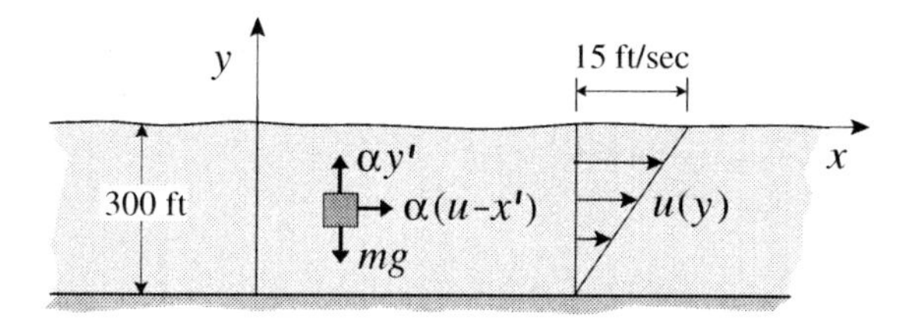

8. Particular Solution Using Operator Factorization Instead. Recall the method of operator factorization that we used, early in this chapter, to help us establish the form of the general solution of linear constant-coefficient homogeneous equations. We can use that method even if the equation is nonhomogeneous. As examples, use that method to find general solutions for each of the following.

(a) $y'' - 3y' + 2y = 24e^{5x}$
(b) $y'' - 4y' + 3y = 120\cosh 2x$
(c) $y'' - y = 4\cos x$
(d) $y'' + y' - 2y = 50e^{-x}$

2.10 PARTICULAR SOLUTION BY VARIATION OF PARAMETERS

The method of undetermined coefficients requires that the differential equation have constant coefficients and that repeated differentiation of the forcing function produce only a finite number of LI terms. In this section we present the more powerful method of **variation of parameters,**[1] which is free of those two restrictions.

2.10.1 First-order equations. This chapter is about equations of order two or greater, since the linear first-order equation

$$y' + p(x)y = q(x) \tag{1}$$

[1]The method of variation of parameters was developed for the first-order linear equation by John Bernoulli in 1697, and for higher-order linear equations by Lagrange in 1774.

was fully treated in Chapter 1. Nevertheless, to explain the method of variation of parameters let us begin by applying it to the first-order equation (1). Assume, as usual, that $p(x)$ and $q(x)$ are continuous on the x interval I under consideration.

First, write the *homogeneous* version of (1),

$$y' + p(x)y = 0, \tag{2}$$

and its general solution $y_h(x) = Ae^{-\int p(x)\,dx}$ or, using the notation in this chapter,

$$y_h(x) = C_1 y_1(x), \tag{3}$$

with $y_1(x) = e^{-\int p(x)\,dx}$.

To find a particular solution of (1) by the method of variation of parameters, let C_1 vary in (3) and seek y_p in the form

Hence the name "variation of parameters," for we let the parameter C_1 vary.

$$\boxed{y_p(x) = C_1(x)y_1(x).} \tag{4}$$

Substituting the latter into (1) gives the equation $C_1'y_1 + C_1y_1' + pC_1y_1 = q$ for $C_1(x)$, or

$$C_1'y_1 + C_1\underbrace{(y_1' + py_1)} = q. \tag{5}$$

The differential equation (5) for C_1 seems no simpler than the original equation (1), so it appears that dubious progress has been made. However, the key to the method is that the underbraced terms in (5) cancel to zero because y_1 is a homogeneous solution, a solution of (2)! Then, (5) becomes $C_1'y_1 = q$, so

$$C_1'(x) = \frac{q(x)}{y_1(x)} = q(x)e^{\int p\,dx}, \tag{6}$$

which can be solved by integration to give

$$C_1(x) = \int e^{\int p\,dx} q(x)\,dx + C. \tag{7}$$

Putting (7) into (4) gives

$$y_p(x) = e^{-\int p(x)\,dx}\left(\int e^{\int p(x)\,dx} q(x)\,dx + C\right). \tag{8}$$

We can set $C = 0$ in (7) and (8) since we don't need the most general $C_1(x)$; any $C_1(x)$ will do. If we do keep C, however, then the additional term $Ce^{-\int p\,dx}$ in (8) is actually the homogeneous solution, so (8) is then the general solution for $y(x)$, not just a particular solution $y_p(x)$.

It is striking that the assumed form (4) works, because the power of the method of assumed form is in assuming the right form, to narrow the search (for instance, to the set of exponential functions in the case of constant-coefficient homogeneous equations and powers of x for Cauchy–Euler equations). Yet, the assumed form (4) does not narrow the search at all, because *any* given function $F(x)$ can be expressed in the form $C_1(x)y_1(x)$ simply by choosing $C_1(x) = F(x)/y_1(x)$.

The reason the method works is that the undifferentiated C_1 terms in (5) (the underbraced terms) *must* cancel to zero because if C_1 were a constant [in which case the $C_1'y_1$ term in (5) would drop out], then $C_1y_1(x)$ would be a solution of the homogeneous equation; that is, its insertion into the left-hand side of (1) would necessarily give zero.

How can the assumed form (4) be useful since it does not narrow the search at all?

2.10.2 Second-order equationa. Now apply the method to the linear *second-order* equation

$$Ly = y'' + p_1(x)y' + p_2(x)y = f(x). \tag{9}$$

L need not be of constant-coefficient or Cauchy–Euler type. We ask only that $p_1(x)$ and $p_2(x)$, in L, be continuous on the interval of interest I.

We know from Theorem 2.7.2 that if $y_1(x)$ and $y_2(x)$ are any two LI solutions of the homogeneous version of (9), $Ly = 0$, then a general solution of the homogeneous equation is of the form

$$y_h(x) = C_1y_1(x) + C_2y_2(x). \tag{10}$$

As above, let the parameters C_1 and C_2 vary and seek a particular solution of $Ly = f$ in the form

$$\boxed{y_p(x) = C_1(x)y_1(x) + C_2(x)y_2(x).} \tag{11}$$

To substitute (11) into (9), we'll need y_p' and y_p'', so differentiate (11). First,

$$y_p' = C_1y_1' + C_2y_2' + \underbrace{C_1'y_1 + C_2'y_2}. \tag{12}$$

Looking ahead to one more differentiation, y_p'' will include $C_1, C_2, C_1', C_2', C_1'', C_2''$ terms, so (9) will become a nonhomogeneous second-order differential equation for $C_1(x)$ and $C_2(x)$, which can hardly be expected to be simpler than the original equation (9) for $y(x)$! However, since it will be only *one equation for the two unknowns* C_1, C_2, we should be able to impose another relation on C_1, C_2.

Remember, y_1 and y_2 are known; the unknowns are C_1 and C_2.

A particularly convenient one to impose is

$$C_1'y_1 + C_2'y_2 = 0, \tag{13}$$

because this will eliminate the underbraced terms in (12), so the next derivative y_p'' will contain only *first*-order derivatives of C_1 and C_2. By adopting (13), (12) reduces to $y_p' = C_1y_1' + C_2y_2'$, and a second differentiation gives

$$y_p'' = C_1y_1'' + C_2y_2'' + C_1'y_1' + C_2'y_2'. \tag{14}$$

Then, substitution of y_p' and y_p'' into (9) gives

$$C_1\underbrace{\left(y_1'' + p_1y_1' + p_2y_1\right)}_{0} + C_2\underbrace{\left(y_2'' + p_1y_2' + p_2y_2\right)}_{0} + C_1'y_1' + C_2'y_2' = f. \tag{15}$$

The cancelations in (15) are analogous to the cancelation of the underbraced terms in (5).

The two underbraced groups vanish because y_1 and y_2 are solutions of the homogeneous equation $Ly = 0$, so (15) simplifies to $C_1'y_1' + C_2'y_2' = f$. That result and (13) give the equations

$$\begin{aligned} y_1C_1' + y_2C_2' &= 0, \\ y_1'C_1' + y_2'C_2' &= f \end{aligned} \tag{16}$$

Equations (16) do not give C_1, C_2 directly, but they do give C_1', C_2', which can be integrated to give C_1, C_2.

for C_1', C_2'. We recognize the determinant of the coefficients as the Wronskian of y_1 and y_2,

$$W[y_1, y_2](x) = \begin{vmatrix} y_1(x) & y_2(x) \\ y_1'(x) & y_2'(x) \end{vmatrix}, \tag{17}$$

and the latter is necessarily nonzero on I because the solutions y_1 and y_2 of $Ly = 0$ are, by assumption, LI. Therefore, (16) admits a unique solution for C_1', C_2', which can be found algebraically, for instance by Cramer's rule (Appendix C), as

$$C_1'(x) = \frac{\begin{vmatrix} 0 & y_2 \\ f & y_2' \end{vmatrix}}{\begin{vmatrix} y_1 & y_2 \\ y_1' & y_2' \end{vmatrix}} \equiv \frac{W_1(x)}{W(x)}, \quad C_2'(x) = \frac{\begin{vmatrix} y_1 & 0 \\ y_1' & f \end{vmatrix}}{\begin{vmatrix} y_1 & y_2 \\ y_1' & y_2' \end{vmatrix}} \equiv \frac{W_2(x)}{W(x)}, \tag{18}$$

in which W_1, W_2 denote the determinants in the numerators: $W_1(x) = -f(x)y_2(x)$ and $W_2(x) = f(x)y_1(x)$. Finally, integrating (18), to obtain $C_1(x)$ and $C_2(x)$, and putting the results into (11) gives the particular solution

$W_1(x)$ is $-f(x)y_2(x)$,
$W_2(x)$ is $f(x)y_1(x)$,
$W(x)$ is given by (17).

$$\boxed{y_p(x) = \left[\int \frac{W_1(x)}{W(x)}\,dx\right] y_1(x) + \left[\int \frac{W_2(x)}{W(x)}\,dx\right] y_2(x),} \tag{19}$$

in which we take the two arbitrary integration constants to be zero, without loss, just as we set $C = 0$ in (8).

EXAMPLE 1. Use variation of parameters to obtain a particular solution of

$$y'' - 4y = 8e^{2x}. \qquad (-\infty < x < \infty) \tag{20}$$

We will do it two ways, first using the formula (19), and then using the "method."

Using the formula (19): To use (19) we need y_1 and y_2. Thus, solve the homogeneous version $y'' - 4y = 0$, obtaining

$$y_h(x) = C_1e^{2x} + C_2e^{-2x}. \tag{21}$$

From (21), identify $y_1(x) = e^{2x}$ and $y_2(x) = e^{-2x}$. Next,

You should be able to use both the method and also the formula (19).

$$W_1(x) = -f(x)y_2(x) = -(8e^{2x})e^{-2x} = -8, \tag{22a}$$
$$W_2(x) = f(x)y_1(x) = (8e^{2x})e^{2x} = 8e^{4x}, \tag{22b}$$
$$W(x) = y_1y_2' - y_1'y_2 = -4. \tag{22c}$$

Then (19) gives

$$y_p(x) = \Big(\int 2\,dx\Big)e^{2x} + \Big(\int -2e^{4x}\,dx\Big)e^{-2x}$$
$$= (2x)e^{2x} + \Big(-\frac{e^{4x}}{2}\Big)e^{-2x} = 2xe^{2x} - \frac{e^{2x}}{2}. \tag{23}$$

We can drop the $-e^{2x}/2$ term at the end of (23) because it is a homogeneous solution and can be absorbed by the C_1e^{2x} term in (21). Doing so, obtain

$$y_p(x) = 2xe^{2x}. \tag{24}$$

Using the method: A general solution of the homogeneous equation $y'' - 4y = 0$ is given by (21), so seek

$$y_p(x) = C_1(x)e^{2x} + C_2(x)e^{-2x}. \tag{25}$$

Then,

$$y_p'(x) = 2C_1e^{2x} - 2C_2e^{-2x} + \underbrace{C_1'e^{2x} + C_2'e^{-2x}}_{\text{set} = 0}, \tag{26a}$$

$$y_p''(x) = 2C_1'e^{2x} + 4C_1e^{2x} - 2C_2'e^{-2x} + 4C_2e^{-2x}. \tag{26b}$$

Putting these expressions for y_p, y_p', and y_p'' into (20) gives, after cancelations, $2e^{2x}C_1' - 2e^{-2x}C_2' = 8e^{2x}$. The latter, plus the equation obtained by setting the underbraced terms in (26a) equal to zero, give

$$e^{2x}C_1' + e^{-2x}C_2' = 0, \tag{27a}$$
$$2e^{2x}C_1' - 2e^{-2x}C_2' = 8e^{2x} \tag{27b}$$

for C_1', C_2'. Solving them by algebra, or Cramer's rule, gives $C_1' = (-8)/(-4) = 2$ and $C_2' = -2e^{4x}$, integration of which gives $C_1(x) = 2x$ and $C_2(x) = -e^{4x}/2$. Finally, putting these into (25) gives $y_p(x) = 2xe^{2x} - e^{2x}/2$, as in (23). ∎

EXAMPLE 2. Using (19), obtain a particular solution of

This equation has nonconstant coefficients, but is a Cauchy–Euler equation, so we will be able to find a general solution of the homogeneous equation.

$$x^2y'' - 2xy' + 2y = 3x^2 \qquad (0 < x < \infty) \tag{28}$$

using the formula (19). First,

$$y_h(x) = C_1x + C_2x^2. \tag{29}$$

From (29), identify y_1 and y_2 as x and x^2, respectively. Next,

$$W_1(x) = -f(x)y_2(x) = -3x^2, \tag{30a}$$
$$W_2(x) = f(x)y_1(x) = 3x, \tag{30b}$$
$$W(x) = y_1y_2' - y_1'y_2 = (x)(2x) - (1)(x^2) = x^2. \tag{30c}$$

We used $f(x) = 3$, rather than $f(x) = 3x^2$, in the expressions for W_1 and W_2, because *to identify $f(x)$ we must first put (28) into the form (9)* by dividing through by x^2.

Finally, put (30a,b,c) into (19) and, setting integration constants to zero, as above, obtain

$$\begin{aligned} y_p(x) &= \left(\int \frac{-3x^2}{x^2}\,dx\right)x + \left(\int \frac{3x}{x^2}\,dx\right)x^2 \\ &= (-3x)x + (3\ln x)x^2 \\ &= 3x^2(\ln x - 1). \end{aligned} \tag{31}$$

We wrote $\ln x$ rather than $\ln|x|$ in (31) because the interval I was stated in (28) to be $x > 0$.

If we also want a general solution, write

$$y(x) = y_h(x) + y_p(x) = C_1 x + C_2 x^2 + 3x^2(\ln x - 1), \tag{32}$$

in which C_1 and C_2 are arbitrary constants. The $-3x^2$ term at the end of (32) can be absorbed into the $C_2 x^2$ term, so we can write a general solution more compactly as

$$y(x) = y_h(x) + y_p(x) = C_1 x + C_2 x^2 + 3x^2 \ln x, \tag{33}$$

in which C_1 and C_2 are arbitrary constants. ▮

Closure. Of the two methods, undetermined coefficients and variation of parameters, the latter is the more powerful since it applies to (9) even if the p_j's are nonconstant, and even if the forcing function does not have a finite number of LI derivatives; we asked only that the p_j's and f be continuous on the interval. Realize, however, that you do need to be able to find two LI solutions of the homogeneous equation, which may not be feasible (in closed form) if (9) is not of constant-coefficient or Cauchy–Euler type. Further, there are two integrations involved, whereas in undetermined coefficients there are none. As a rule of thumb, we suggest using (a) undetermined coefficients if it applies and (b) variation of parameters if it does not.

Application of the method to equations of higher order is considered in the exercises.

EXERCISES 2.10

1. Obtain a general solution, using the method of variation of parameters to obtain a particular solution. If the equation is of constant-coefficient type, state whether the particular solution could have been obtained, instead, by the method of undetermined coefficients. The x interval is $-\infty < x < \infty$ unless specified otherwise.
(a) $y' + 2y = 4e^{2x}$
(b) $y' - y = 6xe^x + 5$
(c) $xy' - y = 6x^3 \quad (x > 0)$
(d) $x^3 y' + x^2 y = 1 \quad (x > 0)$
(e) $y' + (\tan x)y = 2\sec x \quad (-\pi/2 < x < \pi/2)$
(f) $y'' - 2y' + y = 6x^2$
(g) $y'' + y = \sec x \quad (-\pi/2 < x < \pi/2)$
(h) $y'' + y = \tan x \quad (-\pi/2 < x < \pi/2)$
(i) $y'' + 2y' - 3y = 80(e^x + 3)$
(j) $y'' - 2y' = 3\cosh x$
(k) $y'' - 2y' + y = 3 - 4e^x/x \quad (x > 0)$
(l) $y'' + 4y' + 4y = e^{-2x}/x^2 \quad (x > 0)$
(m) $y'' - 2y' + 2y = 4e^{2x}$
(n) $xy'' - 4y' = 36x^3 - 30x \quad (x > 0)$
(o) $x^2 y'' + xy' - 9y = 72 \quad (x > 0)$
(p) $x^2 y'' - 2y = 3x^2 \quad (x > 0)$

(q) $4x^2y'' + 2xy' - 2y = 18x \quad (x > 0)$
(r) $4x^2y'' - 3xy' + 3y = 2x \quad (x > 0)$

2. All equations in Exercise 1 were of constant-coefficient or Cauchy–Euler type, but in this exercise the equations are not of those types. Thus, we can expect them to be much more difficult, but we've designed them to have simple closed-form homogeneous solutions, which are given and which you need not derive. You may use them without derivation. Find a particular solution by variation of parameters, using software for integrations if you wish.

(a) $(\cos x)y'' + (\sin x)y' = \sin x;\ y_1(x) = 1,\ y_2(x) = \sin x$ $(-\pi/2 < x < \pi/2)$
(b) $(\sin x)y'' - (\cos x)y' = 4;\ y_1(x) = 1,\ y_2(x) = \cos x$ $(0 < x < \pi)$
(c) $(x-1)y'' - xy' + y = 2x - 2 - x^2;\ y_1(x) = x,\ y_2(x) = e^x$ $(x > 1)$
(d) $y'' + (\tan x)y' = 20\tan x;\ y_1(x) = 1,\ y_2(x) = \sin x$ $(-\pi/2 < x < \pi/2)$
(e) $y'' - (\cot x)y' = \cot x;\ y_1(x) = 1,\ y_2(x) = \cos x$ $(0 < x < \pi)$

3. Forced Harmonic Oscillator. Consider the displacement $x(t)$ of a body of mass m restrained by a spring of stiffness k, on a frictionless table, as in the figure. Newton's second

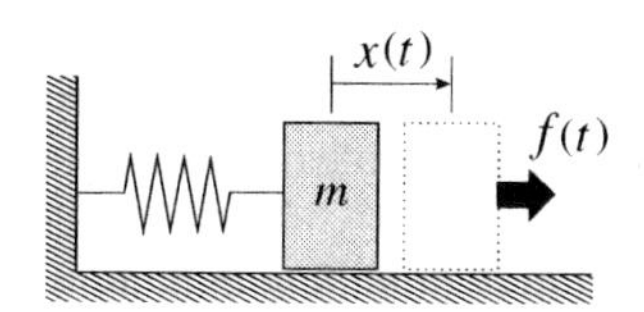

law of motion gives the differential equation

$$mx'' + kx = f(t) \tag{3.1}$$

for $x(t)$. We don't wish to limit ourselves to a specific forcing function, so we will simply call the force $f(t)$. It is convenient to divide (3.1) by m and to write

$$x'' + \omega^2 x = F(t), \tag{3.2}$$

with $\omega = \sqrt{k/m}$ and $F(t) = f(t)/m$. *Since $F(t)$ is not specified, we cannot use the method of undetermined coefficients to determine a particular solution, but we can use variation of parameters.*

(a) With the help of (19), derive the general solution

$$x(t) = C_1 \sin\omega t + C_2 \cos\omega t + \frac{\sin\omega t}{\omega}\int \cos\omega t\, F(t)\, dt - \frac{\cos\omega t}{\omega}\int \sin\omega t\, F(t)\, dt. \tag{3.3}$$

(b) Explain why the following *definite* integral form is equivalent to (3.3):

$$x(t) = C_1 \sin\omega t + C_2 \cos\omega t + \frac{\sin\omega t}{\omega}\int_a^t \cos\omega\tau\, F(\tau)\, d\tau - \frac{\cos\omega t}{\omega}\int_a^t \sin\omega\tau\, F(\tau)\, d\tau. \tag{3.4}$$

in which a is an arbitrary constant.
(c) Complete the solution if $x(t)$ is subjected to the initial conditions $x(0) = x_0$, $x'(0) = x_0'$. HINT: You should find that the form (3.4) is more convenient than (3.3), particularly if you choose a to be the initial t, namely, 0.
(d) Show that (3.4) can be written more compactly as

$$x(t) = C_1 \sin\omega t + C_2 \cos\omega t + \frac{1}{\omega}\int_a^t \sin\omega(t-\tau)\, F(\tau)\, d\tau. \tag{3.5}$$

4. Extension of the Method to Higher-Order Equations. Regarding the extension of the method of variation of parameters to higher-order equations, let it suffice to consider the third-order equation

$$y''' + p_1(x)y'' + p_2(x)y' + p_3(x)y = f(x), \tag{4.1}$$

with the $p_j(x)$'s and $f(x)$ continuous on the interval. Since the homogeneous solution of (4.1) is of the form

$$y_h(x) = C_1y_1(x) + C_2y_2(x) + C_3y_3(x), \tag{4.2}$$

seek $y_p(x)$ as

$$y_p(x) = C_1(x)y_1(x) + C_2(x)y_2(x) + C_3(x)y_3(x). \tag{4.3}$$

To put (4.3) into (4.1) you will need to differentiate (4.3) three times. Proceeding along the same lines as for second-order equations, set the sum of the $C_j'(x)$ terms in the expression for $y_p'(x)$ equal to zero. [This step is analogous to (13).] Then work out $y_p''(x)$ and again set the sum of the $C_j'(x)$ terms equal to zero. Those steps give two equations for the three unknowns C_1', C_2', C_3'. The third equation is obtained by putting your expressions for y_p, y_p', y_p'' and y_p''' into (4.1). Solving those three equations for C_1', C_2', C_3', by Cramer's rule, show that

$$C_1'(x) = \frac{W_1(x)}{W(x)},\quad C_2'(x) = \frac{W_2(x)}{W(x)},\quad C_3'(x) = \frac{W_3(x)}{W(x)}, \tag{4.4}$$

in which W is the Wronskian of y_1, y_2, y_3 and W_j is identical to W but with its column entries y_j, y_j', y_j'' replaced by $0, 0, f$, respectively. Finally, by integrating (4.4), show that

$$y_p(x) = \left[\int \frac{W_1(x)}{W(x)}\,dx\right] y_1(x) + \left[\int \frac{W_2(x)}{W(x)}\,dx\right] y_2(x) + \left[\int \frac{W_3(x)}{W(x)}\,dx\right] y_3(x). \tag{4.5}$$

NOTE: Comparing (4.5) with (19), it should not be hard to see the pattern that is developing for equations of higher order.

5. Application of (4.5). Use (4.5), above, to obtain a particular solution of the given differential equation. LI homogeneous solutions $y_1(x), y_2(x), y_3(x)$ are provided, for your convenience. If you wish, use computer software to evaluate the determinants and integrals.

(a) $y''' - 2y'' - y' + 2y = 24e^{3x} \quad (-\infty < x < \infty)$
$y_1(x) = e^x,\ y_2(x) = e^{2x},\ y_3(x) = e^{-x}$
(b) $y''' + 3y'' - y' - 3y = 36x - 108 \quad (-\infty < x < \infty)$
$y_1(x) = e^x,\ y_2(x) = e^{-x},\ y_3(x) = e^{-3x}$
(c) $x^3y''' - 3x^2y'' + 6xy' - 6y = 4x \quad (x > 0)$
$y_1(x) = x,\ y_2(x) = x^2,\ y_3(x) = x^3$
(d) $x^3y''' + x^2y'' - 2xy' + 2y = 16x^3 \quad (x > 0)$
$y_1(x) = x,\ y_2(x) = x^2,\ y_3(x) = 1/x$

CHAPTER 2 REVIEW

For first-order equations, in Chapter 1, we were able to solve the linear equation $y' + p(x)y = q(x)$ in integral form, and even nonlinear equations if they were separable or if they could be put into exact differential form. However, higher-order equations are substantially more difficult.

For a *linear* equation of order n we were able to establish the form of a general solution as

$$y(x) = C_1y_1(x) + \cdots + C_ny_n(x) + y_p(x), \tag{1}$$

in which the y_j's are n LI solutions of the homogeneous equation, and y_p is any particular solution. And, by the method of variation of parameters we were able to obtain explicit integral formulas for y_p in terms of the y_j's and the forcing function, although we did that only for second-order equations and only outlined the results for higher-order equations in the exercises.

But, in obtaining closed-form solutions for the y_j's, we were successful only for "elementary" equations, namely, those with constant coefficients and those of Cauchy–Euler type. Fortunately, those two are of great importance in applications. Thus, in most of this chapter we focused on those elementary equations. In Chapter 6 we will give infinite series methods that can generate solutions even for nonelementary linear equations.

Nonlinear equations were not considered in this chapter. That case is more difficult, in general, and requires special techniques. Chapter 7 will provide an introduction to such methods, particularly the phase plane, which is a two-dimensional version of the phase line that was met in Chapter 1.

The scope of the theory in Chapter 2 was greater than that in Chapter 1, so we deferred the applications to Chapter 3, so as not to obscure the forest by the trees. Only a handful are contained in this chapter, often among the ADDITIONAL EXERCISES.

Following is an outline of the sections.

Section 2.1. We began with the linearity property of the operator L,

$$L(\alpha u + \beta v) = \alpha L u + \beta L v, \tag{2}$$

which proved to be crucial all through this chapter. From linearity followed the *superposition principle*: If $y_1(x), \ldots, y_k(x)$ are solutions of the homogeneous linear equation $Ly = 0$, then an arbitrary linear combination of them, $C_1 y_1(x) + \cdots + C_k y_k(x)$, is a solution as well.

Sections 2.2 – 2.3. Constant-Coefficient Equations. If the simplest, "bottom line" first-order equation is $y' + py = 0$, with p a constant, the analogous simplest equation of second order is the constant-coefficient homogeneous equation $y'' + p_1 y' + p_2 y = 0$, and that is where we began in Section 2.2. We quickly derived a general solution of the latter by an operator factorization method and studied the various solution forms that arise, including exponentials and hyperbolic and circular functions. To evaluate the complex exponentials that arise when the characteristic equation $r^2 + p_1 r + p_2 = 0$ has complex roots, in Section 2.3, we used Euler's formula $e^{\pm ix} = \cos x \pm i \sin x$.

Section 2.4. To be able to handle nonconstant-coefficient equations we needed to establish the concepts of linear dependence and linear independence of sets of functions, particularly sets of solutions of a given linear, homogeneous differential equation, and we did that in Section 2.4, although still limiting discussion to equations of second order. Theorem 2.4.3 established the form of a general solution of

$$y'' + p_1(x) y' + p_2(x) y = 0 \tag{3}$$

as $y(x) = C_1 y_1(x) + C_2 y_2(x)$, in which y_1 and y_2 are two LI solutions of (3). It is here that the need to understand linear dependence and independence first entered. Theorem 2.4.4(c) gave a convenient Wronskian test for linear independence of a given set of two solutions of (3).

Section 2.5. Sometimes we are able to find a solution $y_1(x)$ of a given homogeneous equation $y'' + p_1(x) y' + p_2(x) y = 0$, but not a second one. To find a second solution, by reduction of order, we can seek it in the form $C(x) y_1(x)$, which amounts to letting the parameter C_1 vary in $y(x) = C_1 y_1(x)$. Putting that form into the differential equation necessarily gives a reduction of order to a simple first-order equation for $C(x)$. The result is the formula (11) for the second LI solution $y_2(x)$. We recommend that you use the method, rather than the formula.

Section 2.6. For constant-coefficient equations the appropriate assumed solution form is the exponential e^{rx}, and for the Cauchy–Euler equation

$$x^2 y'' + a_1 x y' + a_2 y = 0 \tag{4}$$

it is x^r. Using that form, and theorems from Section 2.4, we derived a general solution of (4). In a sense, (4) is a constant-coefficient equation "in disguise," since the substitution

$$x = e^t \tag{5}$$

converts (4) to a constant-coefficient equation. That point was discussed in the exercises.

Section 2.7. The general theory for the nth-order linear equation

$$y^{(n)} + p_1(x)y^{(n-1)} + \cdots + p_{n-1}(x)y' + p_n y = 0. \tag{6}$$

amounted to a natural extension of the Theorems 2.4.2 – 2.4.4 that covered the second-order equation (3). In addition to giving those theorems, we discussed their application to constant-coefficient and Cauchy–Euler equations of any order n.

Section 2.8. In this section we turned to the *nonhomogeneous* linear equation $Ly = f$ and showed that a general solution is of the form of a general solution of the homogeneous equation plus a particular solution of the full equation $Ly = f$. At that point we already knew how to find general solutions of the homogeneous equation, at least for elementary equations, so we devoted Sections 2.9 and 2.10 to two methods for finding particular solutions.

Sections 2.9–2.10. Undetermined coefficients, in Section 2.9, is an algorithm for finding particular solutions for constant-coefficient equations for which the forcing function has only a finite number of LI derivatives. The key is to assume a form that will work, and the method guides us to such a form. Variation of parameters, in Section 2.10, is more powerful in that it is free of the two restrictions that limit the method of undetermined coefficients, but is more tedious to apply, especially as the order n is increased, and involves integrations, whereas undetermined coefficients does not.

Chapter 3

Applications of Higher-Order Equations

3.1 INTRODUCTION

In this chapter we study representative applications of higher-order equations, most involving equations of second order. Besides problems of initial value type, we will include boundary value problems and their applications.

Initial value problems (Sections 3.2–3.4). The initial value problems in these sections will be for "harmonic oscillators," such as a mass restrained by a spring. Besides the obvious importance of oscillators to engineers, in connection with vibrations and controls, electrical oscillators are important in circuit theory, and biological oscillators are prominent as well. The latter involve phenomena such as heart beat and respiration, the periodic firing of the nerve cell, and the periodic variation in the populations of competing species.

Boundary value problems (Sections 3.5–3.6). The problems of boundary value type will be for the steady-state diffusion of heat and mass, and for the buckling of columns. The diffusion problem will lead to constant-coefficient equations in Cartesian geometries and to Cauchy–Euler equations in cylindrical and spherical geometries. The buckling problem will introduce a class of problems known as *eigenvalue problems*, a matrix version of which will be studied in Chapters 4 and 7.

3.2 LINEAR HARMONIC OSCILLATOR; FREE OSCILLATION

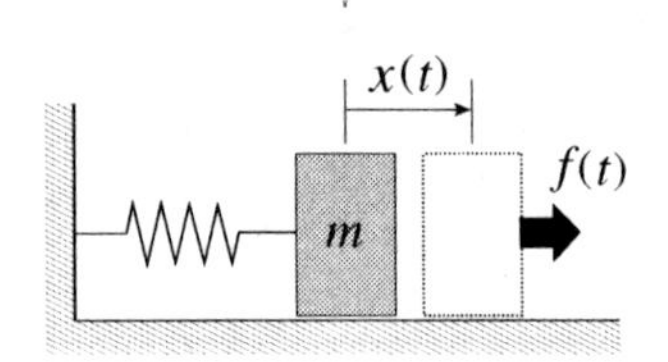

Figure 1. Mechanical oscillator.

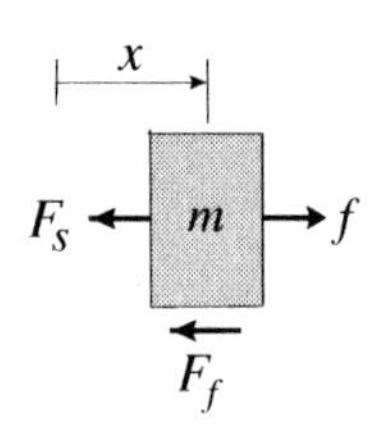

Figure 2. The horizontal forces on the mass if $x > 0$ and $x' > 0$.

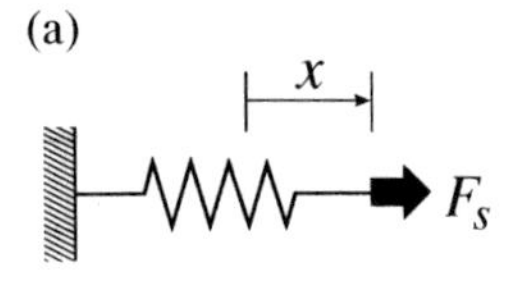

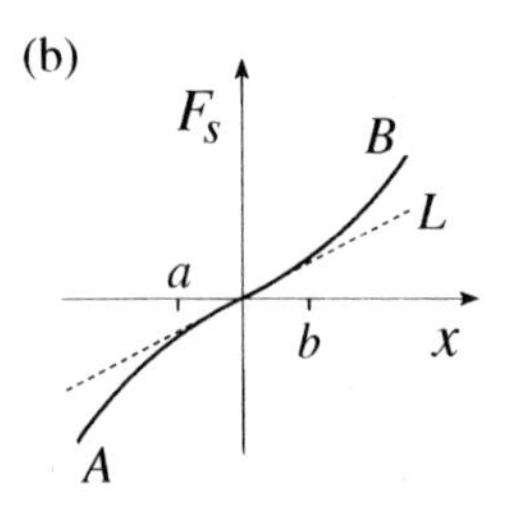

Figure 3. Force–displacement graph for spring, the curve AB.

3.2.1 Mass–spring oscillator. A block of mass m on a tabletop is restrained laterally by an ordinary coil spring, as in Fig. 1. The displacement x of the mass from its equilibrium position (in which the spring is neither stretched nor compressed) is measured as positive to the right. The mass is subjected to a force $f(t)$, where t is the time, and we want the equation governing the displacement $x(t)$.

The relevant physical principle is Newton's second law. Neglecting air resistance, there is the applied force $f(t)$, a force F_s exerted on the mass by the spring, and a horizontal force F_f exerted on the bottom of the mass due to friction. If, for definiteness, both $x(t) > 0$ and $x'(t) > 0$ at the instant under consideration, then F_s and F_f are to the left, as in Fig. 2, and Newton's second law gives

$$mx'' = f - F_s - F_f. \tag{1}$$

Consider F_s. An experimental plot of F_s versus x for a typical coil spring is like the curve AB in Fig. 3b, steepening toward A and B as the coils become compressed and extended, respectively. Such a nonlinear function $F_s(x)$ would make (1) nonlinear, so we will suppose the displacement is small enough, for instance between a and b in the figure, for $F_s(x)$ to be well approximated by the tangent-line L at $x = 0$, with slope k:

$$F_s(x) \approx kx; \tag{2}$$

k is called the spring *stiffness*; the larger k is, the stiffer the spring.[1]

For F_f, suppose the tabletop is lubricated, so the mass rides on a thin film of lubricant. Then F_f will be proportional to the velocity $x'(t)$, so

$$F_f = cx'(t). \tag{3}$$

We consider c to be a known positive constant and call it the *damping coefficient* because, as we will see, the effect of the cx' term in the differential equation will be to cause the motion to "damp out," that is, to die out as $t \to \infty$.

[1]The approximation used for the spring force, $F_s \approx kx$, is well known as **Hooke's law**, after *Robert Hooke* (1635–1703). Hooke published his law of elastic behavior in 1676 as the anagram (the letters scrambled, and in Latin!) *ceiiinosssttuv* and, two years later, the solution *ut tensio sic vis*, meaning "as the force, so is the displacement"; the force is proportional to the displacement. Today, Hooke's law must seem modest indeed, but we should appreciate it within its historical context. In spirit, it followed *Galileo Galilei* (1564–1642) who, in breaking with lines established by the ancient Greeks, established a quantitative science, expressed in formulas and mathematical terms. For example, where Aristotle explained the increasing speed of a falling body as the body moving more and more jubilantly as it approached its natural place (the center of the earth, which was believed to be the center of the universe), Galileo would sidestep the question of cause entirely and instead simply give the formula $v = 9.8t$, where v is the speed (meters per second) and t is the time (seconds). Likewise, Newton's law of gravitation simply gives a formula for the gravitational force of attraction betwen two bodies; it does not attempt explanation. Such departure from the less productive ancient Greek tradition marked the beginning of modern science.

Using (2) and (3), the differential equation (1) becomes

$$\boxed{m\frac{d^2x}{dt^2} + c\frac{dx}{dt} + kx = f(t)} \tag{4}$$

on $0 \le t < \infty$, with m, c, k and the applied force $f(t)$ considered as known.

Equivalently, suppose the tabletop is frictionless and that a *dashpot* is inserted, as in Fig. 4, a piston–cylinder device with fluid such as oil inside the cylinder.[1] As the dashpot is extended or compressed, fluid is forced in or out through small ports. The resistance of the dashpot to that extension or compression is proportional to the velocity x', so the system in Fig. 4 is also modeled by (4).

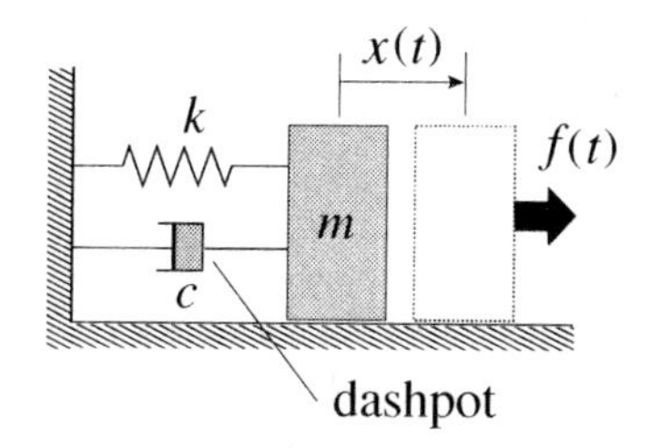

Figure 4. Spring and dashpot; Kelvin–Voigt model. Frictionless tabletop.

In fact, many important materials such as muscle, bone, and polymers respond to forces as though they were spring–dashpot combinations, as in Fig. 4. In such cases the spring–dashpot combination is not literal, but serves as a model of the material behavior and is known as a **Kelvin–Voigt** model.

Besides (4), we prescribe initial conditions

$$x(0) = x_0 \quad \text{and} \quad x'(0) = x_0', \tag{5}$$

and we have an IVP consisting of (4) and (5).

COMMENT 1. To be clear about the directions of F_s and F_f in Fig. 2, we assumed, for definiteness, that $x > 0$ and $x' > 0$ at the instant under consideration. However, (4) holds even if $x < 0$ and/or $x' < 0$. For instance, if $x < 0$, the spring is compressed and the force F_s in Fig. 2 should now be to the right, but its magnitude is now $(k)(-x)$ instead of kx. Thus, there are two sign changes, so the $+kx$ term in (4) is still correct. Similarly for F_f.

COMMENT 2. In Section 1.3.5 we used Kirchhoff's voltage law to derive

$$L\frac{d^2i}{dt^2} + R\frac{di}{dt} + \frac{1}{C}i = \frac{de(t)}{dt} \tag{6}$$

for the current $i(t)$ in the RLC circuit reproduced here in Fig. 5. Observe that equation (6) governing the electrical oscillator in Fig. 5 is *identical* to equation (4) governing the mechanical oscillator in Fig. 4, with these correspondences:

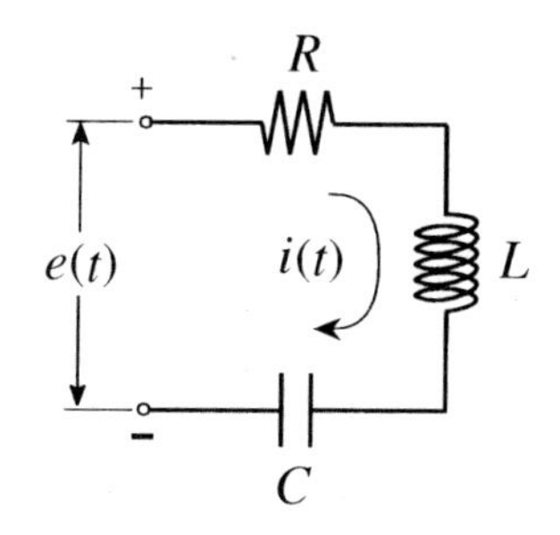

Figure 5. Electric oscillator; RLC circuit. This circuit is an electrical analog of the mechanical system in Fig. 4.

$$L \leftrightarrow m, \quad R \leftrightarrow c, \quad \frac{1}{C} \leftrightarrow k, \quad i(t) \leftrightarrow x(t), \quad \frac{de(t)}{dt} \leftrightarrow f(t). \tag{7}$$

Thus, these mechanical and electrical systems are **analogs** of each other. Since their governing equations are the same, to within name changes, whatever we learn about mechanical oscillators here applies to electrical oscillators as well. This situation, the same differential equation governing diverse physical systems, is more the rule than the exception.

[1]A dashpot is an energy absorbing device. For instance, when a bowling ball is returned to the bowler, it is brought to a stop when it depresses a lever that connects to a dashpot.

3.2.2 Undamped free oscillation. We begin with $f(t) = 0$ in (4), which corresponds to the **unforced** or **free oscillation**. Consider first the **undamped** case, so $c = 0$ as well. That is, the damping is small enough to be neglected.

The IVP for $x(t)$ is

$$mx'' + kx = 0, \quad (0 \le t < \infty) \tag{8a}$$

$$x(0) = x_0, \quad x'(0) = x_0'. \tag{8b}$$

To solve (8a), seek $x(t) = e^{rt}$ and obtain the characteristic equation $mr^2 + k = 0$, with roots $r = \pm i\sqrt{k/m}$. These give the general solution

COMPLEX-EXPONENTIAL FORM. In Chapter 2 we used subscripted C's for the arbitrary constants in general solutions. Here, unsubscripted constants seem simpler.

$$x(t) = Ae^{i\omega_n t} + Be^{-i\omega_n t} \tag{9}$$

of (8a), in which A and B are arbitrary constants and

$$\boxed{\omega_n \equiv \sqrt{k/m}.} \tag{10}$$

Equivalent to (9), we can express $x(t)$ in terms of real-valued functions as

COSINE-AND-SINE FORM.

$$x(t) = C\cos\omega_n t + D\sin\omega_n t. \tag{11}$$

The frequency of the motion is ω_n radians per unit time, such as rad/sec. It is called the **natural frequency** because it is the frequency of the free oscillation, in the absence of any forcing function $f(t)$ and due only to the disturbance introduced by nonzero initial conditions. To convert the frequency to cycles/sec, write

Hz is the symbol for hertz, cycles per second.

$$\text{frequency in Hz} = \left(\omega_n \frac{\text{rad}}{\text{sec}}\right)\left(\frac{1\text{ cycle}}{2\pi\text{ rad}}\right) = \frac{\omega_n}{2\pi}\frac{\text{cycles}}{\text{sec}} = \frac{\omega_n}{2\pi}\text{ Hz},$$

the inverse of which is the **period**, T sec/cycle,

For instance, if ω_n is in rad/sec, then T is in sec.

$$\boxed{T = \frac{2\pi}{\omega_n}.} \tag{12}$$

Here, unless specified otherwise, frequency will be in radians per unit time.

Besides (9) and (11), there is another form of the general solution that is used,[1]

SINE-WITH-PHASESHIFT FORM.

$$x(t) = E\sin(\omega_n t + \phi). \tag{13}$$

To show the equivalence of (11) and (13), apply $\sin(A + B) = \sin A\cos B + \sin B\cos A$ to (13), so $E\sin(\omega_n t + \phi) = E\sin\phi\cos\omega_n t + E\cos\phi\sin\omega_n t$. Then,

[1]Of course, in place of (13) we could use $x(t) = E\cos(\omega_n t + \phi)$, because the sine and cosine functions are identical, to within a phase shift.

to make $C\cos\omega_n t + D\sin\omega_n t = E\sin\phi\cos\omega_n t + E\cos\phi\sin\omega_n t$ an identity, match the coefficients of the $\cos\omega_n t$ terms, and of the $\sin\omega_n t$ terms, so

$$C = E\sin\phi \text{ and } D = E\cos\phi. \tag{14a,b}$$

To solve (14a,b) for E and ϕ in terms of C and D, square and add those equations and use $\cos^2\phi + \sin^2\phi = 1$. Also, divide (14a) by (14b), obtaining

$$\boxed{E = \sqrt{C^2 + D^2} \quad \text{and} \quad \phi = \arctan\frac{C}{D}.} \tag{15a,b}$$

We will take the positive square root in (15a).

In summary, three equivalent forms for the general solution of $mx'' + kx = 0$, or $x'' + \omega_n^2 x = 0$ where $\omega_n = \sqrt{k/m}$, are these:

$$x(t) = \begin{cases} Ae^{i\omega_n t} + Be^{-i\omega_n t}, \\ C\cos\omega_n t + D\sin\omega_n t, \\ E\sin(\omega_n t + \phi). \end{cases} \tag{16a,b,c}$$

Equivalent solution forms.

EXAMPLE 1. Illustrating (16). With $m = 3$, $k = 12$, $x(0) = 5$, and $x'(0) = -6$, solve for $x(t)$ in each of the three forms given above. First, use (16a): $\omega_n = \sqrt{k/m} = 2$, so a general solution of (8) is

$$x(t) = Ae^{i2t} + Be^{-i2t}.$$

Then, $x(0) = 5 = A + B$ and $x'(0) = -6 = i2A - i2B$ give $A = 5/2 + 3i/2$ and $B = 5/2 - 3i/2$, so

$$x(t) = \left(\frac{5}{2} + \frac{3i}{2}\right)e^{i2t} + \left(\frac{5}{2} - \frac{3i}{2}\right)e^{-i2t}. \tag{17}$$

In the form (16a).

Next, use (16b): $\omega_n = 2$, so

$$x(t) = C\cos 2t + D\sin 2t.$$

Then $x(0) = 5 = C$ and $x'(0) = -6 = 2D$ give $C = 5$ and $D = -3$, so

$$x(t) = 5\cos 2t - 3\sin 2t. \tag{18}$$

In the form (16b).

Finally, use (16c): $\omega_n = 2$, so

$$x(t) = E\sin(2t + \phi). \tag{19}$$

Having found C and D, above, we could obtain E and ϕ from (15), but instead we can apply the initial conditions to (19): $x(0) = 5 = E\sin\phi$ and $x'(0) = -6 = 2E\cos\phi$. Squaring and adding $E\sin\phi = 5$ and $E\cos\phi = -3$ gives $E^2 = 34$ so $E = \sqrt{34}$; and dividing $E \sin\phi = 5$ by $E\cos\phi = -3$ gives $\tan\phi = \dfrac{5}{-3}$ so $\phi = \arctan\left(\dfrac{5}{-3}\right)$. However, a calculator or computer does not distinguish between $(+5)/(-3)$ and $(-5)/(+3)$. Assuming it is programmed to give arctangent values in the interval $-\pi/2 < \phi < \pi/2$ (shaded in Fig. 6) it misinterprets $\arctan\left(\dfrac{5}{-3}\right)$, which is in the second quadrant, as $\arctan\left(\dfrac{-5}{3}\right)$ which is in the fourth quadrant. Thus, we must add π (i.e., a counterclockwise rotation through 180°) to obtain ϕ:

$$\phi = \arctan\left(\frac{5}{-3}\right) = \arctan\left(\frac{-5}{3}\right) + \pi = -1.0304 + \pi = 2.111 \text{ rad}, \tag{20}$$

so

In the form (16c).

$$x(t) = \sqrt{34}\,\sin(2t + 2.111). \tag{21}$$

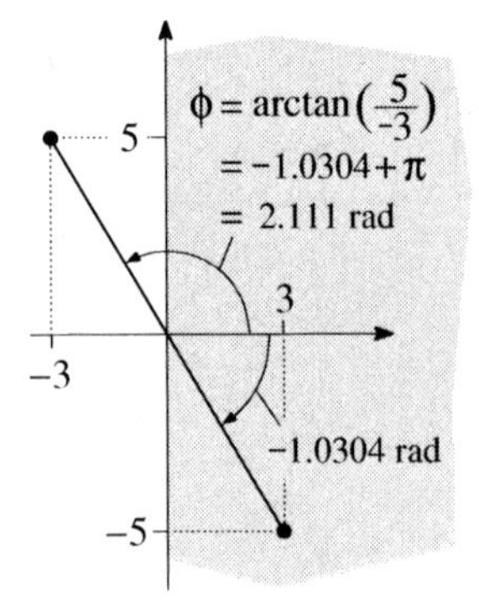

Figure 6. Choosing the correct quadrant for the arctangent.

COMMENT 1. The forms (17), (18), and (21) look different but are the same function. If we imagine adding the graphs of the $5\cos 2t$ and the $-3\sin 2t$, in (18), it is hardly obvious that the result will be a single pure harmonic function. Yet it is, as guaranteed by the equivalence of (16b) and (16c), and as we see in Fig. 7.

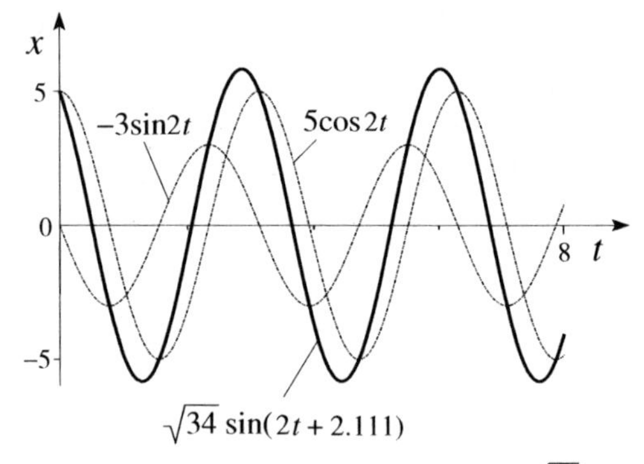

Figure 7. The equivalence of $5\cos 2t - 3\sin 2t$ and $\sqrt{34}\sin(2t + 2.111)$.

COMMENT 2. We've seen that we must be careful in evaluating the arctangent. In general, suppose we wish to evaluate $\arctan(C/D)$. If $D > 0$, then the angle is in the first or fourth quadrants so the computer gives the correct result; but if $D < 0$, then the angle is in the second or third quadrants and we must add π to the computer result. ∎

An advantage of (16c) over (16b) is that whereas C and D in (16b) have no obvious physical significance, E, ω_n, and ϕ in (16c) are these important quantities: the **amplitude**, **frequency**, and **phase angle** of the vibration, respectively, as depicted in Fig. 8.

The sine and cosine are called "harmonic functions," so the motion described by (17) is **harmonic motion**, and the corresponding physical system is a **harmonic oscillator**.

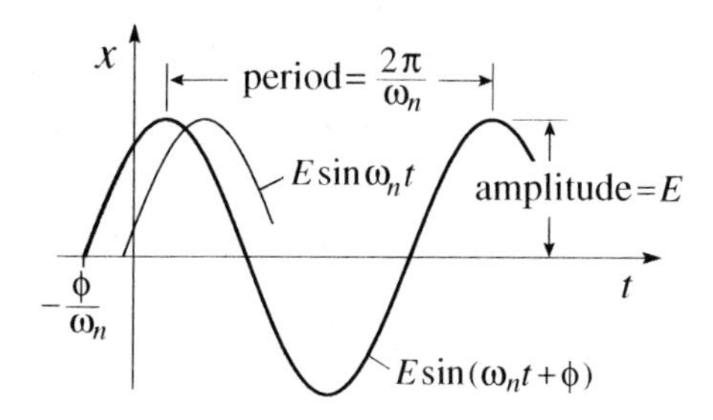

Figure 8. Graphical significance of E, ω_n, ϕ in (16c).

A final comment: The ϕ in $x(t) = E \sin(\omega_n t + \phi)$, like the $\omega_n t$, is in radians. Sometimes the form

$$x(t) = E \sin \omega_n (t - \delta) \tag{22}$$

is used instead. Whereas the phase ϕ is in radians, we see from (22) that δ has units of time. It involves a **time lag**, the time that $\sin \omega_n t$ lags behind $\sin \omega_n (t - \delta)$, as indicated in Fig. 9. Which form to choose, (16c) or (22), depends upon the application. For instance, in the cooling-of-a-house exercise in Section 1.3 it seemed best to emphasize the time lag aspect and to use the form $\sin \omega(t - \delta)$.

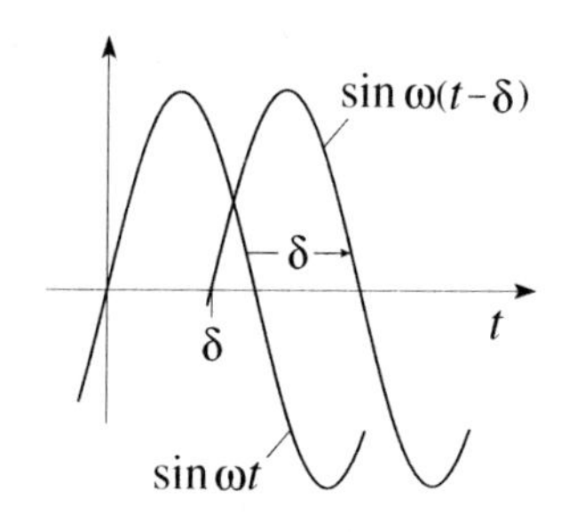

Figure 9. The time lag δ in (22).

3.2.3 Pendulum. Consider the pendulum in Fig. 10. If it is disturbed from its equilibrium position $\theta = 0$, it will undergo an angular oscillation. Thus, it too is an oscillator. We can derive the equation of motion for $\theta(t)$ by recalling from physics that, for angular motion about a point, "$T = I\alpha$:" the torque T or moment of the applied force(s) about the pivot point equals the inertia I of the body about that point times the angular acceleration α.[1] We assume the mass of the string or rod of length l is negligible compared to the mass m of the bob, and that the bob is small enough compared to l to be considered a point mass. Then the inertia of the bob about the pivot point is $I = ml^2$. Further, the angular acceleration (positive counterclockwise) is $d^2\theta/dt^2$ and the counterclockwise torque about the pivot, due to the gravitational force mg, is seen from Fig. 10 to be $-(mg)(l \sin \theta)$. We will neglect friction at the pivot, as well as air resistance, so the only moment about the pivot is due to the weight force mg. Then $T = I\alpha$ becomes

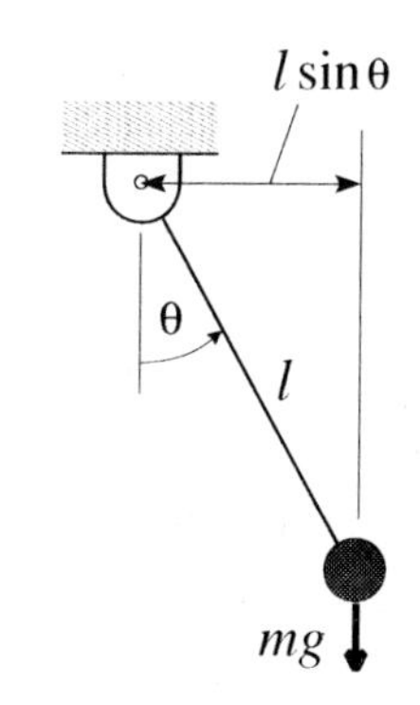

Figure 10. Pendulum.

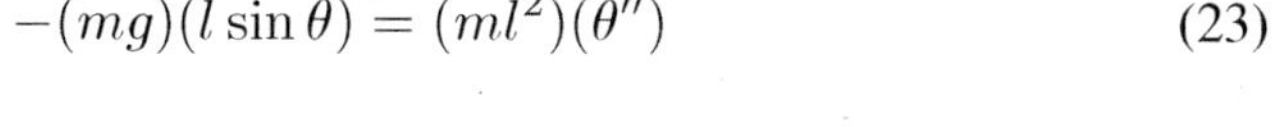

$$-(mg)(l \sin \theta) = (ml^2)(\theta'') \tag{23}$$

or

$$\boxed{\frac{d^2\theta}{dt^2} + \frac{g}{l} \sin \theta = 0.} \tag{24}$$

Noninear pendulum equation.

The latter is *nonlinear* because of the $\sin \theta$ term. However, we know from the

[1]The moment of a force about a point P is the force times the perpendicular distance from P to the line of action of the force. The formula $T = I\alpha$ is not *in addition to* Newton's second law $f = ma$, it can be derived *from* that law.

calculus that

$$\sin\theta = \theta - \frac{1}{3!}\theta^3 + \frac{1}{5!}\theta^5 - \frac{1}{7!}\theta^7 + \cdots \sim \theta \tag{25}$$

as $\theta \to 0$. Thus, if the motion is "small," as it is for the pendulum of a grandfather's clock for instance, then $\sin\theta \approx \theta$, so we can approximate (24) by the *linear* equation

Linearized pendulum equation.

$$\boxed{\theta'' + \frac{g}{l}\theta = 0,} \tag{26}$$

with general solution

$$\theta(t) = C\cos\omega_n t + D\sin\omega_n t, \tag{27}$$

where the natural frequency in this case is

$$\omega_n = \sqrt{\frac{g}{l}}. \tag{28}$$

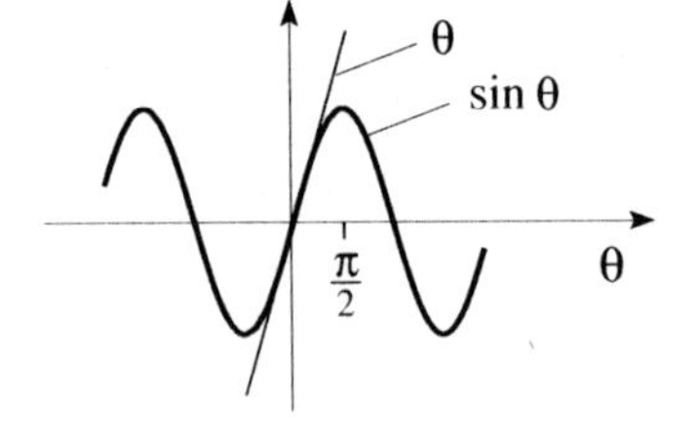

Figure 12. Graphical tangent-line significance of the linearization $\sin\theta \approx \theta$.

In graphical terms, the approximation $\sin\theta \approx \theta$ amounts to replacing the graph of $\sin\theta$ by its tangent line at $\theta = 0$, as in Fig. 12. If $|\theta|$ is less than around $20°$, for instance, the error in the approximation $\sin\theta \approx \theta$ is less than 2%.

Closure. We used the approximate equation of motion $mx'' + cx' + kx = f(t)$ for the motion $x(t)$ of the forced, damped mechanical oscillator in Fig. 4. We limited subsequent discussion to the unforced and undamped case where $f(t) = 0$ and $c = 0$. In that case the governing equation of the oscillator is $x'' + \omega_n^2 x = 0$. That is, the square root of the coefficient of x is the natural frequency, in radians per unit time. For instance, $mx'' + kx = 0$ gives $\omega_n = \sqrt{k/m}$ and $\theta'' + (g/l)\theta = 0$ gives $\omega_n = \sqrt{g/l}$.

The solution of the linear harmonic oscillator equation is familiar from Chapter 2. The only new mathematical point in this section was the three different but equivalent forms of the general solution — using complex exponentials, a cosine and sine, and a sine with phase angle. The cosine-and-sine form (16b) is convenient if initial conditions are prescribed at $t = 0$, and the sine-with-phase-angle form (16c) is attactive because E, ω_n, and ϕ have clear physical significances as the amplitude, frequency, and phase, respectively.

EXERCISES 3.2

1. Given the values of m, k, and the initial conditions, solve (8) for $x(t)$. Obtain the solution in each of the three forms given in (16).
(a) $m = 1,\ k = 1,\ x_0 = 4,\ x_0' = 3$
(b) $m = 1,\ k = 1,\ x_0 = -4,\ x_0' = -3$
(c) $m = 1,\ k = 4,\ x_0 = -4,\ x_0' = 2$
(d) $m = 2,\ k = 4,\ x_0 = -1,\ x_0' = 2$
(e) $m = 2,\ k = 4,\ x_0 = -4,\ x_0' = 5$
(f) $m = 1,\ k = 4,\ x_0 = 4,\ x_0' = 4$
(g) $m = 1,\ k = 9,\ x_0 = -8,\ x_0' = -1$

(h) $m = 4,\ k = 1,\ x_0 = 4,\ x'_0 = -7$
(i) $m = 4,\ k = 9,\ x_0 = 10,\ x'_0 = -1$

2. We solved $mx'' + kx = 0$ by seeking $x(t)$ in the form e^{rt}, which readily led to the solution forms given in (16). A lengthier but interesting alternative is to try to integrate the equation twice. As we've noted earlier, integrating with respect to t doesn't work because we get $mx' + k\int x(t)\,dt = C$ but can't evaluate the integral because we don't yet know $x(t)$. Instead, try integrating with respect to x, rather than t:

$$m\int x''\,dx + k\int x\,dx = C. \tag{2.1}$$

(a) To evaluate the $\int x''\,dx$ term, follow the steps used in equation (9) of Section 1.5, and show that (2.1) gives

$$\underbrace{\frac{1}{2}mx'^2}_{\text{KE}} + \underbrace{\frac{1}{2}kx^2}_{\text{PE}} = C, \tag{2.2}$$

which is a statement of **conservation of energy**: the total energy (kinetic energy of the mass plus potential energy stored in the spring) is constant; it does not change with time. NOTE: $mx''+kx = 0$ was a *force* equation, but multiplying it through by dx converted it to an *energy* equation because force times distance (dx) is work, and work is manifested as energy.
(b) It follows from (2.2) that

$$x' = \omega_n\sqrt{B^2 - x^2}, \tag{2.3}$$

with $2C/k$ renamed as B^2 for convenience. Solve (2.3) by separation of variables and show that

$$x(t) = B\sin(\omega_n t + A), \tag{2.4}$$

which is the same as (16c).

3. Electrical Oscillator. Let R and the forcing function dE/dt be zero in (6), so $Li'' + (1/C)i = 0$. What is the natural frequency, in rad/sec, and in cycles/sec, in terms of L and C? What is the period, in seconds, in terms of L and C?

4. Grandfather's Clock. (a) Determine the length L of the pendulum in a grandfather's clock, in feet, so that each half-period is one second. Take $g = 32.174$ ft/sec^2.
(b) How much error in L would cause the clock to run slow by one minute per week?

5. Ballistic Walking. "Gait analysis" of humans and animals is an active research area in biomechanics. In this exercise we want to estimate the leisurely walking speed of an average human being. The basic idea is that as you plant your left foot, your right leg swings forward as a pendulum, hinged at the hip. We can move faster by using muscle power (that would be a *forced* vibration), but by leisurely walking, called *ballistic walking*, we mean simply letting our leg swing freely as a pendulum (i.e., a *free* vibration), under the sole action of gravity. A leg is not like the point-mass in Fig. 10; it is, roughly, more like a rod, and the inertia of a rod of mass m and length L, about one end, is $I = mL^2/3$.

(a) Using the physics formula $T = I\alpha$, as we did in deriving equation (28), show that the (approximate) governing equation, using the same small-angle approximation as we did in approximating (28) by (30), is

$$\theta'' + \frac{3g}{2L}\theta = 0. \tag{5.1}$$

(b) From (5.1) write down the frequency, in cycles per second. Taking $L \approx 3$ ft, and the distance moved per cycle as 4 ft, obtain an estimate of leisurely walking speed, in ft/sec and in mi/hr.

6. Beam Oscillator. The uniform rigid beam of length L and weight $W = mg$ is constrained by a frictionless pin at its left end and a vertical spring of stiffness k at its right end. Its angular deflection relative to the horizontal is θ, and the spring is unstretched when $\theta = 0$.

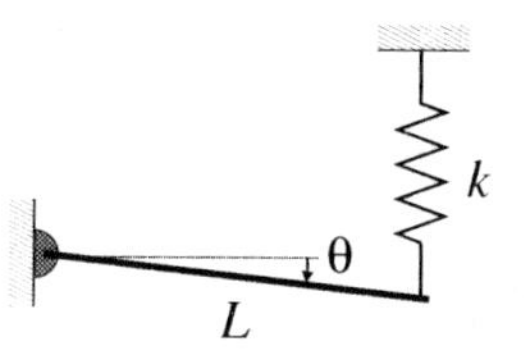

(a) Using the physics formula $T = I\alpha$, as we did in deriving (24), derive the equation of motion for $\theta(t)$, assuming θ is small, and show that

$$m\theta'' + 3k\theta = \frac{3mg}{2L}. \tag{6.1}$$

HINT: Take T, in $T = I\alpha$, as the moment (counted positive clockwise, like θ) about the pin, and realize that the weight force W acts at the midpoint of the beam. The inertia of the rod about the pin is $I = mL^2/3$.
(b) Give an expression for the natural frequency ω_n.
(c) Solve (6.1) subject to the initial conditions $\theta(0) = 0$ and $\theta'(0) = 0$, and show that

$$\theta(t) = \frac{W}{2kL}\left(1 - \cos\sqrt{\frac{3k}{m}}\,t\right). \tag{6.2}$$

(d) Show that the solution (6.2) amounts to a vibration about the rest position of the beam, that is, the angular position at which the beam is in static equilibrium.

7. Mass in a Rotating Tube. The mass shown in the figure moves in the frictionless tube and is restrained by a spring of stiffness k that is neither stretched nor compressed when $x = 0$. The cylinder rotates in a horizontal plane (so gravity is not relevant) at constant angular velocity Ω about P.

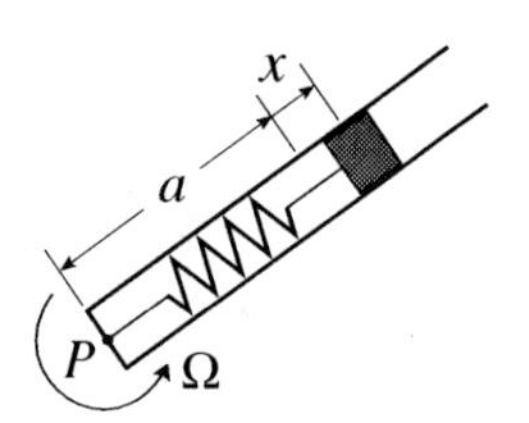

(a) Derive the equation of motion,

$$x'' + \left(\frac{k}{m} - \Omega^2\right)x = a\Omega^2. \tag{7.1}$$

HINT: The centripetal force on a mass m in circular motion at constant angular velocity ω and constant radius R is $mR\omega^2$.
(b) If $x(0) = 0$ and $x'(0) = 0$, solve for $x(t)$ for the case where $\Omega < \sqrt{k/m}$.
(c) Same as part (b), this time for the case where $\Omega > \sqrt{k/m}$.
(d) Describe and discuss your results found in parts (b) and (c).

8. Restrained Inverted Pendulum. The inverted pendulum in the figure is restrained laterally by a spring of stiffness k that is neither stretched nor compressed when $\theta = 0$. The rod of length L is weightless and is attached at its base by a frictionless pin.

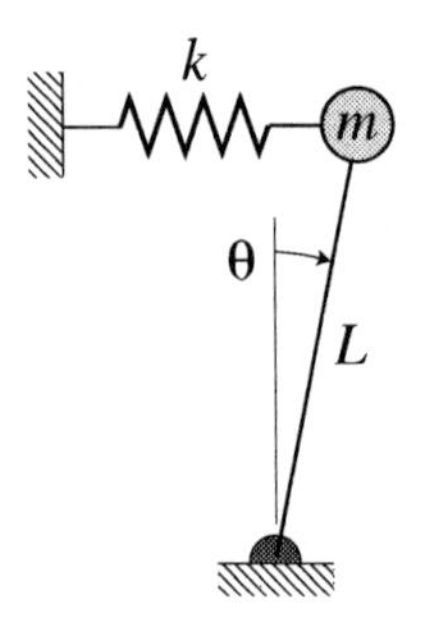

(a) Using the physics formula $T = I\alpha$, as we did in deriving (24), derive the equation of motion governing $\theta(t)$, assuming that θ is small, and show that

$$\theta'' + \left(\frac{k}{m} - \frac{g}{L}\right)\theta = 0. \tag{8.1}$$

(b) If $\theta(0) = 0$ and $\theta'(0) = \theta_0'$, solve for $\theta(t)$ for each of the two cases $k < mg/L$ and $k > mg/L$,
(c) Describe and discuss your results found in part (b).

ABOUT EXERCISES 9 AND 10: The systems in Exercises 9 and 10 don't "look" like mass–spring harmonic oscillators, but each is governed by a harmonic-oscillator equation; one is exactly of that form, and the other is of that form after it is linearized for small motions.

9. Piston Oscillator. Let a piston of mass m be placed at the midpoint of a closed cylinder of cross-sectional area A, as shown in the figure. Assume that the pressure p on either side of the piston satisfies Boyle's law (namely, that the pressure times the volume is constant), and let p_0 be the pressure on both sides when $x = 0$. Neglect friction between the piston and the cylinder.

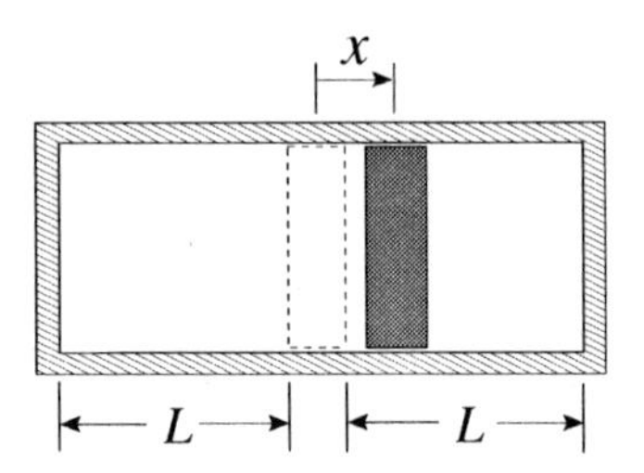

(a) Use Boyle's law to show that the pressure forces on the left and right sides of the piston are $p_0AL/(L + x)$ and $p_0AL/(L - x)$, respectively. Then, show that if the piston is disturbed from its equilibrium position $x = 0$, the governing equation of motion is

$$mx'' + 2p_0AL\frac{x}{L^2 - x^2} = 0. \tag{9.1}$$

(b) Is (9.1) linear or nonlinear? Explain.
(c) Expand the $x/(L^2 - x^2)$ term in a Taylor series about $x = 0$, up to the third-order term, say. Keeping only the leading term of that series, derive the linearized version

$$mx'' + \frac{2p_0A}{L}x = 0 \tag{9.2}$$

of (9.1), for the case of small oscillations, namely, where the amplitude of oscillation is small compared to L: $|x/L| << 1$.
(d) To better understand the approximation of (9.1) by (9.2), sketch the graph of $x/(L^2 - x^2)$ and its linearized version x, on $-L < x < L$, and label any key values.
(e) From (9.2), determine the natural frequency of oscillation, ω_n, in rad/sec and in cycles/sec.
(f) Is the resulting linearized model equivalent to the vibration

of a mass–spring system, with an equivalent spring stiffness $k_{eq} = 2p_0A/L$? Explain.

10. Oscillating Platform. A uniform horizontal platform of mass m is supported by oppositely rotating cylinders a distance L apart (see figure). The friction force f exerted on the platform by each cylinder is proportional to the

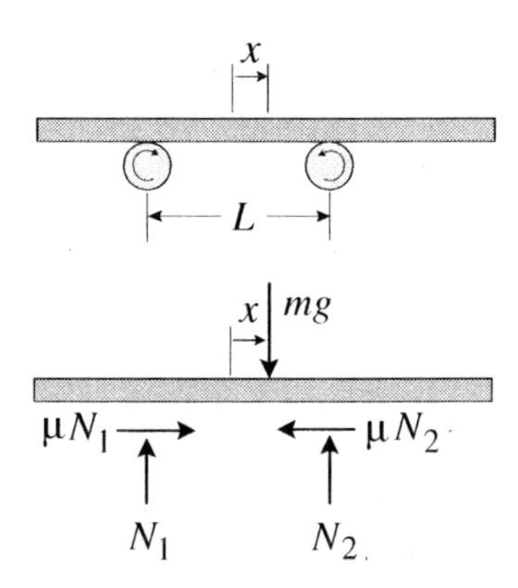

normal force N between the platform and the cylinder, with constant of proportionality (called the *coefficient of sliding friction*) μ; that is, $f = \mu N$. Show that if the platform is disturbed from its equilibrium position (which is $x = 0$) by nonzero initial conditions, then it will undergo a lateral oscillation of frequency $\omega_n = \sqrt{2\mu g/L}$ rad/sec, where g is the gravitational acceleration. NOTE: x in the figure is the lateral displacement of the midpoint of the platform relative to a point midway between the cylinders. HINT: Derive the equation of motion governing $x(t)$. To do that, write the following three equations: (1) mx'' equals the sum of the x-forces, counted positive to the right because x is measured to the right; (2) the sum of the vertical forces equals zero (because there is no vertical motion and hence no vertical acceleration); (3) the sum of the moments about any point (i.e., pick your favorite point) equals zero because there is no angular acceleration of the platform. From those three equations for N_1, N_2, and x, eliminate N_1 and N_2 (because we're not interested in them) and obtain an equation for x alone.

11. Released from rest at a height a above a spring of stiffness k, the mass m falls under the sole action of gravity, until it contacts and compresses the spring, and rebounds.

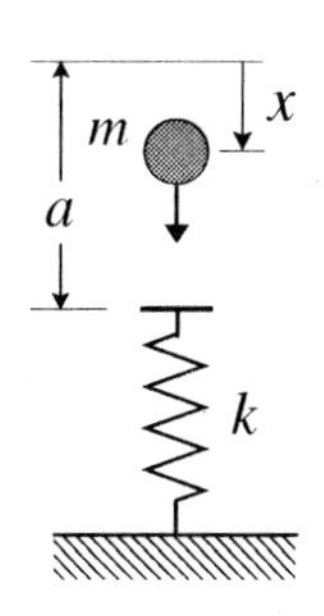

(a) Show that the equation of motion is

$$mx'' = \begin{cases} mg & , \ x < a \\ mg - k(x - a) & , \ x > a. \end{cases} \tag{11.1}$$

(b) You need not solve (11.1); we simply ask if it is linear or nonlinear. Explain.

12. Bead in a Bowl; Energy Equation Approach. A bead of mass m slides on a frictionless parabolic surface $y = x^2$, and we wish to know the frequency of small oscillations about $x = 0$. Applying Newton's second law to derive the differential equation of motion for $x(t)$ is not so simple. It is simpler to observe that the energy of the mass (kinetic energy plus gravitational potential energy) is conserved because the surface is frictionless, to write the energy equation, and to differentiate it with respect to t to obtain the equation of motion.

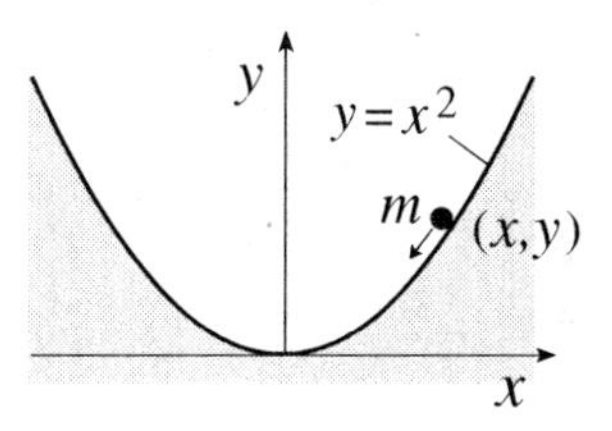

(a) Do that, and thus derive the equation of motion

$$(1 + 4x^2)x'' + 4xx'^2 + 2gx = 0. \tag{12.1}$$

HINT: The kinetic energy, and the potential energy (relative to the x axis), are $(1/2)m(x'^2 + y'^2)$ and mgy, respectively. Then, use the relation $y = x^2$ to eliminate y in favor of x, and thus obtain (12.1).
(b) (12.1) is nonlinear and looks very difficult! However, for small motions you can neglect the higher-order terms $4x^2x''$ and $4xx'^2$ compared to the linear terms x'' and $2gx$. From that linearized equation of motion, derive the frequency of small-amplitude oscillations, in rad/sec.

13. A Slow-Food Delivery Business. Looking ahead to summer job opportunities, an enterprising group of students at State University discover a gap in the marketplace — in the area of slow-food delivery. They know Newton's inverse-square law of gravitation, namely, that the mutual force of attraction exerted on point masses m_1 and m_2 a distance d apart is

$$F = G\frac{m_1 m_2}{d^2}, \tag{13.1}$$

in which G is a universal gravitational constant. They also know the result, derivable from (13.1), that the force of attraction at any point at a radius r *inside* a homogeneous sphere is due only to the mass within the "subsphere" of radius r, and is the same as if that subsphere were compressed to a point mass at the origin. (This result was derived by Newton, and is an exercise in multiiple integration.) With these results in mind, and taking the earth to be a perfect sphere of radius $R = 3,950$ mi, they consider drilling a tunnel, in a straight line from State University (SU), through the center of the earth, to Hong Kong University (HKU) at the other end. They intend to drop the slow food into the tunnel and, with the help of gravity and at no energy cost, deliver it to students at HKU. To determine the delivery time, their next step is to derive the equation of motion of the parcel in the tunnel.

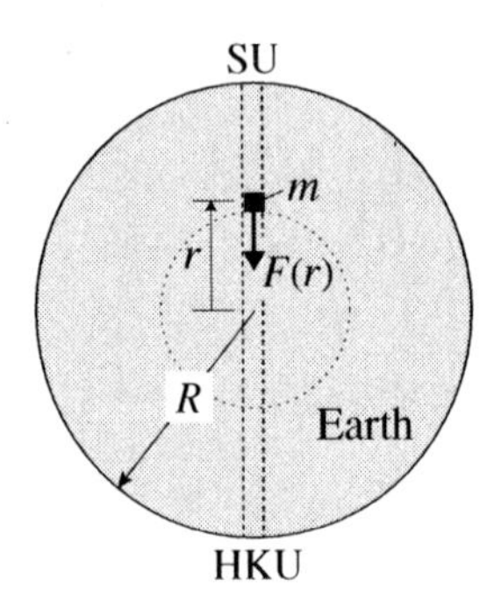

(a) Neglecting any air resistance or friction, show that the result is

$$r'' + \frac{g}{R}r = 0, \tag{13.2}$$

an harmonic oscillator equation! HINT: To obtain the force $F(r)$ that is shown in the figure, use (13.1) with m_1 equal to the earth's mass density γ times the volume of the sphere of radius r (shown as dotted in the figure), and with $m_2 = m$ the mass of the parcel. Then, infer the value of $G\gamma$ by noting that at the earth's surface $F(R)$ equals mg.
(b) With $r(0) = R$, $r'(0) = 0$, and $R = 3,950$ mi, solve for the delivery time, from SU to HKU, and also for the speed of the parcel as it passes through the center of the earth. What is its velocity when it arrives at HKU?
NOTE: As naive as the idea may seem, earth tube travel has been the subject of research, on and off, for over a hundred years.

ADDITIONAL EXERCISES

14. Tube Travel One Step Further. In fact, the tunnel contemplated in Exercise 13 can just as well be used to link any two points on the earth's surface. Let the x, y plane in the figure pass through the line of travel, which is from $(0, R)$ to (a, b), and the center of the earth.

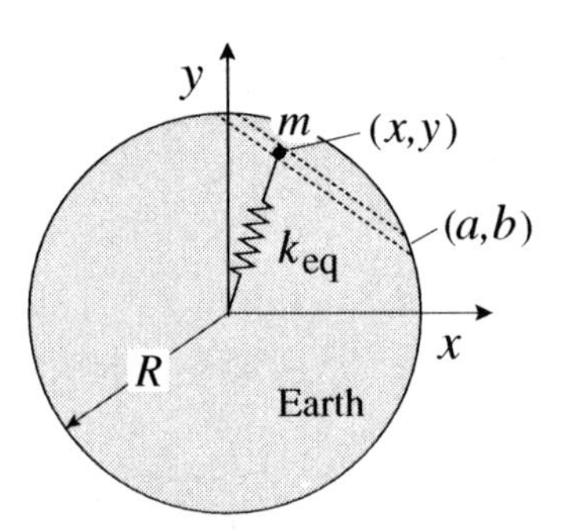

(a) To derive the equation of motion of the mass m along that line it is simpler to write down a statement of the conservation of energy (which holds because we are ignoring any friction and air resistance) and to differentiate it with respect to t than to use Newton's second law. From equation (13.2), above, we can think of the mass m as connected to a spring, of stiffness $k_{\text{eq}} = mg/R$, from the origin to m, stretched by the amount $r = \sqrt{x^2 + y^2}$. Thus, write expressions for the kinetic and potential energies of m (in terms of x, y, x', y'), differentiate, and show that

$$\frac{g}{R}(xx' + yy') + (x''x' + y''y') = 0. \tag{14.1}$$

(b) Next, realize that $x(t)$ and $y(t)$ are related by the equation of the line, $y = R - \alpha x$, say. Put the latter in (14.1) to eliminate y, and obtain the desired equation of motion (in terms of x) as

$$x'' + \frac{g}{R}x = \frac{\alpha g}{1 + \alpha^2}. \tag{14.2}$$

(c) With $x(0) = 0$ and $x'(0) = 0$, solve for $x(t)$ and show that the transit time from $(0, R)$ to (a, b) is the same as obtained in Exercise 13, 42 minutes, independent of the slope α!

15. Including the Kinetic Energy of a Spring. Notice that in this section we've ignored the mass, and hence the kinetic energy, of the spring. For instance, observe that (2.2) in Exercise 2 gives the total energy as the KE of the mass alone, plus the PE of the spring. To the extent that (2.2) is thus defective, the equation $mx'' + kx = 0$ from which it came must also be defective. To correct this defect, suppose the mass of the spring is m_s and its unstretched length is l, and assume the velocity within the spring varies linearly, from 0 at its fixed end to x' at the end attached to the mass m.

(a) With that assumption, write an expression for the kinetic energy "dKE" of a typical element of the spring, of length $d\xi$, integrate on ξ from 0 to $l + x$, and show that the spring's kinetic energy, at any time t, is $m_s x'^2/6$, so in place of (2.2) we have

$$\frac{1}{2}\left(m + \frac{m_s}{3}\right)x'^2 + \frac{1}{2}kx^2 = C. \tag{15.1}$$

That is, the "effective mass," taking the KE of the spring into account, is $m + m_s/3$. If $m_s << m$, the $m_s/3$ correction can be omitted.
(b) Differentiate (15.1), to obtain the corrected equation of motion, and find the corrected natural frequency of the mass–spring system.
(c) Do you expect the assumption of a linear velocity distribution within the spring to be exact? Discuss.

16. Nonlinear Pendulum; Computer. In the text we emphasized the *linearized* pendulum equation (26), for small motions, which has the form of an harmonic oscillator equation, with frequency $\omega_n = \sqrt{g/l}$. We complement that discussion now by asking you to use computer software to solve the full nonlinear equation (24). You will find that the motion is still oscillatory even for motions of larger amplitude, and from your numerical results you will be able to investigate two interesting aspects in particular: how the period varies with the amplitude of the oscillation, and whether the motion remains harmonic as the amplitude is increased. With $g/l = 1$, say, consider the IVP

$$\theta'' + \sin\theta = 0; \qquad \theta(0) = \theta_0, \ \theta'(0) = 0. \tag{16.1}$$

You can control the amplitude of the motion by assigning various values to θ_0, from very small values to values approaching $\theta_0 = \pi$.

(a) Use computer software to obtain a plot of $\theta(t)$ versus t. Do this for $\theta_0 = 1, 10, 30, 60, 110, 170,$ and 179 degrees (converted to radians).
(b) From those plots, measure the period T of the oscillation for each of those seven cases, and plot those values of T versus θ_0. What limiting value do you expect for T as $\theta_0 \to 0$? Do your results seem consistent with that asymptotic value? What do you think T tends to as $\theta_0 \to \pi$?
(c) Does the oscillation remain *harmonic* even as θ_0 is increased?

17. Piston Oscillator; Computer. This exercise follows upon Exercise 9, in which chief interest was in deriving the equation (9.2) governing the small oscillations of the piston, and the natural frequency of those oscillations. We complement that discussion now by asking you to use computer software to solve the full nonlinear equation (9.1). You will find that the motion is still oscillatory even for motions of larger amplitude, and from your numerical results you will be able to investigate two interesting aspects: the variation of the period, and the deviation of the solution from an harmonic oscillation as the amplitude is increased. For definiteness, let $m = L = 2p_0 A = 1$, and hence consider the IVP

$$x'' + \frac{x}{1 - x^2} = 0; \qquad x(0) = x_0, \ x'(0) = 0. \tag{17.1}$$

You can control the amplitude of the motion by assigning various values to x_0, from very small values to values approaching $x_0 = 1$ (which is the maximum possible because the cylinder is closed at its ends and $L = 1$).

(a) Use computer software to obtain a plot of $x(t)$ versus t. Do this for $x_0 = 0.1, 0.2, 0.3, 0.5, 0.8, 0.9, 0.95,$ and 0.99.
(b) From those plots, measure the period T of the oscillation for each of those eight cases, and plot those values of T versus x_0. As $x_0 \to 0$ the period behaves as $T \to 2\pi$, and as $x_0 \to 1$ the period behaves as $T \sim 4/\sqrt{-\ln(1 - x_0)}$ (which we state without proof). Do your eight points seem consistent with those two asymptotic results?
(c) Is the oscillation *harmonic*? Explain. NOTE: Recall from the text that by harmonic we mean a pure sinusoid, give or take a phase shift.

18. An Ecological Oscillator. This exercise is to further indicate the diversity of harmonic oscillators. We will find that in certain cases an ecological system containing predator and prey species can behave as an harmonic oscillator. Imagine an island, inhabited by mice and hawks, and suppose the mice (the prey) are the sole food for the hawks (the predator). If the mouse population is $x(t)$ and the hawk population is $y(t)$, then one possible model is given by the differential equations

$$x' = (a - by)x, \tag{18.1a}$$

$$y' = (-c + dx)y, \tag{18.1b}$$

in which a, b, c, d are positive constants. That is, in the absence of hawks (18.1a) gives the simple exponential growth model $x' = ax$, but predation by the hawks decreases the net birth/death rate of mice so the constant a is reduced in proportion to the population y of hawks, to $a - by$. And in the absence of mice, (18.1b) gives the simple exponential decay model $y' = -cy$, decay because the hawks then have no food. For definiteness, take $a = b = c = d = 1$, so

$$x' = (1-y)x, \tag{18.2a}$$

$$y' = (-1+x)y. \tag{18.2b}$$

Notice that there is an equilibrium state defined by $x(t) = 1$ and $y(t) = 1$, because if $x(t) = 1$ and $y(t) = 1$, then (18.2) becomes $x' = 0$ and $y' = 0$, so both $x(t)$ and $y(t)$ are constant and the system is in equilibrium. [Let the units be in hundreds, for instance, so $x(t) = 1$ means there are 100 mice, not 1 mouse.] Suppose we limit our attention to solutions that are close to that equilibrium state. Show that if we define the deviations from equilibrium as $X(t) = x(t) - 1$ and $Y(t) = y(t) - 1$, and assume that $X(t)$ and $Y(t)$ are small (compared to unity), then (18.2) becomes

$$X' \approx -Y, \tag{18.3a}$$

$$Y' \approx X. \tag{18.3b}$$

Now eliminate Y from (18.3) and show that X satisfies the *harmonic oscillator equation*

$$X'' + X = 0. \tag{18.4a}$$

In the same manner, eliminate X from (18.3) and show that Y satisfies the harmonic oscillator equation

$$Y'' + Y = 0. \tag{18.4b}$$

Thus, the mouse and hawk populations will undergo harmonic variations about their equilibrium values. To illustrate, let $x(0) = 1.1$ and $y(0) = 1$, so $X(0) = 0.1$ and $Y(0) = 0$. Then (18.3) gives $X'(0) = 0$ and $Y'(0) = 0$. With these initial conditions, solve for $X(t)$ and $Y(t)$ and, from them, for $x(t)$ and $y(t)$. Sketch the graphs of $x(t)$ and $y(t)$. Does the out-of-phase nature of the two solutions seem reasonable? Explain.

NOTE: Actually, (18.2) and (18.3) are "systems" of two first-order differential equations, and systems are not studied until Chapter 4, but we used elimination to obtain from (18.3) second-order equations in only one variable [(18.4a) for X and (18.4b) for Y], so we were then able to use the methods studied in Chapter 2.

3.3 FREE OSCILLATION WITH DAMPING

We continue to consider the free oscillation of the linear harmonic oscillator, but now include a damping term cx', so

$$\boxed{mx'' + cx' + kx = 0,} \tag{1}$$

Seek $x(t) = e^{rt}$.

in which m, c, and k are positive constants. Seeking solutions in the form $x(t) = e^{rt}$ gives the characteristic equation $mr^2 + cr + k = 0$, with roots

$$r = -\frac{c}{2m} \pm \sqrt{\left(\frac{c}{2m}\right)^2 - \frac{k}{m}}. \tag{2}$$

With $c = 0$, the latter gives $r = \sqrt{-\frac{k}{m}} = \pm i\sqrt{\frac{k}{m}}$, and as we increase c from zero, the discriminant $(c/2m)^2 - k/m$ in (2) remains negative until c reaches the value

$$\boxed{c_{\text{cr}} = \sqrt{4mk},} \tag{3}$$

which is the **critical damping coefficient**. We will distinguish the three cases $c < c_{cr}$, $c = c_{cr}$, and $c > c_{cr}$ as **underdamped**, **critically damped**, and **overdamped** cases, respectively.

3.3.1 Underdamped. If we have "$\sqrt{a-b}$" in which $a - b < 0$, it is good to rewrite the square root as $i\sqrt{b-a}$ so the discriminant $b - a$ is positive and the imaginariness of the square root is made explicit by the i. In the same way, if $c < c_{cr}$, then the discriminant in (2) is negative, so in that case we re-express (2) as

$$r = -\frac{c}{2m} \pm i\sqrt{\frac{k}{m} - \left(\frac{c}{2m}\right)^2} = -\frac{c}{2m} \pm i\sqrt{\omega_n^2 - \left(\frac{c}{2m}\right)^2}, \tag{4}$$

in which we continue (from Section 3.2) to denote $\omega_n = \sqrt{k/m}$ and to call it the natural frequency. If we define the square root in (4) as

$$\omega \equiv \sqrt{\omega_n^2 - \left(\frac{c}{2m}\right)^2} \tag{5}$$

for brevity, we can write a general solution of (1) as

$$\begin{aligned} x(t) &= C_1 e^{(-c/2m+i\omega)t} + C_2 e^{(-c/2m-i\omega)t} \\ &= e^{-(c/2m)t}\left(C_1 e^{i\omega t} + C_2 e^{-i\omega t}\right) \end{aligned}$$

or, equivalently, as

$$\boxed{x(t) = e^{-(c/2m)t}\left(A\cos\omega t + B\sin\omega t\right),} \tag{6}$$

in which the constants A and B can be determined from initial conditions $x(0) = x_0$ and $x'(0) = x_0'$ if those are specified.

Observe from (6) that the damping c has two effects. First, it introduces the $e^{-(c/2m)t}$ factor, which causes the oscillation to *damp out*; that is, $x(t) \to 0$ as $t \to \infty$, as we see in Fig. 1 (in which we've used representative values of m, c, k, x_0, and x_0' to obtain the computer plot). The upper and lower dotted curves are the envelope of the graph of $x(t)$. Second, the damping *reduces the frequency* of the oscillation from the natural frequency ω_n to the frequency $\omega = \sqrt{\omega_n^2 - (c/2m)^2}$; that is, the damping makes the system more sluggish.

Be clear about ω_n and ω. The frequency ω of the underdamped oscillation, that appears in (6), is $\omega = \sqrt{k/m-(c/2m)^2} = \sqrt{\omega_n^2-(c/2m)^2}$. The latter is less than the natural frequency $\omega_n = \sqrt{k/m}$ that corresponds to the undamped oscillation, and that merely serves as a reference value for the damped oscillation.

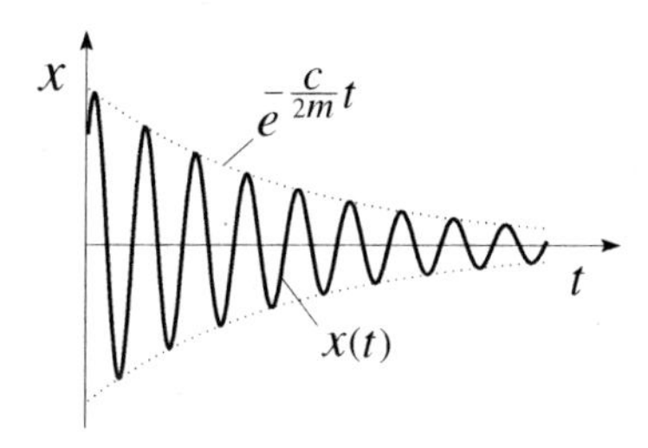

Figure 1. A representative underdamped free oscillation [if $x(0)$ and $x'(0)$ are positive].

3.3.2 Critically damped. If $c = c_{\text{cr}}$, then (2) gives the repeated roots $r = -c/2m, -c/2m$, so a general solution of (1) is

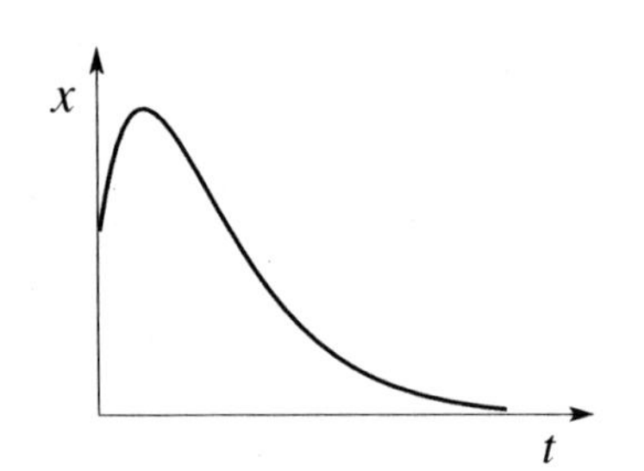

Figure 2. A representative critically damped case [if $x(0)$ and $x'(0)$ are positive].

$$\boxed{x(t) = (A + Bt)\, e^{-(c/2m)t}.} \tag{7}$$

Now the frequency ω has been reduced so much, by the damping, that it has become zero and there is no oscillation at all. A representative solution is plotted in Fig. 2.

Observe that the t in the $A+Bt$ factor becomes infinite as $t \to \infty$, whereas the $\exp(-c/2m)t \to 0$. Which one wins? We leave it for the exercises to show that exponential decay always dominates algebraic growth. That is,

$$t^{\alpha} e^{-\beta t} \to 0 \qquad (\alpha > 0, \beta < 0) \tag{8}$$

as $t \to \infty$, *no matter how large* α *is, and no matter how small* β *is.*

3.3.3 Overdamped. Finally, if we increase the damping beyond c_{cr}, then the discriminant in (2) becomes positive and (2) gives two distinct real roots,

Both r_1 and r_2 are negative.

$$r_1 = \frac{-c + \sqrt{c^2 - 4mk}}{2m} \quad \text{and} \quad r_2 = \frac{-c - \sqrt{c^2 - 4mk}}{2m}, \tag{9}$$

each of which is negative because $\sqrt{c^2 - 4mk} < c$. Thus, in the overdamped case we can write a general solution of (1) as

$$\boxed{x(t) = Ae^{r_1 t} + Be^{r_2 t},} \tag{10}$$

in which $r_1 < 0$ and $r_2 < 0$. We have not shown a representative plot for this case, as we did for the underdamped and critically damped cases in Figs. 1 and 2. Such a plot would be similar to the one in Fig. 2. We will compare the various cases in the following example.

EXAMPLE 1. Comparison of the Four Cases: Undamped, Underdamped, Critically Damped, and Overdamped. To compare the four cases in one example, let $m = 1$, $k = 4$ (so $\omega_n = 2$ and $c_{\text{cr}} = 4$), $x_0 = 2$, and $x_0' = 0$. Let us find the solution $x(t)$ of the IVP and plot it for these cases: $c = 0$ (no damping), $c = 0.5$ (underdamped), $c = 4$ (critically damped), and $c = 8$ (overdamped). Applying the initial conditions $x(0) = 2$ and $x'(0) = 0$ to $x(t) = C\cos\omega_n t + D\sin\omega_n t$ for the undamped case $c = 0$, to (6) for the underdamped case $c = 0.5$, to (7) for the critically damped case $c = 4$, and to (10) for the overdamped case $c = 8$, we obtain these results (Exercise 1).

$c = 0$:

$$x(t) = 2\cos 2t, \tag{11a}$$

$c = 0.5 < c_{\text{cr}}$:

$$x(t) = e^{-0.25t}\left[2\cos\left(\frac{\sqrt{63}}{4}t\right) + \frac{2}{\sqrt{63}}\sin\left(\frac{\sqrt{63}}{4}t\right)\right], \tag{11b}$$

$c = 4 = c_{\text{cr}}$:

$$x(t) = (2 + 4t)e^{-2t}, \tag{11c}$$

$c = 8 > c_{\text{cr}}$:

$$x(t) = \left(1 + \frac{2}{\sqrt{3}}\right)e^{-(4-2\sqrt{3})t} + \left(1 - \frac{2}{\sqrt{3}}\right)e^{-(4+2\sqrt{3})t}, \tag{11d}$$

The frequency $\omega = \sqrt{63}/4 \approx 1.98$ is slightly less than $\omega_n = 2$ because of the damping. As c is increased ω continues to decrease, and becomes zero when $c = c_{\text{cr}} = 4$.

which are plotted in Fig. 3. As c is increased from 0 to 0.5 the frequency decreases slightly, from $\omega = \omega_n = 2$ to $\omega = \sqrt{63}/4 = 1.984$, and as c is increased further to the value $c_{\text{cr}} = 4$ the frequency decreases to zero and the oscillation disappears entirely.

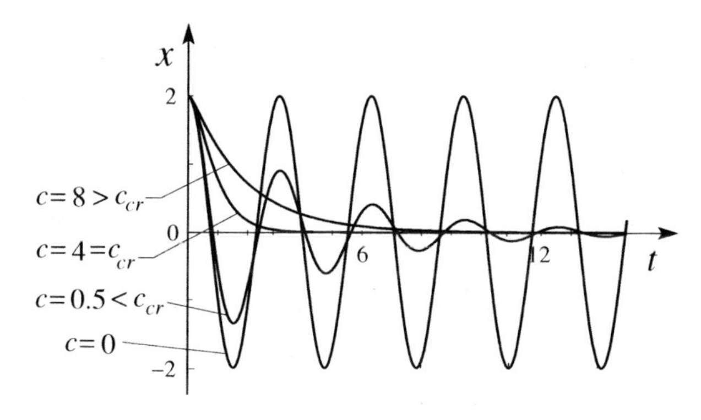

Figure 3. Varying the damping c, with $c_{\text{cr}} = 4$.

COMMENT. Besides the traditional plots of x versus t, as in Fig. 3, it is also informative to plot the displacement x versus the velocity x', which we denote as y. We've done so in Fig. 4 for representative undamped, underdamped, and overdamped cases. In each of those cases the solution curve, or *trajectory*, begins at $(x, y) = (2, 0)$, which corresponds to the initial condition $x(0) = 2$, $x'(0) = y(0) = 0$. The time t does not appear explicitly in Fig. 4, as it did in Fig. 3. Rather, since the trajectories in the x, y **phase plane** are defined by $x = x(t)$, $y = y(t)$, think of the time t as the parameter of those curves; as t increases the representative point $(x(t), y(t))$ generates the trajectory. In Fig. 3, the arrows indicate the direction of increasing t.

For the case $c = 0$, the trajectory closes on itself and the representative point continues to traverse that path, clockwise, over and over. That is, the motion is periodic, as we saw for that case in Fig. 3. [What is the equation of that closed path? From (11a), $x(t) = 2\cos 2t$ and $y(t) = x'(t) = -4\sin 2t$, so $4x^2 + y^2 = 4$, which is an ellipse.]

For $c = 0.05$ the system is underdamped. The trajectory spirals inward, approaching the origin as $t \to \infty$; the motion is a damped oscillation. And for $c = 8$ the system is overdamped, the trajectory approaches the origin without encircling it, and the motion dies out without oscillation.

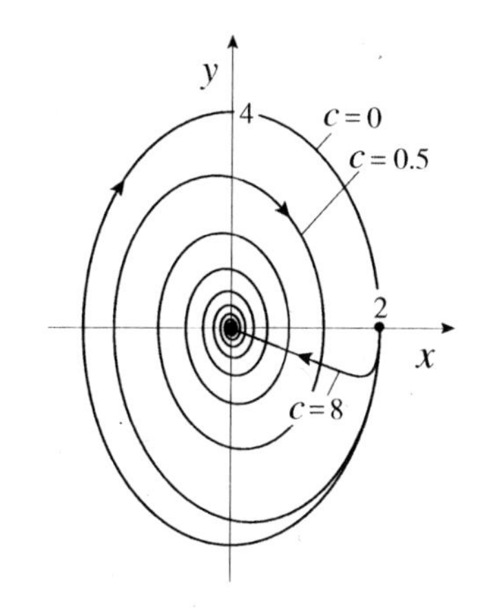

Figure 4. Representative trajectories in the x, x' phase plane. $m = 1$, $k = 4$, $x(0) = x_0 = 2$, $y(0) = x'_0 = 0$, $c = 0$, 0.5, and 8. The heavy dot denotes the initial point.

Although Fig. 4 does not give $x(t)$ versus t directly, as Fig. 3 does, the latter can be seen indirectly, as the projection of the representative point $(x(t), y(t))$ onto the x axis. For $c = 0.5$, for instance, see how that projection is a damped oscillation.

The x, y phase plane is primarily of interest for the more difficult case of nonlinear equations, and will be developed in Chapter 7. ▮

Closure. We've considered the effects of damping on the free oscillation. Whereas the damping coefficient c was zero in Section 3.2, here it is positive. The effects of the damping can be seen from the characteristic roots that are given in (4): If $c = 0$, then $r = \pm i\omega_n$, where $\omega_n = \sqrt{k/m}$ is the natural frequency of the undamped oscillator, and the motion is harmonic. Increasing c from zero has two effects: First, the frequency $\omega = \sqrt{\omega_n^2 - (c/2m)^2}$ decreases. Second, the real part of r, which is $-c/2m$, results in a negative exponential factor in the solution [the $e^{-(c/2m)t}$ in (6)], which causes the solution to damp out as $t \to \infty$, as seen in Fig. 1. When c is increased further, to its critical value $c_{cr} = \sqrt{4mk}$, the frequency is reduced to zero and the oscillation disappears altogether [see (7) and Fig. 2]. If c is increased still further, the damping rate continues to increase but there is no further qualitative change. The four cases were illustrated in Fig. 3.

Remember that in a given application the kx and cx' terms in (1) may be modeling a literal spring–dashpot system, or perhaps a material such as cartilage that *behaves* like a spring–dashpot combination. For if cartilage is compressed it behaves not only elastically as a spring, but also exhibits a damping behavior that can be modeled by the cx' term in (1), because as it is compressed fluid is squeezed out through very small passageways, and this effect causes a resistance proportional to the velocity x'.

Thus, the discs in one's spinal column are, in effect, spring/damper combinations that act as shock absorbers between adjacent vertebrae.

EXERCISES 3.3

1. Example 1. Derive the solutions given in (11a)–(11d).

2. Consider the mass–spring–dashpot system shown, with $x(0) = 0$ and $x'(0) = 0$. Take $g = 32.2$ ft/sec^2.

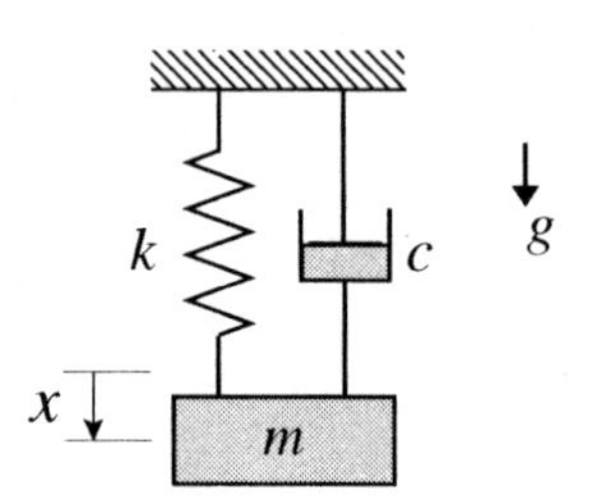

(a) Write down the equation of motion for $x(t)$.
(b) Determine the limit of $x(t)$ as $t \to \infty$.
(c) Let $m = 2$, $c = 1$, $k = 8$. Obtain a computer plot of both the displacement $x(t)$ and the force $F(t)$ exerted on the ceiling. Of course, choose a reasonable t interval for the plot. Is the oscillation underdamped? Overdamped?
(d) The same as (c), but with $m = 2$, $c = 5$, $k = 2$.

3. Computer. (a) To see the effect of increasing the damping coefficient c, while keeping the other constants fixed, obtain computer plots of the solution to (1) subject to the initial conditions $x(0) = 1$ and $x'(0) = 0$, say, with $m = 1$, $k = 9$, and $c = 0, 1, 3, 6, 20$, in turn. In this case $c_{cr} = \sqrt{4mk} = 6$, so these c values proceed from undamped through overdamped. Plot the five graphs together, and on a long enough t interval to obtain the desired comparison.
(b) You should have found that as c is increased, from zero, the solutions die out faster because of the damping. Yet, eventually, as you continue to increase c, the solutions begin to die out more slowly. Explain why this is so. HINT: Examine (6), (7), (9), and (10).

(c) The same as part (a), but using values of m, c, k supplied by your instructor.

4. Nonconservative Due to the Damping. (a) In Exercise 2 of Section 3.2 we found that if the (undamped) equation $mx'' + kx = 0$ is multiplied by dx and integrated, one obtains

$$\frac{1}{2}mx'^2 + \frac{1}{2}kx^2 = C, \tag{4.1}$$

which is a statement of conservation of energy. Now show that multiplying $mx'' + cx' + kx = 0$ by dx and integrating from 0 to t gives

$$\frac{1}{2}mx'^2 + \frac{1}{2}kx^2 = C - \int_0^t c[x'(\tau)]^2 \, d\tau, \tag{4.2}$$

in which C is $[mx'(0)^2 + kx(0)^2]/2$; (4.2) shows that the kinetic energy plus the potential energy is *not* conserved but is a decreasing function of t — nonincreasing, to be more precise, since $x'(t)$ could be identically zero.
(b) Where did that lost energy go?

5. For the underdamped free oscillation, $x(t) \to 0$ as $t \to \infty$. It might not seem unreasonable to also expect the *frequency* to tend to zero as $t \to \infty$. Does it? How do you know?

6. For the critically damped case, $c = c_{\text{cr}}$, show that the graph of $x(t)$ crosses the positive t axis either once or not at all. Specifically, show that there will be one crossing if

$$-\frac{x(0)}{x'(0) + cx(0)/2m} > 0; \tag{6.1}$$

otherwise there will be none.

7. Experimental Determination of c for Subcritical Damping. It is simple to measure the mass m, and even to measure the spring stiffness k from an experimental plot of force versus displacement, but determining c is more difficult. One way to do so is by comparing the analytical solution with a plot of $x(t)$ versus t obtained by experiment. Suppose the oscillation is observed to be a damped oscillation, so $0 < c < c_{\text{cr}}$ and (6) applies.

(a) $T = 2\pi/\omega$ is the period of the $\sin \omega t$ and $\cos \omega t$ terms in (6). Show that

$$\frac{x(t+T)}{x(t)} = e^{-(c/2m)T}. \tag{7.1}$$

Thus, if from our experimental plot of $x(t)$ versus t we measure T, $x(t)$, and $x(t+T)$, and we know m, then (7.1) can be solved for c. Equation (7.1) applies at any time t, but it is convenient to take t to be a time at which $x(t)$ has a local maximum, so t and $t + T$ correspond to successive peaks.
(b) In fact, show that

$$\frac{x(t+NT)}{x(t)} = e^{-(c/2m)NT} \tag{7.2}$$

for any positive integer N. NOTE: If the damping is light, then (7.2) is better than (7.1) because $x(t+NT)/x(t)$ can be determined from the graph more accurately than $x(t+T)/x(t)$.
(c) Solving (7.2) for c, show that

$$c = \frac{2m}{NT} \ln\left(\frac{x(t)}{x(t+NT)}\right). \tag{7.3}$$

8. Application of (7.3). (a) To illustrate the results obtained in Exercise 7, think of the plots in Fig. 3 as experimental data, and consider the one corresponding to $c = 0.5$ — but pretend you don't know that $c = 0.5$. Use (7.2) or (7.3), together with the plot, to evaluate c. (Of course, you should obtain $c \approx 0.5$.)

9. Algebraic Growth Versus Exponential Decay; Who Wins? We stated in (8) that

$$t^{\alpha}e^{-\beta t} \to 0 \qquad (\alpha > 0, \beta > 0) \tag{9.1}$$

as $t \to \infty$, no matter how large α is and no matter how small β is. Prove that claim. HINT: Recall from the calculus that to use l'Hôpital's rule the indeterminate form must be of the type $0/0$ or ∞/∞, not $(\infty)(0)$. Thus, to use l'Hôpital's rule, first re-express the product $t^{\alpha}e^{-\beta t}$ either as $t^{\alpha}/e^{\beta t}$ (giving ∞/∞) or as $e^{-\beta t}/t^{-\alpha}$ (giving $0/0$). Alternatively, using Taylor series you could write

$$t^{\alpha}e^{-\beta t} = \frac{t^{\alpha}}{e^{\beta t}} = \frac{t^{\alpha}}{1 + \beta t + \beta^2 t^2/2! + \cdots} \tag{9.2}$$

and draw conclusions from the form of the right-hand side of (9.2).

ADDITIONAL EXERCISES

10. The Kelvin–Voigt and Maxwell Viscoelastic Models. Thus far the only model we've used for a system with damping is the Kelvin–Voigt model, consisting of a spring and dashpot in parallel. It is called a **viscoelastic** model, "visco" referring to the dashpot, which may contain viscous fluid and behaves somewhat like a hypodermic syringe. This and other spring-dashpot combinations are commonly used to model the behavior of materials that exhibit some combination of springiness and "syringe" behavior, such as most biomaterials, polymers, and rubber. For instance, to study head injuries one might model the head as two masses connected by at least one spring and at least one dashpot. One mass is the skull, which receives the blow, and one is the soft material inside. As another example, to design a high-tech running shoe a viscoelastic model can be used for the shoe bottom in trying to optimize the support system that it provides. In this exercise we first review the force–displacement characteristics of the Kelvin–Voigt model and then introduce another one, the "Maxwell model." The Kelvin–Voigt model is shown below. The spring and dashpots have the *same displacement* x but different forces: $f_s = kx$

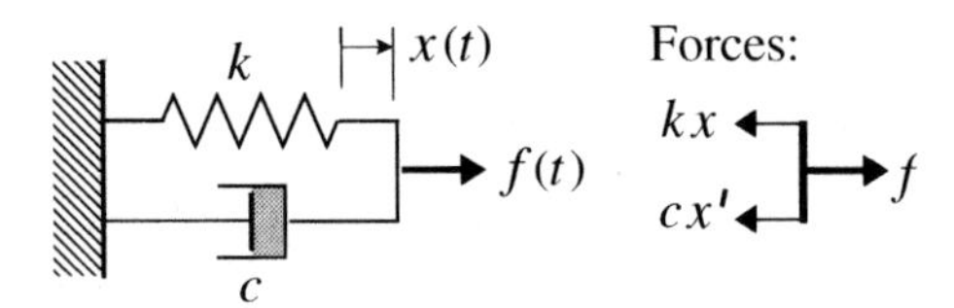

and $f_d = cx'$, and since $f = f_s + f_d$, we have the key relationship

$$f = kx + cx', \tag{10.1}$$

for Kelvin–Voigt, between the force $f(t)$ and displacement $x(t)$.

(a) **Maxwell Model.** Instead of a spring and dashpot in parallel, as above, in the **Maxwell model** they are in series, as shown below, where x and y are absolute, not relative, displacements. Thus, the stretch in the dashpot is $x - y$. This

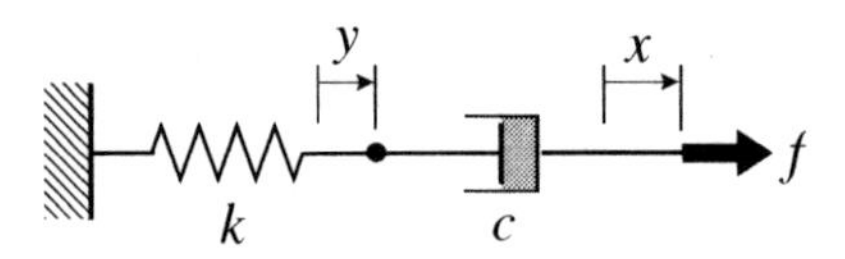

time the elements do not have the same displacement, they have the same *force*, for the force in each element is the applied force f. Write expressions for the force in the spring and in the dashpot, eliminate the variable y, and thus derive the force-displacement relationship

$$x' = \frac{1}{c}f + \frac{1}{k}f', \tag{10.2}$$

in contrast with the Kelvin–Voigt formula (10.1). Two key properties of a Maxwell model, extensional creep and force relaxation, are revealed by two complementary experiments:

(b) **Creep, for Maxwell Model.** In the first experiment we seek the response $x(t)$ due to a "step" increase in $f(t)$, with initial condition $x(0) = 0$. Unfortunately, the derivative f' in (10.2) does not exist at $t = 0$ because of the step (see the left-hand figure, below). We do work with step functions

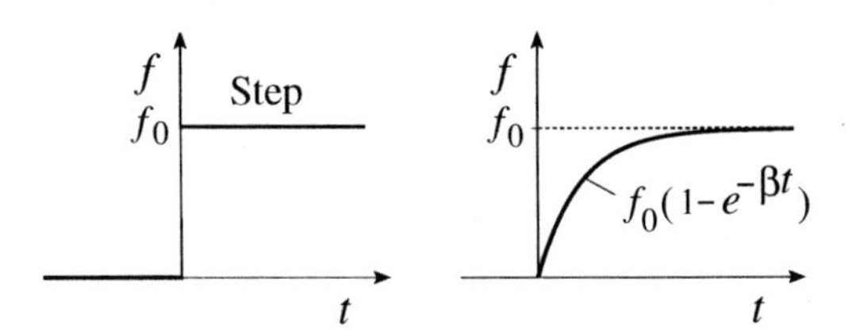

in Chapter 5, but for now we can survive by taking $f(t) = f_0(1 - e^{-\beta t})$ where β is a large positive constant and, at the end, letting $\beta \to \infty$ [because in that limit $f_0(1 - e^{-\beta t})$ tends to the desired step]. Do that: Solve (10.2) for $x(t)$, let $\beta \to \infty$ in your result, and show that

$$x(t) = \frac{f_0}{k} + \frac{f_0}{c}t, \tag{10.3}$$

as sketched below. That is, the spring extension increases instantaneously to f_0/k and remains constant, while the dashpot extends continuously according to $x(t) = f_0 t/c$. This result for x is called **creep**: continuous extension.

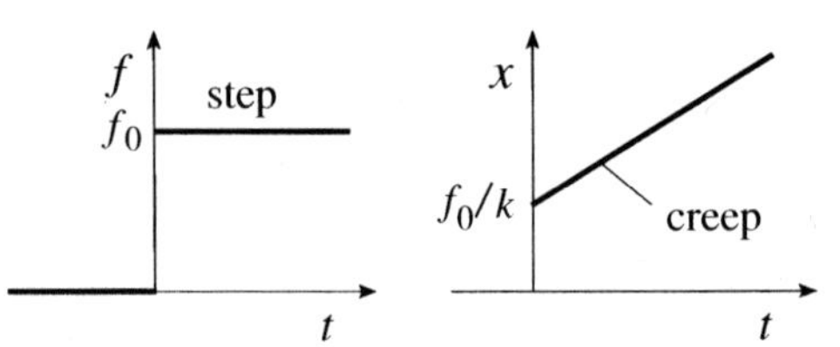

(c) **Relaxation, for Maxwell Model.** The complementary experiment is to determine the force history $f(t)$ that corresponds to a step in *displacment*. In the same spirit as above, take $x(t) = x_0(1 - e^{-\beta t})$, solve (10.2) for $f(t)$ with initial condition $f(0) = 0$, let $\beta \to \infty$ in the result, and show that

$$f(t) = kx_0 e^{-kt/c}, \tag{10.4}$$

as sketched below. Determine the forces $f_s(t)$ in the spring and $f_d(t)$ in the dashpot. NOTE: The force $f(t)$ instantaneously jumps to kx_0 and then relaxes to zero over time. This behavior of the force is called **relaxation**.

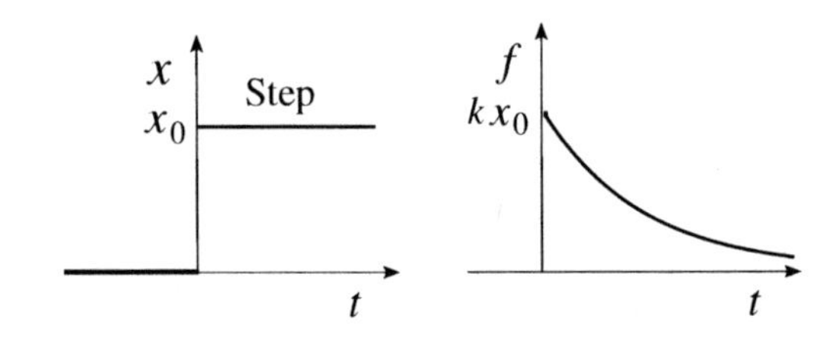

11. Application of Maxwell Model. In Exercise 10 there was no mass, just the "massless" spring and dashpot. Now attach a mass m to the Maxwell spring–dashpot model described in that exercise, as indicated below, where x and y are absolute displacements.

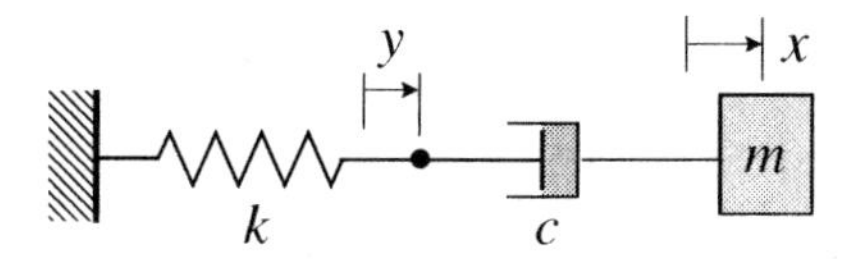

(a) Using Newton's second law, show that the equation of motion for the mass is

$$mx''' + \frac{mk}{c}x'' + kx' = 0. \tag{11.1}$$

(b) Since (11.1) is of *third order* we will need three initial conditions. It is natural to prescribe $x(0)$ and $x'(0)$, but how can we know $x''(0)$? Experimentally, we can expect to know $y(0)$, not $x''(0)$. Show that if $y(0)$ is indeed prescribed, then a third initial condition for (11.1) is $x''(0) = -ky(0)/m$.

(c) Since the coefficients in (11.1) are all positive constants, its solution is of the form

$$x(t) = C_1 + C_2x_2(t) + C_3x_3(t), \tag{11.2}$$

in which the $C_2x_2(t) + C_3x_3(t)$ part is an underdamped, critically damped, or overdamped oscillation. Show that these cases are obtained if $c > \sqrt{mk/4}$, $c = \sqrt{mk/4}$, and $c < \sqrt{mk/4}$, respectively.

(d) It is striking that for the Maxwell model c must be sufficiently *large* for an underdamped oscillation whereas for the Kelvin–Voigt model it must be sufficiently *small*. Explain why that result does make sense.

3.4 FORCED OSCILLATION

Last, we consider the case of forced oscillations (Fig. 1), governed by

$$mx'' + cx' + kx = f(t). \tag{1}$$

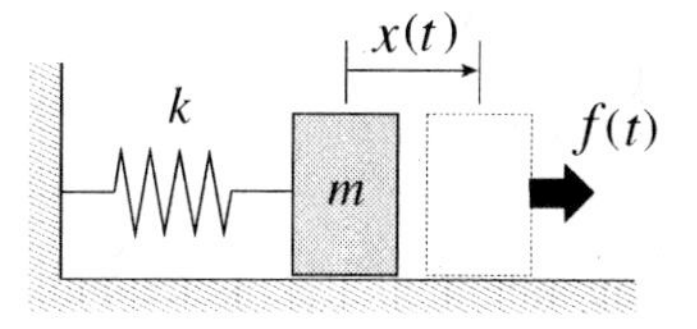

Figure 1. Mechanical oscillator.

What sort of forcing functions $f(t)$ are of interest? In applications, the case where $f(t)$ is an harmonic function is particularly important. As representative of that case, let

$$f(t) = F_0 \cos \Omega t, \tag{2}$$

with prescribed amplitude F_0 and frequency Ω, so (1) becomes

$$\boxed{mx'' + cx' + kx = F_0 \cos \Omega t.} \tag{3}$$

Harmonic excitation.

3.4.1 Undamped, $c = 0$.

To begin, consider the undamped case $c = 0$,

$$mx'' + kx = F_0 \cos \Omega t. \tag{4}$$

The solution of the homogeneous equation $mx'' + kx = 0$ is

$$x_h(t) = C_1 \cos \omega_n t + C_2 \sin \omega_n t, \tag{5}$$

in which $\omega_n = \sqrt{k/m}$ is the natural frequency (i.e., of the free oscillation). To find a particular solution of (4) by the method of undetermined coefficients, note that the forcing function $F_0 \cos \Omega t$ generates the family $\{\cos \Omega t, \sin \Omega t\}$. Hence, seek a particular solution of (4) in the form

Finding $x_p(t)$ by the method of undetermined coefficients.

$$x_p(t) = A \cos \Omega t + B \sin \Omega t. \tag{6}$$

Typically, of course, the driving frequency Ω is not identical to the natural frequency ω_n, so the terms in (6) do not duplicate any of those in (5) and we can accept (6) without modification.

Nonresonant oscillation, $\Omega \neq \omega_n$. Put (6) into the left side of (4) and obtain

$$\left(\omega_n^2 - \Omega^2\right) A \cos \Omega t + \left(\omega_n^2 - \Omega^2\right) B \sin \Omega t = \frac{F_0}{m} \cos \Omega t. \tag{7}$$

If this step, matching coefficients, is unclear see Exercise 8 in Section 2.4.

It follows from (7), by equating the coefficients of the LI functions $\cos \Omega t$ and $\sin \Omega t$ on the left and right sides, that $A = (F_0/m)/(\omega_n^2 - \Omega^2)$ and $B = 0$. Thus,

$$x_p(t) = \frac{F_0/m}{\omega_n^2 - \Omega^2} \cos \Omega t, \tag{8}$$

so a general solution of (4) is

$$\begin{aligned} x(t) &= x_h(t) + x_p(t) \\ &= C_1 \cos \omega_n t + C_2 \sin \omega_n t + \frac{F_0/m}{\omega_n^2 - \Omega^2} \cos \Omega t. \end{aligned} \tag{9}$$

If we impose initial conditions $x(0)$ and $x'(0)$, we can evaluate C_1 and C_2 in (9). Then, for any desired numerical values of m, k, F_0, and Ω we can plot $x(t)$ versus t to see what the solution looks like. However, we are interested not only in obtaining answers, but also in understanding phenomena, so we ask: What can we learn, from (9), about the phenomenon?

The homogeneous solution, the "free vibration," was discussed in Section 3.2. Thus, turn to the particular solution, or "forced response," which is given by (8) and which is the last term in (9). It is natural to regard m and k (and hence ω_n) as fixed, and F_0 and Ω as controllable constants or parameters. The dependence of (8) on F_0 is not very interesting: The forced response (8) simply scales with F_0. That is not surprising because the differential operator in (4) is linear, so [Theorem 2.7.2(a)] if we double the input we double the output, and so on. That is, the response to the forcing function $F_0 \cos \Omega t$ should be proportional to F_0, and it is.

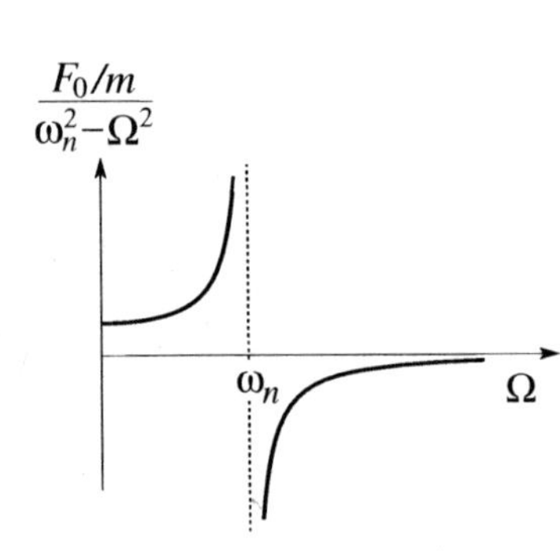

Figure 2. The coefficient of $\cos \Omega t$ in (9).

Particularly important, and not as obvious, is the variation of the amplitude $(F_0/m)/(\omega_n^2 - \Omega^2)$ with Ω, which is plotted in Fig. 2. The change in sign as Ω increases through ω_n prevents us from interpreting the plotted quantity as the amplitude of the response (8), for an amplitude is a magnitude and is positive. However, we can express (8) in the equivalent form

$$\boxed{x_p(t) = \frac{F_0/m}{|\omega_n^2 - \Omega^2|} \cos(\Omega t + \phi),} \tag{10}$$

in which the phase angle ϕ is 0 for $\Omega < \omega_n$ and π for $\Omega > \omega_n$.

Let us verify the claimed equivalence between (10) and (8): For $\Omega < \omega_n$, the $\omega_n^2 - \Omega^2$ in (10) is positive so we can drop the absolute magnitude braces; further, $\cos(\Omega t + \phi) = \cos(\Omega t + 0) = \cos \Omega t$, so for $\Omega < \omega_n$ the right-hand side of (10) agrees with the right-hand side of (8). For $\Omega > \omega_n$, we can express the $|\omega_n^2 - \Omega^2|$ as $\Omega^2 - \omega_n^2$ and the $\cos(\Omega t + \phi)$ as $\cos(\Omega t + \pi) = \cos \Omega t \cos \pi - \sin \Omega t \sin \pi = -\cos \Omega\, t$ so, once again, the right-hand side of (10) agrees with the right-hand side of (8).

The upshot is that $(F_0/m)/|\omega_n^2 - \Omega^2|$ in (10) is the amplitude, and ϕ is the phase angle. *Both* are functions of the "driving frequency" Ω, and are plotted versus Ω in Fig. 3. These are the **amplitude response** and **phase response** curves, respectively.

Observe from Fig. 3a that as the driving frequency Ω increases and approaches the natural frequency ω_n the amplitude tends to infinity! Further, as Ω increases beyond ω_n the amplitude decreases, and as $\Omega \to \infty$ the amplitude tends to zero. Finally, we see from Fig. 3b that the response $x_p(t)$ is *in-phase* (i.e., $\phi = 0$) with the forcing function if $\Omega < \omega_n$, but is 180° *out-of-phase* with it if $\Omega > \omega_n$.

This is a striking result, because imagine varying Ω very slowly, beginning with a value less than ω_n. As we do so, the response $x_p(t)$ remains in-phase with the force. However, when Ω increases from being slightly less than ω_n to being slightly greater than ω_n, the response undergoes a change so that it is then opposing (i.e., 180° out-of-phase) the force for all $\Omega > \omega_n$.

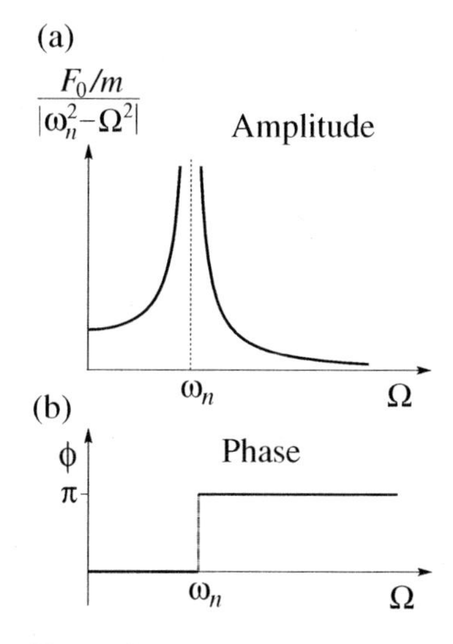

Figure 3. Amplitude response and phase response curves: amplitude and phase of $x_p(t)$ versus driving frequency Ω.

The response $x_p(t)$ being 180° out-of-phase with the force $F_0 \cos \omega t$ (i.e., $\phi = \pi$ in Fig. 3) means that when the force is rightward (i.e., positive) the mass is moving leftward, and vice versa.

Resonant oscillation, $\Omega = \omega_n$. For the special case $\Omega = \omega_n$ (that is, if we force the system precisely at its natural frequency), then the terms in (6) duplicate those in (5) so, according to the method of undetermined coefficients, revise x_p by multiplying the terms by t:

$$x_p(t) = t\,(A \cos \omega_n t + B \sin \omega_n t)\,. \tag{11}$$

Since the duplication has thereby been removed, accept (11) and put it into (4), obtaining $A = 0$ and $B = F_0/(2m\omega_n)$. Thus,

$$\boxed{x_p(t) = \frac{F_0}{2m\omega_n} t \sin \omega_n t,} \tag{12}$$

which is plotted in Fig. 4.

In this special case the response is not a harmonic oscillation but a harmonic function times t, and the t factor causes the magnitude to tend to infinity as $t \to \infty$. Striking is the unbounded growth of the amplitude, with time, which is called **resonance**. Is the resonance predicted by (12) for the case where $\Omega = \omega_n$ actually observed in the behavior of the physical system? Up to a point, because as the oscillation grows larger the idealized equation (4) no doubt becomes inadequate as a model of the physical system. For instance, for large deflections we hardly expect the linearized Hooke's law $F_s = kx$ to accurately describe the spring force. Also, when the motion becomes large the system may simply break.

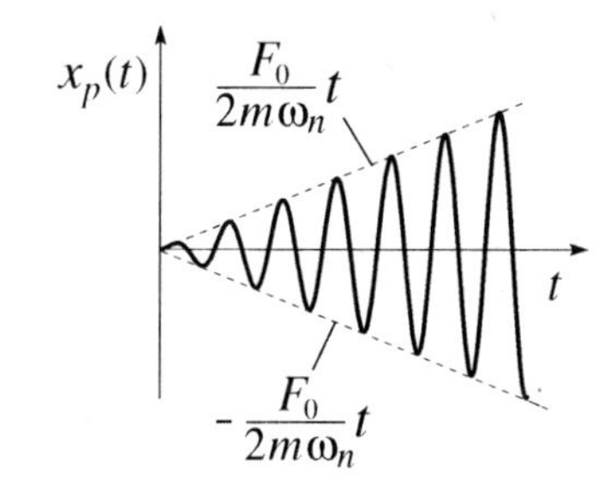

Figure 4. Resonant oscillation, where $\Omega = \omega_n$. The "envelope" is given by $\pm(F_0/2m\omega_n)t$.

3.4.2 Damped, $c > 0$.

Next, include a cx' damping term:

$$mx'' + cx' + kx = F_0 \cos \Omega t. \tag{13}$$

The homogeneous solution $x_h(t)$ was found in Section 3.3 and was given there by equations (6), (7), and (10), for the cases $c < c_{\text{cr}}$, $c = c_{\text{cr}}$, and $c > c_{\text{cr}}$, respectively. We won't repeat those results here.

To find a particular solution, use undetermined coefficients again, and seek

$$x_p(t) = A \cos \Omega t + B \sin \Omega t. \tag{14}$$

Leaving those steps for the exercises, the result is

$$\boxed{x_p(t) = \frac{F_0/m}{(\omega_n^2 - \Omega^2)^2 + (c\Omega/m)^2}\left[(\omega_n^2 - \Omega^2)\cos \Omega t + \frac{c\Omega}{m}\sin \Omega t\right].} \tag{15}$$

We've stressed that the damping term cx' results in $x_h(t)$ tending to zero as $t \to \infty$. That leaves $x_p(t)$, given by (15), as the *steady-state oscillation*. As in the undamped case, above, we are particularly interested in the amplitude of that oscillation. Since it is of the form $M \cos \Omega t + N \sin \Omega t$, where M and N are constants, its amplitude is $\sqrt{M^2 + N^2}$. Thus, the amplitude of the steady-state oscillation (15) is

Recall (15a) in Section 3.2.

$$\text{Amplitude} = \frac{F_0/m}{\sqrt{(\omega_n^2 - \Omega^2)^2 + (c\Omega/m)^2}}. \tag{16}$$

Of course, if $c = 0$, the latter reduces to (10), with resonance at $\Omega = \omega_n$. However, if $c > 0$, the $(c\,\Omega/m)^2$ term keeps the denominator in (16) from vanishing, so the peak amplitude is reduced from infinity to a finite value. To illustrate, let $m = 1$ and $k = 4$. The amplitude (16) is plotted in Fig. 5 for several values of c from $c = 0$ to $c = c_{\text{cr}} = 4$. Observe that as c increases the peak amplitude decreases and moves to the left, approaching $\Omega = 0$ as $c \to c_{\text{cr}}$.

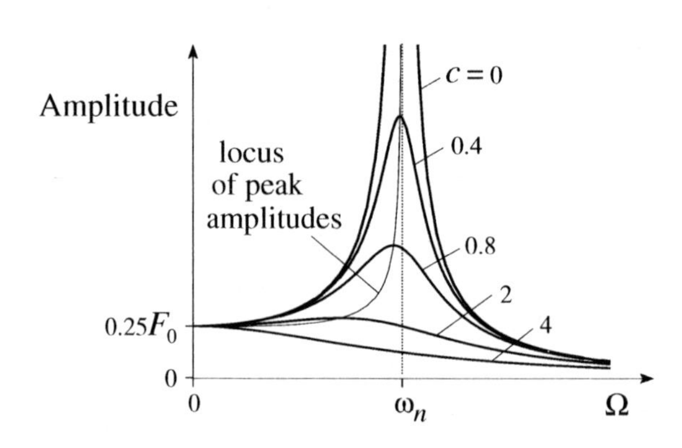

Figure 5. Amplitude response curves for $m = 1$, $k = 4$, and $c = 0, 0.4, 0.8, 2, 4$; $c_{\text{cr}} = 4$.

Besides the amplitude response, in Fig. 5, let us see what $x(t)$ looks like. With $m = 1$ and $k = 4$, as above, let $c = 0.4$, $F_0 = 2$, $\Omega = 0.5$, $x(0) = 1$, and $x'(0) = 0$. Omitting the details, we've plotted $x_h(t)$ and $x_p(t)$ separately in Fig. 6a, and their sum $x(t) = x_h(t) + x_p(t)$ in Fig. 6b. In this example, $x_h(t)$ is negligible by around $t = 25$, after which $x(t) \approx x_p(t)$. Realize that *the steady-state solution $x_p(t)$ is completely insensitve to the initial conditions because those conditions serve only to determine the constants C_1 and C_2 within $x_h(t)$, and $x_h(t) \to 0$ as $t \to \infty$.*

We mentioned that resonance is sometimes desirable and sometimes undesirable. For instance, consider the harmonic vertical forcing function that results when an automobile is driven over a rippled roadway. In that case we want the suspension system to minimize the amplitude of the resulting vertical vibration of the chassis. If we tune the suspension system so that its natural frequency ω_n is much smaller than the anticipated forcing function frequency Ω, then we see from Fig. 5 that the response amplitude will be small. We see the same thing from (16): The amplitude is asymptotic to $F_0/m)/\Omega^2$ as $\Omega \to \infty$, so it tends to zero proportional to $1/\Omega^2$. Thus, if $\Omega >> \omega_n$ then the response amplitude is small. In this application resonance is undesirable, and we want to keep the forcing frequency Ω and the natural frequency ω as far apart as we can.

However, suppose the automobile becomes stuck in a ditch. Someone gets behind it and pushes, not steadily but harmonically and *at* the natural frequency (that is, at which the automobile rocks in the ditch), to generate a large enough response to get the vehicle out of the ditch.

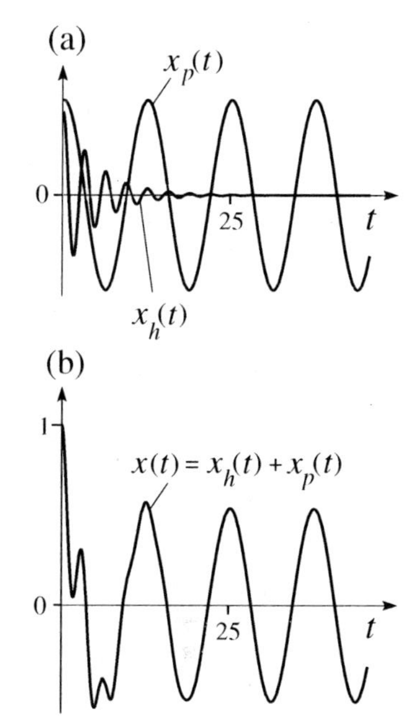

Figure 6. Homogeneous, particular, and total solutions of (13) for $F_0 = 2$, $\Omega = 0.5$, $x(0) = 1$, $m = 1$, $k = 4$, $c = 0.4$, $x(0) = 1$, and $x'(0) = 0$.

Closure. We've studied the oscillator equation $mx'' + cx' + kx = f(t)$ with harmonic forcing function $f(t) = F_0 \cos \Omega t$. To find particular solutions, we used the method of undetermined coefficients. The particular solution is especially important because even an infinitesimal amount of damping (i.e., with c arbitrarily small and positive) introduces an exponential factor $e^{-(c/2m)t}$ in the homogeneous solution, which causes it to tend to zero as $t \to \infty$, leaving the particular solution as the steady-state response. That is, after sufficient time the particular solution is essentially all that is left, as we see in Fig. 6.

Our discussion covered only the important case of harmonic excitation [i.e., in which $f(t)$ is sinusoidal or cosinusoidal]. Other cases are covered in the exercises, and in Chapter 5 we will show how to use the Laplace transform method to obtain solutions for a wide range of forcing functions.

EXERCISES 3.4

1. Derive the particular solution (15) of (13).

2. Show that if we change the $\cos \Omega t$ to $\sin \Omega t$ in (13), we obtain (16) once again, and hence we get the same amplitude response curves as in Fig. 5.

3. Amplitude Response Curves. For the damped harmonically driven oscillator modeled by (13), the amplitude of the steady-state oscillation $x_p(t)$ is given by (16). If we plot that amplitude versus Ω for various c's, we obtain a family of amplitude response curves such as those in Fig. 5. Show that the locus of peak amplitudes is given by

$$\text{Amp.} = \frac{F_0}{m}\frac{1}{\sqrt{\omega_n^4 - \Omega^4}}, \tag{4.1}$$

which tends to F_0/k as $\Omega \to 0$, and to ∞ as $\Omega \to \omega_n$.

4. Sketch the amplitude response curve if the spring is removed — that is, if $k = 0$ in (13). Explain, in physical terms, why it makes sense that the amplitude tends to infinity as $\Omega \to 0$.

5. We see, from (16), that the amplitude tends to F_0/k as $\Omega \to 0$, which is insensitive to c and m, and that it dies out as $F_0/m\Omega^2$ as $\Omega \to \infty$, which is insensitve to c and k. Explain, as best you can, why the low driving frequency behavior is dominated by the spring stiffness, while the high driving frequency behavior is dominated by the mass.

6. Energy Equation. If we multiply the force equation $mx'' + cx' + kx = F_0 \cos \Omega t$ by dx and integrate, we get the energy equation

$$\int mx''\,dx + \int cx'\,dx + \int kx\,dx = \int F_0 \cos \Omega t\,dx. \tag{6.1}$$

(a) Show that the latter gives

$$\frac{1}{2}mx'^2\Big|_{t_1}^{t_2} + c\int_{t_1}^{t_2} x'^2 dt + \frac{1}{2}kx^2\Big|_{t_1}^{t_2} = \int_{t_1}^{t_2} (F_0 \cos \Omega t)x'\,dt, \tag{6.2}$$

in which the terms on the left are kinetic energy of the mass, work absorbed by the dashpot, and potential energy of the spring, respectively, and the term on the right is the work done by the applied force $F_0 \cos \Omega t$.
(b) Once the steady-state oscillation has been attained, is it true that the work done per cycle by the applied force is equal to the work absorbed by the dashpot? Explain. NOTE: $t_2 = t_1 + 2\pi/\Omega$ in (6.2).
(c) Is the latter true *before* steady state is attained? Explain.

EXERCISES 7–9 COVER OTHER FORCING FUNCTIONS, WITH NO DAMPING

7. Constant Force. Solve the IVP

$$mx'' + kx = F_0; \quad x(0) = 0,\, x'(0) = 0, \tag{7.1}$$

and give a labeled sketch of the graph of $x(t)$.

8. Piecewise-Constant Force. Solve the IVP

$$mx'' + kx = f(t); \quad x(0) = 0,\, x'(0) = 0, \tag{8.1}$$

in which $f(t)$ is piecewise constant, being F_0 for $0 < t < T$ and 0 for $T < t < \infty$. NOTE: Piecewise-defined forcing functions will be dealt with most effectively by the Laplace transform, in Chapter 5. Meanwhile, we can solve such problems piece-by-piece: solve for $x(t)$ on $0 < t < T$, then use the final values $x(T)$ and $x'(T)$, from that solution, as initial conditions for $T < t < \infty$.

9. Piecewise-Constant Force. (a) First, read Exercise 8, but you need not solve it. This exercise is the same as Exercise 8, but this time let $f(t)$ be 0 on $0 < t < T$, and F_0 on $T < t < \infty$. Solve for $x(t)$.
(b) Call the solutions of Exercises 8 and 9(a) $x_1(t)$ and $x_2(t)$, respectively. Now consider the problem

$$mx'' + kx = F_0; \quad x(0) = 0,\, x'(0) = 0 \tag{9.1}$$

on $0 < t < \infty$. Is $x(t) = x_1(t) + x_2(t)$? Why (not)?
(c) Here is a variation on part (b). Consider these problems for $x_1(t)$ and $x_2(t)$,

$$mx_1'' + kx_1 = F_1(t); \quad x_1(0) = x_0,\, x_1'(0) = 0 \tag{9.2a}$$

and

$$mx_2'' + kx_2 = F_2(t); \quad x_2(0) = 0,\, x_2'(0) = x_0', \tag{9.2b}$$

where $F_1(t)$ is F_0 on $0 < t < T$ and 0 on $T < t, \infty$, and $F_2(t)$ is 0 on $0 < t < T$ and F_0 on $T < t < \infty$. Then, if

$$mx'' + kx = F_0; \quad x(0) = x_0,\, x'(0) = x_0', \tag{9.2c}$$

is $x(t) = x_1(t) + x_2(t)$? Explain (but you need *not* solve for $x_1(t)$, $x_2(t)$ or $x(t)$).

10. General Forcing Function; Variation of Parameters. In this section we've emphasized the important case of harmonic forcing functions, and some others are covered in Exercises 7–9. Particularly nice would be a formula for $x_p(t)$ for *any* force $f(t)$. Consider the undamped oscillator equation

$$mx'' + kx = f(t), \tag{10.1}$$

and use the method of variation of parameters to derive the formula

$$\boxed{x_p(t) = \frac{1}{\sqrt{mk}}\int_0^t \sin \omega_n (t-\tau) f(\tau)\,d\tau.} \tag{10.2}$$

11. Vibrating Base. An instrument of mass m sits on a laboratory platform, supported by pads having a total stiffness k and damping coefficient c, as indicated schematically in the figure. A disturbance causes the platform to undergo a vertical displacement $y(t) = y_0 \sin \Omega t$.

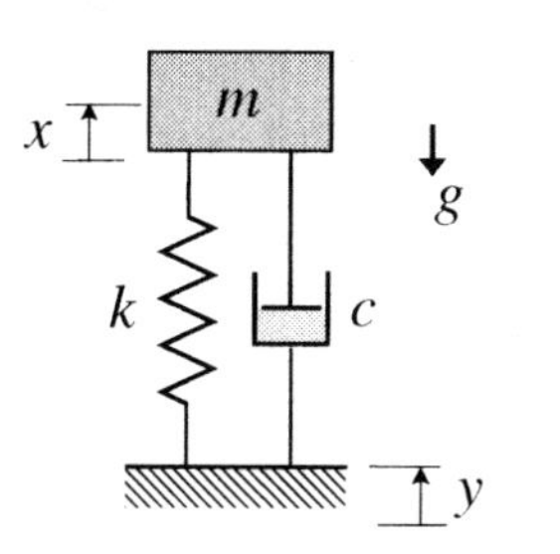

(a) Derive the equation of motion

$$mx'' + cx' + kx = -mg + ky(t) + cy'(t). \tag{11.1}$$

For parts (b) and (c), take $c = 0$:
(b) Determine the amplitude A of the instrument's steady-state oscillation (about the equilibrium position $x = -mg/k$), in terms of m, k, y_0, Ω.
(c) Suppose the anticipated values of y_0 and Ω fall in the ranges $0 \leq y_0 \leq y_{\max}$, $0 \leq \Omega \leq \Omega_{\max}$, and that the instrument must be protected so that its steady-state vibrational amplitude does not exceed some value $A_{\max}$ which, we suppose, is greater than $y_{\max}$. Show that the stiffness k must satisfy

$$k > \frac{m\Omega_{\max}^2 A_{\max}}{A_{\max} - y_{\max}} \tag{11.1}$$

if the instrument is to be protected.

12. Computer. In Fig. 6 we can see the approach of the solution $x(t)$ to the steady-state oscillation $x_p(t)$. To see the effect of increasing the damping on the speed of the approach to the steady-state oscillation, consider the same example as that one [$m = 1$, $k = 4$, $F_0 = 2$, $\Omega = 0.5$, $x(0) = 1$, and $x'(0) = 0$], and use computer software to generate the solution $x(t)$ for several damping coefficients: $c = 0.1$, 0.4, and 1 (and any additional ones if you like.) In this case $c_{cr} = 4$.

13. Vibrating Wall. Suppose we apply an harmonic displacement $X(t) = X_0 \cos \Omega t$ to the wall in the figure below. Neglecting any damping, derive the equation of motion for $x(t)$,

$$mx'' + kx = kX_0 \cos \Omega t. \tag{13.1}$$

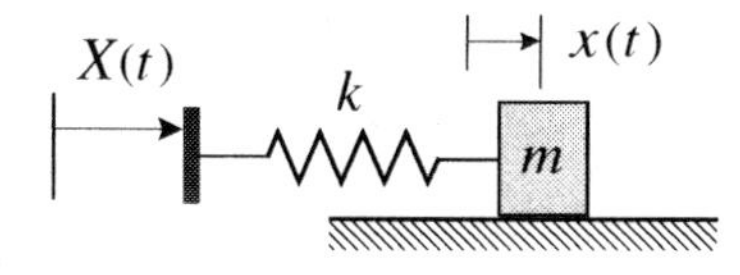

Both $x(t)$ and $X(t)$ are absolute displacements, that is, with respect to fixed points in space. NOTE: This example is interesting from an experimental point of view, because we can see a simple way of applying an harmonic *force* to the mass, namely, by giving the endpoint an harmonic *displacement*, the point being that harmonic displacements are readily generated, using a motor and a "Scotch yoke" mechanism.

14. Tuning a Circuit, Such as a Radio Circuit. Recall from Section 1.3.5 that the equations governing the current $i(t)$ in the circuit shown below, and the charge $Q(t)$ on the capacitor are

$$L\frac{d^2i}{dt^2} + R\frac{di}{dt} + \frac{1}{C}i = \frac{de(t)}{dt}, \tag{14.1}$$

and

$$L\frac{d^2Q}{dt^2} + R\frac{dQ}{dt} + \frac{1}{C}Q = e(t), \tag{14.2}$$

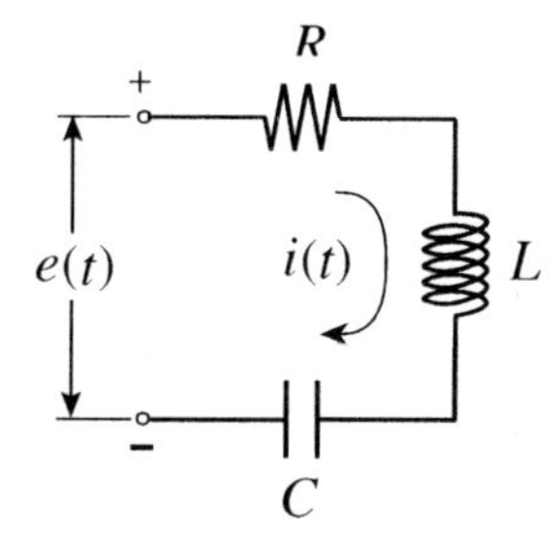

respectively, where L, R, C, e, i, and Q are measured in henrys, ohms, farads, volts, amperes, and coulombs, respectively. In this application the impressed voltage $e(t)$ is the sum of the voltage signals picked up from several input sources, such as broadcasting radio stations. For definiteness, suppose $L = R = 1$, and that we can vary the capacitance C continuously (for instance, by turning a knob). Suppose there are two input voltage signals, $a_1 \cos t$ and $a_2 \cos 2t$, so (14.2) becomes

$$Q'' + Q' + \frac{1}{C}Q = a_1 \cos t + a_2 \cos 2t, \tag{14.3}$$

and that we wish to amplify the response to the $a_1 \cos t$ forcing function, for instance, as much as possible. (That is, we wish to "tune in" to that station.) Using computer software if you wish, determine the best value of C to do that.

15. Beats. For the undamped forced oscillation, governed by the differential equation

$$mx'' + kx = F_0 \cos \Omega t, \tag{15.1}$$

we found that the general solution is

$$x(t) = C_1 \cos \omega_n t + C_2 \sin \omega_n t + \frac{F_0/m}{\omega_n^2 - \Omega^2} \cos \Omega t \tag{15.2}$$

if $\Omega \neq \omega_n$.

(a) Let $x(0) = 0$ and $x'(0) = 0$ for definiteness. Then show that (15.2) gives

$$x(t) = \frac{F_0/m}{\omega_n^2 - \Omega^2}(\cos \Omega t - \cos \omega_n t) \tag{15.3}$$

(b) Use the trigonometric identity

$$\cos A - \cos B = 2 \sin \frac{B+A}{2} \sin \frac{B-A}{2} \tag{15.4}$$

to re-express (15.3) as

$$x(t) = \frac{2F_0/m}{\omega_n^2 - \Omega^2} \sin(\frac{\omega_n + \Omega}{2})t \sin(\frac{\omega_n - \Omega}{2})t. \tag{15.5}$$

Now, suppose Ω is close to but not equal to ω_n. Then the frequency of the second sinusoid in (15.5) is very small compared to that of the first, so the factor $\sin(\frac{\omega_n - \Omega}{2})t$ amounts to a slow amplitude modulation of the relatively high frequency factor $\sin(\frac{\omega_n + \Omega}{2})t$, as shown in the figure below. This phenomenon is known as **beats**.

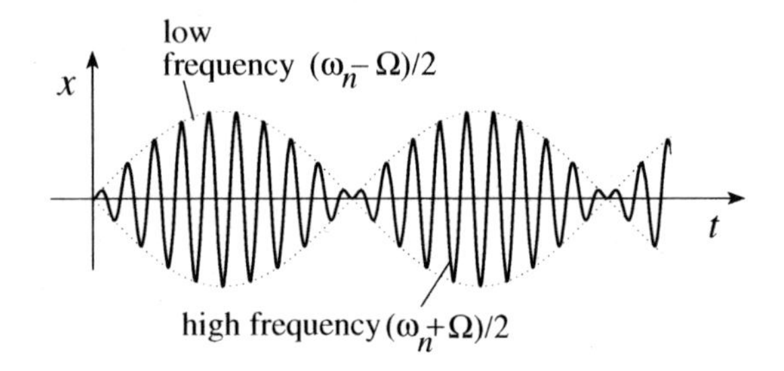

(c) **Using Computer Graphics.** We saw, in (b), that beats will occur if Ω is close to but not equal to ω_n in (15.3). To see it graphically, let $F_0/m = 1$ and $\omega_n = 10$, say, and obtain computer plots of the solution (15.5) for $\Omega = 5$ and 9. Do you observe the beat phenomenon from these plots?

16. Approach to Resonance is Through a Beat Phenomenon. If $\Omega = \omega_n$, then in place of (15.2), given above, the solution is

$$x(t) = C_1 \cos \omega_n t + C_2 \sin \omega_n t + \frac{F_0}{2m\omega_n} t \sin \omega_n t. \tag{16.1}$$

The latter exhibits resonance whereas (15.2) does not. How, we wonder, can the solution behavior be so dramatically different for *one* specific value of the driving frequency, being bounded if $\Omega \neq \omega_n$ and unbounded (as $t \to \infty$) if $\Omega = \omega_n$? We claim that the nonresonant solution (15.2) blends right into the resonant solution (16.1) by developing beats, and the purpose of this exercise is to see how that happens.

(a) Continue to set $x(0) = 0$ and $x'(0) = 0$, just for definiteness, as in Exercise 15. Then show that (16.1) gives

$$x(t) = \frac{F_0/m}{2\omega_n} t \sin \omega_n t. \tag{16.2}$$

(b) Next, show that if we let $\Omega \to \omega_n$ in (15.2) we do (with the help of l'Hôpital's rule) obtain (16.2). Thus, the nonresonant solution (15.2) does blend smoothly into the resonant solution (16.2).

(c) Finally, show how that blending occurs, by using computer graphics. HINT: The key is that as $\Omega \to \omega_n$ the solution (15.2) develops a beat phenomenon.

ADDITIONAL EXERCISES

17. Nonlinear Oscillator: Duffing's Equation and its Jump Phenomenon. The linear Hooke's law $F_s = kx$ for the spring force is generally accurate only if x is sufficiently small. For larger x's one can often improve the model by including a cubic term as well:

$$F_s(x) \approx \alpha x + \beta x^3, \tag{17.1}$$

where α and β are constants. For instance, for the pendulum equation $x'' + (g/l) \sin x = 0$ we approximated the nonlinear $\sin x$ function by just the first term of its Taylor series about $x = 0$, giving $\sin x \approx x$ and hence the linearized equation $x'' + (g/l)x = 0$. For larger motions we can keep not just the first term but the first two. Then $\sin x \approx x - x^3/3!$, so $\alpha = 1$ and $\beta = -1/6$ in (17.1). If $\beta > 0$, we say that the spring is a "hardening spring" because it becomes stiffer as $|x|$ increases; and if $\beta < 0$, it is a "softening spring"; the corresponding graphs are sketched below. With the spring force

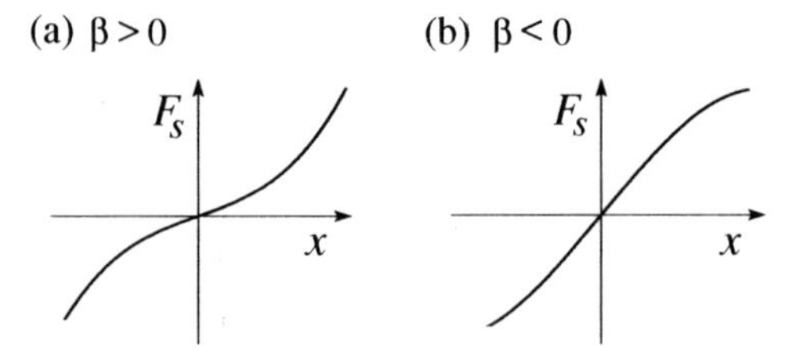

given by (17.1), the equation of motion becomes

$$\boxed{mx'' + cx' + \alpha x + \beta x^3 = F_0 \cos \Omega t,} \tag{17.2}$$

which was studied by *G. Duffing* around 1918 and is known as **Duffing's equation**. *It is essential to understand* that the idea of finding a general solution of (17.2) in the form $x(t) = C_1x_1(t) + C_2x_2(t) + x_p(t)$, in which $x_1(t)$ and $x_2(t)$ are homogeneous solutions and $x_p(t)$ is a particular solution, is incorrect; *it would be valid if (17.2) were linear, but it is not, because of the* x^3 *term.* The theory and solution methods presented in Chapter 2 *do not apply* because (17.2) is nonlinear!

Using methods that are beyond our present scope one can obtain amplitude response curves that are analogous to those given in Fig. 5. However, we find that they *bend* — to the right if $\beta > 0$ and to the left if $\beta < 0$, as shown below for $\beta > 0$. For the moment, pay no attention to the arrows. To see the

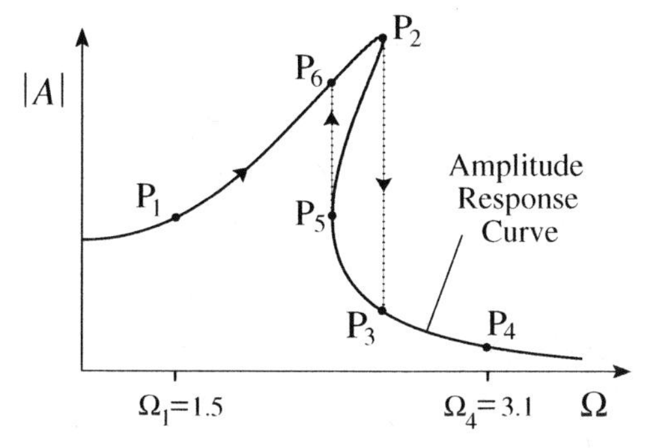

implications of the bending of the response curve, suppose we carry out a numerical experiment and begin with Ω equal to Ω_1. Then the response amplitude is at P_1. If we increase the forcing frequency Ω very slowly (remember that Ω is regarded here as a constant, so we must increase it *very* slowly) then the representative point moves to the right along the response curve, as indicated by the arrow. What happens when it reaches P_2, at which point the response curve has a vertical tangent? Numerical simulation shows that the point jumps from P_2 down to P_3 and continues moving rightward along the lower branch of the response curve if we continue to increase Ω. Suppose we stop at P_4. If we now *decrease* Ω the representative point does *not* jump up from P_3 to P_2. Rather, it continues to the point of vertical tangency at P_5, *then* jumps up to P_6 and continues leftward.

This is the **jump phenomenon** associated with the Duffing equation, whereby a *continuous* variation of a system parameter (Ω in this case) can lead to a *discontinuous* change in the output, namely, the jumps in amplitude from P_2 to P_3 and from P_5 to P_6. We call $P_2P_3P_5P_6P_2$ a **hysteresis loop**.

Finally, here is the problem:

(a) We state without derivation that an approximate equation for the amplitude response curves is

$$F_0^2 = \left[\left(\alpha - m\Omega^2\right) A + \frac{3}{4}\beta A^3\right]^2 + \left(c\Omega A\right)^2. \tag{17.4}$$

For definiteness, let $F_0 = 2$, $\alpha = m = 1$, $\beta = 0.4$, and $c = 0.3$. Then (17.4) becomes

$$4 = \left[\left(1 - \Omega^2\right) A + 0.3A^3\right]^2 + \left(0.3\Omega A\right)^2., \tag{17.5}$$

and (17.2) becomes

$$x'' + 0.3x' + x + 0.4x^3 = 2\cos\Omega t. \tag{17.6}$$

Use computer software to obtain from (17.5) a graph of the amplitude response curve shown above. From that graph, measure the (Ω, A) coordinates of points P_2, P_3, P_5, and P_6.

(b) Now carry out the numerical experiment suggested above. Namely, let Ω vary very slowly, from Ω_1 to Ω_4 and see if the response amplitude follows the path $P_1P_2P_3P_4$ with downward jump from P_2 to P_3. To do that, use the slow variation $\Omega = 1.5 + 0.0002t$, say, from $t = 0$ to $t = 3500$. That is, let Ω in (17.6) be $1.5 + 0.0002t$ and use computer software to solve (17.6) for $x(t)$. You will need initial conditions, so take $x(0) = 0$ and $x'(0) = 0$. Plot the solution $x(t)$ versus t and discuss whether your results exhibit the downward jump in amplitude predicted by the discussion in the beginning of this exercise. You should be able to check both the magnitude of the amplitude jump and the Ω at which it occurs.

(c) Explain, qualitatively, why the amplitude response curves bend to the right for $\beta > 0$ (as in this exercise) and to the left for $\beta < 0$.

3.5 STEADY-STATE DIFFUSION; A BOUNDARY VALUE PROBLEM

The applications in Sections 3.2–3.4 were of initial value type because the governing second-order differential equations were augmented by two conditions at the same point on the independent variable axis, $t = 0$. In fact, problems in which the independent variable is the time are, typically, of initial value type.

On the other hand, problems in which the independent variable is a space variable are often not of initial value type, but of **boundary value type**, which means that conditions are specifed not at a single point but at more than one point, typically at the two endpoints of the interval under consideration. Boundary value problems (BVPs) did not arise in Chapter 1 because only one condition is appropriate for a first-order equation, so one would not have conditions specified at more than one point. Thus, the problems in Chapter 1 were of initial value type. A few BVPs were encountered in Chapter 2, but discussion of BVPs was delayed until now. We begin with a brief general discussion, and then we turn to applications in the remainder of this section and the next.

BVP abbreviation for boundary value problem.

3.5.1 Boundary value problems; existence and uniqueness. To show the implications of initial value versus boundary value, consider a second-order linear differential equation $Ly = f$ with general solution

$$y(x) = y_p(x) + C_1 y_1(x) + C_2 y_2(x), \tag{1}$$

in which y_1 and y_2 are LI homogeneous solutions.

Suppose first that two *initial* conditions are specified at some point x, $y(a) = b_1$ and $y'(a) = b_2$, so we have the IVP

IVP version.

$$Ly = f; \quad y(a) = b_1,\ y'(a) = b_2 \tag{2}$$

on $a < x < \infty$, for instance. Applying the initial conditions to (1) gives

$$y(a) = y_p(a) + C_1 y_1(a) + C_2 y_2(a), \tag{3a}$$
$$y'(a) = y_p'(a) + C_1 y_1'(a) + C_2 y_2'(a), \tag{3b}$$

which are the two equations

$$y_1(a)C_1 + y_2(a)C_2 = y(a) - y_p(a) \tag{4a}$$
$$y_1'(a)C_1 + y_2'(a)C_2 = y'(a) - y_p'(a). \tag{4b}$$

for C_1 and C_2. Observe that the existence and uniqueness of solutions of the IVP (2) amounts to the existence and uniqueness of the solution of the system of linear *algebraic* equations (4) for C_1, C_2. That, in turn, depends on the determinant of the coefficients in (4). The key point is this: that determinant is the Wronskian of the two solutions y_1 and y_2 (evaluated at a), and since those solutions are LI, for (1) to be a general solution, that determinant must be nonzero. And since the determinant of the coefficient matrix is nonzero, there does exist a unique solution of (4) for C_1, C_2. Hence, there exists a unique solution (1) of (2).

Now consider a boundary value version of the problem,

$$Ly = f; \quad y(a) = b_1,\ y(b) = b_2 \tag{5}$$

BVP version; conditions given at $x = a$ and $x = b$.

on $a < x < b$. If we apply the boundary conditions to the general solution (1) we obtain, in place of (4),

$$y_1(a)C_1 + y_2(a)C_2 = y(a) - y_p(a) \tag{6a}$$

$$y_1(b)C_1 + y_2(b)C_2 = y(b) - y_p(b). \tag{6b}$$

This time, the determinant of the coefficient matrix is not the Wronskian of y_1 and y_2, so it may be nonzero or zero. If it is nonzero there is a unique solution for C_1, C_2 and hence for $y(x)$, but if it is nonzero there will not be a unique solution; there will be either no solution, or an infinity of solutions (that is, a nonunique solution).

The foregoing is readily generalized to equations of order greater than two, and the upshot is that **BVPs can have a unique solution, no solution, or a nonunique solution**. The three cases will be illustrated in the application to heat conduction that follows.

IMPORTANT. Three possibilities for a BVP.

3.5.2 Steady-state heat conduction in a rod. Consider the temperature distribution in a homogeneous rod of length L, shown in Fig. 1. By homogeneous we mean it is of the same material throughout, such as aluminum or glass. We assume the temperature u is uniform on each cross section, and that the rod is in thermal equilibrium (steady state), so that u is a function only of x, not of t. To derive the differential equation for $u(x)$, isolate an arbitrary element of the rod, between any x and $x+\Delta x$ (Fig. 2), and write down the relevant physics for that element, namely, a heat balance:

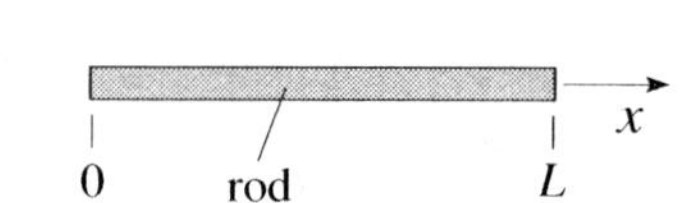

Figure 1. Heat conduction in a rod.

$$\begin{array}{c}\text{Rate of increase}\\ \text{of heat in}\\ \text{the element}\end{array} = \underbrace{\text{Rate in}}_{Q_1} - \underbrace{\text{Rate out}}_{Q_2+Q_3} + \underbrace{\text{Rate generated}}_{q}. \tag{7}$$

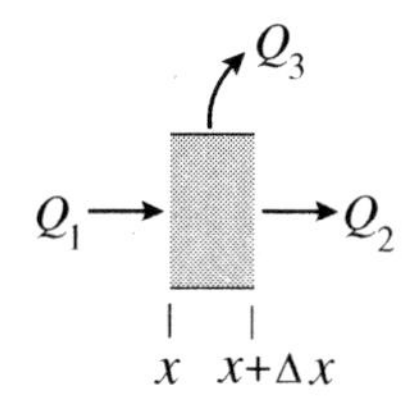

Figure 2. Heat flow into and out of a typical element of the rod.

We want to obtain expressions for the terms on the right-hand side of (7). Consider first the rate at which heat flows into the element at x, denoted in (7) as Q_1. Take the heat flow (calories per second) at x and $x + \Delta x$ to be to the right, as our sign convention; if the heat flow at x is to the left, then Q_1 will be negative. Similarly at $x + \Delta x$.

The **Fourier law of heat conduction** states that Q_1 is proportional to the cross-sectional area A of the rod and also to the **temperature gradient** du/dx at x.[1] Thus, Q_1 is proportional to Adu/dx and is therefore a constant, say k, times

[1]That Q_1 is proportional to A should be no surprise; after all, in the winter we lose heat through a window in proportion to its area. That Q_1 is also proportional to the slope du/dx is also reasonable because the heat flow is "driven" by a temperature difference; the larger du/dx is, the larger is the temperature difference between two points.

Adu/dx. Hence,

$$Q_1 = -kA\frac{du}{dx}(x), \tag{8}$$

in which k is a positive constant called the **thermal conductivity** of the material; the conductivities of some common materials are given in the margin.

Values of k for some common materials, in kcal/(hr m °C):

Silver	213
Copper	200
Aluminum	125
Glass	0.52
Water	0.29
Cast iron	0.16

In engineering units heat is measured in Btu's (British thermal units), and 1 kcal/ (hr m °C) = 0.6721 Btu/(hr ft °F).

To understand the minus sign in (8), note first that both k and A are positive. Now, if du/dx is positive then the right-hand side of (8) is negative, indicating that the heat flow is leftward, as it should be, because heat flows from hot to cold. On the other hand, if du/dx is negative, then (8) gives a positive value of Q_1, as it should because the heat flow will then be rightward.

Similarly, the heat flow Q_2 at $x+\Delta x$ is

$$Q_2 = -kA\frac{du}{dx}(x+\Delta x). \tag{9}$$

Next, the Q_3 term accounts for a heat exchanged between the lateral surface of the rod and the environment, which we assume is at a uniform constant temperature u_∞. If, for definiteness, $u(x)$ is greater than u_∞, then the element will lose heat to the environment at a rate that, according to **Newton's law of cooling**, is proportional to the temperature difference $u-u_\infty$, and to the element's surface area $(2\pi R\Delta x)(u-u_\infty)$, where R is the radius of the rod. Thus,

$$Q_3 = h(2\pi R\Delta x)[u(\xi)-u_\infty], \tag{10}$$

in which ξ is a point (between x and $x+\Delta x$) at which u takes on its average value on that interval, and the constant of proportionality h is a *heat transfer coefficient*, which we assume is known.

Finally, the last term in (7), q, is associated with any internal heating (or cooling) that may be present within the rod. Heat can be generated within the body in various ways. For instance, suppose the rod is made of paraffin and sunlight impinges on the end of the rod at $x=0$. The sunlight penetrates the paraffin and heats it from within, somewhat as a potato is heated internally in a microwave oven.

Thus, let us allow for internal heating, and let $f(x)$ be the volumetric heat generation rate (cal/sec per unit volume). Let f be a continuous function, and suppose its average value within the element under consideration occurs at some point μ in that interval. Multiplying $f(\mu)$ by the volume of the element, $A\Delta x$, we can express q in (7) as

$$q = f(\mu)A\Delta x. \tag{11}$$

Let us check the units in $f(\mu)A\Delta x$: $f(\mu)$ has units of heat per unit volume per unit time, and $A\Delta x$ has units of volume, so their product has units of heat per unit time, which are the correct units for q.

Next, put (8)–(11) into the right-hand side of (7), and realize that the left-hand side is zero because, by the assumption of steady state, all time rates of change are zero. Thus,

$$0 = -kA\frac{du}{dx}(x) + kA\frac{du}{dx}(x+\Delta x) - h(2\pi R\Delta x)[u(\xi) - u_\infty] + f(\mu)A\Delta x \tag{12}$$

or

$$k\,\frac{\dfrac{du}{dx}(x+\Delta x) - \dfrac{du}{dx}(x)}{\Delta x} - \frac{2\pi Rh}{A}[u(\xi)-u_\infty] = -f(\mu). \tag{13}$$

The latter holds for Δx arbitrarily small, so we can take the limit as $\Delta x \to 0$. Doing so, and realizing that both ξ and μ are forced to x as $\Delta x \to 0$, the difference quotient becomes $\frac{d^2u}{dx^2}(x)$, $u(\xi)$ becomes $u(x)$, and $f(\mu)$ becomes $f(x)$, so we obtain the differential equation

Both ξ and μ are between x and $x + \Delta x$, so they are forced to x as $\Delta x \to 0$.

$$\boxed{\frac{d^2u}{dx^2} - \beta(u - u_\infty) = -\frac{1}{k}f(x),} \tag{14}$$

for the steady-state temperature distribution $u(x)$ in the rod shown in Fig. 1. We've used $A = \pi R^2$, and have set $\beta = 2h/kR$ for brevity. We could write (14) as $u'' - \beta u = -\beta u_\infty - f(x)/k$, to have the u's on the left and known terms on the right, but it seems good to keep $\beta(u - u_\infty)$ intact since that group of terms, taken together, models the rate of heat loss to the environment.

The derivation of the governing differential equation (14), together with decisions as to the relevant boundary conditions, discussed below, is the *modeling* part of our analysis. The procedure was to isolate a typical element of the system, in this case a piece of the rod, between x and $x + \Delta x$, and to apply whatever law(s) are relevant. In this case, those laws were the heat balance in (7), the Fourier law of heat conduction expressed by (8) and (9), and Newton's law of cooling expressed by (10). This general procedure is typical in engineering science, and is found elsewhere in this text, as for instance in Exercise 33 of Section 1.3.

EXAMPLE 1. Boundary Conditions, Existence, and Uniqueness. Suppose there is no internal heat generation, so $f(x) = 0$. Further, suppose the lateral face (but not the two end faces) is perfectly insulated so that h, and hence β, is zero. In that case, (14) is simply

$$u'' = 0, \tag{15}$$

with general solution

$$u(x) = C_1 + C_2 x. \tag{16}$$

That is, the steady-state temperature distribution is simply linear.

Incidentally, we've considered the one-dimensional diffusion of heat in the rod shown in Fig. 1, but if $h = 0$ the resulting equation $u'' = 0$ also models the one-dimensional heat conduction through a slab such as a concrete outside wall of a house. Let the x axis be normal to the wall as in Fig. 3, with Cartesian y and z axes normal to it; let y be into the paper and z be upward. If it is warm inside the house and cold outside, for instance, then the temperature distribution u in the wall will vary appreciably with x, on $0 < x < L$, but hardly at all with y and z. Thus, to a good approximation the heat transfer is still essentially one-dimensional, being nonnegligible only in the x direction. The upshot is that equation (15) governs the steady-state temperature distribution $u(x)$ not only within an insulated rod, but also within the slab in Fig. 3.

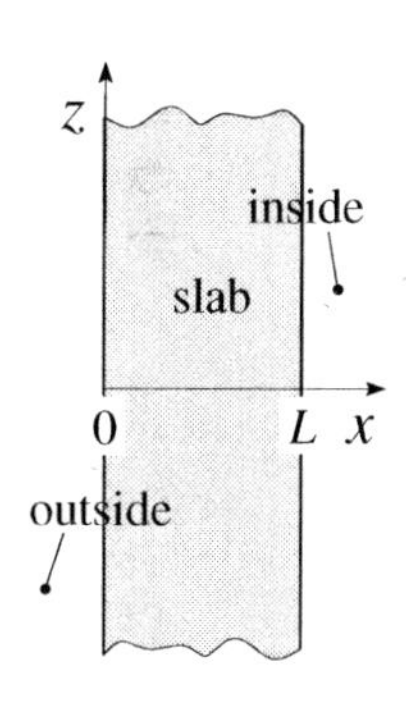

Figure 3. One-dimensional heat conduction in a slab, such as the wall of a house.

Next, consider various boundary conditions, to go along with (15). Three types of conditions, at an end of the rod, are as follows:

$$u \text{ given}: \quad \textbf{first kind}, \text{ Dirichlet}, \tag{17a}$$

$$u' \text{ given : } \textbf{second kind}, \text{ Neumann,} \tag{17b}$$

$$\alpha_1 u + \alpha_2 u' \text{ given : } \textbf{third kind}, \text{ Robin,} \tag{17c}$$

Boundary conditions $u = 0$, $u' = 0$, and $\alpha_1 u + \alpha_2 u' = 0$ are **homogeneous** Dirichlet, Neumann, and Robin conditions, respectively.

in which α_1 and α_2 are nonzero constants.[1]

In physical terms, in the present context, a Dirichlet boundary condition amounts to the temperature u being prescribed at that end. To understand the Neumann boundary condition, recall that the rate of heat transfer through an end face is, by Fourier's law of heat conduction, $-kAu'$. Thus, prescribing u' at an end of the rod amounts to prescribing the heat flow by conduction through that end; it is a *heat flux* condition. In particular, if that condition is homogeneous (that is, if $u' = 0$), then there is no heat flux through that end; it is perfectly insulated. We will defer comment on the Robin-type boundary condition for a moment.

With three possible types of boundary conditions at each of two rod ends, we could consider six combinations. Rather, we will consider just three — that will illustrate the cases of unique solution, no solution, and nonunique solution:

$\mathbf{u(0) = u_1}$, $\mathbf{u(L) = u_2}$: If we apply these Dirichlet conditions to the general solution (16) of the differential equation (15), we can solve for C_1, C_2 uniquely, and obtain the *unique solution*

$$u(x) = u_1 + \frac{u_1 - u_1}{L} x, \tag{18}$$

which is simply the linear variation in Fig. 4, in which we nominally took $u_2 > u_1 > 0$.

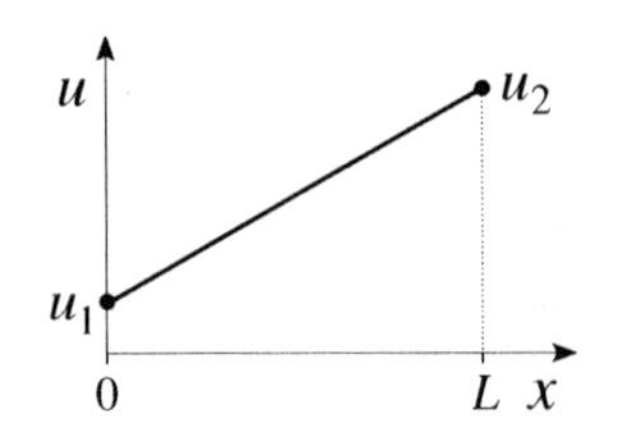

Figure 4. Steady-state solution, for $u(0) = u_1$ and $u(L) = u_2$. Unique solution given by (18).

$\mathbf{u'(0) = u'_1}$, $\mathbf{u'(L) = u'_2}$: Now consider Neumann conditions at both ends. Applying them to (16) gives this simple system of algebraic equations for C_1, C_2:

$$u'_1 = C_2, \quad [\text{i.e., } 0C_1 + C_2] \tag{19a}$$

$$u'_2 = C_2, \quad [\text{i.e., } 0C_1 + C_2]. \tag{19b}$$

Two possibilities exist. If $u'_1 \neq u'_2$ then (19a,b) are inconsistent and there is *no solution*, but if $u'_1 = u'_2$ then they are not only consistent, but there now exist an infinite number of solutions, because $C_2 = u'_1$, and C_1 is arbitrary! To understand that result in physical terms, suppose for instance that $u'_1 < 0$ and $u'_2 > 0$. Those conditions correspond to a heat flux into the rod at both ends. Now, if heat is entering continually at both ends, and the rod's lateral surface is insulated, then there is a continual buildup of heat in the rod and it cannot be in thermal equilibrium, as we supposed in deriving (14). Thus, the significance of the mathematical nonexistence of a solution of the BVP is that such a steady state is physically impossible! However, if $u'_1 = u'_2$, then the heat in at one end equals the heat out at the other, and thermal equilibrium is obtained. In this case the solution is $u(x) = C_1 + (u'_1)x$, with C_1 arbitrary. This linear solution can be shifted up or down arbitrarily by the choice of C_1. It does satisfy $u'' = 0$ and the two boundary conditions, for any value of C_1.

COMMENT 1. What is the physical meaning of a boundary condition of the third kind, at $x = L$ for instance,

$$\alpha_1 u(L) + \alpha_2 u'(L) = C, \tag{20}$$

[1] *Peter Gustav Lejeune-Dirichlet* (1805–1855), *Carl Neumann* (1832–1925), and *Victor Robin* (1855–1897) were German, German, and French mathematicians, respectively.

in which α_1, α_2, and C are prescribed, that is, a Robin condition? Consider the end face of the rod at $x = L$. The heat crossing that face from the left must equal the heat crossing from that face to the environment. By the Fourier law of conduction, the former is $-kAu'(L)$, and by Newton's law of cooling the latter is $hA[u(L) - u_\infty]$. Equating these gives

$$-kAu'(L) = hA[u(L) - u_\infty] \tag{21}$$

or $hu(L) + ku'(L) = hu_\infty$, which is indeed of the form (17c), with $\alpha_1 = h$ and $\alpha_2 = k$. Note these two extreme cases: First, in the limit as $h \to \infty$, (21) gives $u(L) - u_\infty = 0$; that is $u(L) = u_\infty$. The ambient temperature y_∞ simply gets impressed at the end of the rod. Second, in the limit as $h \to 0$, (21) becomes the homogeneous Neumann condition $u'(L) = 0$, which means the end of the rod is perfectly insulated from the environment. Thus, in the context of heat transfer, think of the Dirichlet and Neumann boundary conditions as approximations of a probably-more-accurate Robin condition.

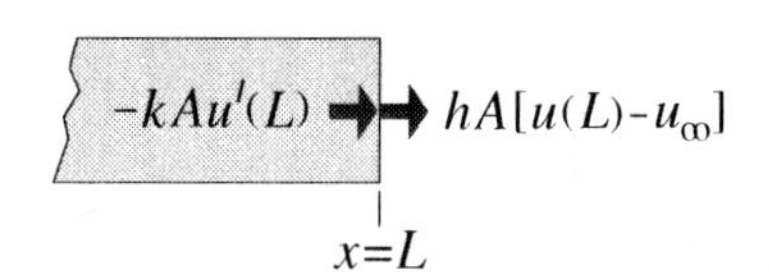

Figure 5. A Robin boundary condition at $x = L$.

If it is not clear that in the limit $h \to \infty$ (21) becomes $u(L) - u_\infty = 0$, try it this way: divide both sides of (21) by h and *then* let $h \to \infty$.

COMMENT 2. We simplified (14) to the form (15) because our aim was to illustrate the idea of the existence and uniqueness of solutions to BVPs with a simple example, but (14) is not a difficult equation to solve. We leave it for the exercises. ▮

Closure. In Section 3.5.1 we stressed the difference between initial value problems and boundary value problems: When *initial* conditions are applied to a general solution, the determinant of the coefficients is $W[y_1, y_2]$, at the initial point, which is nonzero because y_1 and y_2 are LI solutions. Thus, there is a unique solution for C_1 and C_2 in (1) and hence a unique solution of the IVP (2) for $y(x)$. For a BVP, however, that determinant *is not the Wronskian*, so it can be zero or nonzero. Hence, the solution of the BVP (5) can be unique, not exist, or exist and be nonunique.

We derived the equation (14) governing the steady-state temperature distribution $u(x)$ in a rod, allowing for heat generation within the rod and for Newton cooling to the environment. In Example 1 we let both the generation term $f(x)$ and the heat transfer coefficient h be zero so (14) reduced to $u'' = 0$, and illustrated the three possibilities (unique solution, no solution, and nonunique solution) by adopting different sets of boundary conditions. Bear in mind that by "Newton cooling" we mean cooling or heating: If the temperature u in the rod is higher than u_∞, then Q_3 in (10) is positive and we have Newton *cooling*; and if u is less than u_∞, then Q_3 is negative and we have Newton *heating*.

The cases of heat conduction in cylinders and spheres are left for the exercises. Whereas for the one-dimensional Cartesian geometry of a rod the differential equation (14) was of constant-coefficient type, for cylinders and spheres it will be found to be of Cauchy–Euler type.

The diffusion of mass, rather than heat, is also left for the exercises.

EXERCISES 3.5

BOUNDARY VALUE PROBLEMS

1. To illustrate that a BVP can have a unique solution, no solution, or an infinity of solutions, consider the simple BVP

$$u'' = 0; \qquad (0 < x < L) \tag{1.1}$$

subject to the given boundary conditions. Solve that BVP and cite conditions, if any, on b_1, b_2 that result in each of these three cases: no solution, unique solution, infinity of solutions.

(a) $y(0)=b_1,\ y(L)=b_2$ (b) $y(0)=b_1,\ y'(L)=b_2$
(c) $y'(0)=b_1,\ y'(L)=b_2$ (d) $y'(0)=b_1,\ y(L)=b_2$

2. To illustrate that a BVP can have a unique solution, no solution, or an infinity of solutions, consider the BVP

$$y''+y=0, \qquad (0<x<\pi) \tag{2.1}$$

subject to the given boundary conditions. Solve that BVP and state whether the solution is unique or nonunique, or whether a solution does not exist.

(a) $y(0)=50,\ y(\pi)=100$ (b) $y(0)=10,\ y(\pi)=-10$
(c) $y(0)=0,\ y'(\pi)=0$ (d) $y'(0)=0,\ y(\pi)=0$
(e) $y(0)=50,\ y'(\pi)=0$ (f) $y'(0)=0,\ y'(\pi)=0$
(g) $y'(0)=25,\ y'(\pi)=-25$ (h) $y'(0)=5,\ y'(\pi)=9$

3. Introduction to the Eigenvalue Problem. (The eigenvalue problem will be covered in the next section, but is introduced here in the event that your syllabus does not include that section.) It is tempting to think that if there are no inputs then the output must be zero. That is, if both the differential equation and the boundary conditions are homogeneous, then the solution must be the "trivial" solution $y(x)=0$. However, we will see in this exercise that **a homogeneous BVP can have nontrivial solutions**. To illustrate, consider

$$y''+\lambda y=0, \qquad (0<x<L) \tag{3.1}$$

in which λ is a positive parameter that is not specified in advance, subject to the homogeneous boundary conditions given below. Solve for $y(x)$, and show that the trivial solution is the only solution — *unless* λ is assigned one of an infinite set of values, in which case there will be an infinity of nontrivial solutions besides the trivial solution. NOTE: This is an example of an **eigenvalue problem**, and the λ's that lead to the existence of nontrivial solutions are called the **eigenvalues** for that problem. More about the eigenvalue problem in Section 3.6 and, for matrices, in Chapter 4.

(a) $y(0)=0,\ y(L)=0$ (b) $y'(0)=0,\ y(L)=0$
(c) $y(0)=0,\ y'(L)=0$ (d) $y'(0)=0,\ y'(L)=0$

HEAT CONDUCTION IN RODS

4. In (14), let the rod be insulated, so $\beta=0$, and let $f(x)=f_0$ be a constant. Solve the resulting BVP comprised of $u''=-f_0/k$ and the given boundary conditions. Does a solution exist? If so, is it unique? Is it nonunique?

(a) $u(0)=0,\ u(L)=0$ (b) $u(0)=0,\ u'(L)=0$
(c) $u(0)=20,\ u(L)=50$ (d) $u(0)=0,\ u'(L)=-5$
(e) $u'(0)=10,\ u'(L)=10$ (f) $u'(0)=0,\ u(L)=200$

5. As in Exercise 4, let $\beta=0$ and $f(x)=f_0$ in (14). Let the boundary conditions be $u'(0)=u_1'$ and $u'(L)=u_2'$, where the constants u_1' and u_2' can be positive, negative, or zero.
(a) Solve the BVP,

$$u''=-f_0/k; \quad u'(0)=u_1',\ u'(L)=u_2' \tag{5.1}$$

for $u(x)$. Show that unless u_1' and u_2' are related according to

$$k(u_2'-u_1')=-f_0L \tag{5.2}$$

there is no solution for $u(x)$. Show further that if (5.2) *is* satisfied, then a solution exists, but is nonunique in that it can be determined only to within an arbitrary additive constant.
(b) Show that the condition (5.2) can also be obtained from the differential equation itself, $u''=-f_0/k$, by integrating it from $x=0$ to $x=L$.
(c) Give a physical explanation of (5.2). HINT: Multiply (5.2) by the rod's cross-sectional area A, and remember that $u''=-f_0/k$ for steady-state conduction.

6. Piecewise-Constant $f(x)$. Let $\beta=0$ in (14), so the rod is insulated. Suppose $f(x)$ is defined piecewise, as 0 on $0<x<L/2$ and f_0 on $L/2<x<L$, and let $u(0)=0$ and $u(L)=0$.

(a) Solve for $u(x)$. HINT: Solve for $u(x)$ in the two halves separately, and call those solutions $u_L(x)$ and $u_R(x)$. To solve for the four arbitrary constants that arise, apply the boundary conditions $u_L(0)=0$ and $u_R(L)=0$. In addition, to blend $u_L(x)$ and $u_R(x)$ suitably at $x=L/2$ set $u_L(L/2)=u_R(L/2)$ and $u_L'(L/2)=u_R'(L/2)$.
(b) Regarding the foregoing hint, it was probably obvious that

we should match u_L and u_R at $x = L/2$, but it may not be so obvious that we should match their derivatives as well. Explain the logic behind that matching condition. HINT: Consider the rod element in Fig. 2 as straddling the point $x = L/2$, and consider a heat balance for that element. Here, $Q_3 = 0$ since the rod is insulated.

7. (a) Let $\beta = 0$ in (14), so the rod is insulated, and obtain a general solution of $u'' = -f(x)/k$. HINT: Integrate the equation twice, obtaining

$$u(x) = -\frac{1}{k}\int_0^x \int_0^\eta f(\xi)\, d\xi\, d\eta + C_1 + C_2 x, \tag{7.1}$$

in which ξ and η are dummy variables and x is the "fixed" point at which we are computing $u(x)$. We cannot carry out the ξ integration because $f(\xi)$ has not been specified. However, if we interchange the order of integration then we can carry out the η integration, thereby at least reducing the double integral to a single integral. To obtain the new limits of integration, first infer from the limits in (7.1) the region of integration in the ξ, η plane. Once you see that that region is as shown below, you can then deduce

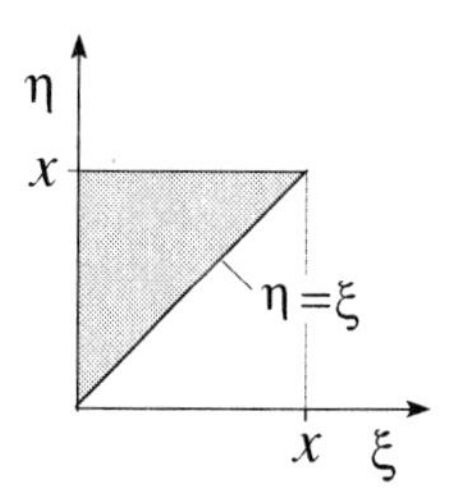

the new integration limits from the figure. Obtain, as your final result,

$$u(x) = -\frac{1}{k}\int_0^x (x-\xi) f(\xi)\, d\xi + C_1 + C_2 x. \tag{7.2}$$

(b) If the boundary conditions are $u(0) = u_1$, $u(L) = u_2$, solve for C_1 and C_2 and show that

$$u(x) = u_1 + (u_2 - u_1)\frac{x}{L} + \frac{x}{kL}\int_0^L (L-\xi) f(\xi)\, d\xi$$
$$- \frac{1}{k}\int_0^x (x-\xi) f(\xi)\, d\xi. \tag{7.3}$$

8. Inclusion of Lateral Heat Loss by Newton Cooling. In Example 1 we found that if there is neither Newton cooling (i.e., $\beta = 0$) nor internal heating [i.e., $f(x) = 0$], and there are Neumann boundary conditions $u'(0) = u_1'$ and $u'(L) = u_2'$, then there is no solution if $u_1' \neq u_2'$, and a nonunique solution if $u_1' = u_2'$. Show that if we now allow for Newton cooling to the environment, so the BVP is

$$u'' - \beta(u - u_\infty) = 0; \quad u'(0) = u_1',\ u'(L) = u_2' \tag{8.1}$$

(with $\beta > 0$), then a unique solution does exist, no matter what values are assigned to u_1' and u_2'.

HEAT CONDUCTION IN CYLINDERS

9. The Governing Differential Equation. Now consider steady-state heat conduction in a cylindrical space, for instance in a hollow pipe with the annular cross section shown in the figure. Consider the temperature distribution to be symmetric about the centerline normal to the paper, so it is

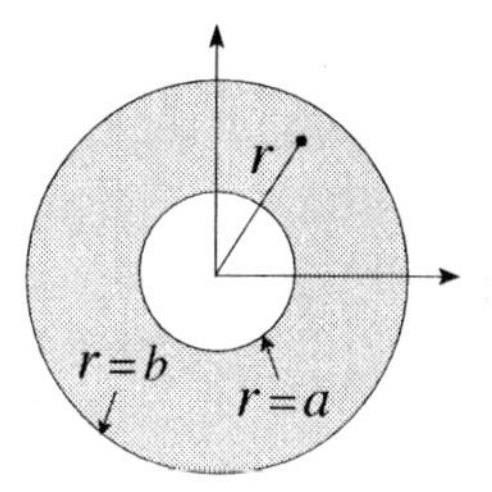

a function only of r, $u(r)$. Include an internal heat generation $f(r)$, in units of heat per unit volume per unit time, with f assumed to be continuous. Derive the differential equation

$$\boxed{k\frac{1}{r}\frac{d}{dr}\left(r\frac{du}{dr}\right) = -f(r).} \tag{9.1}$$

for $u(r)$. HINT: Instead of an axial element, as in Fig. 2, consider a ring of material between r and $r + \Delta r$, of unit depth

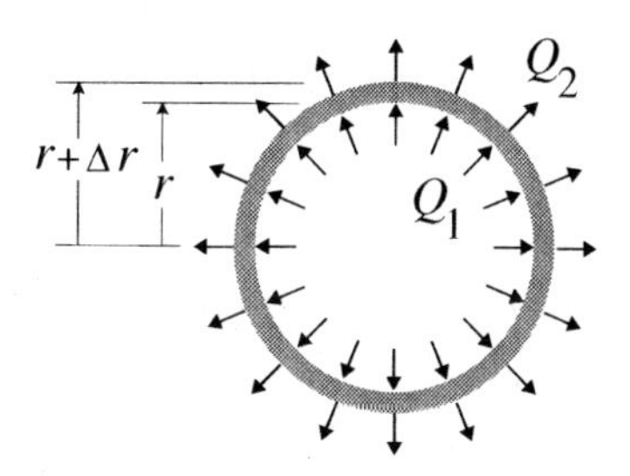

normal to the paper. If μ is the radius at which f takes on its average value over the interval from r to $r + \Delta r$, show that

$$k\,\frac{(r+\Delta r)\dfrac{du}{dr}(r+\Delta r) - r\dfrac{du}{dr}(r)}{\Delta r} = -f(\mu) r, \tag{9.2}$$

and let $\Delta r \to 0$.

10. With $f(r) = 0$, solve (9.1) above, subject to the given boundary conditions. State whether the solution exists, is unique or nonunique, or does not exist. NOTE: $b > a$.
(a) $u(a) = u_1,\ u(b) = u_2$ (b) $u(a) = u_1,\ u'(b) = 0$
(c) $u'(a) = 5,\ u'(b) = 5$ (d) $u'(a) = 5,\ u'(b) = 30$
(e) $u(a) = 50,\ u(b) + 2u'(b) = 100$

HEAT CONDUCTION IN SPHERES

11. The Governing Differential Equation. Finally, consider steady-state heat conduction in a spherical region such as a hollow sphere of inner radius a and outer radius b. The picture is the same as the first one in Exercise 9, but with the three cylindrical coordinate r's changed to spherical coordinate ρ's. Consider the temperature to be a function only of ρ, $u(\rho)$. Include an internal heat generation $f(\rho)$, in units of heat per unit volume per unit time, with f assumed to be continuous. Derive the differential equation

$$\boxed{k\frac{1}{\rho^2}\frac{d}{d\rho}\left(\rho^2\frac{du}{d\rho}\right) = -f(\rho),} \tag{11.1}$$

for $u(\rho)$. HINT: This time consider a thin spherical shell element, between ρ and $\rho + \Delta\rho$.

NOTE: Be careful not to confuse the variable r in cylindrical coordinates with the variable ρ in spherical coordinates. The two are shown below; ρ is the distance from the origin to the point, but r is not. It is the perpendicular distance from the point to the z axis.

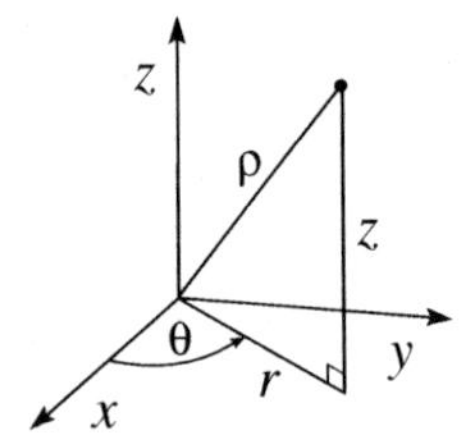

12. With $f(\rho) = 0$, solve (11.1), above, subject to the given boundary conditions. State whether the solution exists, is unique or nonunique, or does not exist. NOTE: $b > a$.
(a) $u(a) = u_1,\ u(b) = u_2$ (b) $u(a) = u_1,\ u'(b) = 0$
(c) $u'(a) = 5,\ u'(b) = 5$ (d) $u'(a) = 5,\ u'(b) = 30$
(e) $u(a) = 50,\ u(b) + 2u'(b) = 100$

ADDITIONAL EXERCISES

13. Nutrient Delivery by Capillaries. The diffusion of mass is central to many biological processes. Here, consider the delivery of the various vital nutrients, such as oxygen, through the walls of capillaries and into the surrounding tissue, by diffusion. (This delivery takes place not through the larger thick-walled vessels, but through the thin-walled capillaries.) Of course, tissue is not homogeneous; it is comprised of discrete cells with intercellular fluid within them and interstitial fluid between them. Nevertheless, according to the **Krogh cylinder model** the tissue surrounding a capillary can be *modeled* as a homogeneous tube with an effective diffusivity D and outer radius r_0 — for the particular nutrient under consideration. The geometry is shown below, and the nutrient concentration $c(r)$ is to be in steady state, symmetric about the z axis, and not varying with z. The equation governing $c(r)$ in $r_i < r < r_o$ is of the same form as (9.1), above, namely,

$$D\frac{1}{r}\frac{d}{dr}\left(r\frac{dc}{dr}\right) = -f(r). \tag{13.1}$$

Here, $f(r)$ will be negative [so the right-hand side of (13.1) will be positive] because it corresponds not to a *generation* of nutrient, but to the metabolic *consumption*, by the tissue, of the nutrient under consideration. (That is the whole point, delivery of the nutrient to the "tissue space," as biologists

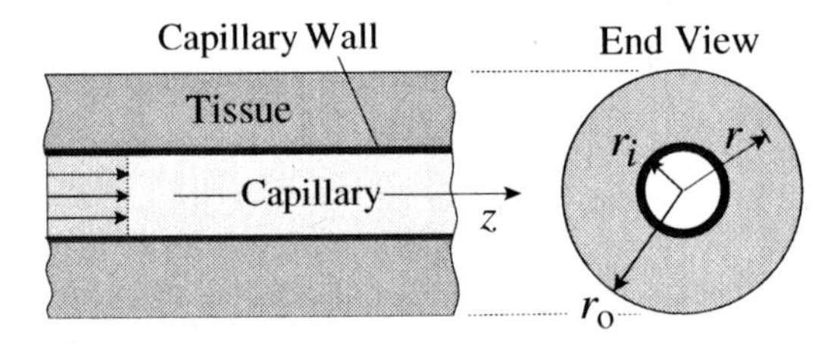

call it, for consumption.) The consumption rate can be modeled approximately by a **Michaelis–Menten**-type relation

$$-f(r) = P\frac{c(r)}{Q + c(r)}. \tag{13.2}$$

in which the right-hand side is designed to be algebraically simple, and to have the following behaviors. First, as $c \to 0$ it should be proportional to c, which it is: It tends to $(P/Q)c$ as $c \to 0$. Second, as $c \to \infty$ it should level off and approach a constant, which it does: It tends to P as $c \to \infty$. That is, the tissue's ability to consume is limited (just as one's rate of consumption of pizza is limited, no matter how much pizza is presented), and levels off as $c(r) \to \infty$. Note that if we put (13.2) into (13.1), the resulting equation for $c(r)$ will be nonlinear and difficult, so let us be content to suppose $c(r)$ is large enough over most of the tissue space to let $-f(r) \approx P$, a positive constant. Thus, the differential equation is

$$D\frac{1}{r}\frac{d}{dr}\left(r\frac{dc}{dr}\right) = P. \tag{13.3}$$

Next, the boundary conditions at r_i and r_o: According to the Krogh model, no nutrient escapes at r_o so, by Fick's law, $\frac{dc}{dr}(r_o) = 0$. At r_i, the boundary condition is that the rate (per unit z length) at which the nutrient crosses the capillary wall, say q, is equal to the rate of diffusion into the tissue:

$$q = -D(2\pi r_i)\frac{dc}{dr}\bigg|_{r=r_i}. \tag{13.4}$$

Thus, the BVP is this:

$$D\frac{1}{r}\frac{d}{dr}\left(r\frac{dc}{dr}\right) = P; \quad -2\pi r_i Dc'(r_i) = q, \; c'(r_o) = 0. \tag{13.5}$$

Actually, let us add a third boundary condition, that $c(r_o) = 0$, so that r_0 is as far as the nutrient penetrates. Having this third boundary condition will enable you to not only solve for the two integration constants, but also to solve for q in terms of P, for they must be related for the system to be in equilibrium.

The problem is this: Solve for $c(r)$ and show that

$$c(r) = \frac{P}{4D}\left[r^2 - 2r_0^2 \ln r + r_0^2(2\ln r_0 - 1)\right]. \tag{13.6}$$

NOTE: Our purpose in this exercise was only to explore key issues in modeling the nutrient uptake. These include: (a) the Krogh cylinder idea of assuming the recipient tissue is a finite annulus with uniform diffusion constant and (b) modeling both the consumption function and the transport across the capillary wall by Michaelis–Menten-type functions. We've *not* taken into account that c is not only a function of r but also of z, since as nutrient is lost through the capillary wall c must decrease as the blood moves downstream. Inclusion of that effect would yield a partial differential equation in place of the ODE considered above.

3.6 INTRODUCTION TO THE EIGENVALUE PROBLEM; COLUMN BUCKLING

3.6.1 An eigenvalue problem.

Consider the BVP

$$\begin{aligned} &y'' + \lambda y = 0 \qquad (0 \le x \le L) \\ &y(0) = 0, \quad y(L) = 0, \end{aligned} \tag{1}$$

in which λ is a constant. Although λ is a constant, its value is not known in advance, so it is a "parameter." The BVP is homogeneous because both the differential equation and boundary conditions are homogeneous. Thus, it surely admits a solution $y(x) = 0$, which we call the *trivial solution*. However, our interest in (1) is only in nontrivial solutions, so the question is this: Are there values of λ such that (1) will admit not only the trivial solution but nontrivial solutions as well? The problem of finding values (if any) of λ that permit nontrivial solutions for $y(x)$, and then finding those corresponding nontrivial solutions, is called an **eigenvalue problem**. Those special λ's and nontrivial solutions $y(x)$ are called the **eigenvalues** and **eigenfunctions** of (1), respectively. We will illustrate how such a problem can arise in a moment, but first let us show how to solve it.

The solution $y(x) = 0$ is not necessarily the only solution because (1) is a *boundary* value problem.

A general solution of the differential equation is

$$y(x) = \begin{cases} A\cos\sqrt{\lambda}x + B\sin\sqrt{\lambda}x, & \lambda > 0, \\ C + Dx, & \lambda = 0, \\ E\cosh\sqrt{|\lambda|}x + F\sinh\sqrt{|\lambda|}x, & \lambda < 0. \end{cases} \tag{2}$$

If the solution for $\lambda < 0$ is unclear, write the equation as $y'' - |\lambda|y = 0$, to put the negativeness of λ in evidence. A general solution is then a linear combination of $\cosh\sqrt{|\lambda|}x$ and $\sinh\sqrt{|\lambda|}x$.

We will consider all three cases because we don't know λ in advance. Consider them, in turn:

$\boldsymbol{\lambda = 0}$: Application of the boundary conditions to the general solution $C + Dx$ gives

$$y(0) = 0 = C + (0)D, \tag{3a}$$
$$y(L) = 0 = C + (L)D. \tag{3b}$$

These give the unique solution $C = D = 0$, so for $\lambda = 0$ we obtain only the trivial solution $y(x) = 0$. Thus, $\lambda = 0$ is not an eigenvalue of (1).

$\boldsymbol{\lambda < 0}$: Application of the boundary conditions to $E\cosh\sqrt{|\lambda|}x + F\sinh\sqrt{|\lambda|}x$ gives

$$y(0) = 0 = E(1) + F(0), \tag{4a}$$
$$y(L) = 0 = E\cosh\sqrt{|\lambda|}L + F\sinh\sqrt{|\lambda|}L. \tag{4b}$$

The first gives $E = 0$, and the second becomes $0 = F\sinh\sqrt{|\lambda|}L$. The sinh is zero only if its argument is zero, and its argument is nonzero because $\lambda \neq 0$ and $L \neq 0$. Thus, we obtain $E = F = 0$ and hence only $y(x) = 0$, so there are no eigenvalues in $\lambda < 0$. Finally, consider the case $\lambda > 0$.

$\boldsymbol{\lambda > 0}$: Application of the boundary conditions to the general solution $A\cos\sqrt{\lambda}x$ $+B\sin\sqrt{\lambda}x$ in (2) gives

$$y(0) = 0 = A(1) + B(0), \tag{5a}$$
$$y(L) = 0 = A(\cos\sqrt{\lambda}L) + B(\sin\sqrt{\lambda}L). \tag{5b}$$

Equation (5a) gives $A = 0$, and then (5b) gives

$$(\sin\sqrt{\lambda}L)B = 0. \tag{6}$$

To satisfy (6) we need $B = 0$, $\sin\sqrt{\lambda}L = 0$, or both. We cannot accept $B = 0$ because we already know that $A = 0$, and if $B = 0$ as well, then we obtain only the trivial solution $y(x) = 0$, which is unacceptable because the point of the eigenvalue problem is to find *nontrivial* solutions.

Thus, we must set

$$\boxed{\sin\sqrt{\lambda}L = 0.} \tag{7}$$

In that case, (6) becomes $0B = 0$, which is satisfied by *any* value of B. Thus, if λ is any value satisfying (7), we have

$$y(x) = B\sin\sqrt{\lambda}x, \tag{8}$$

in which B is arbitrary. After all, if $y(x)$ is a solution of (1), then so is an arbitrary scalar multiple of it; hence, the arbitrary scale factor B is no surprise.

Next, solve (7) for λ. The sine function is zero when its argument is $n\pi$, for any integer n, so it follows from (7) that

$$\sqrt{\lambda}L = n\pi, \text{ or } \lambda_n = \left(\frac{n\pi}{L}\right)^2, \tag{9}$$

for $n = 0, \pm 1, \pm 2, \ldots$. We can omit $n = 0$ because that n gives $\lambda = 0$, which was treated separately, above. We can also omit the negative integers $-1, -2, \ldots$, because $\lambda_n = (n\pi/L)^2$ in (8) is insensitive to the sign of n, and so is the eigenfunction

$$y(x) = B\sin\frac{n\pi x}{L}. \tag{10}$$

For instance, $n = +3$ gives $y(x) = B\sin(3\pi x/L)$ and $n = -3$ gives $y(x) = B\sin(-3\pi x/L) = -B\sin(3\pi x/L)$. These differ only by a factor of -1, which can be absorbed by the arbitrary scale factor B.

The upshot is that, for $\lambda > 0$, we have found the infinite set of eigenvalues given by (9), each one with corresponding eigenfunction given by (10). We can set $B = 1$ in (10), remembering that we can scale that solution by any nonzero constant. Thus,

$$\boxed{\lambda_n = \left(\frac{n\pi}{L}\right)^2, \quad y_n(x) = \sin\frac{n\pi x}{L}} \tag{11}$$

The eigenvalues and eigenfunctions of (1).

for $n = 1, 2, \ldots$.

Equation (7) for λ is called the **characteristic equation** for the eigenvalue problem (1), but that name should not be confused with the equation obtained for r when we seek solutions of a constant-coefficient equation in the form e^{rx}.

Consider a physical application.

3.6.2 Application to column buckling. By a column we mean a stiff slender vertical structural member that supports a vertical load, denoted as P in Fig. 1. Consider the ends of the rod to be secured by frictionless pins, and the upper end to be constrained so that it can move vertically between frictionless rollers, downward, when the rod begins to buckle, but not laterally.

The column supports an end load, a beam is horizontal and supports lateral loads. Euler beam theory applies to both.

It is natural to expect that the column, which is straight when $P = 0$, will bend in proportion to the load P as P is increased. But it does not; it remains straight as

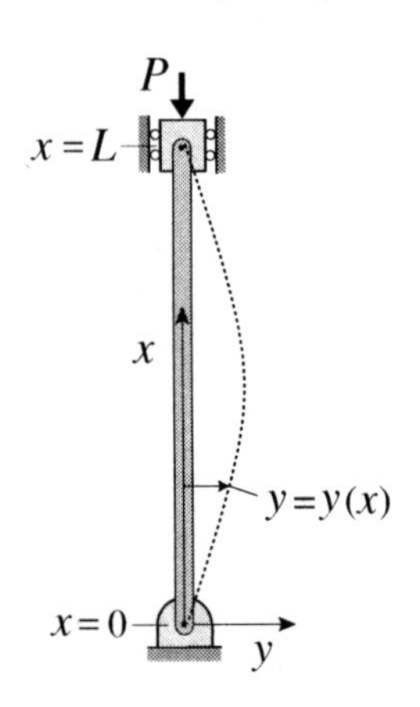

Figure 1. Column buckling.

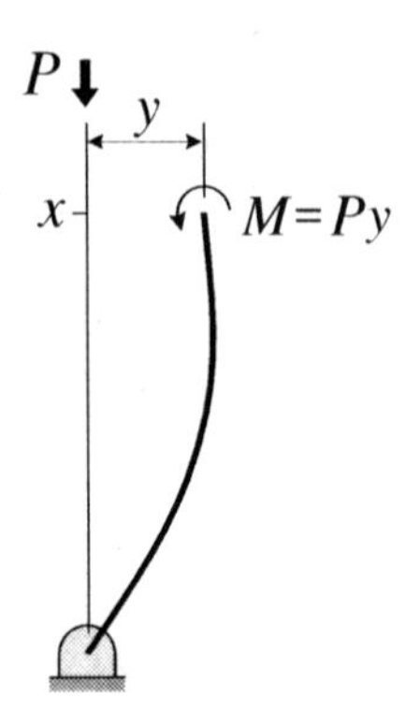

Figure 2. The bending moment M at x is the force P times its moment arm y.

P is increased — until P reaches a critical value P_{cr}, under which load it suddenly bends and collapses. That event is known as **buckling**.

Since buckling amounts to a failure of the load-bearing column, it is important to structural engineers to be able to predict the critical load P_{cr}, called the **buckling load**, and that is our objective here. We will neglect the weight of the column under the assumption that it is negligible compared to the load P.

According to the *Euler beam theory*, studied in sophomore mechanical and civil engineering courses, the curvature of a bent column at any point x is proportional to the bending moment at that point, that is, the moment causing it to bend. In the present example the bending moment M at any point x (between $x = 0$ and $x = L$) is seen from Fig. 2 to be the load P times its "moment arm" y: $M = Py$. Further, it is shown in the calculus that if R is the radius of curvature of a plane curve $y = y(x)$ at x, then the **curvature** κ, which is the numerical inverse of the **radius of curvature** R, is

$$\kappa = \frac{1}{R} = \frac{y''(x)}{[1 + y'^2(x)]^{3/2}}. \tag{12}$$

Since we seek the *initiation* of buckling, we can treat the deflection $y(x)$ as infinitesimal, so the y'^2 term in (12) can be neglected compared to the 1 next to it. In that case, (12) gives

$$\kappa = y''. \tag{13}$$

Thus, since the radius of curvature $\kappa = y''$ is proportional to the bending moment $M = Py$, y'' is proportional to Py. The Euler analysis, which we omit, gives the proportionality constant as $1/EI$, where E is the Young's modulus of the column material and I is the moment of inertia of the beam's cross section, so the final differential equation for the deflection curve $y(x)$ is

$$y'' = -\frac{1}{EI}(Py) \quad \text{or,} \quad y'' + \frac{P}{EI}y = 0. \tag{14}$$

Actually, we changed the proportionality constant to $-1/EI$ rather than $1/EI$, in (14), to get the signs to come out right, because if y is positive (as in Fig. 2) then y'' should be negative. The constant EI is called the *stiffness* of the column because as it increases, the curvature y'' decreases in magnitude, as we can see from (14).

Do not be intimidated by the "Euler beam theory"; (14) simply says that, at each point, the column's curvature is proportional to the moment causing that curvature, which should seem reasonable. In any case, (14) is our differential equation for $y(x)$.

In Fig. 1, we see that $y(0) = 0$ and $y(L) = 0$. If we append these boundary conditions to (14), we have the BVP

$$\begin{aligned} y'' + \lambda y &= 0 \qquad (0 \le x \le L) \\ y(0) = 0&, \quad y(L) = 0, \end{aligned} \tag{15}$$

in which $\lambda = P/EI$; λ is a parameter because it contains P, which is not yet known. Since P, E, and I are all positive, $\lambda = P/EI$ is too.

Surely $y(x) = 0$ satisfies the homogeneous BVP (15), but that is the case where the column does *not* buckle. Buckling corresponds to *nontrivial* solutions of (15), and these will occur only for certain special values of the parameter λ. Thus, (15) is the same as the eigenvalue problem (1), and we can use the solution found above. But note that if we did not already have that solution, we would *not* need to consider the cases $\lambda = 0$ and $\lambda < 0$, because in this application we can see in advance that $\lambda = P/EI$ is positive. (Surely, a negative P cannot produce buckling; we need compression, not tension.)

To understand the results already found, in terms of buckling, imagine beginning our experiment with $P = 0$ and slowly increasing it, and hence λ because $\lambda = P/EI$. The solution $y(x)$ remains the unique trivial solution $y(x) = 0$ for all P's, until λ reaches the first special value $\lambda_1 = (\pi/L)^2$ in (9). When that happens, the nontrivial solution $y(x) = B \sin(\pi x/L)$ occurs, and the column buckles in that shape (Fig. 3). Since we never evaluated B, we cannot predict the direction of the collapse from this analysis. If $B > 0$ it will be to the right as in Fig. 3, and if $B < 0$ it will be to the left.

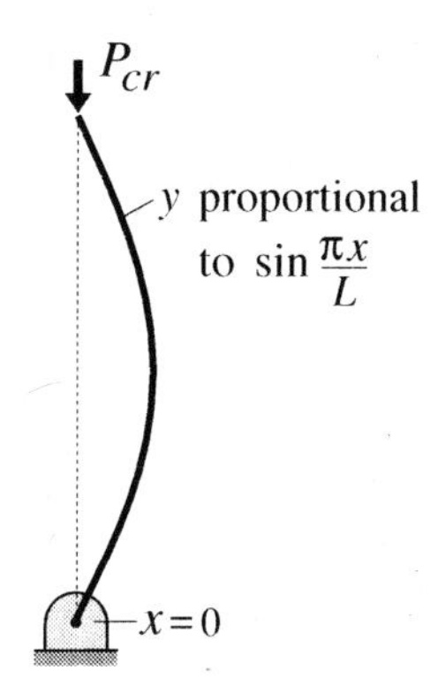

Figure 3. The buckled shape at the onset of buckling.

Keep in mind that our purpose here has not been to model the *process* of collapse, but only to predict its *onset*, the P at which buckling begins. To follow the subsequent process of collapse we would need a dynamic analysis, whereas the present model is static; the time t is not present. Since $\lambda = P/EI$, we can set $\lambda_1 = (\pi/L)^2 = P/EI$ to obtain the critical buckling load P_{cr}. That step gives

$$\boxed{P_{\text{cr}} = \frac{\pi^2 EI}{L^2}.} \tag{16}$$

Here, we are using (11) for λ, with $n = 1$, to obtain the smallest λ, and hence the smallest P, since $\lambda = P/EI$.

It makes sense that P_{cr} decreases rapidly as L increases, because we know from experience that it is much easier to buckle a long stick than a short one.

We've plotted the first several eigenfunctions in Fig. 4. The first, $y_1(x) = \sin(\pi x/L)$, is the buckling mode shown in Fig. 3 (where it was displayed vertically rather than horizontally).

In eigenvalue problems that arise in physical applications, expect both the eigenvalues and the eigenfunctions to have physical significance. In the present application the eigenvalues give the buckling loads $P_n = n^2\pi^2 EI/L^2$ and the eigenfunctions give the corresponding buckling "mode" shapes, $y_n(x) = \sin(n\pi x/L)$. However, we cannot expect to observe any of the "higher buckling modes" (i.e., for $n = 2, 3, \ldots$) because as we increase P, from zero, the column will collapse when the lowest buckling load is reached, the critical buckling load P_{cr} corresponding to $n = 1$. Thus, the higher modes cannot be reached, and will not be observed.

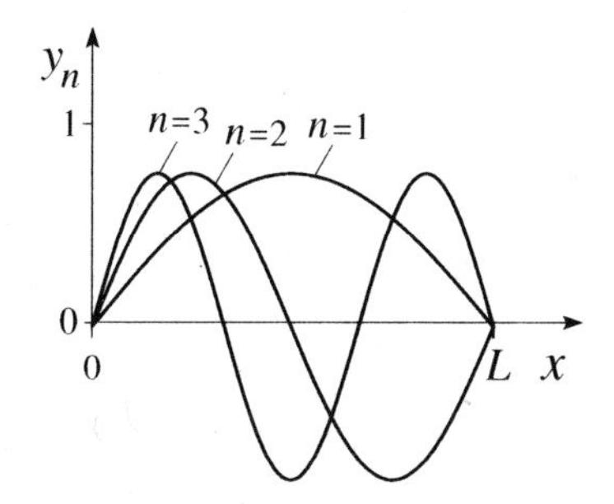

Figure 4. The first several eigenfunctions from (9).

Closure. We have used (1) to introduce a class of eigenvalue problems, of which (1) is but one example. The boundary conditions could be different, and so could the differential equation. For instance, at each end of the interval the boundary condition could be $y = 0$ or $y' = 0$ or $\alpha_1 y + \alpha_2 y' = 0$ with both α's nonzero. Other examples will be given in the exercises, and in Chapters 4 and 7 we will encounter eigenvalue problems for matrix equations rather than for differential equations.

EXERCISES 3.6

1. Additional Eigenvalue Problems. In each case the x interval is implied by the boundary conditions. For example, the conditions $y'(0) = 0$, $y(\pi) = 0$ imply that the interval is $0 < x < \pi$. The problem is to determine the eigenvalues and eigenfunctions. Allow for the three possibilities: $\lambda = 0$, $\lambda < 0$, and $\lambda > 0$.

(a) $y'' + \lambda y = 0$ with $y'(0) = 0$ and $y(\pi) = 0$. Show that

$$\lambda_n = \frac{n^2}{4}, \quad y_n(x) = \cos\frac{nx}{2},$$

for $n = 1, 3, 5, \ldots$.

(b) $y'' + \lambda y = 0$ with $y(0) = 0$ and $y'(\pi) = 0$

(c) $y'' + \lambda y = 0$ with $y'(0) = 0$ and $y'(\pi) = 0$

(d) $y'' + \lambda y = 0$ with $y(0) = 0$ and $y(4) = 0$

(e) $y'' + 2y' + \lambda y = 0$ with $y(0) = 0$ and $y(\pi) = 0$ HINT: This time consider the cases $\lambda < 1$, $\lambda = 1$, $\lambda > 1$.

(f) For $y'' - 4y' + \lambda y = 0$ with $y(0) = 0$ and $y(\pi) = 0$ show that $\lambda_n = 4 + n^2$ and $y_n(x) = e^{2x} \sin nx$ for $n = 1, 2, \ldots$.

ADDITIONAL EXERCISES

2. Buckling of a Flagpole; a Third-Order Equation. Consider the buckling of a "flagpole." Unlike the column shown in Fig. 1, the flagpole is set in concrete at its base, $x = 0$, so besides $y(0) = 0$ we also have $y'(0) = 0$, and the pole is not restrained laterally at $x = L$ as was the column in Fig. 1.

(a) Show that, in place of (14), the governing differential equation is

$$y'' = \frac{P}{EI}[y(L) - y(x)]. \tag{2.1}$$

HINT: As in the text, y'' equals a constant of proportionality $1/EI$ times the moment at x. Show that this time the moment is $M = P[y(L) - y(x)]$.

(b) Unfortunately, we don't know the $y(L)$ in (2.1), but we can eliminate it by differentiating (2.1) with respect to x. Doing so gives $EIy''' = -Py'$. Besides the boundary conditions $y(0) = 0$ and $y'(0) = 0$ at $x = 0$, there is a boundary condition $y''(L) = 0$ that follows from (2.1) by setting $x = L$. [This step recovers information that was lost when we differentiated (2.1) to eliminate the $y(L)$ term.] Thus, the eigenvalue problem is

$$y''' + \lambda y' = 0; \quad y(0) = 0,\ y'(0) = 0,\ y''(L) = 0, \tag{2.2}$$

with $\lambda = P/EI$. Solve (2.2) and show that the critical buckling load is

$$P_{\text{cr}} = \frac{\pi^2 EI}{4L^2}. \tag{2.3}$$

(c) Use (2.3) to compute the buckling load of a flagpole that is 50 feet tall, is made of hollow aluminum, and has a circular cross section with an outside diameter $d_o = 4$ inches and an inside diameter of $d_i = 3.5$ inches. Use $E = 10^7$ lbs/in^2, and $I = \pi(d_o^4 - d_i^4)/64$ in^4.

3. Vibrating String Eigenvalue Problem. Consider a guitar string stretched under tension τ lbs between $x = 0$ and $x = L$, as sketched below, and set in motion in the plane of the paper — for instance, by plucking it. The lateral deflection $y(x, t)$ is governed by the *partial differential equation*

$$\boxed{\tau \frac{\partial^2 y}{\partial x^2} = \rho \frac{\partial^2 y}{\partial t^2},} \tag{3.1}$$

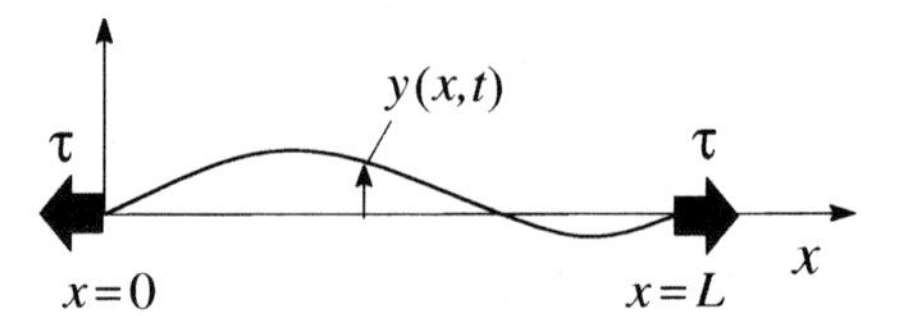

which will be derived in part (c), below; ρ is the string's mass per unit length. We don't cover partial differential equations in this text, but our approach here should seem reasonable: To find vibratory solutions, we will use the method of assumed form, and seek solutions in the form

$$y(x, t) = X(x) \sin \omega t \tag{3.2}$$

in which the temporal frequency ω and the "shape function" $X(x)$ are to be determined so as to satisfy the differential equation (3.1) and the boundary conditions, namely, that $y(0, t) = 0$ and $y(L, t) = 0$ for all t.

(a) Show that it follows that $X(x)$ must satisfy the BVP

$$X'' + \lambda X = 0, \quad X(0) = 0,\ X(L) = 0, \tag{3.3}$$

in which $\lambda = \rho\omega^2/\tau$; λ is not known in advance because the frequency ω is not known in advance, but we know that λ is positive because ρ, τ, and ω are all positive. The BVP (3.3) is homogeneous because both the forcing function and boundary conditions are zero, so it admits the trivial solution $X(x) = 0$. However, we're not interested in that solution because it amounts to the string not moving! Rather, we seek nontrivial solutions, so (3.3) is an eigenvalue problem.

(b) Show that the eigenvalues of (3.3) give the allowable temporal frequencies

$$\omega_n = \frac{n\pi}{L}\sqrt{\frac{\tau}{\rho}} = \frac{n\pi c}{L}, \tag{3.4}$$

in which we've abbreviated $\sqrt{\tau/L}$ as c, and the corresponding standing wave mode shapes

$$X_n(x) = \sin\frac{n\pi}{L}x, \tag{3.5}$$

for $n = 1, 2, \ldots$. Thus, we've been successful in finding not just one, but an infinite number of such solutions:

$$y(x,t) = \sin\frac{\pi}{L}x\,\sin\frac{\pi c}{L}t,\ \ \sin\frac{2\pi}{L}x\,\sin\frac{2\pi c}{L}t.\ \ \ldots, \tag{3.6}$$

The first two are shown below.

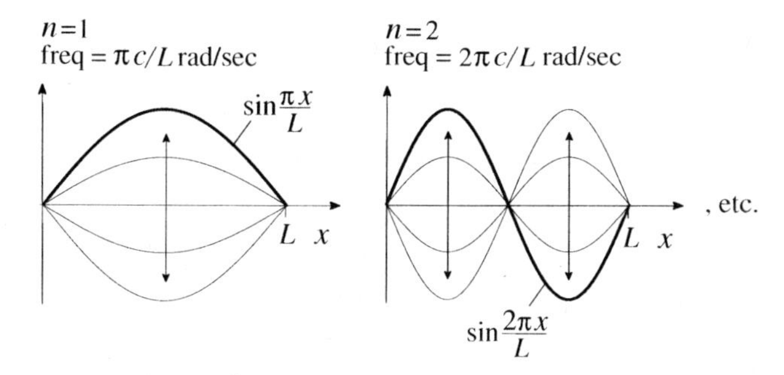

NOTE: These are an infinite number of different "modes" of vibration, and the lowest temporal frequency $\omega_1 = \pi c/L$ is called the *fundamental* frequency. If you are tuning your guitar string to middle A (i.e., middle on a piano), for instance, which is at a frequency of 27.5 cycles/sec, the idea is to adjust the tension τ (remember that $c = \sqrt{\tau/L}$) so the fundamental frequency $1\pi c/L$ equals 27.5 cycles/sec times 2π rad/cycle, or, 172.8 rad/sec. The higher frequency modes in (3.5) provide "overtones," and what we hear, when the string is set in motion, is some combination of all these modes — in different amounts, depending on how the string is set in motion.

(c) Derive (3.1). HINT: A force diagram for a typical element of the string, between x and $x + \Delta x$, is shown below, neglecting the gravitational force on the element compared

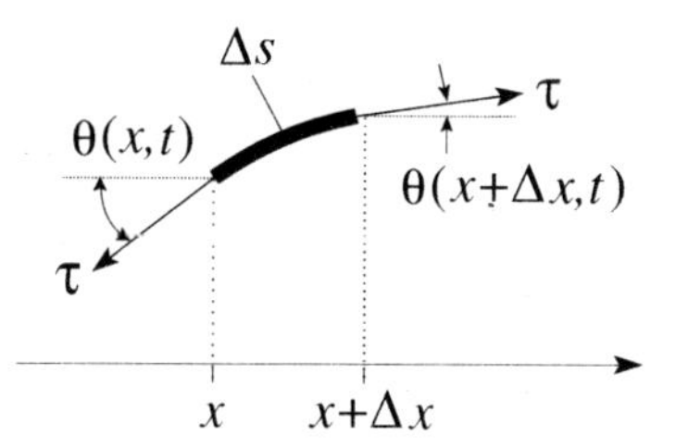

to the effects of the tension τ. (After all, doesn't a guitar sound the same if it is played horizontally or vertically?) Apply Newton's second law in the y direction and obtain

$$\rho\Delta s\frac{\partial^2 y}{\partial t^2}(x+\alpha\Delta x,t) = \tau\left[\sin\theta(x+\Delta x,t) - \sin\theta(x,t)\right], \tag{3.7}$$

in which Δs is the arc length of the element and $x+\alpha\Delta x$ is the x location of the mass center of the element (so $0 < \alpha < 1$). Next, assume that the string deflection is small; specifically, suppose $|y| \ll L$ and $|\theta| \ll 1$. In that case $\Delta s \approx \Delta x$ and, because θ is small,

$$\sin\theta = \theta - \frac{1}{3!}\theta^3 + \frac{1}{5!}\theta^5 - \cdots \approx \theta \tag{3.8}$$

and

$$\tan\theta = \theta + \frac{1}{3}\theta^3 + \frac{2}{15}\theta^5 + \cdots \approx \theta. \tag{3.9}$$

Hence, we can replace

$$\sin\theta \approx \tan\theta = \frac{\partial y}{\partial x}, \tag{3.10}$$

which step enables us to eliminate the temporary variable θ from (3.7), in favor of y. With these approximations in (3.7), let $\Delta x \to 0$ and obtain (3.1).

CHAPTER 3 REVIEW

Chapter 3 is primarily about specific applications, so this review will be brief and limited to the only new mathematical concept that we met, the eigenvalue problem.

BVP, not IVP.

The *eigenvalue problem* for a differential equation consists of a homogeneous BVP that contains a parameter (usually designated as λ). By the BVP being homogeneous we mean that both the differential equation and the boundary conditions are homogeneous. Specifically, the eigenvalue problem that we encountered in Section 3.6 arose in connection with the determination of the buckling load of a column, and consisted of the BVP

$$y'' + \lambda y = 0 \quad (0 \le x \le L) \tag{1a}$$
$$y(0) = 0, \quad y(L) = 0. \tag{1b}$$

Because (1) is homogeneous, $y(x) = 0$ is of course a solution, the trivial solution. But the point of an eigenvalue problem is to find *nontrivial* solutions. In this application, the trivial solution $y(x) = 0$ is the unique solution of (1) in the event that buckling does *not* occur, whereas we are interested in the case where it *does* occur. To proceed, we simply solved (1a) and applied the boundary conditions (1b). We found that for A and B in

$$y(x) = A\cos\sqrt{\lambda}x + B\sin\sqrt{\lambda}x \tag{2}$$

to not both be zero, λ had to be a root of the *characteristic equation*

The characteristic equation for an eigenvalue problem is unrelated to the characteristic equation for r, when we sought solutions in the form e^{rx} in Chapter 2.

$$\sin\sqrt{\lambda}L = 0. \tag{3}$$

Those roots gave the eigenvalues $\lambda_n = (n\pi/L)^2$, for $n = 1, 2, \ldots$. With $A = 0$, B arbitrary, and $\lambda = \lambda_n$ in (2), the corresponding nontrivial solutions were the eigenfunctions. In terms of buckling, we showed that the smallest eigenvalue gave the critical buckling load, and the corresponding eigenfuction gave the shape of the buckled column, the "buckling mode."

Chapter 4

Systems of Linear Differential Equations

4.1 INTRODUCTION, AND SOLUTION BY ELIMINATION

4.1.1 Introduction. We began, in Chapter 1, with a single first-order equation. From there, one can proceed in either of two different directions: to higher-order equations in a single unknown, as we did in Chapter 2, or to systems of more than one first-order equation, as we do now in this chapter.

In Chapter 2 we considered only higher-order *linear* equations. Likewise here, as we now consider systems of more than one equation in more than one unknown, we again consider equations that are linear. A **linear first-order system** of two differential equations is of the form (or can be put in the form)

$$\boxed{\begin{aligned} \frac{dx_1}{dt} &= a_{11}(t)x_1 + a_{12}(t)x_2 + f_1(t), \\ \frac{dx_2}{dt} &= a_{21}(t)x_1 + a_{22}(t)x_2 + f_2(t). \end{aligned}} \tag{1a,b}$$

In (1), t is a generic independent variable which may or may not be the time, and x_1 and x_2 are generic dependent variables.

The **coefficients** $a_{11}(t)$, $a_{12}(t)$, $a_{21}(t)$, $a_{22}(t)$, and the **forcing functions** $f_1(t)$, $f_2(t)$ are prescribed functions of the independent variable t on the interval I of interest, and x_1 and x_2 are dependent variables — the unknowns. If *both* $f_1(t)$ and $f_2(t)$ are zero, then (1) is *homogeneous*; otherwise it is *nonhomogeneous*. The double subscript notation for the coefficients $a_{ij}(t)$ is as follows: The first subscript indicates the equation in which the coefficient occurs, and the second indicates the unknown that it multiplies. For instance, $a_{21}(t)$ is in the second equation (i.e., the x_2 differential equation) and it multiplies the first unknown x_1.

In Chapter 7 we will study systems of two *nonlinear* differential equations, for which other methods will be used.

4.1.2 Physical examples.

Following are examples of (1), involving mixing tanks and electrical circuits.

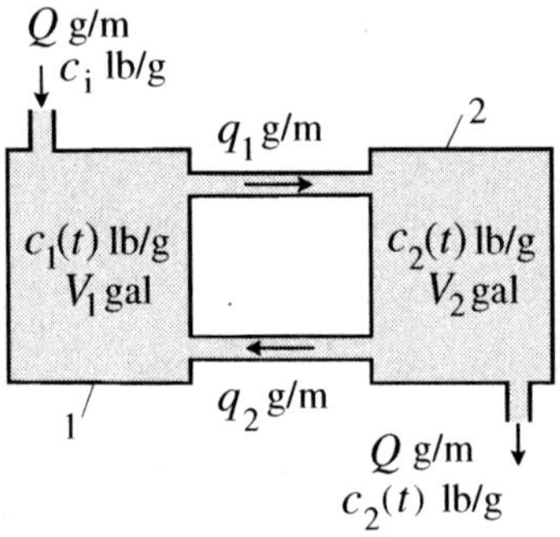

Figure 1. Two mixing tanks.

EXAMPLE 1. Two Mixing Tanks. In Section 1.3.3 we studied the concentration of a solute such as salt, as a function of time, in a single mixing tank. Now suppose there are *two* tanks, connected as in Fig. 4.1. They are constantly stirred so the concentrations c_1 and c_2 (lb/gal) in them are spatially uniform and depend only on the time t. The flow rates Q, q_1, and q_2 (gal/min) are constant, as is the inflow concentration c_i. The liquid volumes V_1 and V_2 (gal) in the tanks are to be constant, so the flow rates Q, q_1, and q_2 must be related according to

$$Q + q_2 = q_1; \tag{2}$$

that is, the inflow $Q+q_2$ to tank 1 must equal the outflow q_1 from tank 1. Or, in terms of tank 2, its inflow q_1 must equal its outflow $Q+q_2$ so, once again, we get $Q+q_2 = q_1$.

To derive differential equations for $c_1(t)$ and $c_2(t)$, carry out a mass balance for each tank, as we did in Section 1.3.3 for a single tank. Equating the rate of increase of solute to the rate in minus the rate out gives, for tank 1,

The unit of each term is lb/min. For instance, consider the Qc_i term: $Q\frac{\text{gal}}{\text{min}}c_i\frac{\text{lb}}{\text{gal}} = Qc_i\frac{\text{lb}}{\text{min}}$.

$$\frac{d}{dt}[c_1(t)V_1] = Qc_i + q_2c_2(t) - q_1c_1(t), \tag{3a}$$

and for tank 2

$$\frac{d}{dt}[c_2(t)V_2] = q_1c_1(t) - q_2c_2(t) - Qc_2(t). \tag{3b}$$

Since V_1 and V_2 are constants, these give

$$\frac{dc_1}{dt} = -(\frac{q_1}{V_1})c_1 + (\frac{q_2}{V_1})c_2 + \frac{Q}{V_1}c_i, \tag{4a}$$

$$\frac{dc_2}{dt} = (\frac{q_1}{V_2})c_1 - (\frac{Q+q_2}{V_2})c_2 \tag{4b}$$

for $c_1(t)$ and $c_2(t)$.

These are of the form (1), with c_1, c_2 in place of the generic variables x_1, x_2, and with $a_{11} = -q_1/V_1$, $a_{12} = q_2/V_1$, $f_1 = Qc_i/V_1$, $a_{21} = q_1/V_2$, $a_{22} = -(Q + q_2)/V_2$, and $f_2 = 0$. Although $f_2 = 0$, (4) is nonhomogeneous because $f_1 \neq 0$. In this example, the four coefficients and the two forcing functions happen to be constants. ▮

Prerequisite for this example is Section 1.3.5, on electrical circuits, without which Example 2 could be omitted.

EXAMPLE 2. An *RL* Circuit. Consider the RL electrical circuit shown in Fig. 2. The resistances and inductances are constants and the voltage source $E(t)$ is known. Let the currents flowing through the two inductors be i_1 and i_2, respectively, and let them be positive if they are in the directions shown in the figure. We wish to obtain the differential equations for the currents in the circuit.

From Kirchhoff's current law (namely, that the algebraic sum of the currents approaching or leaving any point of a circuit is zero) we can determine the current in cf

in terms of i_1 and i_2: The current approaching the junction c from the "south" is i_1 and the current leaving to the north is i_2, so from Kirchhoff's current law the current to the east is $i_1 - i_2$. Or, equivalently, the current to the west in cf is $i_2 - i_1$, as indicated in Fig. 3.

Next, apply Kirchhoff's voltage law (that the algebraic sum of the voltage drops around each loop of a circuit is zero) to the lower loop $bcfab$ and the upper loop $cdefc$, and recall from Section 1.3.5 that the voltage drops across resistors and inductors are Ri and $L\frac{di}{dt}$, respectively. For the lower loop that step gives $L_1\frac{di_1}{dt} + (i_1 - i_2)R_1 - E(t) = 0$, the minus sign in $-E(t)$ included because $E(t)$ amounts to a voltage rise (according to the polarity denoted by the + and − signs in Fig. 2) rather than a voltage drop. Similarly for the upper loop, Kirchhoff's voltage law gives $L_2\frac{di_2}{dt} + i_2R_2 + (i_2 - i_1)R_1 = 0$, so we have

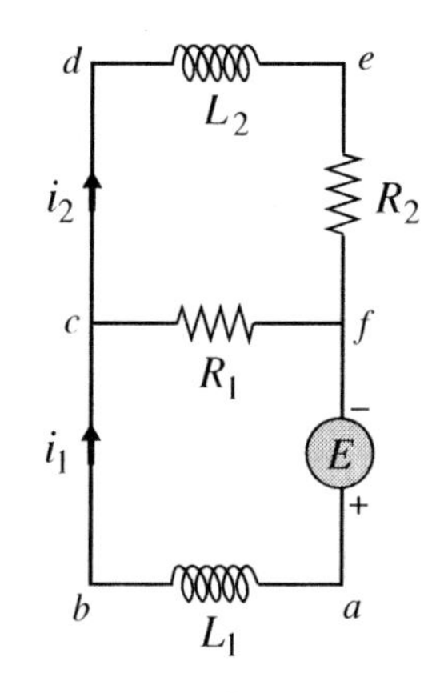

Figure 2. The RL circuit.

$$L_1 i_1' + (i_1 - i_2)R_1 - E(t) = 0, \tag{5a}$$
$$L_2 i_2' + i_2 R_2 + (i_2 - i_1)R_1 = 0 \tag{5b}$$

or, rearranging terms,

$$i_1' = -\frac{R_1}{L_1}i_1 + \frac{R_1}{L_1}i_2 + \frac{1}{L_1}E(t), \tag{6a}$$
$$i_2' = \frac{R_1}{L_2}i_1 - \frac{R_1 + R_2}{L_2}i_2. \tag{6b}$$

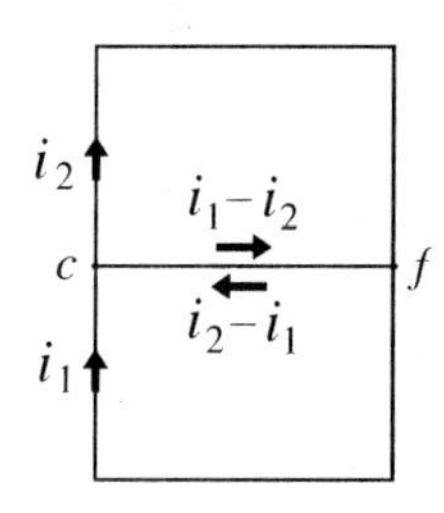

Figure 3. The RL circuit again, but with the circuit elements not shown, in order to focus on the current directions: The current in cf is $i_1 - i_2$ to the right or, equivalently, $i_2 - i_1$ to the left.

The system (6) is nonhomogeneous because of the $E(t)/L_1$ forcing function in (6a).

Since there are two unknown currents $i_1(t)$ and $i_2(t)$, it seems appropriate that we should have two differential equations, which we do in (6a,b). However, we could also apply Kirchhoff's voltage law to the outer loop $bcdefab$, and we wonder if doing that will give a third equation that must be included: If we apply the voltage law to the outer loop, we obtain

$$L_1 i_1' + L_2 i_2' + i_2 R_2 - E(t) = 0, \tag{7}$$

but the latter is actually the sum of equations (5a) and (5b) so it is redundant and can be discarded without loss. ▮

4.1.3 Solutions, existence, and uniqueness.

A linear first-order system does not necessarily contain only two equations and two unknowns. Such a system of n equations in n unknowns is

$$\begin{aligned} x_1' &= a_{11}(t)x_1 + a_{12}(t)x_2 + \cdots + a_{1n}(t)x_n + f_1(t), \\ x_2' &= a_{21}(t)x_1 + a_{22}(t)x_2 + \cdots + a_{2n}(t)x_n + f_2(t), \\ &\vdots \\ x_n' &= a_{n1}(t)x_1 + a_{n2}(t)x_2 + \cdots + a_{nn}(t)x_n + f_n(t). \end{aligned} \tag{8}$$

By functions $x_1(t), \ldots, x_n(t)$ being a **solution** of (8), on a t interval I, we mean that their substitution into the differential equations reduces each to an identity on I. Suppose we augment (8) by appending initial conditions to it at some point $t = a$. That is, append $x_1(a) = b_1$ to the first equation in (8), $x_2(a) = b_2$ to the second, and so on, where the b_j's are prescribed numbers. Then, the system (8), augmented by those initial conditions, is an **initial value problem**.

Realize that we cannot begin by solving the first equation in (8) using results from Section 1.2 even though it is simply a linear first-order equation for $x_1(t)$, because the $x_2(t), \ldots, x_n(t)$ that appear in it are not yet known. Similarly for each of the other equations, so we say the equations are **coupled**. In an application, the mathematical coupling is a reflection of the physical coupling. For instance, (4) is coupled mathematically because of the c_2 in (4a) and the c_1 in (4b), and physically because the tanks are connected by the two pipes, through which there is a flow from one tank to the other.

We wish to begin looking at methods of solution immediately, but how do we know if the given system even *has* a solution and, if it does, if that solution is unique? Thus, although we will give the general theory of the first-order system (8) later, in Section 4.5, we will state here an existence and uniqueness theorem to get us started, because it will give conditions under which we can know in advance that a given linear first-order system of initial value problems does have a unique solution.

THEOREM 4.1.1 *Existence and Uniqueness for the IVP*
Let the $a_{ij}(t)$'s and $f_j(t)$'s in (8) be continuous on an open t interval I, let a be a point in I, and let $b_1, \ldots, b_n$ be prescribed numbers. Then the first-order linear system (8), subject to initial conditions $x_1(a) = b_1, \ldots, x_n(a) = b_n$, has a unique solution on I.

The word "has" corresponds to the existence part of the theorem.

Theorem 4.1.1 should not be surprising since it is essentially a generalization of Theorem 1.2.1 in Section 1.2 for a single first-order linear equation. However, whereas Theorem 1.2.1 contains a *formula* for the solution, the system (8), if $n \geq 2$, is more difficult and an analogous formula for its solution *is not known*. Thus, rather than aim at solution formulas, we are particularly interested now in solution *methods*. The simplest is an elimination procedure similar to the familiar method of elimination used to solve a system of linear *algebraic* equations.

4.1.4 Solution by elimination.

The idea should be clear from an example.

EXAMPLE 3. Solve the IVP

$$x' = 4x - 3y + 4t, \tag{9a}$$
$$y' = 2x - y, \tag{9b}$$

with initial conditions $x(0) = 1$ and $y(0) = 0$, for $x(t)$ and $y(t)$ on $-\infty < t < \infty$.[1]

[1]Sometimes we use a subscript notation for the dependent variables, such as $x_1(t)$ and $x_2(t)$; here

First, note that the four a_{ij} coefficients are constants, and hence continuous for all t, and $f_1(t) = 4t$ and $f_2(t) = 0$ are also continuous for all t, so the IVP does have a unique solution on $-\infty < t < \infty$. To find it, by elimination, we can use (9a) to eliminate y in favor of x [or we could use (9b) to eliminate x in favor of y]. That is, solve (9a) for y by algebra and obtain

$$y = \frac{1}{3}(-x' + 4x + 4t). \tag{10a}$$

Then substitute (10a) into (9b), obtaining $(-x'' + 4x' + 4)/3 = 2x - (-x' + 4x + 4t)/3$ or,

$$x'' - 3x' + 2x = 4t + 4. \tag{10b}$$

Having eliminated y, (10b) is a second-order equation on $x(t)$ *alone*, which can now be solved by the methods of Chapter 2, using undetermined coefficients to obtain a particular solution. Doing so gives a general solution of (10b) as

$$x(t) = C_1e^t + C_2e^{2t} + 2t + 5, \tag{11}$$

in which C_1 and C_2 are arbitrary constants. Then, substitute (11) into (10a) and obtain $y(t) = C_1e^t + (2/3)C_2e^{2t} + 4t + 6$. Thus,

$$x(t) = C_1e^t + C_2e^{2t} + 2t + 5, \tag{12a}$$

$$y(t) = C_1e^t + \frac{2}{3}C_2e^{2t} + 4t + 6. \tag{12b}$$

Finally, apply the initial conditions: $x(0) = 1 = C_1 + C_2 + 5$ and $y(0) = 0 = C_1 + 2C_2/3 + 6$. These give $C_1 = -10$ and $C_2 = 6$, so

$$x(t) = -10e^t + 6e^{2t} + 2t + 5, \tag{13a}$$

$$y(t) = -10e^t + 4e^{2t} + 4t + 6. \tag{13b}$$

That (13) satisfies (9) on the stated interval I is readily checked by substitution and, as noted above, Theorem 4.1.1 assures us that it is the *only* solution of the IVP consisting of (9) and the initial conditions $x(0) = 1$ and $y(0) = 0$. ▮

General and particular solutions of (8).

We can say that (12a) and (12b) comprise a **general solution** of the system (9) because they satisfy (9) for every choice of the constants C_1, C_2 in the first place, and because *every* solution of (9) is contained within (12) by suitable choice of C_1, C_2. After all, (9a) and (9b) implied (10a) and (10b) and, reversing our steps, (10a) and (10b) imply (9a) and (9b). Thus, every solution pair $x(t), y(t)$ of (9) is also a solution pair of (10), and conversely. Finally, we know from Chapter 2 that (11) is a general solution of (10b), and then putting (11) into (10a) gives $y(t)$,

Assigning specific values to C_1 and C_2 in (12) then gave the **particular solution** (13) of (9). That is, any single member of the set of solutions within the general solution is a particular solution.

We will give the general theory of linear first-order systems in Section 4.5 but, meanwhile, we can rely on the existence and uniqueness theorem (Theorem 4.1.1)

we are using $x(t)$ and $y(t)$ instead. If there are many variables, the subscript notation may be better.

and the fact that the elementary elimination procedure leads to general solutions, as it did in Example 3.

EXAMPLE 4. The Mixing Tanks from Example 1. Let us solve the mixing tank problem formulated in Example 1. The result there was the system (4) of differential equations for $c_1(t)$ and $c_2(t)$, on $0 < t < \infty$. Let $V_1 = V_2 = 100$ gal, $q_1 = 20$ gal/min, $q_2 = 5$ gal/min, $Q = q_1 - q_2 = 15$ gal/min [required by (2)], $c_i = 0.3$ lb/gal, $c_1(0) = 0$, and $c_2(0) = 0.4$ lb/gal. Then the IVP is

$$c_1' = -0.2c_1 + 0.05c_2 + 0.045, \tag{14a}$$

$$c_2' = 0.2c_1 - 0.2c_2, \tag{14b}$$

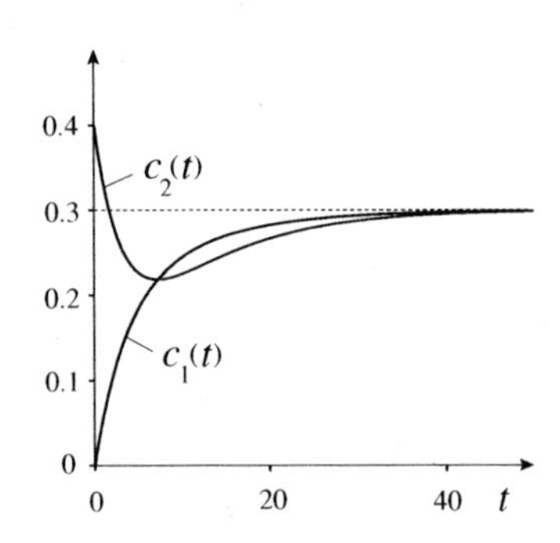

Figure 4. Solution of the IVP.

with initial conditions $c_1(0) = 0$ and $c_2(0) = 0.4$. To solve, use the same elimination procedure as above: Solve (14a) by algebra for c_2, giving $c_2 = 20c_1' + 4c_1 - 9$, and put that expression into (14b), obtaining

$$20c_1'' + 8c_1' + 0.6c_1 = 0.18, \tag{15}$$

with general solution

$$c_1(t) = Ae^{-0.3t} + Be^{-0.1t} + 0.3. \tag{16}$$

Then, putting (16) into (14a) and solving for $c_2(t)$ by algebra gives

$$c_2(t) = -2Ae^{-0.3t} + 2Be^{-0.1t} + 0.3. \tag{17}$$

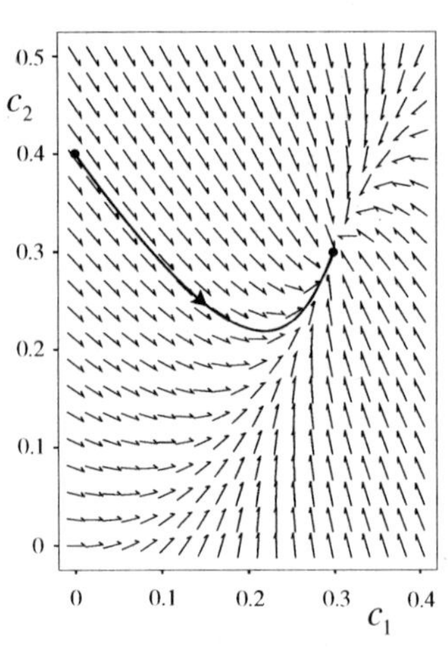

Figure 5. Solution in the c_1, c_2 phase plane. Note that c_2 undershoots its steady-state value 0.3 and then approaches it from below, while c_1 approaches it monotonely from the left, as seen also in Fig. 4.

Finally, applying the initial conditions $c_1(0) = 0$ and $c_2(0) = 0.4$ to (16) and (17) gives $A = -0.175$ and $B = -0.125$. With these values of A and B, we've plotted $c_1(t)$ and $c_2(t)$ in Fig. 4.

COMMENT 1. Steady State. In this application there exists a steady-state solution for $c_1(t)$ and $c_2(t)$, which can be obtained by letting $t \to \infty$ in (16) and (17). Doing so shows that $c_1(t)$ and $c_2(t)$ both tend to the same steady-state value, 0.3, which is also consistent with the plots in Figs.4 and 5. Alternatively, we can obtain those steady-state values directly from the differential equations in (14), without solving them, by setting c_1' and c_2' equal to zero (because, by the definition of steady state, c_1' and c_2' must be zero at steady state) and solving the resulting algebraic equations for c_1 and c_2.

COMMENT 2. The specified t interval was $0 \le t < \infty$ and the initial point was $t = 0$. However, if we wish to apply Theorem 4.1.1 the initial point a must be an *interior* point of I. To apply the theorem, we can simply let the interval I be $-25 < t < \infty$, for instance, instead. The latter is then an open interval containing the initial point $t = 0$ and the a_{ij} coefficients and the f_j forcing functions in (14) are continuous on I. Thus, there exists a unique solution of the IVP on $-25 < t < \infty$, and hence on the actual interval of interest, $0 \le t < \infty$.

COMMENT 3. Phase Plane. To put Fig. 5 into perspective, recall Fig. 7 in Section 1.3. When we compress the c, t plot, in that figure, we end up with the phase line at the left. Analogously, here we can imagine plotting both c_1 and c_2 versus t, with the t axis normal to the c_1, c_2 plane, so the solution curves begin in the plane of the paper and proceed in three-dimensional space toward the reader. If we now compress that plot — "squash it" onto the c_1, c_2 plane — we end up with the c_1, c_2 *phase plane* shown in Fig. 5. We will study the phase plane in Chapter 7. For the present, consider it as simply another useful way of plotting the results. ∎

4.1.5 Auxiliary variables. Theorem 4.1.1 is more general than it appears. For instance, it seems not applicable to the IVP

$$x'' + tx' + x = e^t, \quad x(0) = 4,\ x'(0) = 3, \tag{18a}$$
$$y''' + x' + y = 5t, \quad y(0) = 7,\ y'(0) = 0,\ y''(0) = -1, \tag{18b}$$

on $-\infty < t < \infty$ because of the higher-order derivatives. However, it *is* applicable if we define **auxiliary variables** $u(x)$, $v(x)$, and $w(x)$, according to

We can think of u, v, w as "artificial" variables.

$$x' = u,\ y' = v,\ y'' = w, \tag{19}$$

because then we can re-express (18) as the *first-order* system

$$x' = u, \qquad x(0) = 4, \tag{20a}$$
$$u' = -x - tu + e^t, \qquad u(0) = 3, \tag{20b}$$
$$y' = v, \qquad y(0) = 7, \tag{20c}$$
$$v' = w, \qquad v(0) = 0, \tag{20d}$$
$$w' = -u - y + 5t, \qquad w(0) = -1. \tag{20e}$$

That is, x, u, v, w are our "x_1, x_2, x_3, x_4," but in case the correspondence between (20) and (5) is not clear, let us rewrite (20) fully as

$$x' = 0x + u + 0y + 0v + 0w, \qquad x(0) = 4, \tag{21a}$$
$$u' = -x - tu + 0y + 0v + 0w + e^t, \qquad u(0) = 3, \tag{21b}$$
$$y' = 0x + 0u + 0y + v + 0w, \qquad y(0) = 7, \tag{21c}$$
$$v' = 0x + 0u + 0y + 0v + w, \qquad v(0) = 0, \tag{21d}$$
$$w' = 0x - u - y + 0v + 0w + 5t, \qquad w(0) = -1. \tag{21e}$$

For instance, $a_{21} = -1$, $a_{22} = -t$, $f_2(t) = e^t$, and $f_3(t) = 0$. All the coefficients and forcing functions are continuous on I $(-\infty, \infty)$, so Theorem 4.1.1 assures us that a unique solution of (20) exists on I. The IVPs (18) and (20) are equivalent since we can pass from (18) to (20) by putting $x' = u$, $y' = v$, and $y'' = w$, and from (20) back to (18) by putting $u = x'$, $v = y'$, and $w = y''$. Thus, there also exists a unique solution of (18) for $x(t)$ and $y(t)$ on I.

Of course, that assurance is separate from the matter of *finding* that solution, which would be difficult — chiefly because of the nonconstant coefficient t in (18a) [and (20b)]. For solution techniques, we will confine our attention, in subsequent sections, to the case of constant coefficients. Only in the final section of this chapter will we discuss the numerical solution of such difficult IVPs as (18).

Closure. We showed how linear first-order systems can arise in applications, gave the existence and uniqueness Theorem 4.1.1, and showed how to apply elimination to a system of two equations to obtain a second-order equation in only one unknown. However, not only does the method of elimination grow unwieldy for systems of more than two equations, it may not even work, as is illustrated in the exercises. We will continue to use elimination in the next section, when we study a two-mass oscillator, but after that we will develop matrix methods that are generally preferable.

EXERCISES 4.1

1. Solution by Elimination. For parts (a)–(h) append initial conditions $x(0) = 0, y(0) = 5$. For (i)–(n) use $x(2) = 4, y(2) = 5$. First, does Theorem 4.1.1 assure us of a unique solution on $-\infty < t < \infty$? Next, use the method of elimination to obtain a general solution of the system of two differential equations. Finally, apply the initial conditions to obtain a particular solution.

(a) $x' = -y + 10\sin t$, $y' = -9x + 36$
(b) $x' = x - 8y$, $y' = -x - y - 162t^2$
(c) $x' = 2x + 6y - 16t + 8$, $y' = 2x - 2y$
(d) $x' = x + y$, $y' = x + y + 4$
(e) $x' = x + y + 50$, $y' = -x - y - 50$
(f) $x' = y + e^t$, $y' = x - e^t$
(g) $x' = -x + y - 8$, $y' = 3x + y - 4t$
(h) $x' = 3x - 2y + t$, $y' = 4x - 3y + 12t$
(i) $x' = y$, $y' = x + 3t$
(j) $x' = 3x - y$, $y' = 4x - y$
(k) $x' = 2x + y$, $y' = x + 2y - 27$
(l) $x' = x + 2y$, $y' = 2x + y - 27$
(m) $x' = x + 2y$, $y' = 2x + y + 2e^t$
(n) $x' = 3x - y + 16t$, $y' = -x + 3y - 16t$

2. Example of the Failure of Elimination. It was stated in the Closure that for three or more equations elimination may fail. To illustrate, consider the example

$$x' = x + ay + z, \tag{2.1a}$$

$$y' = x + y + z, \tag{2.1b}$$

$$z' = x + y + z, \tag{2.1c}$$

with a a constant.

(a) Show that elimination fails unless $a = 0$, 1, or -1, and state the symptom(s) of the failure.
(b) Obtain a general solution of (2.1) by elimination for the case where $a = 1$.
(c) Obtain a general solution of (2.1) by elimination for the case where $a = -1$.
(d) Obtain a general solution of (2.1) by elimination for the case where $a = 0$. You need only solve for $x(t)$.
NOTE: A modified version of elimination that *does* work for any value of a is given in Exercise 11.

3. More by Elimination. Find a general solution by elimination. NOTE: These are not first-order systems, but elimination works nonetheless. In (a), for instance, solve the second equation for x and put it into the first equation, to obtain an equation for y alone. (Or, solve the first for y and put it into the second, to obtain an equation for x alone.)

(a) $x'' + 3x' + x + y = 18e^t,\ y' + x + y = 0$
(b) $x'' + 4x - y = 0,\ y'' - x + 4y = 0$
(c) $x' = x + y + 4,\ y'' = x' + y'$
(d) $x'' = -2x + 3y,\ x' + y' = -2x + 4y$
(e) $x'' = x - y + e^t,\ y'' = x - y$
(f) $x'' = -2x + y,\ y'' = x - 2y$

4. Consider two well-stirred tanks containing brine at concentrations $c_1(t)$ and $c_2(t)$. Inflows of brine are maintained at 10 gal/min and 5 gal/min, at concentrations c_{1i} and c_{2i}, respectively, and there are cross-flows and exit flows at the rate of 2, 12, and 15 gal/min, as indicated in the figure. The volumes of brine in the tanks are V_1 and V_2, respectively.

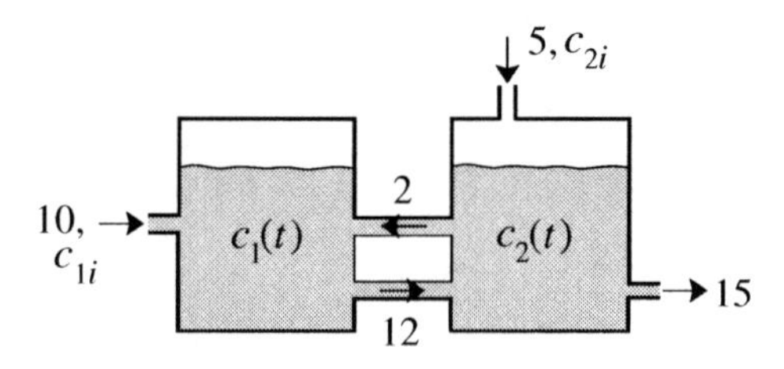

(a) Use a mass balance for each tank to derive the coupled differential equations for $c_1(t)$ and $c_2(t)$.

(b) Looking at those differential equations, will the steady-state values c_{1s} and c_{2s} depend on V_1 and V_2? Why (not)?
(c) Solve for $c_1(t)$ and $c_2(t)$, in terms of c_{1i} and c_{2i}, subject to the initial conditions $c_1(0) = 0$, $c_2(0) = 0$. Take $V_1 = V_2 = 100$ gals.
(d) Determine the steady-state concentrations in each tank, two ways: first, by letting $t \to \infty$ in your solutions, and also directly from the differential equations. If a steady state does not exist, state that.

5. Consider two tanks in series, as shown below, with an inflow of Q gal/min of solution containing c_i lb/gal of solute and an equal outflow rate of Q gal/min. Let the liquid volume in each be v gal, and let the initial conditions be $c_1(0) = c_2(0) = 0$.

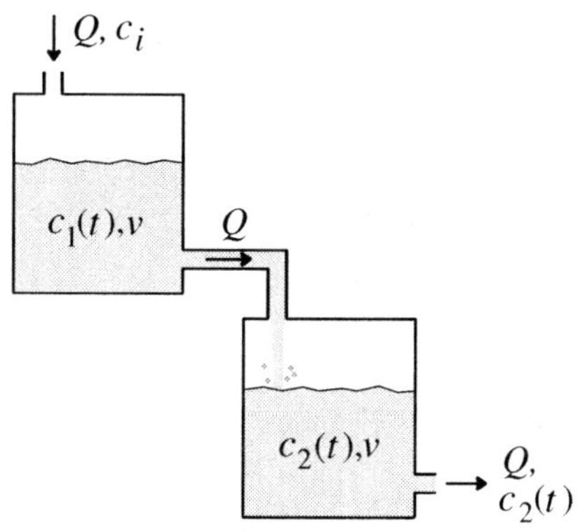

(a) Use a mass balance for each tank to show that the IVPs are

$$c_1'(t) + \frac{Q}{v}c_1(t) = \frac{Q}{v}c_i; \qquad c_1(0) = 0, \tag{5.1a}$$

$$c_2'(t) + \frac{Q}{v}c_2(t) - \frac{Q}{v}c_1(t) = 0; \quad c_2(0) = 0. \tag{5.1b}$$

(b) Solve for $c_1(t)$ and $c_2(t)$.
(c) Determine the steady-state concentrations in each tank, two ways: first, by letting $t \to \infty$ in your solutions, and also directly from the differential equations (5.1a,b).

6. (a) Apply Kirchhoff's laws to the left- and right-hand loops, and show that the currents $i_1(t)$ and $i_2(t)$ in the circuit shown below are modeled by the linear system

$$i_1' = -\frac{R}{L_1}i_1 + \frac{R}{L_1}i_2, \quad i_2' = \frac{R}{L_2}i_1 - \frac{R}{L_2}i_2. \tag{6.1a,b}$$

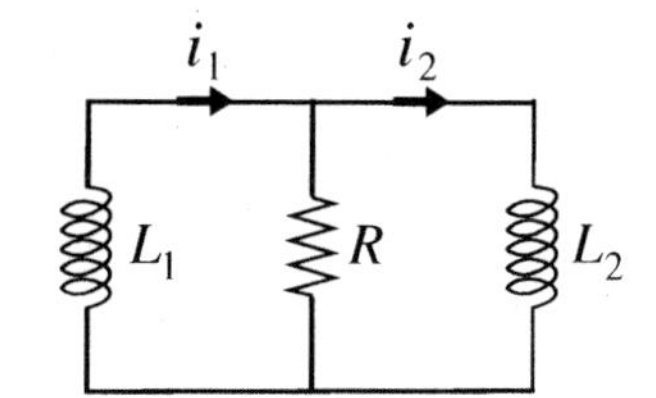

(b) Alternatively, obtain equations (6.1a,b) by using either (i) the left-hand loop and the outer loop or (ii) the right-hand loop and the outer loop.

7. Show that the currents $i_1(t)$ and $i_2(t)$ in the circuit shown below are modeled by the linear system

$$i_1' = -\frac{R_1}{L_1}i_1 + \left(\frac{R_1}{L_1} - \frac{R_2}{L_2}\right)i_2 + \left(\frac{1}{L_1} + \frac{1}{L_2}\right)E(t), \tag{7.1a}$$

$$i_2' = -\frac{R_2}{L_2}i_2 + \frac{1}{L_2}E(t). \tag{7.1b}$$

To do so, apply Kirchhoff's laws to the left-hand and outer loops, and solve by algebra for i_1' and i_2'.

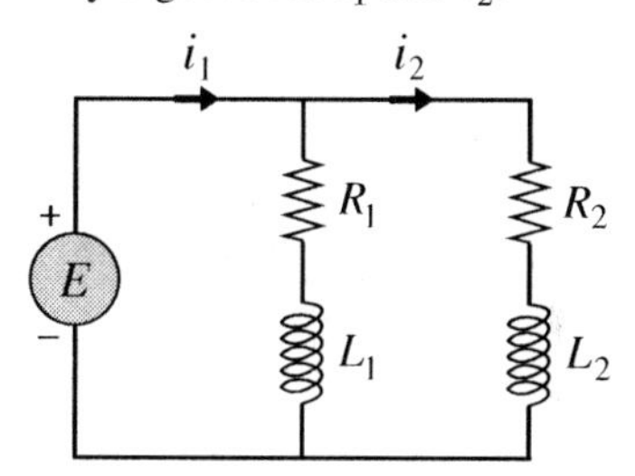

NOTE: Alternatively, you can obtain equations (7.1a,b) by using either (i) the left- and right-hand loops or (ii) the right-hand loop and the outer loop, although additional algebraic steps may be needed.

8. Use Kirchhoff's laws to derive a set of differential equations for the currents $i_1(t)$, $i_2(t)$, $i_3(t)$ in the circuit shown below. State which loops you are using.

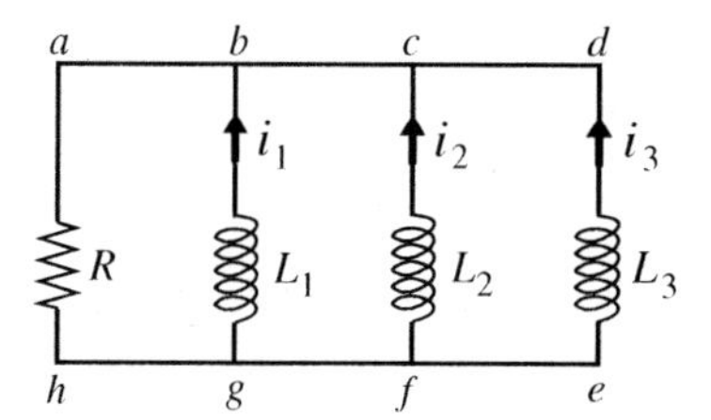

9. Vortex Drift. The motion of a vortex, such as a tornado, as it interacts with solid boundaries, is of important interest. Consider an infinitely long (idealized) counterclockwise vortex, normal to the page, in the presence of rigid walls along the x and y axes. Because of the walls, the vortex will move, with x and y velocity components

$$x' = \frac{\Gamma}{2\pi}\left(\frac{1}{y} - \frac{y}{x^2+y^2}\right), \quad y' = \frac{\Gamma}{2\pi}\left(-\frac{1}{x} + \frac{x}{x^2+y^2}\right), \tag{9.1a,b}$$

in which the positive constant Γ is a measure of the vortex's strength. We wish to determine the vortex's motion, but the system (9.1) is nonlinear and appears to be too difficult to solve for $x(t)$ and $y(t)$.

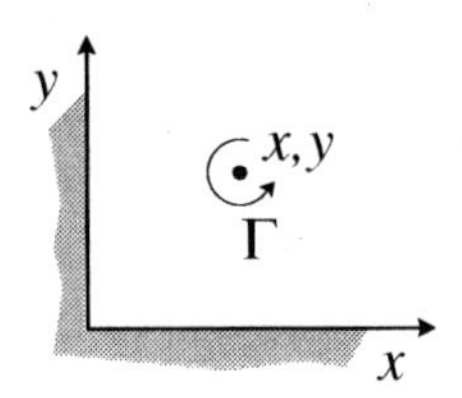

(a) Nevertheless, by dividing (9.1b) by (9.1a), show that

$$\frac{dy}{dx} = -\frac{y^3}{x^3}, \tag{9.2}$$

which is a simple separable equation for the path $y(x)$.
(b) Solve (9.2) subject to initial conditions $x(0) = a$, $y(0) = b$ (with $a > 0$ and $b > 0$) or, in terms of $y(x)$, $y(a) = b$, and show that the path is given by

$$x^2 + y^2 = \frac{a^2+b^2}{a^2b^2}x^2y^2. \tag{9.3}$$

(c) Give a careful sketch, or computer plot, of the path in the x, y plane, and determine any asymptotes. (For purposes of the sketch, choose b several times as large as a.)
NOTE: Realize that the tremendous simplification achieved, in passing from (9.1) to (9.2), was at the expense of information content, because if we had been able to solve (9.1) we would have found $x(t)$ and $y(t)$ and hence both the vortex's path and its motion, *in time*, along that path, but the solution of (9.2) gave only the path; the t dependence was lost.
(d) Evaluate the asymptotic x velocity, along the asymptote that you found in part (c).

ADDITIONAL EXERCISES

10. Compartmental Analysis. The anaesthetic gas halothane is to be put into the arterial blood system, from there to be delivered to tissue sites in the body. Let the system be simplified to consist of just two interacting "compartments:" an arterial blood compartment and a tissue compartment, containing masses of halothane $x(t)$ and $y(t)$, respectively. An exchange of halothane takes place between the two compartments, that can be modeled by the system

$$x' = -\alpha x + \beta y + f(t), \tag{10.1a}$$

$$y' = \alpha x - \beta y, \tag{10.1b}$$

in which α, β are positive empirically known constants, and $f(t)$ is the *dose schedule* of halothane, in mass per unit time, delivered to the arterial blood.

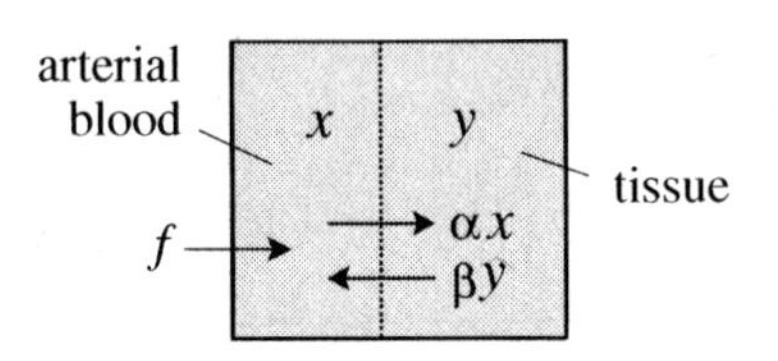

(a) With $x(0) = 0$ and $y(0) = 0$, what dose schedule $f(t)$ will bring the halothane content of the blood compartment up to a desired level x_s quickly, and keep it there? HINT: Model the rise of $x(t)$ to x_s by $x(t) = x_s(1-e^{-\gamma t})$. The larger we make γ, the faster the rise. Put that expression into (10.1) and solve for $f(t)$. Then, in your result, let $\gamma \to \infty$ to obtain the ideal case of an instantaneous rise to x_s. Show that you obtain

$$f(t) = x_s\delta(t) + \alpha x_s e^{-\beta t}, \tag{10.2}$$

in which $\delta(t)$ denotes $\lim_{\gamma\to\infty} \gamma e^{-\gamma t}$.
(b) Actually, the limit of $\gamma e^{-\gamma t}$ does not exist because it gives $\delta(0) = \infty$ and $\delta(t) = 0$ for all $t > 0$, which is not an acceptable function because of the ∞. However, observe that $\int_0^\infty \gamma e^{-\gamma t}\,dt = 1$, so $\delta(t)$ amounts to the delivery of a unit mass of the halothane. Further, sketch or plot $\gamma e^{\gamma t}$ versus t as γ increases. With these ideas in mind, explain why we can interpret the right-hand side of (10.2) as an instantaneous delivery of the full amount x_s at $t = 0$, followed by an additional amount $\alpha x_s e^{-\beta t}$ that diminshes with time. NOTE: This idealized unit "spike" function $\delta(t)$ is called the **Dirac delta function** and will be studied in Section 5.6.2.
(c) If $f(t)$ is given by (10.2), what is the steady-state value of $y(t)$, in terms of x_s?
NOTE: This exercise is based on material in R. F. Brown, *Biomedical Systems Analysis via Compartmental Concept* (Kent, UK: Abacus Press, 1985) which, in turn, is based on the research of I. S. Goldberg et al.

11. A Modified Elimination Method Using Cramer's Rule. In this exercise we outline, as an alternative to elimination, a solution method that will work for any system of the form (5), if the a_{ij} coefficients are constants. We will illustrate it for the system of two equations that was the subject of Example 3, before asking you to use the method to solve other problems: To find a general solution of the system (9), first write it in operator form (i.e., using D in place of d/dt):

$$(D-4)x + 3y = 4t, \tag{11.1a}$$

$$-2x + (D+1)y = 0. \tag{11.1b}$$

Treating the D's formally, as though they were numbers, solve (11.1) using Cramer's rule (Appendix B), and obtain

$$x = \frac{\begin{vmatrix} 4t & 3 \\ 0 & D+1 \end{vmatrix}}{\begin{vmatrix} D-4 & 3 \\ -2 & D+1 \end{vmatrix}} = \frac{4+4t}{D^2-3D+2}, \tag{11.2a}$$

$$y = \frac{\begin{vmatrix} D-4 & 4t \\ -2 & 0 \end{vmatrix}}{\begin{vmatrix} D-4 & 3 \\ -2 & D+1 \end{vmatrix}} = \frac{8t}{D^2-3D+2}. \tag{11.2b}$$

Next, multiply (11.2a,b) through by the $D^2 - 3D + 2$ in the denominators, and obtain

$$(D^2-3D+2)x = 4+4t, \tag{11.3a}$$

$$(D^2-3D+2)y = 8t. \tag{11.3b}$$

General solutions of (11.3a,b) are found, by methods studied in Chapter 2, to be

$$x(t) = C_1e^t + C_2e^{2t} + 2t + 5, \tag{11.4a}$$

$$y(t) = C_3e^t + C_4e^{2t} + 4t + 6. \tag{11.4b}$$

Now, (11.4a) is the same as (12a), but (11.4b) is not quite the same as (12b). The reason is that this **Cramer's rule method** gives too many constants, in the sense that they're not all arbitrary; C_1 and C_2 are arbitrary, but C_3 and C_4 can be expressed in terms of C_1 and C_2. To expose those relationships, substitute (11.4a,b) into the original equations (11.1a,b). Doing that and simplifying gives

$$(3C_3-3C_1)e^t + (3C_4-2C_2)e^{2t} = 0, \tag{11.5a}$$

$$(2C_3-2C_1)e^t + (3C_4-2C_2)e^{2t} = 0. \tag{11.5b}$$

Because of the linear independence of e^t and e^{2t}, we infer from (11.5a) that $3C_3 - 3C_1 = 0$ and $3C_4 - 2C_2 = 0$, so $C_3 = C_1$ and $C_4 = (2/3)C_2$. [The same relations follow from (11.5b).] Thus, C_1 and C_2 are arbitrary constants, but then $C_3 = C_1$ and $C_4 = (2/3)C_2$. Putting these expressions into (11.4b) gives the following general solution of (11.1),

$$x(t) = C_1e^t + C_2e^{2t} + 2t + 5, \tag{11.6a}$$

$$y(t) = C_1e^t + \frac{2}{3}C_2e^{2t} + 4t + 6, \tag{11.6b}$$

which now agrees with (12). NOTE: In this example it is simpler to omit (11.2b) and (11.3b). Once the solution (11.4a) for $x(t)$ is in hand, we can find $y(t)$ simply by substituting that result into (11.1a), which gives $y(t) = [4t + 4x(t) - x'(t)]/3$. That step gives (11.6b).

The problem: Use the Cramer's rule method (with the simplification noted above, if possible) to obtain a general solution for the following systems.

(a) $x' = -x + y - 8,\ y' = 3x + y - 4t$
(b) $x' = 3x - 2y + t,\ y' = 4x - 3y + 12t$
(c) $x' = x + y + 2e^t,\ y' = 2x + 2y + 18e^{3t}$
(d) $x' = 3x + 2y + 12e^{-t},\ y' = 2x + 3y$
(e) $x'' = x - y,\ y'' = x - y$
(f) $x'' = -2x + y + 24,\ y'' = x - 2y$

12. A System of Three Equations. Use the Cramer's rule method outlined in Exercise 11 to find a general solution of the system (2.1) in Exercise 2, for the case where $a = 1$. NOTE: As noted in Exercise 2, elimination does not work if $a = -1, 0,$ or 1, whereas the Cramer's rule method works for any value of a. Nevertheless, Cramer's rule is of little help for large systems because they require the evaluation of large determinants. Thus, for large systems (and indeed even for small ones) the matrix methods introduced later in this chapter are preferred.

13. Charged Particle in Magnetic Field. A charged particle, of mass m and electrical charge q, moves in the presence of a magnetic field $\boldsymbol{B} = B\mathbf{k}$ aligned with the z axis, where $\mathbf{i},\mathbf{j},\mathbf{k}$ are unit vectors in the x, y, z directions and B is a constant. The force on the particle, due to the magnetic field, is the cross product $\mathbf{F} = q\mathbf{v} \times \mathbf{B}$, where $\mathbf{v} = x'(t)\mathbf{i} + y'(t)\mathbf{j} + z'(t)\mathbf{k}$ is the particle's velocity.

(a) Use Newton's law to show that the equations of motion are

$$mx'' = qBy', \quad my'' = -qBx', \quad mz'' = 0. \tag{13.1a,b,c}$$

(b) Solve these, using the Cramer's rule method explained in Exercise 11, subject to initial conditions $x(0) = 0,\ y(0) = 0,\ z(0) = 0,\ x'(0) = V,\ y'(0) = 0,\ z'(0) = 0$. Show that the path of motion is a circle in the x, y plane and determine its center and radius and its direction of traversal (clockwise or counterclockwise). Show that the particle's velocity on the circle is a constant, V.

4.2 APPLICATION TO COUPLED OSCILLATORS

PREREQUISITES: Sections 2.6.2 and 3.2 for 4.2.1–4.2.3, and 3.4 for 4.2.4.

4.2.1 Coupled oscillators. In Chapter 3 we studied the mechanical harmonic oscillator shown here in Fig. 1. Suppose now we have two masses instead of one, as in Fig. 2, with applied forces $F_1(t)$ and $F_2(t)$. Since the masses are connected by the middle spring, surely their motions $x_1(t)$ and $x_2(t)$ will be interdependent, and hence their differential equations will be coupled.

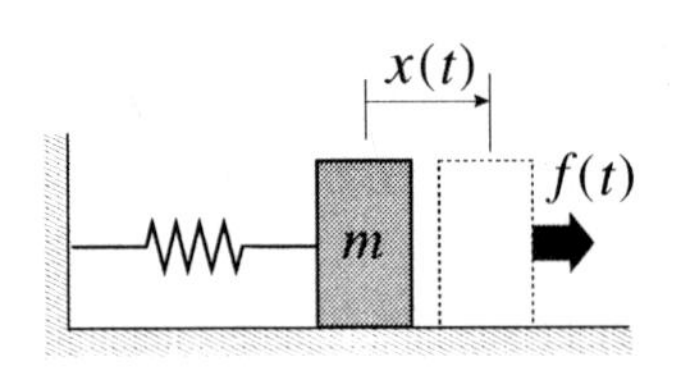

Figure 1. Single-mass oscillator.

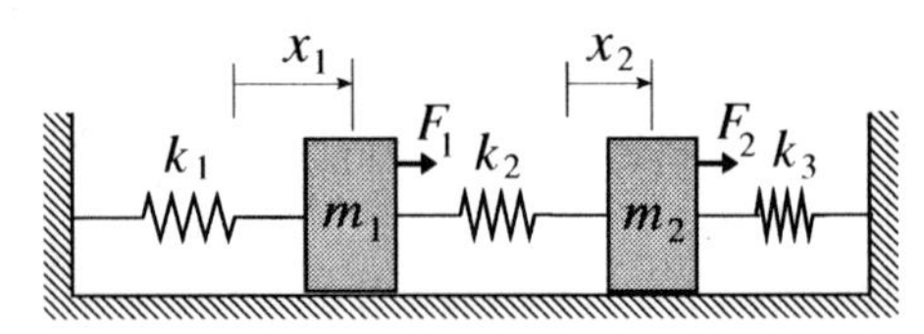

Figure 2. Two-mass oscillator, neglecting friction and air resistance.

To derive those differential equations, apply Newton's second law of motion. Free-body diagrams of the masses (schematic diagrams showing the forces acting on each mass) are given in Fig. 3. We neglect any frictional forces exerted on the masses by the table, and any aerodynamic drag forces on them as well.

We are assuming that the three springs are linear, that is, that they are adequately described by Hooke's law "$f = kx$," with stiffnesses k_1, k_2, k_3.

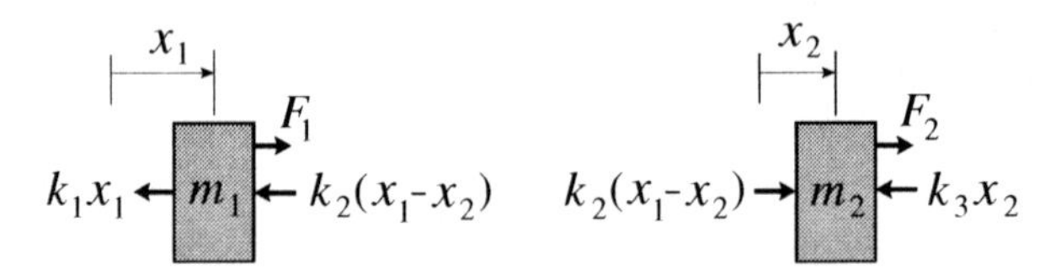

Figure 3. Free-body diagrams of the masses.

There is no need to consider the weight forces and the vertical upward forces exerted by the table on the masses in Fig. 3 because for the horizontal motion the vertical forces are not relevant. How do we know if the spring forces act to the left or to the right? For definiteness, assume the displacement and velocity of each mass is rightward and that $x_1 > x_2 > 0$ at the instant under consideration. (As noted in Section 3.2, the resulting equations of motion are independent of such assumptions.) Then the spring forces will be as shown in Fig. 3, and application of Newton's law to each of the masses gives

$$m_1 x_1'' = -k_1 x_1 - k_2(x_1 - x_2) + F_1(t),$$
$$m_2 x_2'' = k_2(x_1 - x_2) - k_3 x_2 + F_2(t),$$

which are indeed coupled. Or, after rearranging terms,

Equations (1a,b) are coupled by the x_2 in (1a) and the x_1 in (1b).

$$m_1 x_1'' + (k_1 + k_2)x_1 - k_2 x_2 = F_1(t), \tag{1a}$$
$$m_2 x_2'' - k_2 x_1 + (k_2 + k_3)x_2 = F_2(t), \tag{1b}$$

on $0 \leq t < \infty$. Let $m_1 = m_2 = k_1 = k_2 = k_3 = 1$, for definiteness. Then (1) becomes

$$x_1'' + 2x_1 - x_2 = F_1(t), \tag{2a}$$
$$x_2'' - x_1 + 2x_2 = F_2(t). \tag{2b}$$

The latter is a **second-order system** because the highest derivatives are of second order. If we wish to include initial conditions, what sort of conditions are appropriate? We cannot use Theorem 4.1.1 to guide us because it is for first-order systems. But, we can reduce (2) to a first-order system by introducing auxiliary variables, as in Section 4.1.3.

4.2.2 Reduction to first-order system by auxiliary variables. Letting $x_1' = u$ and $x_2' = v$, we can re-express (2) as the first-order system

$$x_1' = u, \tag{3a}$$
$$u' = -2x_1 + x_2 + F_1(t), \tag{3b}$$
$$x_2' = v, \tag{3c}$$
$$v' = x_1 - 2x_2 + F_2(t). \tag{3d}$$

The latter corresponds to Equation (8) of Section 4.1, with $n = 4$. Now we can see from (3) and Theorem 4.1.1 that appropriate initial conditions at $t = 0$ are

$$x_1(0) = b_1,\ x_1'(0) = u(0) = b_2,\ x_2(0) = b_3,\ x_2'(0) = v(0) = b_4, \tag{4}$$

with the b_j's prescribed. If $F_1(t)$ and $F_2(t)$ are continuous, Theorem 4.1.1 applies and assures us that with the latter initial conditions the system (3) [and the equivalent system (2)] has a unique solution.

4.2.3 The free vibration. As in Chapter 3, we begin with the free vibration, where $F_1(t)$ and $F_2(t)$ are zero. Then (2) becomes the homogeneous system

$$\boxed{\begin{aligned} x_1'' + 2x_1 - x_2 &= 0, \\ x_2'' - x_1 + 2x_2 &= 0. \end{aligned}} \tag{5a,b}$$

We converted the second-order equations (2) to the *first-order* equations (3) so we could apply Theorem 4.1.1. But, to *solve* by elimination we proceed in the opposite direction — to uncoupled *higher-order* equations.

We will solve (5) and then apply initial conditions of the form (4). To solve by elimination, should we start with (5) or with the equivalent first-order system (3) (with $F_1 = F_2 = 0$)? Keep in mind that the process of elimination leads from a system of lower-order equations to a single higher-order equation, so beginning with (3) would be a step in the wrong direction. Rather, begin with (5). The latter is a second-order system, but the method of elimination will work: First, solve equation (5a) by algebra for x_2 and obtain

$$x_2 = x_1'' + 2x_1, \tag{6}$$

then put (6) into (5b) and obtain the fourth-order equation

$$x_1'''' + 4x_1'' + 3x_1 = 0 \tag{7}$$

for $x_1(t)$ alone, with general solution

$$x_1(t) = C_1 \cos t + C_2 \sin t + C_3 \cos \sqrt{3}t + C_4 \sin \sqrt{3}t. \tag{8}$$

Next, putting (8) into (6) gives $x_2(t)$, and we have

$$x_1(t) = C_1 \cos t + C_2 \sin t + C_3 \cos \sqrt{3}t + C_4 \sin \sqrt{3}t, \tag{9a}$$
$$x_2(t) = C_1 \cos t + C_2 \sin t - C_3 \cos \sqrt{3}t - C_4 \sin \sqrt{3}t. \tag{9b}$$

The form (10) facilitates physical understanding, but (9) is more convenient for the application of initial conditions.

Regarding the form of (9a) and (9b), recall from Section 3.2 that we can express a trigonometric combination $C \cos \omega t + D \sin \omega t$ in the form $E \sin(\omega t + \phi)$, with E and ϕ related to C and D by equation (15) of that section. That form is not essential, but it facilitates physical understanding because E, ω, ϕ are the *amplitude*, *frequency*, and *phase*, respectively. Thus, to better understand the solution physically, re-express (9) in the equivalent form

$$x_1(t) = G \sin(t + \phi) + H \sin(\sqrt{3}t + \psi), \tag{10a}$$
$$x_2(t) = G \sin(t + \phi) - H \sin(\sqrt{3}t + \psi), \tag{10b}$$

in which G, H, ϕ, ψ are now the arbitrary constants in place of C_1, C_2, C_3, C_4. If, as initial conditions, we prescribe the values of

$$x_1(0),\ x_1'(0),\ x_2(0),\ x_2'(0), \tag{11}$$

then we can determine G, H, ϕ, ψ if we use (10), or C_1, C_2, C_3, C_4 if we use (9).

The solution (10) is a superposition of two modes.

The solution (10) is a superposition of two distinct **modes**: If $H = 0$ and $G \neq 0$ then the solution is a **low-mode** motion; if instead $G = 0$ and $H \neq 0$ then it is a **high-mode** motion. These are called low- and high-mode motions because they are at the lower frequency 1 rad/sec and higher frequency $\sqrt{3}$ rad/sec, respectively.

We see from (10) that in the low mode the two masses move with the same amplitude G, with the same frequency 1, and with the same phase ϕ; they are in-phase. In the high mode they move with the same amplitude H and with the same frequency $\sqrt{3}$, but are $180°$ out-of-phase because of the minus sign in (10b). To see why we say they are $180°$ out-of-phase, use the trigonometric identity $\sin(A+\pi) = \sin A \cos \pi + \sin \pi \cos A = -\sin A$, to re-express the $-\sin(\sqrt{3}t + \psi)$ in (10b) as $\sin(\sqrt{3}t + \psi + \pi)$, which differs in phase from the $\sin(\sqrt{3}t + \psi)$ in (10a) by π (i.e., by $180°$).

The single-mass oscillator in Chapter 3 had one natural frequency; the two-mass oscillator considered here has two.

The modal frequencies are the **natural frequencies** of the system, the frequencies that can occur in the free vibration.

Turning to initial conditions, suppose we wish to excite the system in a purely low-mode motion. What initial conditions will accomplish that? We see from (10) that in the low mode the $x_1(t)$ and $x_2(t)$ motions are identical, so we should be able to excite that mode by starting the system out in that manner, that is, with $x_2(0) = x_1(0)$ and $x_2'(0) = x_1'(0)$. For instance, let $x_1(0) = x_2(0) = a$ and $x_1'(0) = x_2'(0) = 0$, which amounts to moving each mass to the right by a distance a, holding them still (with zero velocity), and letting go at $t = 0$.

Although the form (10) facilitates understanding of the solution, (9) is the more convenient for applying initial conditions, because $\cos 0 = 1$ and $\sin 0 = 0$. Then, (9) gives

$$\begin{aligned} x_1(0) &= a = C_1 + C_3, \\ x_2(0) &= a = C_1 - C_3, \\ x_1'(0) &= 0 = C_2 + \sqrt{3}C_4, \\ x_2'(0) &= 0 = C_2 - \sqrt{3}C_4, \end{aligned} \tag{12}$$

which give $C_1 = a$ and $C_2 = C_3 = C_4 = 0$. Hence,

$$x_1(t) = a\cos t, \tag{13a}$$
$$x_2(t) = a\cos t, \tag{13b}$$

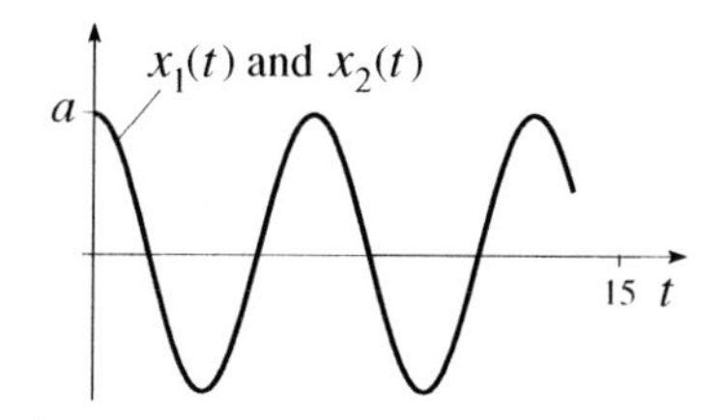

Figure 4. The low-mode response (13). The two masses move in unison, with the same amplitude and the same phase.

which is indeed purely a low-mode motion [because $C_3 = C_4 = 0$ in (9)]. The graph of the motion (13) is given in Fig. 4.

To obtain a high-mode motion instead, choose initial conditions such that $x_2(0) = -x_1(0)$ and $x_2'(0) = -x_1'(0)$. For instance, let $x_1(0) = +b$ and $x_2(0) = -b$, and $x_1'(0) = x_2'(0) = 0$. Then we find that $C_1 = C_2 = C_4 = 0$ and $C_3 = b$, so

$$x_1(t) = b\cos\sqrt{3}t, \tag{14a}$$
$$x_2(t) = -b\cos\sqrt{3}t, \tag{14b}$$

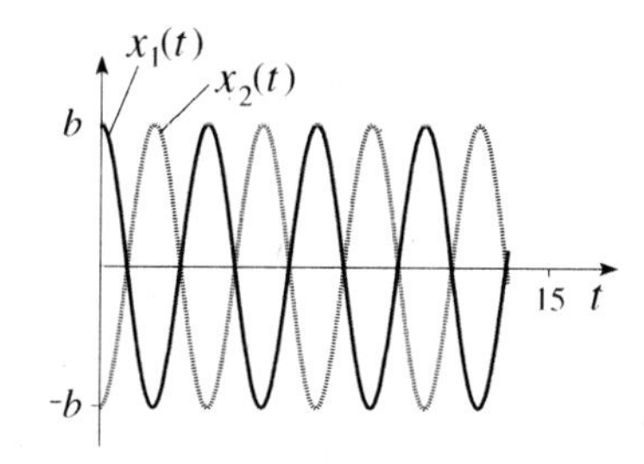

Figure 5. The high-mode response (14); $x_1(t)$ and $x_2(t)$ have the same amplitude but are 180° out-of-phase.

which is a purely high-mode motion; the oscillation is at the higher frequency, $\sqrt{3}$, and the masses are 180° out-of-phase (Fig. 5).

In general, however, the motion will be a mix of the two modes. For instance, if $x_1(0) = 8$ and $x_2(0) = x_1'(0) = x_2'(0) = 0$, we obtain

$$x_1(t) = 4\cos t + 4\cos\sqrt{3}t, \tag{15a}$$
$$x_2(t) = 4\cos t - 4\cos\sqrt{3}t, \tag{15b}$$

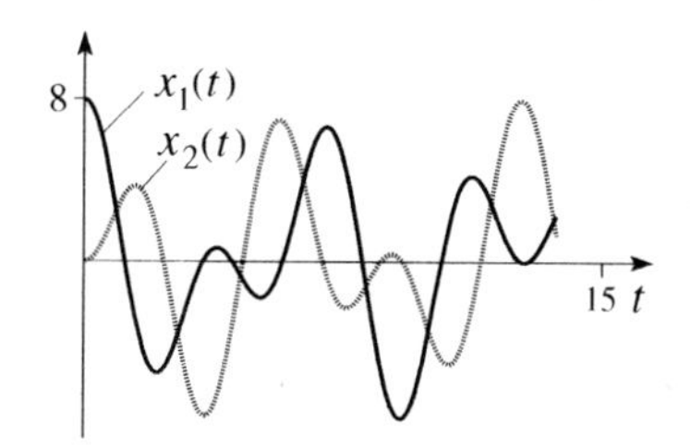

Figure 6. The mixed-mode response (15); no clear pattern is evident.

which includes both modes. Observe from Figs. 4 and 5 that the pure-mode motions are "clean and simple": In the low mode the masses move in-phase at the lower frequency, and in the high mode they move 180° out of phase at the higher frequency. By comparison, the mixed-mode motion is a "mess," with no clear pattern or relationship between $x_1(t)$ and $x_2(t)$. If you use your fists to represent the masses, you should find that to mimic the low- and high-mode motions (Figs. 4 and 5) is simple, but to mimic the mixed-mode motion (Fig. 6) is a challenge.

4.2.4 The forced vibration. Now consider the forced vibration, in which one or both of $F_1(t)$ and $F_2(t)$ is nonzero in (2). As a representative case, let $F_1(t) = F \sin \Omega t$ and $F_2(t) = 0$, so (2) becomes

$$x_1'' + 2x_1 - x_2 = F \sin \Omega t,$$
$$x_2'' - x_1 + 2x_2 = 0. \tag{16a,b}$$

Solution of (16) by elimination gives

$$x_1(t) = \left[C_1 \cos t + C_2 \sin t + C_3 \cos \sqrt{3}t + C_4 \sin \sqrt{3}t\right] + \left\{\frac{(2-\Omega^2)F}{(\Omega^2-1)(\Omega^2-3)} \sin \Omega t\right\}, \tag{17a}$$

$$x_2(t) = \left[C_1 \cos t + C_2 \sin t - C_3 \cos \sqrt{3}t - C_4 \sin \sqrt{3}t\right] + \left\{\frac{F}{(\Omega^2-1)(\Omega^2-3)} \sin \Omega t\right\}. \tag{17b}$$

Note that (17) is valid only if Ω is not equal to either of the natural frequencies 1 and $\sqrt{3}$ because if it is, then the coefficients of the $\sin \Omega t$ terms in (17) are undefined.

That is, solve (16a) by algebra for x_2 and put it into (16b). That gives $x_1'''' + 4x_1'' + x_1 = F(3-\Omega^2) \sin \Omega t$, solution of which is given by (17a). Then, put (17a) into (16a) and solve it by algebra for x_2. That step gives (17b).

The square-bracketed terms on the right-hand sides of (17a,b) are the homogeneous solutions $x_{1h}(t)$ and $x_{2h}(t)$, respectively, and the curly-bracketed terms are particular solutions $x_{1p}(t)$ and $x_{2p}(t)$, respectively. Of course, the homogeneous solutions are the same as the solutions (9a,b) that we discussed for the free vibration, so the only difference in (17) is the additional particular solutions. Thus, the motion of each mass consists of a linear combination of the two natural modes plus a particular solution.

Initial conditions are applied to the *total* solution (17), not just to the homogeneous solutions, to determine the constants $C_1, \ldots, C_4$. We will not pursue those steps. Instead, having already discussed the homogeneous solutions, let us conclude by examining the particular solutions in (17),

$$x_{1p}(t) = \frac{(2-\Omega^2)\,F}{(\Omega^2-1)\,(\Omega^2-3)} \sin \Omega t \equiv I_1(\Omega) \sin \Omega t, \tag{18a}$$

$$x_{2p}(t) = \frac{F}{(\Omega^2-1)\,(\Omega^2-3)} \sin \Omega t \equiv I_2(\Omega) \sin \Omega t, \tag{18b}$$

called the *forced vibration*, or the *forced response*.

Plotting the coefficients of $\sin \Omega t$, in (18), versus Ω, is crucial in understanding the forced response.

The coefficients $I_1(\Omega)$ and $I_2(\Omega)$ in (18) are plotted in Fig. 7a. Their absolute magnitudes $|I_1(\Omega)|$ and $|I_2(\Omega)|$ are the *amplitudes* of the forced vibration and are plotted in Fig. 7b. These curves are called the *amplitude response curves* because they show how the amplitude of the forced vibration varies with the frequency of the forcing function.

Recall from Chapter 3 that for the single-mass oscillator modeled by $mx'' + kx = 0$ there is one natural frequency, $\omega_n = \sqrt{k/m}$, and if the force is $F \sin \Omega t$ so that $mx'' + kx = F \sin \Omega t$, then resonance occurs if the forcing frequency Ω equals the natural frequency ω_n. We saw from Fig. 3a in Section 3.4 that the forced vibrational amplitude tends to infinity as $\Omega \to \omega_n$.

For the *two*-mass oscillator considered in the present section, there are *two* natural frequencies (1 and $\sqrt{3}$), and we see, from Fig. 7b or from (18a,b), that the forced vibration amplitudes $|I_1(\Omega)|$ and $|I_2(\Omega)|$ both "blow up" (tend to infinity) as Ω tends to either of those frequencies. That is, **resonance** occurs if the forcing frequency Ω equals either of the two natural frequencies. This pattern holds

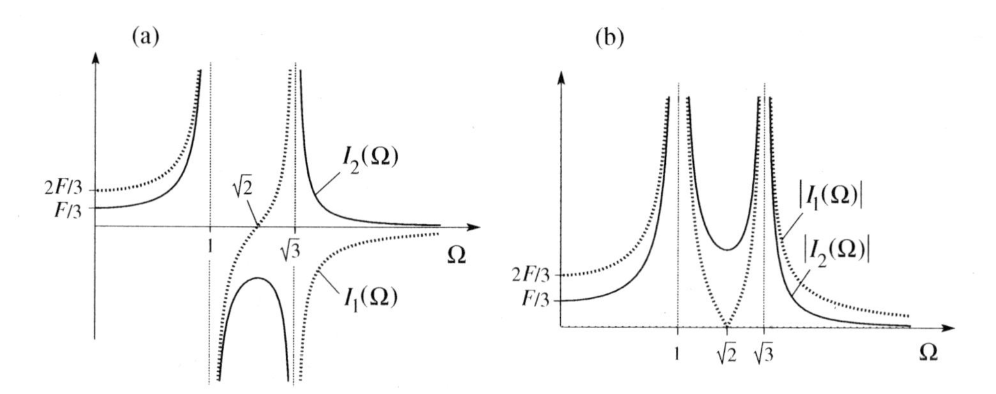

From Fig. 7b we see that resonance occurs when Ω equals either of the two natural frequencies 1 and $\sqrt{3}$.

Figure 7. The amplitude response curves, where $I_1(\Omega)$ and $I_2(\Omega)$ are defined in (18).

for any number of masses: For n coupled masses there will be n natural frequencies, and resonance occurs if the system is forced at any of those natural frequencies (however, see Exercise 12b).

Note that there is a forcing function $F \sin \Omega t$ in the x_1-equation (16a), but *no* forcing function in the x_2-equation (16b), so it might be tempting to conclude that there will be a nonzero particular solution for $x_1(t)$ but not for $x_2(t)$. Yet, we see in (17a,b) that there are nonzero particular solutions (the curly-bracketed terms) for both x_1 and x_2. This is because of the coupling of (16a) and (16b), for if we re-express (1b) as $m_2 x_2'' + (k_2 + k_3)x_2 = k_2 x_1 + F_2(t)$, then even if $F_2(t) = 0$, the $k_2 x_1$ coupling term, on the right, acts as a forcing function for $x_2(t)$.

Closure. For two masses we found the free vibration to be a linear combination of two oscillatory modes, at the natural frequencies. For the forced vibration we let $F_1(t) = F \sin \Omega t$ and $F_2(t) = 0$. As for the free vibration, we solved the system of coupled differential equations by elimination and focused on the particular solutions — the "forced vibration." Figure 7b revealed a resonance phenomenon when the forcing frequency equals either of the two natural frequencies.

EXERCISES 4.2

THE FREE VIBRATION

1. Consider (1), with $F_1(t) = F_2(t) = 0$. Let $m_1 = m_2 = k_1 = k_3 = 2$, and $k_2 = 3$.
(a) Solve for $x_1(t)$ and $x_2(t)$ and give your solution in the sine and cosine form analogous to (9), and identify the natural frequencies.
(b) Also give your solution in the single-sine-with-phase-angle form analogous to (10).
(c) Cite any specific set of initial conditions that will lead to a purely low-mode motion.
(d) Cite any specific set of initial conditions that will lead to a purely high-mode motion.
(e) Cite any specific set of initial conditions that will lead to a mixed-mode motion.

2.(a)–(e) The same as Exercise 1, but using $m_1 = 18$, $m_2 = 8$, $k_1 = 27$, $k_2 = 18$, and $k_3 = 2$.

3.(a)–(e) The same as Exercise 1, but using $m_1 = 36$, $m_2 = 4$, $k_1 = 81$, $k_2 = 9$, and $k_3 = 1$.

4.(a)–(e) The same as Exercise 1, but using parameter values given by your instructor.

5. Application of Initial Conditions. Above (12) we said that it is more convenient to use the solution form (9) than the equivalent form (10) when we apply the initial conditions. We proceeded to use (9) and readily obtained the final result (13). To see why we preferred (9) over (10), here we ask you to use (10) instead. Show that you do obtain the same final result (13), although obtaining it will be trickier. You should find that $G = a$, $H = 0$, and $\phi = \pi/2$.

6. Is (15) Periodic? Recall the solution given by (15),

$$x_1(t) = 4\cos t + 4\cos\sqrt{3}t, \quad x_2(t) = 4\cos t - 4\cos\sqrt{3}t. \tag{6.1a,b}$$

It is tempting to think that these are periodic functions of t because they are linear combinations of periodic functions. [A function $f(t)$ is **periodic** with **period** T if its value repeats as t is incremented by T, that is, if $f(t+T) = f(t)$ for all t.] Let us see. It will suffice to consider just $x_1(t)$. At $t = 0$ we have $x_1(0) = 8$. If $x_1(t)$ is periodic it will have to repeat that value at some time T, at $2T$, $3T$, and so on. For $x_1(t)$ to attain the value 8 at some time T we will need

$$\cos T = 1 \text{ and } \cos\sqrt{3}T = 1. \tag{6.2a,b}$$

Now, (6.2a) holds at $T = 2m\pi$ for every integer m, and (6.2b) holds at $T = 2n\pi$, for every integer n. How does this result show that $x_1(t)$ is, indeed, *not* periodic? HINT: Recall that $\sqrt{3}$ is an irrational number.

7. Beats. Consider again the free vibration of the system shown in Fig. 2, that is, with $F_1(t) = F_2(t) = 0$. Let $m_1 = m_2 = k_2 = 1$ and $k_1 = k_3 = 20$. Since k_2 is small compared to k_1 and k_2 we say that "the coupling is weak." (If we were to let k_2 be zero then there would be no coupling at all and the motions of the two masses would be independent.) In setting up this problem to illustrate the phenomenon of beats we must have k_2 be small compared to k_1 and k_3, which we've done. Any initial conditions that excite both modes will do; for definiteness, let $x_1(0) = 1, x_2(0) = x_1'(0) = x_2'(0) = 0$.

(a) With these choices show that the solution is

$$\begin{aligned} x_1(t) &= \tfrac{1}{2}\left(\cos\sqrt{20}\,t + \cos\sqrt{22}\,t\right), \\ x_2(t) &= \tfrac{1}{2}\left(\cos\sqrt{20}\,t - \cos\sqrt{22}\,t\right). \end{aligned} \tag{7.1}$$

(b) Next, use the trigonometric identities

$$\begin{aligned} \cos A + \cos B &= 2\cos\frac{A+B}{2}\cos\frac{A-B}{2}, \\ \cos A - \cos B &= -2\sin\frac{A+B}{2}\sin\frac{A-B}{2} \end{aligned} \tag{7.2}$$

to show that (to two decimal places) (7.1) can be re-expressed in the form

$$\begin{aligned} x_1(t) &= \cos 4.58t\,\cos 0.11t, \\ x_2(t) &= \sin 4.58t\,\sin 0.11t. \end{aligned} \tag{7.3}$$

(c) Use (7.3) to plot (by hand or by computer) $x_1(t)$ and $x_2(t)$ versus t in separate graphs, one below the other. NOTE: Observe the slow transfer of energy back and forth between the two masses. In vibration theory this phenomenon is known as **beats**. In case this is unclear, let us say it differently: Since $x_1(0) = 1$ and $x_1'(0) = 0$, mass #1 starts out in an oscillation of amplitude 1, whereas mass #2 is initially at rest because $x_2(0) = x_2'(0) = 0$. Subsequently, mass #2 is excited by the force transmitted to it by the middle spring, but only weakly because the middle spring has relatively small stiffness, so the oscillation of mass #2 builds up slowly in time. As the energy is transferred from mass #1 to mass #2 the former loses energy and its oscillation amplitude diminishes. When mass #1 finally almost comes to rest the process repeats — in reverse. Notice that there are two different "time scales" present: the masses oscillate on the faster time scale, but the process of energy transfer occurs at the slower time scale.

8. The Effect of Damping. Thus far we've assumed there is no friction between the masses and the tabletop. We did discuss damping for the single-mass oscillator in Section 3.3, where we considered the friction force to be proportional to the velocity of the mass. Let us do the same here, so in Fig. 3 include a force cx_1' on mass #1 to the left, and cx_2' on mass #2 to the left. Then, in place of (1) we have, for the free vibration,

$$m_1x_1'' + cx_1' + (k_1+k_2)x_1 - k_2x_2 = 0, \tag{8.1a}$$

$$m_2x_2'' + cx_2' - k_2x_1 + (k_2+k_3)x_2 = 0. \tag{8.1b}$$

We saw in Section 3.3 that if the damping coefficient c is "subcritical," then the frequency of the free vibration is less than the natural frequency of the undamped oscillator and the oscillation dies out exponentially; if c is increased to be "supercritical," then the oscillation disappears altogether and the motion dies out exponentially. *Here is the problem*: With $m_1 = m_2 = k_1 = k_2 = k_3 = 1$, for definiteness, discuss the effect on the solution as the damping coefficient c is increased. That is, discuss the effect of c on the two modal frequencies, and identify the critical value of c above which the modes change from underdamped to overdamped. HINT: When you obtain fourth-order equations for $x_1(t)$ and $x_2(t)$, seek solutions in the form e^{rt} and use computer software to solve the quartic characteristic equation for r.

THE FORCED VIBRATION

9. Is it possible to choose initial conditions for (16) so that C_1, C_2, C_3, and C_4 are all zero in (17), so that only the forced vibration is present? If so, give such a set of initial conditions (in terms of F and Ω); if not, explain why not.

10. The Solution *at* Resonance, Without Damping. We noted that *at* resonance the solution (17) fails because the coefficients of the $\sin \Omega t$ terms are undefined. Go back to (16) and let Ω equal either of the resonant frequencies, say $\Omega = 1$, and re-solve the system for $x_1(t)$ and $x_2(t)$. You can use the square-bracketed homogeneous solutions found in (17a,b) without re-deriving them, but you need to find the two particular solutions. As a partial check on your work, here is $x_1(t)$:

$$x_1(t) = C_1 \cos t + C_2 \sin t + C_3 \cos \sqrt{3}t + C_4 \sin \sqrt{3}t - (F/4)t \cos t. \tag{10.1}$$

NOTE: The $t \cos t$ term in (10.1) indicates resonance.

11. With Damping Included. We discussed the inclusion of damping in Exercise 8, for the free vibration. Now let us include it in the forced vibration. Including damping terms in (1) gives

$$m_1x_1'' + c_1x_1' + (k_1+k_2)x_1 - k_2x_2 = F_1(t), \tag{11.1a}$$

$$m_2x_2'' + c_2x_2' - k_2x_1 + (k_3+k_2)x_2 = F_2(t). \tag{11.1b}$$

Let it suffice to consider a specific case; let $m_1 = m_2 = k_2 = k_3 = 1$, $c_1 = c_2 = 0.1$, and $F_1(t) = F \sin \Omega t$ and $F_2(t) = 0$ with $F = 1$ and $\Omega = 2$, in which case (11.1) becomes

$$x_1'' + 0.1x_1' + 2x_1 - x_2 = \sin 2t, \tag{11.2a}$$

$$x_2'' + 0.1x_2' - x_1 + 2x_2 = 0. \tag{11.2b}$$

You need not determine the general solution, just determine the steady-state response, that is, the steady-state oscillation.

12. In Section 4.2.4 we considered, as an illustration, the case where $F_1(t) = F \sin \Omega t$ and $F_2(t) = 0$. Now consider the case where $F_1(t) = F_1 \sin \Omega t$ and $F_2(t) = F_2 \sin \Omega t$, so in place of (16) we have

$$x_1'' + 2x_1 - x_2 = F_1 \sin \Omega t, \tag{12.1a}$$

$$x_2'' - x_1 + 2x_2 = F_2 \sin \Omega t. \tag{12.1b}$$

(a) Use computer software to obtain a general solution of (12.1).

(b) **Forcing at a Natural Frequency Need Not Cause Resonance.** Show that resonance occurs if Ω equals either of the two natural frequencies 1 and $\sqrt{3}$. However, show that if $F_1 = -F_2$ then $\Omega = \sqrt{3}$ still causes resonance, but $\Omega = 1$ does *not*, and that if $F_1 = +F_2$ then $\Omega = 1$ still causes resonance, but $\Omega = \sqrt{3}$ does *not*.

MORE THAN TWO MASSES

13. Three Masses. (a) For the free vibration of the three-mass system shown below, derive the equations

$$m_1x_1'' + (k_1+k_2)x_1 - k_2x_2 = 0, \tag{13.1a}$$

$$m_2x_2'' - k_2x_1 + (k_2+k_3)x_2 - k_3x_3 = 0, \tag{13.1b}$$

$$m_3x_3'' - k_3x_2 + (k_3+k_4)x_3 = 0, \tag{13.1c}$$

neglecting friction and air resistance.

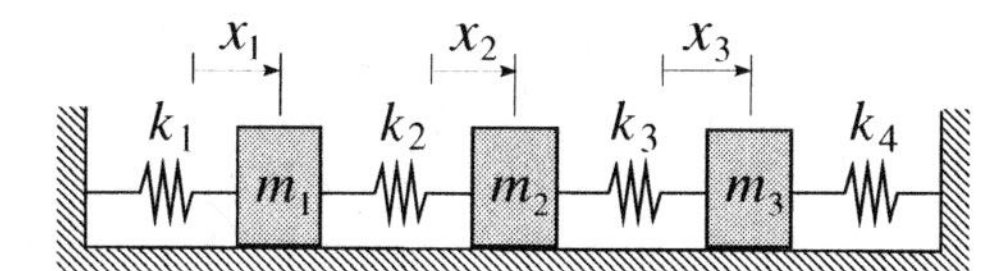

(b) Let all the m's and k's equal 1, say. Using computer software to solve the system of differential equations, show that the natural frequencies are 1.41, 1.85, and 0.765 rad/sec.
(c) In fact, it can be shown that if there are n such masses, each of mass m, and $n+1$ springs, each of stiffness k, then there are the n natural frequencies

$$\omega_j = \left(2-2\cos\frac{j\pi}{n+1}\right)\sqrt{\frac{k}{m}} \tag{13.2}$$

for $j=1,2,\dots,n$. With $n=3$, show that (13.2) agrees with the frequencies found in part (b). Also, show that for n masses the n frequencies are bounded by $0<\omega_j<4\sqrt{k/m}$.

ADDITIONAL APPLICATIONS

14. Coupled *RLC* Circuit. (a) Derive the differential equations for the currents $i_1(t)$ and $i_2(t)$ in the two-loop circuit shown below, stating the physical laws that you use and explaining your steps. HINT: Recall from Section 1.3.5 that the voltage drop (i.e., in the direction of the current) across a capacitor of capacitance C is $Q(t)/C$ where $Q(t)=\int i(t)\,dt$ is the charge on the capacitor and $i(t)$ is the current, and that $dQ/dt=i(t)$. You should obtain this:

$$L_1 i_1'' + \left(\frac{1}{C_1}+\frac{1}{C_{12}}\right)i_1 - \frac{1}{C_{12}}i_2 = E_1'(t), \tag{14.1a}$$

$$L_2 i_2'' - \frac{1}{C_{12}}i_1 + \left(\frac{1}{C_2}+\frac{1}{C_{12}}\right)i_2 = E_2'(t). \tag{14.1b}$$

Observe that equations (14.1a,b) are identical in form to the system (1) that models the two-mass oscillator shown in Fig. 2, subject to the correspondences

$$L_1 \leftrightarrow m_1,\ L_2 \leftrightarrow m_2,\ \frac{1}{C_1}\leftrightarrow k_1,\ \frac{1}{C_{12}}\leftrightarrow k_{12},\ \frac{1}{C_2}\leftrightarrow k_2,$$
$$i_1(t)\leftrightarrow x_1(t),\ i_2(t)\leftrightarrow x_2(t),\ E_1'(t)\leftrightarrow F_1(t),\ E_2'(t)\leftrightarrow F_2(t). \tag{14.2}$$

Thus, we say that the electrical circuit is an **electrical analog** of the two-mass oscillator.
(b) Let $C_1=C_{12}=C_2\equiv C$ and let $L_1=L_2\equiv L$. Determine the natural frequencies of the circuit in terms of L and C.

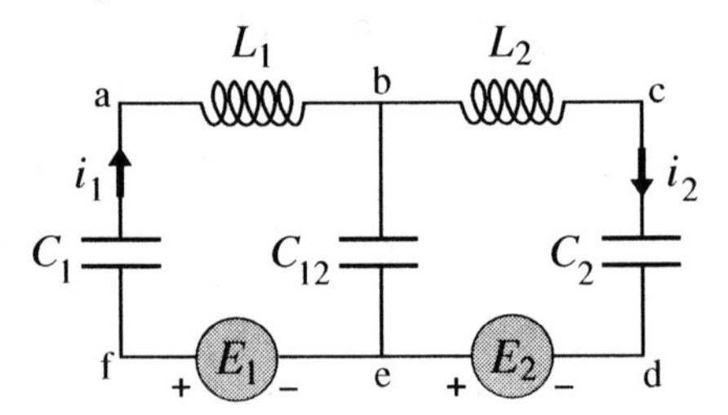

4.3 *N*-SPACE AND MATRICES

The method of elimination used in Sections 4.1 and 4.2 was a good place to start because it enabled us to begin solving systems without needing any new mathematics, but matrix methods are superior and enable us to work conveniently even with systems of many equations. In this section we will introduce enough about vectors in n-space and about matrices to meet our needs in Sections 4.4–4.7.

We will generally use boldface lower case letters such as **u** and **v** for vectors.

4.3.1 Passage from 2-space to n-space. No doubt you are familiar with vectors in two and three dimensions, from the calculus or a course in physics. There, vectors denote quantities such as force and velocity, having both magnitude and direction. A vector is represented as a directed line segment, or "arrow"; a vector of zero length is called the **zero vector** and is denoted as $\mathbf{0}$; the **addition** of (nonzero) vectors $\mathbf{u}$ and $\mathbf{v}$ is defined graphically, and **scalar multiplication** is defined between any (real) scalar α and any vector $\mathbf{u}$. Vector subtraction is defined in terms of vector addition and scalar multiplication by $\mathbf{u} - \mathbf{v} \equiv \mathbf{u} + (-1)\mathbf{v}$. That is, carry out the scalar multiplication -1 times $\mathbf{v}$ and then add $(-1)\mathbf{v}$ to $\mathbf{u}$.

To develop a general theory for systems of differential equations, it is convenient to use a more general concept of vectors known as "n-space." We can introduce n-space as a generalization of 2- and 3-space, as follows. Let us begin in 2-space. Instead of depicting a vector $\mathbf{u}$ in 2-space graphically by an arrow, introduce a reference Cartesian coordinate system (Fig. 1). Then we can express $\mathbf{u}$, alternatively, in terms of its components u_1 and u_2 (which are the orthogonal projections of $\mathbf{u}$ onto the x_1 and x_2 coordinate axes) by the *2-tuple* notation

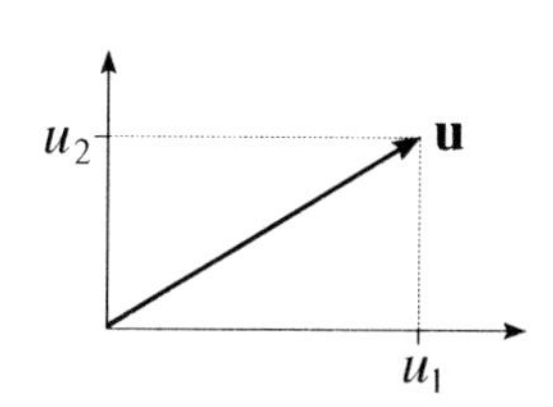

Figure 1. Introducing the idea of coordinate components of $\mathbf{u}$ in 2-space.

$$\mathbf{u} = [u_1, u_2]. \tag{1}$$

We say that vectors $\mathbf{u} = [u_1, u_2]$ and $\mathbf{v} = [v_1, v_2]$ are equal if $u_1 = v_1$ and $u_2 = v_2$; their sum follows from Fig. 2 as

$$\mathbf{u} + \mathbf{v} = [u_1 + v_1, u_2 + v_2]; \tag{2}$$

the scalar multiple $\alpha\mathbf{u}$ is

$$\alpha\mathbf{u} = [\alpha u_1, \alpha u_2]; \tag{3}$$

and the zero vector is $\mathbf{0} = [0, 0]$.

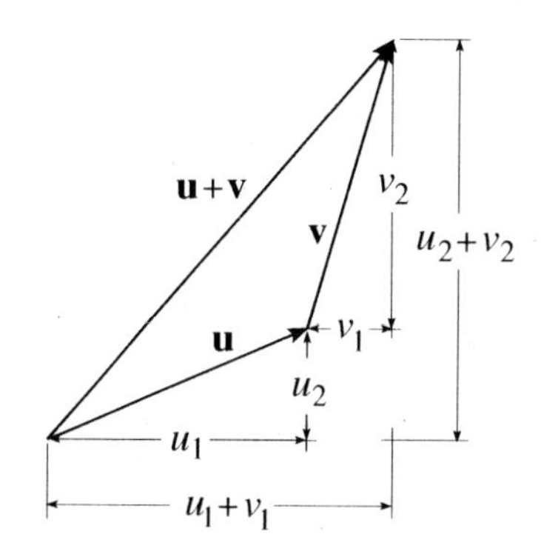

Figure 2. The graphical law of addition, in terms of the components of the vectors.

Similarly for 3-space: If $\mathbf{u} = [u_1, u_2, u_3]$ and $\mathbf{v} = [v_1, v_2, v_3]$, then $\mathbf{u} + \mathbf{v} = [u_1 + v_1, u_2 + v_2, u_3 + v_3]$, $\alpha\mathbf{u} = [\alpha u_1, \alpha u_2, \alpha u_3]$, and so on.

We can define **n-tuple** vectors in this manner even if n is greater than 3, and the collection of all possible n-tuple vectors is called **n-space**. Consider any two vectors $\mathbf{u} = [u_1, \ldots, u_n]$ and $\mathbf{v} = [v_1, \ldots, v_n]$ in n-space. The scalars $u_1, \ldots, u_n$ and $v_1, \ldots, v_n$ are the **components** of $\mathbf{u}$ and $\mathbf{v}$, respectively. As for 2-space and 3-space, $\mathbf{u}$ and $\mathbf{v}$ are **equal** if $u_j = v_j$ for each j, and we define

$$\mathbf{u} + \mathbf{v} \equiv [u_1 + v_1, \ldots, u_n + v_n] \quad \text{(addition)}, \tag{4a}$$
$$\alpha\mathbf{u} \equiv [\alpha u_1, \ldots, \alpha u_n] \quad \text{(scalar multiplication)}, \tag{4b}$$
$$\mathbf{0} \equiv (0, \ldots, 0) \quad \text{(zero vector)}, \tag{4c}$$
$$-\mathbf{u} \equiv (-1)\mathbf{u} \quad \text{(negative inverse)}, \tag{4d}$$
$$\mathbf{u} - \mathbf{v} \equiv \mathbf{u} + (-\mathbf{v}) \quad \text{(subtraction)}. \tag{4e}$$

It follows from these definitions that the familiar arithmetic properties (such as the commutativity property $\mathbf{u}+\mathbf{v} = \mathbf{v}+\mathbf{u}$) hold in n-space, as in 2- and 3-space.

The vector components and the scalar multipliers in n-space can be real or complex numbers, but we will not allow them to be complex until Section 4.6.

EXAMPLE 1. Vector Arithmetic in 4-Space. For instance, if $\mathbf{u} = [2, -1, 4, -8]$ and $\mathbf{v} = [1, 6, 0, 2]$, then $\mathbf{u} + \mathbf{v} = [3, 5, 4, -6]$, $3\mathbf{u} = [6, -3, 12, -24]$, $-\mathbf{u} =$

$(-1)\mathbf{u} = [-2, 1, -4, 8]$, and $\mathbf{u} - 2\mathbf{v} = \mathbf{u} + (-2)\mathbf{v} = [2, -1, 4, -8] + [-2, -12, 0, -4] = [0, -13, 4, -12]$. ▮

The importance of notation!

We cannot display vectors in n-space graphically if $n > 3$, but that will not be a drawback because our methods will be analytical, not graphical. Observe that the bridge from 2- and 3-space to n-space was *notation*; when we expressed vectors in 2- and 3-space in the coordinate form $\mathbf{u} = [u_1, u_2]$ and $\mathbf{u} = [u_1, u_2, u_3]$, respectively, then the notation itself suggested the generalization to vectors $\mathbf{u} = [u_1, \ldots, u_n]$ in "n-space."

4.3.2 Matrix operators on vectors in n-space. Besides vectors, we will need "matrices" that operate on them. A **matrix** is a rectangular array of quantities called the **elements** of the matrix. Here, we restrict the elements to be real numbers unless indicated otherwise. An $m \times n$ matrix $\mathbf{A}$ is written as

$$\mathbf{A} = \begin{bmatrix} a_{11} & a_{12} & \cdots & a_{1n} \\ a_{21} & a_{22} & \cdots & a_{2n} \\ \vdots & \vdots & & \vdots \\ a_{m1} & a_{m2} & \cdots & a_{mn} \end{bmatrix}, \tag{5}$$

with brackets to emphasize that the entire array is considered a single entity. A horizontal line of elements is a **row**, a vertical line of elements is a **column**. Rows are counted from the top and columns from the left, so $a_{21}\ a_{22} \cdots a_{2n}$, for instance, is the second row. The first subscript on a_{ij} is the *row index* and the second is the *column index*; for instance, a_{21} is in the second row and the first column. Sometimes it is convenient to express (5) in the abbreviated form $\mathbf{A} = \{a_{ij}\}$.

We will use boldface capital letters for matrices and boldface lower case letters for vectors.

We will use boldface capital letters to denote matrices and lightface lowercase letters to denote their elements. The matrix $\mathbf{A}$ in (5) has m rows and n columns and is therefore $m \times n$; the **form** of $\mathbf{A}$ is $m \times n$.

For instance,

$$\mathbf{A} = \begin{bmatrix} 3 & -1 & 4 \\ 0 & 2 & 5 \end{bmatrix} \text{ and } \mathbf{B} = \begin{bmatrix} 8 & -6 & 0 \\ 1 & 3 & 27 \\ 2 & 9 & 4 \end{bmatrix} \tag{6}$$

are 2×3 and 3×3, respectively. If we denote $\mathbf{B} = \{b_{ij}\}$, then $b_{11} = 8$, $b_{12} = -6$, $b_{32} = 9$, and so on. Two matrices are **equal** if they are of the same form and their corresponding elements are equal.

Next, the **multiplication** of any $n \times n$ matrix $\mathbf{A}$ into any n-tuple vector $\mathbf{x}$ having components $x_1, \ldots, x_n$ is defined as follows:

$$\mathbf{Ax} = \begin{bmatrix} a_{11} & a_{12} & \cdots & a_{1n} \\ a_{21} & a_{22} & \cdots & a_{2n} \\ \vdots & \vdots & & \vdots \\ a_{n1} & a_{n2} & \cdots & a_{nn} \end{bmatrix} \begin{bmatrix} x_1 \\ x_2 \\ \vdots \\ x_n \end{bmatrix} \equiv \begin{bmatrix} a_{11}x_1 + \cdots + a_{1n}x_n \\ a_{21}x_1 + \cdots + a_{2n}x_n \\ \vdots \\ a_{n1}x_1 + \cdots + a_{nn}x_n \end{bmatrix}. \tag{7}$$

The last member of (7) may look like a matrix with n columns, but it is not; each sum, such as $a_{11}x_1 + \cdots + a_{1n}x_n$, is a *number*, so the final bracketed quantity in (7), like $\mathbf{x}$, is an n-tuple vector.

The n-tuple vector $\mathbf{x}$ is the "input," and the n-tuple vector to the right of the $\equiv$ sign is the "output." Thus, the $n \times n$ matrix $\mathbf{A}$ is an **operator** that sends one n-tuple vector into another, as defined in (7).

When we put the vector $\mathbf{x}$ into (7), we arranged it in vertical format, whereas we used a horizontal format for vectors in Section 4.3.1. The vertical format is more space-efficient in (7), and *we will, henceforth, understand n-tuple vectors to be in column format, as* $\mathbf{x}$ *is in (7).* If we wish to change a vector $\mathbf{x}$ to row format we will "transpose" it and denote its **transpose** as $\mathbf{x}^{\mathrm{T}}$. That is,

$$\text{if} \quad \mathbf{x} = \begin{bmatrix} x_1 \\ \vdots \\ x_n \end{bmatrix}, \quad \text{then} \quad \mathbf{x}^{\mathrm{T}} = [x_1, \ldots, x_n]. \tag{8a}$$

$\mathbf{x}^{\mathrm{T}}$ is not $\mathbf{x}$ to the T power, it is the transpose of $\mathbf{x}$. *Henceforth, we will understand vectors to be in column format, unless they are transposed.*

Similarly, the transpose of a row vector is a column vector,

$$[x_1, \ldots, x_n]^{\mathrm{T}} = \begin{bmatrix} x_1 \\ \vdots \\ x_n \end{bmatrix}. \tag{8b}$$

The vector $\mathbf{x}$ in (8a) is in fact a matrix, an $n \times 1$ matrix; each of its rows has just one element. But we prefer to call it a vector, or **column vector**, a vector in column format. Similarly, the vector $\mathbf{x}^{\mathrm{T}}$ in (8a) is also a $1 \times n$ matrix, but we prefer to call it a **row vector**, a vector in row format.

EXAMPLE 2. Illustrating (7). Let $\mathbf{B}$ be the 3×3 matrix given in (6), and let $\mathbf{x} = [2, -1, 1]^{\mathrm{T}}$. Then,

$$\mathbf{Bx} = \begin{bmatrix} 8 & -6 & 0 \\ 1 & 3 & 27 \\ 2 & 9 & 4 \end{bmatrix} \begin{bmatrix} 2 \\ -1 \\ 1 \end{bmatrix} = \begin{bmatrix} 16+6+0 \\ 2-3+27 \\ 4-9+4 \end{bmatrix} = \begin{bmatrix} 22 \\ 26 \\ -1 \end{bmatrix}. \blacksquare \tag{9}$$

We've written $[2, -1, 1]^{\mathrm{T}}$ rather than as a column simply to conserve vertical space in the text, and will continue to follow that practice.

Multiplication by $\mathbf{A}$ is **linear** because (if $\mathbf{A}$ is $n \times n$ and the $\mathbf{x}$'s are $n \times 1$)

$$\mathbf{A}(\alpha_1\mathbf{x}_1 + \alpha_2\mathbf{x}_2) = \alpha_1\mathbf{A}\mathbf{x}_1 + \alpha_2\mathbf{A}\mathbf{x}_2 \tag{10}$$

That is, the matrix operator $\mathbf{A}$ defined by (7) is a **linear operator**.

for any scalars α_1 and α_2 and for any vectors $\mathbf{x}_1$ and $\mathbf{x}_2$. Verification of (10) follows readily from (4b), (4a), and (7), and the details are left for the exercises. Further, (10) can be extended to *any* finite number of vectors:

$$\boxed{\mathbf{A}(\alpha_1\mathbf{x}_1 + \cdots + \alpha_k\mathbf{x}_k) = \alpha_1\mathbf{A}\mathbf{x}_1 + \cdots + \alpha_k\mathbf{A}\mathbf{x}_k} \tag{11}$$

for any $k \geq 1$.

4.3.3 Identity matrix and zero matrix. The diagonal line of elements from a_{11} to a_{nn}, of an $n \times n$ matrix, is the **main diagonal** of the matrix. A matrix is **diagonal** if all of its elements not on the main diagonal are zero, and a diagonal matrix having each of its diagonal elements equal to 1 (i.e., $a_{jj} = 1$ for $j = 1, \ldots, n$) is called an **identity matrix** and is denoted as **I**. The identity matrix is the matrix analog of the scalar 1, for just as $1x = x$ for all scalar x's, so does it follow from (7) that if **I** is an $n \times n$ identity matrix and **x** is any n-tuple vector, then $\mathbf{Ix} = \mathbf{x}$. For instance, if **I** is 3×3, then

Diagonal matrices.

$$\mathbf{Ix} = \begin{bmatrix} 1 & 0 & 0 \\ 0 & 1 & 0 \\ 0 & 0 & 1 \end{bmatrix} \begin{bmatrix} x_1 \\ x_2 \\ x_3 \end{bmatrix} = \begin{bmatrix} x_1 \\ x_2 \\ x_3 \end{bmatrix} = \mathbf{x}. \tag{12}$$

In addition, an $n \times n$ matrix having *zeros* for *all* of its elements is a **zero matrix**. The zero matrix, denoted as **O**, is the matrix analog of the scalar 0, for just as $0x = 0$ for all scalar x's, so is $\mathbf{Ox} = \mathbf{0}$ for any n-tuple vector **x**. For instance,

$\mathbf{Ix} = \mathbf{x}$ and $\mathbf{Ox} = \mathbf{0}$. NOTE: We use **0** for the zero vector, and **O** for the zero matrix.

$$\mathbf{Ox} = \begin{bmatrix} 0 & 0 & 0 \\ 0 & 0 & 0 \\ 0 & 0 & 0 \end{bmatrix} \begin{bmatrix} x_1 \\ x_2 \\ x_3 \end{bmatrix} = \begin{bmatrix} 0 \\ 0 \\ 0 \end{bmatrix} = \mathbf{0}. \tag{13}$$

4.3.4 Relevance to systems of linear algebraic equations. In studying systems of linear differential equations, we will meet systems of linear *algebraic* equations. A system of n linear algebraic equations for the n unknowns $x_1, \ldots, x_n$ is of the form

$$\begin{aligned} a_{11}x_1 + a_{12}x_2 + \cdots + a_{1n}x_n &= c_1, \\ a_{21}x_1 + a_{22}x_2 + \cdots + a_{2n}x_n &= c_2, \\ &\vdots \\ a_{n1}x_1 + a_{n2}x_2 + \cdots + a_{nn}x_n &= c_n, \end{aligned} \tag{14}$$

in which the a_{ij} coefficients and the c_j's on the right-hand side are given.

Thanks to the way the multiplication of a vector by a matrix has been defined, by (7), we can express (14) compactly, in matrix notation, as

$$\boxed{\mathbf{Ax} = \mathbf{c},} \tag{15}$$

More generally, we could have m equations in n unknowns, but for our purposes it will suffice to restrict attention to the case where $m = n$. Being able to express (14) so compactly, in (15), is the chief motivation for defining the multiplication of **A** into **x** as we did in (7).

in which

$$\mathbf{A} = \begin{bmatrix} a_{11} & a_{12} & \cdots & a_{1n} \\ a_{21} & a_{22} & \cdots & a_{2n} \\ \vdots & \vdots & & \vdots \\ a_{n1} & a_{n2} & \cdots & a_{nn} \end{bmatrix}, \quad \mathbf{x} = \begin{bmatrix} x_1 \\ x_2 \\ \vdots \\ x_n \end{bmatrix}, \quad \text{and } \mathbf{c} = \begin{bmatrix} c_1 \\ c_2 \\ \vdots \\ c_n \end{bmatrix}. \tag{16}$$

$\mathbf{A}$ is called the **coefficient matrix**, and $\mathbf{x}$ is the unknown; its components are the n unknowns $x_1, \ldots, x_n$. To verify the claimed equivalence of (15) and (14), work out the product $\mathbf{Ax}$, using (7), and set the result equal to $\mathbf{c}$. That step gives

$$\begin{bmatrix} a_{11}x_1 + \cdots + a_{1n}x_n \\ a_{21}x_1 + \cdots + a_{2n}x_n \\ \vdots \\ a_{n1}x_1 + \cdots + a_{nn}x_n \end{bmatrix} = \begin{bmatrix} c_1 \\ c_2 \\ \vdots \\ c_n \end{bmatrix}. \tag{17}$$

The vectors in (17) are equal if and only if their corresponding n components are equal. Equating those corresponding components gives the n scalar equations (14), so (14) and (15) are equivalent, as we claimed.

The linear system (15) raises three questions: First, does it *have* any solutions? If it has one or more solutions, it is **consistent**; and if it has no solutions, it is **inconsistent**. Second, if it *is* consistent, then is the solution **unique**? Third, how can we *find* its solutions?

THEOREM 4.3.1 *Existence and Uniqueness for* $\mathbf{Ax} = \mathbf{c}$
Consider the linear equation

$$\mathbf{Ax} = \mathbf{c}, \tag{18}$$

in which $\mathbf{A}$ is $n \times n$.
(a) The system (18) has a *unique* solution for $\mathbf{x}$ if and only if

$$\det\mathbf{A} \neq 0. \tag{19}$$

If $\det\mathbf{A} = 0$, then either there is *no* solution or there is an *infinity* of solutions.
(b) Homogeneous case. If (18) is homogeneous (i.e., $\mathbf{c} = \mathbf{0}$), it has the trivial solution $\mathbf{x} = \mathbf{0}$. If $\det\mathbf{A} \neq 0$, that trivial solution is unique; if $\det\mathbf{A} = 0$, then in addition to the trivial solution there is an infinity of nontrivial solutions (i.e., with the components $x_1, \ldots, x_n$ of $\mathbf{x}$ not all zero).

The homogeneous equation $\mathbf{Ax} = \mathbf{0}$ is necessarily consistent because at the *least* it has the solution $\mathbf{x} = \mathbf{0}$.

To solve (18) we will use *Gauss elimination*. Determinants are reviewed in Appendix B, and Gauss elimination is reviewed in Appendix C.

4.3.5 Vector and matrix functions. If the components of a vector are functions of a variable t, then the vector is a **vector function**. Similarly, if the elements of a matrix are functions of t, then the matrix is a **matrix function**.

The *algebra* of vector and matrix functions is the same as for constant vectors and matrices. For instance, if

$$\mathbf{x}(t) = \begin{bmatrix} 2-t \\ \cos t \end{bmatrix}, \quad \mathbf{y}(t) = \begin{bmatrix} t^2 \\ 5 \end{bmatrix}, \quad \text{and} \quad \mathbf{A}(t) = \begin{bmatrix} t & 4 \\ 0 & e^t \end{bmatrix},$$

then

$$\mathbf{x} + 2\mathbf{y} = \begin{bmatrix} 2-t+2t^2 \\ \cos t + 10 \end{bmatrix} \text{ and } \mathbf{A}\mathbf{x} = \begin{bmatrix} t & 4 \\ 0 & e^t \end{bmatrix} \begin{bmatrix} 2-t \\ \cos t \end{bmatrix} = \begin{bmatrix} 2t-t^2+4\cos t \\ e^t \cos t \end{bmatrix}.$$

The *calculus* of vector and matrix functions mimics the calculus of scalar functions: Vector functions $\mathbf{x}(t) = [x_1(t), \dots, x_n(t)]$ and matrix functions $\mathbf{A}(t)$ are **continuous** at t_0 if all their components $x_j(t)$ and elements $a_{ij}(t)$ are, respectively. For instance, $\mathbf{x}(t) = [3t, \sin t, -2]^{\mathrm{T}}$ is continuous on $-\infty < t < \infty$ because all of $3t$, $\sin t$, and -2 are, and $\mathbf{A} = \begin{bmatrix} 6/t & t^2 \\ \cos t & 0 \end{bmatrix}$ is continuous on $0 < t < \infty$ because all of $6/t$, t^2, $\cos t$, and 0 are.

Differentiation and integration of vector functions are defined in a natural way: A vector function $\mathbf{x}(t)$ is both **differentiable** at t_0, and **integrable**, if its components are, and we then define

$$\frac{d}{dt}\mathbf{x}(t) \equiv \begin{bmatrix} x_1'(t) \\ \vdots \\ x_n'(t) \end{bmatrix} \quad \text{and} \quad \int_a^b \mathbf{x}(t)\,dt \equiv \begin{bmatrix} \int_a^b x_1(t)\,dt \\ \vdots \\ \int_a^b x_n(t)\,dt \end{bmatrix}. \tag{20}$$

For instance, if

$$\mathbf{x}(t) = \begin{bmatrix} 3t^2 \\ e^t \end{bmatrix}, \text{ then } \mathbf{x}'(t) = \begin{bmatrix} 6t \\ e^t \end{bmatrix} \text{ and } \int_0^1 \mathbf{x}(t)\,dt = \begin{bmatrix} \int_0^1 3t^2\,dt \\ \int_0^1 e^t\,dt \end{bmatrix} = \begin{bmatrix} 1 \\ e-1 \end{bmatrix}.$$

Differentiation and integration of matrix functions are defined as well, in similar element-by-element manner, but those operations will not be used in this chapter.

Closure. In this section we introduced vectors in n-space and $n \times n$ matrix operators on those vectors. Note that we defined $\mathbf{A}\mathbf{x}$, not $\mathbf{x}\mathbf{A}$; the two are *not* the same and, indeed, the latter is not defined. This should not be surprising. After all, for differential operators that operate on functions, $\frac{d}{dt}x$ surely is not the same as $x\frac{d}{dt}$.

When we return to systems of differential equations in subsequent sections, we will rely on matrix format. Matrices and n-space are part of the subject of *linear algebra*, which we draw upon in this text only insofar as it is needed. Additional matrix algebra will in fact be needed for Section 4.9 and will be provided beforehand in Section 4.8.

EXERCISES 4.3

1. Consider these vectors in 4-space: $\mathbf{u} = [5, 0, 1, 2]^{\mathrm{T}}$, $\mathbf{v} = [3, -1, 4, 1]^{\mathrm{T}}$, $\mathbf{w} = [4, 3, 2, 1]^{\mathrm{T}}$, and $\mathbf{x} = [1, -1, 2, 0]^{\mathrm{T}}$. Evaluate each of the following.

(a) $\mathbf{u} + 2\mathbf{v}$ (b) $\mathbf{v} - \mathbf{w}$
(c) $-2\mathbf{x} - \mathbf{v}$ (d) $-\mathbf{u} + 4\mathbf{v}$
(e) $\mathbf{u} + \mathbf{v} + \mathbf{w}$ (f) $\mathbf{v} - (\mathbf{w} - \mathbf{x})$
(g) $3(2\mathbf{w}) - 7\mathbf{x}$ (h) $3\mathbf{x} + 2\mathbf{v}$
(i) $2(3\mathbf{u} - \mathbf{v}) + \mathbf{x}$ (j) $2(\mathbf{x} - \mathbf{v}) - 2\mathbf{u}$

2. Algebraic Properties of Vectors in n-Space. We defined the addition, scalar multiplication, and subtraction of vectors in n-space in (4a)–(4e). From those definitions one can prove that the following algebraic properties hold, as they do for "arrow vectors" in 2- and 3-space.
Commutativity:

$$\mathbf{u} + \mathbf{v} = \mathbf{v} + \mathbf{u} \tag{2.1a}$$

Associativity:

$$\begin{aligned} (\mathbf{u} + \mathbf{v}) + \mathbf{w} &= \mathbf{u} + (\mathbf{v} + \mathbf{w}) \\ \alpha(\beta\mathbf{u}) &= (\alpha\beta)\mathbf{u} \end{aligned} \tag{2.1b,c}$$

Distributivity:

$$\begin{aligned} (\alpha + \beta)\mathbf{u} &= \alpha\mathbf{u} + \beta\mathbf{u} \\ \alpha(\mathbf{u} + \mathbf{v}) &= \alpha\mathbf{u} + \alpha\mathbf{v} \end{aligned} \tag{2.1d,e}$$

From (2.1b) it follows that we can write $\mathbf{u} + \mathbf{v} + \mathbf{w}$ without ambiguity. That is, whether we add **u** and **v** and then add **w** to their sum, or whether we add **v** and **w** and then add **u** to their sum, doesn't matter for, according to (2.1b), the results are the same. Similarly, from (2.1e) we see that we can write $\alpha\beta\mathbf{u}$ without ambiguity. There are no surprises in (2.1a)–(2.1e).

(a) Prove (2.1a) (b) Prove (2.1b) (c) Prove (2.1c)
(d) Prove (2.1d) (e) Prove (2.1e)

3. Let $\mathbf{x}_1, \dots, \mathbf{x}_k$ be vectors in n-space and let $\alpha_1, \dots, \alpha_k$ be scalars. From the definitions of vector addition and scalar multiplication, namely,

$$\mathbf{u} + \mathbf{v} = \begin{bmatrix} u_1 \\ \vdots \\ u_n \end{bmatrix} + \begin{bmatrix} v_1 \\ \vdots \\ v_n \end{bmatrix} \equiv \begin{bmatrix} u_1 + v_1 \\ \vdots \\ u_n + v_n \end{bmatrix} \tag{3.1}$$

and

$$\alpha\mathbf{u} = \alpha \begin{bmatrix} u_1 \\ \vdots \\ u_n \end{bmatrix} \equiv \begin{bmatrix} \alpha u_1 \\ \vdots \\ \alpha u_n \end{bmatrix}, \tag{3.2}$$

it follows that

$$\boxed{\alpha_1\mathbf{x}_1 + \cdots + \alpha_k\mathbf{x}_k = \begin{bmatrix} \alpha_1 x_{11} + \cdots + \alpha_k x_{k1} \\ \vdots \\ \alpha_1 x_{1n} + \cdots + \alpha_k x_{kn} \end{bmatrix}.} \tag{3.3}$$

It can be shown that (3.3) holds for *any* $k \geq 1$, but let it suffice, for this exercise, to show that it holds for $k = 3$: with $k = 3$, show that (3.3) follows from the definitions (3.1) and (3.2).

4. If $\mathbf{u} = [1, 2, 3, 1]^{\mathrm{T}}$, $\mathbf{v} = [3, -1, 4, 0]^{\mathrm{T}}$, and $\mathbf{w} = [4, 0, 0, 1]^{\mathrm{T}}$, solve the following for **x**.

(a) $3\mathbf{u} - \mathbf{x} = \mathbf{w} + 2\mathbf{x}$ (b) $\mathbf{u} + \mathbf{v} + 2\mathbf{w} + \mathbf{x} = \mathbf{0}$
(c) $\mathbf{u} + 2\mathbf{x} = 3(\mathbf{w} + \mathbf{x})$ (d) $3(\mathbf{x} + \mathbf{v}) + 2\mathbf{w} = \mathbf{0}$
(e) $2\mathbf{u} + \mathbf{x} = \mathbf{w} + 2\mathbf{x}$ (f) $\mathbf{u} + \mathbf{x} = 2(\mathbf{x} - \mathbf{w})$
(g) $3\mathbf{v} = 2(\mathbf{w} - 4\mathbf{x})$ (h) $3(\mathbf{x} - \mathbf{u}) = 4(\mathbf{x} + \mathbf{v})$

5. Let $\mathbf{A} = \begin{bmatrix} 1 & 2 & 0 \\ 1 & -1 & 1 \\ 2 & 1 & 0 \end{bmatrix}$, $\mathbf{B} = \begin{bmatrix} 2 & 2 & 2 \\ 1 & 1 & 1 \\ 1 & 1 & 1 \end{bmatrix}$

$\mathbf{C} = \begin{bmatrix} 1 & 0 & 0 \\ 2 & 0 & 0 \\ 3 & 0 & 0 \end{bmatrix}$, $\mathbf{D} = \begin{bmatrix} 1 & 2 & 0 \\ 1 & -1 & 1 \\ 2 & 1 & -2 \end{bmatrix}$

$\mathbf{E} = \begin{bmatrix} 3 & 0 & 0 \\ 0 & 3 & 0 \\ 0 & 0 & 3 \end{bmatrix}$, and $\mathbf{F} = \begin{bmatrix} 0 & 0 & 1 \\ 0 & 1 & 0 \\ 1 & 0 & 0 \end{bmatrix}$,

and let $\mathbf{u} = [1, 2, 3]^{\mathrm{T}}$ and $\mathbf{v} = [3, -1, 1]^{\mathrm{T}}$. Evaluate

(a) **Au** (b) **Av** (c) **Bu** (d) **Bv**
(e) **Cu** (f) **Cv** (g) **Du** (h) **Dv**
(i) **Eu** (j) **Ev** (k) **Fu** (l) **Fv**

6. Inferring A. If, for any given vector $\mathbf{x} = [x_1, x_2, x_3]^{\mathrm{T}}$, the product **Ax** is the vector given below, determine the 3×3 matrix **A**.

(a) $\begin{bmatrix} x_1 + 4x_2 - x_3 \\ 6x_1 + 2x_2 + x_3 \\ 3x_1 - x_2 - x_3 \end{bmatrix}$ (b) $\begin{bmatrix} 3x_1 + x_2 - x_3 \\ x_1 + x_3 \\ 2x_1 - x_2 \end{bmatrix}$

(c) $\begin{bmatrix} x_2 \\ x_1 + 2x_2 + x_3 \\ 3x_1 - x_2 \end{bmatrix}$ (d) $\begin{bmatrix} x_3 \\ x_2 \\ x_1 \end{bmatrix}$

(e) $\begin{bmatrix} x_1 + x_2 + x_3 \\ x_2 + x_3 \\ x_3 \end{bmatrix}$ (f) $\begin{bmatrix} x_1 \\ x_1 + x_2 \\ x_1 + x_2 + x_3 \end{bmatrix}$

(g) $\begin{bmatrix} x_1 + x_3 \\ 0 \\ x_1 + x_3 \end{bmatrix}$ (h) $\begin{bmatrix} 0 \\ 0 \\ x_1 - 2x_3 \end{bmatrix}$

(i) $\begin{bmatrix} x_1 \\ x_2 \\ x_3 \end{bmatrix}$ (j) $\begin{bmatrix} 5x_3 \\ 5x_3 \\ 5x_3 \end{bmatrix}$

7. Linearity of Matrix Multiplication; Equations (10) and (11). From the definitions (4a), (4b), and (7),
(a) show that

$$\mathbf{A}(\alpha\mathbf{x}) = \alpha\mathbf{A}\mathbf{x}, \tag{7.1}$$

(b) and that

$$\mathbf{A}(\mathbf{x} + \mathbf{y}) = \mathbf{A}\mathbf{x} + \mathbf{A}\mathbf{y}. \tag{7.2}$$

(c) Now use (7.1) and (7.2) to show that

$$\mathbf{A}(\alpha\mathbf{x} + \beta\mathbf{y}) = \alpha\mathbf{A}\mathbf{x} + \beta\mathbf{A}\mathbf{y}, \tag{7.3}$$

which is really (10) with different letters used.
(d) Next, use (7.3) to show that

$$\mathbf{A}(\alpha\mathbf{x} + \beta\mathbf{y} + \gamma\mathbf{z}) = \alpha\mathbf{A}\mathbf{x} + \beta\mathbf{A}\mathbf{y} + \gamma\mathbf{A}\mathbf{z}, \tag{7.4}$$

which is really (11) for $k = 3$. HINT: First, write the left-hand side of (7.4) as $\mathbf{A}[1(\alpha\mathbf{x} + \beta\mathbf{y}) + \gamma\mathbf{z}]$ and use (7.3). NOTE: Continuing in this manner, or by using induction instead, we can show that (11) holds for *any* $k \geq 1$.

8. Consider $\mathbf{A}\mathbf{x} = \mathbf{c}$ where $\mathbf{A}$ is $n \times n$. What, if anything, can be inferred about the determinant of $\mathbf{A}$, if the system

(a) is inconsistent?
(b) has a unique solution?
(c) has a nonunique solution?

9. Continuity of Vector and Matrix Functions. What is the broadest open interval, containing $t = 2$, on which $\boldsymbol{x}(t)$ is continuous? On which $\boldsymbol{A}(t)$ is continuous?

(a) $\mathbf{x}(t) = \begin{bmatrix} e^t \\ e^{2t} \end{bmatrix}$, $\mathbf{A}(t) = \begin{bmatrix} 1/t & t \\ 3t & e^{-t} \end{bmatrix}$

(b) $\mathbf{x}(t) = \begin{bmatrix} 1/(t+5) \\ 1/(t+6) \end{bmatrix}$, $\mathbf{A}(t) = \begin{bmatrix} 4t & \ln t \\ 5 & e^{7t} \end{bmatrix}$

(c) $\mathbf{x}(t) = \begin{bmatrix} \ln(1+t^2) \\ t \sin t \end{bmatrix}$, $\mathbf{A}(t) = \begin{bmatrix} 4 & \ln(t-1) \\ 1 & e^{-t} \end{bmatrix}$

(d) $\mathbf{x}(t) = \begin{bmatrix} 1/\sinh t \\ t \cos t \end{bmatrix}$, $\mathbf{A}(t) = \begin{bmatrix} 0 & 3/\cosh t \\ 1 & 1/e^t \end{bmatrix}$

(e) $\mathbf{x}(t) = \begin{bmatrix} 1/(t+3) \\ 1/(t-8) \end{bmatrix}$, $\mathbf{A}(t) = \begin{bmatrix} 1/e^t & 0 \\ 0 & 1/\ln t \end{bmatrix}$

(f) $\mathbf{x}(t) = \begin{bmatrix} 1/(t^2-5t+4) \\ 0 \end{bmatrix}$, $\mathbf{A}(t) = \begin{bmatrix} 1/(t-7)^3 & 0 \\ 1/(t-7) & 0 \end{bmatrix}$

(g) $\mathbf{x}(t) = \begin{bmatrix} 1/(t^2-9) \\ 1/(t+2) \end{bmatrix}$, $\mathbf{A}(t) = \begin{bmatrix} t^4 & 1/t^4 \\ 1/(5-t)^2 & \sin t \end{bmatrix}$

(h) $\mathbf{x}(t) = \begin{bmatrix} \ln(4t-t^2-3) \\ \ln t \end{bmatrix}$, $\mathbf{A}(t) = \begin{bmatrix} 4/(t^2-1) & e^{-t} \\ 8/(t^2-9) & e^{-2t} \end{bmatrix}$

10. Differentiation of Vector Functions. Given $\mathbf{x}(t)$, work out its derivative $\mathbf{x}'(t)$.

(a) $[t, t^4, t^7]^{\mathrm{T}}$ (b) $[5, \cos 2t]^{\mathrm{T}}$ (c) $[t + e^{2t}, t^3, 5]^{\mathrm{T}}$
(d) $[2-t, 8\cos 4t]^{\mathrm{T}}$ (e) $[\sinh 5t, 0]^{\mathrm{T}}$ (f) $[\sinh t, \cosh 9t]^{\mathrm{T}}$
(g) $[e^t + e^{-t}, 0]^{\mathrm{T}}$ (h) $[\tan t, t]^{\mathrm{T}}$ (i) $[\ln(3-4t), t^2]^{\mathrm{T}}$

11. Integration of Vector Functions. Given the vectors $\mathbf{f}(t)$ and $\mathbf{x}(0)$, integrate $\mathbf{x}'(t) = \mathbf{f}(t)$ from 0 to t and solve for $\mathbf{x}(t)$. On what largest interval, containing $t = 0$, is $\mathbf{x}(t)$ continuous?

(a) $\mathbf{f}(t) = [4t, 1]^{\mathrm{T}}$, $\mathbf{x}(0) = [-1, 0]^{\mathrm{T}}$
(b) $\mathbf{f}(t) = [\sinh t, 6\cosh 3t]^{\mathrm{T}}$, $\mathbf{x}(0) = [1, 2]^{\mathrm{T}}$
(c) $\mathbf{f}(t) = [2/(t+3), 1]^{\mathrm{T}}$, $\mathbf{x}(0) = [3, 4]^{\mathrm{T}}$
(d) $\mathbf{f}(t) = [8t^3, 5\sinh 5t]^{\mathrm{T}}$, $\mathbf{x}(0) = [-1, 1]^{\mathrm{T}}$
(e) $\mathbf{f}(t) = [2t, \tan t]^{\mathrm{T}}$, $\mathbf{x}(0) = [2, 21]^{\mathrm{T}}$
(f) $\mathbf{f}(t) = [2, 1/(t-1)^2]^{\mathrm{T}}$, $\mathbf{x}(0) = [0, 6]^{\mathrm{T}}$
(g) $\mathbf{f}(t) = [2, 6\sin 3t]^{\mathrm{T}}$, $\mathbf{x}(0) = [0, 0]^{\mathrm{T}}$
(h) $\mathbf{f}(t) = [4\tan 2t, 1/(t-1)^2]^{\mathrm{T}}$, $\mathbf{x}(0) = [0, 6]^{\mathrm{T}}$

12. In retail, what pays off is "location, location, location." In mathematics, it's ____, ____, ____.

4.4 LINEAR DEPENDENCE AND INDEPENDENCE OF VECTORS

In Chapter 2, the linear independence of sets of *functions* arose in connection with the form of a general solution of an nth-order linear homogeneous differential equation. For an arbitrary linear combination of n solutions to be a general solution, that set of n functions must be LI. In this chapter, on *systems* of linear differential equations, the corresponding discussion of the form of a general solution will involve the linear dependence or independence of sets of *vectors*. Discussion here for vectors will parallel that in Section 2.4 for functions.

We will continue to use the abbreviations LI and LD, from Chapter 2.

4.4.1 Linear dependence of a set of constant vectors in n-space. We begin with the following definition, for vectors that are constants, not functions of t:

Definition 4.4.1 *Linear Dependence of a Set of Constant Vectors*
A finite set of n-tuple vectors $\{\mathbf{u}_1, \ldots, \mathbf{u}_k\}$ is **linearly dependent** (LD) if there exist constants $a_1, \ldots, a_k$, not all zero, such that

$$a_1\mathbf{u}_1 + a_2\mathbf{u}_2 + \cdots + a_k\mathbf{u}_k = \mathbf{0}. \tag{1}$$

If the set is not linearly dependent, it is **linearly independent** (LI).

There are k vectors, each with n components; k may be less than, equal to, or greater than n. The set being finite means k is finite.

The words "not all zero" are crucial because *of course* (1) will be satisfied if all the a_j's are zero. Only if (1) is satisfied with a_j's that are *not* all zero does (1) express a nontrivial relation among the $\mathbf{u}_j$'s, in which case the $\mathbf{u}_j$'s are LD.

Normally, we are interested in sets of two or more vectors ($k \geq 2$), but Definition 4.4.1 applies even if $k = 1$. If $k = 1$, it follows from the definition that $\{\mathbf{u}_1\}$ is LD if and only if $\mathbf{u}_1$ is the zero vector.

If $k \geq 2$, it follows from Definition 4.4.1 that a set is LD if and only if *at least one of the* $\mathbf{u}_j$*'s can be expressed as a linear combination of the others.* After all, if the a_j's in (1) are not all zero then at least one of them must be nonzero, for instance a_2. In that case we can divide (1) by a_2 and solve the resulting equation for $\mathbf{u}_2$, thus giving $\mathbf{u}_2$ as a linear combination of the other $\mathbf{u}_j$'s.

EXAMPLE 1. Testing for LD/LI by Inspection. Sometimes we can see that a set is LD merely by inspection, namely, by noticing a linear relationship among its members. For instance, if $\mathbf{u}_1 = [1, 2, 1]^{\mathrm{T}}$, $\mathbf{u}_2 = [1, 0, 0]^{\mathrm{T}}$, and $\mathbf{u}_3 = [4, 8, 4]^{\mathrm{T}}$, we can see that $\mathbf{u}_3 = 4\mathbf{u}_1$, so the set $\{\mathbf{u}_1, \mathbf{u}_2, \mathbf{u}_3\}$ is LD. That is, $4\mathbf{u}_1 + 0\mathbf{u}_2 - 1\mathbf{u}_3 = \mathbf{0}$, in which the coefficients are not all zero. We happen not to be able to solve the latter for $\mathbf{u}_2$ in terms of $\mathbf{u}_1$ and $\mathbf{u}_3$ because its coefficient is 0, but we can solve either for $\mathbf{u}_1$ or $\mathbf{u}_3$: $\mathbf{u}_1 = 0\mathbf{u}_2 + (1/4)\mathbf{u}_3$ and

We've written column vectors as transposed row vectors to save vertical space.

$\mathbf{u}_3 = 4\mathbf{u}_1 + 0\mathbf{u}_2$. ∎

EXAMPLE 2. Using Definition 4.4.1 to Test for LD/LI. The method of inspection is not always feasible, so let us now indicate a systematic way to test for linear dependence using Definition 4.4.1. To illustrate, consider the vectors

$$\mathbf{u}_1 = [1, 1, -1]^{\mathrm{T}}, \quad \mathbf{u}_2 = [2, 1, 1]^{\mathrm{T}}, \quad \mathbf{u}_3 = [2, 3, 4]^{\mathrm{T}}. \tag{2}$$

Is $\{\mathbf{u}_1, \mathbf{u}_2, \mathbf{u}_3\}$ LD or LI? Write out (1), and see if it can be satisfied by a_j's that are not all zero:

$$a_1 \begin{bmatrix} 1 \\ 1 \\ -1 \end{bmatrix} + a_2 \begin{bmatrix} 2 \\ 1 \\ 1 \end{bmatrix} + a_3 \begin{bmatrix} 2 \\ 3 \\ 4 \end{bmatrix} = \begin{bmatrix} 0 \\ 0 \\ 0 \end{bmatrix} \tag{3}$$

or, according to the definition of the scalar multiplication of vectors,

$$\begin{bmatrix} a_1 \\ a_1 \\ -a_1 \end{bmatrix} + \begin{bmatrix} 2a_2 \\ a_2 \\ a_2 \end{bmatrix} + \begin{bmatrix} 2a_3 \\ 3a_3 \\ 4a_3 \end{bmatrix} = \begin{bmatrix} 0 \\ 0 \\ 0 \end{bmatrix}. \tag{4}$$

Then, by the definition of vector addition,

$$\begin{bmatrix} a_1 + 2a_2 + 2a_3 \\ a_1 + a_2 + 3a_3 \\ -a_1 + a_2 + 4a_3 \end{bmatrix} = \begin{bmatrix} 0 \\ 0 \\ 0 \end{bmatrix}. \tag{5}$$

For the vectors on the left- and right-hand sides of (5) to be equal, their respective components must be equal, so we obtain the system

$$a_1 + 2a_2 + 2a_3 = 0, \tag{6a}$$
$$a_1 + a_2 + 3a_3 = 0, \tag{6b}$$
$$-a_1 + a_2 + 4a_3 = 0 \tag{6c}$$

(6a,b,c) are equivalent to the single vector equation (3). The question now is whether (6) admits nontrivial solutions (i.e., not all of the a_j's being zero). If so, conclude that $\{\mathbf{u}_1, \mathbf{u}_2, \mathbf{u}_3\}$ is LD; if not, conclude that it is LI.

of linear algebraic equations for a_1, a_2, a_3. Omitting the steps, application of Gauss elimination to (6) gives the unique trivial solution $a_1 = a_2 = a_3 = 0$. Since (3) holds *only* if the three a_j's are zero, the vector set $\{\mathbf{u}_1, \mathbf{u}_2, \mathbf{u}_3\}$ is LI. ∎

EXAMPLE 3. Using Definition 4.4.1 to Test for LD/LI. As a second illustration, consider

$$\mathbf{u}_1 = [2, 0, 1]^{\mathrm{T}}, \quad \mathbf{u}_2 = [0, 1, 1]^{\mathrm{T}}, \quad \mathbf{u}_3 = [2, 2, 3]^{\mathrm{T}}. \tag{7}$$

Is $\{\mathbf{u}_1, \mathbf{u}_2, \mathbf{u}_1\}$ LD or LI? Proceeding as in Example 2 [this time using the more compact transpose notation in (8) and (9)], write out (1):

$$a_1[2, 0, 1]^{\mathrm{T}} + a_2[0, 1, 1]^{\mathrm{T}} + a_3[2, 2, 3]^{\mathrm{T}} = [0, 0, 0]^{\mathrm{T}}, \tag{8}$$

or

$$[2a_1 + 2a_3,\ a_2 + 2a_3,\ a_1 + a_2 + 3a_3]^{\mathrm{T}} = [0, 0, 0]^{\mathrm{T}}. \tag{9}$$

Thus,

$$\begin{aligned} 2a_1 + 0a_2 + 2a_3 &= 0, && (10a)\\ 0a_1 + a_2 + 2a_3 &= 0, && (10b)\\ a_1 + a_2 + 3a_3 &= 0, && (10c) \end{aligned}$$

Gauss elimination of which gives

$$\begin{aligned} a_1 \quad + a_3 &= 0, && (11a)\\ a_2 + 2a_3 &= 0, && (11b)\\ 0 &= 0. && (11c) \end{aligned}$$

Thus, $a_3 = \alpha$, $a_2 = -2\alpha$, and $a_1 = -\alpha$, with α arbitrary. Since, by choosing any nonzero value for α, the a_j's are not all zero, the set $\{\mathbf{u}_1, \mathbf{u}_2, \mathbf{u}_3\}$ is LD. Further, the nontrivial linear relationship among the vectors is $-\alpha\mathbf{u}_1 - 2\alpha\mathbf{u}_2 + \alpha\mathbf{u}_3 = \mathbf{0}$ or, canceling the α's, $-\mathbf{u}_1 - 2\mathbf{u}_2 + \mathbf{u}_3 = \mathbf{0}$. ▮

Both examples involved 3 vectors in 3-space. More generally, we can consider the linear dependence or independence of k vectors in n-space, with $k > n$, $k = n$, or $k < n$. But, the case $k = n$, as in Examples 1 and 2, will be the most important in this chapter. In that case, we can bypass the Gauss elimination of the system of linear algebraic equations for the a_j's, and go right to the determinant of the coefficient matrix $\mathbf{A}$: If $\det\mathbf{A} \neq 0$, there is only the trivial solution for the a_j's and the vector set is LI, and if $\det\mathbf{A} = 0$ there are nontrivial solutions as well and the vector set is LD.

In fact, notice that *the columns of the coefficient matrix* $\mathbf{A}$ *are the given* $\mathbf{u}_j$ *vectors*. For instance, consider (6) in matrix form:

$$\begin{bmatrix} 1 & 2 & 2 \\ 1 & 1 & 3 \\ -1 & 1 & 4 \end{bmatrix} \begin{bmatrix} a_1 \\ a_2 \\ a_3 \end{bmatrix} = \begin{bmatrix} 0 \\ 0 \\ 0 \end{bmatrix}. \tag{12}$$

The columns of the coefficient matrix are indeed the $\mathbf{u}_j$ vectors given in (2). Similarly for (10), the columns of the coefficient matrix in (10) are the vectors given in (7). That pattern always holds. Since there will be nontrivial solutions for the a_j's if and only if the determinant of that matrix is zero, we have this useful result:

THEOREM 4.4.1 *LI Test for a Set of n Constant Vectors in n-Space*
Given n vectors in n-space, $\mathbf{u}_1, \dots, \mathbf{u}_n$, let $[\mathbf{u}_1, \dots, \mathbf{u}_n]$ denote an $n \times n$ matrix having those vectors as its respective columns (or rows). Then, $\{\mathbf{u}_1, \dots, \mathbf{u}_n\}$ is LI if $\det[\mathbf{u}_1, \dots, \mathbf{u}_n] \neq 0$, and LD if $\det[\mathbf{u}_1, \dots, \mathbf{u}_n] = 0$.

Theorem 4.4.1 applies only if the number of vectors in the set of n-tuples is n, no more and no less.

EXAMPLE 4. Examples 2 and 3 Again, This Time Using Theorem 4.4.1. Consider again the vectors $\mathbf{u}_1 = [1, 1, -1]^{\mathrm{T}}$, $\mathbf{u}_2 = [2, 1, 1]^{\mathrm{T}}$, and $\mathbf{u}_3 = [2, 3, 4]^{\mathrm{T}}$ in Example

2. Since the set consists of three vectors in 3-space, Theorem 4.4.1 applies, and we can use it as follows: Construct the determinant $\det[\mathbf{u}_1, \mathbf{u}_2, \mathbf{u}_3]$ and evaluate it, finding that it equals -9. Since the latter is nonzero, $\{\mathbf{u}_1, \mathbf{u}_2, \mathbf{u}_3\}$ is LI, as found in Example 2. For the vectors in Example 3, $\det[\mathbf{u}_1, \mathbf{u}_2, \mathbf{u}_3] = 0$, so that set is LD, as found in Example 3.

COMMENT. The sequencing of the vectors, as the columns of $[\mathbf{u}_1, \dots, \mathbf{u}_n]$, does not affect the conclusion of the LD or LI of the set because the interchanging of two columns of a determinant merely changes the sign of the determinant. Thus, it doesn't affect the zeroness or nonzeroness of the determinant, which is all that matters regarding the LI or LD of the set of column vectors. ▮

4.4.2 Linear dependence of vector functions in n-space. In Section 4.4.1 the vectors were constants, that is, their components were constants. We will also need the concept of linear dependence for sets of vectors that are functions of the (generic) independent variable t. For instance, the 3-tuple $\mathbf{u}(t) = [e^t, 3, 1+5t]^{\mathrm{T}}$ is a vector function of t.

Regarding notation, if the components of such a vector have a common factor, we can scale it out. For instance, we may prefer to write $[6e^t, e^t, e^t]^{\mathrm{T}}$, more compactly, as $e^t[6, 1, 1]^{\mathrm{T}}$ or $[6, 1, 1]^{\mathrm{T}}e^t$.

The concept of linear dependence and independence for vector functions is essentially the same as for constant vectors, but we urge you to read this subsection carefully, paying attention to the role of the t dependence in the definition, theorem, and examples.

Definition 4.4.2 *Linear Dependence of a Set of Vector Functions in n-Space*
A finite set of n-tuple vector functions $\{\mathbf{u}_1(t), \dots, \mathbf{u}_k(t)\}$ is **linearly dependent** on an interval I if there exist *constants* $a_1, \dots, a_k$, not all zero, such that

For linear dependence *on an interval I*, the a_j's in (13) must be *constants*, not functions of t.

$$a_1\mathbf{u}_1 + a_2\mathbf{u}_2 + \cdots + a_k\mathbf{u}_k = \mathbf{0} \tag{13}$$

for all t in I. If it is not linearly dependent on I, it is **linearly independent** on I.

To determine if a given set of vector functions is LD or LI, we can proceed exactly as in Examples 2 and 3, by writing out (1) as a set of scalar equations and solving for the a_j's, for instance by Gauss elimination. *If there exist solutions for them that are constants and not all zero, then the set is LD; if not, it is LI.*

Typically, in subsequent sections, k will equal n, in which case we would prefer to use a convenient determinant condition, like the one given in Theorem 4.4.1 for constant vectors, rather than having to solve for the a_j's. We do have this analog of Theorem 4.4.1:

THEOREM 4.4.2 *LI Test for a Set of n Vector Functions in n-Space*
Given n vector functions, $\mathbf{u}_1(t), \dots, \mathbf{u}_n(t)$ in n-space, on a t interval I, let

$[\mathbf{u}_1(t), \ldots, \mathbf{u}_n(t)]$ denote an $n \times n$ matrix having those vectors as its respective columns. Then, $\{\mathbf{u}_1(t), \ldots, \mathbf{u}_n(t)\}$ is LI on I if $\det[\mathbf{u}_1(t), \ldots, \mathbf{u}_n(t)] \neq 0$ *anywhere* in I.

EXAMPLE 5. Vector Functions. Determine if the set $[1-t, 1]^{\mathrm{T}}$, $[1+t, 2+t]^{\mathrm{T}}$, on $-\infty < t < \infty$, is LD or LI. We can always fall back on the Definition 4.4.2. Write $a_1[1-t, 1]^{\mathrm{T}} + a_2[1+t, 2+t]^{\mathrm{T}} = [0, 0]^{\mathrm{T}}$ or, equivalently,

$$(1-t)a_1 + (1+t)a_2 = 0, \tag{14a}$$

$$a_1 + (2+t)a_2 = 0. \tag{14b}$$

To solve for a_1 and a_2, (14b) gives $a_1 = -(2+t)a_2$ and putting that into (14a) gives $(t^2 + 2t - 1)a_2 = 0$ for a_2. The only solution of $(t^2 + 2t - 1)a_2 = 0$ for *all* values of t in I is $a_2 = 0$, and then $a_1 = -(2+t)a_2$ gives $a_1 = 0$ too. Thus, the only constants a_1 and a_2 for which (13) is satisfied for all t in I are $a_1 = a_2 = 0$, so the given set is LI on I.

COMMENT 1. A quicker way to obtain the same result is to note that since (14) is to hold for all t in I, it must hold at $t = 0$, for instance. There, it becomes $a_1 + a_2 = 0$ and $a_1 + 2a_2 = 0$, which give $a_1 = a_2 = 0$. Further, those values of a_1 and a_2 do satisfy (14) for all t, so the only constants a_1 and a_2 that satisfy (14) for all t on I are $a_1 = a_2 = 0$. Hence, the set is LI on I.

COMMENT 2. In this example we used Definition 4.4.2, showing that the only constant coefficients satisfying (13) for all t in I are $a_1 = a_2 = 0$. That approach has the advantage of "keeping our feet on the ground" by appealing directly to the definition of linear dependence. However, since the set consists of two vectors in 2-space, Theorem 4.4.2 applies. To apply it, form and compute $\det[\mathbf{u}_1(t), \mathbf{u}_2(t)]$:

$$\det[\mathbf{u}_1(t), \mathbf{u}_2(t)] = \begin{vmatrix} 1-t & 1+t \\ 1 & 2+t \end{vmatrix} = -t^2 - 2t + 1. \tag{15}$$

The latter is indeed nonzero at points in I (in fact, it is zero only at the two points $-1 \pm \sqrt{2}$), so the set is LI on I. ▮

Note that if the set consists of n vector functions in n-space, so that Theorem 4.4.2 applies, if the $\det[\mathbf{u}_1, \ldots, \mathbf{u}_n]$ is zero for all t in I, then the theorem gives *no information*, and the set can be LD *or* LI, as illustrated in the next example.

EXAMPLE 6. Let $\mathbf{u}_1 = [1, 3]^{\mathrm{T}}$ and $\mathbf{u}_2 = [t, 3t]^{\mathrm{T}}$, and consider the set $\{\mathbf{u}_1(t), \mathbf{u}_2(t)\}$ on $-\infty < t < \infty$. It is tempting to think the set is LD because $\mathbf{u}_2$ is a multiple of $\mathbf{u}_1$: $\mathbf{u}_2 = t\mathbf{u}_1$. That is, $a_1\mathbf{u}_1 + a_2\mathbf{u}_2 = \mathbf{0}$ on I with $a_1 = t$ and $a_2 = -1$. However, remember from Definition 4.4.2 that the a_j's in (13) must be *constants*, whereas $a_1 = t$ is not. Perhaps surprisingly then, $\{\mathbf{u}_1, \mathbf{u}_2\}$ is LI, even though $\mathbf{u}_1(t)$ and $\mathbf{u}_2(t)$ are aligned. In case this point is unclear, let us proceed more slowly: First, try Theorem 4.4.2. We have $\det[\mathbf{u}_1(t), \mathbf{u}_2(t)] = \begin{vmatrix} 1 & t \\ 3 & 3t \end{vmatrix} = 0$ for all t, so the theorem gives no information. Falling back on Definition 4.4.2 then, write $a_1[1, 3]^{\mathrm{T}} + a_2[t, 3t]^{\mathrm{T}} = \mathbf{0} = [0, 0]^{\mathrm{T}}$. Thus, $a_1 + ta_2 = 0$ and $3a_1 + 3ta_2 = 0$, and solving these by Gauss elimination gives $a_2 = \alpha$

Are you concerned that $\mathbf{u}_1$ is not really a vector function since both its components are constants? Vector functions can be constant on their intervals of definition, just as scalar functions can be constant on their intervals of definition.

(arbitrary) and $a_1 = -t\alpha$. For these to not both be zero, we need $\alpha \neq 0$, but then a_1 and a_2 cannot both be constants. Then, it follows from Definition 4.4.2 that $\{\mathbf{u}_1(t), \mathbf{u}_2(t)\}$ is LI, not LD, on the interval. ▮

Closure. Definition 4.4.1 of the linear dependence and independence of a set of constant vectors is essentially the same as the corresponding definition in Section 2.4 for a set of functions. To test a given set of constant vectors for linear dependence, we can write out equation (1) and solve for the a_j's, for instance by Gauss elimination; the set is LI if and only if the a_j's are necessarily all zero. For the important case in which $k = n$, testing is simpler: Construct an $n \times n$ matrix with the given vectors as its columns. The set is LI if the determinant of that matrix is not zero, and LD if the determinant is zero.

For a set of vector *functions* the issue is now the linear dependence or independence on a *t interval* I. We can use Gauss elimination to solve for the a_j's. The set is LD on I if and only if we obtain solutions for the a_j's that are not all zero, *and are constants* on I. For the important case where $k = n$ we can, alternatively, use Theorem 4.4.2: Form a matrix having the vectors as its columns. If its determinant is nonzero anywhere in I, then the set is LI on I. But, if the determinant is zero on I (i.e., everywhere in I), the theorem gives no information. In that case we can fall back on writing (13) and solving for the a_j's.

Another important result, regarding linear dependence of vector functions, will be given in Theorem 4.5.4.

EXERCISES 4.4

CONSTANT VECTORS

1. (a) Can a set of vectors be neither LD nor LI? Explain.
(b) Can it be both LD and LI? Explain.

2. If $\{\mathbf{u}_1, \mathbf{u}_2, \mathbf{u}_3\}$ is LD, can $\mathbf{u}_3$ be expressed as a linear combination of $\mathbf{u}_1$ and $\mathbf{u}_2$? Explain.

3. Using Inspection. Show by inspection that the following vector sets are LD, by giving a nontrivial linear relation among the vectors, of the form (1) with the a_j's not all zero. Call the vectors $\mathbf{u}_1, \mathbf{u}_2, \ldots,$ in turn.

(a) $[1, 4]^{\mathrm{T}}, [-3, -12]^{\mathrm{T}}$
(b) $[1, 1, 3]^{\mathrm{T}}, [1, 2, 3]^{\mathrm{T}}, [2, 3, 6]^{\mathrm{T}}$
(c) $[2, 1, 5]^{\mathrm{T}}, [0, 0, 0]^{\mathrm{T}}$
(d) $[0, 0]^{\mathrm{T}}, [0, 0]^{\mathrm{T}}, [0, 0]^{\mathrm{T}}$
(e) $[1, 1]^{\mathrm{T}}, [1, 2]^{\mathrm{T}}, [3, 4]^{\mathrm{T}}$
(f) $[1, 2, 3]^{\mathrm{T}}, [3, 2, 1]^{\mathrm{T}}, [8, 8, 8]^{\mathrm{T}}$
(g) $[1, 1, 1]^{\mathrm{T}}, [1, 1, 1]^{\mathrm{T}}, [3, 4, 5]^{\mathrm{T}}$
(h) $[1, 4, 0]^{\mathrm{T}}, [2, 8, 0]^{\mathrm{T}}, [3, -1, 0]^{\mathrm{T}}$
(i) $[1, 0, 1]^{\mathrm{T}}, [4, 5, 4]^{\mathrm{T}}, [1, 1, 1]^{\mathrm{T}}$
(j) $[1, 0, 2]^{\mathrm{T}}, [1, 0, 3]^{\mathrm{T}}, [4, 0, 10]^{\mathrm{T}}$
(k) $[1, 0, 0]^{\mathrm{T}}, [0, 1, 0]^{\mathrm{T}}, [0, 0, 1]^{\mathrm{T}}, [6, 4, 1]^{\mathrm{T}}$

4. LD or LI? Determine whether the given set is LI or LD. If it is LD, give any nontrivial linear relation among the vectors, that is, of the form (1) with the a_j's not all zero. SUGGESTION: If the set consists of n vectors in n-space, then use Theorem 4.4.1,which will answer the question fully: if the determinant is nonzero the set is LI, and if it is zero the set is LD. If Theorem 4.4.1 does not apply, fall back on the Definition 4.4.1, as we did in Examples 2 and 3.

(a) $[1, 3]^{\mathrm{T}}, [2, 0]^{\mathrm{T}}, [-1, 3]^{\mathrm{T}}, [7, 3]^{\mathrm{T}}$
(b) $[1, 3]^{\mathrm{T}}, [2, 0]^{\mathrm{T}}, [1, 2]^{\mathrm{T}}$
(c) $[2, 3, 0]^{\mathrm{T}}, [1, -2, 3]^{\mathrm{T}}$
(d) $[2, 3, 0]^{\mathrm{T}}, [1, -2, 4]^{\mathrm{T}}, [1, 1, 0]^{\mathrm{T}}$
(e) $[0, 0, 2]^{\mathrm{T}}, [0, 0, 3]^{\mathrm{T}}, [2, -1, 5]^{\mathrm{T}}, [1, 2, 4]^{\mathrm{T}}$
(f) $[2, 3, 0, 0]^{\mathrm{T}}, [1, -5, 0, 2]^{\mathrm{T}}, [3, 1, 2, 2]^{\mathrm{T}}$
(g) $[1, 3, 2, 0]^{\mathrm{T}}, [4, 1, -2, -2]^{\mathrm{T}}, [0, 2, 0, 3]^{\mathrm{T}}$
(h) $[2, 0, 1, -1, 0]^{\mathrm{T}}, [1, 2, 0, 3, 1]^{\mathrm{T}}, [4, -4, 3, -9, -2]^{\mathrm{T}}$

(i) $[1,3,0]^{\mathrm{T}}$, $[0,1,-1]^{\mathrm{T}}$, $[0,0,0]^{\mathrm{T}}$
(j) $[1,1,0,1]^{\mathrm{T}}$, $[2,-1,1,3]^{\mathrm{T}}$, $[4,1,1,5]^{\mathrm{T}}$
(k) $[1,-3,0,2,1]^{\mathrm{T}}$, $[-2,6,0,-4,-2]^{\mathrm{T}}$
(l) $[1,2,3,4]^{\mathrm{T}}$, $[2,4,6,8]^{\mathrm{T}}$, $[0,0,0,0]^{\mathrm{T}}$
(m) $[1,2,3,4]^{\mathrm{T}}$, $[2,3,4,5]^{\mathrm{T}}$
(n) $[2,1,-1]^{\mathrm{T}}$, $[1,4,2]^{\mathrm{T}}$, $[3,-2,-4]^{\mathrm{T}}$
(o) $[7,1,0]^{\mathrm{T}}$, $[-1,1,4]^{\mathrm{T}}$, $[2,3,5]^{\mathrm{T}}$
(p) A vector set supplied by your instructor.

5. Make up any LI vector set of the specified type and verify its linear independence.

(a) three 3-tuples (b) three 4-tuples (c) three 5-tuples
(d) four 4-tuples (e) four 5-tuples (f) four 6-tuples

6. More Than n Constant Vectors in n-Space. A set of more than n constant n-tuple vectors is necessarily LD. Prove that for the case where the set consists of three 2-tuple vectors.

7. Less Than n Constant Vectors in n-Space. Show that a set of m constant n-tuples can be LD *or* LI if $m < n$.

8. Sets of *Two* Constant Vectors. Show that *if a set contains only two constant n-tuple vectors, then it is LI if and only if neither one is a scalar multiple of the other.*

VECTOR FUNCTIONS

9. LD or LI? Determine whether the set is LD or LI on $-\infty < t < \infty$. SUGGESTION: It may be that you can see a linear relationship (13) by inspection. If so, simply state that the set is LD and, to verify your claim, give that linear relationship. If inspection does not suffice, see if the set consists of n vectors in n-space; if so, use Theorem 4.4.2. Finally, if that theorem proves inconclusive (namely, if the determinant is zero everywhere on I), fall back on the Definition 4.4.2, and see if there exist constants a_j, not all zero, such that a linear relationship (13) is satisfied on I.

(a) $[2,t]^{\mathrm{T}}$, $[3,4]^{\mathrm{T}}$
(b) $[1,t]^{\mathrm{T}}$, $[2,t]^{\mathrm{T}}$
(c) $e^t[1,2]^{\mathrm{T}}$, $e^t[-3,-6]^{\mathrm{T}}$
(d) $\cos t[1,1]^{\mathrm{T}}$, $\sin t[1,1]^{\mathrm{T}}$
(e) $[t,0]^{\mathrm{T}}$, $[0,t]^{\mathrm{T}}$, $[t,t]^{\mathrm{T}}$
(f) $[t,0]^{\mathrm{T}}$, $[0,t]^{\mathrm{T}}$, $[1,1]^{\mathrm{T}}$
(g) $[1,1]^{\mathrm{T}}$, $[t,t]^{\mathrm{T}}$, $[t^2,t^2]^{\mathrm{T}}$
(h) $[1,1]^{\mathrm{T}}$, $[2,2]^{\mathrm{T}}$, $[0,\sin t]$
(i) $[1,e^t]^{\mathrm{T}}$, $[e^t,1]^{\mathrm{T}}$, $[5,5]^{\mathrm{T}}$
(j) $[t,0]^{\mathrm{T}}$, $[2t,0]^{\mathrm{T}}$, $[3t,0]^{\mathrm{T}}$
(k) $[1,2,1]^{\mathrm{T}}$, $[2e^t,4e^t,2e^t]^{\mathrm{T}}$
(l) $[1,2,1]^{\mathrm{T}}$, $e^{-t}[3,6,3]^{\mathrm{T}}$
(m) $[1,2,3,4]^{\mathrm{T}}$, $[0,0,0,t]^{\mathrm{T}}$
(n) $[1,2,3,4t]^{\mathrm{T}}$, $[0,0,0,0]^{\mathrm{T}}$
(o) $[1,1,1]^{\mathrm{T}}$, $[2,2,2]^{\mathrm{T}}$, $[t,t,t]^{\mathrm{T}}$
(p) $[3,3,3]^{\mathrm{T}}$, $[1,1,1]^{\mathrm{T}}$, $[0,t,0]^{\mathrm{T}}$
(q) $[1,0,0]^{\mathrm{T}}$, $[t,0,0]^{\mathrm{T}}$, $[0,0,t]^{\mathrm{T}}$
(r) $e^t[1,0]^{\mathrm{T}}$, $e^{2t}[1,0]^{\mathrm{T}}$, $e^{3t}[1,0]^{\mathrm{T}}$
(s) $[e^t,0,0]^{\mathrm{T}}$, $[0,e^t,0]^{\mathrm{T}}$, $[e^t,2e^t,e^{3t}]^{\mathrm{T}}$, $[3,4,-2]^{\mathrm{T}}$
(t) A vector set supplied by your instructor.

10. Sets of *Two* Vector Functions. Show that *if a set contains only two n-tuple vector functions, then it is LI if and only if neither one is a constant multiple of the other.* [NOTE: This result provides a simple way to see that the set in part (a) of Exercise 9, above, is LI, and that the one in part (c) is LD.]

4.5 EXISTENCE, UNIQUENESS, AND GENERAL SOLUTION

4.5.1 The key theorems. Using vectors, matrices, and the concept of the linear independence of sets of vector functions, we can now study the existence and uniqueness of solutions for systems of initial value problems and establish the form of general solutions of such systems. We continue to study systems of n *linear* differential equations in n unknowns $x_1(t), \dots, x_n(t)$,

$$\begin{aligned} x_1' &= a_{11}(t)x_1 + \cdots + a_{1n}(t)x_n + f_1(t), \\ &\vdots \\ x_n' &= a_{n1}(t)x_1 + \cdots + a_{nn}(t)x_n + f_n(t), \end{aligned} \tag{1}$$

with the primes denoting differentiation with respect to t. The system is linear because the equations are linear in the x_j's and their derivatives.

Recall that (1) is homogeneous if *all* of $f_1(t), \ldots, f_n(t)$ are identically zero on the interval, and it is nonhomogeneous if *at least one* of them is not. Likewise, (3) is homogeneous if $\mathbf{f}(t) = \mathbf{0}$ on the interval, and nonhomogeneous otherwise.

We can express (1) in matrix form as

$$\begin{bmatrix} x_1' \\ \vdots \\ x_n' \end{bmatrix} = \begin{bmatrix} a_{11}(t) & \cdots & a_{1n}(t) \\ \vdots & & \vdots \\ a_{n1}(t) & \cdots & a_{nn}(t) \end{bmatrix} \begin{bmatrix} x_1 \\ \vdots \\ x_n \end{bmatrix} + \begin{bmatrix} f_1(t) \\ \vdots \\ f_n(t) \end{bmatrix}, \tag{2}$$

or,

$$\boxed{\mathbf{x}' = \mathbf{A}(t)\mathbf{x} + \mathbf{f}(t),} \tag{3}$$

in which $\mathbf{A}(t)$ is the **coefficient matrix**, with $a_{ij}(t)$ as its i, j element.

EXAMPLE 1. Example of the Matrix Equation (3). The system

$$x_1' = 2e^t x_1 + 3x_2 + 25 \sin t, \tag{4a}$$
$$x_2' = (2 + 3t)x_1 - x_2 - t + 4 \tag{4b}$$

can be expressed in the matrix form (3), with

$$\mathbf{x}(t) = \begin{bmatrix} x_1(t) \\ x_2(t) \end{bmatrix}, \quad \mathbf{A}(t) = \begin{bmatrix} 2e^t & 3 \\ 2+3t & -1 \end{bmatrix}, \quad \text{and} \quad \mathbf{f}(t) = \begin{bmatrix} 25 \sin t \\ -t+4 \end{bmatrix}. \quad ▮$$

By $\{x_1(t), \ldots, x_n(t)\}$ being a **solution** of (1) on I we mean that if those functions are substituted into (1), then each equation in (1) is reduced to an identity on I. Similarly for the matrix version (3); $\mathbf{x}(t)$ being a solution of (3) on I means that if it is substituted into (3) it reduces (3) to an identity on I.

If initial conditions

$$x_1(a) = b_1, \; x_2(a) = b_2, \; \ldots, \; x_n(a) = b_n \tag{5}$$

are specified at some point $t = a$ in I, then (1) and (5) constitute an **initial value problem**. Express (5) as $\mathbf{x}(a) = \mathbf{b}$, in which $\mathbf{x}(a) = [x_1(a), \ldots, x_n(a)]^{\mathrm{T}}$ and $\mathbf{b} = [b_1, \ldots, b_n]^{\mathrm{T}}$.

We begin with the following foundational result.

The key condition here is the continuity of $\mathbf{A}(t)$ and $\mathbf{f}(t)$. The existence and uniqueness of solutions is the foundation upon which the rest of the theory will be built.

THEOREM 4.5.1 *Existence and Uniqueness for the IVP*
Let $\mathbf{A}(t)$ be an $n \times n$ matrix and let $\mathbf{x}(t)$ and $\mathbf{f}(t)$ be n-tuple vectors. If $\mathbf{A}(t)$ and $\mathbf{f}(t)$ are continuous on an open interval I, and a is in I, then the IVP

$$\boxed{\mathbf{x}' = \mathbf{A}(t)\mathbf{x} + \mathbf{f}(t); \quad \mathbf{x}(a) = \mathbf{b}} \tag{6}$$

has a unique solution for $\mathbf{x}(t)$ on I.

Recall that $\mathbf{A}(t)$ being continuous on I means that each $a_{ij}(t)$ element of $\mathbf{A}(t)$ is continuous on I; similarly for $\mathbf{f}(t)$. For instance, $\mathbf{A}(t)$ and $\mathbf{f}(t)$ in Example 1 are continuous on every interval I because $2e^t$, 3, $2+3t$, -1, $25\sin t$, and $-t+4$ are continuous for all t.

If $n = 1$, (6) reduces (though in "Chapter 1 notation") to the IVP

$$y' + p(x)y = q(x), \quad y(a) = b \tag{7}$$

studied in Chapter 1. Sure enough, for the case $n = 1$ Theorem 4.5.1 is identical to Theorem 1.2.1, except that Theorem 1.2.1 went much further by giving a *formula* for the solution. For $n > 1$, the problem is more difficult, and such a solution formula is not known, as we've mentioned before.

Next, we say that a family of solutions of (3) is a **general solution** of (3) if it contains every solution of (3). As a first step toward establishing the form of a general solution, we have the following superposition principle for the *homogeneous* equation $\mathbf{x}' = \mathbf{A}(t)\mathbf{x}$:

THEOREM 4.5.2 *Superposition Principle for* $\mathbf{x}' = \mathbf{A}(t)\mathbf{x}$
Let $\mathbf{A}(t)$ be an $n \times n$ matrix and let the unknown $\mathbf{x}(t)$ be an n-tuple vector. If $\mathbf{x}_1(t), \ldots, \mathbf{x}_k(t)$ are solutions of the homogeneous equation $\mathbf{x}' = \mathbf{A}(t)\mathbf{x}$, then so is any linear combination of them, $\mathbf{x}(t) = c_1\mathbf{x}_1(t) + \cdots + c_k\mathbf{x}_k(t)$.

Proof: The linear combination $\mathbf{x}(t) = c_1\mathbf{x}_1(t)+\cdots+c_k\mathbf{x}_k(t)$ satisfies $\mathbf{x}' = \mathbf{A}(t)\mathbf{x}$, for any set of values of the c_j's, because

$$\begin{aligned}\frac{d}{dt}(c_1\mathbf{x}_1 + \cdots + c_k\mathbf{x}_k) &= c_1\mathbf{x}_1' + \cdots + c_k\mathbf{x}_k' \\ &= c_1\mathbf{A}\mathbf{x}_1 + \cdots + c_k\mathbf{A}\mathbf{x}_k \\ &= \mathbf{A}\left(c_1\mathbf{x}_1 + \cdots + c_k\mathbf{x}_k\right).\end{aligned} \tag{8}$$

That is, $c_1\mathbf{x}_1(t)+\cdots+c_k\mathbf{x}_k(t)$ does satisfy the homogeneous equation $\mathbf{x}' = \mathbf{A}(t)\mathbf{x}$. The steps in (8) rest on linearity: The first equality follows from the linearity of the derivative operator, and the last one follows from the linearity of matrix multiplication [equation (11) in Section 4.3]. ∎

We can now state the form of a general solution:

THEOREM 4.5.3 *General Solution of the Homogeneous Equation* $\mathbf{x}' = \mathbf{A}(t)\mathbf{x}$, *and of the Nonhomogeneous Equation* $\mathbf{x}' = \mathbf{A}(t)\mathbf{x} + \mathbf{f}(t)$
Let $\mathbf{A}(t)$ be an $n \times n$ matrix and let $\mathbf{x}(t)$ and $\mathbf{f}(t)$ be n-tuple vectors, with $\mathbf{A}(t)$ and $\mathbf{f}(t)$ continuous on an open interval I.

(**a**) *Homogeneous equation*: There exist n LI solutions of the homogeneous equation

$$\boxed{\mathbf{x}' = \mathbf{A}(t)\mathbf{x}} \tag{9}$$

on I, no more and no less. If $\mathbf{x}_1(t), \dots, \mathbf{x}_n(t)$ are n such LI solutions, then

$$\boxed{\mathbf{x}(t) = c_1\mathbf{x}_1(t) + \cdots + c_n\mathbf{x}_n(t)} \tag{10}$$

is a general solution of (9) on I, in which the c_j's are arbitrary constants.
(**b**) *Nonhomogeneous equation*: If $\mathbf{x}_1(t), \dots, \mathbf{x}_n(t)$ are n LI solutions of (9) on I, and $\mathbf{x}_p(t)$ is any solution (i.e., a "particular solution") of the nonhomogeneous equation

$$\boxed{\mathbf{x}' = \mathbf{A}(t)\mathbf{x} + \mathbf{f}(t)} \tag{11}$$

on I, then

$$\boxed{\mathbf{x}(t) = c_1\mathbf{x}_1(t) + \cdots + c_n\mathbf{x}_n(t) + \mathbf{x}_p(t)} \tag{12}$$

is a general solution of (11) on I, in which the c_j's are arbitrary constants.[1]

Any such set $\mathbf{x}_1(t), \dots, \mathbf{x}_n(t)$ is called a **fundamental set** (or **basis**) of solutions of the system $\mathbf{x}' = \mathbf{A}(t)\mathbf{x}$ for the interval I.

Fundamental set and fundamental matrix for (9). Do you see that (13) is equivalent to (10)? If not, do the matrix multiplication and examine the result, remembering that the columns of the matrix in (13) are the homogeneous solutions $\mathbf{x}_1(t), \dots, \mathbf{x}_n(t)$.

Observe that if we let the c_j's be the components of a column vector $\mathbf{c}$, and let $\mathbf{X}(t) = [\mathbf{x}_1(t), \dots, \mathbf{x}_n(t)]$ be a matrix having the n LI solutions $\mathbf{x}_j(t)$ as its columns, then the general solution (10) can be expressed compactly as

$$\mathbf{x}(t) = \begin{bmatrix} x_{11}(t) & \cdots & x_{n1}(t) \\ \vdots & \cdots & \vdots \\ x_{1n}(t) & \cdots & x_{nn}(t) \end{bmatrix} \begin{bmatrix} c_1 \\ \vdots \\ c_n \end{bmatrix} = \mathbf{X}(t)\mathbf{c}. \tag{13}$$

The matrix $\mathbf{X}(t)$, having a fundamental set of solutions of (9) as its columns, is called a **fundamental matrix** of (9).

To apply Theorem 4.5.3 we need to be able to test a set of solutions $\mathbf{x}_1(t), \dots, \mathbf{x}_n(t)$ for linear independence. The following theorem provides a convenient test.

THEOREM 4.5.4 *Linear Independence of Solutions of (9)*
Let the $n \times n$ matrix $\mathbf{A}(t)$ in (9) be continuous on I, and let $\mathbf{x}_1(t), \dots, \mathbf{x}_n(t)$ be n solutions of (9) on I. Then
(a) $\det[\mathbf{x}_1(t), \dots, \mathbf{x}_n(t)]$ is either zero everywhere on I, or nonzero everywhere on I.

[1] The existence of an $\mathbf{x}_p(t)$ follows from Theorem 4.5.1, for if we specify *any* initial point a in I and *any* vector $\mathbf{b}$, then, according to Theorem 4.5.1, there exists a particular solution of (11) satisfying that initial condition. That particular solution can be used as $\mathbf{x}_p(t)$ in (12).

(b) If it is nonzero on I, then $\{\mathbf{x}_1(t), \dots, \mathbf{x}_n(t)\}$ is LI on I, and if it is zero on I then $\{\mathbf{x}_1(t), \dots, \mathbf{x}_n(t)\}$ is LD on I.

4.5.2 Illustrating the theorems.

Let us illustrate the foregoing theorems.

EXAMPLE 2. Homogeneous Equation, General Solution, and Fundamental Set. Consider the homogeneous system

$$x_1' = 4x_1 + x_2, \tag{14a}$$

$$x_2' = -6x_1 - x_2 \tag{14b}$$

on $-\infty < t < \infty$. The elimination method (Section 4.1) gives the general solution

$$x_1(t) = c_1 e^t + c_2 e^{2t}, \tag{15a}$$

$$x_2(t) = -3c_1 e^t - 2c_2 e^{2t}, \tag{15b}$$

or, in vector form,

$$\mathbf{x}(t) = c_1 \begin{bmatrix} e^t \\ -3e^t \end{bmatrix} + c_2 \begin{bmatrix} e^{2t} \\ -2e^{2t} \end{bmatrix}. \tag{16}$$

When writing $\mathbf{x}_1(t)$ and $\mathbf{x}_2(t)$, use a vector symbol such as an overhead arrow or a wavy underline, to distinguish them from the scalar functions $x_1(t)$ and $x_2(t)$.

The latter is indeed of the form (10), with

$$\mathbf{x}_1(t) = \begin{bmatrix} e^t \\ -3e^t \end{bmatrix} = \begin{bmatrix} 1 \\ -3 \end{bmatrix} e^t \quad \text{and} \quad \mathbf{x}_2(t) = \begin{bmatrix} e^{2t} \\ -2e^{2t} \end{bmatrix} = \begin{bmatrix} 1 \\ -2 \end{bmatrix} e^{2t}. \tag{17}$$

To verify that the solutions $\mathbf{x}_1(t)$ and $\mathbf{x}_2(t)$ are LI, evaluate

$$\det\left[\mathbf{x}_1(t), \mathbf{x}_2(t)\right] = \begin{vmatrix} e^t & e^{2t} \\ -3e^t & -2e^{2t} \end{vmatrix} = e^{3t}, \tag{18}$$

which is nonzero. Thus, by Theorem 4.5.4, $\mathbf{x}_1(t)$ and $\mathbf{x}_2(t)$ are LI on I. Hence, by Theorem 4.5.3(a), (16) is a general solution of the system (14), and $\{\mathbf{x}_1(t), \mathbf{x}_2(t)\}$ is a fundamental set of solutions of (14) on I. ▮

EXAMPLE 3. An IVP. Solve

$$\begin{aligned} x' &= 8x + 6y + 8z, & x(0) &= -2 \\ y' &= -8x - 6y - 8z, & y(0) &= 6 \\ z' &= 4x + 3y + 4z + 5e^t, & z(0) &= -5. \end{aligned} \tag{19}$$

To solve the IVP, first find a general solution, then apply the initial conditions to evaluate the arbitrary constants. As we noted in Section 4.1, elimination may not work if there are more than two differential equations. In fact it does not work for the set given in (19), but the latter can be solved by matrix methods that will be given in Section 4.7. To continue with our illustration of the foregoing theorems, we will give those solutions without derivation. Homogeneous and particular solutions are

$$\mathbf{x}_1(t) = \begin{bmatrix} 0 \\ 4 \\ -3 \end{bmatrix}, \ \mathbf{x}_2(t) = \begin{bmatrix} 4 \\ -4 \\ -1 \end{bmatrix}, \ \mathbf{x}_3(t) = \begin{bmatrix} 2 \\ -2 \\ 1 \end{bmatrix} e^{6t}, \ \mathbf{x}_p(t) = \begin{bmatrix} -8 \\ 8 \\ 1 \end{bmatrix} e^t, \quad (20)$$

respectively. To check for linear independence of the homogeneous solutions, evaluate

$$\det\left[\mathbf{x}_1(t), \mathbf{x}_2(t), \mathbf{x}_3(t)\right] = \begin{vmatrix} 0 & 4 & 2e^{6t} \\ 4 & -4 & -2e^{6t} \\ -3 & -1 & e^{6t} \end{vmatrix} = e^{6t} \begin{vmatrix} 0 & 4 & 2 \\ 4 & -4 & -2 \\ -3 & -1 & 1 \end{vmatrix} = -24e^{6t}, \quad (21)$$

which is nonzero. Thus, by Theorem 4.5.4, the homogeneous solutions $\mathbf{x}_1(t), \mathbf{x}_2(t), \mathbf{x}_3(t)$ are LI on I. Hence, by Theorem 4.5.3(a), a general solution of the system of three differential equations in (20) is

$$\begin{aligned} \mathbf{x}(t) &= c_1\mathbf{x}_1(t) + c_2\mathbf{x}_2(t) + c_3\mathbf{x}_3(t) + \mathbf{x}_p(t) \\ &= c_1 \begin{bmatrix} 0 \\ 4 \\ -3 \end{bmatrix} + c_2 \begin{bmatrix} 4 \\ -4 \\ -1 \end{bmatrix} + c_3 \begin{bmatrix} 2 \\ -2 \\ 1 \end{bmatrix} e^{6t} + \begin{bmatrix} -8 \\ 8 \\ 1 \end{bmatrix} e^t. \end{aligned} \quad (22)$$

Next, the initial conditions

$$\begin{aligned} x(0) &= -2 = 0c_1 + 4c_2 + 2c_3 - 8, \\ y(0) &= \ \ 6 = 4c_1 - 4c_2 - 2c_3 + 8, \\ z(0) &= -5 = -3c_1 - c_2 + c_3 + 1, \end{aligned} \quad (23)$$

give $c_1 = 1, c_2 = 2, c_3 = -1$. With those c_j's, (22) is the solution of the IVP (20) and, by Theorem 4.5.1, it is unique. ▮

Closure. Theorem 4.5.1 gives conditions for the existence and uniqueness of solutions of the IVP (6), and the Theorem 4.5.3 tells us what we must do to find general solutions of the homogeneous system (9), and the nonhomogeneous system (11): A general solution of (9) is given by an arbitrary linear combination of n LI solutions of (9), as shown in (10). To obtain a general solution of (11), we must augment that general solution of the homogeneous system by adding any particular solution $\mathbf{x}_p(t)$ of (11), as shown in (12). To tie these theorems and results together, pay careful attention to Examples 2 and 3.

To be sure that $\mathbf{x}_1(t), \ldots, \mathbf{x}_n(t)$ are LI on I, we can use the determinant condition in Theorem 4.5.4.

We continue our discussion of how to *find* those solutions in Section 4.7, after introducing the "eigenvalue problem" in Section 4.6.

EXERCISES 4.5

1. Matrix Form. Write each system in the matrix form (3), and identify $\mathbf{A}(t)$ and $\mathbf{f}(t)$. What is the broadest interval, containing $t = 0$, on which both $\mathbf{A}(t)$ and $\mathbf{f}(t)$ are continuous?

(a) $x_1' = tx_1 + 2x_2, \ x_2' = x_1 + e^t x_2 + \sin t$
(b) $x_1' + x_1 - x_2 - 3t = 0, \ x_2' + e^{-t}x_1 - 4 = 0$
(c) $x_1' = (\tan t)x_1 + x_2 + 3, \ x_2' = x_1 + 5x_2 + 6t - 1$
(d) $(t+1)x_1' = tx_2 + (t+1)^2, \ x_1' = x_1 - x_2$
(e) $(\cos t)x_1' = tx_1 + x_2 + 10\sin t, \ (\cot 2t)x_2' = x_1 - x_2 + 20$
(f) $(t^2 + t - 6)x_1' = (t-2)x_1 + x_2, \ (t+5)x_1' = tx_1 - x_2 + 8$

2. Example 2. (a) Use the method of elimination to derive the solution (15) of the system (14).
(b) Verify, by substitution, that $\mathbf{x}_1(t)$ and $\mathbf{x}_2(t)$ given by (17) are solutions of (14), as claimed.

3. Evaluate c_1 and c_2 in the general solution (16) so that the following initial conditions are satisfied. Four significant figures will suffice.

(a) $x_1(0) = 5, \ x_2(0) = -4$ (b) $x_1(0) = 5, \ x_2(0) = 0$
(c) $x_1(0) = 0, \ x_2(0) = -4$ (d) $x_1(0.5) = 1, \ x_2(0.5) = 3$
(e) $x_1(1) = 1, \ x_2(1) = 4$ (f) $x_1(2) = 0, \ x_2(2) = 8$

4. Example 3. Verify, by substitution, that $\mathbf{x}_1(t), \mathbf{x}_2(t), \mathbf{x}_3(t)$, and $\mathbf{x}_p(t)$ given by (20) are solutions of the homogeneous and nonhomogeneous systems, respectively, as claimed.

5. Fundamental Matrix. (a) Verify that the general solution (10) can be expressed in the form $\mathbf{X}(t)\mathbf{c}$, as in (13). That is, carry out the matrix multiplication in (13) and show that the result is the same as the right-hand side of (10).
(b) Show that a fundamental matrix $\mathbf{X}(t)$ satisfies (9), namely, that

$$\boxed{\mathbf{X}'(t) = \mathbf{A}(t)\mathbf{X}(t).} \tag{5.1}$$

6. Systems of Two Equations. Find a general solution $x_1(t)$ and $x_2(t)$ of the given system by the elimination method given in Section 4.1. From that general solution, identify LI homogeneous solutions $\mathbf{x}_1(t)$ and $\mathbf{x}_2(t)$, and the particular solution $\mathbf{x}_p(t)$ if there is one. Finally, find the solution $\mathbf{x}(t)$ satisfying the given initial conditions. The interval is $-\infty < t < \infty$.

(a) System: $x_1' = x_1 + x_2, x_2' = x_1 + x_2$
Initial conditions: $x_1(0) = 5, x_2(0) = -3$
(b) System: $x_1' = 4x_1 + 4x_2, x_2' = x_1 + x_2 + 125t$
Initial conditions: $x_1(0) = 0, x_2(0) = 0$
(c) System: $x_1' = 4x_2 + 144(1+t), x_2' = 3x_1 + x_2$
Initial conditions: $x_1(-1) = 0, x_2(-1) = 16$
(d) System: $x_1' = 4x_1 + 2x_2, x_2' = 2x_1 + x_2 - 12e^t$
Initial conditions: $x_1(4) = 0, x_2(4) = 0$
(e) System: $x_1' = 2x_1 + x_2 + 2 - 3t, x_2' = x_1 + 2x_2 - 1$
Initial conditions: $x_1(0) = 2, x_2(0) = 2$
(f) System: $x_1' = 3x_1 + x_2 - e^t, x_2' = x_1 + 3x_2 - 2e^t$
Initial conditions: $x_1(0) = 3, x_2(0) = 4$
(g) System: $x_1' = 3x_1 + 2x_2 + 2 - 5e^{2t}, x_2' = 4x_1 + 5x_2 + 5$
Initial conditions: $x_1(0) = -6, x_2(0) = 6$
(h) System: $x_1' = 5x_1 + 2x_2 - 6 - 3t, x_2' = x_1 + 4x_2 - 6 + 3t$
Initial conditions: $x_1(0) = 3, x_2(0) = 2$

7. Systems of Three Equations. Verify that $\mathbf{x}_1(t)$, $\mathbf{x}_2(t)$, and $\mathbf{x}_3(t)$ are LI solutions of the homogeneous system and that $\mathbf{x}_p(t)$ is a particular solution, on $-\infty < t < \infty$. From them, obtain a general solution. Then, find the solution $\mathbf{x}(t)$ satisfying the given initial conditions.

(a) System: $x_1' = x_1 + x_2 + x_3, \ x_2' = x_1 + x_2 + x_3, \ x_3' = x_1 + x_2 + x_3$.
Solutions:

$$\mathbf{x}_1 = \begin{bmatrix} 2 \\ -1 \\ -1 \end{bmatrix}, \mathbf{x}_2 = \begin{bmatrix} 1 \\ -2 \\ 1 \end{bmatrix}, \mathbf{x}_3 = \begin{bmatrix} e^{3t} \\ e^{3t} \\ e^{3t} \end{bmatrix}.$$

Initial conditions: $x_1(0) = 15, x_2(0) = 21, x_3(0) = -12$.

(b) System: $x_1' = 2x_1 + x_2 + x_3, x_2' = x_1 + 2x_2 + x_3, x_3' = x_1 + x_2 + 2x_3$.
Solutions:

$$\mathbf{x}_1 = \begin{bmatrix} e^{4t} \\ e^{4t} \\ e^{4t} \end{bmatrix}, \mathbf{x}_2 = \begin{bmatrix} 0 \\ e^{t} \\ -e^{t} \end{bmatrix}, \mathbf{x}_3 = \begin{bmatrix} e^{t} \\ 0 \\ -e^{t} \end{bmatrix}.$$

Initial conditions: $x_1(0) = 3, x_2(0) = 0, x_3(0) = 0$.

(c) System: $x_1' = 3x_3 + 9e^{-t}, x_2' = x_3 - 9e^{-t}, x_3' = x_1 - 13x_2 + 7x_3$.
Solutions:

$$\mathbf{x}_1 = \begin{bmatrix} 13 \\ 1 \\ 0 \end{bmatrix}, \mathbf{x}_2 = \begin{bmatrix} 3e^{5t} \\ e^{5t} \\ 5e^{5t} \end{bmatrix}, \mathbf{x}_3 = \begin{bmatrix} 3e^{2t} \\ e^{2t} \\ 2e^{2t} \end{bmatrix},$$

$$\mathbf{x}_p = \begin{bmatrix} -30e^{-t} \\ 2e^{-t} \\ 7e^{-t} \end{bmatrix}.$$

Initial conditions: $x_1(0) = 3, x_2(0) = 0, x_3(0) = 0$.

(d) System: $x_1' = 2x_1 + 2x_2 + x_3 + 3$, $x_2' = x_1 + 3x_2 + x_3 + 2$, $x_3' = x_1 + 2x_2 + 2x_3 + 3$.
Solutions:

$$\mathbf{x}_1 = \begin{bmatrix} e^{5t} \\ e^{5t} \\ e^{5t} \end{bmatrix}, \mathbf{x}_2 = \begin{bmatrix} -e^t \\ 0 \\ e^t \end{bmatrix}, \mathbf{x}_3 = \begin{bmatrix} -2e^t \\ e^t \\ 0 \end{bmatrix}, \mathbf{x}_p = \begin{bmatrix} -1 \\ 0 \\ -1 \end{bmatrix}.$$

Initial conditions: $x_1(0) = 0$, $x_2(0) = 24$, $x_3(0) = 0$.

(e) System: $x_1' = x_2 + x_3 + 4 - 16t$, $x_2' = x_1 + x_3 + 16t$, $x_3' = x_1 + x_2$.

Solutions: $\mathbf{x}_1 = \begin{bmatrix} e^{-t} \\ e^{-t} \\ -2e^{-t} \end{bmatrix}$, $\mathbf{x}_2 = \begin{bmatrix} e^{2t} \\ e^{2t} \\ e^{2t} \end{bmatrix}$, $\mathbf{x}_3 = \begin{bmatrix} 0 \\ e^{-t} \\ -e^{-t} \end{bmatrix}$,

and $\mathbf{x}_p = \begin{bmatrix} 18 - 16t \\ -18 + 16t \\ -2 \end{bmatrix}$. Initial conditions: $x_1(0) = 1$, $x_2(0) = 0$, $x_3(0) = 0$.

8. Consider the system $x_1' = x_1 + x_2 - 1 - t$, $x_2' = 3x_1 + 3x_2 - 6 - 3t$. Of the following vector functions, which are homogeneous solutions, which are particular solutions, and which are neither? From them, put forward a fundamental set of solutions of the homogeneous version of the system, along with a general solution of the full (nonhomogeneous) system.

$$\begin{bmatrix} e^{4t} - 3 \\ 3e^{4t} + 3 \end{bmatrix}, \begin{bmatrix} e^{4t} \\ 1 \end{bmatrix}, \begin{bmatrix} e^{4t} \\ 3e^{4t} \end{bmatrix}, \begin{bmatrix} t + 1 \\ 1 \end{bmatrix}, \begin{bmatrix} t \\ 2 \end{bmatrix}, \begin{bmatrix} 1 \\ -1 \end{bmatrix}$$

9. Consider the system $x_1' = x_3 - 1$, $x_2' = x_3 - 1$, $x_3' = x_1 + x_2 + x_3 - 7$. Of the following vector functions, which are homogeneous solutions, which are particular solutions, and which are neither? From them, put forward a fundamental set of solutions of the homogeneous version of the system, along with a general solution of the full (nonhomogeneous) system.

$$\begin{bmatrix} 1 \\ -1 \\ 0 \end{bmatrix}, \begin{bmatrix} 6 \\ 0 \\ 1 \end{bmatrix}, \begin{bmatrix} 6 \\ 1 \\ 2 \end{bmatrix}, \begin{bmatrix} 5 \\ 1 \\ 1 \end{bmatrix}, \begin{bmatrix} e^{-t} \\ e^{-t} \\ -e^{-t} \end{bmatrix}, \begin{bmatrix} e^{2t} \\ e^{2t} \\ 2e^{2t} \end{bmatrix}, \begin{bmatrix} e^t \\ e^t \\ -e^t \end{bmatrix}$$

10. Inferring the Matrix from Solutions. Here is an "inverse problem." Suppose we know a general solution of $\mathbf{x}' = \mathbf{A}\mathbf{x}$ but don't know the $\mathbf{A}$ matrix. Can we infer $\mathbf{A}$ from the solution? Let $n = 2$, and suppose we know two LI solutions, $\mathbf{u}(t) = [u_1(t), u_2(t)]^{\mathrm{T}}$ and $\mathbf{v}(t) = [v_1(t), v_2(t)]^{\mathrm{T}}$. Avoiding the double subscript notation, for simplicity, denote $\mathbf{A} = \begin{bmatrix} a(t) & b(t) \\ c(t) & d(t) \end{bmatrix}$. If possible, infer $a(t)$, $b(t)$, $c(t)$, $d(t)$ in terms of the components $u_1(t)$, $u_2(t)$, $v_1(t)$, and $v_2(t)$.

11. A Linear Combination of Inputs Results in the Same Linear Combination of the Corresponding Outputs.
(a) Suppose the IVPs

$$\mathbf{x}' = \mathbf{A}\mathbf{x}; \quad \mathbf{x}(a) = \mathbf{b} \tag{11.1a}$$

and

$$\mathbf{x}' = \mathbf{A}\mathbf{x}; \quad \mathbf{x}(a) = \mathbf{c} \tag{11.1b}$$

have solutions $\mathbf{u}(t)$ and $\mathbf{v}(t)$, respectively. Then show that the IVP

$$\mathbf{x}' = \mathbf{A}\mathbf{x}; \quad \mathbf{x}(a) = \alpha\mathbf{b} + \beta\mathbf{c} \tag{11.2}$$

has the solution $\mathbf{x}(t) = \alpha\mathbf{u}(t) + \beta\mathbf{v}(t)$.

(b) Now consider inputs in the form of forcing functions. If

$$\mathbf{x}' = \mathbf{A}\mathbf{x} + \mathbf{f}_1(t) \tag{11.3a}$$

and

$$\mathbf{x}' = \mathbf{A}\mathbf{x} + \mathbf{f}_2(t) \tag{11.3b}$$

have particular solutions $\mathbf{x}_{p1}(t)$ and $\mathbf{x}_{p2}(t)$, respectively, show that

$$\mathbf{x}' = \mathbf{A}\mathbf{x} + \alpha\mathbf{f}_1(t) + \beta\mathbf{f}_2(t) \tag{11.4}$$

has a particular solution $\mathbf{x}(t) = \alpha\mathbf{x}_{p1}(t) + \beta\mathbf{x}_{p2}(t)$.

(c) Suppose $\mathbf{x}' = \mathbf{A}\mathbf{x}$, where $\mathbf{A}$ is a certain constant 2×2 matrix, has solutions $\mathbf{u}(t)$ and $\mathbf{v}(t)$ corresponding to the initial conditions $\mathbf{x}(0) = [1, 0]^{\mathrm{T}}$ and $\mathbf{x}(0) = [0, 1]^{\mathrm{T}}$, respectively. Then, what is the solution corresponding to an initial condition $\mathbf{x}(0) = [12, -6]^{\mathrm{T}}$?

(d) Suppose, for a certain constant 2×2 matrix $\mathbf{A}$, that the problem $\mathbf{x}' = \mathbf{A}\mathbf{x} + [\sin t, 0]^{\mathrm{T}}$ has a particular solution $\mathbf{x}_p(t) = [-(3\cos t + 4\sin t)/10, (2\cos t + \sin t)/10]^{\mathrm{T}}$ and that $\mathbf{x}' = \mathbf{A}\mathbf{x} + [0, e^t]^{\mathrm{T}}$ has a particular solution $\mathbf{x}_p(t) = [-(1 + t)e^t/2, te^t/2]^{\mathrm{T}}$. Then, find a particular solution if $\mathbf{x}' = \mathbf{A}\mathbf{x} + [100\sin t, -8e^t]^{\mathrm{T}}$.

4.6 MATRIX EIGENVALUE PROBLEM

We will need more linear algebra, the "eigenvalue problem" in particular, to continue our study of systems of differential equations in the next section, and also in Chapter 7 on the phase plane.

4.6.1 The eigenvalue problem. By the eigenvalue problem we mean an algebraic problem of the form

$$\boxed{\mathbf{Ax} = \lambda \mathbf{x},} \tag{1}$$

This is an algebraic equation, not a differential equation.

in which $\mathbf{A}$ is a given $n \times n$ matrix, $\mathbf{x}$ is an unknown n-tuple vector, and λ is an unknown scalar. Perhaps we should call (1) a "matrix eigenvalue problem" because eigenvalue problems arise in other formats as well, such as the differential equation eigenvalue problems in Section 3.6. In this chapter, we are concerned exclusively with the matrix eigenvalue problem (1).

We can get (1) into the form $\mathbf{Ax} = \mathbf{c}$ that was already studied in Section 4.3, as follows. Recall that if $\mathbf{I}$ is an $n \times n$ identity matrix then $\mathbf{Ix} = \mathbf{x}$. Thus, if we replace the $\mathbf{x}$ in (1) by $\mathbf{Ix}$, and move that term to the left, we can write

$$\boxed{(\mathbf{A} - \lambda \mathbf{I})\mathbf{x} = \mathbf{0}.} \tag{2}$$

If we don't insert the $\mathbf{I}$ in the right-hand side of (1), and write $(\mathbf{A} - \lambda)\mathbf{x} = \mathbf{0}$ instead, then the latter is incorrect because the matrix $\mathbf{A}$ minus the scalar λ is undefined; we can subtract only an $n \times n$ matrix from an $n \times n$ matrix.

To be sure the foregoing steps are clear, let us write out (1) and (2) in scalar form for $n = 3$, for instance. Then (1) is the system

$$\begin{aligned} a_{11}x_1 + a_{12}x_2 + a_{13}x_3 &= \lambda x_1, \\ a_{21}x_1 + a_{22}x_2 + a_{23}x_3 &= \lambda x_2, \\ a_{31}x_1 + a_{32}x_2 + a_{33}x_3 &= \lambda x_3, \end{aligned}$$

and if we subtract the terms on the right from those on the left, we obtain

$$\begin{aligned} (a_{11} - \lambda)x_1 + \quad a_{12}x_2 + \quad a_{13}x_3 &= 0, \\ a_{21}x_1 + (a_{22} - \lambda)x_2 + \quad a_{23}x_3 &= 0, \\ a_{31}x_1 + \quad a_{32}x_2 + (a_{33} - \lambda)x_3 &= 0, \end{aligned}$$

which, in matrix form, is (2). That is, $\mathbf{A} - \lambda\mathbf{I}$ in (2) means the matrix $\mathbf{A}$ with λ subtracted from each element on its main diagonal.

Since (2) is homogeneous, it surely admits the trivial solution $\mathbf{x} = \mathbf{0}$. However, the point of an eigenvalue problem is to find *nontrivial* solutions for $\mathbf{x}$. Thus, the eigenvalue problem can be stated as follows: *Given an $n \times n$ matrix* $\mathbf{A}$*, find values of the parameter λ such that (2) admits nontrivial solutions for* $\mathbf{x}$*; then, find those corresponding nontrivial solutions.* The λ's that lead to nontrivial solutions are called the **eigenvalues** of $\mathbf{A}$, and the corresponding nontrivial solutions for $\mathbf{x}$ are called the **eigenvectors** of $\mathbf{A}$. Thus, the unknowns in (2) [or, equivalently, in (1)] are not only the vector $\mathbf{x}$, but the scalar λ as well.

Statement of the eigenvalue problem for a given matrix $\mathbf{A}$.

The eigenvalue problem occurs in a wide variety of applications such as vibration theory, chemical kinetics, stability of equilibria, buckling of structures, and the convergence of iterative techniques, but out interest in it in this text is due to its application to the solution of systems of coupled ordinary differential equations.

4.6.2 Solving an eigenvalue problem.

As mentioned above, the problem (2) is of the type "$\mathbf{Ax} = \mathbf{c}$" that was the subject of Theorem 4.3.1, where the coefficient matrix "$\mathbf{A}$" in (2) is $\mathbf{A} - \lambda\mathbf{I}$ and the "$\mathbf{c}$" is $\mathbf{0}$. It follows from part (b) of that theorem that (2) admits nontrivial solutions (in addition to the obvious trivial solution $\mathbf{x} = \mathbf{0}$) if and only if

$$\boxed{\det(\mathbf{A} - \lambda\mathbf{I}) = 0.} \tag{3}$$

Expansion of the determinant in (3) gives an nth-degree polynomial in λ, so (3) is an nth-degree polynomial equation, known as the **characteristic equation** corresponding to the matrix $\mathbf{A}$. The latter has exactly n roots in the complex plane and these may include repeated roots. Thus, it follows that there is *at least one* eigenvalue λ, and *at most* n distinct eigenvalues λ, corresponding to any given $n \times n$ matrix $\mathbf{A}$.

The first step in solving the eigenvalue problem (1) is to expand the determinant in (3) and to solve the resulting nth-degree characteristic equation for the eigenvalues λ. Having done that, number the distinct eigenvalues as $\lambda_1, \dots, \lambda_k$; if there are no repeated roots then $k = n$, and if there are repeated roots then $k < n$.

We don't solve an eigenvalue problem for $\mathbf{x}$ and λ simultaneously. Rather, first solve (3) for the λ's; then, for each λ thus found, solve (2) for $\mathbf{x}$.

Next, set $\lambda = \lambda_1$ in (2). Since $\det(\mathbf{A} - \lambda_1\mathbf{I}) = 0$, $(\mathbf{A} - \lambda_1\mathbf{I})\mathbf{x} = \mathbf{0}$ is guaranteed to have nontrivial solutions, namely, the eigenvectors corresponding to λ_1, which we now wish to find. We can find them by Gauss elimination (Appendix C) and, using standard notation, we designate them as $\mathbf{e}_1$ instead of $\mathbf{x}$, e being the first letter of the word eigenvector. The set of all solutions $\mathbf{e}_1$ of $(\mathbf{A} - \lambda_1\mathbf{I})\mathbf{x} = \mathbf{0}$ is called the **eigenspace** corresponding to the eigenvalue λ_1. Next, set $\lambda = \lambda_2$ and find its eigenspace, and so on, until the k eigenspaces have been found.

We will summarize those steps and then give examples.

1. To guarantee nontrivial solutions of (2), set $\det(\mathbf{A} - \lambda\mathbf{I}) = 0$.

2. Expand the determinant to obtain the nth-degree characteristic equation for λ.

3. Solve the characteristic equation for the roots λ (i.e., for the eigenvalues of $\mathbf{A}$), using computer software if necessary, and number them.

4. Set $\lambda = \lambda_1$ in (2), and solve (2) for $\mathbf{x}$ by Gauss elimination. We denote those solutions as $\mathbf{e}_1$, and call the set of all such solutions the **eigenspace** corresponding to the eigenvalue λ_1.

5. Repeat the process for λ_2, λ_3, and so on, in turn.

EXAMPLE 1. Solving an Eigenvalue Problem. Determine all eigenvalues and eigenspaces of

$$\mathbf{A} = \begin{bmatrix} 3 & 4 \\ 2 & 1 \end{bmatrix}. \tag{4}$$

The characteristic equation for this $\mathbf{A}$ is

$$\det(\mathbf{A} - \lambda\mathbf{I}) = \begin{vmatrix} 3-\lambda & 4 \\ 2 & 1-\lambda \end{vmatrix} = \lambda^2 - 4\lambda - 5$$
$$= (\lambda - 5)(\lambda + 1) = 0 \tag{5}$$

Here, $\mathbf{A}$ is 2×2 so its characteristic equation is of second degree.

so the eigenvalues of $\mathbf{A}$ are $\lambda_1 = 5$ and $\lambda_2 = -1$ or vice versa, the order being immaterial.

With the eigenvalues determined, find the corresponding eigenspaces from (2):

$\lambda_1 = 5$: Then $(\mathbf{A} - \lambda_1\mathbf{I})\mathbf{x} = \mathbf{0}$ becomes

$$\begin{bmatrix} 3-5 & 4 \\ 2 & 1-5 \end{bmatrix}\begin{bmatrix} x_1 \\ x_2 \end{bmatrix} = \begin{bmatrix} -2 & 4 \\ 2 & -4 \end{bmatrix}\begin{bmatrix} x_1 \\ x_2 \end{bmatrix} = \begin{bmatrix} 0 \\ 0 \end{bmatrix}, \tag{6}$$

Gauss elimination of which gives (in the "augmented matrix form" discussed in Appendix C)

$$\begin{bmatrix} -2 & 4 & 0 \\ 2 & -4 & 0 \end{bmatrix} \to \begin{bmatrix} -2 & 4 & 0 \\ 0 & 0 & 0 \end{bmatrix}. \tag{7}$$

It follows from (7) that $x_2 = \alpha$ (arbitrary) and $x_1 = 2\alpha$, so the eigenspace corresponding to λ_1 is

$$\mathbf{e}_1 = \begin{bmatrix} 2\alpha \\ \alpha \end{bmatrix} = \alpha\begin{bmatrix} 2 \\ 1 \end{bmatrix}. \tag{8}$$

Since α is arbitrary, the set of vectors given by (8) is the set of all scalar multiples of $[2, 1]^{\mathrm{T}}$, namely, the set of all vectors on the line S_1 in Fig. 1: the line S_1 in Fig. 1 is the eigenspace corresponding to the eigenvalue $\lambda_1 = 5$.

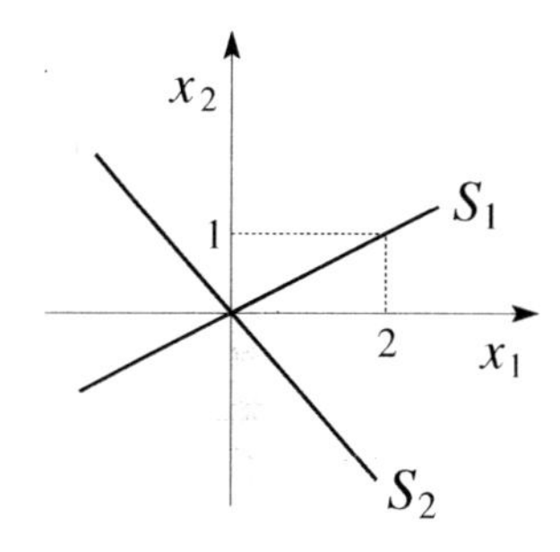

Figure 1. The eigenspaces S_1, S_2.

$\lambda_2 = -1$: Then $(\mathbf{A} - \lambda_2\mathbf{I})\mathbf{x} = \mathbf{0}$ becomes

$$\begin{bmatrix} 3+1 & 4 \\ 2 & 1+1 \end{bmatrix}\begin{bmatrix} x_1 \\ x_2 \end{bmatrix} = \begin{bmatrix} 4 & 4 \\ 2 & 2 \end{bmatrix}\begin{bmatrix} x_1 \\ x_2 \end{bmatrix} = \begin{bmatrix} 0 \\ 0 \end{bmatrix}, \tag{9}$$

Gauss elimination of which gives

$$\begin{bmatrix} 4 & 4 & 0 \\ 2 & 2 & 0 \end{bmatrix} \to \begin{bmatrix} 4 & 4 & 0 \\ 0 & 0 & 0 \end{bmatrix}. \tag{10}$$

The solution of (10) is $x_2 = \beta$ (arbitrary) and $x_1 = -\beta$, so

$$\mathbf{e}_2 = \begin{bmatrix} -\beta \\ \beta \end{bmatrix} = \beta\begin{bmatrix} -1 \\ 1 \end{bmatrix}. \tag{11}$$

Since β is arbitrary, the set of vectors given by (11) is the set of all scalar multiples of $[-1, 1]^{\mathrm{T}}$, namely, the set of all vectors on the line S_2 in Fig. 1: The line S_2 in Fig. 1 is the eigenspace corresponding to the eigenvalue $\lambda_2 = -1$.

COMMENT 1. Let us be more careful about the arbitrariness of α and β, stated above: Eigenvectors are always to be nontrivial (i.e., nonzero), so the *eigenvectors* of $\mathbf{A}$ are given

Distinguishing between the set of eigenvectors and the eigenspace.

by (8) and (11) with α and β arbitrary *but nonzero*. However, the *eigenspace* of a matrix $\mathbf{A}$ corresponding to a given eigenvalue means the set of all eigenvectors corresponding to that eigenvalue, plus the zero vector. Thus, the eigenspace S_1 shown in the figure is the line *including* the point at the origin, and similarly for S_2.

COMMENT 2. Eigenvectors are always nonunique. After all, if $\mathbf{e}$ satisfies $\mathbf{Ax} = \lambda\mathbf{x}$, then so does any scalar multiple of $\mathbf{e}$; hence the arbitrary scale factors α and β in (8) and (11).

COMMENT 3. It would have been equally correct to express $\mathbf{e}_2$ as $\beta\,[1,-1]^{\mathrm{T}}$, if you prefer, because the difference between the latter and (11) is a factor of -1, which can be absorbed by the arbitrary scale factor β.

Checking your eigenvalues and eigenvectors.

COMMENT 4. You can, and should, check your results. For instance, for the pair $\lambda_1 = 5$ and $\mathbf{e}_1$ given by (8), work out $\mathbf{Ae}_1$ and see if you get 5 times $\mathbf{e}_1$, as you should, because the eigenvalue problem is $\mathbf{Ax} = \lambda\mathbf{x}$. In doing so, there's no harm in taking $\alpha = 1$, for instance, in (8). ▮

EXAMPLE 2. Determine the eigenvalues and eigenspaces of

$$\mathbf{A} = \begin{bmatrix} 2 & 2 & 1 \\ 1 & 3 & 1 \\ 1 & 2 & 2 \end{bmatrix}. \tag{12}$$

Here $\mathbf{A}$ is 3×3 and the characteristic equation is of third degree. More generally, an $n \times n$ matrix $\mathbf{A}$ has an nth-degree characteristic equation. We've made up these examples to have integer roots of the characteristic equation, just for convenience.

The characteristic equation is

$$\det(\mathbf{A}-\lambda\mathbf{I}) = \begin{vmatrix} 2-\lambda & 2 & 1 \\ 1 & 3-\lambda & 1 \\ 1 & 2 & 2-\lambda \end{vmatrix} = -\lambda^3 + 7\lambda^2 - 11\lambda + 5$$
$$= -(\lambda - 5)(\lambda - 1)^2 = 0, \tag{13}$$

so the eigenvalues of $\mathbf{A}$ are $\lambda_1 = 5$ and $\lambda_2 = 1$, with $\lambda_2 = 1$ a **repeated eigenvalue**. A repeated eigenvalue λ_j is an eigenvalue of **multiplicity** k if the factor $\lambda - \lambda_j$ in the characteristic equation occurs to the kth power. In this example we see from (13) that the repeated eigenvalue $\lambda_2 = 1$ is of multiplicity 2.

Next, find the eigenspaces.

$\lambda_1 = 5$: Then $(\mathbf{A}-\lambda_1\mathbf{I})\mathbf{x} = \mathbf{0}$ becomes

$$\begin{bmatrix} 2-5 & 2 & 1 \\ 1 & 3-5 & 1 \\ 1 & 2 & 2-5 \end{bmatrix}\begin{bmatrix} x_1 \\ x_2 \\ x_3 \end{bmatrix} = \begin{bmatrix} -3 & 2 & 1 \\ 1 & -2 & 1 \\ 1 & 2 & -3 \end{bmatrix}\begin{bmatrix} x_1 \\ x_2 \\ x_3 \end{bmatrix} = \begin{bmatrix} 0 \\ 0 \\ 0 \end{bmatrix}, \tag{14}$$

Gauss elimination of which gives

$$\begin{bmatrix} -3 & 2 & 1 & 0 \\ 0 & 1 & -1 & 0 \\ 0 & 0 & 0 & 0 \end{bmatrix}. \tag{15}$$

From the eliminated form (15), the solution is $x_3 = \alpha$ (arbitrary), $x_2 = \alpha$, $x_1 = \alpha$, so

$$\mathbf{e}_1 = \begin{bmatrix} \alpha \\ \alpha \\ \alpha \end{bmatrix} = \alpha\begin{bmatrix} 1 \\ 1 \\ 1 \end{bmatrix}. \tag{16}$$

Thus, the eigenspace corresponding to $\lambda_1 = 5$ is the set of all scalar multiples of $[1, 1, 1]^{\mathrm{T}}$, namely, the line S_1 through the origin shown in Fig. 2a.

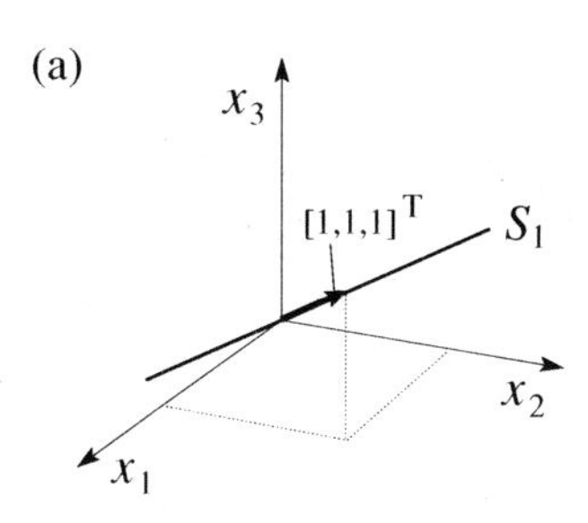

$\lambda_2 = 1$: Then $(\mathbf{A} - \lambda_2 \mathbf{I})\mathbf{x} = \mathbf{0}$ becomes

$$\begin{bmatrix} 2-1 & 2 & 1 \\ 1 & 3-1 & 1 \\ 1 & 2 & 2-1 \end{bmatrix} \begin{bmatrix} x_1 \\ x_2 \\ x_3 \end{bmatrix} = \begin{bmatrix} 1 & 2 & 1 \\ 1 & 2 & 1 \\ 1 & 2 & 1 \end{bmatrix} \begin{bmatrix} x_1 \\ x_2 \\ x_3 \end{bmatrix} = \begin{bmatrix} 0 \\ 0 \\ 0 \end{bmatrix}, \tag{17}$$

Gauss elimination of which gives

$$\begin{bmatrix} 1 & 2 & 1 & 0 \\ 1 & 2 & 1 & 0 \\ 1 & 2 & 1 & 0 \end{bmatrix} \rightarrow \begin{bmatrix} 1 & 2 & 1 & 0 \\ 0 & 0 & 0 & 0 \\ 0 & 0 & 0 & 0 \end{bmatrix}. \tag{18}$$

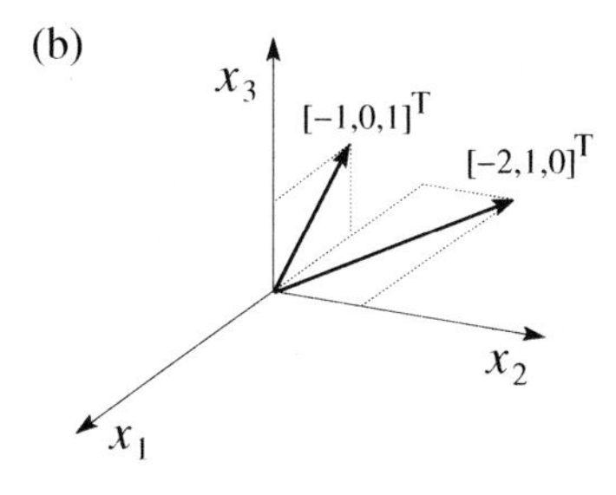

The solution is $x_3 = \beta$ (arbitrary), $x_2 = \gamma$ (arbitrary), $x_1 = -\beta - 2\gamma$, so

$$\mathbf{e}_2 = \begin{bmatrix} -\beta - 2\gamma \\ \gamma \\ \beta \end{bmatrix} = \beta \begin{bmatrix} -1 \\ 0 \\ 1 \end{bmatrix} + \gamma \begin{bmatrix} -2 \\ 1 \\ 0 \end{bmatrix}. \tag{19}$$

Figure 2. (a) The eigenspace corresponding to $\lambda_1 = 5$ is the *line* S_1. (b) The eigenspace corresponding to $\lambda_2 = 1$ is the *plane* through the vectors $[-1, 0, 1]^{\mathrm{T}}$ and $[-2, 1, 0]^{\mathrm{T}}$.

The vectors $[-1, 0, 1]^{\mathrm{T}}$ and $[-2, 1, 0]^{\mathrm{T}}$ in (19) are sketched in Fig. 2b. Since the eigenspace corresponding to $\lambda_2 = 1$ consists of all linear combinations of them, that eigenspace is the infinite *plane* (not shown in Fig. 2b) containing them. In fact, the top scalar equation in the right-hand side of (18), $x_1 + 2x_2 + x_3 = 0$, is the equation of that plane!

COMMENT 1. The number of LI vectors in an eigenspace is the **dimension** of that eigenspace. Since every vector in the eigenspace corresonding to the eigenvalue $\lambda_1 = 5$ is a multiple of $[1, 1, 1]^{\mathrm{T}}$, there is only one LI vector in that eigenspace, so the latter is one-dimensional. And since every vector in the eigenspace corresponding to the eigenvalue $\lambda_2 = 1$ is a linear combination of the two LI vectors on the right-hand side of (19), that eigenspace is two-dimensional. These dimensions should not be surprising since, after all, the eigenspace (16) is the one-dimensional line S_1 in Fig. 2a, and the eigenspace (19) is the two-dimensional plane through the two vectors shown in Fig. 2b.

The number of LI vectors in an eigenspace is the dimension of that eigenspace.

COMMENT 2. Notice that when we are done with the Gauss elimination in (18) there are two rows of zeros at the bottom of the array. At this point in *every* eigenvalue problem there *must be at least one row of zeros at the bottom*, for that is the result of choosing λ so that nontrivial solutions are obtained for $\boldsymbol{x}$. If you do *not* obtain at least one row of zeros you have made a mistake: your λ is incorrect or your Gauss elimination steps are incorrect, or both. Notice the row of zeros in (7) as well. ▮

Eigenspaces will always be arbitrary linear combinations of one or more LI vectors. In Examples 1 and 2 the number of such LI vectors happened to be the same as the multiplicity of the corresponding eigenvalue,[1] but it can be less, as occurs in some of the exercises.

---

[1] In Example 2, for instance, $\lambda_1 = 5$ is of multiplicity 1 and its eigenspace contains one LI vector [see (16)]; $\lambda_2 = 1$ is of multiplicity 2 and its eigenspace contains two LI vectors, as we see in (19).

4.6.3 Complex eigenvalues and eigenvectors.

Although we've considered only real matrices (i.e., with real elements), the eigenvalues of a real matrix $\mathbf{A}$ can be complex. After all, it's true that if $\mathbf{A}$ is real, then the coefficients in the characteristic equation $\det(\mathbf{A}-\lambda\mathbf{I})=0$ will be real, but it is possible for a polynomial equation with real coefficients to have complex roots. For instance, $x^2-4x+5=0$ has the roots $2+i$ and $2-i$. Furthermore, if an eigenvalue λ is complex, we can expect its eigenvectors to be complex as well because we obtain them by solving the equation $(\mathbf{A}-\lambda\mathbf{I})\mathbf{x}=\mathbf{0}$, which contains the complex number λ.

However, when we introduced n-space in Section 4.3 we restricted all components of the vectors and the scalar multipliers of vectors to be real. Now, we drop that restriction and permit those numbers to be complex. We begin with an example of the complex arithmetic that arises.

The set of all n-tuple vectors, with vector addition, scalar multiplication, negative inverse, and zero vector defined as in Section 4.3, is called $\mathbb{R}^n$ (real n-space) if all vector components and scalar multipliers are real. If they are permitted to be complex, the set is called $\mathbb{C}^n$ (complex n-space).

EXAMPLE 3. Complex Vector Arithmetic. Consider the column vectors $\mathbf{u}=[2-i, 1+5i, 4]^{\mathrm{T}}$ and $\mathbf{v}=[3i, 6, -1]^{\mathrm{T}}$. Work out the vector sum $\mathbf{u}+\mathbf{v}$ and the scalar multiple $(1-2i)\mathbf{u}$:

$$\mathbf{u}+\mathbf{v}=\begin{bmatrix}2-i\\1+5i\\4\end{bmatrix}+\begin{bmatrix}3i\\6\\-1\end{bmatrix}=\begin{bmatrix}(2-i)+(3i)\\(1+5i)+(6)\\(4)+(-1)\end{bmatrix}=\begin{bmatrix}2+2i\\7+5i\\3\end{bmatrix} \tag{20}$$

and

$$(1-2i)\mathbf{u}=(1-2i)\begin{bmatrix}2-i\\1+5i\\4\end{bmatrix}=\begin{bmatrix}(1-2i)(2-i)\\(1-2i)(1+5i)\\(1-2i)(4)\end{bmatrix}=\begin{bmatrix}-5i\\11+3i\\4-8i\end{bmatrix}, \tag{21}$$

respectively. ∎

We conclude with an example of an eigenvalue problem with complex eigenvalues.

EXAMPLE 4. Complex Eigenvalues. The matrix

$$\mathbf{A}=\begin{bmatrix}1&-4\\1&1\end{bmatrix} \tag{22}$$

has the characteristic equation

$$\det(\mathbf{A}-\lambda\mathbf{I})=\begin{vmatrix}1-\lambda&-4\\1&1-\lambda\end{vmatrix}=\lambda^2-2\lambda+5=0, \tag{23}$$

which gives $\lambda_1=1+2i$ and $\lambda_2=1-2i$. To find the eigenspace corresponding to $\lambda_1=1+2i$, we must find the nontrivial solutions of $(\mathbf{A}-\lambda_1\mathbf{I})\mathbf{x}=\mathbf{0}$, namely,

$$\begin{bmatrix}-2i&-4\\1&-2i\end{bmatrix}\begin{bmatrix}x_1\\x_2\end{bmatrix}=\begin{bmatrix}0\\0\end{bmatrix}. \tag{24}$$

Rather than solve (24) by Gauss elimination as usual, we can minimize any complex arithmetic by proceeding as follows: Denoting the matrix in (24) as $\begin{bmatrix} a & b \\ c & d \end{bmatrix}$, the two scalar equations are $ax_1 + bx_2 = 0$ and $cx_1 + dx_2 = 0$. Because (by our choice of λ) we've arranged for the determinant $ad - bc$ to be zero, for (24) to have nontrivial solutions, it follows that the two scalar equations are redundant; one is simply a multiple of the other, since a is to c as b is to d. Thus, we can discard either of the two equations, for instance the first, without loss. That leaves the second scalar equation in (24),

This idea, discarding one of the two equations, applies only for 2×2 matrices.

$$x_1 - 2ix_2 = 0,$$

which is one equation in the two unknowns x_1 and x_2, one of which is therefore arbitrary. With $x_2 = \alpha$, an arbitrary constant, we obtain $x_1 = 2i\alpha$ and $x_2 = \alpha$, so

$$\mathbf{e}_1 = \begin{bmatrix} 2i\alpha \\ \alpha \end{bmatrix} = \alpha \begin{bmatrix} 2i \\ 1 \end{bmatrix}. \tag{25}$$

To check the calculations, verify that $\mathbf{Ae}_1$ does equal $\lambda_1\mathbf{e}_1$. Similarly for $\mathbf{e}_2$, below. Remember that eigenvectors can be scaled at your pleasure. If we scale (25) by $-i$, for instance, we obtain $\mathbf{e}_1 = \alpha[2, -i]^{\mathrm{T}}$, which is *equivalent* to the form in (25), and equally correct.

To find the other eigenspace, corresponding to $\lambda_2 = 1{-}2i$, we can proceed in the same manner, but there is a short cut. Because λ_1 and λ_2 are complex conjugates, the system $(\mathbf{A} - \lambda_2\mathbf{I})\mathbf{x} = \mathbf{0}$ will be the same as (24), but with all i's changed to $-i$'s, namely,

$$\begin{bmatrix} 2i & -4 \\ 1 & 2i \end{bmatrix} \begin{bmatrix} x_1 \\ x_2 \end{bmatrix} = \begin{bmatrix} 0 \\ 0 \end{bmatrix}. \tag{26}$$

Thus, its solution will be the same as (25), but with all i's changed to $-i$'s. Thus,

$$\mathbf{e}_2 = \beta \begin{bmatrix} -2i \\ 1 \end{bmatrix}. \ \blacksquare \tag{27}$$

COMMENT. To generalize the foregoing idea, suppose λ is an eigenvalue of $\mathbf{A}$, with eigenvector $\mathbf{e}$, so $\mathbf{Ae} = \lambda\mathbf{e}$. Take the complex conjugate of both sides: On the left, $\overline{\mathbf{Ae}} = \overline{\mathbf{A}}\overline{\mathbf{e}} = \mathbf{A}\overline{\mathbf{e}}$, the last step following because we are considering only real matrices $\mathbf{A}$. And on the right, $\overline{\lambda\mathbf{e}} = \overline{\lambda}\overline{\mathbf{e}}$. Thus, $\mathbf{A}\overline{\mathbf{e}} = \overline{\lambda}\overline{\mathbf{e}}$, so it follows that if λ is an eigenvalue, with eigenvector $\mathbf{e}$, then $\overline{\lambda}$ is an eigenvalue as well, with eigenvector $\overline{\mathbf{e}}$. Of course if λ is real, then we haven't learned anything, but if it is complex, then its eigenvectors are simply the complex conjugates of the eigenvectors of λ.

Closure. The eigenvalue problem consists of the search for *non*trivial solutions $\mathbf{x}$ of the system $\mathbf{Ax} = \lambda\mathbf{x}$ or, equivalently, $(\mathbf{A} - \lambda\mathbf{I})\mathbf{x} = \mathbf{0}$. Not only is $\mathbf{x}$ unknown, but so is the scalar parameter λ. The idea is to set $\det(\mathbf{A} - \lambda\mathbf{I}) = 0$ because, according to Theorem 4.3.1, that condition guarantees the existence of nontrivial solutions of $(\mathbf{A} - \lambda\mathbf{I})\mathbf{x} = \mathbf{0}$. By expanding the determinant, obtain the characteristic equation, which is an nth-degree polynomial equation in λ. Solve that equation for the eigenvalues $\lambda_1, \dots, \lambda_k$, where k will be at least 1 and at most n; in each of Examples 1 and 2, k happened to equal n. Then, setting the λ in $(\mathbf{A} - \lambda\mathbf{I})\mathbf{x} = \mathbf{0}$ equal to λ_1 solve that system by Gauss elimination for the nontrivial solutions that are guaranteed to exist; instead of calling the solutions $\mathbf{x}$, it is traditional to call them $\mathbf{e}_1$ since they are the eigenvectors corresponding to the eigenvalue λ_1. Do that for $\lambda_1, \ldots, \lambda_k$, in turn.

By the eigenspace corresponding to a particular eigenvalue, we mean all of the corresponding eigenvectors together with the trivial solution $\mathbf{0}$, which is not itself an eigenvector.

Note that although both unknowns, $\mathbf{x}$ and λ, appear in the original equation $\mathbf{Ax} = \lambda\mathbf{x}$, we can solve for them separately: first, solve the characteristic equation for the λ's, then put them into $(\mathbf{A} - \lambda\mathbf{I})\mathbf{x} = \mathbf{0}$, in turn, and solve that equation by Gauss elimination for the corresponding eigenvectors.

An eigenvalue can be zero; it is the eigen*vectors* that must be nontrivial. If you obtain only the trivial solution $\mathbf{x} = \mathbf{0}$ when you solve $(\mathbf{A} - \lambda\mathbf{I})\mathbf{x} = \mathbf{0}$, you've made a mistake, either in the calculation of the eigenvalue λ or in the Gauss elimination solution of $(\mathbf{A} - \lambda\mathbf{I})\mathbf{x} = \mathbf{0}$, because if λ is indeed an eigenvalue, then $\det(\mathbf{A} - \lambda\mathbf{I}) = 0$ *guarantees* the existence of nontrivial solutions for $\mathbf{x}$, that is, for $\mathbf{e}$.

EXERCISES 4.6

1. Find the eigenvalues and eigenspaces for the given matrix. If the multiplicity of an eigenvalue is greater than one, then state its multiplicity. Also, state the dimension of each eigenspace.

(a) $\begin{bmatrix} 1 & -3 \\ 0 & 0 \end{bmatrix}$ (b) $\begin{bmatrix} 0 & 0 \\ 0 & 0 \end{bmatrix}$

(c) $\begin{bmatrix} 4 & 1 \\ 2 & 3 \end{bmatrix}$ (d) $\begin{bmatrix} 3 & -1 \\ 1 & 1 \end{bmatrix}$

(e) $\begin{bmatrix} 2 & 3 \\ 2 & 7 \end{bmatrix}$ (f) $\begin{bmatrix} 20 & 9 \\ -35 & -16 \end{bmatrix}$

(g) $\begin{bmatrix} -4 & 3 \\ 1 & -2 \end{bmatrix}$ (h) $\begin{bmatrix} 1 & 2 \\ -2 & 5 \end{bmatrix}$

(i) $\begin{bmatrix} 3 & 3 \\ 3 & 3 \end{bmatrix}$ (j) $\begin{bmatrix} -2 & 2 \\ 3 & 3 \end{bmatrix}$

(k) $\begin{bmatrix} 0 & 1 & -1 \\ -5 & 4 & 1 \\ 1 & 0 & -1 \end{bmatrix}$ (l) $\begin{bmatrix} 4 & 1 & 0 \\ -5 & 0 & 1 \\ 1 & -1 & -1 \end{bmatrix}$

(m) $\begin{bmatrix} 3 & 6 & 9 \\ 0 & 4 & 8 \\ 0 & 0 & 4 \end{bmatrix}$ (n) $\begin{bmatrix} 4 & 4 & 4 \\ 4 & 4 & 4 \\ 4 & 4 & 4 \end{bmatrix}$

(o) $\begin{bmatrix} 1 & 0 & 2 \\ 1 & 0 & 2 \\ 1 & 0 & 2 \end{bmatrix}$ (p) $\begin{bmatrix} 0 & 0 & 1 \\ 0 & 0 & 1 \\ 1 & 1 & 1 \end{bmatrix}$

(q) $\begin{bmatrix} 1 & 0 & 1 \\ 0 & 2 & 0 \\ 4 & 0 & 4 \end{bmatrix}$ (r) $\begin{bmatrix} 0 & 0 & 3 \\ 0 & 0 & 1 \\ 1 & -13 & 7 \end{bmatrix}$

(s) $\begin{bmatrix} 2 & 0 & 2 \\ 2 & 2 & 0 \\ 0 & 0 & 2 \end{bmatrix}$ (t) $\begin{bmatrix} 1 & 0 & 0 \\ 0 & 2 & 0 \\ 0 & 0 & 3 \end{bmatrix}$

(u) $\begin{bmatrix} 1 & 1 & 1 & 1 \\ 0 & 1 & 1 & 1 \\ 0 & 0 & 1 & 1 \\ 0 & 0 & 0 & 1 \end{bmatrix}$ (v) $\begin{bmatrix} 0 & 0 & 0 & 6 \\ 0 & 0 & 0 & 0 \\ 0 & 0 & 0 & 0 \\ 0 & 0 & 0 & 0 \end{bmatrix}$

2. (m),(p),(s),(v) **Computer.** Solve the problem in the corresponding part of Exercise 1, this time using computer software.

3. Trace. For an $n \times n$ matrix $\mathbf{A}$, the sum of its diagonal elements $a_{11}, a_{22}, \ldots, a_{nn}$ is called the **trace** of $\mathbf{A}$ and is written as $\operatorname{tr}\mathbf{A}$: $\operatorname{tr}\mathbf{A} = a_{11} + a_{22} + \cdots + a_{nn}$. If $\mathbf{A}$ is 2×2, show that its chacteristic equation can be written as

$$\boxed{\lambda^2 + (\operatorname{tr}\mathbf{A})\lambda + (\det\mathbf{A}) = 0.} \tag{3.1}$$

NOTE: The latter provides a quick way to write down the characteristic equation for any 2×2 matrix.

4. Is the following vector $\mathbf{x}$ an eigenvector of $\mathbf{A}$? Explain. If so, what is the corresponding eigenvalue?

$$\mathbf{A} = \begin{bmatrix} 1 & 8 & 5 & 3 \\ 2 & 16 & 10 & 6 \\ 5 & -14 & -11 & -3 \\ -1 & -8 & -5 & -3 \end{bmatrix}.$$

(a) $[1,2,-1,3]^{\mathrm{T}}$ (b) $[1,2,-4,-1]^{\mathrm{T}}$ (c) $[1,2,1,1]^{\mathrm{T}}$
(d) $[1,0,1,-2]^{\mathrm{T}}$ (e) $[1,0,1,-1]^{\mathrm{T}}$ (f) $[1,1,0,-3]^{\mathrm{T}}$

5. The given matrix is known to have $\lambda = 2$ among its eigenvalues. Find the eigenspace corresponding to that eigenvalue, and state its dimension.

(a) $\begin{bmatrix} 3 & 2 & 2 & 1 \\ 2 & 3 & 1 & 2 \\ -1 & 1 & 2 & 0 \\ 2 & 4 & 3 & 5 \end{bmatrix}$ (b) $\begin{bmatrix} 3 & 1 & 2 & 1 \\ -1 & 3 & 1 & 2 \\ 0 & 2 & 5 & 3 \\ 1 & 3 & 5 & 6 \end{bmatrix}$

(c) $\begin{bmatrix} 3 & 1 & 1 & 1 \\ 1 & 3 & 1 & 1 \\ 1 & 1 & 3 & 1 \\ 1 & 1 & 1 & 3 \end{bmatrix}$ (d) $\begin{bmatrix} 3 & 0 & 1 & 1 \\ 1 & 3 & 0 & 1 \\ -1 & 1 & 3 & -1 \\ 1 & 2 & 2 & 3 \end{bmatrix}$

(e) $\begin{bmatrix} 2 & 0 & 0 & 0 \\ 3 & 2 & 0 & 0 \\ 0 & 4 & 2 & 0 \\ 1 & 0 & 5 & 1 \end{bmatrix}$ (f) $\begin{bmatrix} 0 & 0 & 0 & 2 \\ 0 & 0 & 1 & 0 \\ 0 & 4 & 0 & 0 \\ 2 & 0 & 0 & 0 \end{bmatrix}$

6. Determine values of a, b, c, d so that

$$\mathbf{A} = \begin{bmatrix} 0 & 0 & 0 & a \\ 0 & 0 & b & b \\ 0 & c & 0 & 0 \\ d & 0 & 0 & 0 \end{bmatrix},$$

has 3 and -4 among its eigenvalues.

7. We saw in Example 2 that a given eigenvalue can have more than one LI eigenvector. Can a given eigenvector correspond to more than one eigenvalue? Explain.

8. Triangular Matrices. A square matrix is **upper triangular** if all its elements below the main diagonal are zero; it is **lower triangular** if all its elements above the main diagonal are zero; and it is **triangular** if it is either upper or lower triangular. Show that if $\mathbf{A}$ is triangular, then its eigenvalues are simply its diagonal elements, with the same multiplicity as their occurence on the main diagonal.

9. Show that the eigenvalues of $k\mathbf{A}$, for any scalar k, are k times those of $\mathbf{A}$. How do the eigenvectors corresponding to an eigenvalue λ of $\mathbf{A}$ compare with the eigenvectors corresponding to the eigenvalue $k\lambda$ of $k\mathbf{A}$? Explain.

10. Complex Eigenvalues and Eigenvectors. Solve for the eigenvalues and corresponding eigenspaces.

(a) $\begin{bmatrix} 0 & 4 \\ -1 & 0 \end{bmatrix}$ (b) $\begin{bmatrix} 2 & 2 \\ -2 & 1 \end{bmatrix}$

(c) $\begin{bmatrix} 3 & 2 \\ -2 & 3 \end{bmatrix}$ (d) $\begin{bmatrix} 1 & -3 \\ 3 & 1 \end{bmatrix}$

(e) $\begin{bmatrix} 0 & 0 & 1 \\ 0 & 0 & 0 \\ -1 & 5 & 0 \end{bmatrix}$ (f) $\begin{bmatrix} 2 & 2 & 1 \\ -2 & 0 & 0 \\ -1 & 0 & 0 \end{bmatrix}$

11. Symmetric Matrices Have Real Eigenvalues. If an $n \times n$ matrix $\mathbf{A} = \{a_{ij}\}$ has $a_{ij} = a_{ji}$ for each i and j, then it is **symmetric**. For instance, $\begin{bmatrix} 2 & 3 & 0 \\ 3 & 1 & -1 \\ 0 & -1 & 5 \end{bmatrix}$ is symmetric, while $\begin{bmatrix} 2 & 3 & 0 \\ 3 & 1 & -1 \\ 6 & -1 & 5 \end{bmatrix}$ is not. It can be shown that if a (real) matrix $\mathbf{A}$ is symmetric, then its eigenvalues are necessarily *real*. Prove that statement for the case where $n = 2$; that is, show that the eigenvalues of $\begin{bmatrix} a & b \\ b & c \end{bmatrix}$ are real for any (real) values of a, b, c.

ADDITIONAL EXERCISES

12. Population Exchange and Markov Matrices. Our interest in the eigenvalue problem, in this text, will be in connection with solving systems of differential equations, which we study in the next section. However, the eigenvalue problem occurs in a wide variety of applications, and our purpose in this exercise is to illustrate just one of these, in connection with population dynamics. Suppose there is a population exchange between the states of Delaware (DE), Maryland (MD), and Pennsylvania (PA) such that each year 20% of DE's residents move to MD and 8% to PA; 12% of MD's residents move to DE and 3% to PA; and 10% of PA's residents move to DE and 3% to MD, as indicated in the figure. Ignore gains due to births and losses due to deaths or, equivalently, as-

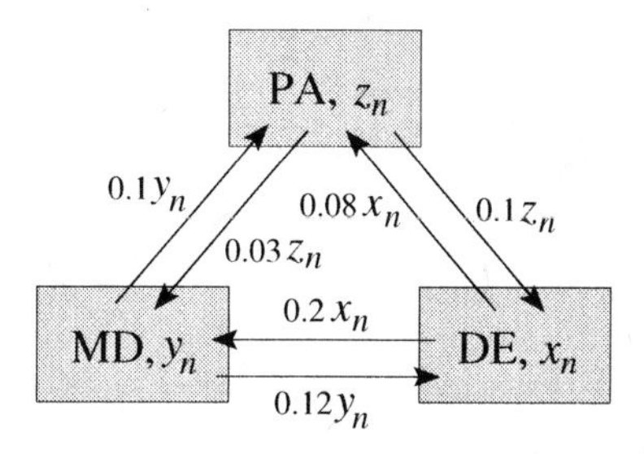

sume these effects are nonzero but cancel. If we denote the populations at the end of the nth year in DE, MD, and PA as x_n, y_n, and z_n, respectively, then it follows from the data given that

$$\begin{aligned} x_{n+1} &= x_n - (0.2 + 0.08)x_n + 0.12y_n + 0.1z_n, \\ y_{n+1} &= y_n + 0.2x_n - (0.12 + 0.1)y_n + 0.03z_n, \\ z_{n+1} &= z_n + 0.08x_n + 0.1y_n - (0.1 + 0.03)z_n, \end{aligned} \tag{12.1}$$

or, in matrix form,

$$\begin{bmatrix} x_{n+1} \\ y_{n+1} \\ z_{n+1} \end{bmatrix} = \begin{bmatrix} 0.72 & 0.12 & 0.1 \\ 0.2 & 0.78 & 0.03 \\ 0.08 & 0.1 & 0.87 \end{bmatrix} \begin{bmatrix} x_n \\ y_n \\ z_n \end{bmatrix} \tag{12.2}$$

or

$$\mathbf{p}_{n+1} = \mathbf{A}\mathbf{p}_n \tag{12.3}$$

in which $\mathbf{p}_n = [x_n, y_n, z_n]^{\mathrm{T}}$ is the "population vector" at the end of the nth year. (In this application the populations are not functions of a continuous time variable but of a discrete time variable, namely, the integer subscript n.) A fundamental question is whether $\mathbf{p}_n$ tends to a steady state, say $\mathbf{p}$, as $t \to \infty$. If there does exist such a steady state, then $\mathbf{p}_{n+1} = \mathbf{p}_n \equiv \mathbf{p}$, so (12.3) becomes

$$\mathbf{p} = \mathbf{A}\mathbf{p}. \tag{12.4}$$

Thus, the existence of any equlibrium populations amounts to the existence of nontrivial solutions of (12.4).

(a) Note that (12.4) can be viewed as an eigenvalue problem ($\mathbf{A}\mathbf{p} = \lambda\mathbf{p}$) in which the eigenvalue is 1. Thus, the question of whether or not the system (12.4) admits any steady states (i.e., other than the obvious one where $\mathbf{p} = \mathbf{0}$) is equivalent to the question of whether 1 is among the eigenvalues of $\mathbf{A}$. If it is, then the corresponding eigenvector is the steady-state population vector and its arbitrary scale factor can be determined from knowledge of the initial populations. Show that 1 is indeed among the eigenvalues of $\mathbf{A}$, and thus determine the steady-state population vector if the total population (which is not changing because there are no births or deaths and which must therefore equal the initial total population) is one million people.

(b) The next important question is this: Is that equilibrium state *stable*? After all, only if it is stable will $\mathbf{p}_n$ *tend* to $\mathbf{p}$ as $n \to \infty$. It can be shown to be stable, but here we ask only for a heuristic approach. Namely, beginning with any initial population vector, such as $\mathbf{p}_0 = [1, 0, 0]^{\mathrm{T}}$ (in which the 1 means one million), use (12.3) and computer software to determine $\mathbf{p}_n$ for a number of successive values of n. From your computer results, does $\mathbf{p}_n$ appear to be approaching $\mathbf{p}$ as $n \to \infty$?

(c) In fact, the matrix $\mathbf{A}$ in problems of this type are **Markov matrices**: $\mathbf{A} = \{a_{ij}\}$ is a Markov (or stochastic) matrix if $a_{ij} \geq 0$ for each i and j, and if the elements of each column, or each row, sum to unity. In this example, the columns sum to unity. Prove that claim for the case of a 2×2 Markov matrix,

$$\mathbf{A} = \begin{bmatrix} a & b \\ 1-a & 1-b \end{bmatrix}, \quad (0 \leq a \leq 1, 0 \leq b \leq 1) \tag{12.5}$$

namely, that 1 is among its eigenvalues.

4.7 HOMOGENEOUS SYSTEMS WITH CONSTANT COEFFICIENTS

In Sections 4.3–4.6 we laid the vector and matrix foundations that we will now use to solve systems of linear homogeneous differential equations with constant coefficients.

4.7.1 Solution by the method of assumed exponential form. We found, in Sections 4.1 and 4.2, that the solution of systems of linear constant-coefficient differential equations by elimination leads to higher-order constant-coefficient equations in a single dependent variable. Since such equations admit exponential solutions, it is reasonable to seek solutions of systems in that form as well. The following example will indicate the main ideas.

In this section, the interval will be $-\infty < t < \infty$ unless stated otherwise.

EXAMPLE 1. Obtain a general solution of the system

$$x_1' = 3x_1 + 4x_2, \tag{1a}$$
$$x_2' = 2x_1 + x_2. \tag{1b}$$

Seek solutions of (1) in the exponential form

$$x_1(t) = q_1 e^{rt}, \quad x_2(t) = q_2 e^{rt}, \tag{2}$$

in which the constants q_1, q_2, and r are to be determined so that (1a) and (1b) are satisfied. Putting (2) into (1) gives

$$rq_1 e^{rt} = 3q_1 e^{rt} + 4q_2 e^{rt}, \tag{3a}$$
$$rq_2 e^{rt} = 2q_1 e^{rt} + q_2 e^{rt}, \tag{3b}$$

or, canceling the e^{rt}'s,

$$3q_1 + 4q_2 = rq_1, \tag{4a}$$
$$2q_1 + q_2 = rq_2. \tag{4b}$$

The unknowns in (4) are q_1, q_2, and r.

Of course, (4) admits the trivial solution $q_1 = q_2 = 0$; but if both q_1 and q_2 are zero, then (2) is only the trivial solution $x_1(t) = x_2(t) = 0$, which is of no help toward finding a general solution of (1). Thus, we insist on nontrivial solutions of (4), for q_1 and q_2, in which case (4) is an *eigenvalue problem* $\mathbf{Aq} = r\mathbf{q}$, with $\mathbf{A} = \begin{bmatrix} 3 & 4 \\ 2 & 1 \end{bmatrix}$. The eigenvalues and eigenvectors of $\mathbf{A}$ are found to be

Here we have $\mathbf{q}$ in place of $\mathbf{x}$ and r in place of λ, but the letters used don't matter.

By a nontrivial solution of (4), we mean a solution in which q_1 and q_2 are not both zero.

$$\lambda_1 = 5, \; \mathbf{e}_1 = \alpha \begin{bmatrix} 2 \\ 1 \end{bmatrix}; \qquad \lambda_2 = -1, \; \mathbf{e}_2 = \beta \begin{bmatrix} -1 \\ 1 \end{bmatrix}. \tag{5}$$

Since r is "λ" and $\mathbf{q}$ is "$\mathbf{e}$" in $\mathbf{Aq} = r\mathbf{q}$, the eigenvalues and eigenvectors in (5) give the r's and $\mathbf{q}$'s, respectively, to use in (2). Thus, we have found the solutions

$$\mathbf{x}(t) = \begin{bmatrix} q_1 e^{rt} \\ q_2 e^{rt} \end{bmatrix} = \begin{bmatrix} 2\alpha e^{5t} \\ \alpha e^{5t} \end{bmatrix} = \alpha \begin{bmatrix} 2 \\ 1 \end{bmatrix} e^{5t} \tag{6}$$

from the first "eigenpair" [i.e., the λ_1 and $\mathbf{e}_1$ in (5)], and

$$\mathbf{x}(t) = \begin{bmatrix} q_1 e^{rt} \\ q_2 e^{rt} \end{bmatrix} = \begin{bmatrix} -\beta e^{-t} \\ \beta e^{-t} \end{bmatrix} = \beta \begin{bmatrix} -1 \\ 1 \end{bmatrix} e^{-t} \tag{7}$$

from the second eigenpair [λ_2 and $\mathbf{e}_2$ in (5)]. Since the system (1) is linear and homogeneous, an arbitrary linear combination of the solutions (6) and (7) will (according to Theorem 4.5.2) also be a solution. Thus,

$$\mathbf{x}(t) = \alpha \begin{bmatrix} 2 \\ 1 \end{bmatrix} e^{5t} + \beta \begin{bmatrix} -1 \\ 1 \end{bmatrix} e^{-t} \tag{8}$$

is also a solution of (1). The latter is of the form

The names of the arbitrary constants don't matter, but we are getting ready to appeal to Theorem 4.5.3, in which our usual c_j notation is used. Thus, we changed α, β to c_1, c_2.

$$\mathbf{x}(t) = c_1\mathbf{x}_1(t) + c_2\mathbf{x}_2(t), \tag{9}$$

in which we've simply renamed the arbitrary constants α and β as c_1 and c_2, respectively, and where

$$\mathbf{x}_1(t) = \begin{bmatrix} 2 \\ 1 \end{bmatrix} e^{5t} \text{ and } \mathbf{x}_2(t) = \begin{bmatrix} -1 \\ 1 \end{bmatrix} e^{-t}. \tag{10}$$

Further, by Theorem 4.5.4 $\mathbf{x}_1(t)$ and $\mathbf{x}_2(t)$ are LI on I because

$$\det[\mathbf{x}_1(t), \mathbf{x}_2(t)] = \begin{vmatrix} 2e^{5t} & -e^{-t} \\ e^{5t} & e^{-t} \end{vmatrix} = e^{5t}e^{-t} \begin{vmatrix} 2 & -1 \\ 1 & 1 \end{vmatrix} = 3e^{4t} \neq 0, \tag{11}$$

so it follows from Theorem 4.5.3(a) that (9) is a general solution of (1). ▮

This name is not standard.

We can organize the method of **assumed exponential form** that was used in Example 1 as follows. To solve

$$\boxed{\mathbf{x}' = \mathbf{A}\mathbf{x},} \tag{12}$$

in which $\mathbf{A}$ is an $n \times n$ constant matrix, seek

$$\boxed{\mathbf{x}(t) = \mathbf{q}e^{rt},} \tag{13}$$

with the constant vector $\mathbf{q}$ and the constant r to be determined so that (13) satisfies (12). Put (13) into (12) and obtain

$$r\mathbf{q}e^{rt} = \mathbf{A}\mathbf{q}e^{rt} \tag{14}$$

which, upon canceling the exponentials, gives the eigenvalue problem

$$\boxed{\mathbf{A}\mathbf{q} = r\mathbf{q}} \tag{15}$$

for $\mathbf{q}$ and r. Solve for the eigenvalues and eigenvectors of $\mathbf{A}$. These give r and $\mathbf{q}$, respectively, to use in (13), enabling us to find solutions to (12) such as the $\mathbf{x}_1(t)$ and $\mathbf{x}_2(t)$ in (10).

Hopefully, we can construct a general solution of the given system as an arbitrary linear combination of these solutions. But, can we count on obtaining enough LI solutions by this method to construct a general solution? The following theorem guarantees success — if the $n \times n$ matrix $\mathbf{A}$ has n *distinct* eigenvalues.

THEOREM 4.7.1 *General Solution if the Eigenvalues of* **A** *are Distinct*
Consider the system (12) on any t interval I, where $\mathbf{A}$ is an $n \times n$ constant matrix. If $\mathbf{A}$ has n distinct eigenvalues $\lambda_1, \dots, \lambda_n$, then the corresponding eigenvectors $\mathbf{e}_1, \dots, \mathbf{e}_n$ are LI, and

$$\boxed{\mathbf{x}(t) = c_1 \mathbf{e}_1 e^{\lambda_1 t} + \cdots + c_n \mathbf{e}_n e^{\lambda_n t}} \tag{16}$$

is a general solution of (12) on I.

Proof: Each of the terms $\mathbf{e}_j e^{\lambda_j t}$ in (16) is a solution of (12). After all, we sought solutions in the form $\mathbf{q}e^{rt}$, and the eigenvalues gave the r's and the eigenvectors gave the $\mathbf{q}$'s. Thus, it remains only to verify that this set of n solutions is LI on I, so that we can apply Theorem 4.5.3(a). It is shown in the exercises that if $\mathbf{A}$ has n distinct eigenvalues $\lambda_1, \dots, \lambda_n$, then the corresponding eigenvectors $\mathbf{e}_1, \dots, \mathbf{e}_n$ are LI. Thus,

$$\det[\mathbf{e}_1 e^{\lambda_1 t}, \dots, \mathbf{e}_n e^{\lambda_n t}] = e^{(\lambda_1 + \cdots + \lambda_n)t} \det[\mathbf{e}_1, \dots, \mathbf{e}_n] \neq 0,$$

so the solutions $\boldsymbol{x}_1(t) = \mathbf{e}_1 e^{\lambda_1 t}, \dots, \mathbf{x}_n(t) = \mathbf{e}_n e^{\lambda_n t}$ in (16) are indeed LI. ■

The first step is by property D6 in Appendix B (with "α's" equal to $e^{\lambda_1 t}, \dots, e^{\lambda_n t}$), and the second step is by the linear independence of the $\mathbf{e}_j$'s.

Let us illustrate the use of Theorem 4.7.1.

EXAMPLE 2. Using Theorem 4.7.1. Obtain a general solution of the system

$$x' = x + z, \tag{17a}$$
$$y' = -x - y, \tag{17b}$$
$$z' = 2x + 2z. \tag{17c}$$

In this case

$$\mathbf{A} = \begin{bmatrix} 1 & 0 & 1 \\ -1 & -1 & 0 \\ 2 & 0 & 2 \end{bmatrix}, \tag{18}$$

and we find its eigenvalues and eigenvectors to be

$$\lambda_1 = 0, \ \mathbf{e}_1 = \alpha \begin{bmatrix} -1 \\ 1 \\ 1 \end{bmatrix}; \ \lambda_2 = -1, \ \mathbf{e}_2 = \beta \begin{bmatrix} 0 \\ 1 \\ 0 \end{bmatrix}; \ \lambda_3 = 3, \ \mathbf{e}_3 = \gamma \begin{bmatrix} 4 \\ -1 \\ 8 \end{bmatrix}. \tag{19}$$

Since the eigenvalues $0, -1$, and 3 are distinct, it follows from Theorem 4.7.1 that

$$\mathbf{x}(t) = c_1 \begin{bmatrix} -1 \\ 1 \\ 1 \end{bmatrix} e^{0t} + c_2 \begin{bmatrix} 0 \\ 1 \\ 0 \end{bmatrix} e^{-t} + c_3 \begin{bmatrix} 4 \\ -1 \\ 8 \end{bmatrix} e^{3t} \tag{20}$$

Sometimes we use subscripted dependent variables, such as the x_1 and x_2 in Example 1, and sometimes we use other names, such as the x, y, and z in the present example. In this case the $\mathbf{x}(t)$ in (20) denotes $[x(t), y(t), z(t)]^{\mathrm{T}}$ rather than $[x_1(t), x_2(t), x_3(t)]^{\mathrm{T}}$.

or, in scalar form,

$$x(t) = -c_1 + 4c_3e^{3t}, \tag{21a}$$
$$y(t) = c_1 + c_2e^{-t} - c_3e^{3t}, \tag{21b}$$
$$z(t) = c_1 + 8c_3e^{3t}, \tag{21c}$$

is a general solution of (17). ▮

Subsection 4.7.2 is not a prerequisite for the remainder of this section.

4.7.2 Application to the two-mass oscillator. We return to the coupled oscillator studied in Section 4.2 by the method of elimination, this time using the method of assumed exponential form to determine the free vibration.

The physical system is shown again in Fig. 1, and the differential equations for the unknown motions $x_1(t)$ and $x_2(t)$ are, from Section 4.2,

$$m_1x_1'' + (k_1 + k_2)x_1 - k_2x_2 = F_1(t), \tag{22a}$$
$$m_2x_2'' - k_2x_1 + (k_3 + k_2)x_2 = F_2(t). \tag{22b}$$

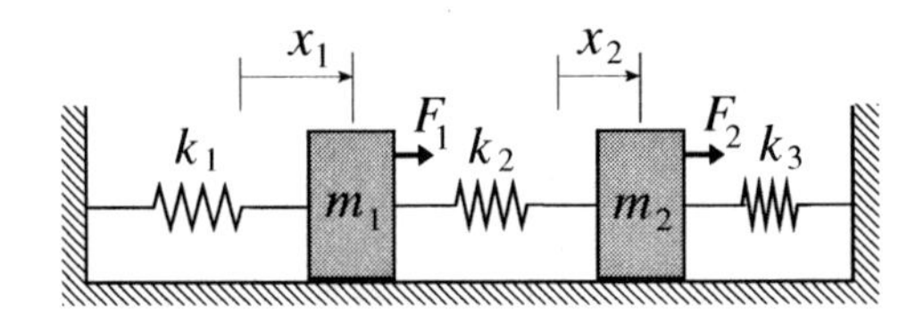

Figure 1. Two-mass oscillator, neglecting friction and air resistance.

We are considering the free vibration, where $F_1(t) = F_2(t) = 0$.

Our method of assumed exponential form is only for homogeneous systems, so we consider only the free vibration, for which $F_1(t) = F_2(t) = 0$ in (22). For definiteness, let $m_1 = m_2 = k_1 = k_2 = k_3 = 1$, so (22) becomes

$$x_1'' + 2x_1 - x_2 = 0, \tag{23a}$$
$$x_2'' - x_1 + 2x_2 = 0. \tag{23b}$$

Merely to get (23) to look more like our standard form $\mathbf{x}' = \mathbf{A}\mathbf{x}$, re-express it as

$$x_1'' = -2x_1 + x_2, \tag{24a}$$
$$x_2'' = x_1 - 2x_2. \tag{24b}$$

Actually, (24) is of the form $\mathbf{x}'' = \mathbf{A}\mathbf{x}$, not $\mathbf{x}' = \mathbf{A}\mathbf{x}$, but let us try our method of assumed exponential form; then we will assess our results. Thus, seek

$$x_1(t) = q_1e^{rt}, \quad x_2(t) = q_2e^{rt}, \tag{25}$$

put the latter into (24), cancel the e^{rt}'s, and obtain the eigenvalue problem

Here, $\mathbf{A} = \begin{bmatrix} -2 & 1 \\ 1 & -2 \end{bmatrix}$, and λ is r^2, not r.

$$\begin{bmatrix} -2 & 1 \\ 1 & -2 \end{bmatrix}\begin{bmatrix} q_1 \\ q_2 \end{bmatrix} = r^2\begin{bmatrix} q_1 \\ q_2 \end{bmatrix}. \tag{26}$$

Since we have $\mathbf{x}'' = \mathbf{A}\mathbf{x}$ rather than $\mathbf{x}' = \mathbf{A}\mathbf{x}$, λ is r^2 in (26) rather than r. The eigenpairs of $\mathbf{A}$ are

$$\lambda_1 = r_1^2 = -1, \ \mathbf{e}_1 = \alpha \begin{bmatrix} 1 \\ 1 \end{bmatrix}; \quad \lambda_2 = r_2^2 = -3, \ \mathbf{e}_2 = \beta \begin{bmatrix} 1 \\ -1 \end{bmatrix}. \tag{27}$$

In this example, each eigenpair gives *two* solutions because $r_1^2 = -1$ gives $r_1 = \pm i$, and $r_2^2 = -3$ gives $r_2 = \pm\sqrt{3}\,i$. If we put these results into (25) and use superposition, we obtain

$$\mathbf{x}(t) = c_1 \begin{bmatrix} 1 \\ 1 \end{bmatrix} e^{it} + c_2 \begin{bmatrix} 1 \\ 1 \end{bmatrix} e^{-it} + c_3 \begin{bmatrix} 1 \\ -1 \end{bmatrix} e^{i\sqrt{3}t} + c_4 \begin{bmatrix} 1 \\ -1 \end{bmatrix} e^{-i\sqrt{3}t}, \tag{28}$$

in which c_1, c_2, c_3, c_4 are arbitrary constants. Alternatively, we can use Euler's formula to express the complex exponentials in terms of cosines and sines:

$$\begin{aligned} \mathbf{x}(t) &= \begin{bmatrix} c_1 e^{it} + c_2 e^{-it} \\ c_1 e^{it} + c_2 e^{-it} \end{bmatrix} + \begin{bmatrix} c_3 e^{i\sqrt{3}t} + c_4 e^{-i\sqrt{3}t} \\ -c_3 e^{i\sqrt{3}t} - c_4 e^{-i\sqrt{3}t} \end{bmatrix} \\ &= \begin{bmatrix} c_5 \cos t + c_6 \sin t \\ c_5 \cos t + c_6 \sin t \end{bmatrix} + \begin{bmatrix} c_7 \cos\sqrt{3}\,t + c_8 \sin\sqrt{3}\,t \\ -c_7 \cos\sqrt{3}\,t - c_8 \sin\sqrt{3}\,t \end{bmatrix}, \end{aligned} \tag{29}$$

Recall Euler's formula, $e^{\pm ix} = \cos x \pm i \sin x$.

or

$$\mathbf{x}(t) = \underbrace{\begin{bmatrix} 1 \\ 1 \end{bmatrix} (c_5 \cos t + c_6 \sin t)}_{\text{low mode}} + \underbrace{\begin{bmatrix} 1 \\ -1 \end{bmatrix} (c_7 \cos\sqrt{3}\,t + c_8 \sin\sqrt{3}\,t)}_{\text{high mode}}. \tag{30}$$

The latter is the same as the result obtained in Section 4.2. If we prefer, we can express it in the equivalent form

$$\mathbf{x}(t) = \underbrace{G \begin{bmatrix} 1 \\ 1 \end{bmatrix} \sin(t+\phi)}_{\text{low mode}} + \underbrace{H \begin{bmatrix} 1 \\ -1 \end{bmatrix} \sin(\sqrt{3}\,t+\psi)}_{\text{high mode}} \tag{31}$$

in which G, H, ϕ, ψ are now the arbitrary constants in place of the c_5, c_6, c_7, and c_8 in (30).

It is important to realize that, since (24) is of the form $\mathbf{x}'' = \mathbf{A}\mathbf{x}$ and not of the form (12), Theorem 4.7.1 *does not apply*, and cannot be used to assure us that (28) is a general solution of (24). Equation (28) does agree with the general solution obtained in Section 4.2, so we are confident that it is indeed a general solution of (24). But, how can we be sure it is a general solution of (24) using the theorems presented in this section? First, introduce auxiliary variables (discussed in Section 4.1.5)

$$x_1' = u, \quad x_2' = v, \tag{32}$$

for then (24) can be re-expressed, equivalently, as the *first*-order system

$$x_1' = u, \tag{33a}$$
$$u' = -2x_1 + x_2, \tag{33b}$$
$$x_2' = v, \tag{33c}$$
$$v' = x_1 - 2x_2, \tag{33d}$$

or, in matrix form,

$$\begin{bmatrix} x_1' \\ u' \\ x_2' \\ v' \end{bmatrix} = \begin{bmatrix} 0 & 1 & 0 & 0 \\ -2 & 0 & 1 & 0 \\ 0 & 0 & 0 & 1 \\ 1 & 0 & -2 & 0 \end{bmatrix} \begin{bmatrix} x_1 \\ u \\ x_2 \\ v \end{bmatrix}, \tag{34}$$

Now we can use Theorem 4.7.1.

which is now of the form (12). We can seek solutions of (34) in the form $\mathbf{x}(t) = \mathbf{q}e^{rt}$ (where $\mathbf{x}(t)$ here denotes $[x_1(t), u(t), x_2(t), v(t)]^{\mathrm{T}}$), put that into (33), obtain an eigenvalue problem, with $\lambda = r$ this time, and proceed according to Theorem 4.7.1 to a general solution of (33). If you do that, the steps of which are left for the exercises, and at the end delete the auxiliary variables $u(t)$ and $v(t)$, you will obtain the same solution as in (28).

That is, if we want to use Theorem 4.7.1 we must first put the system into the form (12) by using auxiliary variables.

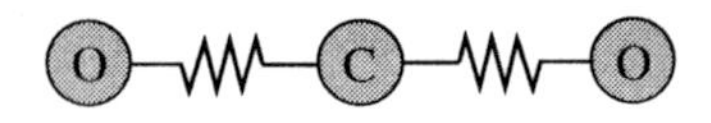

Figure 2. Schematic of a mass-spring model of a CO_2 molecule.

COMMENT. An interesting application of this type of analysis can be found in chemistry and thermodynamics, when a molecule of a perfect gas such as CO_2 is modeled by point masses (representing the atoms) connected by linear springs (representing the interatomic electrical forces), as indicated schematically in Fig. 2. One would expect the eigenfrequencies and eigenmodes to have important consequences with respect to the thermodynamic properties and behavior of the gas, and they do; from them one can determine the specific heat and entropy of the gas.

4.7.3 The case of repeated eigenvalues.

Theorem 4.7.1 tells us that all is well if $\mathbf{A}$ has n distinct eigenvalues. To see what can happen if the eigenvalues are not distinct, consider the following two examples, in which repeated eigenvalues occur.

EXAMPLE 3. Falling Short of a General Solution Because of a Repeated Eigenvalue. Consider the system

$$x' = y, \tag{35a}$$
$$y' = -9x + 6y, \tag{35b}$$

which is of the form (12) with

$$\mathbf{A} = \begin{bmatrix} 0 & 1 \\ -9 & 6 \end{bmatrix}. \tag{36}$$

The characteristic equation for $\mathbf{A}$ is $\lambda^2 - 6\lambda + 9 = (\lambda - 3)^2 = 0$. Thus, $\lambda = 3$ is a repeated root of multiplicity two, and its eigenspace is found to be $\mathbf{e} = \alpha [1, 3]^{\mathrm{T}}$. If we put these results into (13) we have

$$\mathbf{x}(t) = \alpha \begin{bmatrix} 1 \\ 3 \end{bmatrix} e^{3t}, \tag{37}$$

but we are missing a second LI solution. To show where we stand, let us write the foregoing result as

$$\mathbf{x}(t) = c_1 \begin{bmatrix} 1 \\ 3 \end{bmatrix} e^{3t} + c_2\,(?), \tag{38}$$

with the question mark term reminding us that we're missing a second LI solution.

To understand why we came up short in this example, solve (35) by elimination instead, because that method will lead to a general solution no matter what. Omitting the details, elimination gives the general solution

$$\mathbf{x}(t) = c_1 \begin{bmatrix} e^{3t} \\ 3e^{3t} \end{bmatrix} + c_2 \begin{bmatrix} te^{3t} \\ (1+3t)e^{3t} \end{bmatrix}. \tag{39}$$

The first solution in (39) is the one we found in (38), but we did not find the second one. Now we can see why we missed it, because it is not of the assumed form! It includes te^{3t} terms, whereas (13) allows for solutions only of the form e^{rt}. ▮

We cannot conclude from one example that repeated eigenvalues *necessarily* cause us to fall short of obtaining a general solution by the method of assumed exponential form. In the next example there is a repeated eigenvalue, yet we *do* obtain a general solution.

EXAMPLE 4. General Solution Obtained in Spite of a Repeated Eigenvalue. Consider the system

$$\begin{aligned} x' &= 2x + 2y + z, \\ y' &= x + 3y + z, \\ z' &= x + 2y + 2z. \end{aligned} \tag{40}$$

We find that the characteristic equation of the $\mathbf{A}$ matrix can be factored as $-(\lambda - 5)(\lambda - 1)^2 = 0$, so $\lambda = 5, 1, 1$, and $\lambda = 1$ is a repeated eigenvalue, of multiplicity two. The eigenvalues and their eigenspaces are

$$\lambda_1 = 5,\ \mathbf{e}_1 = \alpha \begin{bmatrix} 1 \\ 1 \\ 1 \end{bmatrix}; \qquad \lambda_2 = 1,\ \mathbf{e}_2 = \beta \begin{bmatrix} -1 \\ 0 \\ 1 \end{bmatrix} + \gamma \begin{bmatrix} -2 \\ 1 \\ 0 \end{bmatrix} \tag{41}$$

so

$$\begin{aligned} \mathbf{x}(t) &= \alpha \begin{bmatrix} 1 \\ 1 \\ 1 \end{bmatrix} e^{5t} + \left(\beta \begin{bmatrix} -1 \\ 0 \\ 1 \end{bmatrix} + \gamma \begin{bmatrix} -2 \\ 1 \\ 0 \end{bmatrix} \right) e^{t} \\ &= \alpha \begin{bmatrix} 1 \\ 1 \\ 1 \end{bmatrix} e^{5t} + \beta \begin{bmatrix} -1 \\ 0 \\ 1 \end{bmatrix} e^{t} + \gamma \begin{bmatrix} -2 \\ 1 \\ 0 \end{bmatrix} e^{t} \end{aligned}$$

$$\equiv c_1\mathbf{x}_1(t) + c_2\mathbf{x}_2(t) + c_3\mathbf{x}_3(t), \tag{42}$$

in which we changed α, β, γ to c_1, c_2, c_3, respectively. Finally, $\{\mathbf{x}_1(t), \mathbf{x}_2(t), \mathbf{x}_3(t)\}$ is LI because

$$\det[\mathbf{x}_1(t), \mathbf{x}_2(t), \mathbf{x}_3(t)] = \begin{vmatrix} e^{5t} & -e^t & -2e^t \\ e^{5t} & 0e^t & e^t \\ e^{5t} & e^t & 0e^t \end{vmatrix}$$

$$= e^{5t}e^te^t \begin{vmatrix} 1 & -1 & -2 \\ 1 & 0 & 1 \\ 1 & 1 & 0 \end{vmatrix} = -4e^{7t} \tag{43}$$

is nonzero. Thus, (42) is a general solution of (40) in spite of the repeated eigenvalue. ∎

The **dimension** of an eigenspace is the number of LI vectors in it.

An eigenvalue is defective if it is of multiplicity k but has less than k LI eigenvectors.

It follows from Examples 3 and 4 that repeated eigenvalues may, but need not, cause a defect in the sense of $\mathbf{x} = \mathbf{q}e^{rt}$ not leading to a general solution of the system $\mathbf{x}' = \mathbf{A}\mathbf{x}$. In Example 4 the repeated eigenvalue was of multiplicity *two* and the corresponding eigenspace contained *two* LI eigenvectors, so we were led to a general solution of (43). In Example 3, however, the eigenvalue of multiplicity *two* had an eigenspace that contained only *one* LI eigenvector (i.e., it was only a one-dimensional eigenspace). Thus, we say that the eigenvalue $r = 3$ in Example 3 was **defective**.

If an eigenvalue of multiplicity k has an eigenspace containing l LI eigenvectors (i.e., if that eigenspace is l-dimensional) with $l < k$, that eigenvalue is of **defect** $k-l$ since there results a shortage in the solution of $\mathbf{x}' = \mathbf{A}\mathbf{x}$, by $k-l$ LI solutions.

4.7.4 Modifying the method if there are defective eigenvalues. We can modify the assumed form so that a general solution is obtained even if there are defective eigenvalues:

1. Seek a solution of $\mathbf{x}' = \mathbf{A}\mathbf{x}$ in the form

$$\mathbf{x}(t) = \mathbf{q}e^{rt}. \tag{44}$$

That choice will lead to a general solution (as in Examples 1, 2, and 4) even if $\mathbf{A}$ has repeated eigenvalues, provided that none of its repeated eigenvalues is defective.

The forms (45) and (46) should not be surprising, for we know from Chapter 2 that a repeated root r of multiplicity k, of a linear homogeneous constant-coefficient equation, contributes the terms $(c_1+c_2t+\cdots+c_kt^{k-1})e^{rt}$ to the general solution.

2. If there *are* defective eigenvalues, the solution obtained using (44) will fall short of a general solution. To find the missing solutions corresponding to a defective eigenvalue r_j, proceed as follows. If r_j is of defect one, seek a solution (that is, the contribution to the solution coming from the eigenvalue r_j) in the modified form

$$\boxed{\mathbf{x}(t) = (\mathbf{q} + \mathbf{p}t)\,e^{r_jt},} \tag{45}$$

if it is of defect two seek

$$\boxed{\mathbf{x}(t) = \left(\mathbf{q} + \mathbf{p}t + \mathbf{s}t^2\right) e^{r_j t},} \tag{46}$$

and so on.

3. Do so for each defective eigenvalue, in turn.

To illustrate, we will return to Example 3 and complete it using the foregoing steps. We already found, in Example 3, the eigenvalue $r = 3$ of multiplicity two and its one-dimensional eigenspace $\mathbf{e} = \alpha[1, 3]^{\mathrm{T}}$. Thus $r = 3$ is of defect one, so seek $\mathbf{x}(t)$ in the modified form $(\mathbf{q} + \mathbf{p}t)\, e^{3t}$ to find the missing LI solution associated with $r = 3$. Put the latter into $\mathbf{x}' = \mathbf{A}\mathbf{x}$ and obtain

$$\mathbf{p}e^{3t} + 3\,(\mathbf{q} + \mathbf{p}t)\, e^{3t} = \mathbf{A}\,(\mathbf{q} + \mathbf{p}t)\, e^{3t} \tag{47}$$

or, upon canceling the e^{3t}'s,

$$(\mathbf{p} + 3\mathbf{q}) + 3\mathbf{p}t = \mathbf{A}\mathbf{q} + \mathbf{A}\mathbf{p}t. \tag{48}$$

Regarding equation (48): more generally, if the $\mathbf{u}_j$'s and $\mathbf{v}_j$'s are constant vectors, $\mathbf{u}_1 + \mathbf{u}_2 t + \cdots + \mathbf{u}_N t^N = \mathbf{v}_1 + \mathbf{v}_2 t + \cdots + \mathbf{v}_N t^N$ implies that $\mathbf{u}_j = \mathbf{v}_j$ for each j. See the exercises.

Differentiating (48) with respect to t gives $3\mathbf{p} = \mathbf{A}\mathbf{p}$. It follows that we can cancel the $3\mathbf{p}t$ and $\mathbf{A}\mathbf{p}t$ terms in (48), leaving $\mathbf{p} + 3\mathbf{q} = \mathbf{A}\mathbf{q}$. Thus,

$$\mathbf{A}\mathbf{p} = 3\mathbf{p}, \tag{49a}$$
$$\mathbf{A}\mathbf{q} = \mathbf{p} + 3\mathbf{q}. \tag{49b}$$

Since $r = 3$ is an eigenvalue of $\mathbf{A}$, we don't need to solve (49a) by Gauss elimination; we already know that its solution is $\mathbf{p} = \alpha[1, 3]^{\mathrm{T}}$. Put that into (49b), recall that $\mathbf{A} = \begin{bmatrix} 0 & 1 \\ -9 & 6 \end{bmatrix}$, and solve (49b) for $\mathbf{q}$ by Gauss elimination. That step gives $\mathbf{q} = [\beta, \alpha + 3\beta]^{\mathrm{T}}$ in which β, like α, is arbitrary. Then

$$\begin{aligned} \mathbf{x}(t) = (\mathbf{q} + \mathbf{p}t)\, e^{3t} &= \left(\begin{bmatrix} \beta \\ \alpha + 3\beta \end{bmatrix} + \alpha \begin{bmatrix} 1 \\ 3 \end{bmatrix} t \right) e^{3t} \\ &= \beta \begin{bmatrix} 1 \\ 3 \end{bmatrix} e^{3t} + \alpha \begin{bmatrix} t \\ 1 + 3t \end{bmatrix} e^{3t} \end{aligned} \tag{50}$$

is the desired general solution of (35). The latter is the same as (39), obtained by elimination, with β and α instead of C_1 and C_2, respectively.

4.7.5 Complex eigenvalues. We saw in Section 4.6.3 that even if $\mathbf{A}$ is real (i.e., if all its elements are real numbers) it can have complex eigenvalues and hence eigenvectors with complex components. The preceding results in this section hold whether or not $\mathbf{A}$ has complex eigenvalues, but it will be good to work an example with complex eigenvalues, being careful about the complex arithmetic.

EXAMPLE 5. Use the method of assumed form to obtain a general solution of

$$x_1' = -x_1 + x_2, \tag{51a}$$
$$x_2' = -2x_1 + x_2. \tag{51b}$$

Putting the assumed exponential form $\mathbf{x}(t) = \mathbf{q}e^{rt}$ into (51) gives the eigenvalue problem $\mathbf{Aq} = r\mathbf{q}$ with $\mathbf{A} = \begin{bmatrix} -1 & 1 \\ -2 & 1 \end{bmatrix}$. The characteristic equation is $\lambda^2 + 1 = 0$, so we obtain the complex conjugate roots $\lambda = \pm i$. To find the eigenspace corresponding to $\lambda_1 = +i$, we must find the nontrivial solutions of $(\mathbf{A} - r\mathbf{I})\mathbf{q} = \mathbf{0}$ in which r is i, namely,

$$\begin{bmatrix} -1-i & 1 \\ -2 & 1-i \end{bmatrix} \begin{bmatrix} q_1 \\ q_2 \end{bmatrix} = \begin{bmatrix} 0 \\ 0 \end{bmatrix}. \tag{52}$$

By the same reasoning as in Example 4 of Section 4.6, the two scalar equations within (52) are redundant, so we can discard the second, for instance, without loss. That leaves the first,

$$(-1-i)q_1 + q_2 = 0,$$

so $q_2 = (1+i)q_1$ in which q_1 is arbitrary, say α. Thus, $q_1 = \alpha$ and $q_2 = (1+i)\alpha$, so

$$\mathbf{e}_1 = \begin{bmatrix} \alpha \\ (1+i)\alpha \end{bmatrix} = \alpha \begin{bmatrix} 1 \\ 1+i \end{bmatrix}. \tag{53}$$

As in Example 4 of Section 4.6, we can obtain $\mathbf{e}_2$ by changing the i's in $\mathbf{e}_2$ to $-i$'s:

$$\mathbf{e}_2 = \beta \begin{bmatrix} 1 \\ 1-i \end{bmatrix}. \tag{54}$$

From these eigenpairs we can write a general solution of (51) as

Of course, we could use Euler's formulas to re-express (55) in terms of $\cos t$ and $\sin t$ if we prefer. We'll do that at the end.

$$\mathbf{x}(t) = \alpha \begin{bmatrix} 1 \\ 1+i \end{bmatrix} e^{it} + \beta \begin{bmatrix} 1 \\ 1-i \end{bmatrix} e^{-it}, \tag{55}$$

in which α and β are arbitrary constants (real or complex).

Suppose initial conditions are given, say

$$x_1(0) = 2 \text{ and } x_2(0) = 12. \tag{56}$$

To apply these, let us first write (55) in scalar form:

$$x_1(t) = \alpha e^{it} + \beta e^{-it}, \tag{57a}$$
$$x_2(t) = \alpha(1+i)e^{it} + \beta(1-i)e^{-it}. \tag{57b}$$

Then,

$$x_1(0) = \ \ 2 = \alpha + \beta, \tag{58a}$$
$$x_2(0) = 12 = \alpha(1+i) + \beta(1-i), \tag{58b}$$

which give $\alpha = 1-5i$ and $\beta = 1+5i$. If we put these into (57) and use Euler's formulas for e^{it} and e^{-it}, we finally obtain

$$x_1(t) = 2\cos t + 10\sin t, \tag{59a}$$
$$x_2(t) = 12\cos t + 8\sin t \tag{59b}$$

as the solution of (51) satisfying the initial conditions (56).

COMMENT. Realize that (53) can take on different appearances. For instance, scaling both components by $1-i$ gives $\mathbf{e}_1 = \alpha \begin{bmatrix} 1-i \\ 2 \end{bmatrix}$, which is equivalent to (53) since the scale factor $1-i$ can be absorbed by the arbitrary factor α. ▮

Closure. In this section we discussed the solution of linear homogeneous systems of the form $\mathbf{x}' = \mathbf{A}\mathbf{x}$ where $\mathbf{A}$ is an $n \times n$ constant matrix. Using the method of assumed form, we looked for exponential solutions in the form $\mathbf{x}(t) = \mathbf{q}e^{rt}$. Substituting the latter into the system gave the eigenvalue problem $\mathbf{A}\mathbf{q} = r\mathbf{q}$ for r and $\mathbf{q}$. If there are n distinct eigenvalues, they give n LI solutions of the system, and a general solution (Theorem 4.7.1)

$$\mathbf{x}(t) = c_1\mathbf{e}_1 e^{\lambda_1 t} + \cdots + c_n\mathbf{e}_n e^{\lambda_n t}. \tag{60}$$

Even if there are repeated eigenvalues the method still leads to a general solution if each repeated eigenvalue gives as many LI eigenvectors as its multiplicity. But if any repeated eigenvalue gives less than that number of LI eigenvectors then it is "defective," and we come up short of a general solution unless we modify the method to allow not only for exponentials but also powers of t times exponentials, as explained in Section 4.7.4.

EXERCISES 4.7

NOTE ABOUT EQUATION NUMBERING: In Exercise (2), for instance, "(17)" means equation (17) within the text, and in Exercise 7, for instance, "(7.1)" means the first numbered equation within Exercise 7.

1. Example 1. Solve (1) by elimination instead, and verify that your general solution agrees with that in (8).

2. Example 2. Solve (17) by elimination instead, and show that your solution agrees with (21).

3. Example 3. Solve (35) by elimination instead, and show that your general solution agrees with that given in (39).

4. Example 4. For the $\mathbf{A}$ matrix in (40), accept the eigenvalues 5 and 1 as given, and just solve for their eigenvectors, showing that your results agree with (41).

5. Example 5. Verify that if you put $\alpha = 1-5i$ and $\beta = 1+5i$ into (57), and use Euler's formulas, then you do obtain (59).

6. Since we used q_1 and q_2 in the assumed form (2), why did we not also use r_1 and r_2, and seek $x_1(t) = q_1 e^{r_1 t}$ and $x_2(t) = q_2 e^{r_2 t}$? Try it, and see. You should find that you are *forced* to have $r_2 = r_1$. Explain.

7. The Step From (48) to (49). (a) When there is a defective eigenvalue, the solution method outlined in Section 4.7.4 inevitably leads to an equation like (48), so we want to be sure to understand the step that took us from (48) to (49). More generally than (48), consider the equation

$$\mathbf{u}_1 + \mathbf{u}_2 t + \cdots + \mathbf{u}_N t^N = \mathbf{v}_1 + \mathbf{v}_2 t + \cdots + \mathbf{v}_N t^N, \tag{7.1}$$

in which the $\mathbf{u}_j$'s and $\mathbf{v}_j$'s are constant vectors. Show that if (7.1) holds on any open t interval, then we must have $\mathbf{u}_j = \mathbf{v}_j$ for each $j = 1, \ldots, N$. HINT: Differentiate (7.1) with respect to t a sufficient number of times.

(b) What does this result have to do with linear independence?

8. No Defective Eigenvalues. Solve, by seeking $\mathbf{x}(t)$ in the form $\mathbf{q}e^{rt}$. Primes denote d/dt. NOTE: You should obtain a general solution in each case because there are no defective eigenvalues in these examples. If the eigenvalues are given, you need not derive them.

(a) $x' = x + 2y$, $y' = 3x + 6y$

(b) $x' = 2x + 3y$, $y' = 2x + 7y$

(c) $x' = 12x + 3y$
$y' = -20x - 5y$

(d) $x' = 2x + y$
$y' = 9x + 2y$

(e) $x' = 4x + y$
$y' = -5x + z$
$z' = x - y - z$

(f) $x' = y - z$
$y' = -5x + 4y + z$
$z' = x - z$

(g) $x' = 2x + 2z$
$y' = x + y + 2z$
$z' = x + 3z$
$\lambda = 1, 1, 4$

(h) $x' = y + z$
$y' = x + z$
$z' = x + y$
$\lambda = -1, -1, 2$

(i) $x' = 4x + y + z$
$y' = x + 4y + z$
$z' = x + y + 4z$
$\lambda = 3, 3, 6$

(j) $x' = z$
$y' = z$
$z' = x + y + z$

(k) $x' = x + 2y + 3z$
$y' = x + 2y + 3z$
$z' = 3x + 6y + 9z$

(l) $x' = x + 3y + 2z$
$y' = 2x + 6y + 4z$
$z' = -x - 3y - 2z$

9. Defective Eigenvalues. Obtain a general solution using the methods given in this section. If the eigenvalues are given, you need not derive them. NOTE: Read Exercise 7 first.

(a) $x' = 2x + 4y$
$y' = -x + 6y$

(b) $x' = 2x - 4y$
$y' = x - 2y$

(c) $x' = 3x + y$
$y' = -4x - y$

(d) $x' = -x + y$
$y' = -4x + 3y$

(e) $x' = x - y$
$y' = 9x - 5y$

(f) $x' = 3x - y$
$y' = 25x - 7y$

(g) $x' = -4x - 5y - 9z$
$y' = -x + 8z$
$z' = x + y - 7z$
$\lambda = 1, -6, -6$

(h) $x' = 2x + y$
$y' = -x + z$
$z' = -3x - 5y + 5z$
$\lambda = 3, 2, 2$

(i) $x' = 2x - y$
$y' = 4x - 2y - z$
$z' = -4x + 3z$
$\lambda = -1, 2, 2$

(j) $x' = -x + y - 3z$
$y' = x + 2z$
$z' = 2x - y + 4z$
$\lambda = 1, 1, 1$

(k) $x' = x + y$
$y' = -x - y + z$
$z' = -2x - 3y + 3z$
$\lambda = 1, 1, 1$

(l) $x' = -x + y$
$y' = -x + y + z$
$z' = 2x - 3y - 3z$
$\lambda = -1, -1, -1$

10. (a) Consider the system

$$x'' = -11x + 12y \tag{10.1a}$$

$$y' = -5x + 6y. \tag{10.1b}$$

The latter is not of the form $\mathbf{x}' = \mathbf{A}\mathbf{x}$ because of the second-order derivative in (10.1a). Nevertheless, try the assumed form $x(t) = q_1 e^{rt}$, $y(t) = q_2 e^{rt}$ and show that

$$r^2 q_1 = -11q_1 + 12q_2 \tag{10.2a}$$

$$rq_2 = -5q_1 + 6q_2. \tag{10.2b}$$

Evidently, the latter cannot be put into eigenvalue problem form because the equations contain different factors, r^2 and r, respectively, so we cannot proceed.
(b) To overcome this difficulty let us re-express (10.1) as an equivalent system of first-order equations by introducing an auxiliary variable. Namely, let $x' = z$, say, and obtain

$$x' = z \tag{10.3a}$$

$$z' = -11x + 12y \tag{10.3b}$$

$$y' = -5x + 6y, \tag{10.3c}$$

which you could rearrange in alphabetical order if you wish. The system obtained is a system of first-order equations and now the method of assumed exponential form can be successfully applied. Do that, and obtain the general solution

$$x(t) = c_1 e^t + 4c_2 e^{2t} + 3c_3 e^{3t} \tag{10.4a}$$

$$y(t) = c_1 e^t + 5c_2 e^{2t} + 5c_3 e^{3t} \tag{10.4b}$$

$$z(t) = c_1 e^t + 8c_2 e^{2t} + 9c_3 e^{3t}, \tag{10.4c}$$

where c_1, c_2, and c_3 are arbitrary constants. NOTE: Of equations (10.4), we can drop (10.4c) because z was only an auxiliary variable. If we do that, we can express the general solution of (10.1) as

$$\mathbf{x}(t) = c_1 \begin{bmatrix} 1 \\ 1 \end{bmatrix} e^t + c_2 \begin{bmatrix} 4 \\ 5 \end{bmatrix} e^{2t} + c_3 \begin{bmatrix} 3 \\ 5 \end{bmatrix} e^{3t}, \tag{10.5}$$

in which $\mathbf{x}(t)$ means $[x(t), y(t)]^{\mathrm{T}}$.

11. (a) Do what was suggested below (34): Seek solutions of (34) in the form $\mathbf{q}r^{rt}$ and thus obtain a general solution of (34). Show that if you then delete the solutions for the auxiliary variables $u(t)$ and $v(t)$ you do obtain (28).
(b) Show that the four vector functions in (28) are LI.

12. Distinct Eigenvalues Give LI Eigenvectors. In the proof of Theorem 4.7.1 we stated, but did not prove, the following: *If an $n \times n$ matrix $\mathbf{A}$ has n distinct eigenvalues $\lambda_1, \dots, \lambda_n$, and $\mathbf{e}_1, \dots, \mathbf{e}_n$ are eigenvectors associated with those eigenvalues, respectively, then $\mathbf{e}_1, \dots, \mathbf{e}_n$ are LI.* Prove that claim. HINT: We need to show that

$$c_1\mathbf{e}_1 + \cdots + c_n\mathbf{e}_n = \mathbf{0} \tag{12.1}$$

holds only if $c_1 = \cdots = c_n = 0$. Multiply (12.1) through by $\mathbf{A}$ repeatedly, $n - 1$ times, each time using the fact that

$\mathbf{A}\mathbf{e}_j = \lambda_j \mathbf{e}_j$. Although the c_j's are the unknowns, it is convenient to keep the products $c_j\mathbf{e}_j$ together. Doing so, obtain the algebraic system

$$\begin{bmatrix} 1 & 1 & \cdots & 1 \\ \lambda_1 & \lambda_2 & \cdots & \lambda_n \\ \vdots & \vdots & & \vdots \\ \lambda_1^{n-1} & \lambda_2^{n-1} & \cdots & \lambda_n^{n-1} \end{bmatrix} \begin{bmatrix} c_1\mathbf{e}_1 \\ c_2\mathbf{e}_2 \\ \vdots \\ c_n\mathbf{e}_n \end{bmatrix} = \begin{bmatrix} \mathbf{0} \\ \mathbf{0} \\ \vdots \\ \mathbf{0} \end{bmatrix} \tag{12.2}$$

for $c_1\mathbf{e}_1, \dots, c_n\mathbf{e}_n$. The key point is that the coefficient matrix is a "Vandermonde matrix." Since the λ's have been assumed to be distinct, show from the properties of determinants given in Appendix B that the determinant of the coefficieint matrix in (12.2) is nonzero. Thus, show that (12.2) has the unique solution $c_1\mathbf{e}_1 = \mathbf{0}, \dots, c_n\mathbf{e}_n = \mathbf{0}$. Finally, show that it follows that all the c_j's must be zero. NOTE: (12.2) may look unusual because the elements of the two column vectors are themselves vectors rather than numbers, but that is permissible in matrix algebra: If you work out the matrix product on the left and equate the first through nth elements of the resulting vector on the left and the vector on the right, you do obtain the original set of n equations.

4.8 DOT PRODUCT AND ADDITIONAL MATRIX ALGEBRA

4.8.1 More about n-space: dot product, norm, and angle For "arrow vectors," the reader is no doubt familiar, probably from the calculus, with the norm or length of a vector, the dot product of two vectors, and the angle between them. We will now define these quantities for real n-tuple vectors — in terms of the components of the vectors. Thus, if $\mathbf{u} = [u_1, \dots, u_n]^{\mathrm{T}}$, we wish to define the **norm** or "length" of $\mathbf{u}$, denoted here as $\|\mathbf{u}\|$; and given another vector $\mathbf{v} = [v_1, \dots, v_n]^{\mathrm{T}}$, we wish to define the **angle** θ between $\mathbf{u}$ and $\mathbf{v}$, and the **dot product** $\mathbf{u} \cdot \mathbf{v}$. Furthermore, if $n = 2$ or 3 we want these definitions to reduce to the familiar definitions used for 2- and 3-space.

Thus far, for vectors in n-space, we've defined only the scalar multiplication of vectors and vector addition, in Section 4.3. Now we define norm, angle between nonzero vectors, and dot product.

Consider first the dot product. We will begin with the arrow vector definition

$$\mathbf{u} \cdot \mathbf{v} = \|\mathbf{u}\| \, \|\mathbf{v}\| \cos\theta, \tag{1}$$

will re-express it in terms of vector components for 2-space and 3-space, and will then generalize the definition of the dot product to n-space.

If $\mathbf{u}$ and $\mathbf{v}$ are vectors in 2-space, as in Fig. 1, (1) may be expressed in terms of the components of $\mathbf{u}$ and $\mathbf{v}$ as follows:

$$\begin{aligned} \mathbf{u} \cdot \mathbf{v} &= \|\mathbf{u}\| \, \|\mathbf{v}\| \cos\theta \\ &= \|\mathbf{u}\| \, \|\mathbf{v}\| \cos(\beta - \alpha) \\ &= \|\mathbf{u}\| \, \|\mathbf{v}\| (\cos\beta\cos\alpha + \sin\beta\sin\alpha) \\ &= (\|\mathbf{u}\| \cos\alpha)(\|\mathbf{v}\| \cos\beta) + (\|\mathbf{u}\| \sin\alpha)(\|\mathbf{v}\| \sin\beta) \\ &= u_1v_1 + u_2v_2. \end{aligned} \tag{2}$$

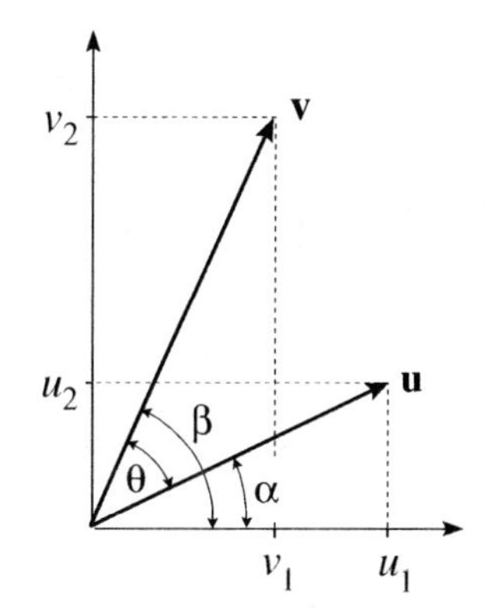

Figure 1. $\mathbf{u} \cdot \mathbf{v}$ in terms of the components of $\mathbf{u}$ and $\mathbf{v}$, for 2-space.

Omitting the steps, the analogous result for 3-space is

$$\mathbf{u} \cdot \mathbf{v} = u_1v_1 + u_2v_2 + u_3v_3. \tag{3}$$

Generalizing (2) and (3) to n-space, it is natural to define the (scalar-valued) dot product of two n-tuple vectors $\mathbf{u} = [u_1, \ldots, u_n]^{\mathrm{T}}$ and $\mathbf{v} = [v_1, \ldots, v_n]^{\mathrm{T}}$ as

$$\mathbf{u} \cdot \mathbf{v} \equiv u_1 v_1 + u_2 v_2 + \cdots + u_n v_n = \sum_{j=1}^{n} u_j v_j. \tag{4}$$

Understand that we have not "derived" (4); it is a *definition* motivated by the formulas (2) and (3) for arrow vectors in 2- and 3-space.

Having defined the dot product is the key, for now $\|\mathbf{u}\|$ and θ follow. We can use the arrow vector definitions $\|\mathbf{u}\| = \sqrt{\mathbf{u} \cdot \mathbf{u}}$ and $\mathbf{u} \cdot \mathbf{v} = \|\mathbf{u}\|\|\mathbf{v}\| \cos \theta$ given above for 2- and 3-space, together with (4), to define $\|\mathbf{u}\|$ and θ for n-space as

$$\|\mathbf{u}\| \equiv \sqrt{\mathbf{u} \cdot \mathbf{u}} = \sqrt{\sum_{j=1}^{n} u_j^2} \tag{5}$$

and

$$\theta \equiv \arccos\left(\frac{\mathbf{u} \cdot \mathbf{v}}{\|\mathbf{u}\|\,\|\mathbf{v}\|}\right), \tag{6}$$

It is understood that $\mathbf{u}$ and $\mathbf{v}$ in (6) are nonzero, because if $\|\mathbf{u}\| = 0$ and/or $\|\mathbf{v}\| = 0$ then the argument of the arccosine is undefined. Of course: The angle between two vectors is not defined if one or both of them are "point vectors," that is, zero vectors.

with the inverse cosine (arccosine) in the interval $[0, \pi]$.

The norm (5) amounts to a generalization of the familiar Pythagorean theorem, which gives $\|\mathbf{u}\| = \sqrt{u_1^2 + u_2^2}$ in 2-space and $\|\mathbf{u}\| = \sqrt{u_1^2 + u_2^2 + u_3^2}$ in 3-space.

EXAMPLE 1. Dot Product, Norm, and Angle in 2-Space. Let $\mathbf{u} = [1, 0]^{\mathrm{T}}$ and $\mathbf{v} = [2, -2]^{\mathrm{T}}$. Then (4)–(6) give

$$\begin{aligned}
\mathbf{u} \cdot \mathbf{v} &= (1)(2) + (0)(-2) = 2, \\
\|\mathbf{u}\| &= \sqrt{(1)^2 + (0)^2} = 1, \\
\|\mathbf{v}\| &= \sqrt{(2)^2 + (-2)^2} = 2\sqrt{2}, \\
\theta &= \arccos\left(\frac{2}{2\sqrt{2}}\right) = \frac{\pi}{4} \text{ rad} \quad (\text{or } 45^\circ),
\end{aligned}$$

as is readily verified if we sketch $\mathbf{u}$ and $\mathbf{v}$ as arrow vectors in a Cartesian plane. ▮

EXAMPLE 2. In 4-Space. Let $\mathbf{u} = [2, -2, 4, -1]^{\mathrm{T}}$ and $\mathbf{v} = [5, 9, -1, 0]^{\mathrm{T}}$. Then,

$$\mathbf{u} \cdot \mathbf{v} = (2)(5) + (-2)(9) + (4)(-1) + (-1)(0) = -12, \tag{7a}$$

$$\|\mathbf{u}\| = \sqrt{(2)^2 + (-2)^2 + (4)^2 + (-1)^2} = 5, \tag{7b}$$

$$\|\mathbf{v}\| = \sqrt{(5)^2 + (9)^2 + (-1)^2 + (0)^2} = \sqrt{107}, \tag{7c}$$

$$\theta = \arccos\left(\frac{-12}{5\sqrt{107}}\right) \approx \arccos(-0.232) \approx 1.805 \text{ rad} \quad (\text{or } 103.4^\circ). \tag{7d}$$

In this case, $n = 4$ is greater than 3 so (7a) through (7d) are not to be understood in a graphical sense, but simply in terms of the definitions (4)–(6). ▮

There is one catch: (6) defines an angle θ only if

$$-1 \leq \frac{\mathbf{u} \cdot \mathbf{v}}{\|\mathbf{u}\|\|\mathbf{v}\|} \leq 1, \tag{8}$$

because $\cos\theta = (\mathbf{u} \cdot \mathbf{v})/(\|\mathbf{u}\|\|\mathbf{v}\|)$ must lie between -1 and $+1$; for instance, there is no (real) θ such that $\cos\theta = 8$ or $\cos\theta = -13$. If there are $\mathbf{u}$,$\mathbf{v}$ pairs that do *not* satisfy (8), then (6) fails as a definition of the angle between vectors. However, it can be shown that (8) is indeed satisfied for *any* vectors $\mathbf{u}$ and $\mathbf{v}$ in n-space, and it is known as the **Schwarz inequality**. Thanks to the Schwarz inequality (8), (6) does define an angle for every possible pair of nonzero vectors $\mathbf{u}$ and $\mathbf{v}$ in n-space.

The Schwarz inequality is usually written in the form $|\mathbf{u} \cdot \mathbf{v}| \leq \|\mathbf{u}\|\|\mathbf{v}\|$.

If $\mathbf{u} \cdot \mathbf{v} = 0$, $\mathbf{u}$ and $\mathbf{v}$ are said to be **orthogonal**. For instance, the vectors $\mathbf{u} = [1, 3, -2]^{\mathrm{T}}$ and $\mathbf{v} = [5, 1, 4]^{\mathrm{T}}$ are orthogonal because their dot product is zero, and they are also perpendicular because θ defined by (6) is $90°$. However, orthogonality does not necessarily imply perpendicularity because if one or both vectors are $\mathbf{0}$ then their dot product is zero and they are orthogonal, but one can hardly say they are perpendicular because the angle between a nonzero vector and a point vector, or between two point vectors, is not defined.

Further, a vector set $\{\mathbf{u}_1, \ldots, \mathbf{u}_k\}$ is an **orthogonal set** if every one of the vectors is orthogonal to every other one. For instance, let $\mathbf{u}_1 = [1, 1, 0, 1]^{\mathrm{T}}$, $\mathbf{u}_2 = [1, 1, 0, -2]^{\mathrm{T}}$, and $\mathbf{u}_3 = [5, -5, 4, 0]^{\mathrm{T}}$. The set $\{\mathbf{u}_1, \mathbf{u}_2, \mathbf{u}_3\}$ is orthogonal because $\mathbf{u}_1 \cdot \mathbf{u}_2 = 0$, $\mathbf{u}_1 \cdot \mathbf{u}_3 = 0$, and $\mathbf{u}_2 \cdot \mathbf{u}_3 = 0$.

4.8.2 Algebra of matrix operators. Recall that there is an algebra of vectors: we can add two vectors to form a new vector, we can scale a vector by a constant factor to form a new vector, and so on. Likewise, there is an algebra of the matrix operators that operate on the vectors.

Matrix Addition. If $\mathbf{A} = \{a_{ij}\}$ and $\mathbf{B} = \{b_{ij}\}$ are $n \times n$ matrix operators (that is, $n \times n$ matrices), then a matrix operator $\mathbf{A} + \mathbf{B}$, called the **sum** of $\mathbf{A}$ and $\mathbf{B}$, is defined by

$$\boxed{\mathbf{A} + \mathbf{B} \equiv \{a_{ij} + b_{ij}\}.} \tag{9}$$

In other words, the sum of $\mathbf{A}$ and $\mathbf{B}$ is the matrix obtained by adding the corresponding elements of $\mathbf{A}$ and $\mathbf{B}$.

EXAMPLE 1. If

$$\mathbf{A} = \begin{bmatrix} 3 & -2 & 0 \\ 1 & 2 & 3 \\ 5 & 0 & 1 \end{bmatrix} \quad \text{and} \quad \mathbf{B} = \begin{bmatrix} 4 & 5 & 6 \\ -3 & 2 & 1 \\ 3 & 2 & -1 \end{bmatrix}, \tag{10}$$

then

$$\mathbf{A}+\mathbf{B} = \begin{bmatrix} 3+4 & -2+5 & 0+6 \\ 1-3 & 2+2 & 3+1 \\ 5+3 & 0+2 & 1-1 \end{bmatrix} = \begin{bmatrix} 7 & 3 & 6 \\ -2 & 4 & 4 \\ 8 & 2 & 0 \end{bmatrix}. \quad \blacksquare \tag{11}$$

It follows from the definition (9), and the definition of vector addition, that

$$(\mathbf{A}+\mathbf{B})\mathbf{x} = \mathbf{A}\mathbf{x} + \mathbf{B}\mathbf{x}. \tag{12}$$

Scalar Multiplication. If $\mathbf{A} = \{a_{ij}\}$ is an $n \times n$ matrix operator and α is a scalar, a new matrix operator $\alpha\mathbf{A}$ is defined by

$$\boxed{\alpha\mathbf{A} \equiv \{\alpha a_{ij}\},} \tag{13}$$

from which it follows that

$$(\alpha\mathbf{A})\mathbf{x} = \alpha(\mathbf{A}\mathbf{x}). \tag{14}$$

EXAMPLE 2. For instance, if $\mathbf{A} = \begin{bmatrix} 2 & -5 \\ 1 & 0 \end{bmatrix}$, then

$$3\mathbf{A} = 3\begin{bmatrix} 2 & -5 \\ 1 & 0 \end{bmatrix} = \begin{bmatrix} 6 & -15 \\ 3 & 0 \end{bmatrix}. \quad \blacksquare$$

No, we will use Cayley's definition.

Matrix Multiplication. Judging from the natural way in which matrix addition and scalar multiplication are defined by (9) and (13), we might expect the multiplication of two matrices $\mathbf{A} = \{a_{ij}\}$ and $\mathbf{B} = \{b_{ij}\}$ to be defined by $\mathbf{AB} \equiv \{a_{ij}b_{ij}\}$. In fact, this definition will *not* be adopted. Rather, the standard definition of matrix multiplication is the one suggested by Cayley,[1] called the **Cayley product**:

$$\boxed{\mathbf{AB} \equiv \left\{\sum_{k=1}^{n} a_{ik}b_{kj}\right\}.} \tag{15}$$

That is, if we denote $\mathbf{AB} \equiv \mathbf{C} = \{c_{ij}\}$, then (15) says that

$$c_{ij} = \sum_{k=1}^{n} a_{ik}b_{kj}. \tag{16}$$

---

[1]*Arthur Cayley* (1821–1895) published around 200 papers in a 15-year period during which he was engaged in the practice of law. In 1863, he accepted a professorship of mathematics at Cambridge.

The motivation behind Cayley's definition is not obvious. We will first show how to use it, before motivating it.

EXAMPLE 3. Let

$$\mathbf{A} = \begin{bmatrix} 2 & 0 & -5 \\ 1 & 3 & 2 \\ 4 & 1 & -1 \end{bmatrix} \quad \text{and} \quad \mathbf{B} = \begin{bmatrix} 5 & 1 & 4 \\ -2 & 3 & 1 \\ 1 & 0 & 2 \end{bmatrix}.$$

Then, according to the definition given by (15) and (16),

$$\mathbf{AB} = \begin{bmatrix} 2 & 0 & -5 \\ 1 & 3 & 2 \\ 4 & 1 & -1 \end{bmatrix} \begin{bmatrix} 5 & 1 & 4 \\ -2 & 3 & 1 \\ 1 & 0 & 2 \end{bmatrix} = \begin{bmatrix} 5 & 2 & -2 \\ 1 & 10 & 11 \\ 17 & 7 & 15 \end{bmatrix} = \mathbf{C}. \tag{17}$$

To compute c_{32}, for example, (16) gives

$$c_{32} = \sum_{k=1}^{3} a_{3k} b_{k2} = a_{31}b_{12} + a_{32}b_{22} + a_{33}b_{32} = (4)(1) + (1)(3) + (-1)(0) = 7,$$

3rd row of **A**   2nd column of **B**

that is, $a_{31}b_{12} + a_{32}b_{22} + a_{33}b_{32}$, as indicated by the arrows in (17). One more:

$$c_{11} = \sum_{k=1}^{3} a_{1k}b_{k1} = (2)(5) + (0)(-2) + (-5)(1) = 5.$$

Move across a row of the first matrix and down a column of the second. ∎

Observe, in (17), that c_{32} is the *dot product* of the third row of **A**, considered as a 3-tuple vector, with the second column of **B**. More generally, if $\mathbf{AB} = \mathbf{C} = \{c_{ij}\}$, then c_{ij} is the dot product of the ith row of **A** with the jth column of **B**, which is simpler to remember than (16).

c_{ij} is the dot product of the ith row of **A** with the jth column of **B**.

Having shown how to implement the Cayley product (15), let us motivate it. It is the matrix analog of *composite functions*. For recall from the calculus that if f and g are functions, then the composite function $f \circ g$ is defined by $(f \circ g)(x) \equiv f(g(x))$. Similarly for matrices; for any n-tuple vector **x**, we want to define a new matrix operator **AB**, called the **product** of **A** and **B**, by

$$(\mathbf{AB})\mathbf{x} \equiv \mathbf{A}(\mathbf{Bx}). \tag{18}$$

We could write $\mathbf{A} \circ \mathbf{B}$ and call it the "composition" of **A** and **B**, but following standard notation we write **AB** and call it the "product" of **A** and **B**.

The right-hand side is meaningful because we've already defined the multiplication of a vector by a matrix. Thus, first compute the vector **Bx**. Then, multiply **A** into that vector. Thus, the right-hand side of (18) serves to define the matrix operator "**AB**" on the left-hand side.

For $\mathbf{AB}$ to have the operational character given in (18), we *must* define its i, j element by (15). Let us verify that claim for the case where $n = 2$. For simplicity, we will use $a, b, \ldots, h$ for the elements of $\mathbf{A}$ and $\mathbf{B}$, rather than double-subscript notation. Written out, the right-hand side of (18) is

$$\begin{aligned} \mathbf{A}(\mathbf{Bx}) = \begin{bmatrix} a & b \\ c & d \end{bmatrix} \left(\begin{bmatrix} e & f \\ g & h \end{bmatrix} \begin{bmatrix} x_1 \\ x_2 \end{bmatrix} \right) &= \begin{bmatrix} a & b \\ c & d \end{bmatrix} \begin{bmatrix} ex_1 + fx_2 \\ gx_1 + hx_2 \end{bmatrix} \\ &= \begin{bmatrix} a(ex_1 + fx_2) + b(gx_1 + hx_2) \\ c(ex_1 + fx_2) + d(gx_1 + hx_2) \end{bmatrix} \\ &= \begin{bmatrix} (ae + bg)x_1 + (af + bh)x_2 \\ (ce + dg)x_1 + (cf + dh)x_2 \end{bmatrix} \\ &= \begin{bmatrix} ae + bg & af + bh \\ ce + dg & cf + dh \end{bmatrix} \begin{bmatrix} x_1 \\ x_2 \end{bmatrix}. \end{aligned} \tag{19}$$

The latter is the right-hand side of (18). For the left-hand side of (18), $(\mathbf{AB})\mathbf{x}$, to be the same as (19) we *must* define the product matrix $\mathbf{AB}$ as

$$\mathbf{AB} = \begin{bmatrix} a & b \\ c & d \end{bmatrix} \begin{bmatrix} e & f \\ g & h \end{bmatrix} \equiv \begin{bmatrix} ae + bg & af + bh \\ ce + dg & cf + dh \end{bmatrix}$$

which is indeed the Cayley product of $\mathbf{A}$ and $\mathbf{B}$!

The point, then, is that if we want the operator $\mathbf{AB}$ to be the composition of $\mathbf{A}$ and $\mathbf{B}$, which we do, then we must use the Cayley definition.

In matrix multiplication, matrices *may* commute, but in general they do *not*!

It is extremely important that *matrix multiplication is not, in general, commutative*; that is, in general,

$$\boxed{\mathbf{AB} \neq \mathbf{BA}.} \tag{20}$$

EXAMPLE 4. In General, AB≠BA. If

$$\mathbf{A} = \begin{bmatrix} 2 & 3 \\ 3 & 0 \end{bmatrix}, \quad \mathbf{B} = \begin{bmatrix} -2 & 3 \\ 3 & -4 \end{bmatrix}, \quad \mathbf{C} = \begin{bmatrix} 2 & 3 \\ 1 & 0 \end{bmatrix},$$

you will find, by working out the various products, that $\mathbf{A}$ and $\mathbf{B}$ happen to commute ($\mathbf{AB} = \mathbf{BA}$), but $\mathbf{A}$ and $\mathbf{C}$ do not ($\mathbf{AC} \neq \mathbf{CA}$), nor do $\mathbf{B}$ and $\mathbf{C}$ ($\mathbf{BC} \neq \mathbf{CB}$). ∎

However, commutativity always holds if one of the matrices is the identity matrix $\mathbf{I}$:

$$\mathbf{IA} = \mathbf{AI}, \tag{21}$$

as can be verified by working out $\mathbf{IA}$ and $\mathbf{AI}$, and it also holds if one of the matrices is a zero matrix: $\mathbf{OA} = \mathbf{AO}$. Further, $\mathbf{IA} = \mathbf{AI} = \mathbf{A}$ and $\mathbf{OA} = \mathbf{AO} = \mathbf{0}$.

Because it matters which factor is first and which is second, in a matrix product $\mathbf{AB}$, we will say that $\mathbf{B}$ is **pre-multiplied** by $\mathbf{A}$, and $\mathbf{A}$ is **post-multiplied** by $\mathbf{B}$, in this text, so as to leave no doubt as to which matrix is first and which is second.

Do any other familiar arithmetic rules, besides commutativity, fail to hold for the multiplication of matrices? Yes, the following rules for real numbers a, b, c do *not* carry over to matrices:

1. $ab = ba$ (commutative).

2. If $ab = ac$ and $a \neq 0$, then $b = c$ (cancelation rule).

3. If $ab = 0$, then $a = 0$ and/or $b = 0$.

4. If $a^2 = 1$, then $a = +1$ or -1.

We collect these results as a theorem, for convenient reference.

THEOREM 4.8.1 *Exceptional ("Singular") Properties of Matrix Multiplication*

(i) $\mathbf{AB} \neq \mathbf{BA}$, in general.

(ii) Even if $\mathbf{A} \neq \mathbf{O}$, $\mathbf{AB} = \mathbf{AC}$ does *not* imply that $\mathbf{B} = \mathbf{C}$.

(iii) $\mathbf{AB} = \mathbf{O}$ does *not* imply that $\mathbf{A} = \mathbf{O}$ and/or $\mathbf{B} = \mathbf{O}$.

(iv) $\mathbf{A}^2 = \mathbf{I}$ does *not* imply that $\mathbf{A} = +\mathbf{I}$ or $-\mathbf{I}$.

In (iv), $\mathbf{A}^2$ means $\mathbf{AA}$. We do not suggest memorization of the regular properties (22a)–(22d). Just keep in mind the singular ones, particularly (i).

The first of these has already been discussed. Proof of the others is left for the exercises. Theorem 4.8.1 notwithstanding, several important properties *do* carry over from the multiplication of real numbers to the multiplication of matrices:

THEOREM 4.8.2 *Ordinary ("Regular") Properties of Matrix Multiplication*
If α, β are scalars, and the matrices $\mathbf{A}$, $\mathbf{B}$, $\mathbf{C}$ are suitably conformable, then

$$(\alpha\mathbf{A})\mathbf{B} = \mathbf{A}(\alpha\mathbf{B}) = \alpha(\mathbf{AB}), \quad \text{(associative)} \tag{22a}$$

$$\mathbf{A}(\mathbf{BC}) = (\mathbf{AB})\mathbf{C}, \quad \text{(associative)} \tag{22b}$$

$$(\mathbf{A} \pm \mathbf{B})\mathbf{C} = \mathbf{AC} \pm \mathbf{BC}, \quad \text{(distributive)} \tag{22c}$$

$$\mathbf{C}(\mathbf{A} \pm \mathbf{B}) = \mathbf{CA} \pm \mathbf{CB}, \quad \text{(distributive)} \tag{22d}$$

$$\mathbf{A}(\alpha\mathbf{B} \pm \beta\mathbf{C}) = \alpha\mathbf{AB} \pm \beta\mathbf{AC}. \quad \text{(linear)} \tag{22e}$$

4.8.3 Inverse matrix. There is one more matrix concept that will be needed in Section 4.9, that of the "inverse" of an $n \times n$ matrix. We will introduce the idea by visiting once again the classical problem of linear algebra,

$$\mathbf{Ax} = \mathbf{c}, \tag{23}$$

in which $\mathbf{A}$ is an $n \times n$ matrix and $\mathbf{x}$ and $\mathbf{c}$ are n-tuple vectors. Appendices B and C review the solution of (23) by Cramer's rule and Gauss elimination – the former applying only if $\mathbf{A}$ is square ($n \times n$) and has a nonzero determinant, and the former applying whether $\mathbf{A}$ is square or not.

A different solution strategy is suggested by the compact matrix notation itself, because just as we solve $3x = 12$ by dividing both sides by 3, we wonder if we can solve (23) by dividing both sides by $\mathbf{A}$, giving $\mathbf{x} = \mathbf{c}/\mathbf{A}$. No, because the division of a vector by a matrix is not defined. However, rather than dividing both sides of $3x = 12$ by 3 we could *multiply* both sides by $1/3$, which is the numerical inverse of 3. That idea is more promising for the problem $\mathbf{Ax} = \mathbf{c}$, because matrix multiplication *is* defined.

The idea then is to seek a matrix "$\mathbf{A}^{-1}$" having the property that $\mathbf{A}^{-1}\mathbf{A} = \mathbf{I}$, where $\mathbf{I}$ is the identity matrix, for if we premultiply both sides of $\mathbf{Ax} = \mathbf{c}$ by $\mathbf{A}^{-1}$ we obtain $\mathbf{A}^{-1}\mathbf{Ax} = \mathbf{A}^{-1}\mathbf{c}$ and hence $\mathbf{Ix} = \mathbf{A}^{-1}\mathbf{c}$. Then, since $\mathbf{Ix} = \mathbf{x}$, we have the solution

$$\mathbf{x} = \mathbf{A}^{-1}\mathbf{c} \tag{24}$$

of (23). As a check, let us substitute (24) into (23). That step gives $\mathbf{AA}^{-1}\mathbf{c} = \mathbf{c}$, which is an identity, for any choice of $\mathbf{c}$, if $\mathbf{AA}^{-1} = \mathbf{I}$.

Thus, given an $n \times n$ matrix $\mathbf{A}$ we seek a matrix $\mathbf{A}^{-1}$ such that *both* $\mathbf{A}^{-1}\mathbf{A} = \mathbf{I}$ *and* $\mathbf{AA}^{-1} = \mathbf{I}$, namely, such that

$$\boxed{\mathbf{A}^{-1}\mathbf{A} = \mathbf{AA}^{-1} = \mathbf{I}.} \tag{25}$$

Just as the superscript T in $\mathbf{x}^{\mathrm{T}}$ denotes the transpose of $\mathbf{x}$, not $\mathbf{x}$ to the Tth power, $\mathbf{A}^{-1}$ denotes the inverse of $\mathbf{A}$, not $\mathbf{A}$ to the -1 power or $1/\mathbf{A}$. The latter division is not even a defined operation.

If such a matrix exists, then $\mathbf{A}$ is **invertible** and $\mathbf{A}^{-1}$ is the **inverse** of $\mathbf{A}$, read as "A-inverse." If $\mathbf{A}$ is not invertible it is **singular**.

THEOREM 4.8.3 *Inverse Matrix*

(a) $\mathbf{A}$ is invertible if and only if it is square ($n \times n$) and $\det\mathbf{A} \neq 0$.

(b) If it is invertible, its inverse is unique.

Partial proof: Let us first prove part (a), but in only one direction: for $\mathbf{A}$ to be invertible, it is necessary that it be square, with $\det\mathbf{A} \neq 0$. To begin, consider (25) "dimensionally," supposing that $\mathbf{A}$ is $m \times n$ and $\mathbf{A}^{-1}$ is $p \times q$. In that case the dimensions of the quantities on the left and right of the first equal sign in (25) are

$$(p \times q)(m \times n) = (m \times n)(p \times q). \tag{26}$$

For the matrices on the left to conform under multiplication we need $q = m$, and for those on the right to conform we need $p = n$. In that case the left- and right-hand sides of (26) are $n \times n$ and $m \times m$, so we must also have $m = n$. Thus, both $\mathbf{A}$ and $\mathbf{A}^{-1}$ must be $n \times n$.[1]

Knowing now that $\mathbf{A}^{-1}$, like $\mathbf{A}$, is $n \times n$, take the determinant of both sides of (25), recalling the determinant property $\det(\mathbf{AB}) = (\det\mathbf{A})(\det\mathbf{B})$ from Appendix

[1] For proof of the sufficiency of the two conditions given in part (a), along with further discussion, see for instance M. Greenberg, *Advanced Engineering Mathematics*, 2nd ed. (Upper Saddle River, NJ: Prentice Hall, 1998).

B and realizing that $\det \mathbf{I} = 1$. Thus, from (25) obtain $(\det \mathbf{A}^{-1})(\det \mathbf{A}) = 1$, so $\det \mathbf{A}$ cannot be zero if $\mathbf{A}$ is to be invertible.

To prove part (b), suppose $\mathbf{B}$ and $\mathbf{C}$ are inverses of $\mathbf{A}$. Then $\mathbf{BA} = \mathbf{I}$ and $\mathbf{CA} = \mathbf{I}$. Subtracting these gives $\mathbf{BA} - \mathbf{CA} = \mathbf{O}$ or $(\mathbf{B} - \mathbf{C})\mathbf{A} = \mathbf{O}$ by (22c). Next, post-multiplying the latter equation by $\mathbf{A}^{-1}$ (which exists by assumption) gives $(\mathbf{B} - \mathbf{C})\mathbf{AA}^{-1} = \mathbf{OA}^{-1}$, so $(\mathbf{B} - \mathbf{C})\mathbf{I} = \mathbf{O}$. Hence, $\mathbf{B} - \mathbf{C} = \mathbf{O}$, $\mathbf{B} = \mathbf{C}$, and the inverse matrix is unique. ■

Assuming a given square matrix $\mathbf{A}$ is invertible, that is, $\det \mathbf{A} \neq 0$, how can we *find* its inverse? We will give one method now, and additional methods in the exercises.

Consider once again the problem (23), $\mathbf{Ax} = \mathbf{c}$, in which we assume $\mathbf{A}$ is invertible. On the one hand we know that (23) admits the unique solution $\mathbf{x} = \mathbf{A}^{-1}\mathbf{c}$, and on the other hand we know from our discussion of Gauss elimination in Appendix C that we can solve (23) using elementary operations. Thus, it seems reasonable that we should, likewise, be able to evaluate $\mathbf{A}^{-1}$ using elementary operations. We can, and the idea is as follows. Beginning with $\mathbf{Ax} = \mathbf{c}$ or, equivalently, with $\mathbf{Ax} = \mathbf{Ic}$, elementary operations produce the solution $\mathbf{x} = \mathbf{A}^{-1}\mathbf{c}$ or, equivalently, $\mathbf{Ix} = \mathbf{A}^{-1}\mathbf{c}$. Symbolically then, the sequence of elementary operations effects the transformation

$$\begin{gathered} \mathbf{Ax} = \mathbf{Ic} \\ \downarrow \\ \mathbf{Ix} = \mathbf{A}^{-1}\mathbf{c}. \end{gathered} \tag{27}$$

That is, at the same time that the elementary operations are transforming $\mathbf{A}$ to $\mathbf{I}$ on the left of (27), they are transforming $\mathbf{I}$ to $\mathbf{A}^{-1}$ on the right. Thus, we can omit $\mathbf{x}$ and $\mathbf{c}$ altogether, put $\mathbf{A}$ and $\mathbf{I}$ side by side, as an augmented matrix $\mathbf{A}|\mathbf{I}$, and carry out elementary row operations on $\mathbf{A}|\mathbf{I}$ so as to reduce $\mathbf{A}$, on the left, to $\mathbf{I}$. When that is done, the matrix on the right will be $\mathbf{A}^{-1}$. Symbolically,

$$\boxed{\mathbf{A}|\mathbf{I} \rightarrow \mathbf{I}|\mathbf{A}^{-1}.} \tag{28}$$

We derived the algorithm indicated in (28) by a consideration of the solution of $\mathbf{Ax}=\mathbf{c}$, but (28) stands alone. That is, even if we are given an $\mathbf{A}$ matrix and asked for its inverse, in the absence of a problem $\mathbf{Ax}=\mathbf{c}$, we can use (28), provided that $\det \mathbf{A} \neq 0$ so $\mathbf{A}$ is indeed invertible.

EXAMPLE 5. Invert the matrix

$$\mathbf{A} = \begin{bmatrix} 1 & 1 & 1 \\ 3 & 4 & 6 \\ 4 & 6 & 8 \end{bmatrix}. \tag{29}$$

We find that $\det\mathbf{A} = -2 \neq 0$, so $\mathbf{A}$ is invertible. To find its inverse, convert $\mathbf{A}|\mathbf{I}$ to $\mathbf{I}|\mathbf{A}^{-1}$ by elementary row operations:

Symbolically, the elimination steps in (30) are these, in which rj means the jth row:

1. Add $-3 \times r1$ to $r2$
2. Add $-4 \times r1$ to $r3$
3. Add $-2 \times r2$ to $r3$
4. Multiply $r3$ by $-1/2$
5. Add $-3 \times r3$ to $r2$
6. Add $-1 \times r3$ to $r1$
7. Add$-1 \times r2$ to $r1$.

$$\begin{aligned}
\mathbf{A}|\mathbf{I} &= \left[\begin{array}{ccc|ccc} 1 & 1 & 1 & 1 & 0 & 0 \\ \mathbf{3} & 4 & 6 & 0 & 1 & 0 \\ 4 & 6 & 8 & 0 & 0 & 1 \end{array}\right] \rightarrow \left[\begin{array}{ccc|ccc} 1 & 1 & 1 & 1 & 0 & 0 \\ 0 & 1 & 3 & -3 & 1 & 0 \\ \mathbf{4} & 6 & 8 & 0 & 0 & 1 \end{array}\right] \\
&\rightarrow \left[\begin{array}{ccc|ccc} 1 & 1 & 1 & 1 & 0 & 0 \\ 0 & 1 & 3 & -3 & 1 & 0 \\ 0 & \mathbf{2} & 4 & -4 & 0 & 1 \end{array}\right] \rightarrow \left[\begin{array}{ccc|ccc} 1 & 1 & 1 & 1 & 0 & 0 \\ 0 & 1 & 3 & -3 & 1 & 0 \\ 0 & 0 & -2 & 2 & -2 & 1 \end{array}\right] \\
&\rightarrow \left[\begin{array}{ccc|ccc} \underline{1} & 1 & 1 & 1 & 0 & 0 \\ 0 & \underline{1} & 3 & -3 & 1 & 0 \\ 0 & 0 & \underline{1} & -1 & 1 & -1/2 \end{array}\right] \rightarrow \left[\begin{array}{ccc|ccc} 1 & 1 & 1 & 1 & 0 & 0 \\ 0 & 1 & 0 & 0 & -2 & 3/2 \\ 0 & 0 & 1 & -1 & 1 & -1/2 \end{array}\right] \\
&\rightarrow \left[\begin{array}{ccc|ccc} 1 & 1 & 0 & 2 & -1 & 1/2 \\ 0 & 1 & 0 & 0 & -2 & 3/2 \\ 0 & 0 & 1 & -1 & 1 & -1/2 \end{array}\right] \rightarrow \left[\begin{array}{ccc|ccc} 1 & 0 & 0 & 2 & 1 & -1 \\ 0 & 1 & 0 & 0 & -2 & 3/2 \\ 0 & 0 & 1 & -1 & 1 & -1/2 \end{array}\right] \\
&= \mathbf{I}|\mathbf{A}^{-1},
\end{aligned} \tag{30}$$

from which it follows that

$$\mathbf{A}^{-1} = \begin{bmatrix} 2 & 1 & -1 \\ 0 & -2 & 3/2 \\ -1 & 1 & -1/2 \end{bmatrix}. \tag{31}$$

As a check on our arithmetic in (30) we can (and should) verify by matrix multiplication that $\mathbf{A}$ times $\mathbf{A}^{-1}$ (or $\mathbf{A}^{-1}$ times $\mathbf{A}$) gives $\mathbf{I}$.

COMMENT 1. Regarding the Gauss elimination steps, observe that we worked "from the top down" to obtain zeros below the main diagonal of the first matrix. That is, in the first three steps we added -3 times the first row to the second to "knock out" the boldface 3, and we added -4 times the first row to the third to knock out the boldface 4. We added -2 times the second row to the third to knock out the boldface 2. Then we scaled the rows to obtain **leading 1**'s in each row; in this example only the third row had to be scaled, namely, by $-1/2$. Once we obtained 1's on the main diagonal [highlighted by boldface and underlining] and 0's below it, the Gauss elimination part was done. In the remaining three steps we continued to use elementary row operations, but now working "from the bottom up," to obtain 0's *above* the main diagonal as well. Namely, we used the leading 1 in the third row to obtain 0's above it, and the leading 1 in the second row to obtain a 0 above it, finally obtaining $\mathbf{I}$ as the left-hand matrix.

COMMENT 2. The algorithm indicated symbolically by (28) can be used to invert any invertible matrix. If, besides simply desiring the inverse of $\mathbf{A}$, we are trying to solve a problem $\mathbf{Ax} = \mathbf{c}$, we can use $\mathbf{A}^{-1}$ to do so, for the (unique) solution is $\mathbf{x} = \mathbf{A}^{-1}\mathbf{c}$. For instance, suppose we wish to solve

$$\begin{bmatrix} 1 & 1 & 1 \\ 3 & 4 & 6 \\ 4 & 6 & 8 \end{bmatrix} \begin{bmatrix} x_1 \\ x_2 \\ x_3 \end{bmatrix} = \begin{bmatrix} 8 \\ -6 \\ 0 \end{bmatrix}, \tag{32}$$

namely, $\mathbf{Ax} = \mathbf{c}$ in which $\mathbf{A}$ is given by (29) and $\mathbf{c} = [8, -6, 0]^{\mathrm{T}}$. The solution is

$$\mathbf{x} = \mathbf{A}^{-1}\mathbf{c} = \begin{bmatrix} 2 & 1 & -1 \\ 0 & -2 & 3/2 \\ -1 & 1 & -1/2 \end{bmatrix} \begin{bmatrix} 8 \\ -6 \\ 0 \end{bmatrix} = \begin{bmatrix} 10 \\ 12 \\ -14 \end{bmatrix}. \quad \blacksquare \tag{33}$$

It may be hard to see how the procedure (28) can *fail*. That is, we know that $\mathbf{A}$ is invertible only if $\det\mathbf{A} \neq 0$, yet it may appear that (28) will work no matter what, since it makes no explicit reference to $\det\mathbf{A}$. You will find that if $\det\mathbf{A} = 0$, then when you try to convert the left-hand matrix in $\mathbf{A}|\mathbf{I}$ to an identity matrix, one or more rows of zeros will result at the bottom, so the desired identity matrix cannot be obtained. For instance, try re-working Example 5 with the bottom row of $\mathbf{A}$ changed from $[4, 6, 8]$ to $[4, 5, 7]$ (which will cause $\det\mathbf{A}$ to be zero).

Closure. We began this section by adding three definitions to n-space: a dot product, a norm, and the angle between two nonzero vectors. These were designed to build upon the familiar dot product, norm, and angle in 2- and 3-space. Turning to matrices, we defined the addition of matrices, their scalar multiplication, and matrix multiplication. Of these, the Cayley definition of matrix multiplication was not obvious. If we call the product $\mathbf{AB} = \mathbf{C}$, say, the Cayley definition amounts to the i, j component of $\mathbf{C}$ being the dot product of the ith row of $\mathbf{A}$ with the jth column of $\mathbf{B}$. Keep in mind that in general matrix multiplication is not commutative! We are so accustomed to commutativity holding in the ordinary multiplication of scalars that it is easy to forget that it does not hold, in general, for matrices.

The other chief topic in this section was that of the inverse of a matrix. Necessary and sufficient conditions for a matrix $\mathbf{A}$ to be invertible are that it be square ($n \times n$), with $\det\mathbf{A} \neq 0$. If a matrix *is* invertible, its inverse can be computed by row elimination, according to the algorithm given symbolically by $\mathbf{A}|\mathbf{I} \to \mathbf{I}|\mathbf{A}^{-1}$.

EXERCISES 4.8

1. Norm, Dot Product, and Angle. Given the following vectors $\mathbf{u}$ and $\mathbf{v}$, determine $\mathbf{u} \cdot \mathbf{v}$, $\|\mathbf{u}\|$, $\|\mathbf{v}\|$, and θ (in radians and degrees). If $\mathbf{u}$ and $\mathbf{v}$ are orthogonal, state that.

(a) $\mathbf{u} = [4, 3]^{\mathrm{T}}, \ \mathbf{v} = [1, 2]^{\mathrm{T}}$
(b) $\mathbf{u} = [5, 3]^{\mathrm{T}}, \ \mathbf{v} = [-4, 1]^{\mathrm{T}}$
(c) $\mathbf{u} = [1, 2, 3]^{\mathrm{T}}, \ \mathbf{v} = [-4, -3, -2]^{\mathrm{T}}$
(d) $\mathbf{u} = [0, 2, 7]^{\mathrm{T}}, \ \mathbf{v} = [3, -4, 1]^{\mathrm{T}}$
(e) $\mathbf{u} = [-2, 1, 5]^{\mathrm{T}}, \ \mathbf{v} = [2, -1, 1]^{\mathrm{T}}$
(f) $\mathbf{u} = [8, 2, 0]^{\mathrm{T}}, \ \mathbf{v} = [1, -4, 3]^{\mathrm{T}}$
(g) $\mathbf{u} = [1, 3, 5]^{\mathrm{T}}, \ \mathbf{v} = [-3, 5, 6]^{\mathrm{T}}$
(h) $\mathbf{u} = [2, 2, 2, 2]^{\mathrm{T}}, \ \mathbf{v} = [-4, -5, -6, 0]^{\mathrm{T}}$
(i) $\mathbf{u} = [1, 2, 3, 4]^{\mathrm{T}}, \ \mathbf{v} = [-4, -3, -2, -1]^{\mathrm{T}}$
(j) $\mathbf{u} = [1, 2, 3, 4]^{\mathrm{T}}, \ \mathbf{v} = [4, 3, 2, 1]^{\mathrm{T}}$
(k) $\mathbf{u} = [3, 2, 0, -1]^{\mathrm{T}}, \ \mathbf{v} = [-5, 0, 0, 2]^{\mathrm{T}}$
(l) $\mathbf{u} = [1, 2, 0, 0]^{\mathrm{T}}, \ \mathbf{v} = [0, 1, 1, 1]^{\mathrm{T}}$
(m) $\mathbf{u} = [1, 3, 5, 7, 2]^{\mathrm{T}}, \ \mathbf{v} = [-3, 5, 6, 2, 1]^{\mathrm{T}}$
(n) $\mathbf{u} = [5, 4, 3, 2, 1]^{\mathrm{T}}, \ \mathbf{v} = [1, 0, -2, 0, 1]^{\mathrm{T}}$

2. Orthogonal Sets. Determine if the given vector set is orthogonal.

(a) $\mathbf{u} = [4, 1]^{\mathrm{T}}, \ \mathbf{v} = [-2, 8]^{\mathrm{T}}$

(b) $\mathbf{u} = [4, 1]^{\mathrm{T}}$, $\mathbf{v} = [-2, 8]^{\mathrm{T}}$, $\mathbf{w} = [0, 0]^{\mathrm{T}}$
(c) $\mathbf{u} = [3, 4]^{\mathrm{T}}$, $\mathbf{v} = [1, 1]^{\mathrm{T}}$, $\mathbf{w} = [8, -6]^{\mathrm{T}}$
(d) $\mathbf{u} = [0, 2, 7]^{\mathrm{T}}$, $\mathbf{v} = [3, -14, 4]^{\mathrm{T}}$, $\mathbf{w} = [0, 0, 1]^{\mathrm{T}}$
(e) $\mathbf{u} = [-2, 1, 1]^{\mathrm{T}}$, $\mathbf{v} = [2, -1, 5]^{\mathrm{T}}$, $\mathbf{w} = [3, 6, 1]^{\mathrm{T}}$
(f) $\mathbf{u} = [5, -1, 2]^{\mathrm{T}}$, $\mathbf{v} = [0, 6, 3]^{\mathrm{T}}$, $\mathbf{w} = [-1, -1, 2]^{\mathrm{T}}$, $\mathbf{x} = [2, 2, 1]^{\mathrm{T}}$
(g) $\mathbf{u} = [1, 3, 5, 7]^{\mathrm{T}}$, $\mathbf{v} = [-4, 2, 1, -1]^{\mathrm{T}}$
(h) $\mathbf{u} = [2, 2, 2, 2]^{\mathrm{T}}$, $\mathbf{v} = [2, 1, -1, -2]^{\mathrm{T}}$
(i) $\mathbf{u} = [1, 2, 3, 4]^{\mathrm{T}}$, $\mathbf{v} = [4, 3, -2, -1]^{\mathrm{T}}$, $\mathbf{w} = [0, 0, 0, 0]^{\mathrm{T}}$
(j) $\mathbf{u} = [1, 0, 1, 0]^{\mathrm{T}}$, $\mathbf{v} = [0, 3, 0, 1]^{\mathrm{T}}$, $\mathbf{w} = [2, 0, -2, 0]^{\mathrm{T}}$, $\mathbf{x} = [0, 2, 0, -6]^{\mathrm{T}}$

3. Show that **if a set of nonzero vectors $\{\mathbf{u}_1, \dots, \mathbf{u}_k\}$ is orthogonal, then it is LI**. HINT: Write

$$a_1\mathbf{u}_1 + a_2\mathbf{u}_2 + \cdots + a_k\mathbf{u}_k = \mathbf{0} \tag{3.1}$$

and take suitable dot products. NOTE: Indeed, we can think of orthogonality as the extreme case of linear independence. For instance, consider the vectors $\mathbf{u}$ and $\mathbf{v}$ shown below. When $\theta = 0$ they are LD. If we rotate $\mathbf{u}$ through the angle θ they are LI for all $\theta > 0$, until θ is $180°$, when they are LD again. In a sense, they are the most misaligned, "the most linearly independent," when $\theta = 90°$. (This idea is only heuristic because we don't distinguish degrees of linear independence; vectors are either LI or not.)

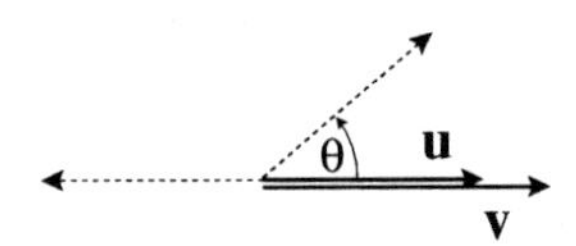

4. Schwarz Inequality. Make up any pair of nonzero vectors $\mathbf{u}$ and $\mathbf{v}$ in 5-space, say, and verify that they satisfy the Schwarz inequality $|\mathbf{u} \cdot \mathbf{v}| \leq \|\mathbf{u}\|\|\mathbf{v}\|$.

5. Show that (12) follows from the definition of $\mathbf{A} + \mathbf{B}$ given by (9), and the definition of $\mathbf{Ax}$ given by (7) in Section 4.3.

6. Composition. (a) Recall from (18) that the product of $\mathbf{A}$ and $\mathbf{B}$ was defined as their composition. Let

$$\mathbf{A} = \begin{bmatrix} 2 & 3 & -1 \\ 1 & 0 & 4 \\ 5 & -2 & 1 \end{bmatrix} \text{ and } \mathbf{B} = \begin{bmatrix} 1 & 1 & 1 \\ 2 & 1 & 0 \\ 3 & 1 & -1 \end{bmatrix}$$

and let $\mathbf{x} = [x_1, x_2, x_3]^{\mathrm{T}}$. Verify (18); that is, show that the left- and right-hand sides of (18) are equal.
(b) Having defined the product of two matrices as their composition, it should not be surprising that, in general, $\mathbf{AB} \neq \mathbf{BA}$. After all, the same is true for the composition of functions. For instance, let $f(x) = x^2$ and $g(x) = e^x$. Obtain $f(g(x))$ and $g(f(x))$ and show they are not the same.

7. Example 4. In Example 4, verify that $\mathbf{A}$ and $\mathbf{B}$ commute, but $\mathbf{A}$ and $\mathbf{C}$ do not, nor do $\mathbf{B}$ and $\mathbf{C}$, as claimed there.

8. Verify, by working out the products, that $\mathbf{IA} = \mathbf{AI} = \mathbf{A}$ for any $n \times n$ matrix $\mathbf{A}$, where $\mathbf{I}$ is an $n \times n$ identity matrix.

9. If $\mathbf{A}$ and $\mathbf{B}$ are any $n \times n$ matrices, show that the following formulas are corrrect if $\mathbf{A}$ and $\mathbf{B}$ commute, that is, if $\mathbf{AB} = \mathbf{BA}$.

(a) $(\mathbf{A}+\mathbf{B})^2 = \mathbf{A}^2 + 2\mathbf{AB} + \mathbf{B}^2$
(b) $(\mathbf{A}+\mathbf{B})(\mathbf{A}-\mathbf{B}) = \mathbf{A}^2 - \mathbf{B}^2$
(c) $(\mathbf{AB})^2 = \mathbf{A}^2\mathbf{B}^2$
(d) $(\mathbf{AB})^3 = \mathbf{A}^3\mathbf{B}^3$

10. Singular Properties of Matrix Multiplication.
(a) Make up 2×2 matrices $\mathbf{A}$, $\mathbf{B}$, and $\mathbf{C}$ that illustrate Theorem 4.8.1 part (ii).
(b) Make up 2×2 matrices $\mathbf{A}$ and $\mathbf{B}$ that illustrate Theorem 4.8.1 part (iii).
(c) Make up a 2×2 matrix $\mathbf{A}$ that illustrates Theorem 4.8.1 part (iv).

11. Show that the most general matrix that commutes with $\begin{bmatrix} 1 & 2 \\ 3 & 4 \end{bmatrix}$ is $\begin{bmatrix} \alpha - 3\beta/2 & \beta \\ 3\beta/2 & \alpha \end{bmatrix}$, where α and β are arbitrary.

12. Given $\mathbf{A}$, find the most general matrix $\mathbf{B}$ such that $\mathbf{AB} = \mathbf{O}$. HINT: Write $\mathbf{B} = \begin{bmatrix} a & b \\ c & d \end{bmatrix}$ and see what $\mathbf{AB} = \mathbf{O}$ requires of a, b, c, d.

(a) $\mathbf{A} = \begin{bmatrix} 2 & 3 \\ 0 & 0 \end{bmatrix}$ (b) $\mathbf{A} = \begin{bmatrix} 1 & 2 \\ 3 & 4 \end{bmatrix}$
(c) $\mathbf{A} = \begin{bmatrix} 0 & 0 \\ 5 & 0 \end{bmatrix}$ (d) $\mathbf{A} = \begin{bmatrix} 1 & 2 \\ 3 & 6 \end{bmatrix}$

13. Regular Properties of Matrix Multiplication. Let

$$\mathbf{A} = \begin{bmatrix} 1 & 2 \\ 3 & 4 \end{bmatrix}, \quad \mathbf{B} = \begin{bmatrix} 5 & 0 \\ 1 & -2 \end{bmatrix}, \quad \mathbf{C} = \begin{bmatrix} 6 & 1 \\ 3 & 3 \end{bmatrix}.$$

For these matrices, verify (by working out both sides of the equation) the regular property expressed in
(a) equation (22b)
(b) equation (22c)
(c) equation (22d)
(d) equation (22e) with $\alpha = 3$, $\beta = 2$.

14. Exponentiation. Given $\mathbf{A} = \begin{bmatrix} 1 & 1 \\ 1 & 1 \end{bmatrix}$, $\mathbf{B} = \begin{bmatrix} 1 & 1 \\ 0 & 1 \end{bmatrix}$,

$$\mathbf{C} = \begin{bmatrix} 0 & 3 \\ 0 & 0 \end{bmatrix}, \text{ and } \mathbf{D} = \begin{bmatrix} 0 & 1 & 2 & 3 \\ 0 & 0 & 4 & 5 \\ 0 & 0 & 0 & 6 \\ 0 & 0 & 0 & 0 \end{bmatrix}.$$

(a) $\mathbf{A}^{100}$ HINT: It should suffice to work out the first few powers $\mathbf{A}^2$, $\mathbf{A}^3$, ..., and to observe the pattern that develops.

(b) $\mathbf{B}^{100}$ (c) $\mathbf{C}^{100}$
(d) $\mathbf{D}^2$, $\mathbf{D}^3$, $\mathbf{D}^4$, $\mathbf{D}^{50}$ (e) $(\mathbf{ABC})^3$
(f) $(\mathbf{CBA})^3$ (g) $\mathbf{A}^2\mathbf{B}^2$ and $(\mathbf{AB})^2$

15. Nonsquare Matrices. (a) Given the matrices

$$\mathbf{A} = \begin{bmatrix} 0 & 3 \\ 2 & -5 \\ 1 & 10 \end{bmatrix}, \mathbf{B} = \begin{bmatrix} 5 & -1 \\ 0 & 2 \end{bmatrix}, \mathbf{x} = \begin{bmatrix} 4 \\ 3 \end{bmatrix}, \mathbf{y} = [-1, 2],$$

work out whichever of the products $\mathbf{AB}$, $\mathbf{BA}$, $\mathbf{Ax}$, $\mathbf{xA}$, $\mathbf{Bx}$, $\mathbf{xB}$, $\mathbf{yB}$, $\mathbf{A}^2$, $\mathbf{B}^2$, $\mathbf{x}^2$, $\mathbf{xy}$, and $\mathbf{yx}$ are defined.
(b) Make up any 3×2 matrix $\mathbf{A}$, any 2×4 matrix $\mathbf{B}$, and any 4×1 matrix (i.e., a vector in 4-space) $\mathbf{x}$. Evaluate $\mathbf{AB}$ and then $(\mathbf{AB})\mathbf{x}$. Alternatively, evaluate $\mathbf{Bx}$ and then $\mathbf{A}(\mathbf{Bx})$, and verify that your two results are the same, as they must be; that is, $(\mathbf{AB})\mathbf{x} = \mathbf{A}(\mathbf{Bx})$.

16. Inferring A. If, with $\mathbf{x} = [x_1, \ldots, x_4]^{\mathrm{T}}$, the product $\mathbf{Ax}$ is the column vector given below, determine $\mathbf{A}$.

(a) $\begin{bmatrix} x_1 - 3x_4 \\ 2x_2 + x_3 \end{bmatrix}$ (b) $\begin{bmatrix} x_2 + x_3 - x_4 \\ x_1 + 5x_3 \end{bmatrix}$

(c) $\begin{bmatrix} x_1 + x_2 \\ x_2 + x_3 \\ x_3 + x_4 \end{bmatrix}$ (d) $\begin{bmatrix} 2x_1 - x_3 - x_4 \\ -2x_1 + x_2 \\ x_2 + x_4 \end{bmatrix}$

(e) $\begin{bmatrix} x_4 \\ x_3 \\ x_2 \\ x_1 \end{bmatrix}$ (f) $\begin{bmatrix} x_1 + 3x_4 \\ x_2 - x_4 \\ x_3 + x_4 \\ 2x_4 \end{bmatrix}$

(g) $\begin{bmatrix} x_1 \\ x_4 \end{bmatrix}$ (h) $\begin{bmatrix} x_1 + x_2 \\ x_3 + x_4 \end{bmatrix}$

17. Computer. Let

$$\mathbf{A} = \begin{bmatrix} 4 & 1 \\ 0 & 5 \\ -2 & 3 \\ 1 & 1 \end{bmatrix}, \quad \mathbf{B} = \begin{bmatrix} 1 & -4 & 2 & 1 \\ 8 & 1 & 4 & 0 \end{bmatrix}, \quad \mathbf{x} = \begin{bmatrix} 6 \\ 5 \\ 2 \\ 1 \end{bmatrix},$$

Use computer software to evaluate

(a) $\mathbf{ABx}$ (b) $3\mathbf{BA}$
(c) $(\mathbf{AB})^3 - 2\mathbf{AB}$ (d) $2(\mathbf{AB})^2$
(e) $5(\mathbf{BA})^2$ (f) $(\mathbf{AB} - (\mathbf{AB})^2)\mathbf{x}$

18. Symmetric Matrices Give Orthogonal Eigenvectors. An $n \times n$ matrix $\mathbf{A} = \{a_{ij}\}$ is **symmetric** if $a_{ij} = a_{ji}$ for each i and j, that is, if $\mathbf{A}^{\mathrm{T}} = \mathbf{A}$. It was indicated in Exercise 11 of Section 4.6 that all eigenvalues of a real symmetric matrix are real. Also important is that its eigenvectors corresponding to different eigenvalues are *orthogonal*. That is, if λ_1 and λ_2 denote any two eigenvalues of $\mathbf{A}$ and $\lambda_1 \neq \lambda_2$, then all eigenvectors corresponding to λ_1 are orthogonal to all eigenvectors corresponding to λ_2. For the given symmetric matrix, work out its eignvalues and eigenvectors, and verify that claim for each pair of distinct eigenvalues.

(a) $\begin{bmatrix} 0 & 2 \\ 2 & 3 \end{bmatrix}$ (b) $\begin{bmatrix} 5 & 0 \\ 0 & -2 \end{bmatrix}$

(c) $\begin{bmatrix} -1 & 2 \\ 2 & 2 \end{bmatrix}$ (d) $\begin{bmatrix} 1 & 2 \\ 2 & 1 \end{bmatrix}$

(e) $\begin{bmatrix} 1 & 1 & 1 \\ 1 & 1 & 1 \\ 1 & 1 & 1 \end{bmatrix}$ (f) $\begin{bmatrix} 1 & 2 & 2 \\ 2 & 1 & 2 \\ 2 & 2 & 1 \end{bmatrix}$

(g) $\begin{bmatrix} 0 & 0 & 1 \\ 0 & 0 & 1 \\ 1 & 1 & 1 \end{bmatrix}$ (h) $\begin{bmatrix} 3 & 0 & 0 & 0 \\ 0 & 2 & 2 & 2 \\ 0 & 2 & 2 & 2 \\ 0 & 2 & 2 & 2 \end{bmatrix}$

NOTE: Recall that the two modes of vibration, given by the eigenvectors $[1, 1]^{\mathrm{T}}$ and $[1, -1]^{\mathrm{T}}$ in equation (31) of Section 4.7, were orthogonal. Not that the two modes are spatially perpendicular — they are actually collinear, but they are orthogonal in the sense that their dot product is zero. That orthogonality was not an accident, but followed from the symmetry of the $\mathbf{A}$ matrix $\begin{bmatrix} -2 & 1 \\ 1 & -2 \end{bmatrix}$ in (26).

19. Exercise 18, Continued; Proof for 2 × 2 Matrices. Let $\mathbf{A} = \begin{bmatrix} a & b \\ b & c \end{bmatrix}$, in which a, b, c are real. Show that the eigenvectors corresponding to distinct eigenvalues are necessarily orthogonal.

20. Inverse Matrix. Use the method of elementary row operations to evaluate the inverse matrix, and verify that $\mathbf{A}^{-1}\mathbf{A} = \mathbf{I}$. If the matrix is *not* invertible, state that.

(a) $\begin{bmatrix} 1 & 4 \\ 3 & 2 \end{bmatrix}$ (b) $\begin{bmatrix} 0 & 1 \\ 1 & 0 \end{bmatrix}$

(c) $\begin{bmatrix} 2 & 1 \\ 2 & -1 \end{bmatrix}$ (d) $\begin{bmatrix} 3 & 0 \\ 0 & 4 \end{bmatrix}$

(e) $\begin{bmatrix} 1 & 2 & 1 \\ 2 & 1 & 1 \\ 4 & 3 & 1 \end{bmatrix}$ (f) $\begin{bmatrix} 0 & 1 & 0 \\ 2 & 0 & 5 \\ 0 & 0 & 3 \end{bmatrix}$

(g) $\begin{bmatrix} 1 & 1 & 1 \\ 0 & 1 & 1 \\ 0 & 0 & 1 \end{bmatrix}$ (h) $\begin{bmatrix} 0 & 0 & 2 \\ 0 & -1 & 0 \\ 1 & 0 & 0 \end{bmatrix}$

(i) $\begin{bmatrix} 3 & 1 & -2 \\ 1 & 2 & 1 \\ 4 & 3 & -1 \end{bmatrix}$ (j) $\begin{bmatrix} 2 & 1 & 1 \\ 1 & 2 & 1 \\ 1 & 1 & 2 \end{bmatrix}$

(k) $\begin{bmatrix} 2 & 0 & 2 \\ 0 & 3 & 0 \\ 4 & 0 & 5 \end{bmatrix}$ (l) $\begin{bmatrix} 2 & 4 & -6 \\ 2 & 1 & 0 \\ 4 & 4 & 2 \end{bmatrix}$

(m) $\begin{bmatrix} 1 & 1 & 1 & 1 \\ 2 & 2 & 2 & 0 \\ 3 & 3 & 0 & 0 \\ 4 & 0 & 0 & 0 \end{bmatrix}$ (n) $\begin{bmatrix} 5 & 0 & 0 & 0 \\ 0 & 1 & 2 & 0 \\ 0 & 3 & 4 & 0 \\ 0 & 0 & 0 & 6 \end{bmatrix}$

(o) A matrix supplied by your instructor.

21. Formula for the Inverse of a 2 × 2 Matrix. (a) If $\mathbf{A} = \begin{bmatrix} a & b \\ c & d \end{bmatrix}$ and $ad - bc \neq 0$, verify that

$$\mathbf{A}^{-1} = \frac{1}{ad-bc}\begin{bmatrix} d & -b \\ -c & a \end{bmatrix}. \tag{21.1}$$

(b) Find all 2 × 2 matrices **A** that are their own inverse; that is, such that $\mathbf{A}^{-1} = \mathbf{A}$. That is, determine their elements a, b, c, d.

22. (a)–(n) **Computer.** Use computer software to evaluate the inverse of the matrix given in the corresponding part of Exercise 20, if it exists.

23. Prove that if an $n \times n$ matrix **A** is invertible, then

$$\det(\mathbf{A}^{-1}) = \frac{1}{\det\mathbf{A}}. \tag{23.1}$$

HINT: Take the determinant of both sides of $\mathbf{A}\mathbf{A}^{-1} = \mathbf{I}$ and cite any properties of determinants that you use, from Appendix B.

24. More About the Eigenvalue Problem. If **A** has an eigenvalue λ and corresponding eigenvector **e**, show that

(a) $\alpha\mathbf{A}$ has eigenvalue $\alpha\lambda$ and corresponding eigenvector **e**.
(b) $\mathbf{A}^2$ has eigenvalue λ^2 and corresponding eigenvector **e**. Verify that result for the case where $\mathbf{A} = \begin{bmatrix} 3 & 1 \\ 1 & 3 \end{bmatrix}$.
(c) $\mathbf{A}^n$ has eigenvalue λ^n and corresponding eigenvector **e**, for any $n = 2, 3, 4, \dots$
(d) $\mathbf{A}^{-1}$, if **A** is invertible, has eigenvalue $1/\lambda$ and corresponding eigenvector **e**. Verify that result for the case where $\mathbf{A} = \begin{bmatrix} 1 & 0 \\ 2 & 5 \end{bmatrix}$. You may use the formula given above in Exercise 21.

ADDITIONAL EXERCISES

25. Inverse by Cayley–Hamilton. In the text we presented the row elimination method of inverting a matrix. A different and interesting method is afforded by the **Cayley–Hamilton theorem**:

Cayley–Hamilton Theorem. Let **A** be an $n \times n$ matrix, and let its characteristic equation be

$$\det(\mathbf{A} - \lambda\mathbf{I}) = \alpha_0\lambda^n + \alpha_1\lambda^{n-1} + \cdots + \alpha_{n-1}\lambda + \alpha_n = 0, \tag{25.1}$$

in which α_0 is $+1$ if n is even and -1 if n is odd. Then,

$$\alpha_0\mathbf{A}^n + \alpha_1\mathbf{A}^{n-1} + \cdots + \alpha_{n-1}\mathbf{A} + \alpha_n\mathbf{I} = \mathbf{O}; \tag{25.2}$$

that is, "**A** satisfies its own characteristic equation."

A proof of this theorem will be outlined in the exercises for Section 4.9. The theorem holds for *every* $n \times n$ matrix **A**, whether it is invertible or not; likewise, (25.2) holds no matter what. Now suppose that $\det\mathbf{A} \neq 0$ so **A** is invertible. To find its inverse, premultiply both sides of (25.2) by $\mathbf{A}^{-1}$ and solve by algebra, obtaining

$$\mathbf{A}^{-1} = -\frac{1}{\alpha_n}\left(\alpha_0\mathbf{A}^{n-1} + \alpha_1\mathbf{A}^{n-2} + \cdots + \alpha_{n-1}\mathbf{I}\right). \tag{25.3}$$

The only way (25.3) can fail is if $\alpha_n = 0$, but if α_n were zero then zero would be an eigenvalue, a root of (25.1). In turn, that would imply that $\det\mathbf{A} = 0$, which cannot be if **A** is invertible, as we assumed.

(a) For the matrix given in (29), show that (25.1) and (25.2) become

$$-\lambda^3 + 13\lambda^2 - \lambda - 2 = 0, \tag{25.4}$$

and

$$-\mathbf{A}^3 + 13\mathbf{A}^2 - \mathbf{A} - 2\mathbf{I} = \mathbf{O}. \tag{25.5}$$

Verify that the left side of (25.5) does sum to zero.
(b) Next, use (25.3) to evaluate $\mathbf{A}^{-1}$. Or, equivalently, multiply (25.5) by $\mathbf{A}^{-1}$, which gives $-\mathbf{A}^2+13\mathbf{A}-\mathbf{I}-2\mathbf{A}^{-1}=\mathbf{O}$, and solve the latter by algebra for $\mathbf{A}^{-1}$. Your result should agree with (31).

26. Use the method explained above to find the inverse of the matrix given in

(a) Exercise 20, part (a) (b) Exercise 20, part (c)
(c) Exercise 20, part (e) (d) Exercise 20, part (g)

4.9 EXPLICIT SOLUTION OF $\mathbf{x}' = \mathbf{A}\mathbf{x}$ AND THE MATRIX EXPONENTIAL FUNCTION

4.9.1 Matrix exponential solution. In this section we consider once again the problem

$$\mathbf{x}' = \mathbf{A}\mathbf{x}, \tag{1}$$

in which the prime denotes differentiation with respect to t, and in which $\mathbf{A}$ is an $n \times n$ constant matrix. From Theorem 4.5.3 we know the form of a general solution of (1). Further, two solution methods have been discussed: elimination (Section 4.1), and seeking $\mathbf{x}(t)$ in exponential form and solving the eigenvalue problem that results (Section 4.7).

However, we did not obtain an explicit expression, a formula, for that solution. For the *scalar* case

$$x' = Ax, \tag{2}$$

we did find the explicit solution formula

$$x(t) = ce^{At}. \tag{3}$$

Since the matrix system (1) is a generalization of the single scalar equation (2), we wonder if we can find an explicit solution formula for the solution of (1) that is a generalization of the solution (3) of (2). Perhaps we need merely change x, c, and A in (3) to $\mathbf{x}$, $\mathbf{c}$, and $\mathbf{A}$, respectively, so that

$$\mathbf{x}(t) = \mathbf{c}e^{\mathbf{A}t}? \tag{4}$$

No, we will see that (4) is INCORRECT. The corrected solution will be given below, by (7).

If (4) is to make sense, how are we to understand the $e^{\mathbf{A}t}$? The only matrix-valued functions of $n \times n$ matrices that we've met so far have been integer powers of $\mathbf{A}$. In (3), the *scalar* exponential function can be expressed by its familiar Taylor series

$$e^{At} = 1 + At + \frac{1}{2!}A^2t^2 + \frac{1}{3!}A^3t^3 + \cdots \tag{5}$$

so, as a natural generalization of (5), the **matrix exponential function** $e^{\mathbf{A}t}$ is *defined* as[1]

$$\boxed{e^{\mathbf{A}t} \equiv \mathbf{I} + \mathbf{A}t + \frac{1}{2!}\mathbf{A}^2t^2 + \frac{1}{3!}\mathbf{A}^3t^3 + \cdots .} \tag{6}$$

We know from the calculus that the scalar series (5) converges for all t. It can be shown that the matrix series (6) does as well.

[1]If we add the terms of the series on the right-hand side of (6), each element of the resulting matrix will be an infinite series of scalars. The matrix series is **convergent** if each of those scalar series is convergent.

Since $\mathbf{A}, \mathbf{A}^2, \ldots$ in (6) are all $n \times n$, the identity matrix $\mathbf{I}$ in (6) must be $n \times n$ as well. The sum of $n \times n$ matrices is an $n \times n$ matrix, so (6) defines $e^{\mathbf{A}t}$ as an $n \times n$ matrix. In that case, the product $\mathbf{c}e^{\mathbf{A}t}$ in (4) is not meaningful, because the $n \times 1$ vector $\mathbf{c}$ and an $n \times n$ matrix are not conformable for multiplication. Thus, (4) cannot be correct as it stands.

To see if we might repair (4), note that we patterned it after (3), and since (3) is a scalar equation it can just as well be written as $x(t) = e^{At}c$. Thus, let us try modifying (4) as

$$\mathbf{x}(t) = e^{\mathbf{A}t}\mathbf{c}. \tag{7}$$

Now the matrix multiplication on the right *is* meaningful because it is an $n \times n$ matrix times an $n \times 1$ matrix. The result is $n \times 1$, as is the $\mathbf{x}(t)$ on the left-hand side, so (7) is at *least* dimensionally correct.

To see if (7) satisfies the differential equation $\mathbf{x}' = \mathbf{A}\mathbf{x}$ we will need to know the t-derivative of $e^{\mathbf{A}t}$. Proceeding formally, differentiation of (6) gives

$$\begin{aligned}\frac{d}{dt}e^{\mathbf{A}t} &= \mathbf{0} + \mathbf{A} + \mathbf{A}^2 t + \frac{1}{2!}\mathbf{A}^3 t^2 + \cdots \\ &= \mathbf{A}\Big(\mathbf{I} + \mathbf{A}t + \frac{1}{2!}\mathbf{A}^2 t^2 + \cdots\Big) \\ &= \mathbf{A}e^{\mathbf{A}t},\end{aligned} \tag{8}$$

so the familiar scalar formula $\frac{d}{dt}e^{at} = ae^{at}$ carries over to the matrix case as

$$\boxed{\frac{d}{dt}e^{\mathbf{A}t} = \mathbf{A}e^{\mathbf{A}t},} \tag{9}$$

for all t. With the help of (9) we can now substitute (7) into the differential equation $\mathbf{x}' = \mathbf{A}\mathbf{x}$, in (1), obtaining

That is, $\mathbf{x} = e^{\mathbf{A}t}\mathbf{c}$ does satisfy $\mathbf{x}' = \mathbf{A}\mathbf{x}$.

$$\frac{d\mathbf{x}}{dt} = \frac{d}{dt}\left(e^{\mathbf{A}t}\mathbf{c}\right) = \left(\frac{d}{dt}e^{\mathbf{A}t}\right)\mathbf{c} = \mathbf{A}e^{\mathbf{A}t}\mathbf{c} = \mathbf{A}(e^{\mathbf{A}t}\mathbf{c}) = \mathbf{A}\mathbf{x}. \tag{10}$$

Thus, (7) does indeed satisfy (1). Is it a *general* solution? Let us denote the columns of $e^{\mathbf{A}t}$ as $\mathbf{u}_1(t), \ldots, \mathbf{u}_n(t)$, respectively. Since $\mathbf{c}$ is arbitrary, we can set $\mathbf{c} = [1, 0, \ldots, 0]^{\mathrm{T}}$, in which case the right-hand side of (7) is $\mathbf{u}_1(t)$. Thus, the column vector $\mathbf{u}_1(t)$ is a solution of (1). Similarly, letting $\mathbf{c} = [0, 1, 0, \ldots, 0]^{\mathrm{T}}$ reveals that $\mathbf{u}_2(t)$ is a solution of (1), and so on. Thus, the columns of $e^{\mathbf{A}t}$ are solutions of (1), on $-\infty < t < \infty$ because $\mathbf{A}$ is a constant matrix and is therefore continuous for all t. Further, they are LI by the determinant test (Theorem 4.5.4) because, at the point $t = 0$ in the interval, $\det(e^{\mathbf{A}t}) = \det(e^{\mathbf{A}0}) = \det(e^{\mathbf{0}}) = \det(\mathbf{I}) = 1 \neq 0$.

$e^{\mathbf{0}} = \mathbf{I}$ by setting $t = 0$ in (6).

We conclude that the n columns of $e^{\mathbf{A}t}$ are LI solutions of (1), so $e^{\mathbf{A}t}$ is a fundamental matrix for (1), and (7) is a general solution of (1) on $-\infty < t < \infty$.

Remarkably, just as the explicit solution of (2) is given by the scalar exponential form (3), the explicit solution of (1) is given by the matrix exponential form (7)!

Finally, if an initial condition $\mathbf{x}(0) = \mathbf{b}$ is appended to (1), then (7) gives $\mathbf{x}(0) = \mathbf{b} = e^{\mathbf{0}}\mathbf{c} = \mathbf{I}\mathbf{c} = \mathbf{c}$, so $\mathbf{c} = \mathbf{b}$, and we have this result:

THEOREM 4.9.1 *Matrix Exponential Explicit Solution*
Let $\mathbf{A}$ in (1) be an $n \times n$ constant matrix, and let I be $-\infty < t < \infty$. Then
(a) A general solution of (1), on I, is

$$\boxed{\mathbf{x}(t) = e^{\mathbf{A}t}\mathbf{c},} \tag{11}$$

in which the components of $\mathbf{c}$ are arbitrary constants.
(b) The solution of (1), with initial condition $\mathbf{x}(0) = \mathbf{b}$, is

$$\boxed{\mathbf{x}(t) = e^{\mathbf{A}t}\mathbf{b},} \tag{12}$$

which exists and is unique on I.
(c) $e^{\mathbf{A}t}$ is a fundamental matrix for (1).

Regarding part (c), the columns of $e^{\mathbf{A}t}$ comprise a fundamental set, or basis, of solutions of $\mathbf{x}' = \mathbf{A}\mathbf{x}$.

However, realize that to use (11) or (12) we must evaluate the matrix $e^{\mathbf{A}t}$, and that evaluation involves summing the infinite matrix series (6)! A simplification occurs if $\mathbf{A}$ is **nilpotent**, that is, if there is a positive integer K such that $\mathbf{A}^k = \mathbf{O}$ for all $k \geq K$, because then the series terminates. $\mathbf{A}$ *will be nilpotent if all its elements below and on (or above and on) the main diagonal are zero.*

For mnemonic purposes, think of nilpotent as "potentially nil."

EXAMPLE 1. Use the foregoing results to solve the IVP

$$x_1' = 3x_2 + 2x_3, \quad x_1(0) = 8, \tag{13a}$$
$$x_2' = 4x_3, \quad x_2(0) = 3, \tag{13b}$$
$$x_3' = 0, \quad x_3(0) = 6. \tag{13c}$$

In this case, the coefficient matrix $\mathbf{A}$ is nilpotent because

$$\mathbf{A} = \begin{bmatrix} 0 & 3 & 2 \\ 0 & 0 & 4 \\ 0 & 0 & 0 \end{bmatrix}, \quad \mathbf{A}^2 = \begin{bmatrix} 0 & 0 & 12 \\ 0 & 0 & 0 \\ 0 & 0 & 0 \end{bmatrix}, \tag{14}$$

All elements above and on the main diagonal of $\mathbf{A}$ are zero, so $\mathbf{A}$ is nilpotent.

and $\mathbf{A}^k = \mathbf{O}$ for all $k \geq 3$. Using these results, (6) gives

$$e^{\mathbf{A}t} = \begin{bmatrix} 1 & 0 & 0 \\ 0 & 1 & 0 \\ 0 & 0 & 1 \end{bmatrix} + \begin{bmatrix} 0 & 3 & 2 \\ 0 & 0 & 4 \\ 0 & 0 & 0 \end{bmatrix} t + \frac{1}{2}\begin{bmatrix} 0 & 0 & 12 \\ 0 & 0 & 0 \\ 0 & 0 & 0 \end{bmatrix} t^2$$

$$= \begin{bmatrix} 1 & 3t & 2t+6t^2 \\ 0 & 1 & 4t \\ 0 & 0 & 1 \end{bmatrix}, \tag{15}$$

and the solution of (13) is given by (12) as

$$\begin{bmatrix} x_1(t) \\ x_2(t) \\ x_3(t) \end{bmatrix} = \begin{bmatrix} 1 & 3t & 2t+6t^2 \\ 0 & 1 & 4t \\ 0 & 0 & 1 \end{bmatrix} \begin{bmatrix} 8 \\ 3 \\ 6 \end{bmatrix} = \begin{bmatrix} 8+21t+36t^2 \\ 3+24t \\ 6 \end{bmatrix}. \tag{16}$$

That is, $x_1(t) = 8 + 21t + 36t^2$, $x_2(t) = 3 + 24t$, and $x_3(t) = 6$. ∎

Of course, the system (13) could have been solved without using the methods of this section, for it would have been simple to solve (13c), (13b), and then (13a), in that order. However, the case in which $\mathbf{A}$ is nilpotent makes a nice first example, and what we want to do next is to find a way to sum the exponential series in (6) into closed form even if $\mathbf{A}$ is *not* nilpotent.

4.9.2 Getting the exponential matrix series into closed form. The matrix exponential series in (6) can indeed be summed in closed form, whether the $n \times n$ coefficient matrix $\mathbf{A}$ is nilpotent or not.

Recall from Section 4.7 that solving (1) is complicated slightly if $\mathbf{A}$ has any defective eigenvalues, because then we must allow for solutions that are powers of t times exponentials. *We will limit our attention here to the case in which* $\mathbf{A}$ *has no defective eigenvalues.* That is, *we assume that* $\mathbf{A}$ *has* n *LI eigenvectors.*

Assuming that $\mathbf{A}$ has n LI eigenvectors is equivalent to assuming that none of its eigenvalues is defective.

The key will be the following theorem regarding the "diagonalization" of $\mathbf{A}$.

THEOREM 4.9.2 *Diagonalization of* $\mathbf{A}$
Let $\mathbf{A}$ be $n \times n$, with n LI eigenvectors $\mathbf{e}_1, \dots, \mathbf{e}_n$, and make these the columns of a $\mathbf{Q}$ matrix, so

$$\mathbf{Q} = [\mathbf{e}_1, \dots, \mathbf{e}_n]. \tag{17}$$

Then

$$\boxed{\mathbf{Q}^{-1}\mathbf{A}\mathbf{Q} = \mathbf{D}} \tag{18}$$

is diagonal, and the jth diagonal element of $\mathbf{D}$ is the jth eigenvalue of $\mathbf{A}$:

$$\mathbf{D} = \begin{bmatrix} \lambda_1 & 0 & \cdots & 0 \\ 0 & \lambda_2 & & \vdots \\ \vdots & & \ddots & 0 \\ 0 & 0 & \cdots & \lambda_n \end{bmatrix}. \tag{19}$$

Expressing $\mathbf{Q}$ as $[\mathbf{e}_1, \dots, \mathbf{e}_n]$ is an example of the "partitioning" of a matrix into submatrices, in this case the submatrices being the n columns of $\mathbf{Q}$, each of which is an $n \times 1$ matrix.

The eigenvalues are not necessarily distinct; all we ask here is that there are n LI eigenvectors.

Proof: First, note that if we form a matrix $\mathbf{Q}$ with n LI eigenvectors of the given matrix $\mathbf{A}$ as its columns, then the assumed linear independence of its columns assures us, by Theorem 4.4.1, that $\det\mathbf{Q} \neq 0$, so $\mathbf{Q}$ is invertible. Thus, the $\mathbf{Q}^{-1}$ in (18) does exist. Now to prove (18), begin with the product $\mathbf{AQ}$ therein:

$$\begin{aligned}\mathbf{AQ} &= \mathbf{A}\left[\mathbf{e}_1, \dots, \mathbf{e}_n\right] \\ &= \left[\mathbf{A}\mathbf{e}_1, \dots, \mathbf{A}\mathbf{e}_n\right] = \left[\lambda_1\mathbf{e}_1, \dots, \lambda_n\mathbf{e}_n\right] = \begin{bmatrix} \lambda_1 e_{11} & \cdots & \lambda_n e_{n1} \\ \vdots & & \vdots \\ \lambda_1 e_{1n} & \cdots & \lambda_n e_{nn} \end{bmatrix} \\ &= \begin{bmatrix} e_{11} & \cdots & e_{n1} \\ \vdots & & \vdots \\ e_{1n} & \cdots & e_{nn} \end{bmatrix} \begin{bmatrix} \lambda_1 & 0 & \cdots & 0 \\ 0 & \lambda_2 & & 0 \\ \vdots & & \ddots & \vdots \\ 0 & & \cdots & \lambda_n \end{bmatrix} = \mathbf{QD}. \end{aligned} \tag{20}$$

In (20), the second equality follows from the definition of matrix multiplication, and the third from the fact that each $\mathbf{e}_j$ is an eigenvector of $\mathbf{A}$, corresponding to the eigenvalue λ_j. The fifth equality is not obvious, but if you work out the matrix product to the right of the equality you will see that the result is the same as the matrix given on the left side of that equality.

Finally, pre-multiplying both sides of (20) by $\mathbf{Q}^{-1}$ gives $\mathbf{Q}^{-1}\mathbf{AQ} = \mathbf{D}$ and completes the proof. ∎

Before we show how to apply the foregoing theorem to the summing of the matrix exponential series, let us illustrate equation (18).

EXAMPLE 2. Illustrating (18). Let

$$\mathbf{A} = \begin{bmatrix} 3 & 4 \\ 2 & 1 \end{bmatrix}. \tag{21}$$

We find that

$$\lambda_1 = 5, \ \mathbf{e}_1 = \alpha \begin{bmatrix} 2 \\ 1 \end{bmatrix}; \quad \lambda_2 = -1, \ \mathbf{e}_2 = \beta \begin{bmatrix} -1 \\ 1 \end{bmatrix}. \tag{22}$$

With $\alpha = \beta = 1$, say, construct $\mathbf{Q}$ according to (17) as

$$\mathbf{Q} = \left[\mathbf{e}_1, \mathbf{e}_2\right] = \begin{bmatrix} 2 & -1 \\ 1 & 1 \end{bmatrix}. \tag{23}$$

The $\mathbf{e}_1$ and $\mathbf{e}_2$ column vectors that make up $\mathbf{Q}$ are indeed LI because $\det\left[\mathbf{e}_1, \mathbf{e}_2\right] = 3 \neq 0$. Next, compute the inverse of $\mathbf{Q}$,

$$\mathbf{Q}^{-1} = \begin{bmatrix} 1/3 & 1/3 \\ -1/3 & 2/3 \end{bmatrix}, \tag{24}$$

so

$$\mathbf{Q}^{-1}\mathbf{AQ} = \begin{bmatrix} 1/3 & 1/3 \\ -1/3 & 2/3 \end{bmatrix} \begin{bmatrix} 3 & 4 \\ 2 & 1 \end{bmatrix} \begin{bmatrix} 2 & -1 \\ 1 & 1 \end{bmatrix} = \begin{bmatrix} 5 & 0 \\ 0 & -1 \end{bmatrix} = \begin{bmatrix} \lambda_1 & 0 \\ 0 & \lambda_2 \end{bmatrix},$$

as promised by (18).

COMMENT. How do we know how to number the eigenvalues and eigenvectors? That is, which is λ_1 and which is λ_2? It doesn't matter; if we had called $\lambda_1 = -1, \mathbf{e}_1 = \beta\left[-1, 1\right]^{\mathrm{T}}$

and $\lambda_2 = 5$, $\mathbf{e}_2 = \alpha\,[2,1]^{\mathrm{T}}$ instead, then we would have used $\mathbf{Q} = \begin{bmatrix} -1 & 2 \\ 1 & 1 \end{bmatrix}$, and would have obtained $\mathbf{Q}^{-1}\mathbf{AQ} = \begin{bmatrix} -1 & 0 \\ 0 & 5 \end{bmatrix}$, which is the diagonal $\mathbf{D}$ matrix again, but with the diagonal elements in a different order. ▮

Now to get $e^{\mathbf{A}t}$ into closed form, we begin by solving (18) for $\mathbf{A}$. Pre-multiplication of both sides by $\mathbf{Q}$ and then post-multiplication of both sides by $\mathbf{Q}^{-1}$ gives

$$\mathbf{A} = \mathbf{QDQ}^{-1}. \tag{25}$$

Then,

$$\mathbf{A}^2 = \mathbf{QDQ}^{-1}\mathbf{QDQ}^{-1} = \mathbf{QDIDQ}^{-1} = \mathbf{QD}^2\mathbf{Q}^{-1}, \tag{26a}$$
$$\mathbf{A}^3 = \mathbf{QDQ}^{-1}\mathbf{QD}^2\mathbf{Q}^{-1} = \mathbf{QDID}^2\mathbf{Q}^{-1} = \mathbf{QD}^3\mathbf{Q}^{-1}, \tag{26b}$$

and so on. Thus,

$$\boxed{\mathbf{A}^k = \mathbf{QD}^k\mathbf{Q}^{-1}.} \tag{27}$$

Normally, exponentiation of a matrix is increasingly laborious as the exponent is increased. However, (27) makes exponentiation simple because $\mathbf{D}$ is diagonal, and the exponentiation of a *diagonal* matrix is simple:

If (28) is not obvious, take $\mathbf{D}$ to be 3×3, for instance, and work out $\mathbf{D}^2$ and $\mathbf{D}^3$, for instance.

$$\mathbf{D}^k = \begin{bmatrix} \lambda_1^k & 0 & \cdots & 0 \\ 0 & \lambda_2^k & & 0 \\ \vdots & & \ddots & \vdots \\ 0 & 0 & \cdots & \lambda_n^k \end{bmatrix}. \tag{28}$$

Thus, if we begin by computing the eigenvalues and eigenvectors of $\mathbf{A}$, once and for all, then (27) is a closed-form expression for raising $\mathbf{A}$ to any integer power k, no matter how large k is.

To accomplish the closed-form evaluation of the infinite matrix series for $e^{\mathbf{A}t}$, given in (6), (27) is now the key because the terms in (6) are powers of $\mathbf{A}$:

$$\begin{aligned} e^{\mathbf{A}t} &= \mathbf{I} + \mathbf{A}t + \frac{1}{2!}\mathbf{A}^2t^2 + \cdots = \mathbf{I} + \mathbf{QDQ}^{-1}t + \frac{1}{2!}\mathbf{QD}^2\mathbf{Q}^{-1}t^2 + \cdots \\ &= \mathbf{Q}\left[\mathbf{I} + \mathbf{D}t + \frac{1}{2!}\mathbf{D}^2t^2 + \cdots\right]\mathbf{Q}^{-1} \\ &= \mathbf{Q}\begin{bmatrix} (1 + \lambda_1 t + \frac{1}{2!}\lambda_1^2 t^2 + \cdots) & \cdots & 0 \\ \vdots & \ddots & \vdots \\ 0 & \cdots & (1 + \lambda_n t + \frac{1}{2!}\lambda_n^2 t^2 + \cdots) \end{bmatrix}\mathbf{Q}^{-1} \end{aligned}$$

$$= \mathbf{Q}\begin{bmatrix} e^{\lambda_1 t} & \cdots & 0 \\ \vdots & \ddots & \vdots \\ 0 & \cdots & e^{\lambda_n t} \end{bmatrix}\mathbf{Q}^{-1}$$

$$= \mathbf{Q}e^{\mathbf{D}t}\mathbf{Q}^{-1}, \tag{29}$$

where the $e^{\mathbf{D}t}$ in the final expression denotes

$$e^{\mathbf{D}t} = \begin{bmatrix} e^{\lambda_1 t} & \cdots & 0 \\ \vdots & \ddots & \vdots \\ 0 & \cdots & e^{\lambda_n t} \end{bmatrix}. \tag{30}$$

Let us review the steps in (29). We began with a series in powers of $\mathbf{A}$. Equation (27) enabled us to convert it to a series in powers of the *diagonal* matrix $\mathbf{D}$. Finally, summing the series into closed form was accomplished when the *scalar* series on the main diagonal were recognized to be the Taylor series of $e^{\lambda_1 t}, \ldots, e^{\lambda_n t}$.

The result is this:

THEOREM 4.9.3 *Closed Form Expression of the Matrix Exponential Function*
Let $\mathbf{A}$ be $n \times n$ with eigenvalues $\lambda_1, \ldots, \lambda_n$ and n LI eigenvectors $\mathbf{e}_1, \ldots, \mathbf{e}_n$, and let $\mathbf{Q} = [\mathbf{e}_1, \ldots, \mathbf{e}_n]$. Then

$$\boxed{e^{\mathbf{A}t} = \mathbf{Q}e^{\mathbf{D}t}\mathbf{Q}^{-1},} \tag{31}$$

with $e^{\mathbf{D}t}$ given by (30).

EXAMPLE 3. Using (31) in (11). Obtain a general solution of

$$x' = x + 4y \tag{32a}$$
$$y' = x + y. \tag{32b}$$

In this case

$$\mathbf{A} = \begin{bmatrix} 1 & 4 \\ 1 & 1 \end{bmatrix} \tag{33}$$

and its eigenvalues and eigenvectors are found to be

$$\lambda_1 = 3,\ \mathbf{e}_1 = \alpha\begin{bmatrix} 2 \\ 1 \end{bmatrix}; \qquad \lambda_2 = -1,\ \mathbf{e}_2 = \beta\begin{bmatrix} -2 \\ 1 \end{bmatrix}. \tag{34}$$

With

In (32) we used $x(t)$ and $y(t)$ for the dependent variables, instead of $x_1(t)$ and $x_2(t)$. Of course, the names don't matter. Thus, $\mathbf{x}(t)$ in (36) means $[x(t), y(t)]^{\mathrm{T}}$.

$$\mathbf{Q} = [\mathbf{e}_1, \mathbf{e}_2] = \begin{bmatrix} 2 & -2 \\ 1 & 1 \end{bmatrix}, \ \mathbf{Q}^{-1} = \begin{bmatrix} 1/4 & 1/2 \\ -1/4 & 1/2 \end{bmatrix}, \text{ and } e^{\mathbf{D}t} = \begin{bmatrix} e^{3t} & 0 \\ 0 & e^{-t} \end{bmatrix}, \tag{35}$$

(11) and (31) give

$$\begin{aligned} \mathbf{x}(t) &= e^{\mathbf{A}t}\mathbf{c} = \mathbf{Q}e^{\mathbf{D}t}\mathbf{Q}^{-1}\mathbf{c} \\ &= \begin{bmatrix} 2 & -2 \\ 1 & 1 \end{bmatrix}\begin{bmatrix} e^{3t} & 0 \\ 0 & e^{-t} \end{bmatrix}\begin{bmatrix} 1/4 & 1/2 \\ -1/4 & 1/2 \end{bmatrix}\begin{bmatrix} c_1 \\ c_2 \end{bmatrix} \\ &= \frac{c_1}{4}\begin{bmatrix} 2e^{3t} + 2e^{-t} \\ e^{3t} - e^{-t} \end{bmatrix} + \frac{c_2}{2}\begin{bmatrix} 2e^{3t} - 2e^{-t} \\ e^{3t} + e^{-t} \end{bmatrix} \end{aligned} \tag{36}$$

Since c_1 and c_2 are arbitrary, we could "clean up" (36) by renaming $c_1/4$ as C_1 and $c_2/2$ as C_2.

or, in scalar form, $x(t) = c_1(e^{3t} + 2e^{-t})/2 + c_2(e^{3t} - e^{-t})$ and $y(t) = c_1(e^{3t} - e^{-t})/4 + c_2(e^{3t} + e^{-t})/2$. ▮

EXAMPLE 4. Using (31) in (12). Solve

$$x' = 2x + 2y + z, \quad x(0) = 4, \tag{37a}$$
$$y' = x + 3y + z, \quad y(0) = -2, \tag{37b}$$
$$z' = x + 2y + 2z, \quad z(0) = 4. \tag{37c}$$

The eigenvalue $\lambda = 1$ is of multiplicity two, but it is not defective because its eigenspace is two-dimensional, containing two LI eigenvectors.

We find that $\lambda = 5$ is an eigenvalue of multiplicity one and that $\lambda = 1$ is an eigenvalue of multiplicity two:

$$\lambda_1 = 5, \ \mathbf{e}_1 = \alpha\begin{bmatrix} 1 \\ 1 \\ 1 \end{bmatrix}; \quad \lambda_2 = 1, \ \mathbf{e}_2 = \beta\begin{bmatrix} -1 \\ 0 \\ 1 \end{bmatrix} + \gamma\begin{bmatrix} -2 \\ 1 \\ 0 \end{bmatrix}. \tag{38}$$

We will need $\lambda_1, \lambda_2, \lambda_3$ and $\mathbf{e}_1, \mathbf{e}_2, \mathbf{e}_3$ for $\mathbf{Q}$ and $\mathbf{D}$, so let us re-number the eigenvalues and eigenvectors as follows: $\lambda_1 = 5$, $\mathbf{e}_1 = [1, 1, 1]^{\mathrm{T}}$; $\lambda_2 = 1$, $\mathbf{e}_2 = [-1, 0, 1]^{\mathrm{T}}$; $\lambda_3 = 1$, $\mathbf{e}_3 = [-2, 1, 0]^{\mathrm{T}}$. Then,

$$\mathbf{Q} = [\mathbf{e}_1, \mathbf{e}_2, \mathbf{e}_3] = \begin{bmatrix} 1 & -1 & -2 \\ 1 & 0 & 1 \\ 1 & 1 & 0 \end{bmatrix}, \tag{39}$$

from which we obtain

$$\mathbf{Q}^{-1} = \begin{bmatrix} 1/4 & 1/2 & 1/4 \\ -1/4 & -1/2 & 3/4 \\ -1/4 & 1/2 & -1/4 \end{bmatrix}. \tag{40}$$

Then, (12) and (31) give

$$\begin{aligned} \mathbf{x}(t) &= e^{\mathbf{A}t}\mathbf{b} = \mathbf{Q}e^{\mathbf{D}t}\mathbf{Q}^{-1}\mathbf{b} \\ &= \begin{bmatrix} 1 & -1 & -2 \\ 1 & 0 & 1 \\ 1 & 1 & 0 \end{bmatrix}\begin{bmatrix} e^{5t} & 0 & 0 \\ 0 & e^{t} & 0 \\ 0 & 0 & e^{t} \end{bmatrix}\begin{bmatrix} 1/4 & 1/2 & 1/4 \\ -1/4 & -1/2 & 3/4 \\ -1/4 & 1/2 & -1/4 \end{bmatrix}\begin{bmatrix} 4 \\ -2 \\ 4 \end{bmatrix} \end{aligned}$$

$$= \begin{bmatrix} e^{5t} + 3e^t \\ e^{5t} - 3e^t \\ e^{5t} + 3e^t \end{bmatrix} \tag{41}$$

or, in scalar form, $x(t) = e^{5t} + 3e^t$, $y(t) = e^{5t} - 3e^t$, and $z(t) = e^{5t} + 3e^t$. ∎

Closure. The material in this section demonstrates the power of matrix methods in solving systems of homogeneous linear differential equations with constant coefficients. We showed that the exponential solution $x(t) = ce^{at}$ of the scalar equation $x' = ax$ can be generalized to the matrix equation $\mathbf{x}' = \mathbf{A}\mathbf{x}$, the general solution being $\mathbf{x}(t) = e^{\mathbf{A}t}\mathbf{c}$, in which the $\mathbf{c}$ must follow, not precede, the exponential. We used a matrix Taylor series to define the matrix exponential function $e^{\mathbf{A}t}$. Although it was striking to see that the simple result from Section 1.2 could be generalized so elegantly, the explicit solution $\mathbf{x}(t) = e^{\mathbf{A}t}\mathbf{c}$ left much to be desired in that the evaluation of $e^{\mathbf{A}t}$ required the summing of an infinite matrix series. To get that series into closed form, we established the Diagonalization Theorem 4.9.1, subject to the condition that $\mathbf{A}$ has n LI eigenvectors, and arrived at the exponentiation formula (27) that enabled us to sum the matrix series. The result was the formula

$$\boxed{\mathbf{x}(t) = \mathbf{Q}e^{\mathbf{D}t}\mathbf{Q}^{-1}\mathbf{c}} \tag{42}$$

General solution of $\mathbf{x}' = \mathbf{A}\mathbf{x}$, in closed form. The components of $\mathbf{c}$ are arbitrary constants $c_1, \dots, c_k$.

for the general solution. That is, we do not need to evaluate the exponential matrix $e^{\mathbf{A}t}$ in the solution $\mathbf{x}(t) = e^{\mathbf{A}t}\mathbf{c}$ by summing an infinite matrix series, because the right-hand side of (42) contains the simpler exponential $e^{\mathbf{D}t}$, which is given in closed form by (30). If an initial condition $\mathbf{x}(0) = \mathbf{b}$ is prescribed, then the solution is given by (42) again, with $\mathbf{c}$ changed to $\mathbf{b}$:

$$\boxed{\mathbf{x}(t) = \mathbf{Q}e^{\mathbf{D}t}\mathbf{Q}^{-1}\mathbf{b}} \tag{43}$$

Solution of the IVP $\mathbf{x}' = \mathbf{A}\mathbf{x}$; $\mathbf{x}(0) = \mathbf{b}$, in closed form.

If $\mathbf{A}$ has n LI eigenvectors it is **diagonalizable**, and is **diagonalized** according to the formula $\mathbf{Q}^{-1}\mathbf{A}\mathbf{Q} = \mathbf{D}$. Further, $\mathbf{Q}$ is called a **modal matrix** for $\mathbf{A}$.

We did not treat the nongeneric case in which one or more of $\mathbf{A}$'s eigenvalues is defective. In that case we are not able to make up the modal matrix $\mathbf{Q}$ because we have less than n LI eigenvectors to use for its columns. However, instead of diagonalizing $\mathbf{A}$, one is able to do almost as well, namely, "triangularizing" it and, once again, summing the exponential matrix series into closed form.[1]

[1] See, for instance, M. Greenberg, *Differential Equations & Linear Algebra* (Upper Saddle River, NJ: Prentice Hall, 2001).

EXERCISES 4.9

1. Nilpotent Matrices. Verify that each is nilpotent, by showing that repeated exponentiation of the matrix gives $\mathbf{O}$.

(a) $\mathbf{A} = \begin{bmatrix} 0 & a \\ 0 & 0 \end{bmatrix}$, $\mathbf{B} = \begin{bmatrix} 0 & a & b \\ 0 & 0 & c \\ 0 & 0 & 0 \end{bmatrix}$

(b) $\mathbf{C} = \begin{bmatrix} 0 & 0 \\ a & 0 \end{bmatrix}$, $\mathbf{D} = \begin{bmatrix} 0 & 0 & 0 \\ a & 0 & 0 \\ b & c & 0 \end{bmatrix}$

2. Similar to Example 1. Given $\mathbf{A}$ and $\mathbf{x}(0) = \mathbf{b}$, solve $\mathbf{x}' = \mathbf{Ax}$ for $\mathbf{x}(t)$, following the same lines as in Example 1.

(a) $\mathbf{A} = \begin{bmatrix} 0 & 1 & -1 \\ 0 & 0 & 4 \\ 0 & 0 & 0 \end{bmatrix}$, $\mathbf{b} = \begin{bmatrix} 12 \\ 0 \\ 24 \end{bmatrix}$

(b) $\mathbf{A} = \begin{bmatrix} 0 & 0 & 0 & 0 \\ 4 & 0 & 0 & 0 \\ 1 & 2 & 0 & 0 \\ 8 & 4 & 1 & 0 \end{bmatrix}$, $\mathbf{b} = \begin{bmatrix} 6 \\ 1 \\ 0 \\ 3 \end{bmatrix}$

3. If $\mathbf{P} = \{p_{ij}\}$ is *any* $n \times n$ diagonal matrix, then $e^{\mathbf{P}}$ is also diagonal, with $e^{p_{jj}}$ as its jth diagonal element. Verify that claim for the case $n = 3$, That is, show that if

$$\mathbf{P} = \begin{bmatrix} a & 0 & 0 \\ 0 & b & 0 \\ 0 & 0 & c \end{bmatrix}, \text{ then } e^{\mathbf{P}} = \begin{bmatrix} e^a & 0 & 0 \\ 0 & e^b & 0 \\ 0 & 0 & e^c \end{bmatrix}. \tag{3.1}$$

HINT: Use $e^{\mathbf{P}} = \mathbf{I} + \mathbf{P} + \frac{1}{2!}\mathbf{P}^2 + \cdots$.

4. Properties of the Exponential Matrix. (a) Show that the property $e^{(a+b)t} = e^{at}e^{bt}$ for the scalar exponential function does *not* carry over to the matrix exponential function. Specifically, show that

$$e^{(\mathbf{A}+\mathbf{B})t} = e^{\mathbf{A}t}e^{\mathbf{B}t} \tag{4.1}$$

holds if and only if $\mathbf{A}$ *and* $\mathbf{B}$ *commute*, that is, only if and only if $\mathbf{AB} = \mathbf{BA}$. HINT: Use (6) for each of the three exponentials in (4.1), multiply the two series on the right, and arrange their product as a power series in t, that is, in increasing powers of t. Then show that the coefficients of like powers of t are equal, on both sides of the equation, if and only if $\mathbf{AB} = \mathbf{BA}$. Let it suffice to do this only through terms of order t^2.
(b) Show that, for any $n \times n$ matrix $\mathbf{A}$, the inverse of the exponential matrix $e^{\mathbf{A}t}$ is simply

$$\boxed{(e^{\mathbf{A}t})^{-1} = e^{-\mathbf{A}t}.} \tag{4.2}$$

HINT: Use (4.1) and (6) to show that $e^{-\mathbf{A}t}e^{\mathbf{A}t} = e^{\mathbf{A}t}e^{-\mathbf{A}t} = \mathbf{I}$.

5. Recall from part (c) of Theorem 4.9.1 that a fundamental matrix for the system $\mathbf{x}' = \mathbf{Ax}$ is $e^{\mathbf{A}t}$, that is, its columns are n LI solutions of $\mathbf{x}' = \mathbf{Ax}$ and are therefore a fundamental set of solutions.

(a) Use that fact to find a fundamental set for the system given in Example 1, that is, for the differential equations given in (13) (i.e., not including the initial conditions). Verify that the column vectors are indeed a fundamental set. You may use any of the results given in that example without reproducing them.
(b) The same for the system given in Example 3.
(c) The same for the system given in Example 4.

6. Because $\mathbf{A}$ in Example 1 was nilpotent, we were able to solve readily by using the expansion (6), since the series terminated. Why could we *not* have solved Example 1 using Theorem 4.9.2 and equation (31) instead?

7. Similar to Examples 3 and 4. Use (31) to obtain $e^{\mathbf{A}t}$, which is a fundamental matrix for $\mathbf{x}' = \mathbf{Ax}$. Then, use the latter and (12) to solve the initial value problem $\mathbf{x}' = \mathbf{Ax}$, with $\mathbf{x}(0) = \mathbf{b}$. You may use computer software for the steps in part (h) if you wish.

(a) $\mathbf{A} = \begin{bmatrix} 3 & 1 \\ 1 & 3 \end{bmatrix}$, $\mathbf{b} = \begin{bmatrix} 1 \\ 0 \end{bmatrix}$

(b) $\mathbf{A} = \begin{bmatrix} 2 & 2 \\ 1 & 3 \end{bmatrix}$, $\mathbf{b} = \begin{bmatrix} 5 \\ 8 \end{bmatrix}$

(c) $\mathbf{A} = \begin{bmatrix} 1 & 1 \\ 1 & 1 \end{bmatrix}$, $\mathbf{b} = \begin{bmatrix} 0 \\ -3 \end{bmatrix}$

(d) $\mathbf{A} = \begin{bmatrix} 2 & 4 \\ 0 & 4 \end{bmatrix}$, $\mathbf{b} = \begin{bmatrix} 1 \\ 1 \end{bmatrix}$

(e) $\mathbf{A} = \begin{bmatrix} 0 & 0 & 2 \\ 0 & 0 & 2 \\ 2 & 2 & 2 \end{bmatrix}$, $\mathbf{b} = \begin{bmatrix} 8 \\ 0 \\ 0 \end{bmatrix}$

(f) $\mathbf{A} = \begin{bmatrix} 1 & 0 & 1 \\ 1 & 0 & 1 \\ 1 & 0 & 1 \end{bmatrix}$, $\mathbf{b} = \begin{bmatrix} 1 \\ 2 \\ 3 \end{bmatrix}$

(g) $\mathbf{A} = \begin{bmatrix} 1 & 1 & 1 \\ 2 & 2 & 2 \\ 3 & 3 & 3 \end{bmatrix}$, $\mathbf{b} = \begin{bmatrix} 0 \\ 0 \\ -5 \end{bmatrix}$

(h) $\mathbf{A} = \begin{bmatrix} 1 & 0 & 3 \\ 0 & 1 & 1 \\ 1 & -13 & 8 \end{bmatrix}, \quad \mathbf{b} = \begin{bmatrix} 10 \\ 5 \\ 0 \end{bmatrix}$

8. What if the Initial Point is Not $t = 0$? Derive a modified version of (43) if the initial point is t_0 instead of 0. That is, let the initial condition be $\mathbf{x}(t_0) = \mathbf{b}$. HINT: Use the general solution of $\mathbf{x}' = \mathbf{A}\mathbf{x}$ given by (42), apply the initial condition, and solve for $\mathbf{c}$.

ADDITIONAL EXERCISES

9. A Limited Proof of Cayley–Hamilton Theorem. The Cayley–Hamilton Theorem was stated without proof in the Additional Exercises for Section 4.8. It holds for *any* $n \times n$ matrix $\mathbf{A}$, but here we ask you to prove it only for the case in which $\mathbf{A}$ is diagonalizable, that is, if $\mathbf{A}$ has n LI eigenvectors. HINT: Let the characteristic polynomial

$$\alpha_0\lambda^n + \alpha_1\lambda^{n-1} + \cdots + \alpha_{n-1}\lambda + \alpha_n$$

of $\mathbf{A}$ be denoted as $p(\lambda)$. We wish to prove that

$$p(\mathbf{A}) = \alpha_0\mathbf{A}^n + \alpha_1\mathbf{A}^{n-1} + \cdots + \alpha_{n-1}\mathbf{A} + \alpha_n\mathbf{I} = \mathbf{O}. \quad (9.1)$$

To do that, use the diagonalization formula $\mathbf{D} = \mathbf{Q}^{-1}\mathbf{A}\mathbf{Q}$ and show that

$$p(\mathbf{A}) = \mathbf{Q}\,[\alpha_0\mathbf{D}^n + \alpha_1\mathbf{D}^{n-1} + \cdots + \alpha_{n-1}\mathbf{D} + \alpha_n\mathbf{I}]\,\mathbf{Q}^{-1}, \quad (9.2)$$

and that the matrix sum within the brackets is $\mathbf{O}$.

4.10 NONHOMOGENEOUS SYSTEMS

4.10.1 Solution by variation of parameters. In Section 2.9.1 we first solved the homogeneous equation $y' + p(x)y = 0$, obtaining the general solution $y(x) = Ce^{-\int p\,dx}$, and then solved the nonhomogeneous equation $y' + p(x)y = q(x)$ by varying the parameter C and seeking $y(x)$ in the form $C(x)e^{-\int p\,dx}$.

We will do the same here, as we now turn to the nonhomogeneous equation

$$\mathbf{x}' = \mathbf{A}(t)\mathbf{x} + \mathbf{f}(t). \quad (1)$$

Later in this section we will restrict the $n \times n$ matrix $\mathbf{A}$ to be constant, but initially we permit it to be a nonconstant function of t.

Assume that we know a fundamental matrix $\mathbf{X}(t)$ for the homogeneous equation $\mathbf{x}' = \mathbf{A}(t)\mathbf{x}$, so a general solution of the latter is $\mathbf{x}(t) = \mathbf{X}(t)\mathbf{c}$, in which $\mathbf{c}$ is arbitrary. Then, by "variation of parameters," seek a solution of the nonhomogeneous equation (1) by varying the vector parameter $\mathbf{c}$ and taking

$$\mathbf{x}(t) = \mathbf{X}(t)\mathbf{c}(t). \quad (2)$$

To determine $\mathbf{c}(t)$ so that (2) satisfies (1), substitute (2) into (1). For the left-hand side, we will need to know how to differentiate the product $\mathbf{X}(t)\mathbf{c}(t)$. Recalling the product rule $[f(t)g(t)]' = f'g + fg'$ from the scalar calculus, it is not surprising that the derivative of a product of matrices is

$$[\mathbf{A}(t)\mathbf{B}(t)]' = \mathbf{A}'(t)\mathbf{B}(t) + \mathbf{A}(t)\mathbf{B}'(t), \quad (3)$$

assuming, of course, that both matrices are differentiable and that they are conformable for multiplication, as are the $n \times n$ matrix $\mathbf{X}$ and $n \times 1$ matrix $\mathbf{c}$ in (2).

Equation (3) is a product rule for matrices, not necessarily square ($n \times n$), but conformable for multiplication.

Thus, substituting (2) into (1), with the help of (3), gives

$$\mathbf{X}'\mathbf{c} + \mathbf{X}\mathbf{c}' = \mathbf{A}\mathbf{X}\mathbf{c} + \mathbf{f}. \tag{4}$$

Since $\mathbf{X}' = \mathbf{A}\mathbf{X}$, the first terms on the left- and right-hand sides cancel, leaving $\mathbf{X}\mathbf{c}' = \mathbf{f}$. Now, $\mathbf{X}$ is invertible because its columns are LI, so we can pre-multiply both sides of $\mathbf{X}\mathbf{c}' = \mathbf{f}$ by the inverse $\mathbf{X}^{-1}(t)$, and obtain

The columns of $\mathbf{X}$ are LI, so $\det[\mathbf{X}(t)] \neq 0$ by Theorem 4.5.4. Then, since its determinant is nonzero, $\mathbf{X}(t)$ is invertible.

$$\mathbf{c}' = \mathbf{X}^{-1}(t)\mathbf{f}(t), \tag{5}$$

which can be integrated to give $\mathbf{c}(t)$. We can use an indefinite integral or a definite integral. If we are aiming at a general solution of (1) it is convenient to use an indefinite integral, obtaining

$$\mathbf{c}(t) = \int \mathbf{X}^{-1}(t)\mathbf{f}(t)\, dt + \mathbf{c}_0, \tag{6}$$

in which $\mathbf{c}_0$ is arbitrary. Finally, putting (6) into (2) gives the general solution

General solution of (1).

$$\boxed{\mathbf{x}(t) = \mathbf{X}(t)\Big(\int \mathbf{X}^{-1}(t)\mathbf{f}(t)\, dt + \mathbf{c}_0\Big),} \tag{7}$$

which is the matrix system generalization of the corresponding general solution of the scalar equation $y' + p(x)y = q(x)$ that was obtained in Section 1.2.

In (7), the $\mathbf{X}(t)\mathbf{c}_0$ part, with $\mathbf{c}_0$ arbitrary, is a general solution of the homogeneous equation $\mathbf{x}' = \mathbf{A}\mathbf{x}$, and the $\mathbf{X}(t)\int \mathbf{X}^{-1}(t)\mathbf{f}(t)\, dt$ part is a particular solution of (1). Their sum is a general solution of (1).

If, instead, we have an IVP version of (1), with initial condition $\mathbf{x}(t_0) = \mathbf{b}$, then it is convenient to re-express (7) as

In (8), τ is simply a dummy variable of integration.

$$\mathbf{x}(t) = \mathbf{X}(t)\Big(\int_{t_0}^{t} \mathbf{X}^{-1}(\tau)\mathbf{f}(\tau)\, d\tau + \mathbf{c}_0\Big), \tag{8}$$

which is equivalent to (7) because the difference between the indefinite integral in (7) and the definite integral in (8) is only an additive constant vector — which can be absorbed by the arbitrary $\mathbf{c}_0$ anyhow. Now apply the initial condition to (8) and obtain

$$\mathbf{b} = \mathbf{X}(t_0)(\mathbf{0} + \mathbf{c}_0), \tag{9}$$

so $\mathbf{c}_0 = \mathbf{X}^{-1}(t_0)\mathbf{b}$. Putting the latter into (8) gives the solution of the IVP as

Solution of IVP version of (1).

$$\boxed{\mathbf{x}(t) = \mathbf{X}(t)\Big(\int_{t_0}^{t} \mathbf{X}^{-1}(\tau)\mathbf{f}(\tau)\, d\tau + \mathbf{X}^{-1}(t_0)\mathbf{b}\Big).} \tag{10}$$

EXAMPLE 1. General Solution. Use (7) to obtain a general solution of the system $x_1' = x_2$, $x_2' = (1/t)x_2 + 3t$ on $0 < t < \infty$, or, in the matrix form (1),

$$\mathbf{x}' = \begin{bmatrix} 0 & 1 \\ 0 & 1/t \end{bmatrix} \begin{bmatrix} x_1 \\ x_2 \end{bmatrix} + \begin{bmatrix} 0 \\ 3t \end{bmatrix}. \tag{11}$$

We made up this example with a nonconstant matrix $\mathbf{A}(t)$ to highlight the fact that the foregoing derivation of (7) and (10) does not require $\mathbf{A}$ to be constant.

The latter was designed to have the simple LI homogeneous solutions $\mathbf{x}(t) = [1, 0]^{\mathrm{T}}$ and $\mathbf{x}(t) = [t^2, 2t]^{\mathrm{T}}$, so a fundamental matrix is $\mathbf{X}(t) = \begin{bmatrix} 1 & t^2 \\ 0 & 2t \end{bmatrix}$. Using the formula given in Exercise 21 of Section 4.8, its inverse is $\mathbf{X}^{-1}(t) = \dfrac{1}{2t} \begin{bmatrix} 2t & -t^2 \\ 0 & 1 \end{bmatrix}$. With these, (7) gives

If $\mathbf{A} = \begin{bmatrix} a & b \\ c & d \end{bmatrix}$ and $\det\mathbf{A} = ad - bc \neq 0$, then $\mathbf{A}^{-1} = \dfrac{1}{ad-bc} \begin{bmatrix} d & -b \\ -c & a \end{bmatrix}$.

$$\begin{aligned} \mathbf{x}(t) = \begin{bmatrix} x_1(t) \\ x_2(t) \end{bmatrix} &= \begin{bmatrix} 1 & t^2 \\ 0 & 2t \end{bmatrix} \left(\int \frac{1}{2t} \begin{bmatrix} 2t & -t^2 \\ 0 & 1 \end{bmatrix} \begin{bmatrix} 0 \\ 3t \end{bmatrix} dt + \mathbf{c}_0 \right) \\ &= \begin{bmatrix} 1 & t^2 \\ 0 & 2t \end{bmatrix} \left(\int \begin{bmatrix} -3t^2/2 \\ 3/2 \end{bmatrix} dt + \mathbf{c}_0 \right) \\ &= \begin{bmatrix} 1 & t^2 \\ 0 & 2t \end{bmatrix} \left(\begin{bmatrix} -t^3/2 \\ 3t/2 \end{bmatrix} + \mathbf{c}_0 \right) \\ &= \begin{bmatrix} 1 & t^2 \\ 0 & 2t \end{bmatrix} \mathbf{c}_0 + \begin{bmatrix} t^3 \\ 3t^2 \end{bmatrix}, \end{aligned} \tag{12}$$

in which the final right-hand side is the general solution of the homogeneous equation and a particular solution of the nonhomogeneous equation, respectively. Or, if we express the arbitrary vector $\mathbf{c}_0$ as $[c_1, c_2]^{\mathrm{T}}$,

$$\mathbf{x}(t) = c_1 \begin{bmatrix} 1 \\ 0 \end{bmatrix} + c_2 \begin{bmatrix} t^2 \\ 2t \end{bmatrix} + \begin{bmatrix} t^3 \\ 3t^2 \end{bmatrix}. \tag{13}$$

In scalar form, $x_1(t) = c_1 + c_2 t^2 + t^3$ and $x_2(t) = 2c_2 t + 3t^2$.

COMMENT. Both the matrix inversion formula given above in the margin, that we used to invert $\mathbf{X}(t)$, and also the elementary row operation method given in Section 4.8.3, hold even if the elements of the matrix are functions of t, for matrix inversion is an arithmetic operation involving no calculus-based steps. But, as review, let us repeat the inversion, this time using the elementary row operation which, unlike the formula in the margin for a 2×2 matrix $\mathbf{A}$, can be used for any $n \times n$ invertible matrix.

$$\begin{aligned} \mathbf{X}|\mathbf{I} &= \left[\begin{array}{cc|cc} 1 & t^2 & 1 & 0 \\ 0 & 2t & 0 & 1 \end{array}\right] \rightarrow \left[\begin{array}{cc|cc} 1 & t^2 & 1 & 0 \\ 0 & 1 & 0 & 1/2t \end{array}\right] \\ &\rightarrow \left[\begin{array}{cc|cc} 1 & 0 & 1 & -t/2 \\ 0 & 1 & 0 & 1/2t \end{array}\right] \rightarrow \mathbf{X}^{-1}(t) = \frac{1}{2t} \begin{bmatrix} 2t & -t^2 \\ 0 & 1 \end{bmatrix}. \quad \blacksquare \end{aligned} \tag{14}$$

EXAMPLE 2. IVP Version. Consider the same system (11), but this time with an initial condition $\mathbf{x}(1) = [7, 5]^{\mathrm{T}}$. Of course, we could simply apply that initial condition to the general solution (13). That would give $c_1 = 5$ and $c_2 = 1$, and hence

$$\mathbf{x}(t) = 5 \begin{bmatrix} 1 \\ 0 \end{bmatrix} + \begin{bmatrix} t^2 \\ 2t \end{bmatrix} + \begin{bmatrix} t^3 \\ 3t^2 \end{bmatrix} = \begin{bmatrix} 5 + t^2 + t^3 \\ 2t + 3t^2 \end{bmatrix}. \tag{15}$$

But, let us use (10) instead, to illustrate the use of that formula. It gives

$$\begin{aligned}\mathbf{x}(t) &= \begin{bmatrix} 1 & t^2 \\ 0 & 2t \end{bmatrix} \left(\int_1^t \frac{1}{2\tau} \begin{bmatrix} 2\tau & -\tau^2 \\ 0 & 1 \end{bmatrix} \begin{bmatrix} 0 \\ 3\tau \end{bmatrix} d\tau + \frac{1}{2} \begin{bmatrix} 2 & -1 \\ 0 & 1 \end{bmatrix} \begin{bmatrix} 7 \\ 5 \end{bmatrix} \right) \\ &= \begin{bmatrix} 1 & t^2 \\ 0 & 2t \end{bmatrix} \left(\int_1^t \begin{bmatrix} -3\tau^2/2 \\ 3/2 \end{bmatrix} d\tau + \begin{bmatrix} 9/2 \\ 5/2 \end{bmatrix} \right) \\ &= \begin{bmatrix} 1 & t^2 \\ 0 & 2t \end{bmatrix} \left(\begin{bmatrix} -t^3/2 + 1/2 \\ 3t/2 - 3/2 \end{bmatrix} + \begin{bmatrix} 9/2 \\ 5/2 \end{bmatrix} \right) \\ &= \begin{bmatrix} 5 + t^2 + t^3 \\ 2t + 3t^2 \end{bmatrix}, \end{aligned} \tag{16}$$

as in (15). That is, $x_1(t) = 5 + t^2 + t^3$ and $x_2(t) = 2t + 3t^2$. ∎

4.10.2 Constant coefficient matrix. The foregoing discussion did not require the coefficient matrix $\mathbf{A}$ to be constant, but realize that to implement (7) or (10) we need to determine a fundamental matrix $\mathbf{X}(t)$; and for nonconstant coefficient matrices, this can be a formidable task. We survived in Examples 1 and 2 because we made up the systems to admit simple closed-form solutions.

Thus, let us now require $\mathbf{A}$ to be constant, in which case we know how to find n LI solutions of the homogeneous equation, and hence a fundamental matrix of them. Let us illustrate.

EXAMPLE 3. Use (10) to solve the IVP

$$\mathbf{x}' = \begin{bmatrix} 1 & 4 \\ 1 & 1 \end{bmatrix} \begin{bmatrix} x_1 \\ x_2 \end{bmatrix} + \begin{bmatrix} 8e^t \\ 18t \end{bmatrix}; \quad \mathbf{x}(0) = \begin{bmatrix} 4 \\ 0 \end{bmatrix}. \tag{17}$$

The eigenvalues and eigenvectors of $\mathbf{A}$ are found to be $\lambda_1 = 3$, $\mathbf{e}_1 = \alpha[2,1]^{\mathrm{T}}$ and $\lambda_2 = -1$, $\mathbf{e}_2 = \beta[2,-1]^{\mathrm{T}}$, so a fundamental matrix for $\mathbf{x}' = \mathbf{A}\mathbf{x}$ is

$$\mathbf{X}(t) = \begin{bmatrix} 2e^{3t} & 2e^{-t} \\ e^{3t} & -e^{-t} \end{bmatrix}, \quad \text{with} \quad \mathbf{X}^{-1}(t) = \frac{1}{4e^{2t}} \begin{bmatrix} e^{-t} & 2e^{-t} \\ e^{3t} & -2e^{3t} \end{bmatrix}. \tag{18}$$

Then, (10) gives the solution to (17) as

$$\int_0^t \tau e^{a\tau}\, d\tau = (a\tau - 1)\frac{e^{a\tau}}{a^2}\bigg|_0^t$$

$$\begin{aligned}\mathbf{x}(t) &= \begin{bmatrix} 2e^{3t} & 2e^{-t} \\ e^{3t} & -e^{-t} \end{bmatrix} \left(\int_0^t \frac{1}{4e^{2\tau}} \begin{bmatrix} e^{-\tau} & 2e^{-\tau} \\ e^{3\tau} & -2e^{3\tau} \end{bmatrix} \begin{bmatrix} 8e^{\tau} \\ 18\tau \end{bmatrix} d\tau + \frac{1}{4} \begin{bmatrix} 1 & 2 \\ 1 & -2 \end{bmatrix} \begin{bmatrix} 4 \\ 0 \end{bmatrix} \right) \\ &= \begin{bmatrix} 2e^{3t} & 2e^{-t} \\ e^{3t} & -e^{-t} \end{bmatrix} \left(\int_0^t \begin{bmatrix} 2e^{-2\tau} + 9\tau e^{-3\tau} \\ 2e^{2\tau} - 9\tau e^{\tau} \end{bmatrix} d\tau + \begin{bmatrix} 1 \\ 1 \end{bmatrix} \right) \\ &= \begin{bmatrix} 2e^{3t} & 2e^{-t} \\ e^{3t} & -e^{-t} \end{bmatrix} \left(\begin{bmatrix} 2 - e^{-2t} - 3te^{-3t} - e^{-3t} \\ -10 + e^{2t} - 9te^{t} + 9e^{t} \end{bmatrix} + \begin{bmatrix} 1 \\ 1 \end{bmatrix} \right) \\ &= \begin{bmatrix} 6e^{3t} - 18e^{-t} - 24t + 16 \\ 3e^{3t} + 9e^{-t} - 2e^{t} + 6t - 10 \end{bmatrix}. \end{aligned} \tag{19}$$

Thus, $x_1(t) = 6e^{3t} - 18e^{-t} - 24t + 16$ and $x_2(t) = 3e^{3t} + 9e^{-t} - 2e^t + 6t - 10$. ▮

4.10.3 Particular solution by undetermined coefficients. In Examples 1–3, we relied on the explicit solution formulae (7) and (10). Alternatively, if $\mathbf{A}$ is constant, we can find a homogeneous solution (as in Section 4.7) and (subject to the same sort of limitations as were cited in Section 2.8) a particular solution by undetermined coefficients.

EXAMPLE 4. Undetermined Coefficients. To illustrate, let us re-solve the IVP (17) in that manner. As noted in Example 3, the eigenvalues and eigenvectors of $\mathbf{A}$ are $\lambda_1 = 3$, $\mathbf{e}_1 = \alpha[2,1]^{\mathrm{T}}$, and $\lambda_2 = -1$, $\mathbf{e}_2 = \beta[2,-1]^{\mathrm{T}}$, so a fundamental matrix is the matrix $\mathbf{X}(t)$ given in (18), and a homogeneous solution of the system is

$$\mathbf{x}_h(t) = \mathbf{X}(t)\mathbf{c} = \begin{bmatrix} 2e^{3t} & 2e^{-t} \\ e^{3t} & -e^{-t} \end{bmatrix} \begin{bmatrix} c_1 \\ c_2 \end{bmatrix}, \tag{20}$$

Realize that (20) can also be expressed as $\mathbf{x}_h(t) = c_1 \begin{bmatrix} 2e^{3t} \\ e^{3t} \end{bmatrix} + c_2 \begin{bmatrix} 2e^{-t} \\ -e^{-t} \end{bmatrix}$.

in which the constant vector $\mathbf{c}$ is arbitrary.

Next, we wish to find a particular solution $\mathbf{x}_p(t)$. To do so by undetermined coefficients, first write the system as

$$\mathbf{x}' = \mathbf{A}\mathbf{x} + \begin{bmatrix} 8 \\ 0 \end{bmatrix} e^t + \begin{bmatrix} 0 \\ 18 \end{bmatrix} t. \tag{21}$$

The e^t term generates (by repeated differentiations) only the single term e^t, and the t term generates the two terms t and 1, so tentatively seek $\mathbf{x}_p$ in the form

$$\mathbf{x}_p(t) = (\mathbf{p}e^t) + (\mathbf{q}t + \mathbf{r}), \tag{22}$$

Of course the $\mathbf{r}$ in (22) is really $\mathbf{r}$ times 1, so the right-hand side of (22) is a linear combination of e^t, t, and 1.

in which $\mathbf{p}$, $\mathbf{q}$, $\mathbf{r}$ are the undetermined (vector) coefficients. The e^t in the first term does not duplicate the e^{3t} or the e^{-t} in the homogeneous solution so leave the $\mathbf{p}e^t$ intact. Similarly, neither the t nor the 1 in the second term duplicates terms in the homogeneous solution. Thus, accept (22) and proceed.

Substituting (22) into (21) gives

$$\begin{aligned} \mathbf{p}e^t + \mathbf{q} &= \mathbf{A}(\mathbf{p}e^t + \mathbf{q}t + \mathbf{r}) + \begin{bmatrix} 8 \\ 0 \end{bmatrix} e^t + \begin{bmatrix} 0 \\ 18 \end{bmatrix} t \\ &= (\mathbf{A}\mathbf{p})e^t + (\mathbf{A}\mathbf{q})t + \mathbf{A}\mathbf{r} + \begin{bmatrix} 8 \\ 0 \end{bmatrix} e^t + \begin{bmatrix} 0 \\ 18 \end{bmatrix} t. \end{aligned} \tag{23}$$

Next, to render (23) an identity, match the coefficients of the e^t, t, and constant terms, respectively, obtaining the three equations

$$e^t: \quad \mathbf{p} = \mathbf{A}\mathbf{p} + \begin{bmatrix} 8 \\ 0 \end{bmatrix}, \tag{24a}$$

$$t: \quad \mathbf{0} = \mathbf{A}\mathbf{q} + \begin{bmatrix} 0 \\ 18 \end{bmatrix}, \tag{24b}$$

$$1: \quad \mathbf{q} = \mathbf{A}\mathbf{r} \tag{24c}$$

In (24a,b,c) we match coefficients of the LI functions $e^t, t, 1$, respectively, in (23).

for **p**, **q**, and **r**. Written out, these are

$$\begin{bmatrix} 0 & 4 \\ 1 & 0 \end{bmatrix}\begin{bmatrix} p_1 \\ p_2 \end{bmatrix} = \begin{bmatrix} -8 \\ 0 \end{bmatrix}, \quad \begin{bmatrix} 1 & 4 \\ 1 & 1 \end{bmatrix}\begin{bmatrix} q_1 \\ q_2 \end{bmatrix} = \begin{bmatrix} 0 \\ -18 \end{bmatrix}, \quad \begin{bmatrix} 1 & 4 \\ 1 & 1 \end{bmatrix}\begin{bmatrix} r_1 \\ r_2 \end{bmatrix} = \begin{bmatrix} q_1 \\ q_2 \end{bmatrix}, \tag{25}$$

which give $p_1 = 0$, $p_2 = -2$, $q_1 = -24$, $q_2 = 6$, $r_1 = 16$, and $r_2 = -10$. Thus, (22) becomes

$$\mathbf{x}_p(t) = \begin{bmatrix} 0 \\ -2 \end{bmatrix} e^t + \begin{bmatrix} -24 \\ 6 \end{bmatrix} t + \begin{bmatrix} 16 \\ -10 \end{bmatrix} \tag{26}$$

and a general solution of (21) is

$$\mathbf{x}(t) = \mathbf{x}_h(t) + \mathbf{x}_p(t) = \begin{bmatrix} 2e^{3t} & 2e^{-t} \\ e^{3t} & -e^{-t} \end{bmatrix}\begin{bmatrix} c_1 \\ c_2 \end{bmatrix} + \begin{bmatrix} 0 \\ -2 \end{bmatrix} e^t + \begin{bmatrix} -24 \\ 6 \end{bmatrix} t + \begin{bmatrix} 16 \\ -10 \end{bmatrix}. \tag{27}$$

Finally, imposing the initial condition $\mathbf{x}(0) = [4, 0]^{\mathrm{T}}$ gives two linear algebraic equations for c_1 and c_2, with solution $c_1 = 2$ and $c_2 = -9$. With these values, (27) is identical to the solution (19) found in Example 3. ▮

Closure. If we know a fundamental matrix $\mathbf{X}(t)$ for $\mathbf{x}' = \mathbf{A}(t)\mathbf{x}$, we can use variation of parameters and seek a solution of the nonhomogeneous equation $\mathbf{x}' = \mathbf{A}(t)\mathbf{x} + \mathbf{f}(t)$ in the form $\mathbf{x}(t) = \mathbf{X}(t)\mathbf{c}(t)$. We did that, and obtained the explicit formula (7) for the general solution, with $\mathbf{c}_0$ an arbitrary constant vector. If an initial condition $\mathbf{x}(t_0) = \mathbf{b}$ is prescribed, we can solve for $\mathbf{c}_0$ and, doing that, we obtained the solution (10).

We emphasized that if the coefficient matrix is not constant, then determining a fundamental matrix is problably difficult. If, however, a fundamental matrix *is* known [that is, if a general solution of the homogeneous version of (1) is known], then it is relatively simple to find, from that, a particular solution of (1), this step "merely" involving the integrations in (7) and (10). This result, for systems, echoes the same result for single equations of higher order. For instance, we saw in Section 2.10.2 that *if* we know two LI solutions of $y'' + p(x)y' + q(x)y = 0$, then we can use them, by means of variation of parameters, to obtain a particular solution of the nonhomogeneous equation $y'' + p(x)y' + q(x)y = f(x)$ by integration; see (19) in that section.

Not surprisingly, if the coefficient matrix is *constant*, then the situation is much simpler. We know how to find a general solution of the homogeneous equation and can probably use the method of undetermined coefficients to find a particular solution. With a general solution of the homogeneous equation and a particular solution of the nonhomogeneous equation in hand, a general solution is then given by their sum.

EXERCISES 4.10

1. Product Rule for Matrices. Derive the product rule for matrices, given by (3). HINT: The derivation mimics that for scalar functions. Write the derivative as a difference quotient and rewrite the numerator $\mathbf{A}(t+\Delta t)\mathbf{B}(t+\Delta t)-\mathbf{A}(t)\mathbf{B}(t)$ as $[\mathbf{A}(t+\Delta t)\mathbf{B}(t+\Delta t)-\mathbf{A}(t+\Delta t)\mathbf{B}(t)]+[\mathbf{A}(t+\Delta t)\mathbf{B}(t)-\mathbf{A}(t)\mathbf{B}(t)]$.

2. Using the Product Rule. Let $\mathbf{A}(t)$, $\mathbf{B}(t)$, $\mathbf{x}(t)$, and $\mathbf{y}(t)$ be $\begin{bmatrix} 2 & t^3 \\ 4 & e^t \end{bmatrix}$, $\begin{bmatrix} e^t & 5 \\ \sin t & -t \end{bmatrix}$, $\begin{bmatrix} t \\ 0 \end{bmatrix}$, and $\begin{bmatrix} 1 \\ t^2 \end{bmatrix}$, respectively. Work out the t-derivative of the following product, with the help of the product rule, (3). Then rework it without the product rule (i.e., by doing the matrix multiplication first and *then* the derivative), and verify that the results agree.

(a) $\mathbf{AB}$ (b) $\mathbf{BA}$ (c) $\mathbf{Ax}$ (d) $\mathbf{Ay}$ (e) $\mathbf{Bx}$ (f) $\mathbf{By}$

3. (a) Verify, by substitution, that (7) satisfies (1).
(b) Verify, by substitution, that (10) satisfies both (1) and the initial condition $\mathbf{x}(t_0)=\mathbf{b}$.

4. Verify, by substitution, that (16) satisfies (11) and the initial condition $\mathbf{x}(1)=[7,5]^{\mathrm{T}}$.

5. Verify, by substitution, that (19) satisfies the IVP (17).

6. About the Application of Linear Independence to (23). Surely we *can* make (23) an identity (on any t interval I) by equating the coefficients of the e^t, t, and "1" terms on the left- and right-hand sides, as we did in (24). More than that, show that we *must* do that because $\{e^t,t,1\}$ is LI on I. That is, show that if a scalar function set $\{u(t),v(t),w(t)\}$ is LI on I, then

$$\mathbf{a}u(t)+\mathbf{b}v(t)+\mathbf{c}w(t)=\mathbf{d}u(t)+\mathbf{e}v(t)+\mathbf{f}w(t), \qquad (6.1)$$

where $\mathbf{a}$ through $\mathbf{f}$ are constant n-tuple vectors, holds on I if and only if $\mathbf{a}=\mathbf{d}$, $\mathbf{b}=\mathbf{e}$, and $\mathbf{c}=\mathbf{f}$. [Similarly for sets of more or less than three functions, but let it suffice to consider (6.1).]

7. Use (7) to find a general solution of the given system. The interval I is $-\infty<t<\infty$, but $0<t<\infty$ in parts (i) and (j). NOTE: You will first need to find LI solutions of the homogeneous system. In parts (i)–(l), such LI solutions are given, and you may use them without derivation. For parts (a)–(j) you can compute $\mathbf{X}^{-1}(t)$ by using the inverse matrix formula given as a margin note for Example 1, and for parts (k) and (l) you may use computer software for that step if you wish.

(a) $x_1'=x_1+2x_2+6e^t,\ x_2'=3x_1+6x_2+7$
(b) $x_1'=2x_1+3x_2+6e^{2t},\ x_2'=2x_1+7x_2$
(c) $x_1'=7x_1+2x_2,\ x_2'=3x_1+2x_2-64t$
(d) $x_1'=2x_1+x_2,\ x_2'=9x_1+2x_2+100$
(e) $x_1'=5x_1+x_2+15e^t,\ x_2'=x_1+5x_2$
(f) $x_1'=12x_1+3x_2,\ x_2'=-20x_1-5x_2+6e^t$
(g) $x_1'=3x_1+4x_2,\ x_2'=2x_1+x_2+4e^t-9e^{2t}$
(h) $x_1'=2x_1+x_2,\ x_2'=x_1+2x_2+10e^t$
(i) $x_1'=x_2+6t^2,\ x_2'=(2/t)x_2;\ \ [1,0]^{\mathrm{T}},[t^3,3t^2]^{\mathrm{T}}$
(j) $x_1'=(1/t^2)x_2+8t^3,\ x_2'=(2/t)x_2-2t^2;$ $[1,0]^{\mathrm{T}},[t,t^2]^{\mathrm{T}}$
(k) $x_1'=x_3,\ x_2'=x_3,\ x_3'=x_1+x_2+x_3+4e^t$ $[1,-1,0]^{\mathrm{T}},e^{-t}[1,1,-1]^{\mathrm{T}},e^{2t}[1,1,2]^{\mathrm{T}}$
(l) $x_1'=4x_1+x_2+x_3,\ x_2'=x_1+4x_2+x_3,\ x_3'=x_1+x_2+4x_3+36t;\ e^{3t}[1,-1,0]^{\mathrm{T}},e^{6t}[1,1,1]^{\mathrm{T}},e^{3t}[2,-1,-1]^{\mathrm{T}}$

8. Computing the Inverse of $\mathbf{X}(t)$. In this section we must be able to compute the inverse matrix $\mathbf{X}^{-1}(t)$ matrix. For 2×2 matrices there is the convenient formula given in the Exercises for Section 4.8, and again here as a margin note for Example 1. For matrices larger than 2×2, one can use the *same* elementary row operation procedure that was explained in Section 4.8 [see (28) and Example 1 there], even though the matrix elements are now functions of t.
(a) Use that method to derive the margin formula. That is, use elementary row operations to derive

$$\begin{bmatrix} a(t) & b(t) \\ c(t) & d(t) \end{bmatrix}^{-1}=\frac{1}{a(t)d(t)-b(t)c(t)}\begin{bmatrix} d(t) & -b(t) \\ -c(t) & a(t) \end{bmatrix}. \qquad (8.1)$$

(b) Use the method of elementary row operations to derive the inverse

$$\begin{bmatrix} 1 & e^{-t} & e^{2t} \\ -1 & e^{-t} & e^{2t} \\ 0 & -e^{-t} & 2e^{2t} \end{bmatrix}^{-1}=\frac{1}{6}\begin{bmatrix} 3 & -3 & 0 \\ 2e^t & 2e^t & -2e^t \\ e^{-2t} & e^{-2t} & 2e^{-2t} \end{bmatrix}.$$

(c) Use the method of elementary row operations to derive the inverse

$$\begin{bmatrix} e^{3t} & e^{6t} & 2e^{3t} \\ -e^{3t} & e^{6t} & -e^{3t} \\ 0 & -e^{6t} & -e^{3t} \end{bmatrix}^{-1}=\begin{bmatrix} 2e^{-3t} & e^{-3t} & 3e^{-3t} \\ e^{-6t} & e^{-6t} & e^{-6t} \\ -e^{-3t} & -e^{-3t} & -2e^{-3t} \end{bmatrix}.$$

9. Verify that $\mathbf{x}(t)$ given by (10) satisfies the initial condition $\mathbf{x}(t_0)=\mathbf{b}$.

10. Undetermined Coefficients. (b)–(g) For the corresponding part of Exercise 7, use undetermined coefficients to obtain a particular solution, as we did in Example 4.

ADDITIONAL EXERCISES

11. Undetermined Coefficients Again, This Time with Redundancy. In this section on *systems*, we've used the method of variation of parameters in Section 4.10.1, and the method of undetermined coefficients in Section 4.10.3. When we studied undetermined coefficients for a *single* differential equation, in Section 2.9, we found that certain "redundancies" can occur, in which case the trial form for the particular solution had to be modified accordingly. Similarly, analogous redundancies can occur when we apply undetermined coefficients to systems. For instance, consider the system

$$\mathbf{x}' = \begin{bmatrix} 2 & 3 \\ 2 & 7 \end{bmatrix} \begin{bmatrix} x_1 \\ x_2 \end{bmatrix} + \begin{bmatrix} 7e^t \\ 0 \end{bmatrix}. \tag{11.1}$$

(a) To find a particular solution, note that the forcing function is $\mathbf{f}(t) = \begin{bmatrix} 7 \\ 0 \end{bmatrix} e^t$, so tentatively seek a particular solution in the form $\mathbf{x}_p(t) = \mathbf{p}e^t$, where the constant vector $\mathbf{p}$ is to be determined. Try that form and show that it does not work.
(b) The reason the latter did not work is the "redundancy," the forcing function and one of the homgeneous solutions both being proportional to e^t. Try the modified form $\mathbf{x}_p(t) = \mathbf{p}te^t + \mathbf{q}e^t$, and show that it does work. Use it to derive the particular solution $\mathbf{x}_p(t) = \begin{bmatrix} (6t-1)e^t \\ -2te^t \end{bmatrix}$. NOTE: You will need to think your way through some of the steps.

CHAPTER 4 REVIEW

This chapter dealt only with *linear* systems.

Section 4.1 introduced the idea of linear systems of coupled first-order differential equations, including physical examples. For systems of *two* equations we showed how to solve them using elimination to obtain instead a single second-order equation in only one unknown.

Continuing to use elimination, **Section 4.2** was devoted to an important application, both the free and forced vibrations of a coupled harmonic oscillator, and followed up on the harmonic oscillator material in Chapter 3. Whereas a single-mass oscillator has a single natural frequency, a two-mass oscillator has two natural frequencies, and a general solution for the free vibration is a linear combination of two modes, each of which is an oscillation at one of the natural frequencies. If the system is forced harmonically, resonance occurs if the frequency of the forcing function matches either of those natural frequencies. Similarly for N-mass oscillators. Although the oscillators studied were mechanical, the same theory and results apply if they are electrical, biological, or of any other type.

For systems of two equations, elimination is simple, but that method is not well suited to systems of many equations, and it may not even work if there are more than two. Thus, we began, in Section 4.3, to develop vector and matrix methods that would prove particularly convenient for systems. In **Sections 4.3** and **4.4**, we introduced vectors in n-space, matrix operators on those vectors, and the linear dependence and linear independence of sets of n-tuple vectors.

With that preparation completed, **Section 4.5** gave theorems for the form of a general solution of a system of linear differential equations, and for the existence and uniqueness of solutions of a system of linear differential equations with initial conditions. These were similar to the corresponding theorems for a single linear higher-order equation, from Chapter 2. Fundamental throughout was linear independence: of sets of functions in Chapter 2 and of sets of vectors in Chapter 4. We considered only linear systems, but did allow for nonconstant coefficient matrices. Particularly important were Theorems 4.5.1 on existence and uniqueness of solutions and 4.5.3 on the form of the general solution, both for homogeneous and nonhomogeneous systems. Theorem 4.5.4 gave a convenient *test* for the linear independence of a set of solutions $\mathbf{x}_1(t), \ldots, \mathbf{x}_n(t)$ of the homogeneous system $\mathbf{x}' = \mathbf{A}(t)\mathbf{x}$. Namely, the set is LI on I if and only if $\det[\mathbf{x}_1(t), \ldots, \mathbf{x}_n(t)]$ is nonzero in I.

However, that section did not give *methods* for generating those n LI solutions, and if the coefficient matrix is not constant, then obtaining solutions can be quite difficult. Thus, to proceed, we restricted the coefficient matrix $\mathbf{A}$ to be constant, rather than a nonconstant function of t, and introduced additional linear algebra material on the eigenvalue problem in **Section 4.6**.

Treating the homogeneous equation $\mathbf{x}' = \mathbf{A}\mathbf{x}$ first, we used a solution method of assumed exponential form in **Section 4.7**. The latter is a version of the method of assumed exponential form that was successfully used in Chapter 2 for single linear higher-order equations with constant coefficients. That is, we assumed a solution form $\mathbf{x}(t) = \mathbf{q}e^{rt}$, which is a vector generalization of the form $x(t) = e^{rt}$ used in Chapter 2. Putting the latter into the system $\mathbf{x}' = \mathbf{A}\mathbf{x}$ gave an eigenvalue problem $\mathbf{A}\mathbf{q} = r\mathbf{q}$ for r and $\mathbf{q}$. The only complication occurred if the coefficient matrix $\mathbf{A}$ had one or more defective eigenvalues — eigenvalues having less LI eigenvectors than their multiplicity. To find the "missing" solutions corresponding to a defective eigenvalue, we modified the assumed solution form to include powers of t times the exponentials, the situation being essentially the same as occurred in Chapter 2 when the characteristic equation had repeated roots.

Besides the method of assumed exponential form, for finding n LI solutions of the homogeneous equation, we also aimed at finding an *explicit solution formula*, if possible, as we did for the single first-order linear equation in Chapter 1. To prepare for that, we paused again, in **Section 4.8**, to present additional linear algebra concepts: matrix operators, matrix multiplication, and the inverse matrix.

Then, in **Section 4.9** we obtained an exponential matrix solution for the IVP

$$\mathbf{x}' = \mathbf{A}\mathbf{x}; \quad \mathbf{x}(0) = \mathbf{b}$$

and obtained the explicit solution $\mathbf{x}(t) = e^{\mathbf{A}t}\mathbf{b}$, but had to deal with the infinite series defining the exponential matrix $e^{\mathbf{A}t}$. We showed how to get that series into closed form for the generic case in which $\mathbf{A}$ has no defective eigenvalues, the key being the diagonalization formula $\mathbf{Q}^{-1}\mathbf{A}\mathbf{Q} = \mathbf{D}$ for the $\mathbf{A}$ matrix.

It remained to obtain particular solutions due to a forcing function $\mathbf{f}(t)$ in the system $\mathbf{x}' = \mathbf{A}\mathbf{x} + \mathbf{f}(t)$, and we did that in **Section 4.10**. Assuming a homogeneous solution to be known, we used the method of variation of parameters to derive an explicit formula for the general solution, including both homogeneous and particular solutions. We also showed how to find particular solutions by a method of undetermined coefficients, like that used in Chapter 2 for a single higher-order equation with a forcing function, that is useful in simple cases.

Chapter 5

LAPLACE TRANSFORM

5.1 INTRODUCTION

The Laplace transform method is quite different from the methods developed in the preceding chapters. It will be useful for initial value problems governed by differential equations with constant coefficients, particularly with forcing functions that are defined piecewise such as the one in Fig. 1 or that are periodic such as the one in Fig. 2.

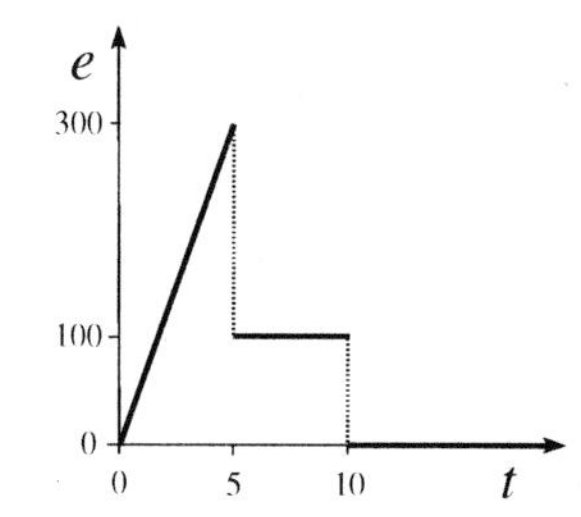

Figure 1. A piecewise-defined forcing function.

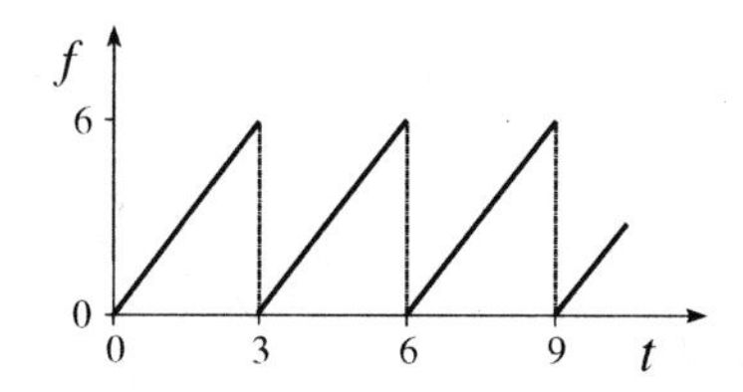

Figure 2. A periodic forcing function.

In this chapter we will use t as the independent variable because, typically, the Laplace transform is used for IVPs on $0 \leq t < \infty$, in which t is the time.

We will begin by defining the Laplace transform and its inverse in Section 5.2. The concepts of transform and inverse will become clear once we begin to use the transform to solve differential equations in Section 5.3 but, until then, Section 5.2 may seem unmotivated and hence unclear. Thus, we will try to motivate the basic idea behind transform methods, in this introduction, by explaining a very simple transform, which we will call the "logarithmic transform."

Imagine that this is before the advent of calculators and computers, so numerical calculations must be done by hand. Then, even elementary operations such as multiplication can be quite tedious. (Try multiplying 3.715649 and 8.699237!)

But, suppose we are familiar with the natural logarithm function $\ln x$ defined on $0 < x < \infty$ (Fig. 3), as well as with its various properties, in particular

$$\ln(uv) = \ln u + \ln v. \tag{1}$$

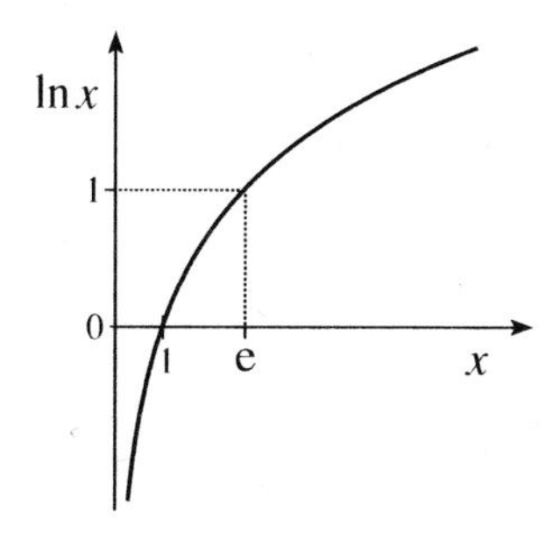

Figure 3. Graph of $\ln x$.

Suppose further that we have a logarithm table, with values of x in the first column and the corresponding values of $\ln x$ in the second. In the other direction, we say that the "antilog" of $\ln x$ is x. Then, to evaluate a product uv we can proceed as follows: Write $z = uv$, where z is to be determined. Take the natural log of both sides, obtaining $\ln z = \ln(uv)$. Then use the property (1), obtaining $\ln z = \ln u + \ln v$. Look up $\ln u$ and $\ln v$ in the table, and *add* them. Finally, take the antilog of their sum to obtain z.

To illustrate, suppose we want to compute 1.7356×6.8102. The steps are these:

By taking the transform of (2a) we mean taking the logarithm of both sides of (2a). In (2f) we are denoting the inverse logarithm, or "antilog," as "$\ln^{-1}$."

$$\begin{aligned}
z &= (1.7356)(6.8102) && \text{(2a)}\\
\ln z &= \ln\left[(1.7356)(6.8102)\right] \quad \text{[taking the "logarithmic transform" of (2a)]} && \text{(2b)}\\
&= \ln(1.7356) + \ln(6.8102) \quad \text{[by transform property (1)]} && \text{(2c)}\\
&= 0.55135 + 1.91842 \quad \text{[by looking up } \ln(1.7356) \text{ and } \ln(6.8102) \\
&\qquad\qquad \text{in the right-hand column of the table]} && \text{(2d)}\\
&= 2.46977 \quad \text{[by addition]} && \text{(2e)}\\
z &= \ln^{-1}(2.46977) \quad \text{[taking the inverse transform]}\\
&= 11.8201 \quad \text{[looking up the inverse in the left- hand column of table].} && \text{(2f)}
\end{aligned}$$

"Logarithmic Transform" Table

x	$\ln x$
⋮	⋮
1.7356 →	0.55135
⋮	+
6.8102 →	1.91842
⋮	↓
11.8201 ←	2.46977

Figure 4. Logarithmic transform calculation.

Taking the transform of both sides of (2a) moved us to the "transform domain." What was a multiplication in the original domain [i.e., in (2a)] became a simpler operation, addition, in the transform domain [i.e., in (2d)]. We carried out the addition, then returned to the original domain by the inverse transform, the antilog. The steps are indicated in Fig. 4.

For the procedure to work, we needed both the transform and the inverse transform to *exist* and to be *unique*, so we could indeed obtain the transforms in (2d) and the inverse transform in (2f), each in a unique way. These requirements were satisfied because, as we can see from Fig. 3, the logarithm function is a one-to-one function on $0 < x < \infty$.

The Laplace transform, on the other hand, is designed not to facilitate algebraic steps such as multiplication but, primarily, for the solution of linear differential equation initial value problems. Somewhat as the "logarithmic transform" converted the *product* operation in (2a) into the simple *sum* operation in the transform domain, the Laplace transform converts a constant-coefficient linear *differential* equation in the t domain into a linear *algebraic* equation in the transform domain.

Of course, the logarithm is not normally presented, in pre-calculus, in terms of transforms. But, as mentioned above, the foregoing logarithmic transform analogy should help to motivate the material that follows in Section 5.2.

Besides the Laplace transform there is also a **Fourier transform**, which is of comparable importance, but the latter is outside the scope of this text. Fortunately, the transform methodology is so similar, from one transform to another, that an understanding of one is of great help in studying others.

5.2 THE TRANSFORM AND ITS INVERSE

This section is to introduce the transform and to give just enough about the transform and inversion processes so we can get started with differential equation applications in the next section. Additional material on transforms and inverses is given in subsequent sections.

5.2.1 Laplace transform. If a function $f(t)$ defined on $0 \leq t < \infty$ is multiplied by e^{-st}, in which s is a constant, and integrated from zero to infinity, the result is a function of the parameter s. Known as the **Laplace transform** of $f(t)$, it is denoted as $\mathcal{L}\{f(t)\}$ or as $F(s)$:[1]

The $\mathcal{L}$ is a script L, for Laplace. Denote the transform of $f(t)$ either as $\mathcal{L}\{f(t)\}$ or as $F(s)$.

$$\boxed{\mathcal{L}\{f(t)\} = F(s) = \int_0^\infty f(t)e^{-st}\,dt.} \tag{1}$$

In some contexts it is important to allow the parameter s to be a complex number, but in this chapter it will be simpler and sufficient to restrict it to be real.

Thus, for the Laplace transform the input is a function of t and the output, its transform, is a function of s. It is an example of an **integral transform** because it is defined by the *integral* in (1).

The integral is *improper* because of its infinite upper limit, and is defined as the following limit of a sequence of proper integrals (i.e., with finite limits),

$$F(s) = \int_0^\infty f(t)e^{-st}\,dt \equiv \lim_{B\to\infty}\int_0^B f(t)e^{-st}\,dt. \tag{2}$$

The definition (2) is in the same spirit as the definition of an infinite series $\sum_0^\infty a_n$ as the limit of the sequence of partial sums, $\lim_{N\to\infty}\sum_0^N a_n$, that one meets in the calculus.

The whole Laplace transform method rests upon the definition (1). As we indicated in Section 5.1, it is designed to convert linear constant-coefficient *differential* equations to simple linear *algebraic* equations, but we won't apply the transform to differential equations until Section 5.3. Here, to get started, let us begin by evaluating the integral (1) for several familiar functions, that is, by working out their Laplace transforms.

EXAMPLE 1. $f(t) = 1$. Evaluate the Laplace transform of the simple function $f(t) = 1$. By the definition of the transform, we have

$$\begin{aligned}\mathcal{L}\{1\} &= \int_0^\infty 1e^{-st}\,dt = \lim_{B\to\infty}\int_0^B e^{-st}\,dt \\ &= \lim_{B\to\infty}\left(\frac{e^{-st}}{-s}\Big|_0^B\right) = \lim_{B\to\infty}\left(\frac{1}{s} - \frac{e^{-sB}}{s}\right).\end{aligned} \tag{3}$$

As $B \to \infty$, s remains fixed.

Now take the limit. If $s > 0$, $\lim_{B\to\infty} e^{-sB} = 0$, but if $s < 0$ the limit does not exist (it diverges to infinity). Thus, the right-hand side of (3) is $1/s$ if $s > 0$ and does not exist if $s < 0$. Actually, for the borderline case $s = 0$ equation (3) is not valid, because if $s = 0$, then $\int e^{-st}\,dt = \int dt = t$, not $e^{-st}/(-s)$. Treating that case separately, for $s = 0$ we have $\mathcal{L}\{1\} = \int_0^\infty 1e^0\,dt = \lim_{B\to\infty}\int_0^B dt = \lim_{B\to\infty} B$, which does not exist.

[1]The Laplace transform is named after the great French applied mathematician *Pierre-Simon de Laplace* (1749 – 1827), who contributed chiefly to celestial mechanics but also to fluid mechanics and other branches of science. Although Laplace used integrals of this type in his study of probability, they had already appeared in the work of Euler, and the Laplace transform "method" was developed only later, by the British electrical engineer *Oliver Heaviside* (1850 – 1925). For an historical account see J. L. B. Cooper, "Heaviside and the Operational Calculus," *Math. Gazette*, Vol. 36 (1952), pp. 5–19.

Thus,

$$\boxed{\mathcal{L}\{1\} = \frac{1}{s}} \tag{4}$$

Remember, the transform of a function $f(t)$ is a function of s; it is *not* a function of t.

for $s > 0$. To reiterate, the Laplace transform of $f(t) = 1$ is $1/s$, and it exists for $s > 0$; that is, the domain of definition of $F(s) = 1/s$ is $s > 0$.

COMMENT 1. The transform $F(s)$ of a given function $f(t)$ is a function of s, and any definition of a function includes a statement as to the interval on which it is defined. In this example we found that the s interval on which the transform (4) is defined is $s > 0$, but the s interval of definition differs, in general, from one transform to another. Looking ahead, however, we will not be concerned with what the specific s interval of definition is, as long as there is *some* such interval, an s interval on which the transform integral (1) converges.

COMMENT 2. We examined the borderline case $s = 0$ separately in this example, but the situation at such values of s will not be important and we will ignore them in subsequent examples. ▮

EXAMPLE 2. $f(t) = t^n$. Consider t^n, for any $n = 1, 2, \ldots$, and use integration by parts:

$$\mathcal{L}\{t^n\} = \int_0^\infty \underbrace{t^n}_{u} \underbrace{e^{-st}\,dt}_{dv} = \lim_{B\to\infty}\left(t^n \frac{e^{-st}}{-s}\Big|_0^B - \int_0^B \frac{e^{-st}}{-s} n t^{n-1}\,dt \right)$$

$$= \lim_{B\to\infty}\left(\frac{B^n e^{-sB}}{-s} \right) + \frac{n}{s}\mathcal{L}\{t^{n-1}\}. \tag{5}$$

The limit in (5) does not exist if $s < 0$. If $s > 0$, it is indeterminate because $B^n \to \infty$ and $e^{-sB} \to 0$ as $B \to \infty$. Which one wins? Repeated application of l'Hôpital's rule shows that the limit exists and is zero. For instance, if $n = 2$, then

$$\lim_{B\to\infty} B^2 e^{-sB} = \lim_{B\to\infty} \frac{B^2}{e^{sB}} = \lim_{B\to\infty} \frac{2B}{s\,e^{sB}} = \lim_{B\to\infty} \frac{2}{s^2\,e^{sB}} = 0.$$

To find the transform of t^5, for instance, we could integrate by parts five times, eventually "knocking down" the t^5 in the integrand to t^0 and thus obtaining a simple integral. But it is more convenient to use the formula (6) recursively, as we've done. The result is given by (7).

Thus,

$$\mathcal{L}\{t^n\} = \frac{n}{s}\mathcal{L}\{t^{n-1}\}. \tag{6}$$

Putting $n = 1, 2, \ldots$ in (6), in turn, gives

$$\mathcal{L}\{t\} = \frac{1}{s}\mathcal{L}\{t^0\} = \frac{1}{s}\mathcal{L}\{1\} = \frac{1}{s^2} \text{ from (4)},$$

$$\mathcal{L}\{t^2\} = \frac{2}{s}\mathcal{L}\{t\} = \frac{2}{s}\frac{1}{s^2} = \frac{2}{s^3},$$

$$\mathcal{L}\{t^3\} = \frac{3}{s}\mathcal{L}\{t^2\} = \frac{3}{s}\frac{2}{s^3} = \frac{3\cdot 2}{s^4},$$

and so on. The result is

$$\boxed{\mathcal{L}\{t^n\} = \frac{n!}{s^{n+1}} \quad (n = 1, 2, \ldots)} \tag{7}$$

for $s > 0$. Since $0! = 1$, (7) agrees with (4) for the case $n = 0$, so (7) is valid for $n = 0, 1, 2, \ldots$. ∎

EXAMPLE 3. $\boldsymbol{f(t) = e^{at}}$**.** Next, consider $f(t) = e^{at}$ in which a is real.

$$\begin{aligned} \mathcal{L}\{e^{at}\} &= \int_0^\infty e^{at}e^{-st}\,dt = \lim_{B\to\infty}\int_0^B e^{-(s-a)t}\,dt \\ &= \lim_{B\to\infty}\frac{e^{-(s-a)t}}{-(s-a)}\bigg|_0^B = \lim_{B\to\infty}\left(\frac{1}{s-a} - \frac{e^{-(s-a)B}}{s-a}\right). \end{aligned} \tag{8}$$

If $s < a$, the $e^{-(s-a)B} = e^{(a-s)B} \to \infty$ as $B \to \infty$, and the limit does not exist. But, if $s > a$, then $e^{-(s-a)B} \to 0$, the limit exists, and we obtain

$$\boxed{\mathcal{L}\{e^{at}\} = \frac{1}{s-a}.} \tag{9}$$

COMMENT. It will also be useful to know the transform of e^{at} if a is complex, say $a = \alpha + i\beta$ in which α and β are real. The result (8) is still valid, but we must re-examine the limit of $e^{-(s-a)B}$ therein as $B \to \infty$. Write

$$e^{-(s-a)B} = e^{-(s-\alpha-i\beta)B} = e^{-(s-\alpha)B}e^{i\beta B}.$$

Since $|e^{i\beta B}| = |\cos\beta B + i\sin\beta B| = \sqrt{\cos^2\beta B + \sin^2\beta B} = \sqrt{1} = 1$, everything hinges on the $e^{-(s-\alpha)B}$. The latter diverges to infinity if $s - \alpha < 0$ $(s < \alpha)$, and converges to zero if $s > \alpha$, that is, if $s > \text{Re}\,a$. Thus, (9) holds if $s > \text{Re}\,a$. ∎

Conclusion: If a is real, (9) holds for $s > a$; if a is complex, it holds for $s > \text{Re}\,a$.

5.2.2 Linearity property of the transform. Since the Laplace transform is defined by an integral, it satisfies the linearity property of integrals, which will help us evaluate transforms just as it helped us evaluate integrals in the calculus:

THEOREM 5.2.1 *Linearity of the Transform*
For any constants α and β,

$$\boxed{\mathcal{L}\{\alpha f(t) + \beta g(t)\} = \alpha\mathcal{L}\{f(t)\} + \beta\mathcal{L}\{g(t)\},} \tag{10}$$

That is, $\int_0^\infty [\alpha f(t) + \beta g(t)]e^{-st}\,dt = \alpha\int_0^\infty f(t)e^{-st}\,dt + \beta\int_0^\infty g(t)e^{-st}\,dt.$

for all s for which $\mathcal{L}\{f(t)\}$ and $\mathcal{L}\{g(t)\}$ exist.

EXAMPLE 4. Using the Linearity Property. To evaluate the transform of $\sin at$, in which a is real, we could evaluate $\int_0^\infty (\sin at)e^{-st}\,dt$, but it is simpler to use the definition of the sine, together with the transform (9) and the linearity property (10):

$$\begin{aligned}\mathcal{L}\{\sin at\} &= \mathcal{L}\{\frac{e^{iat}-e^{-iat}}{2i}\} = \mathcal{L}\{\frac{1}{2i}e^{iat} - \frac{1}{2i}e^{-iat}\}\\ &= \frac{1}{2i}\mathcal{L}\{e^{iat}\} - \frac{1}{2i}\mathcal{L}\{e^{-iat}\}\\ &= \frac{1}{2i}\frac{1}{s-ia} - \frac{1}{2i}\frac{1}{s+ia}\\ &= \frac{1}{2i}\frac{1}{s-ia}\frac{s+ia}{s+ia} - \frac{1}{2i}\frac{1}{s+ia}\frac{s-ia}{s-ia}\\ &= \frac{a}{s^2+a^2}.\end{aligned} \tag{11}$$

The third equality follows from the linearity property (10), and the fourth from (9) with ia in place of "a." In the last two lines we convert $1/(s-ia)$ and $1/(s+ia)$ to the standard Cartesian form "$a+ib$." (Appendix D)

For what s interval? Recall the margin note for Example 3: $\mathcal{L}\{e^{iat}\} = 1/(s-ia)$ holds for $s > \text{Re}(ia)$, which is zero because ia is purely imaginary, and $\mathcal{L}\{e^{-iat}\}$ holds for $s > \text{Re}(-ia)$, which is also zero. Thus,

$$\boxed{\mathcal{L}\{\sin at\} = \frac{a}{s^2+a^2}} \tag{12}$$

for $s > 0$. ▮

Evaluating $\mathcal{L}\{\cos at\}$ is similar. Begin with

$$\mathcal{L}\{\cos at\} = \mathcal{L}\Big\{\frac{e^{iat}+e^{-iat}}{2}\Big\} \tag{13}$$

and follow steps analogous to those in (11). The result is

$$\boxed{\mathcal{L}\{\cos at\} = \frac{s}{s^2+a^2},} \tag{14}$$

again for $s > 0$. The steps are left for the exercises.

Likewise, we can evaluate the transforms of $\sinh at$ and $\cosh at$ by recalling their definitions $(e^{at}-e^{-at})/2$ and $(e^{at}+e^{-at})/2$, respectively, and using the linearity property and (9); we leave these as exercises as well.

As we continue to evaluate the transforms of various elementary functions, we can begin to generate a *Laplace transform table*. Thus far we have these entries:

	$f(t)$	$\mathcal{L}\{f(t)\} = \int_0^\infty f(t)e^{-st}\,dt$
1.	1	$\frac{1}{s}$
2.	t^n	$\frac{n!}{s^{n+1}}$
3.	e^{at}	$\frac{1}{s-a}$
4.	$\sin at$	$\frac{a}{s^2+a^2}$
5.	$\cos at$	$\frac{s}{s^2+a^2}$

A longer table is given in the endpapers. When we speak of "the table," in this chapter, that is the one we mean.

EXAMPLE 5. Using the Table and the Linearity Property. Evaluate

$$\mathcal{L}\left\{6e^{-2t} + 5\sin 3t\right\}.$$

Life is simple if we find $f(t) = 6e^{-2t} + 5\sin 3t$ in the table (among the end papers). We do not, but we do find e^{at} and $\sin at$, so we can use the linearity property (10) and those two transforms and write

The linearity property (10) also holds for more than two functions; see Exercise 6.

$$\begin{aligned}\mathcal{L}\left\{6e^{-2t} + 5\sin 3t\right\} &= 6\mathcal{L}\left\{e^{-2t}\right\} + 5\mathcal{L}\left\{\sin 3t\right\}\\ &= 6\frac{1}{s+2} + 5\frac{3}{s^2+9}.\end{aligned} \tag{15}$$

For what s interval? The transform $\mathcal{L}\{e^{-2t}\} = 1/(s+2)$ holds for $s > -2$ and $\mathcal{L}\{\sin 3t\} = 3/(s^2+9)$ holds for $s > 0$. *Both* are valid on the overlap of $s > -2$ and $s > 0$, namely, on $s > 0$. Thus, (15) is valid for $s > 0$. ▮

5.2.3 Exponential order, piecewise continuity, and conditions for existence of the transform. We need the transform to *exist* in the first place. For the Laplace transform, the existence of the transform of a given function $f(t)$ defined on $0 \le t < \infty$ amounts to the convergence of its transform integral, for some s interval. Each of the transform integrals of the functions considered thus far has indeed converged for some s interval.

We will now give a set of sufficient conditions on $f(t)$ for its transform to exist, but must first define two concepts, exponential order and piecewise continuity.

Exponential order. A function $f(t)$ is of **exponential order** as $t \to \infty$ if there exist nonnegative constants K, c, and T, such that

$$\boxed{|f(t)| \le Ke^{ct}} \tag{16}$$

If $f(t)$ is real valued, then $|f(t)|$ in (16) means its absolute value; if it is complex valued (as e^{it} is, for instance), then $|f(t)|$ means the modulus of the complex number.

for all $t \geq T$.[1]

We will call c the **exponential coefficient**. In place of the phrase "for all $t \geq T$" we could say that (16) holds "eventually," that is, for sufficiently large t. *The set of functions of exponential order is the set of functions that do not grow (in magnitude) faster than exponentially*, and that set includes the vast majority of functions of interest to scientists and engineers.

Certainly, every bounded function is of exponential order, because if $|f(t)| \leq M$, say, for all $t \geq 0$, then (16) holds with $K = M$, $c = 0$, and $T = 0$. For instance, $f(t) = 6\cos t$ is of exponential order because $|6\cos t| \leq 6$, so (16) holds with $K = 6$, $c = 0$, and $T = 0$, These values are by no means unique: For $f(t) = 6\cos t$, (16) holds for any $K \geq 6$, for any $c \geq 0$, and for any $T \geq 0$. The numerical values of K, c, T will not be important. Only the existence of such numbers, so that $f(t)$ is of exponential order, will be important.

As one more example, consider $f(t) = t^3$. Showing that $f(t)$ satisfies (16) is equivalent to showing that $|f(t)/e^{ct}| \leq K$ for some $c \geq 0$ and for all sufficiently large T. With $c = 1$, for instance, $t^3/e^t \to 0$ as $t \to \infty$ (by three applications of l'Hôpital's rule), so surely $t^3/e^t \leq 0.01$, for instance, for all sufficiently large t. Thus, t^3 is of exponential order. Similarly, t^N is of exponential order — no matter how large N is.

Piecewise continuity. A function $f(t)$ is **piecewise continuous** on $a \leq t \leq b$ if there exist a finite number of points $t_1, \ldots, t_N$ (with $a < t_1 < \cdots < t_N < b$) such that

(i) $f(t)$ is continuous on each open subinterval $a < t < t_1$, $t_1 < t < t_2, \ldots,$ $t_N < t < b$, and

(ii) on each subinterval $f(t)$ has finite limits as t approaches the left and right endpoints from within that subinterval.

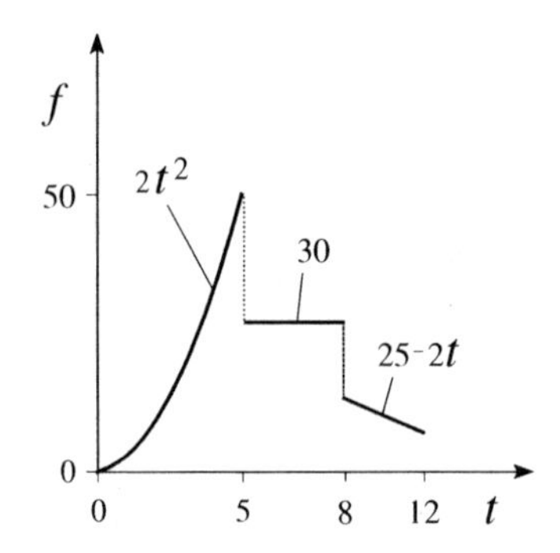

Figure 1. $f(t)$ is piecewise continuous on $0 \leq t \leq 12$.

For instance, consider the function $f(t)$ defined on $0 \leq t \leq 12$, the graph of which is given in Fig. 1. In this case $a = 0$, $b = 12$, $t_1 = 5$, and $t_2 = 8$. The values of $f(t)$ *at* the endpoints $0, 5, 8$, and 12, of the subintervals, are irrelevant insofar as the piecewise continuity of f is concerned; assign any (finite) values you like [such as $f(0) = 837$, $f(5) = 60$, $f(8) = -34$, and $f(12) = \pi$] and $f(t)$ is still piecewise continuous on $0 \leq t \leq 12$. After all, condition (i) involves only the open subintervals, and condition (ii) involves only the limits as the endpoints are approached; neither one involves the values of $f(t)$ *at* the endpoints.

Besides piecewise continuity on a closed interval, we say that $f(t)$ is piecewise continuous **on $0 \leq t < \infty$** *if it is piecewise continuous on $0 \leq t \leq t_0$ for every $t_0 > 0$.*

We now give sufficient conditions on $f(t)$ for its Laplace transform to exist.

[1]For brevity, we will say that $f(t)$ is of exponential order, rather than saying that it is of exponential order "as $t \to \infty$."

THEOREM 5.2.2 *Existence of the Laplace Transform*
Let $f(t)$ be

(i) *piecewise continuous* on $0 \leq t < \infty$, and

(ii) of *exponential order* as $t \to \infty$, with exponential coefficient c.

Then the Laplace transform of $f(t)$, defined by (1), exists for all $s > c$.

In intuitive language, if $f(t)$ is of exponential order then it grows no faster than Ke^{ct}, for some K and c, so the e^{-st} in $\int_0^\infty f(t)e^{-st}\,dt$ can be made to "clobber" $f(t)$ as $t \to \infty$, ensuring convergence of the integral, by choosing $s > c$.

Proof: Since $f(t)$ is of exponential order, there exist nonnegative constants K, c, and T, such that $|f(t)| \leq Ke^{ct}$ for all $t \geq T$. Using that T, break up the transform integral as

$$\mathcal{L}\{f(t)\} = \int_0^T f(t)e^{-st}\,dt + \int_T^\infty f(t)e^{-st}\,dt \equiv I_1 + I_2, \tag{17}$$

respectively. I_1 exists because it can be written as a finite sum of integrals over intervals on each of which $f(t)e^{-st}$ is continuous, with finite limits at the endpoints. For I_2, use the following comparison test from the calculus: If $|g(t)| \leq h(t)$ for $t \geq a$ and $\int_a^\infty h(t)\,dt$ exists, then $\int_a^\infty g(t)\,dt$ does too. Since $|f(t)| \leq Ke^{ct}$ on $T \leq t < \infty$, by assumption,

$$|f(t)e^{-st}| \leq Ke^{ct}e^{-st} = Ke^{-(s-c)t}. \tag{18}$$

To apply the comparison test to I_2, take "$|g(t)|$" and "$h(t)$" to be the $|f(t)e^{-st}|$ and the $Ke^{-(s-c)t}$ in (18), respectively. Since $\int_T^\infty Ke^{-(s-c)t}\,dt$ exists for all $s > c$, it follows from the comparison test cited above that I_2 does too. Since both I_1 and I_2 exist, it follows from (17) that $\mathcal{L}\{f(t)\}$ exists. ■

Theorem 5.2.2 is reassuring since the class of functions satisfying conditions (i) and (ii) cover the functions typically encountered in engineering-science applications. Those conditions are sufficient, not necessary. For instance, consider $f(t) = 1/\sqrt{t}$ (Fig. 2). The latter does not satisfy condition (i) because its limit as $t \to 0$ does not exist; thus, $1/\sqrt{t}$ is not piecewise continuous on $0 \leq t < \infty$. Nevertheless, its transform integral does exist and is

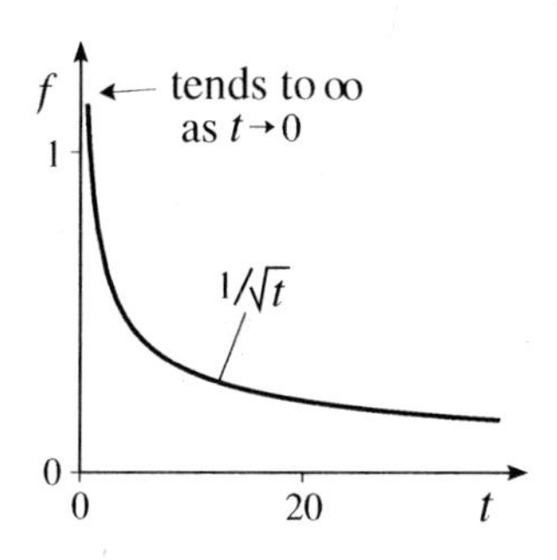

Figure 2. $1/\sqrt{t}$ is *not* piecewise continuous on $0 \leq t < \infty$.

$$\mathcal{L}\left\{\frac{1}{\sqrt{t}}\right\} = \sqrt{\frac{\pi}{s}} \tag{19}$$

for $s > 0$, as is shown in the exercises.

One of the main advantages of the Laplace transform method of solving differential equations is the ease with which it deals with piecewise-continuous forcing functions, as we begin to see in the next example.

EXAMPLE 6. A Piecewise-Defined Function. Consider the piecewise-defined function indicated in Fig. 3. To determine its transform, break up the integral:

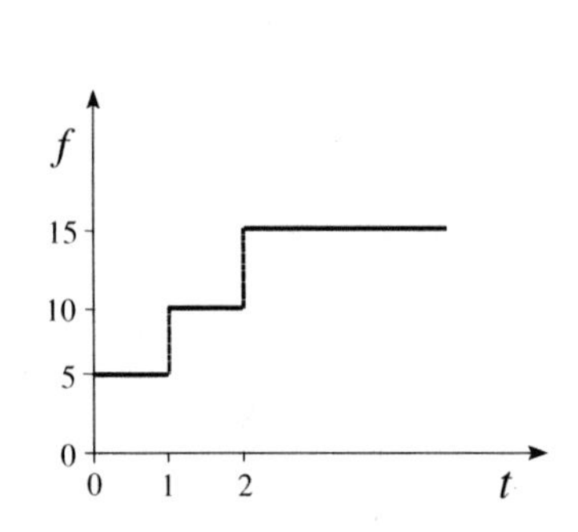

Figure 3. Example of a piecewise-defined function.

$$\begin{aligned}\mathcal{L}\{f(t)\} &= \int_0^\infty f(t)e^{-st}\,dt = \int_0^1 5e^{-st}\,dt + \int_1^2 10e^{-st}\,dt + \int_2^\infty 15e^{-st}\,dt \\ &= 5\frac{e^{-st}}{-s}\bigg|_0^1 + 10\frac{e^{-st}}{-s}\bigg|_1^2 + 15\lim_{B\to\infty}\frac{e^{-st}}{-s}\bigg|_2^B \\ &= \frac{5}{s}\left(1-e^{-s}\right) + \frac{10}{s}\left(e^{-s}-e^{-2s}\right) - \frac{15}{s}\left(\lim_{B\to\infty} e^{-Bs} - e^{-2s}\right) \\ &= \frac{5}{s}\left(1+e^{-s}+e^{-2s}\right) \quad (s>0). \end{aligned} \tag{20}$$

COMMENT. This example illustrates how conveniently the Laplace transform handles piecewise-continuous functions: the input $f(t)$ was the "three-tier" piecewise-defined function

$$f(t) = \begin{cases} 5, & 0<t<1, \\ 10, & 1<t<2, \\ 15, & 2<t<\infty, \end{cases} \tag{21}$$

yet the output was the single expression $5\left(1+e^{-s}+e^{-2s}\right)/s$. In fact, piecewise-defined functions will be even simpler to deal with once we introduce the Heaviside step function in Section 5.4. ▮

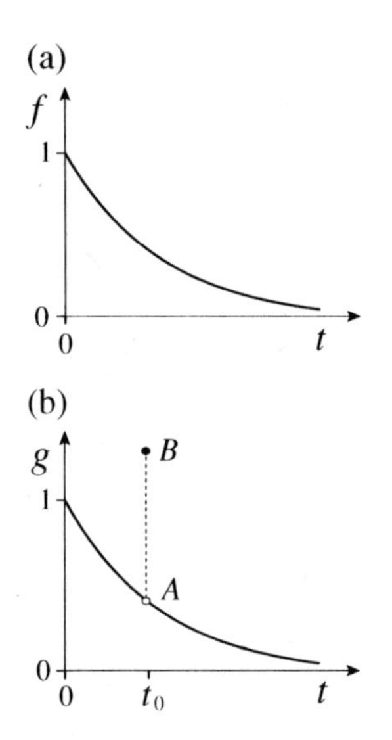

Figure 4. Different functions having the same transform.

5.2.4 Inverse transform. When we use the Laplace transform to solve differential equations in the next section we will need to proceed in both directions: from a given function $f(t)$ to its transform $F(s)$, and from a transform $F(s)$ to the function $f(t)$ "from whence it came." We call the latter the **inverse transform** and denote it as $\mathcal{L}^{-1}\{F(s)\} = f(t)$. Thus, we have the **transform pair**

$$\boxed{\mathcal{L}\{f(t)\} = F(s), \text{ and } \mathcal{L}^{-1}\{F(s)\} = f(t).} \tag{22}$$

For instance, $\mathcal{L}\left\{e^{at}\right\} = 1/(s-a)$ and $\mathcal{L}^{-1}\left\{1/(s-a)\right\} = e^{at}$.

However, the idea fails if inverses are not uniquely determined. That is, if more than one function can have the same transform $F(s)$, then the inverse of $F(s)$ is not defined because it is not *uniquely* defined. In fact, inverses are *not* uniquely defined. To illustrate, let $f(t) = e^{-t}$ (Fig. 4a), and let $g(t)$ be the same as $f(t)$ but with its value at t_0 moved from A to B (Fig. 4b). Surely, the transform of $g(t)$ is the same as that of $f(t)$, namely $1/(s+1)$, because the integrands of their transform integrals differ by a finite value at the point t_0 and therefore have the same areas under their graphs.

But, the good news is twofold: First, such pointwise differences will be of no concern in applications. Second, it turns out that functions satisfying the two conditions of the existence Theorem 5.2.2 and having the same transform can have pointwise differences at most:

THEOREM 5.2.3 *Uniqueness of the Inverse Transform*
Let $f(t)$ and $g(t)$ satisfy the two conditions in Theorem 5.2.2, so their transforms $F(s)$ and $G(s)$ both exist (on $s > c$ for some constant c). If $F(s) = G(s)$, then $f(t) = g(t)$ for all $t \geq 0$ at which both are continuous.

Note the words "at which both are continuous."

Thus reassured that the idea of the inverse transform is on solid ground, we devote the remainder of this section to the *evaluation* of inverse transforms. That step will rely on the transform table and various helpful properties of the transform and inverse transform.

Linearity of the inverse transform. Recall that the transform is linear:

$$\mathcal{L}\{\alpha f(t) + \beta g(t)\} = \alpha F(s) + \beta G(s). \tag{23}$$

If we express (10) in the reverse direction, we have

$$\boxed{\mathcal{L}^{-1}\{\alpha F(s) + \beta G(s)\} = \alpha f(t) + \beta g(t),} \tag{24}$$

Just as the linearity property (23) holds for any finite number of functions (24) does as well.

so the inverse transform is linear too. Both (23) and (24) hold not just for two functions, but for any finite number of them (Exercise 6). The linearity of the inverse will be a key property in evaluating inverse transforms.

5.2.5 Introduction to the determination of inverse transforms. Recall from the calculus that functions that we wish to integrate may not be found in our integral table. But, we learned how to get more mileage out of the table by using various properties of integration such as integration by parts, partial fraction expansions, and substitutions.

Similarly for the evaluation of inverse Laplace transforms, we will use various properties of Laplace inverses, along with various techniques, to get more mileage out of our transform table. Particularly indispensible will be the linearity property of the inverse, stated in (24), and partial fractions. We will illustrate these two in the following examples, just to get started so we can begin to solve differential equations by the Laplace transform method in the next section. Additional useful inversion properties and techniques will be introduced as we go along.

EXAMPLE 7. Using Linearity and the Table. Evaluate $\mathcal{L}^{-1}\left\{\frac{3}{s^5}\right\}$. We could use item 7 of the table, with $n = 4$, but to do that we need a 4! in the numerator and we don't have it, because the numerator is 3. So put it there, and compensate by multiplying by $1/4!$. Thus,

$$\begin{aligned}\mathcal{L}^{-1}\left\{\frac{3}{s^5}\right\} = \mathcal{L}^{-1}\left\{\frac{3}{4!}\frac{4!}{s^5}\right\} = \frac{3}{4!}\mathcal{L}^{-1}\left\{\frac{4!}{s^5}\right\} &= \frac{3}{4!}t^4 \quad \text{(by item 7)}\\ &= \frac{1}{8}t^4.\end{aligned} \tag{25}$$

The second equality in (25) is, by the linearity property (24) with $\beta = 0$: $\mathcal{L}^{-1}\{\alpha F(s)\} = \alpha\mathcal{L}^{-1}\{F(s)\}$.

COMMENT. More generally, it follows from the linearity of the inverse transform that

$$\mathcal{L}^{-1}\{F(s)\} = \frac{1}{\alpha}\mathcal{L}^{-1}\{\alpha F(s)\} \tag{26}$$

for any nonzero constant α. That is, *if we would like a (nonzero) constant α introduced into the function of s that is being inverted, just put it there and, to compensate, put $1/\alpha$ out in front.* ▮

EXAMPLE 8. Using Linearity and the Table. Evaluate

$$\mathcal{L}^{-1}\left\{\frac{5}{s+4} + \frac{2}{s^2+9}\right\}.$$

By linearity,

$$\mathcal{L}^{-1}\left\{\frac{5}{s+4} + \frac{2}{s^2+9}\right\} = 5\mathcal{L}^{-1}\left\{\frac{1}{s+4}\right\} + 2\mathcal{L}^{-1}\left\{\frac{1}{s^2+9}\right\}. \tag{27}$$

The first inverse on the right-hand side is e^{-4t} by item 2 with $a = -4$. For the second we could use item 3, with $a = 3$, but we need an a in the numerator and don't have it. However, we can use (26) and invert as follows:

$$\mathcal{L}^{-1}\left\{\frac{1}{s^2+9}\right\} == \frac{1}{3}\mathcal{L}^{-1}\left\{\frac{3}{s^2+9}\right\} = \frac{1}{3}\sin 3t. \quad \text{(by item 3)} \tag{28}$$

Thus,

$$\mathcal{L}^{-1}\left\{\frac{5}{s+4} + \frac{2}{s^2+9}\right\} = 5e^{-4t} + \frac{2}{3}\sin 3t. \quad ▮ \tag{29}$$

EXAMPLE 9. Linearity and Item 17. Evaluate

$$\mathcal{L}^{-1}\left\{\frac{6s+3}{(s-5)^4}\right\}. \tag{30}$$

We could use item 17, if not for the s in the numerator. However, we can write that s as $(s-5)+5$ and proceed as follows:

$$\begin{aligned}\mathcal{L}^{-1}\left\{\frac{6s+3}{(s-5)^4}\right\} &= \mathcal{L}^{-1}\left\{\frac{6(s-5)+33}{(s-5)^4}\right\} = \mathcal{L}^{-1}\left\{\frac{6}{(s-5)^3} + \frac{33}{(s-5)^4}\right\} \\ &= \mathcal{L}^{-1}\left\{\frac{6}{2!}\frac{2!}{(s-5)^3} + \frac{33}{3!}\frac{3!}{(s-5)^4}\right\} = \frac{6}{2!}\mathcal{L}^{-1}\left\{\frac{2!}{(s-5)^3}\right\} \\ &\quad + \frac{33}{3!}\mathcal{L}^{-1}\left\{\frac{3!}{(s-5)^4}\right\} = 3t^2e^{5t} + \frac{11}{2}t^3e^{5t}.\end{aligned} \tag{31}$$

COMMENT. Note that the simple step $6s+3 = 33+6(s-5)$ is actually a Taylor series expansion of $6s+3$ about $s=5$, which series terminates after only two terms. If, instead, the numerator in (30) were $6s^2 + 3$, for instance, then its Taylor expansion about 5 would give $153 + 60(s-5) + 6(s-5)^2$. ∎

Using partial fractions. Also of great help will be the technique of partial fractions, as we now illustrate.

EXAMPLE 10. Evaluate

$$\mathcal{L}^{-1}\left\{\frac{7s-5}{s^2-s-2}\right\}. \tag{32}$$

First, factor the denominator: $s^2 - s - 2 = (s-2)(s+1)$. Thus (Appendix A), we can express

$$\frac{7s-5}{s^2-s-2} = \frac{7s-5}{(s-2)(s+1)} = \frac{A}{s-2} + \frac{B}{s+1}, \tag{33}$$

and find that $A = 3$ and $B = 4$. Next,

$$\begin{aligned}\mathcal{L}^{-1}\left\{\frac{7s-5}{s^2-s-2}\right\} &= \mathcal{L}^{-1}\left\{\frac{3}{s-2} + \frac{4}{s+1}\right\}\\ &= 3\mathcal{L}^{-1}\left\{\frac{1}{s-2}\right\} + 4\mathcal{L}^{-1}\left\{\frac{1}{s+1}\right\} \quad \text{(by linearity)}\\ &= 3e^{2t} + 4e^{-t}. \quad \text{(each by item 2)}\end{aligned} \tag{34}$$

This is only a modest introductory example. The use of partial fractions to obtain transform inverses will be explained more fully in subsequent sections.

In summary, the form $(7s-5)/(s^2-s+2)$ was not found in the table, so we used partial fractions, then the linearity property (24) and, finally, item 2 in the table.

COMMENT. The technique used in this example works also if the roots for s are complex. However, in this section and in the exercises we consider only examples in which the roots are real. The complex case is also important, and will be included in the next section. ∎

Closure. The Laplace transform of a function $f(t)$ defined on $0 \le t < \infty$ is given by the integral in (1). The integral is improper because of the infinite upper integration limit. For the transform to exist, the integral must converge. If $f(t)$ is piecewise continuous on $0 \le t < t_0$ for every t_0, and $|f(t)|$ does not grow faster than some constant times an exponential e^{ct}, then the integral will indeed converge if s is sufficiently positive. The vast majority of functions likely to arise in applications satisfy those conditions and therefore have Laplace transforms; they are "transformable."

We evaluated the transforms of several elementary functions, constructed a mini table of them, and called attention to the longer table given in the endpapers, upon which we will rely.

We can get more mileage out of the table by using it in conjunction with a number of properties such as the linearity of the transform and inverse transform, along with algebraic techniques such as partial fractions. The same is true in the calculus, in using such properties of integration as linearity and integration by parts to integrate many more functions than are given in the text's integral table.

Note that the linearity property (23) is equivalent to the reverse statement (24). Linearity is so important that we wrote it both ways. However, as we proceed we will establish many more such properties of the transform, and it seems "overkill" to state each in both directions. Rather, we leave it to the reader to understand that they hold for transforms and also for inverses.

In the next section we begin using the Laplace transform to solve differential equations — which is our purpose in this chapter.

EXERCISES 5.2

NOTE: *We refer to items in the table as "item such and such," and equations in the text are referred to in parentheses. For instance, "(13) and (14)" in Exercise 1, below, means equations (13) and (14) in the text, not items 13 and 14 in the table.*

1. Fill in the steps between (13) and (14).

2. Using the exponential definitions of the hyperbolic sine and cosine, and the integral definition (2), derive these transforms, where a is positive.

(a) $\mathcal{L}\{\sinh at\} = a/(s^2 - a^2)$ for $s > a$
(b) $\mathcal{L}\{\cosh at\} = s/(s^2 - a^2)$ for $s > a$

3. Evaluate the Laplace transform of the given function $f(t)$ by working out the right-hand side of (2), and give the s interval on which the transform exists. HINT: Express sines or cosines in terms of complex exponentials, and hyperbolic sines or cosines in terms of real exponentials. Then, combine those terms with the e^{-st} before integrating.

(a) t (b) t^2
(c) e^{2t-3} (d) $5e^{4-3t}$
(e) $\cos(t-2)$ (f) $3\sin(t+5)$
(g) $\sinh(t+1)$ (h) $e^t \cos t$
(i) $e^{-t}\cosh 2t$ (j) $e^{3t}\cos 3t$

4. Piecewise-Defined Functions. Evaluate the transform of the given piecewise-defined function $f(t)$. Is $f(t)$ piecewise continuous on $0 \le t < \infty$?

(a) 0 on $0 \le t < 3$, 50 on $3 \le t < 5$, 0 on $t \ge 5$
(b) 25 on $0 \le t < 5$, 0 on $t \ge 5$
(c) 0 on $0 \le t < 5$, 100 on $t \ge 5$
(d) 50 on $0 < t \le 1$, 25 on $1 < t \le 2$, 0 on $t > 2$
(e) e^t on $0 \le t < 2$, 0 on $t \ge 2$
(f) e^{-t} on $0 \le t < 2$, e^{2-t} on $t \ge 2$
(g) 0 on $0 \le t < 3$, e^{3-t} on $t \ge 3$
(h) t on $0 \le t < 1$, 1 on $t \ge 1$

5. Piecewise Continuous? Let $f(t)$ be the periodic function shown in Fig. 2. Is it piecewise continuous on $0 \le t < \infty$? That is, does it satisfy condition (i) in Theorem 5.2.2? Explain.

6. Linearity. Show that it follows from the linearity property (10) that

(a) $\mathcal{L}\{\alpha_1 f_1(t) + \alpha_2 f_2(t) + \alpha_3 f_3(t)\}$
$= \alpha_1 F_1(s) + \alpha_2 F_2(s) + \alpha_3 F_3(s)$

(b) $\mathcal{L}\{\alpha_1 f_1(t) + \cdots + \alpha_4 f_k(t)\}$
$= \alpha_1 F_1(s) + \cdots + \alpha_4 F_4(s)$

In fact, (10) holds not just for two functions but for any finite number of them.

7. Transform, Using Linearity and the Table. Use the linearity property and the Laplace transform table to find the transform of the given function. NOTE: You may use the results stated in Exercise 6.

(a) $6\sin 2t$ (b) $-3e^{-2t}$
(c) $e^t + \cos 4t$ (d) $1 + t^3 e^{-t}$
(e) $\sin t - 3\cos t$ (f) $1 + 7t\cosh 3t$
(g) $t(\sin t + \sinh t)$ (h) $7t^{10} - 4t + 3e^t$
(i) $3(1 - t + t^2)$ (j) $(1 - 4t)\cos 5t$
(k) $1 + 2t + 3t^2$ (l) $(1 - t - 3t^2)\, e^{2t} - 5$
(m) $e^t - e^{2t} - 4t$ (n) $1 + t\,(1 - e^t - \sin 3t)$

(o) $\cos^2 at$ HINT: $\cos^2 A = (1 - \cos 2A)/2$
(p) $\sin^2 at$ HINT: See part (o).

8. Exponential Order. Show whether or not the given function $f(t)$ is of exponential order. If it is, give any suitable values for K, c, and T.

(a) $5e^{4t}$ (b) $-10e^{-5t}$ (c) $\sinh 2t$
(d) $3 + 2t$ (e) $\sinh t^2$ (f) $e^{4t}\sin t$
(g) $\cos t^3$ (h) $\sin t + 3\cos t$ (i) $(t+1)/(t+2)$
(j) $6t + e^t\cos t$ (k) $4e^t - 5e^{2t}$ (l) $1 + t + t^2$

9. Inversion Using Linearity and the Table. Evaluate the inverse using linearity and one or more of items 1–7 and 17 in the table. Identify any items used.

(a) $\dfrac{1-5s}{s^3}$ (b) $\dfrac{2}{(s+2)^4}$ (c) $\dfrac{5}{s}-\dfrac{6}{s^4}$

(d) $\dfrac{3s}{2s^2+10}$ (e) $\dfrac{5}{(9s-1)^4}$ (f) $\dfrac{2-5s}{(s+6)^3}$

(g) $\dfrac{1}{(s+1)^5}$ (h) $\dfrac{3}{4-2s}-\dfrac{2}{s^5}$ (i) $\dfrac{6-7s}{2s^2-8}$

(j) $\dfrac{12}{(s-1)^4}$ (k) $\dfrac{2}{s^3}-\dfrac{12}{3-s^2}$ (l) $\dfrac{2s+3}{s^2+6}$

(m) $\dfrac{6s^2+3}{(s-5)^4}$ (n) $\dfrac{10}{s^2-4}$ (o) $\dfrac{5}{2s^2+6}$

10. Using Partial Fractions. By partial fractions, express the given transform in the form $A/(s-a)+B/(s-b)$ and invert.

(a) $\dfrac{1}{s^2-2s-3}$ (b) $\dfrac{8-s}{s^2-s-2}$ (c) $\dfrac{4}{s^2-3s+2}$

(d) $\dfrac{2s}{s^2+4s+3}$ (e) $\dfrac{6}{s(s+6)}$ (f) $\dfrac{6}{3s-s^2}$

(g) $\dfrac{2s+7}{s^2+s-2}$ (h) $\dfrac{7s+8}{3s(s+4)}$ (i) $\dfrac{3s-2}{(2-s)(2+s)}$

11. Multiplication by e^{at}; s-shift. (a) If $\mathcal{L}\{f(t)\}=F(s)$ for $s>c$, and a is any real number, show that

$$\boxed{\mathcal{L}\{e^{at}f(t)\}=F(s-a),} \tag{11.1}$$

which is called the **s-shift formula**.
(b) Use (11.1) and table item 3 to obtain table item 9.
(c) Use (11.1) and table item 4 to obtain table item 10.
(d) Use (11.1) and table item 5 to obtain table item 11.
(e) Use (11.1) and table item 6 to obtain table item 12.

ADDITIONAL EXERCISES

12. The Gamma Function. The transform $\mathcal{L}\{t^n\}$ can be evaluated by integrating $\int_0^\infty t^n e^{-st}\,dt$ by parts n times because by doing so we can "knock down" the t^n to t^0, and $\int_0^\infty t^0 e^{-st}\,dt=\int_0^\infty e^{-st}\,dt$ is readily evaluated. This strategy doesn't work for $\mathcal{L}\{t^p\}$ if p is not an integer, but we *can* evaluate

$$\mathcal{L}\{t^p\}=\int_0^\infty t^p e^{-st}\,dt \qquad (p>-1) \tag{12.1}$$

in terms of the **gamma function** $\Gamma(x)$, which is defined by the formula

$$\boxed{\Gamma(x)=\int_0^\infty t^{x-1}e^{-t}\,dt \qquad (x>0).} \tag{12.2}$$

The integral in (12.2) is nonelementary in that it cannot be evaluated in closed form in terms of elementary functions, but it arises often enough in applications for it to have been given a name, the gamma function, and to have been studied extensively. In this exercise we study it, and in the next exercise we use it to evaluate $\mathcal{L}\{t^p\}$ where p is not an integer. To begin, observe that the integral in (12.2) is improper for two reasons: first, the upper limit is ∞ and, second, the integrand is unbounded as $t\to 0$ if $x-1<0$, because $t^{x-1}e^{-t}=t^{x-1}(1-t+t^2/2-\cdots)\sim t^{x-1}$ as $t\to 0$. The more negative the exponent $x-1$, the stronger the "blow-up" of t^{x-1} as $t\to 0$. Nevertheless, it can be shown from the theory of improper integrals that the integral in (12.2) does converge if $x-1$ is not "too" negative, namely, if $x-1>-1$, i.e., if $x>0$. That is why we included the stipulation $x>0$ in (12.2).

(a) Integrating by parts, use (12.2) to show that

$$\boxed{\Gamma(x)=(x-1)\Gamma(x-1) \qquad (x>1).} \tag{12.3}$$

NOTE: The latter *recursion formula* is the most important property of the gamma function. If we know (e.g., by numerical integration) the values of $\Gamma(x)$ on a unit interval such as $0<x\le 1$, then we can use (12.3) to evaluate $\Gamma(x)$ for any $x>1$, by making steps to get into interval on which $\Gamma(x)$ is tabulated. For example,

$$\begin{aligned}\Gamma(3.2)=2.2\Gamma(2.2)&=(2.2)(1.2)\Gamma(1.2)\\&=(2.2)(1.2)(0.2)\Gamma(0.2),\end{aligned}$$

where $\Gamma(0.2)$ is known if $\Gamma(x)$ is indeed known on $0<x\le 1$.

(b) Show, by direct integration, that

$$\Gamma(1)=1. \tag{12.4}$$

(c) Using (12.3) and (12.4), show that if x is a positive integer n, then

$$\boxed{\Gamma(n)=(n-1)!} \tag{12.5}$$

where $0!\equiv 1$.

(d) Besides being able to evaluate $\Gamma(x)$ analytically at $x=1,2,3,\ldots$, we can also evaluate it at $x=\frac{1}{2},\frac{3}{2},\frac{5}{2},\ldots$. In particular, show that

$$\Gamma\left(\frac{1}{2}\right)=\sqrt{\pi}. \tag{12.6}$$

HINT: With $x = 1/2$ in (12.2), change variables from t to u according to $t = u^2$ and show that

$$\Gamma\left(\frac{1}{2}\right) = 2\int_0^\infty e^{-u^2}\,du.$$

Then

$$\left[\Gamma\left(\frac{1}{2}\right)\right]^2 = 4\int_0^\infty e^{-u^2}\,du \int_0^\infty e^{-v^2}\,dv$$
$$= 4\int_0^\infty \int_0^\infty e^{-(u^2+v^2)}\,du\,dv.$$

Regard the latter as a double integral in a Cartesian u, v plane, change from u, v to polar variables r, θ, and remember that in place of the Cartesian area element $du\,dv$ the polar area element is $r dr d\theta$. The resulting double integral should be simpler to evaluate, thanks to the r in the $r dr d\theta$.

(e) Using (12.3) and (12.6), show that

$$\Gamma\left(\frac{3}{2}\right) = \frac{\sqrt{\pi}}{2} \quad \text{and} \quad \Gamma\left(\frac{5}{2}\right) = \frac{3\sqrt{\pi}}{4}. \tag{12.7}$$

NOTE: Similarly, one can evaluate $\Gamma(\frac{7}{2})$, $\Gamma(\frac{9}{2})$, and so on.

13. Finally, the Laplace Transform of t^p. Using (12.2), show that if $p > -1$, then

$$\boxed{\mathcal{L}\left\{t^p\right\} = \frac{\Gamma(p+1)}{s^{p+1}},} \tag{13.1}$$

which is item 8 in the table. For instance, it follows from (12.1) and (12.6) that $\mathcal{L}\left\{t^{-1/2}\right\} = \Gamma(1/2)/s^{1/2} = \sqrt{\pi/s}$.

14. Transforms Tend to Zero as $s \to 0$. Notice that all transforms in our transform table tend to zero as $s \to \infty$. In fact, prove that if $f(t)$ is piecewise continuous on $0 \le t < \infty$ and of exponential order, and its transform is $F(s)$, then

$$\boxed{\lim_{s\to\infty} F(s) = 0.} \tag{14.1}$$

HINT: Recall from the calculus the bound

$$\left|\int_a^b g(t)\,dt\right| \le \int_a^b |g(t)|\,dt. \tag{14.2}$$

Now, break up the transform integral as we did in (17). Show that $I_1 \to 0$ as $s \to \infty$ by noting that $|f(t)|$ is bounded on $0 \le t \le T$ by some positive number M and using (14.2). Then, show that $I_2 \to 0$ as $s \to \infty$ by using (14.2) and $|f(t)| \le Ke^{ct}$.

NOTE: One can obtain (14.1) more simply as follows:

$$\lim_{s\to\infty} F(s) = \lim_{s\to\infty} \int_0^\infty f(t)e^{-st}\,dt$$
$$= \int_0^\infty \lim_{s\to\infty} \left[f(t)e^{-st}\right] dt = \int_0^\infty 0\,dt = 0, \tag{14.3}$$

but that derivation is only formal, not rigorous, because we inverted the order of integration and the limit, in the second equality, without justification.

15. Inverting as a Power Series. If, to invert a given transform, we are willing to end up with a power series rather than a closed-form expression, we can proceed as follows. To illustrate, let us seek the inverse of $F(s) = 1/(s-a)$. Expanding the latter in a Taylor series in s, about $s = 0$, gives

$$F(s) = \frac{1}{s-a} = -\frac{1}{a} - \frac{1}{a^2}s - \frac{1}{a^3}s^2 - \cdots. \tag{15.1}$$

The individual terms are of the simple form s^n, so we might expect to find their inverses in the table, and hence to invert the series term by term. But s^n is not to be found in the table. In fact, as follows from the property (14.1) in Exercise 14, all transforms in our table tend to zero as $s \to \infty$. The positive integer powers of s in (15.1) do not tend to zero as $s \to \infty$, so it appears that they are not invertible, and that (15.1) is not helpful. (More precisely, they are not invertible in terms of functions that satisfy our usual conditions of piecewise continuity and exponential order.) However, suppose we re-express $F(s)$ as

$$F(s) = \frac{1}{s}\frac{1}{1-\dfrac{a}{s}}. \tag{15.2}$$

If we let $a/s \equiv z$, say, then when we expand

$$\frac{1}{1-\dfrac{a}{s}} = \frac{1}{1-z} = 1 + z + z^2 + \cdots$$
$$= 1 + \frac{a}{s} + \frac{a^2}{s^2} + \cdots \tag{15.3}$$

we obtain *inverse* powers of s, which *are* invertible. Thus, (15.2) and (15.3) give

$$f(t) = \mathcal{L}^{-1}\left\{\frac{1}{s} + \frac{a}{s^2} + \frac{a^2}{s^3} + \cdots\right\}$$
$$= \mathcal{L}^{-1}\left\{\frac{1}{s}\right\} + a\mathcal{L}^{-1}\left\{\frac{1}{s^2}\right\} + a^2\mathcal{L}^{-1}\left\{\frac{1}{s^3}\right\} + \cdots$$
$$= 1 + at + \frac{1}{2!}a^2t^2 + \frac{1}{3!}a^3t^3 + \cdots, \tag{15.4}$$

where, in the second equality, we assumed the infinite series could be inverted term by term. We know the result (15.4) is correct because we know the inverse of $1/(s-a)$ is e^{at} and the final member of (15.4) is indeed the Taylor series of e^{at}. This inversion method is of interest for functions $F(s)$ that are difficult to invert by conventional means, but the drawback is that the inverse obtained is in the form of an infinite series.
The problem: Use this method in each case, obtaining the inverse in the form of a power series; three or four terms of the series will suffice. Then, use the table to invert the given transform in closed form and verify that the power series that you obtained is correct. HINT: In part (a), for instance, factor an s^2 out of the denominator just as we factored an s out of the denominator in (15.2).

(a) $\dfrac{1}{s^2+a^2}$ (b) $\dfrac{1}{s^2-a^2}$

(c) $\dfrac{2s}{(s^2+1)^2}$ (d) $\dfrac{s^2-4}{(s^2+4)^2}$

16. Continuation of Exercise 15. Invert $1/\sqrt{s^2+1}$ using the idea outlined in Exercise 15. That is, write

$$\frac{1}{\sqrt{s^2+1}} = \frac{1}{s\sqrt{1+\dfrac{1}{s^2}}} = \frac{1}{s}\left(1+\frac{1}{s^2}\right)^{1/2}, \tag{16.1}$$

let $1/s^2 = z$, say, expand $(1+z)^{-1/2}$ in a Taylor series about $z=0$, replace z by $1/s^2$, invert term by term, and thus show that

$$\mathcal{L}^{-1}\left\{\frac{1}{\sqrt{s^2+1}}\right\} = 1-\frac{1}{4}t^2+\frac{1}{64}t^4-\frac{1}{2304}t^6+\cdots. \tag{16.2}$$

The function defined by the series in (16.2) is the **Bessel function** $J_0(t)$, which will be the subject of Section 6.4.

17. Transform of Periodic Functions. A function $f(t)$ is **periodic with period T** if

$$f(t+T) = f(t) \tag{17.1}$$

for all t's in the domain of t. For instance, the *sawtooth wave*, below, is periodic with period $T=3$, and the segment from $t=0$ to $t=3$, say, is one period. If we repeat that basic unit indefinitely to the right we generate the sawtooth wave.

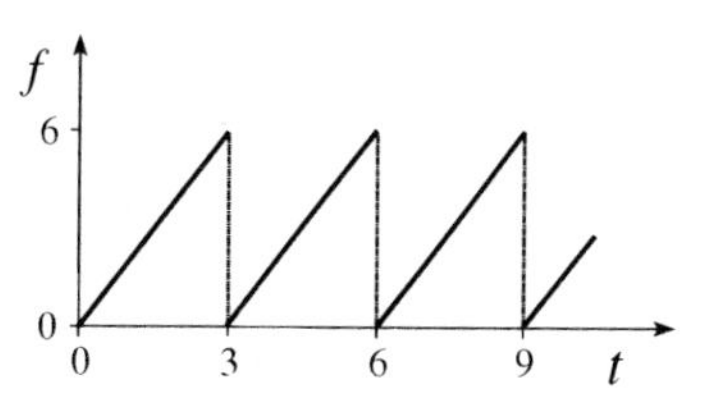

(a) Let $f(t)$ be periodic, of period T. Show that its transform can be simplified to an integration over only one period, as

$$\mathcal{L}\{f(t)\} = \frac{1}{1-e^{-st}}\int_0^T f(t)e^{-st}\,dt. \tag{17.2}$$

HINT: Write

$$\mathcal{L}\{f(t)\} = \int_0^\infty f(t)e^{-st}\,dt = \int_0^T f(t)e^{-st}\,dt + \int_T^{2T} f(t)e^{-st}\,dt + \cdots. \tag{17.3}$$

Make changes of variables in the integrals so that each has, as new limits, 0 to T, obtaining

$$\mathcal{L}\{f(t)\} = \left(1+e^{-sT}+e^{-2sT}+\cdots\right)\int_0^T f(t)e^{-st}\,dt, \tag{17.4}$$

and use the *geometric series* formula

$$\frac{1}{1-z} = 1+z+z^2+z^3+\cdots \qquad (|z|<1) \tag{17.5}$$

to sum the series in (17.4) and obtain (17.2).
(b) For instance, use (17.2) to show that the transform of the sawtooth wave shown above is

$$\mathcal{L}\{f(t)\} = \frac{2}{s^2}-\frac{6}{s}\frac{e^{-3s}}{1-e^{-3s}}. \tag{17.6}$$

(c) For fun, show that the inverse of (17.6) does give us back the sawtooth wave. HINT: This time use the geometric series formula not to sum the series into the closed form $1/(1-z)$, but to expand the $1/(1-e^{-3s})$ into a geometric series in powers of e^{-3s}.

5.3 APPLICATION TO THE SOLUTION OF DIFFERENTIAL EQUATIONS

5.3.1 First-order equations. We are ready to use the Laplace transform to solve linear differential equations with constant coefficients. To do that, we will need to be able to transform whatever derivatives appear in the given equation, so we will address that first.

THEOREM 5.3.1 *Transform of a Derivative*
Let $f(t)$ and $f'(t)$ satisfy the conditions of Theorem 5.2.2; that is, let them be piecewise continuous on $0 \leq t < \infty$ and of exponential order as $t \to \infty$, with exponential coefficient c. Then $\mathcal{L}\{f'(t)\}$ exists for all $s > c$, and

This will be the key.

$$\boxed{\mathcal{L}\{f'(t)\} = s\mathcal{L}\{f(t)\} - f(0).} \tag{1}$$

Proof: Let it suffice to suppose that f and f' are not only piecewise continuous, but even continuous. Use integration by parts,

Understand $\int_0^\infty (\ )\, dt$ and $(\ )\Big|_0^\infty$ to mean $\lim_{B\to\infty} \int_0^B (\ )\, dt$ and $\lim_{B\to\infty} (\ )\Big|_0^B$, respectively.

$$\mathcal{L}\{f'(t)\} = \int_0^\infty f'(t)e^{-st}\, dt = \int_0^\infty \underbrace{e^{-st}}_{u} \underbrace{f'(t)\, dt}_{dv}$$

$$= \left[f(t)e^{-st}\right]\Big|_0^\infty + s\int_0^\infty f(t)e^{-st}\, dt. \tag{2}$$

Since, by assumption, $|f(t)e^{-st}| \leq Ke^{(c-s)t}$ for all sufficiently large t, and $Ke^{(c-s)t} \to 0$ as $t \to \infty$ for all $s > c$, it follows that $|f(t)e^{-st}| \to 0$ as well. Thus, the $\left[f(t)e^{-st}\right]|^\infty$ part of (2) is zero. Further, the last integral in (2) is $\mathcal{L}\{f(t)\}$, for all $s > c$, so (2) gives

$$\mathcal{L}\{f'(t)\} = 0 - f(0)e^0 + s\mathcal{L}\{f(t)\}, \tag{3}$$

for $s > c$, which is (1). ■

EXAMPLE 1. Application to an IVP. Solve the IVP

The steps:
0. The problem given in the "t domain."

$$x' + 2x = 10e^{3t}, \tag{4a}$$

$$x(0) = 5 \tag{4b}$$

for $x(t)$. First, multiply both sides of (4a) by e^{-st} and integrate from 0 to ∞. That is, "take the transform" of both sides:

1. Transform the equation.

$$\int_0^\infty (x' + 2x)e^{-st}\, dt = \int_0^\infty (10e^{3t})e^{-st} dt, \tag{5}$$

or, in Laplace transform notation,

$$\mathcal{L}\{x' + 2x\} = \mathcal{L}\{10e^{3t}\}. \tag{6}$$

By linearity,

$$\mathcal{L}\{x'\} + 2\mathcal{L}\{x\} = 10\mathcal{L}\{e^{3t}\} \tag{7}$$

2. Use the linearity of $\mathcal{L}$.

and, using (1) above, item 2 of the table, and denoting $\mathcal{L}\{x(t)\}$ as $X(s)$, we obtain

$$\big[sX(s) - x(0)\big] + 2X(s) = \frac{10}{s-3}, \tag{8}$$

3. Evaluate the transforms.

in which $x(0) = 5$ from (4b).

Note carefully, and this is the key regarding the solution of linear differential equations with constant coefficients by the Laplace transform, that whereas (4a) was a linear *differential* equation for $x(t)$, its transformed version (8) is merely a linear *algebraic* equation for $X(s)$. Solving it for $X(s)$ by simple algebra gives

$$X(s) = \frac{5}{s+2} + \frac{10}{(s+2)(s-3)},$$

4. In the "s domain," solve by algebra for $X(s)$.

which must now be inverted to obtain $x(t)$. For the second term, partial fractions gives

$$\frac{10}{(s+2)(s-3)} = -\frac{2}{s+2} + \frac{2}{s-3},$$

so

$$X(s) = \frac{5}{s+2} + \left(-\frac{2}{s+2} + \frac{2}{s-3}\right) = \frac{3}{s+2} + \frac{2}{s-3} \tag{9}$$

and, again using linearity,

$$\begin{aligned} x(t) &= \mathcal{L}^{-1}\Big\{\frac{3}{s+2} + \frac{2}{s-3}\Big\} = 3\mathcal{L}^{-1}\Big\{\frac{1}{s+2}\Big\} + 2\mathcal{L}^{-1}\Big\{\frac{1}{s-3}\Big\} \\ &= 3e^{-2t} + 2e^{3t}. \end{aligned} \tag{10}$$

5. Invert $X(s)$ to get $x(t)$.

COMMENT 1. When we used Theorem 5.3.1 to write $\mathcal{L}\{x'\} = sX(s) - x(0)$, we did not know if $x(t)$ satisfied the conditions of that theorem because $x(t)$ was not yet known. However, with the solution (10) now in hand, we see that $x(t)$ does satisfy the conditions of the theorem. Of course, if we do wish to do any checking, it would be more to the point simply to check that (10) satisfies (4a) and (4b)!

COMMENT 2. Normally, the Laplace transform is used to solve initial value problems such as (4). However, we *could* use it to obtain a general solution of (4a) alone, if we wish, by letting $x(0) \equiv$ "A" be an arbitrary constant in (8). Then, in place of (10) we would have obtained $x(t) = (A - 2)e^{-2t} + 2e^{3t}$ or, equivalently,

$$x(t) = Ce^{-2t} + 2e^{3t} \tag{11}$$

in which C is arbitrary; (11) is indeed a general solution of (4a).

COMMENT 3. Notice that the initial condition was not applied at the end; it became "built in" when we used (1) and wrote $\mathcal{L}\{x'(t)\} = sX(s) - x(0) = sX(s) - 5$ in (8). ▮

We could have solved (4) by methods developed in earlier chapters, but this first example was only to illustrate the steps in the Laplace transform method.

5.3.2 Higher-order equations. To solve higher-order equations we will need the transforms of higher-order derivatives. Just as we evaluated $\mathcal{L}\{f'\}$ using one integration by parts, in (2), we can evaluate $\mathcal{L}\{f''\}$ using two integrations by parts, and so on. However, it is simpler to use (1) recursively. That is, if we replace the function "f" in (1) by f', then (1) gives

$$\begin{aligned}\mathcal{L}\{f''\} &= \mathcal{L}\{(f')'\} = s\mathcal{L}\{f'\} - f'(0) \quad \text{[by (1)]}\\ &= s[s\mathcal{L}\{f\} - f(0)] - f'(0) \quad \text{[by (1) again]},\end{aligned}$$

so

$$\boxed{\mathcal{L}\{f''(t)\} = s^2F(s) - sf(0) - f'(0).} \tag{12}$$

Similarly, if we replace the function f in (1) by "f''," then (1) gives

$$\begin{aligned}\mathcal{L}\{f'''\} &= s\mathcal{L}\{f''\} - f''(0) \quad \text{[by (1)]}\\ &= s\left[s^2F(s) - sf(0) - f'(0)\right] - f''(0) \quad \text{[by (14)]}\\ &= s^3F(s) - s^2f(0) - sf'(0) - f''(0),\end{aligned} \tag{13}$$

and so on, for higher derivatives.

Sufficient conditions for (12) are that f, f', and f'' be piecewise continuous on $0 \leq t < \infty$ and of exponential order. Similarly, sufficient conditions for (13) are that f, f', f'', and f''' be piecewise continuous on $0 \leq t < \infty$ and of exponential order, and so on.

We see that *differentiation in the t domain corresponds to multiplication by s in the s domain*. For instance, the transform of $f'(t)$ is s times the transform $F(s)$ of $f(t)$ [plus a constant coming from the initial value $f(0)$] and the transform of $f''(t)$ is s^2 times $F(s)$ (plus constant terms coming from initial values). Consequently, a linear constant-coefficient differential equation for $x(t)$ is converted to a linear algebraic equation for $X(s)$ in the s domain, and such an equation can be solved for $X(s)$ by simple algebra.

EXAMPLE 2. A Second-Order Equation. Solve

$$x'' + 9x = 24\sin t; \qquad x(0) = 2,\ x'(0) = -1. \tag{14}$$

The steps are the same as in Example 1:

$$\begin{aligned}\mathcal{L}\{x'' + 9x\} &= \mathcal{L}\{24\sin t\},\\ \mathcal{L}\{x''\} + 9\mathcal{L}\{x\} &= 24\mathcal{L}\{\sin t\},\\ \left[s^2X(s) - sx(0) - x'(0)\right] + 9X(s) &= 24\frac{1}{s^2+1},\\ (s^2+9)X(s) - 2s + 1 &= 24\frac{1}{s^2+1}.\end{aligned} \tag{15}$$

Solve (15) by algebra for $X(s)$:

$$X(s) = \frac{2s-1}{s^2+9} + 24\frac{1}{(s^2+9)(s^2+1)}. \tag{16}$$

To invert the latter we could use partial fractions and expand the last term as

$$\frac{1}{(s^2+9)(s^2+1)} = \frac{A}{s+3i} + \frac{B}{s-3i} + \frac{C}{s+i} + \frac{D}{s-i}. \tag{17}$$

The form (17) is fine, but it will be more convenient to use (18) instead.

However, inverting the first two of these four terms will give an e^{-i3t} and e^{i3t}, which can be combined into a sine and a cosine. Similarly for the last two terms. Thus, it is more convenient to leave the s^2+9 and s^2+1 factors intact and only *partially expand* as

$$\frac{1}{(s^2+9)(s^2+1)} = \frac{E+Fs}{s^2+9} + \frac{G+Hs}{s^2+1}, \tag{18}$$

If this step is not clear, see Example 2 in Appendix A.

as we look ahead to using items 3 and 4 in the table to invert.

To evaluate E, F, G, H in (18), combine the terms over a common denominator as

$$\begin{aligned}\frac{1}{(s^2+9)(s^2+1)} &= \frac{(E+Fs)(s^2+1)+(G+Hs)(s^2+9)}{(s^2+9)(s^2+1)} \\ &= \frac{(F+H)s^3+(E+G)s^2+(F+9H)s+(E+9G)}{(s^2+9)(s^2+1)}\end{aligned} \tag{19}$$

and equate coefficients of like powers of s in the numerators of the left- and right-hand sides:

$$\begin{aligned} s^0 &: \quad 1 = E+9G \\ s^1 &: \quad 0 = F+9H \\ s^2 &: \quad 0 = E+G \\ s^3 &: \quad 0 = F+H, \end{aligned} \tag{20}$$

Notice that obtaining $X(s)$, by (15) and (16), is short and simple. Most of our effort is in inverting $X(s)$. This is typical.

which give $E=-1/8$, $F=0$, $G=1/8$, and $H=0$. Thus, (18) becomes

$$\frac{1}{(s^2+9)(s^2+1)} = -\frac{1}{8}\frac{1}{s^2+9} + \frac{1}{8}\frac{1}{s^2+1}, \tag{21}$$

so

$$X(s) = \frac{2s-4}{s^2+9} + \frac{3}{s^2+1}, \tag{22}$$

with inverse

$$\begin{aligned} x(t) &= \mathcal{L}^{-1}\left\{\frac{2s-4}{s^2+9} + \frac{3}{s^2+1}\right\} \\ &= 2\mathcal{L}^{-1}\left\{\frac{s}{s^2+9}\right\} - 4\mathcal{L}^{-1}\left\{\frac{1}{s^2+9}\right\} + 3\mathcal{L}^{-1}\left\{\frac{1}{s^2+1}\right\} \\ &= 2\cos 3t - \frac{4}{\sqrt{9}}\mathcal{L}^{-1}\left\{\frac{\sqrt{9}}{s^2+9}\right\} + 3\sin t \\ &= 2\cos 3t - \frac{4}{3}\sin 3t + 3\sin t. \quad \blacksquare \end{aligned} \tag{23}$$

EXAMPLE 3. Solve

$$x'' + 3x' + 2x = 10e^{-t}; \qquad x(0) = 7,\ x'(0) = 0. \tag{24}$$

Transforming the equation, with the help of (1) and (12) and using the initial conditions, gives

$$\left[s^2X(s) - 7s - 0\right] + 3\left[sX(s) - 7\right] + 2X(s) = \frac{10}{s+1}, \tag{25}$$

so

$$X(s) = \frac{7(s+3)}{(s+1)(s+2)} + \frac{10}{(s+1)^2(s+2)}. \tag{26}$$

By combining the terms in (26) we will need only one partial fraction expansion instead of two.

We could expand each of the two terms in (26) in partial fractions, but it is more efficient to combine them first, as

$$X(s) = \frac{7(s+3)(s+1) + 10}{(s+1)^2(s+2)} = \frac{7s^2 + 28s + 31}{(s+1)^2(s+2)}, \tag{27}$$

for then only one partial fraction expansion is needed. Next, following the steps described in Appendix A, expand

Occurrence of a repeated factor in the denominator.

$$\begin{aligned}\frac{7s^2 + 28s + 31}{(s+1)^2(s+2)} &= \frac{A}{s+1} + \frac{B}{(s+1)^2} + \frac{C}{s+2} \\ &= \frac{A(s+1)(s+2) + B(s+2) + C(s+1)^2}{(s+1)^2(s+2)}.\end{aligned} \tag{28}$$

Then, match the coefficients of the s^0, s^1, and s^2 terms in the numerators of the left- and right-hand sides:

$$\begin{aligned}s^0 &: 2A + 2B + C = 31, \\ s^1 &: 3A + B + 2C = 28, \\ s^2 &: A + C = 7,\end{aligned} \tag{29}$$

which give $A = 4$, $B = 10$, and $C = 3$. Thus,

$$X(s) = \frac{4}{s+1} + \frac{10}{(s+1)^2} + \frac{3}{s+2} \tag{30}$$

and, using items 2 and 17, obtain

$$x(t) = 4e^{-t} + 10te^{-t} + 3e^{-2t}. \ \blacksquare \tag{31}$$

EXAMPLE 4. Solve

$$x'' - 6x' + 13x = 0, \qquad x(0) = 2,\ x'(0) = 0. \tag{32}$$

Transforming the equation gives

$$\left[s^2 X(s) - 2s\right] - 6\left[sX(s) - 2\right] + 13X(s) = 0,$$

so

$$X(s) = \frac{2s - 12}{s^2 - 6s + 13}. \tag{33}$$

We can expand the right-hand side in partial fractions, as in Examples 1–3. However, this time the roots of the denominator are complex, $s = 2 \pm 3i$. If we wish to avoid the resulting complex arithmetic, we can proceed as follows. *Complete the square* in the denominator:

$$s^2 - 6s + 13 = \left[(s-3)^2 - 9\right] + 13 = (s-3)^2 + 4,$$

For a quadratic denominator, completing the square provides an alternative to partial fractions.

which is of the same form as the denominators in items 9 and 10, with $a = 3$ and $b = \sqrt{4} = 2$. Next, re-express the numerator in (33) as

$$2s - 12 = 2(s-3) - 6.$$

Then (33) becomes

$$X(s) = \frac{2(s-3) - 6}{(s-3)^2 + 4} = 2\frac{s-3}{(s-3)^2 + 4} - \frac{6}{2}\frac{2}{(s-3)^2 + 4}, \tag{34}$$

and items 9 and 10 (*not* 11 and 12) give

$$x(t) = 2e^{3t} \cos 2t - 3e^{3t} \sin 2t. \quad \blacksquare \tag{35}$$

5.3.3 Systems.

Extension of the method to *systems* of linear constant-coefficient equations is straightforward.

EXAMPLE 5. Two First-Order Equations. Solve

$$x' = x + y - 9t, \qquad x(0) = 1, \tag{36a}$$

$$y' = 4x + y, \qquad y(0) = 4 \tag{36b}$$

for $x(t)$ and $y(t)$. Transforming (36a) and (36b) gives

$$sX(s) - x(0) = X(s) + Y(s) - \frac{9}{s^2}, \tag{37a}$$

$$sY(s) - y(0) = 4X(s) + Y(s) \tag{37b}$$

By transforming (36a) and (36b) we mean, as usual, multiplying them through by e^{-st} and integrating from 0 to ∞. That is, we take the transform of both sides of those equations.

or, since $x(0) = 1$ and $y(0) = 4$,

$$(s-1)X(s) - Y(s) = 1 - \frac{9}{s^2}, \tag{38a}$$

$$-4X(s) + (s-1)Y(s) = 4. \tag{38b}$$

Thus, just as a single differential equation in the t domain leads to a single linear algebraic equation in the s domain, a system of differential equations in the t domain leads to a system of linear algebraic equations in the s domain.

Solving (38), by elimination or by Cramer's rule, gives

$$X(s) = \frac{s^3 + 3s^2 - 9s + 9}{s^2(s^2 - 2s - 3)}, \qquad Y(s) = \frac{4s^3 - 36}{s^2(s^2 - 2s - 3)}. \tag{39a,b}$$

In this case the denominators $s^2(s+1)(s-3)$ have the repeated root $s = 0$; that is, the denominators are $(s-0)(s-0)(s+1)(s-3)$. Thus (see the Repeated Roots part of Appendix A), the appropriate partial fraction expansions are of the form

$$X(s) = \frac{A}{s} + \frac{B}{s^2} + \frac{C}{s+1} + \frac{D}{s-3} \text{ and } Y(s) = \frac{E}{s} + \frac{F}{s^2} + \frac{G}{s+1} + \frac{H}{s-3}.$$

Omitting the algebra this time, we can solve for A through H, and obtain

$$X(s) = -\frac{3}{s^2} + \frac{5}{s} - \frac{5}{s+1} + \frac{1}{s-3}, \tag{40a}$$

$$Y(s) = \frac{12}{s^2} - \frac{8}{s} + \frac{10}{s+1} + \frac{2}{s-3}, \tag{40b}$$

Actually, we did not need to solve for both $x(t)$ *and* $y(t)$. We could have omitted (39b) and (40b) and solved only for $x(t)$. With $x(t)$ in hand, (36a) then gives $y(t)$ from $y(t) = x'(t) - x(t) + 9t$.

inversion of which gives

$$x(t) = -3t + 5 - 5e^{-t} + e^{3t}, \tag{41a}$$

$$y(t) = 12t - 8 + 10e^{-t} + 2e^{3t}. \quad \blacksquare \tag{41b}$$

EXAMPLE 6. Coupled Mechanical Oscillator. In Section 4.2 we discussed the motion of the two-mass oscillator shown below in Fig. 1. For the free vibration [i.e., with $F_1(t) = F_2(t) = 0$], with m_1, m_2, k_1, k_2, and k_3 all equal to 1, the coupled differential equations are

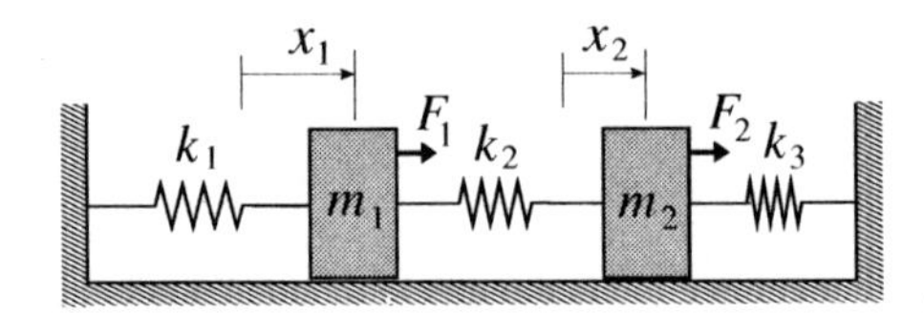

Figure 1. Two-mass oscillator.

$$x_1'' + 2x_1 - x_2 = 0, \tag{42a}$$

$$x_2'' - x_1 + 2x_2 = 0. \tag{42b}$$

Let the initial conditions be $x_1(0) = 1$ and $x_1'(0) = x_2(0) = x_2'(0) = 0$, so the disturbance from rest is the initial displacement of mass #1. Transforming the two equations and using the initial conditions gives

$$(s^2 + 2)X_1(s) - X_2(s) = s, \tag{43a}$$

$$-X_1(s) + (s^2 + 2)X_2(s) = 0. \tag{43b}$$

Let us solve only for $x_1(t)$, and then put that result into (42a) and solve by algebra for $x_2(t)$. Solving (43) for $X_1(s)$ gives

$$X_1(s) = \frac{s(s^2+2)}{s^4+4s^2+3}. \tag{44}$$

Factor the denominator as s^4+4s^2+3 as $(s^2+1)(s^2+3)$, and expand

$$X_1(s) = \frac{s^3+2s}{(s^2+1)(s^2+3)} = \frac{A+Bs}{s^2+1} + \frac{C+Ds}{s^2+3}. \tag{45}$$

The step (45) is like (18); recall that discussion in Example 2.

Solving for the coefficients in the usual way gives $A = C = 0$ and $B = D = 1/2$, so

$$x_1(t) = \mathcal{L}^{-1}\left\{\frac{1}{2}\frac{s}{s^2+1} + \frac{1}{2}\frac{s}{s^2+3}\right\} = \frac{1}{2}\left(\cos t + \cos\sqrt{3}t\right). \tag{46a}$$

Finally, if we put the latter into (42a) and solve for $x_2(t)$, we obtain

$$x_2(t) = x_1''(t) + 2x_1(t) = \frac{1}{2}\left(\cos t - \cos\sqrt{3}t\right); \tag{46b}$$

$x_1(t)$ and $x_2(t)$ are plotted in Fig. 2. If you studied Section 4.2, you will recognize the presence, in (46a,b), of both vibrational modes, the low mode (at frequency 1) and the high mode (at frequency $\sqrt{3}$). ∎

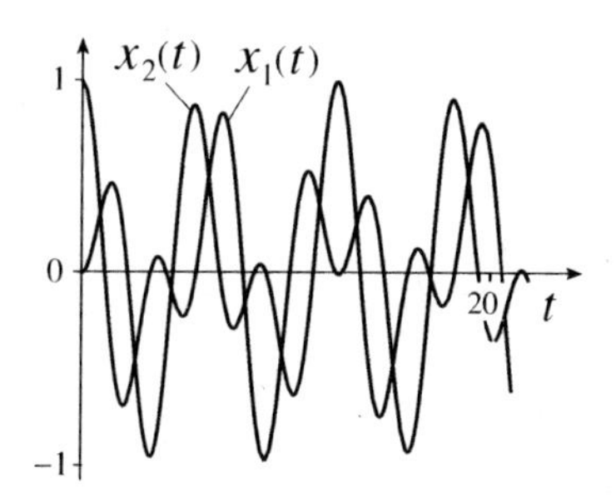

Figure 2. The motions of the two masses.

5.3.4 Application to a nonconstant-coefficient equation; Bessel's equation.

In general, equations with nonconstant coefficients are not solvable by the Laplace transform method. However, if the nonconstant coefficients are low-degree powers of t such as t ot t^2 then there is hope, thanks to the following two theorems:

Subsection 5.3.4 can be omitted without loss of continuity.

THEOREM 5.3.2 *Multiplication by t*
Let $f(t)$ satisfy the conditions of Theorem 5.2.2 with exponential coefficient c, so $\mathcal{L}\{f(t)\} = F(s)$ for $s > c$. Then, for $s > c$,

$$\boxed{\mathcal{L}\{tf(t)\} = -\frac{dF}{ds}.} \tag{47}$$

That is, multiplication by t in the t domain corresponds to $-d/ds$ in the s domain.

In fact,

$$\boxed{\mathcal{L}\{t^n f(t)\} = (-1)^n \frac{d^n F}{ds^n} \quad (n = 1, 2, \ldots).} \tag{48}$$

For $n = 1$, (48) reduces to (47), so (48) is a generalization of (47).

We will give only a formal derivation of these results. If we differentiate

$$\mathcal{L}\{f(t)\} = F(s) = \int_0^\infty f(t)e^{-st}\,dt$$

with respect to s, we obtain

d/ds because the integral is a function only of s, and $\partial/\partial s$ because the integrand is a function of s *and* t.

$$F'(s) = \frac{d}{ds}\int_0^\infty f(t)e^{-st}\,dt = \int_0^\infty \frac{\partial}{\partial s}[f(t)e^{-st}]\,dt$$
$$= -\int_0^\infty tf(t)e^{-st}\,dt = -\mathcal{L}\{tf(t)\}, \tag{49}$$

which is (47). We said the steps were "formal" because the interchange in the order of the integration and the differentiation in (49) is permissible only if $f(t)$ is sufficiently well behaved. Continuing formally, differentiate again and obtain

$$F''(s) = \int_0^\infty t^2 f(t)e^{-st}\,dt = \mathcal{L}\{t^2 f(t)\},$$

which is (48) for $n = 2$; and so on.

The second theorem we will need is this:

THEOREM 5.3.3 *Large-s Behavior*
(a) If $f(t)$ satisfies the conditions of Theorem 5.2.2, then

$$\boxed{\lim_{s\to\infty} F(s) = 0.} \tag{50}$$

(b) If $f(t)$ and $f'(t)$ satisfy the conditions of Theorem 5.2.2, then

The initial value formula.

$$\boxed{\lim_{s\to\infty}[sF(s)] = f(0),} \tag{51}$$

known as the **initial value formula.**

Proof: Proof of (50) was outlined in Exercise 14 of Section 5.2. To prove (51), let $s \to \infty$ in the formula $\mathcal{L}\{f'(t)\} = sF(s) - f(0)$. Since $f'(t)$ satisfies the conditions of Theorem 5.2.2, by hypothesis, its transform tends to zero as $s \to \infty$. Thus, the limit of $sF(s) - f(0)$ is zero, which gives (51). ■

EXAMPLE 7. Bessel's Equation. The equation

$$\boxed{tx'' + x' + tx = 0,} \tag{52}$$

known as **Bessel's equation of order zero**, is nonelementary, its solutions not being expressible in closed form in terms of elementary functions. It is one of the most important nonelementary differential equations in engineering and science, and we will study it further in Chapter 6. Because it has nonconstant coefficients we cannot expect the Laplace transform solution method to work, but since the coefficients happen to be t's to only the first power we will see that the method will, nevertheless, enable us to find the transform $X(s)$. Consider the IVP

$$tx'' + x' + tx = 0, \qquad x(0) = 1, \ x'(0) = 0. \tag{53}$$

To transform the tx'' term use (47), with $x''(t)$ as "$f(t)$," and to transform the tx term use (47) again, this time with $x(t)$ as "$f(t)$." Thus,

$$\begin{aligned} \mathcal{L}\{tx''(t)\} &= -\frac{d}{ds}\mathcal{L}\{x''(t)\} \\ &= -\frac{d}{ds}[s^2X(s) - sx(0) - x'(0)] \end{aligned} \tag{54}$$

and

$$\mathcal{L}\{tx(t)\} = -\frac{d}{ds}\mathcal{L}\{x(t)\} = -\frac{d}{ds}X(s). \tag{55}$$

Thus, transforming the differential equation gives

$$-\frac{d}{ds}[s^2X(s) - sx(0) - x'(0)] + [sX(s) - x(0)] - \frac{d}{ds}X(s) = 0 \tag{56}$$

(57) is a first-order equation because the coefficients are t to the first power; t^2 would have led to a second-order equation, and so on. It is best to write dX/ds rather than X' because the prime might be mistaken for d/dt. Obtain $\dfrac{dX}{X} + \dfrac{s\,ds}{s^2+1} = 0$, integrate, and obtain (58).

or, since $x(0) = 1$ and $x'(0) = 0$,

$$(s^2 + 1)\frac{dX}{ds} + sX = 0, \tag{57}$$

which is *not* a linear algebraic equation (as obtained for constant-coefficient equations) but a *differential equation* for $X(s)$! However, it is a simple one because it is separable, and its solution is

$$X(s) = \frac{A}{\sqrt{s^2 + 1}}, \tag{58}$$

with A an arbitrary constant of integration. To evaluate A, use (51):

$$\lim_{s\to\infty} sX(s) = \lim_{s\to\infty} \frac{sA}{\sqrt{s^2+1}} = x(0).$$

The limit of the middle member is A, and $x(0)$ is 1, so $A = 1$. Thus,

$$X(s) = \frac{1}{\sqrt{s^2 + 1}}. \tag{59}$$

Inverting the latter is difficult; but if we give up on obtaining a closed-form result, we can obtain the inverse in power series form as

$$x(t) = \mathcal{L}^{-1}\left\{\frac{1}{\sqrt{s^2+1}}\right\} = 1 - \frac{1}{4}t^2 + \frac{1}{64}t^4 - \frac{1}{2304}t^6 + \cdots; \tag{60}$$

we did that in Exercise 16 of Section 5.2. It can be shown that the series in (60) converges for all t, and its sum is the **Bessel function** of the first kind and of order zero, $J_0(t)$. We will obtain that series again in Chapter 6, when we develop methods for obtaining solutions in series form. ▮

Closure. The point of Laplace transforming a linear differential equation with constant coefficients is that what was a linear differential equation in the t domain becomes a linear algebraic equation in the s domain, so solving for $X(s)$ in the s domain is simple. Expect the most difficult part of the method to be the last step, inverting the transform $X(s)$ to obtain $x(t)$. Understand, especially, the use of partial fractions in (16)–(22), in (27)–(30), and in (39)–(40), and completing the square in (33)–(35).

In general, nonconstant-coefficient equations are not solvable by the Laplace transform. However, if the equation has nonconstant coefficients that are constants times integer powers of t, then the transform method gives a linear *differential* equation for $X(s)$ instead of a linear *algebraic* one. In the case of the Bessel equation we were able to solve that equation for $X(s)$, and to obtain $x(t)$ in series form.

Extension to systems of constant-coefficient differential equations is straightforward. Instead of obtaining a single linear algebraic equation for $X(s)$, we obtain a system of linear algebraic equations for the transforms $X_1(s), \ldots, X_n(s)$ of the unknowns $x_1(t), \ldots, x_n(t)$. Solve for $X_1(s), \ldots, X_n(s)$ by algebra and invert them.

EXERCISES 5.3

1. We derived (12) and (13) by applying (1) two and three times, respectively. Instead, derive (12) by integrating $\int_0^\infty f''(t)e^{-st}\,dt$ by parts twice. [Similarly, you could derive (13) by integrating $\int_0^\infty f'''(t)e^{-st}\,dt$ by parts three times, but you need not do that.]

2. Initial Value Problems. Solve the given IVP by the Laplace transform.

(a) $x' = \sin t;\quad x(0) = 2$
(b) $x' + 2x = 4t^2;\quad x(0) = -12$
(c) $x' - 6x = 10e^t;\quad x(0) = 3$
(d) $x' - x = 24\sinh t;\quad x(0) = 0$
(e) $x'' = 6t;\quad x(0) = 2,\ x'(0) = -1$
(f) $x'' + 5x' = 25;\quad x(0) = 6,\ x'(0) = 0$
(g) $x'' + x = 2t - 1;\quad x(0) = x'(0) = 0$
(h) $x'' + 9x = 18\cos 3t;\quad x(0) = 0,\ x'(0) = 150$
(i) $x'' + 4x' + 3x = 8e^{-t};\quad x(0) = 0,\ x'(0) = 6$
(j) $x'' + 4x' + 4x = 0;\quad x(0) = 1,\ x'(0) = 0$
(k) $x'' - 4x' + 4x = 12te^{2t};\quad x(0) = x'(0) = 0$
(l) $x'' - 2x' + x = 24e^t;\quad x(0) = 0,\ x'(0) = 0$
(m) $x'' - x = 4te^{-t};\quad x(0) = 0,\ x'(0) = 0$
(n) $x''' - x'' = 6t;\quad x(0) = x'(0) = x''(0) = 1,$
(o) $x'''' - x'' = 6t;\quad x(0) = 2,$
$x'(0) = x''(0) = x'''(0) = 0$
(p) $x'''' - x = 4;\quad x(0) = 8,$
$x'(0) = x''(0) = x'''(0) = 0$

3. (c)–(g) **Obtaining a General Solution.** Use the Laplace transform to find a *general solution* of the differential equation given in the corresponding part of Exercise 2. HINT: For instance, consider the equation $x'' - 9x = 18$. To find a general solution by the Laplace transform, solve the IVP

$$x'' - 9x = 18;\quad x(0) = a,\ x'(0) = b, \tag{3.1}$$

in which a and b are arbitrary. Doing so, one obtains

$$x(t) = -2 + (2 + a)\cosh 3t + \frac{b}{3}\sinh 3t. \tag{3.2}$$

Since a and b are arbitrary, we can write (3.2) in the equivalent and more compact form

$$x(t) = -2 + C_1\cosh 3t + C_2\sinh 3t. \tag{3.3}$$

4. Initial Conditions Not at $t = 0$. Because the transforms of derivatives involve initial conditions at $t = 0$, it is particularly convenient (and typical) if the initial conditions are specified at $t = 0$. In the problems given below they are not. Nevertheless, we can still solve by the Laplace transform, as follows: Ignore the given initial conditions and use $x(0) = a$ and $x'(0) = b$ instead, in which a and b are arbitrary constants. That will give the general solution (as was discussed in Exercise 3). With the general solution in hand, apply the given initial conditions to evaluate a and b and hence to obtain the desired particular solution.

(a) $x'' - x = 0;\quad x(5) = 12,\ x'(5) = 0$
(b) $x'' + 4x = 8t;\quad x(\pi) = x'(\pi) = 0$
(c) $x'' + x' = 12e^{3t};\quad x(-1) = x'(-1) = 0$
(d) $x'' + 9x = 24\sin t;\quad x(\pi) = -2,\ x'(\pi) = 0$

5. Systems. Solve by the Laplace transform. You may use computer software to do the partial fraction expansions if you wish.

(a) $x' = x + 3y - 4t;\quad x(0) = 0$
$y' = x - y - 8;\quad y(0) = 5$
(b) $x' = y;\quad x(0) = 0$
$y' = x + 4e^t;\quad y(0) = 0$
(c) $x' = -y + 10\sin t;\quad x(0) = 5$
$y' = 9(10 - x);\quad y(0) = 0$
(d) $x'' = y;\quad x(0) = a,\ x'(0) = 0$
$y' + x' = 0;\quad y(0) = b$
(e) $x'' = x + y;\quad x(0) = 4,\ x'(0) = 0$
$y'' = x + y;\quad y(0) = y'(0) = 0$
(f) $x'' = 2x + y;\quad x(0) = 2,\ x'(0) = 0$
$y'' = x + 2y;\quad y(0) = 2,\ y'(0) = 0$

6. Two-Mass Oscillator. This time, let the forces applied to the masses (in Fig.1) be the constants $F_1(t) = F_2(t) = 50$, and let all initial conditions be zero. Then the IVP is

$$x_1'' + 2x_1 - x_2 = 50;\quad x_1(0) = x_1'(0) = 0. \tag{6.1a}$$

$$x_2'' - x_1 + 2x_2 = 50;\quad x_2(0) = x_2'(0) = 0. \tag{6.1b}$$

(a) Solve (6.1) for $x_1(t)$ and $x_2(t)$ using the Laplace transform.
(b) Solve using initial conditions and forcing functions prescribed by your instructor.

7. Two-Loop Oscillator. The currents $i_1(t)$, $i_2(t)$ in the circuit shown below are governed by the equations

$$i_1' = -\frac{R}{L_1}(i_1 + i_2);\quad i_1(0) = I \tag{7.1a}$$

$$i_2' = -\frac{R}{L_2}(i_1 + i_2);\quad i_2(0) = 0. \tag{7.1b}$$

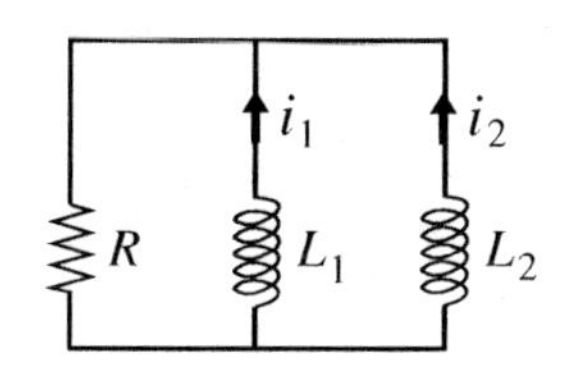

(a) Let $R = L_1 = L_2 = 1$ and $I = 4$, and solve for the two currents by the Laplace transform. As a partial check on your work, you should find that $i_2(t) = 2e^{-2t} - 2$.
(b) Solve again, using values of R, L_1, L_2, I given by your instructor.

8. Three-Loop Oscillator. The currents $i_1(t)$, $i_2(t)$, $i_3(t)$ in the circuit shown below are governed by the equations

$$i_1' = -\frac{R}{L_1}(i_1 + i_2 + i_3);\quad i_1(0) = I \tag{8.1a}$$

$$i_2' = -\frac{R}{L_2}(i_1 + i_2 + i_3);\quad i_2(0) = 0 \tag{8.1b}$$

$$i_3' = -\frac{R}{L_3}(i_1 + i_2 + i_3);\quad i_3(0) = 0. \tag{8.1c}$$

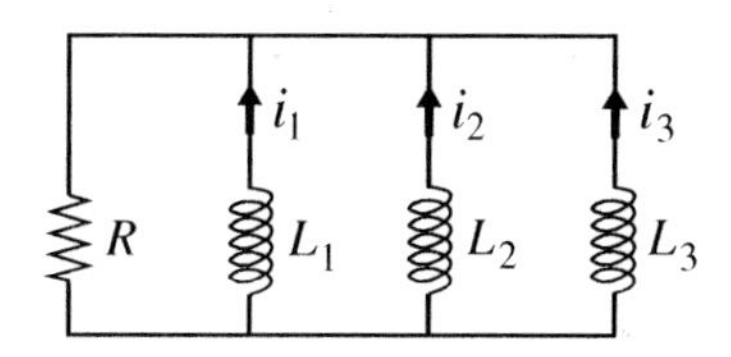

(a) Let $R = L_1 = L_2 = L_3 = 1$ and $I = 6$, and solve for the three currents by the Laplace transform. As a partial check on your work, you should find that $i_3(t) = 2e^{-3t} - 2$.
(b) Solve again, using values of R, L_1, L_2, L_3, I given by your instructor.

9. Motion of Charged Particle. A charged particle, having mass m and electrical charge q, moves in a uniform magnetic field of strength B. The field is in the positive z direction. The equations of motion of the particle are found (you need not derive them) to be

$$\begin{aligned} mx'' &= qBy', \\ my'' &= -qBx', \\ mz'' &= 0, \end{aligned} \tag{9.1}$$

where $x(t), y(t), z(t)$ are the x, y, z displacements as functions of the time t. Let $\alpha \equiv qB/m$ for brevity, and (with no loss of generality) let the initial location be the origin, so $x(0) = y(0) = z(0) = 0$. Further, let $x'(0) = U$, $y'(0) = V$, and $z'(0) = 0$.

(a) Use the Laplace transform to solve for $x(t), y(t), z(t)$.
(b) Show that the resulting motion is a circle in the x, y plane, and give its radius R and center in terms of α, U, and V.

10. Application to Boundary Value Problems on a Finite Subinterval. The Laplace transform method is especially convenient for initial value problems on $0 < t < \infty$, but it can also be used to solve boundary value problems on a subinterval. To illustrate, consider this BVP for $y(x)$:

$$y'' - y' = 2 - 2x;\quad y(0) = 1,\ y(2) = 5, \tag{10.1}$$

in which the independent variable happens to be x instead of t. (In this text we generally take the generic independent variable to be x for boundary value problems and t for initial value problems. The name of the independent variable is not relevant insofar as the Laplace transform procedure is concerned.) To apply the Laplace transform to (10.1) we begin by artificially extending the x domain to $0 < x < \infty$. Having done that, we must define the forcing function [which was given in (10.1) as $2-2x$ on $0 < x < 2$] on the extension, that is, on $2 < x < \infty$. We will find that the extended definition of the forcing function is immaterial, so let us simply let the forcing function be $2-2x$ on the whole interval $0 < x < \infty$. Realize that the Laplace transform is now in terms of x, so the transform $Y(s)$ of $y(x)$ is $\int_0^\infty y(x)e^{-sx}\,dx$, that is, with x's instead of t's. Apply that transform to (10.1) and obtain

$$s^2Y(s) - sy(0) - y'(0) - sY(s) + y(0) = \frac{2}{s} - \frac{2}{s^2}. \quad (10.2)$$

Now, we know that $y(0) = 1$ but we do not know the $y'(0)$ needed in (10.2). Simply call $y'(0) = A$, where the constant A is not yet known, and proceed. Show that (10.2) leads to

$$y(x) = 1 + x^2 + A\,(e^x - 1)\,. \quad (10.3)$$

Now apply the boundary condition $y(2) = 5$ to (10.3), solve for A, and show that

$$y(x) = 1 + x^2. \quad (10.4)$$

NOTE: To see that the extended definition of the forcing function is immaterial, as we claimed, observe that (10.4) does satisfy the differential equation in (10.1) on the *actual* interval $0 < x < 2$, and we forced it to satisfy both the initial condition $y(0) = 1$ and the boundary condition $y(2) = 5$. That's all that is required of the solution of (10.1).

11. Use the idea explained in Exercise 10 to solve the following for $y(x)$.

(a) $y'' - 2y' = e^x;\quad y(0) = 9,\ y(1) = 10 - e$
(b) $y'' + y = 3 + x + x^2;\quad y(0) = 1,\ y(2) = 7$
(c) $y''' - y'' = 0;\quad y(0) = 0,\ y(1) = 2,\ y'(1) = 2$
(d) $y''' = 48x;\quad y(0) = 1,\ y(2) = 37,\ y'(2) = 68$

12. Another BVP: Deflection of a Loaded Beam. Consider a cantilever beam of length L subjected to a uniform load distribution w (downward force per unit x-length), as

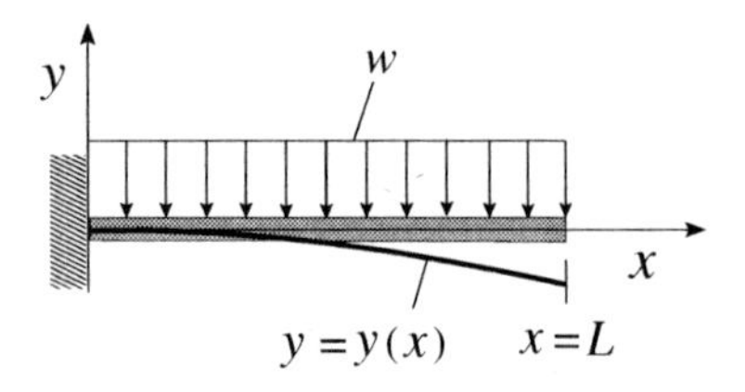

shown in the sketch. We state, without derivation, that *Euler beam theory* gives the BVP

$$EIy'''' = -w;\quad y(0) = y'(0) = 0,\ y''(L) = y'''(L) = 0 \quad (12.1)$$

for the resulting deflection $y(x)$, where the constant EI is called the stiffness of the beam and depends on the beam cross section and the beam material. (In the terminology of beam theory E is the Young's modulus of the beam material and I is the inertia of the beam cross section about a horizontal axis through its centroid.) Using the idea contained in Exercise 10, solve (12.1) for $y(x)$ by the Laplace transform.

ADDITIONAL EXERCISES

13. On the Design of the Laplace Transform. The integral

$$\text{“}T\{f(t)\}\text{”} = \int_\alpha^\beta f(t)K(t,s)\,dt \equiv F(s); \quad (13.1)$$

is called an **integral transform**; the input is $f(t)$ and the output is its transform $F(s)$. For the Laplace transform the **kernel** $K(s,t)$ is e^{-st}, the limits are $\alpha = 0$ and $\beta = \infty$, and we use the letter $\mathcal{L}$, after Laplace, on the left-hand side; in (13.1) we use instead the generic T for "transform." By designing a transform (13.1) we mean choosing α, β, and the kernel $K(t,s)$. Imagining that the Laplace transform were not yet invented, let us think about how to design a transform that will be tailored to constant-coefficient equations on the interval $0 \le t < \infty$. Since the interval is $0 \le t < \infty$, let us choose $\alpha = 0$ and $\beta = \infty$. But how shall we choose $K(t,s)$? The key property that we wish to establish, by suitable choice of the kernel, is that the transform of the derivative $f'(t)$ be linear in $F(s)$,

$$T\{f'(t)\} = a(s)F(s) + b(s), \quad (13.2)$$

because *it is thanks to that form that the Laplace transform converts linear constant-coefficient differential equations to linear algebraic equations*. Show how the form of (13.2) can guide us to the choice $K(t,s) = e^{-st}$ and hence to the Laplace transform. HINT: Use (13.1) to transform $f'(t)$ and insist that the transform be of the form (13.2).

14. Designing a Transform for Cauchy–Euler Equations. First, read Exercise 13.

(a) Just as the Laplace transform is designed for linear constant-coefficient differential equations, now design one for Cauchy–Euler equations—on the interval $1 \leq t < \infty$, say. That is, choose the kernel $K(t, s)$ in

$$\mathcal{C}\{f(t)\} = \int_1^\infty f(t)K(t, s)\, dt \tag{14.1}$$

so that the transform of $tf'(t)$ is of the form

$$\mathcal{C}\{tf'(t)\} = a(s)F(s) + b(s), \tag{14.2}$$

in which $F(s)$ denotes $\mathcal{C}\{f(t)\}$, the Cauchy–Euler transform of $f(t)$. You should find that the kernel is $K(t, s) = t^{-s}$, so

$$\boxed{\mathcal{C}\{f(t)\} = \int_1^\infty f(t)t^{-s}\, dt \equiv F(s).} \tag{14.3}$$

(b) Derive formulas analogous to (1) and (12), namely,

$$\boxed{\mathcal{C}\{tf'(t)\} = (s-1)F(s) - f(1)} \tag{14.4}$$

and

$$\boxed{\mathcal{C}\{t^2 f''(t)\} = (s-1)(s-2)F(s) - (s-2)f(1) - f'(1).} \tag{14.5}$$

(c) One can also work out the transforms of various elementary functions to develop a transform table, but we ask you to evaluate just one: $\mathcal{C}\{t^n\}$, for any integer n.

(d) Use the transform formulas you obtained in parts (a)–(c) to solve

$$t^2x'' + tx' - 4x = 0; \quad x(1) = 6,\ x'(1) = 0.$$

(e) Use them to solve

$$t^2x'' + tx' - x = 12t^2; \quad x(1) = 4,\ x'(1) = -10.$$

(f) Use them to solve

$$t^2x'' - 2tx' + 2x = 2t^3; \quad x(1) = x'(1) = 0.$$

5.4 DISCONTINUOUS FORCING FUNCTIONS; HEAVISIDE STEP FUNCTION

5.4.1 Motivation. The Laplace transform is particularly well suited to equations with *discontinuous forcing functions*.

5.4.2 Heaviside step function and piecewise-defined functions. To represent functions that are defined piecewise, it is convenient to use the Heaviside **step function** $H(t)$, defined as[1]

$$\boxed{H(t) = \begin{cases} 1, & t > 0 \\ 0, & t < 0. \end{cases}} \tag{1}$$

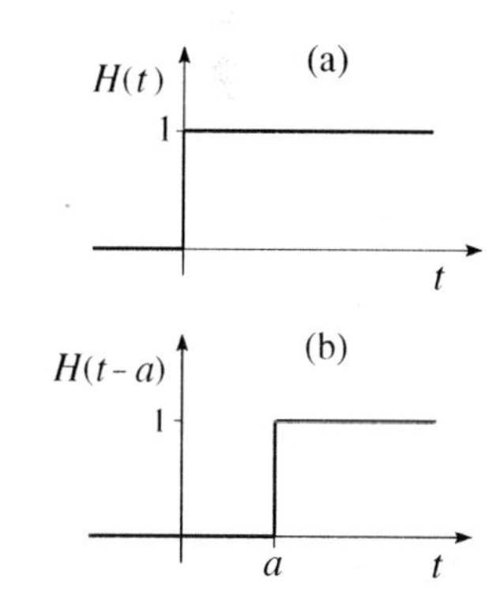

Figure 1. The Heaviside unit step function. It is 1 when its argument is positive, and 0 when its argument is negative.

[1] *Oliver Heaviside* (1850–1925), initially a telegraph and telephone engineer, is best known for his contributions to vector field theory and to the development of a systematic Laplace transform methodology for the solution of differential equations. Note the spelling: Heaviside, not Heavyside. For an intersesting biography of his life and work, see Paul J. Nahin's *Oliver Heaviside: Sage in Solitude* (NY: IEEE Press, 1987).

The value assigned to $H(t)$ *at* the point of discontinuity $t = 0$ will be of no importance here, so we've omitted it in (1). Since $H(t)$ is a unit step at $t = 0$ (Fig. 1a), $H(t-a)$ is a unit step shifted to $t = a$, as in Fig. 1b.

The Heaviside function can be used to build up piecewise-defined functions. We begin by forming a unit **rectangular pulse** $P(t;a,b)$, which is unity on $a < t < b$ and zero outside that interval. It is expressible as the difference of two Heaviside functions:

$$\boxed{P(t;a,b) = H(t-a) - H(t-b),} \tag{2}$$

as indicated in Fig. 2.

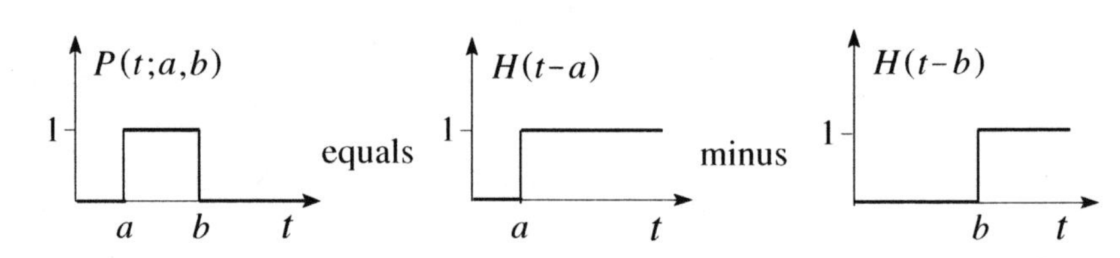

Figure 2. The rectangular pulse function is a difference of Heaviside functions.

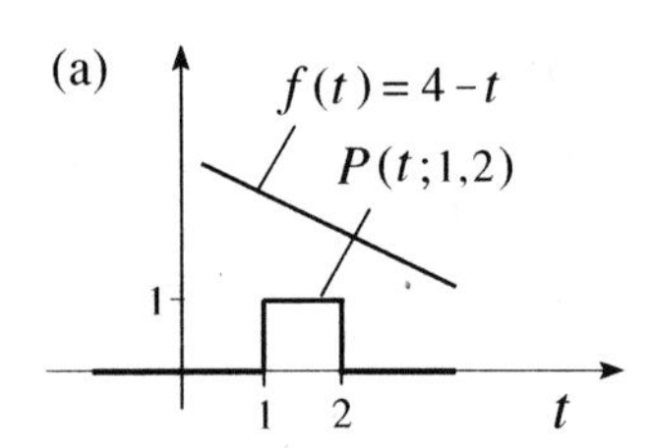

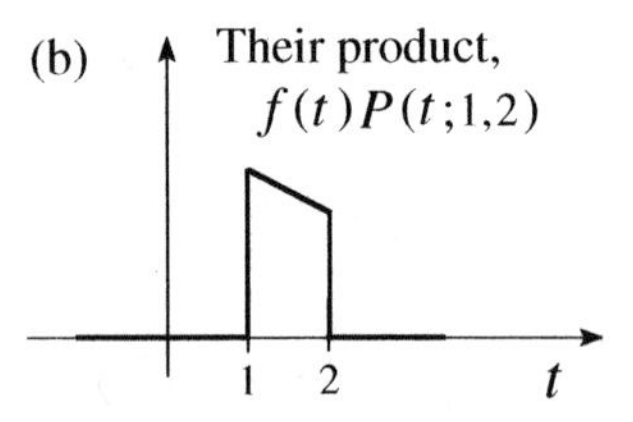

Figure 3. Picking out a "piece" of the function $f(t) = 4-t$.

We can use the rectangular pulse function to pick out a "piece" of a function. For instance, if $f(t) = 4 - t$ is multiplied by $P(t;1,2)$, graphs of which are shown in Fig. 3a, then their product is the function shown in Fig. 3b.

EXAMPLE 1. A Piecewise-Defined Function. To illustrate, let us use pulse functions to obtain a single expression for the function

$$f(t) = \begin{cases} 4e^t, & 0 < t < 1 \\ e^t, & 1 < t < 2 \\ 10/t^2, & 2 < t < \infty \end{cases} \tag{3}$$

shown in Fig. 4:

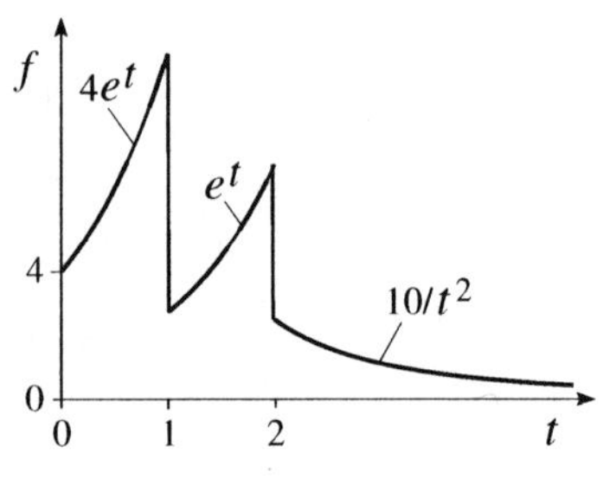

Figure 4. Graph of the function $f(t)$ defined by (3).

$$\begin{aligned} f(t) &= 4e^t P(t;0,1) + e^t P(t;1,2) + \frac{10}{t^2} P(t;2,\infty) \\ &= 4e^t\,[H(t-0) - H(t-1)] + e^t\,[H(t-1) - H(t-2)] \\ &\quad + \frac{10}{t^2}\,[H(t-2) - H(t-\infty)] \end{aligned} \tag{4}$$

or, since $H(t-0) = H(t) = 1$ and $H(t-\infty) = 0$ on the interval of interest $(0 < t < \infty)$,

$$\begin{aligned} f(t) &= 4e^t\,[1 - H(t-1)] + e^t\,[H(t-1) - H(t-2)] + \frac{10}{t^2} H(t-2) \\ &= 4e^t - 3e^t H(t-1) + \left(\frac{10}{t^2} - e^t\right) H(t-2). \end{aligned} \tag{5}$$

As a check on (5), observe that it does give $4e^t$ on $0 < t < 1$, e^t on $1 < t < 2$, and $10/t^2$ on $2 < t < \infty$.

The breaking down of f into three parts, in the first line of equation (4), is shown graphically in Fig. 5:

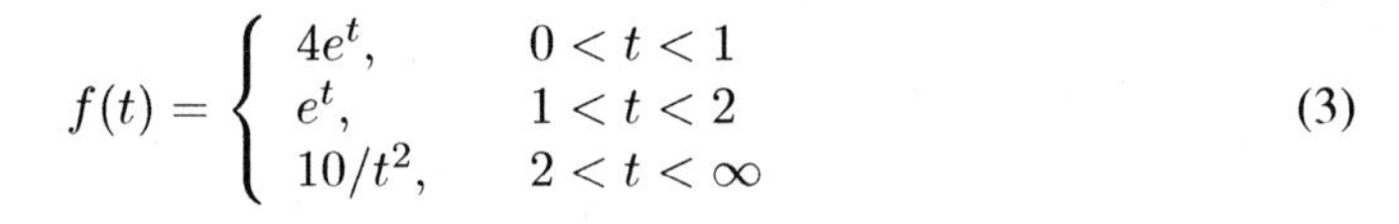

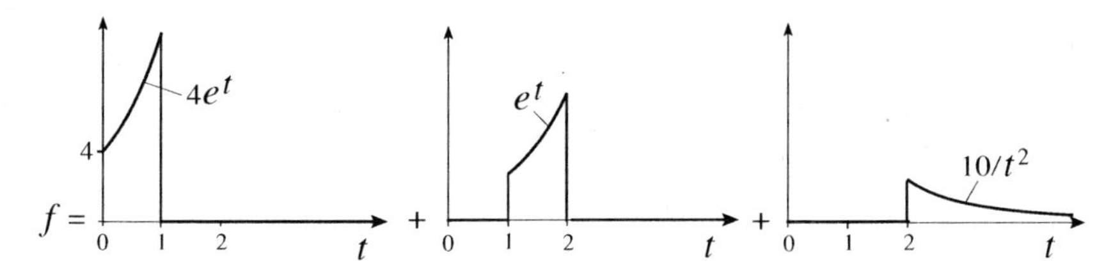

Figure 5. $f(t) = 4e^t P(t; 0, 1) + e^t P(t; 1, 2) + (10/t^2)P(t; 2, \infty)$.

Adding the three graphs in Fig. 5 gives the graph of f shown in Fig. 4. ∎

Time delays. Besides using Heaviside functions to express piecewise-defined functions, we can also use them to create "time delays." To explain delays, first consider the combination

$$H(t)f(t) = \begin{cases} 0, & t < 0 \\ f(t), & t > 0. \end{cases} \tag{6}$$

Graphically, $H(t)f(t)$ is the function $f(t)$ "cut off" for $t < 0$. For instance, $f(t) = \sin t$ is shown in Fig. 6a, and $H(t)\sin t$ is the cut-off version in Fig. 6b. Next, recall that the graph of a function $f(t-a)$, for any $a > 0$, is the same as the graph of $f(t)$, but shifted rightward by a. Putting these ideas together (i.e., cutting and shifting), it follows that

$$\boxed{H(t-a)f(t-a)} \tag{7}$$

is a **time-delayed function**: $H(t)f(t)$ cuts off $f(t)$ for $t < 0$, and then changing the t's to $(t-a)$'s shifts the cut-off graph to the right by a, as illustrated in Fig. 6c for $H(t-a)\sin(t-a)$.

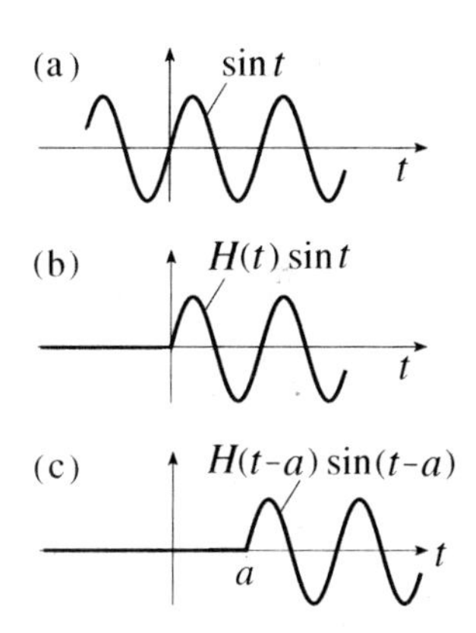

Figure 6. Example of a time-delayed function.

5.4.3 Transforms of Heaviside and time-delayed functions. We will need to know the transform of the Heaviside function. The transform of $H(t-a)$, if $a \geq 0$, is

$$\begin{aligned} \mathcal{L}\{H(t-a)\} &= \int_0^\infty H(t-a)e^{-st}\,dt = \int_a^\infty e^{-st}\,dt \\ &= \frac{e^{-st}}{-s}\bigg|_a^\infty = \frac{e^{-as}}{s} \end{aligned} \tag{8}$$

if $s > 0$:

$$\boxed{\mathcal{L}\{H(t-a)\} = \frac{e^{-as}}{s}.} \tag{9}$$

As a partial check of (9), if $a = 0$, then $H(t-a) = H(t) = 1$ (on $0 < t < \infty$), so (9) reduces to $\mathcal{L}\{1\} = 1/s$, which agrees with item 1 in the table.

Second, we will need to know how to transform time-delayed functions. We have the following important result:

THEOREM 5.4.1 *Transform of Time-Delayed Functions*
Let $\mathcal{L}\{f(t)\} = F(s)$ exist for $s > c$ and let a be any positive number. Then

$$\boxed{\mathcal{L}\{H(t-a)f(t-a)\} = e^{-as}F(s)} \tag{10}$$

for $s > c$.

For the special case $f(t) = 1$, its transform is $F(s) = 1/s$, and (10) reduces to (9).

Proof: From the definition of the transform,

$$\begin{aligned}\mathcal{L}\{H(t-a)f(t-a)\} &= \int_0^\infty H(t-a)f(t-a)e^{-st}\,dt \\ &= \int_a^\infty f(t-a)e^{-st}\,dt \quad \text{(Now let } t-a=\tau) \\ &= \int_0^\infty f(\tau)e^{-s(\tau+a)}\,d\tau \\ &= e^{-as}\int_0^\infty f(\tau)e^{-s\tau}\,d\tau = e^{-as}F(s). \quad \blacksquare\end{aligned} \tag{11}$$

Remember, the $F(s)$ in (10) is the transform of $f(t)$, *not* of $f(t-a)$. Equation (10) will be the focal point of this section, and we will use it in both directions: the transform of $H(t-a)f(t-a)$ is $e^{-as}F(s)$ and, in the reverse direction, the inverse of $e^{-as}F(s)$ is $H(t-a)f(t-a)$, which we now illustrate.

EXAMPLE 2. Using (10). Evaluate $\mathcal{L}\{H(t-3)\sin(t-3)\}$. The function to be transformed is of the form $H(t-a)f(t-a)$ so identify a and $f(t)$: $a = 3$ and $f(t-a)$ is $f(t-3)$ is $\sin(t-3)$, so $f(t)$ is $\sin t$, *not* $\sin(t-3)$. Then (10) gives

It is essential to understand that if $f(t-3)$ is $\sin(t-3)$, then $f(t)$ is $\sin t$.

$$\begin{aligned}\mathcal{L}\{H(t-3)\sin(t-3)\} &= e^{-as}\mathcal{L}\{f(t)\} = e^{-3s}\mathcal{L}\{\sin t\} \\ &= e^{-3s}\frac{1}{s^2+1}.\end{aligned} \tag{12}$$

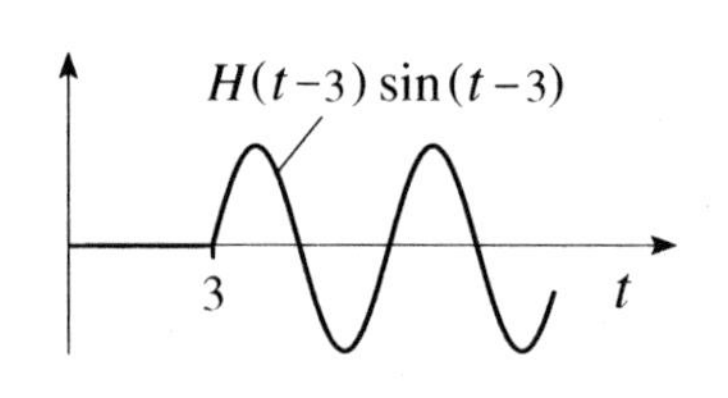

Figure 7. Graph of $H(t-3)\sin(t-3)$.

We must also know how to use (10) in the reverse direction. To illustrate, now evaluate $\mathcal{L}^{-1}\left\{\frac{e^{-3s}}{s^2+1}\right\}$. [Of course, we should get back the original function $H(t-3)\sin(t-3)$.] The function of s to be inverted is indeed of the form $e^{-as}F(s)$, as in (10), with $a = 3$ and $F(s) = \frac{1}{s^2+1}$. Then, $f(t) = \mathcal{L}^{-1}\{F(s)\} = \sin t$, and (10) gives

$$\mathcal{L}^{-1}\left\{e^{-3s}\frac{1}{s^2+1}\right\} = H(t-3)\sin(t-3), \tag{13}$$

the graph of which is given in Fig. 7. ▮

EXAMPLE 3. Using (10). Evaluate $\mathcal{L}\{H(t-2)t^2\}$. We see from the argument of the Heaviside function that $a=2$, but this time the t^2 spoils the $H(t-2)f(t-2)$ delay pattern that we need if we are to use (10). That difficulty is overcome by rewriting the t^2 as $[(t-2)+2]^2$. Then,

$$\mathcal{L}\{H(t-2)t^2\} = \mathcal{L}\{H(t-2)\underbrace{[(t-2)+2]^2}_{f(t-2)}\} = e^{-2s}\mathcal{L}\{\underbrace{(t+2)^2}_{f(t)}\} \quad \text{by (10)}$$
$$= e^{-2s}\mathcal{L}\{t^2+4t+4\}$$
$$= e^{-2s}\left(\frac{2}{s^3}+\frac{4}{s^2}+\frac{4}{s}\right). \tag{14}$$

Do you see that $f(t)=(t+2)^2$? Let us call the argument "arg." Then $f(t-2)=[(t-2)+2]^2$ gives $f(\text{arg})=(\text{arg}+2)^2$. Now let arg be t and obtain $f(t)=(t+2)^2$.

COMMENT 1. We don't *have to* use (10); we can always fall back on the integral definition, in this case $\mathcal{L}\{H(t-2)t^2\} = \int_0^\infty H(t-2)t^2e^{-st}\,dt = \int_2^\infty t^2e^{-st}\,dt$, which can be evaluated by integrating by parts twice.

COMMENT 2. Generalizing this example, if we have the form $H(t-a)f(t)$ instead of $H(t-a)f(t-a)$, we can always re-express this as $H(t-a)f((t-a)+a)$, as we did following the first equal sign in (14). ▮

5.4.4 Differential equations with piecewise-defined forcing functions. We are ready to solve differential equations with piecewise-defined forcing functions.

EXAMPLE 4. A First-Order Equation. Let $p(t)$ be the function whose graph is shown in Fig. 8. Solve

$$x'+x = p(t)$$
$$= 4t[H(t-0)-H(t-2)] + 2[H(t-2)-H(t-\infty)]$$
$$= 4t + 2H(t-2) - 4H(t-2)t, \tag{15}$$

subject to the initial condition $x(0)=3$. [We've used the name $p(t)$ for the forcing function so as not to confuse it with the generic $f(t)$ in (10).] The transforms of the first and second terms on the right-hand side of (15) (omitting the scale factors 4 and 2 for the moment) are simple:

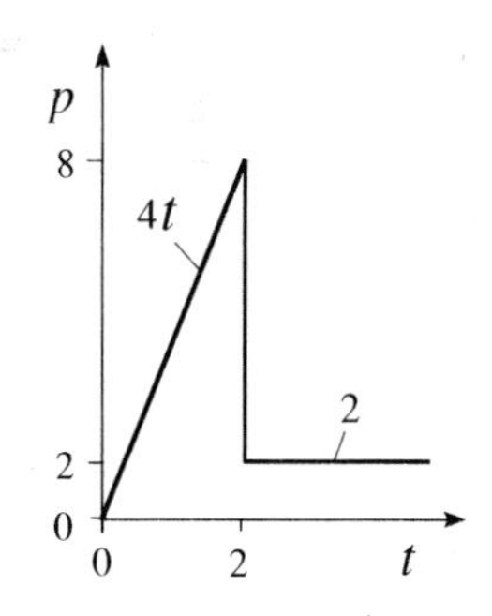

Figure 8. The forcing function $p(t)$ in (15).

$$\mathcal{L}\{t\} = \frac{1}{s^2} \qquad \text{(item 7 in the table)} \tag{16a}$$

$$\mathcal{L}\{H(t-2)\} = \frac{e^{-2s}}{s} \qquad \text{(item 19 in the table).} \tag{16b}$$

The third term on the right-hand side of (15) is not in the table but, following the same lines as in Example 3,

$$\mathcal{L}\{H(t-2)t\} = \mathcal{L}\{H(t-2)\underbrace{[(t-2)+2]}_{f(t-2)}\} = e^{-2s}\mathcal{L}\{\underbrace{t+2}_{f(t)}\} \quad \text{by (10)}$$

$$= e^{-2s}\left(\frac{1}{s^2} + \frac{2}{s}\right). \tag{17}$$

Thus, the transform of the IVP consisting of (15) and the initial condition $x(0) = 3$ is

$$[sX(s) - 3] + X(s) = \frac{4}{s^2} + 2\frac{e^{-2s}}{s} - 4\left(\frac{1}{s^2} + \frac{2}{s}\right)e^{-2s}. \tag{18}$$

Solving the latter for $X(s)$ gives

$$X(s) = \frac{3}{s+1} + \frac{4}{s^2(s+1)} - \frac{6}{s(s+1)}e^{-2s} - \frac{4}{s^2(s+1)}e^{-2s}. \tag{19}$$

Using partial fractions, express

$$\frac{1}{s^2(s+1)} = -\frac{1}{s} + \frac{1}{s^2} + \frac{1}{s+1} \quad \text{and} \quad \frac{1}{s(s+1)} = \frac{1}{s} - \frac{1}{s-1}, \tag{20}$$

and obtain

$$X(s) = \frac{7}{s+1} + \frac{4}{s^2} - \frac{4}{s} + e^{-2s}\left(-\frac{4}{s^2} - \frac{2}{s} + \frac{2}{s+1}\right). \tag{21}$$

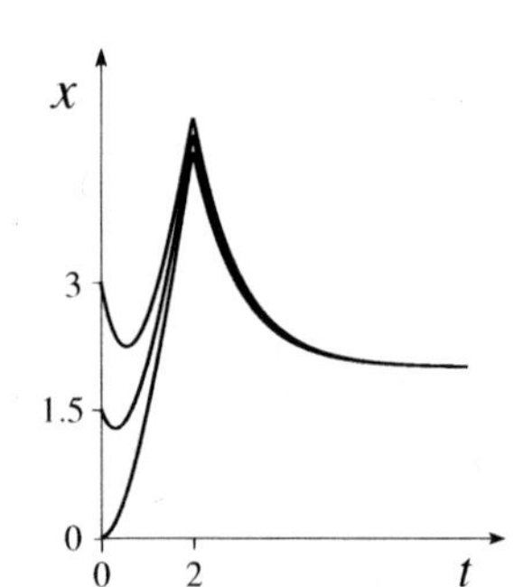

Figure 9. Solutions of (15) for $x(0) = 0,\ 1.5,\ 3$.

The first three terms in (21) are readily inverted using the table. To invert the last term, use (10) *in the reverse direction*: The inverse of the square-bracketed terms is $-4t - 2 + 2e^{-t}$, so the inverse of e^{-2s} times the bracketed terms is, according to (10), that function with each t replaced by $t - 2$, times $H(t-2)$. Thus,

$$\begin{aligned} x(t) &= 7e^{-t} + 4t - 4 + H(t-2)[-4(t-2) - 2 + 2e^{-(t-2)}] \\ &= 7e^{-t} + 4t - 4 + H(t-2)(2e^{2-t} - 4t + 6). \end{aligned} \tag{22}$$

The solution (22) is plotted in Fig. 9, along with the solutions corresponding to two other initial conditions, $x(0) = 1.5$ and $x(0) = 0$. ∎

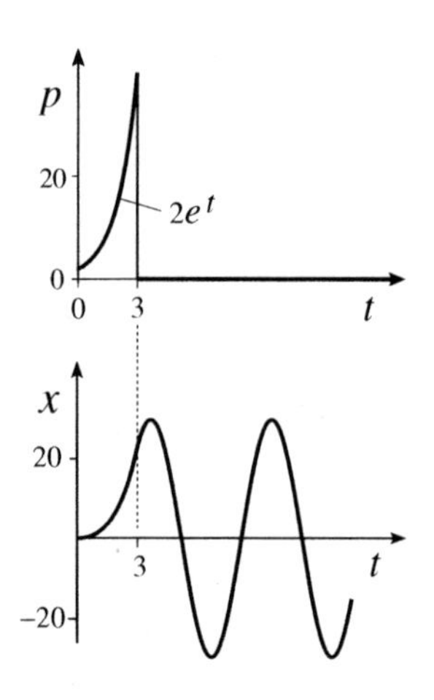

Figure 10. The forcing function $p(t)$ and the response $x(t)$.

EXAMPLE 5. A Second-Order Equation. Let $p(t)$ be the function whose graph is given in Fig. 10. Solve the problem

$$\begin{aligned} x'' + x &= p(t) \\ &= 2e^t[H(t-0) - H(t-3)] \\ &= 2e^t[1 - H(t-3)] \end{aligned} \tag{23}$$

subject to the initial conditions $x(0) = 0$ and $x'(0) = 0$. To transform the $H(t-3)e^t$ term in the forcing function, write

$$\begin{aligned} \mathcal{L}\{H(t-3)e^t\} &= \mathcal{L}\{H(t-3)e^{(t-3)+3}\} = e^3\mathcal{L}\{H(t-3)\underbrace{e^{(t-3)}}_{f(t-3)}\} \\ &= e^3e^{-3s}\mathcal{L}\{\underbrace{e^t}_{f(t)}\} \quad \text{by (10)} \\ &= e^3e^{-3s}\frac{1}{s-1}. \end{aligned} \tag{24}$$

Then, transforming (23) with $x(0) = x'(0) = 0$ gives

$$(s^2+1)X(s) = \frac{2}{s-1} - 2e^3 e^{-3s}\frac{1}{s-1} \tag{25}$$

or, solving for $X(s)$,

$$X(s) = \frac{2}{(s-1)(s^2+1)} - 2e^3 e^{-3s}\frac{1}{(s-1)(s^2+1)}. \tag{26}$$

Typically, there is more than one way to invert a given transform, as we've seen. Here, let us use partial fractions and expand

$$\frac{1}{(s-1)(s^2+1)} = \frac{A}{s-1} + \frac{Bs+C}{s^2+1}. \tag{27}$$

Doing so, we find that $A = 1/2$ and $B = C = -1/2$, so we can write (26) as

$$X(s) = \frac{1}{s-1} - \frac{s}{s^2+1} - \frac{1}{s^2+1} - e^3 e^{-3s}\left(\frac{1}{s-1} - \frac{s}{s^2+1} - \frac{1}{s^2+1}\right), \tag{28}$$

with inverse

$$x(t) = e^t - \cos t - \sin t - e^3 H(t-3)\left[e^{t-3} - \cos(t-3) - \sin(t-3)\right]. \tag{29}$$

The graph of the solution (29) is given in Fig. 10.

COMMENT. It may seem strange that there is a Heaviside function $H(t-3)$ in (29), yet the graph of $x(t)$ in Fig. 10 shows no discontinuity at $t = 3$. To understand this result, call the square-bracketed factor in (29) $g(t)$. Even though the $H(t-3)$ jumps at $t = 3$, $H(t-3)g(t)$ does not because $g(3) = 0$; thus, $x(t)$ is continuous at $t = 3$. Similarly, $g'(3) = 0$ so $x'(t)$ is continuous there as well. But $g''(3) = 2 \neq 0$, so there *is* a jump in $x''(t)$, and hence the curvature of the graph of $x(t)$, at $t = 3$. ▮

If $g(t)$ is continuous and $g(3) = 0$, then $H(t-3)g(t)$ has no jump at $t = 3$.

5.4.5 Periodic forcing functions.

To illustrate the case of periodic forcing functions, consider the following example.

EXAMPLE 6. A Periodic Forcing Function. Consider the IVP

$$x' + x = f(t); \qquad x(0) = x_0, \tag{30}$$

in which $f(t)$ is the periodic *square wave* in Fig. 11, with period $T = 2$. Transforming (30) gives

$$X(s) = \frac{x_0}{s+1} + \frac{1}{s+1}F(s) \tag{31}$$

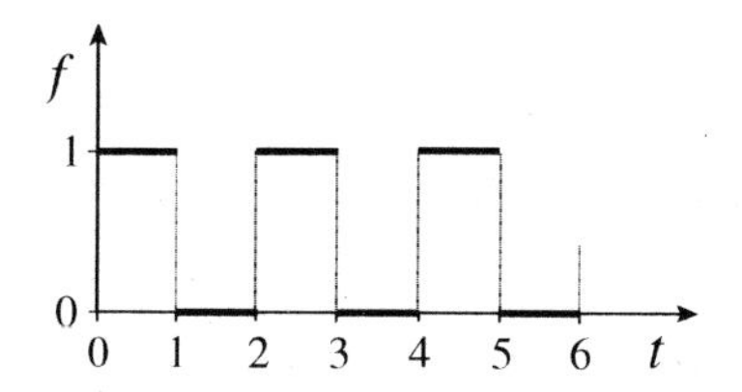

Figure 11. The square wave $f(t)$ in (30).

in which

$$\begin{aligned} F(s) &= \int_0^\infty f(t)e^{-st}\,dt \qquad [f(t) \text{ given in Fig. 11}] \\ &= \int_0^1 1e^{-st}\,dt + \int_2^3 1e^{-st}\,dt + \int_4^5 1e^{-st}\,dt + \cdots \\ &= \frac{1}{s}\left(1 - e^{-s} + e^{-2s} - e^{-3s} + e^{-4s} - \cdots\right). \end{aligned} \tag{32}$$

Putting the latter into (31) gives

$$X(s) = \frac{x_0}{s+1} + \frac{1}{s(s+1)}\left(1 - e^{-s} + e^{-2s} - e^{-3s} + e^{-4s} - \cdots\right). \tag{33}$$

To invert the series part of (33), begin with

$$\mathcal{L}^{-1}\left\{\frac{1}{s(s+1)}\right\} = \mathcal{L}^{-1}\left\{\frac{1}{s} - \frac{1}{s+1}\right\} = 1 - e^{-t}. \tag{34}$$

Here, a is $0, 1, 2, \ldots$, and $F(s)$ is $1/[s(s+1)]$.

Then, the typical term in the series in (33) is inverted with the help of the time-delay formula $\mathcal{L}^{-1}\{e^{-as}F(s)\} = H(t-a)f(t-a)$ (which is Theorem 5.4.1, and also table item 29), and we obtain

$$\begin{aligned} x(t) &= x_0 e^{-t} + (1-e^{-t}) - H(t-1)\left(1-e^{-(t-1)}\right) + H(t-2)\left(1-e^{-(t-2)}\right) \\ &\quad - H(t-3)\left(1-e^{-(t-3)}\right) + \cdots \\ &= x_0 e^{-t} + \sum_{n=0}^{\infty}(-1)^n H(t-n)\left(1-e^{-(t-n)}\right), \end{aligned} \tag{35}$$

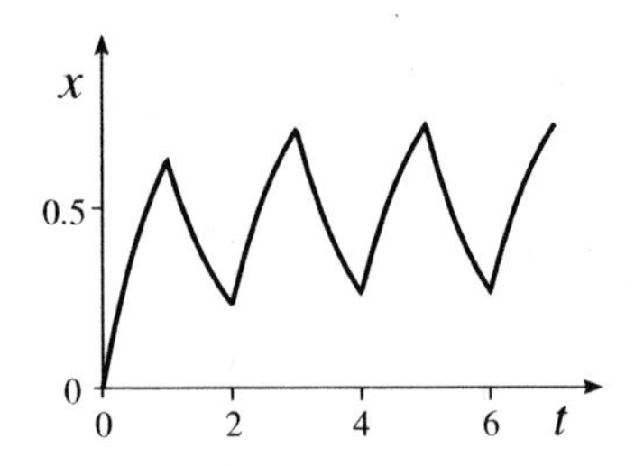

Figure 12. The solution (35) of the IVP (30), for the case $x_0 = 0$.

the graph of which is given in Fig. 12 for the case where $x_0 = 0$.

COMMENT. It's true that the result (35) is an infinite series, but for any given value of t the series *terminates* after a finite number of terms! For instance, if we are interested in the interval $0 < t < 7$, as in Fig. 12, then we need sum only the terms up to and including the $H(t-6)$ term because $H(t-7), H(t-8), \ldots$ are zero for $0 < t < 7$. ▮

Closure. We defined the Heaviside step function $H(t)$ and its translated version $H(t-a)$, we used two Heaviside functions to form rectangular pulses $P(t; a, b)$, and showed how to express a given piecewise-defined function as the sum of unit rectangular pulse functions times the individual pieces of the given function.

Then we defined time-delayed functions, of the form $H(t-a)f(t-a)$. The transform of such a function was given by (10) as

$$\mathcal{L}\{H(t-a)f(t-a)\} = e^{-as}F(s). \tag{36}$$

When we turned to the solution of differential equations having piecewise-defined forcing functions, that formula was the key: *when we transformed the differential equation we used it in the forward direction [as in (36)] because the piecewise-defined forcing function contained Heaviside functions, and when we inverted the transform $X(s)$ we used it again, this time in the reverse direction*, namely,

$$\mathcal{L}^{-1}\{e^{-as}F(s)\} = H(t-a)f(t-a)\}. \tag{37}$$

We also found, in Section 5.4.5, that Heaviside functions can be expected to arise in the Laplace transform solution of equations with periodic forcing functions.

EXERCISES 5.4

1. Express $f(t)$ in Heaviside notation. RECALL: (a, b) is standard notation for the open interval $a < t < b$.

(a) $f(t) = 50t$ on $(0, 20)$, 1000 on $(20, \infty)$
(b) $f(t) = 10$ on $(0, 1)$, 20 on $(1, 2)$, 30 on $(2, 3)$, 0 on $(3, \infty)$
(c) $f(t) = 10t^2$ on $(0, 2)$, 25 on $(2, 5)$, 30 on $(5, \infty)$
(d) $f(t) = \sin t$ on $(0, \pi)$, 0 on $(\pi, 2\pi)$, $\sin t$ on $(2\pi, 3\pi)$, 0 on $(3\pi, \infty)$
(e) $f(t) = f_1(t)$ on $(0, a)$, $f_2(t)$ on (a, b), and $f_3(t)$ on (b, ∞)

2. Give a labeled hand sketch of the graph of the given function f, on $0 < t < \infty$. NOTE: Recall the definition of the Heaviside function: it is 1 when its argument is positive and 0 when its argument is negative.

(a) $H(t - \pi/2) \sin t$
(b) $H(t-1)t$
(c) $H(t) \cosh t$
(d) $H(t)e^{-t}$
(e) $H(t-\pi) \sin(t-\pi)$
(f) $H(t-\pi) \cos(t-\pi)$
(g) $H(t-1)e^{t-1}$
(h) $\cos(t-2\pi)H(t-2\pi)$
(i) $H(t-2)H(t-3)$
(j) $H(t^2-4)$
(k) $H(t^2-4t+2)$
(l) $H(\sin t)$
(m) $1 - H(t-1) + H(t-2)$
(n) $[H(t-1) - H(t-2)]t^2$
(o) $H(t-1)50t + (25-50t)H(t-2)$
(p) $[H(t-1) - H(t-2) + H(t-3)]t$

3. In each case evaluate $x'(2)$, $x'(5)$, and $x''(5)$.

(a) $x(t) = t^2 + H(t-3)e^{-2t}$
(b) $x(t) = 6t + [H(t-3) - H(t-4)] \cos t$
(c) $x(t) = 1 + tH(t-3) + t^2H(t-4) + t^4H(t-7)$
(d) $x(t) = e^{-t}[H(t-1) + 2H(t-4) + 5H(t-6)]$

4. Evaluate the Laplace transform.

(a) $H(t-1) \sinh(t-1)$
(b) $H(t-\pi) \sin(t-\pi)$
(c) $4H(t-2)e^{3t}$
(d) $H(t-1)(e^{3t} - 5e^t)$
(e) $H(t-2) \cos t$
(f) $H(t-3) \sin t$
(g) $H(t-1)t^2$
(h) $H(t-3)(t+5)$
(i) $H(t-2)(5t+7)$
(j) $H(t-5)(1+t)$

5. Evaluate the inverse.

(a) e^{-4s}/s^3
(b) $se^{-s}/(s^2-1)$
(c) $\dfrac{e^{-5s}}{s^2-3s}$
(d) $e^{-s}\left(\dfrac{1+s}{1+s^2}\right)$
(e) $e^{-3s}\left(\dfrac{4}{s} - \dfrac{1}{s^2}\right)$
(f) $\dfrac{e^{-s} + 3e^{-2s}}{s^2(s+1)}$
(g) $\dfrac{e^{-s}}{s^2} - \dfrac{4e^{-2s}}{s^3}$
(h) $\dfrac{e^{-s}}{(s+3)^4}$

6. Show, from (29), that the solution $x(t)$ is both continuous and smooth (i.e., its derivative x' is continuous) at $t = 3$, even though the forcing function has a jump discontinuity at that point. Show that the *second* derivative $x''(t)$ suffers a jump discontinuity at $t = 3$. Thus, it is the curvature of the graph of $x(t)$ that jumps at $t = 3$. NOTE: *Recall from the calculus that a function $f(t)$ is continuous at t_0 if the limits of $f(t)$ as $t \to t_0$ from the left and right both exist and equal $f(t_0)$. However, in (29), as elsewhere in this section, we have not specified the value of the Heaviside function at its jump discontinuity, so by continuous here we will simply mean that the limit from the left is equal to the limit from the right.*

7. Simplified Version of Example 4. As a "simplest" version of a first-order equation with a discontinuous forcing function, consider the IVP

$$x' = H(t-a); \quad x(0) = 0, \tag{7.1}$$

with $a > 0$. Solve for $x(t)$ by the Laplace transform, and give a labeled sketch of its graph. Show that whereas the input is discontinuous at $t = a$, $x(t)$ is continuous there, and that $x'(t)$ has a jump discontinuity there.

8. Simplified Version of Example 5. As a "simplest" version of a second-order equation with a discontinuous forcing function, consider the IVP

$$x'' = H(t-a); \quad x(0) = 0,\ x'(0) = 0, \tag{8.1}$$

with $a > 0$. Solve for $x(t)$ by the Laplace transform. Show that whereas the input is discontinuous at $t = a$, $x(t)$ and $x'(t)$ are continuous there, and that $x''(t)$ has a jump discontinuity there. NOTE: This should not be surprising. Think of (8.1) as governing the displacement of an automobile (of unit mass), subjected to a unit force starting at $t = a$. That is, at $t = a$ we depress the accelerator to the floor and keep it there. The automobile's acceleration jumps, at $t = a$, from 0 to unity, but its velocity and displacement increase *continuously* from rest.

9. Solving Example 4 Without the Laplace Transform. Re-solve (15) by breaking the problem into two parts and solving each by hand, that is, without using the Laplace transform. The first is

$$x' + x = 4t; \ x(0) = 3 \tag{9.1}$$

on $0 < t < 2$, and the second is

$$x' + x = 2; \ x(2) \text{ obtained from solution to (9.1)} \tag{9.2}$$

on $2 < t < \infty$. Show that your results agree with (22). NOTE: The point to observe is that for piecewise defined forcing functions the Laplace transform method is much simpler. As the number of "pieces" is increased, the benefit is even more pronounced.

10. First-Order Equations with Discontinuous Forcing Functions. Use the Laplace transform to solve the given IVP for $x(t)$.
(a) $x' + 2x = p(t);\ x(0) = x_0$, where $p(t)$ is 50 for $1 < t < 2$ and 0 elsewhere
(b) $x' + 2x = p(t);\ x(0) = x_0$, where $p(t)$ is $40t$ for $0 < t < 2$ and 20 for $t > 2$
(c) $x' - x = p(t);\ x(0) = x_0$, where $p(t)$ is 40 for $0 < t < 5$, 20 for $5 < t < 10$, and 0 for $t > 10$
(d) $x' + x = p(t);\ x(0) = 0$, where $p(t)$ is $20t$ for $0 < t < 5$, $200 - 20t$ for $5 < t < 10$, and 0 for $t > 10$
(e) $x' - 4x = p(t);\ x(0) = 0$, where $p(t)$ is $\sin t$ for $0 < t < \pi$ and 0 for $t > \pi$
(f) $x' + 2x = p(t);\ x(0) = 6$, where $p(t)$ is 0 for $0 < t < 4$, for $t - 4$ for $4 < t < 8$, and 4 for $t > 8$
(g) $x' - 3x = H(t-1) + H(t-2) + H(t-3);\ x(0) = 0$
(h) $x' + x = H(t-2)e^t;\ x(0) = 12$

11. (a)–(h) **Computer.** As a check against your solution to the corresponding part of Exercise 10, use computer software to solve for $x(t)$.

12. Higher-Order Equations with Discontinuous Forcing Functions. Use the Laplace transform to solve the given IVP for $x(t)$.
(a) $x'' - x = p(t);\ x(0) = 0,\ x'(0) = 0$, where $p(t)$ is 50 for $4 < t < 6$, and 0 elsewhere
(b) $x'' + x' = p(t);\ x(0) = 0,\ x'(0) = 0$, where $p(t)$ is $100 - 25t$ for $0 < t < 2$, and 0 for $t > 2$
(c) $x'' - 2x' = p(t);\ x(0) = 0,\ x'(0) = 6$, where $p(t)$ is 10 for $0 < t < 4$ and 20 for $t > 4$
(d) $x'' + x = p(t);\ x(0) = 0,\ x'(0) = 0$, where $p(t)$ is $3 - t$ for $0 < t < 4$, and 0 for $t > 4$
(e) $x'' - 2x' + x = H(t-1)t;\ x(0) = 0,\ x'(0) = 0$
(f) $x''' - x'' = p(t);\ x(0) = 0,\ x'(0) = 0,\ x''(0) = -2$, where $p(t)$ is $2e^{-t}$ for $0 < t < 4$, and 0 for $t > 4$
(g) $x''' - x' = p(t);\ x(0) = 0,\ x'(0) = 0,\ x''(0) = -10$, where $p(t)$ is $10t$ for $0 < t < 3$, and 30 for $t > 3$

EQUATIONS WITH PERIODIC FORCING FUNCTIONS

13. More About Example 6. Show that the solution (35) *tends to* a periodic function of period 2 as $t \to \infty$. HINT: Write out $x(t + 2)$ and $x(t)$ and show that their difference tends to zero as $t \to \infty$.

14. Let $f(t)$ be the periodic function defined below. Using the Laplace transform, solve the IVP

$$x' + x = f(t); \quad x(0) = 0 \tag{14.1}$$

and plot $x(t)$ over three periods of f. Use a computer algebra system to help with the steps if you wish. [As in Example 6, we've taken the equation in (14.1) to be of first order to reduce the labor, so the calculations are not so long that they obscure the main ideas.] On one period, $f(t)$ is defined as follows:

(a) $\sin t$ on $0 < t < \pi$ NOTE: This function is the same as $|\sin t|$, and is called a "rectified" sine wave.
(b) 1 on $0 < t < 2$, 0 on $2 < t < 3$
(c) t on $0 < t < 1$, 0 on $1 < t < 2$
(d) 100 on $0 < t < \pi$, -100 on $\pi < t < 2\pi$
(e) 0 on $0 < t < 9$, 50 on $9 < t < 10$
(f) $\sin t$ on $0 < t < \pi$, 0 on $\pi < t < 2\pi$
(g) t on $0 < t < 1$

15. Continuation of Exercise 14; Are the Solutions Periodic? A natural question, in solving Exercise 14, is whether the solution $x(t)$ is periodic — like the forcing function $f(t)$. As representative, consider parts (b), (d), and (e):

(b) Show that the solution to Exercise 14(b) is not periodic, but that it *tends to* a periodic function, with the same period as $f(t)$ (namely, 3) as $t \to \infty$. HINT: Write out $x(t) - x(t + 3)$, show that only a few terms fail to cancel, and that the result tends to zero as $t \to \infty$.
(d) Show that the solution to Exercise 14(d) is not periodic, but that it *tends to* a periodic function, with the same period as $f(t)$ (namely, 2π) as $t \to \infty$.
(e) Show that the solution to Exercise 14(e) is not periodic, but that it *tends to* a periodic function, with the same period as $f(t)$ (namely, 10) as $t \to \infty$.

ADDITIONAL EXERCISES

16. Car Following, and a Delay Equation. Since the 1950s there has been extensive differential-equations-based modeling of traffic flow. Consider single-lane driving with no passing, and suppose the velocities of our car, and the car ahead of us, are $v(t)$ and $v_a(t)$, respectively. As a first cut at a simple model, suppose that we accelerate (by depressing the gas pedal) and decelerate (by breaking) proportional to the perceived velocity difference,

$$\frac{dv}{dt} = k[v_a(t) - v(t)], \tag{16.1}$$

in which k is a constant that could be determined empirically, and which we consider as known. To pose a specific problem, suppose the car ahead stops abruptly so $v_a(t) = 0$ for all

$t \geq 0$, and the question is whether the car behind ("our car") can stop in time to avoid a collision. Since $v_a(t) = 0$ for all $t > 0$, (16.1) becomes

$$\frac{dv}{dt} = -kv(t), \tag{16.2}$$

which is readily solved. But, to be more realistic let us build into (16.2) a nonzero driver reaction time, say δ. In that case we have, in place of (16.2),

$$\boxed{\frac{dv}{dt} = -kH(t-\delta)v(t-\delta),} \tag{16.3}$$

that is, the right-hand side of (16.2) is delayed by the response time δ; (16.3) is an example of a **delay-type differential equation**. Thus far in this text we've seen nothing like (16.3) because the argument of the unknown v is different in the two terms of the differential equation: t and $t-\delta$. But perhaps the Laplace transform can help because it can deal with delay-type functions.

(a) With the initial condition $v(0) = v_0$, show that application of the Laplace transform to (16.3) gives

$$V(s) = \frac{v_0}{s + ke^{-\delta s}} \tag{16.4}$$

for the transform of $v(t)$.
(b) The latter is difficult to invert. However, proceeding formally, let us pull out a factor of s from the denominator, and re-express (16.4) as

$$V(s) = \frac{v_0}{s} \frac{1}{1 + k\dfrac{e^{-\delta s}}{s}}. \tag{16.5}$$

Next, let the $ke^{-\delta s}/s$ be a new variable, say z, and use the binomial series

$$\frac{1}{1+z} = 1 - z + z^2 - z^3 + z^4 - \cdots \tag{16.6}$$

to obtain

$$V(s) = v_0\left[\frac{1}{s} - k\frac{e^{-\delta s}}{s^2} + k^2\frac{e^{-2\delta s}}{s^3} - \cdots\right]. \tag{16.7}$$

(c) Show that inversion of (16.7) gives

$$v(t) = v_0\Big[1 - kH(t-\delta)(t-\delta) + k^2H(t-2\delta)\frac{(t-2\delta)^2}{2!}$$
$$-k^3H(t-3\delta)\frac{(t-3\delta)^3}{3!} + \cdots\Big]. \tag{16.8}$$

(d) To find the stopping distance, recall that $dx/dt = v(t)$. Thus, replace the $v(t)$ in (16.8) by dx/dt, integrate from 0 to t, with $x(0) = 0$, and thus show that

$$x(t) = v_0\left[t - kH(t-\delta)\frac{(t-\delta)^2}{2!} + k^2H(t-2\delta)\frac{(t-2\delta)^3}{3!} - \cdots\right]. \tag{16.9}$$

(e) For the case of zero reaction time ($\delta = 0$), show that the series in (16.9) can be summed, giving

$$x(t) = \frac{v_0}{k}\left(1 - e^{-kt}\right). \tag{16.10}$$

(f) For definiteness, let $k = 1$ and $v_0 = 50$, and consider these reaction times: $\delta = 0, 0.3, 0.6, 1$ seconds. Use (16.9) to obtain a computer plot of $x(t)$ versus t, for those δ's. [Of course, for the case $\delta = 0$ you could use (16.10)]. From those plots, determine the safe following distance to avoid hitting the car ahead, for each of those values of δ.
(g) Explain why we pulled out an s from the denominator and re-expressed (16.4) in the form (16.5).
(h) Is the differential equation (16.3) linear? Explain.

COMMENT. You should have found in part (f) that if δ is made large enough then the solution develops a larger and larger damped *oscillation*. It is striking that (16.2) gives only exponential decay, yet when a sufficient time delay is included, in (16.3), oscillations are obtained. This result is important in control systems and biological systems.

5.5 CONVOLUTION

The Laplace convolution theorem is one of the most useful properties of the Laplace transform. It involves an operation called the "Laplace convolution" of two given functions, so we begin by defining that operation. Don't be concerned if the definition seems obscure; its significance will be established by Theorem 5.5.1, and illustrated in the examples.

5.5.1 Definition of Laplace convolution. Given functions $f(t)$ and $g(t)$ defined on $0 \leq t < \infty$, their **Laplace convolution**, or simply their **convolution**, is denoted as $f * g$ and defined as

$$(f * g)(t) \equiv \int_0^t f(\tau)\, g(t - \tau)\, d\tau, \tag{1}$$

in which the notation $(f * g)(t)$ is to indicate that $f * g$, like f and g, is a function of t, because of the t dependence in the integral.

EXAMPLE 1. Let $f(t) = e^t$ and $g(t) = 4e^{-3t}$. Their convolution is

$$\begin{aligned}(f * g)(t) &= \int_0^t e^{\tau} 4\, e^{-3(t-\tau)}\, d\tau \\ &= 4e^{-3t} \int_0^t e^{4\tau}\, d\tau = e^t - e^{-3t}.\end{aligned} \tag{2}$$

The convolution $f(t) * g(t)$ is NOT $f(t)$ *times* $g(t)$, it is the integral in (1).

COMMENT. Be clear that the convolution of $f(t)$ and $g(t)$ is *not* the ordinary product of $f(t)$ and $g(t)$, but is the function defined by (1). In this example the ordinary product of $f(t) = e^t$ and $g(t) = 4e^{-3t}$ is $f(t)g(t) = 4e^{-2t}$, whereas their convolution is $(f * g)(t) = e^t - e^{-3t}$. ∎

Although it is not an ordinary product, the Laplace transform does obey some of the rules of ordinary multiplication:

$$f * g = g * f, \qquad \text{(commutative)} \tag{3a}$$

$$f * (g * h) = (f * g) * h, \qquad \text{(associative)} \tag{3b}$$

$$f * (\alpha g + \beta h) = \alpha f * g + \beta f * h, \qquad \text{(linear)} \tag{3c}$$

$$f * 0 = 0, \tag{3d}$$

where α and β in (3c) are arbitrary constants. To prove commutativity, (3a), make the change of variable $t - \tau = \mu$ in (1), from the dummy integration variable τ to a dummy integration variable μ, with t fixed. Then $\tau = t - \mu$ and $d\tau = -d\mu$, so

$$(f * g)(t) = \int_0^t f(\tau)\, g(t-\tau)\, d\tau = \int_t^0 f(t-\mu)\, g(\mu)\, (-d\mu)$$

$$= \int_0^t g(\mu)\, f(t-\mu)\, d\mu = (g * f)(t). \tag{4}$$

Regarding the final equality in (4), keep in mind that the name of the dummy variable of integration in a definite integral is immaterial: μ, τ, or whatever you like; it does not matter.

For instance, in Example 1 we found that $(f * g)(t) = e^t - e^{-3t}$; if you work out $(g*f)(t)$, you will, according to (3a), obtain the same result: $(g*f)(t) = e^t - e^{-3t}$. Proofs of (3b)–(3d) are left for the exercises.

By the way, in (4) we set $\tau = t - \mu$ and, taking a differential, got $d\tau = -d\mu$. Why did we not get $d\tau = dt - d\mu$? Because in the convolution integral in (1), t is regarded as fixed; the variable is τ. Thus, in $\tau = t - \mu$ the old variable is τ and the new variable is μ. To keep this step clear, it might help to underline the variables and to write $\underline{\tau} = t - \underline{\mu}$. We will do that subsequently, in situations like this.

5.5.2 Convolution theorem. We're ready for the convolution theorem.

THEOREM 5.5.1 *Laplace Convolution Theorem*
Let $f(t)$ and $g(t)$ be piecewise continuous on $0 \le t < \infty$ and of exponential order with exponential constant c, so both $\mathcal{L}\{f(t)\} = F(s)$ and $\mathcal{L}\{g(t)\} = G(s)$ exist for $s > c$. Then

$$\boxed{\mathcal{L}\{(f * g)(t)\} = F(s)G(s).} \tag{5}$$

or, equivalently and in the other direction,

$$\boxed{\mathcal{L}^{-1}\{F(s)G(s)\} = (f * g)(t) = \int_0^t f(\tau)g(t-\tau)\, d\tau,} \tag{6}$$

The inverse of a product, $F(s)G(s)$, is NOT the product of the inverses, $f(t)g(t)$; it is their *convolution* $(f * g)(t)$.

Proof: Since (5) and (6) are equivalent, to prove the theorem it suffices to prove either one. Let us prove (5). The key will be to reverse the order of integration. We will give the steps and then explain them:

$$\mathcal{L}\{(f * g)(t)\} = \mathcal{L}\Big\{ \int_0^t f(\tau)g(t-\tau)\, d\tau \Big\} = \int_0^\infty \Big(\int_0^t f(\tau)g(t-\tau)\, d\tau \Big) e^{-st}\, dt$$

$$= \int_0^\infty \int_\tau^\infty f(\tau)g(t-\tau)e^{-st}\, dt\, d\tau$$

$$= \int_0^\infty \Big(\int_\tau^\infty g(t-\tau)e^{-st}\, dt \Big) f(\tau)\, d\tau \quad (\underline{t} - \tau = \underline{\mu},\ \ dt = d\mu)$$

$$= \int_0^\infty \Big(\int_0^\infty g(\mu)e^{-s(\mu+\tau)}\, d\mu \Big) f(\tau)\, d\tau$$

$$= \int_0^\infty \Big(\int_0^\infty g(\mu)e^{-s\mu}\, d\mu \Big) f(\tau)e^{-s\tau}\, d\tau$$

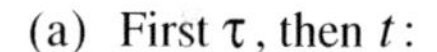

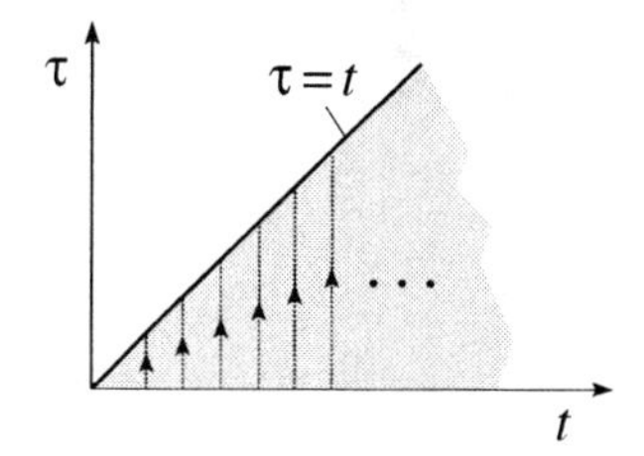

(b) First t, then τ:

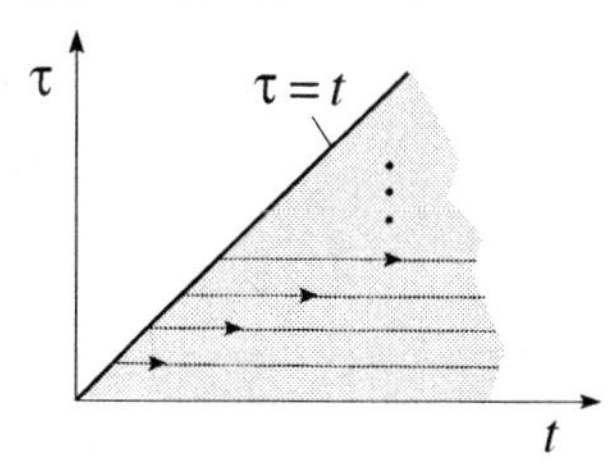

Figure 1. Reversing the order of integration.

$$= \left(\int_0^\infty g(\mu) e^{-s\mu} \, d\mu \right) \left(\int_0^\infty f(\tau) e^{-s\tau} \, d\tau \right) = F(s)G(s). \tag{7}$$

The first and second equalities are by the definitions of the convolution and the Laplace transform, respectively. In the third, which is the key, we reverse the order of integration.[1] To obtain the new integration limits, we must know the region of integration in the t, τ plane. From the limits in $\int_0^\infty \int_0^t (\ \ ) \, d\tau \, dt$ we can *infer* that it is the 45° wedge shown in Fig. 1a. The left and right sides of the third equal sign correspond to the integration sequences indicated in Figs. 1a and 1b, respectively. In the fourth equality we merely moved the $f(\tau)$ out of the t integral because $f(\tau)$ does not depend on t; with respect to t it is a constant. In the fifth we made the change of variables $t - \tau = \mu$ from t to μ, to change the lower integration limit to 0. In the sixth, we moved the $e^{-s\tau}$ out of the μ integral because it does not depend on μ. In the seventh, we noticed that the μ integral has no τ dependence, so what began as an iterated integral has now broken apart into an ordinary product of two integrals. In the final step we identify those integrals as the transforms $F(s)$ and $G(s)$ of $g(t)$ and $f(t)$, respectively. ■

To get the new integration limits, we must know the region of integration. We can *infer* that region from the original integration limits.

5.5.3 Applications. The usefulness of the convolution theorem is illustrated in the following examples.

EXAMPLE 2. Oscillator. Consider the IVP

$$mx'' + kx = f(t); \quad x(0) = 0, \ x'(0) = 0 \tag{8}$$

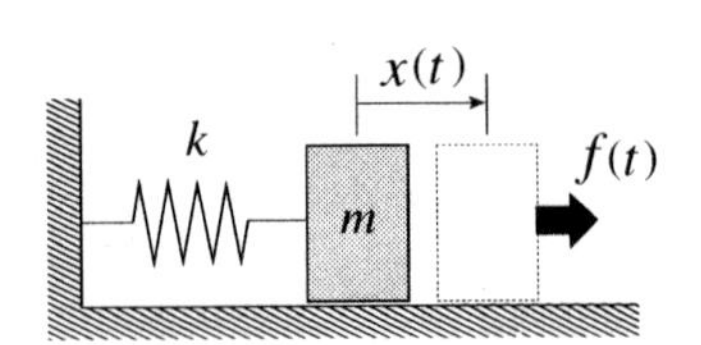

Figure 2. Forced mechanical oscillator.

for the mechanical oscillator in Fig. 2, neglecting any friction or air resistance on the mass. Application of the Laplace transform to (8) gives

$$X(s) = \frac{F(s)}{ms^2 + k}. \tag{9}$$

If we know the force $f(t)$, we can find its transform $F(s)$ and then, hopefully, invert (9) to obtain $x(t)$. But, suppose we do not wish to specify the forcing function in advance; we would like to obtain the solution of (8) *in terms of* $f(t)$. In that case, we cannot compute the $F(s)$ in (9), so we cannot proceed with the solution for $x(t)$. However, if we re-express the right-hand side of (9) as a *product* rather than as a fraction, as

The key is in re-expressing (9) as a product.

$$X(s) = \frac{1}{ms^2 + k} F(s), \tag{10}$$

then it can be inverted using the convolution theorem because we know the inverse of each of the two factors. Denoting the natural frequency $\sqrt{k/m}$ as ω,

$$\mathcal{L}^{-1}\left\{ \frac{1}{ms^2 + k} \right\} = \mathcal{L}^{-1}\left\{ \frac{1}{m} \frac{1}{s^2 + k/m} \right\} = \mathcal{L}^{-1}\left\{ \frac{1}{m\omega} \frac{\omega}{s^2 + \omega^2} \right\}$$
$$= \frac{1}{m\omega} \mathcal{L}^{-1}\left\{ \frac{\omega}{s^2 + \omega^2} \right\} = \frac{1}{\sqrt{mk}} \sin \omega t, \tag{11a}$$

[1]This reversal in order can be justified rigorously, for all $s > c$, thanks to the conditions required of $f(t)$ and $g(t)$ in the theorem.

and of course the inverse of $F(s)$ is simply

$$\mathcal{L}^{-1}\{F(s)\} = f(t). \tag{11b}$$

By (6) then, the solution of (8) is

$$x(t) = \frac{1}{\sqrt{mk}} \sin \omega t * f(t) = \frac{1}{\sqrt{mk}} \int_0^t \sin \omega(t-\tau) f(\tau)\, d\tau. \tag{12}$$

COMMENT. Even if $f(t)$ *is* prescribed, if we use the convolution theorem to invert (10), then there is no point in computing the transform $F(s)$ of $f(t)$ because it will not be needed: To apply the convolution theorem we must invert $F(s)$, and that step simply gives us back $f(t)$! ▮

EXAMPLE 3. Using Convolution for Inversion. Evaluate the inverse of $F(s) = \dfrac{3}{s^2+3s-10}$. We know we can use partial fractions and write

We can invert this $F(s)$ conveniently using additive *or* multiplicative decompositions.

$$F(s) = \frac{3}{(s+5)(s-2)} = -\frac{3}{7}\frac{1}{s+5} + \frac{3}{7}\frac{1}{s-2},$$

$$\mathcal{L}^{-1}\{F(s)\} = -\frac{3}{7}\mathcal{L}^{-1}\{\frac{3}{s+5}\} + \frac{3}{7}\mathcal{L}^{-1}\{\frac{1}{s-2}\} = -\frac{3}{7}e^{-5t} + \frac{3}{7}e^{2t}. \tag{13}$$

Instead of using the foregoing "additive" partial fraction decomposition, we can use the "mulitplicative" decomposition of $F(s)$, and then the convolution theorem:

$$\begin{aligned}\mathcal{L}^{-1}\{\frac{3}{s^2+3s-10}\} &= \mathcal{L}^{-1}\{\frac{3}{(s+5)(s-2)}\} = \mathcal{L}^{-1}\{\frac{3}{s+5}\frac{1}{s-2}\}\\ &= 3e^{-5t} * e^{2t} = 3\int_0^t e^{-5\tau}e^{2(t-\tau)}\, d\tau = 3e^{2t}\int_0^t e^{-7\tau}\, d\tau\\ &= \frac{3}{7}\left(e^{2t} - e^{-5t}\right),\end{aligned} \tag{14}$$

which agrees with (13). ▮

A specific useful result follows from the convolution theorem if we let one of the factors on the right side of (6) be $1/s$, because the inverse of $1/s$ is simply 1:

Let $G(s)$ in (6) be $1/s$. Its inverse is 1.

$$\boxed{\mathcal{L}^{-1}\left\{\frac{1}{s}F(s)\right\} = 1 * f(t) = \int_0^t f(\tau)\, d\tau.} \tag{15}$$

That is, we saw in Section 5.4 that *differentiation in the t domain corresponds to multiplication by* s in the s domain [through the formula $\mathcal{L}\{f'(t)\} = sF(s) - f(0)$].

Now we see that *integration in the t domain corresponds to division by s* (i.e., multiplication by $1/s$) in the s domain.

EXAMPLE 4. Using (15). Evaluate $\mathcal{L}^{-1}\left\{\frac{1}{s(s^2+1)}\right\}$. Knowing that $\mathcal{L}^{-1}\left\{\frac{1}{s^2+1}\right\} = \sin t$, we can use that result and (15) to obtain

$$\mathcal{L}^{-1}\left\{\frac{1}{s(s^2+1)}\right\} = \int_0^t \sin\tau\, d\tau = 1 - \cos t. \quad \blacksquare \tag{16}$$

In Examples 2–4 we used the convolution theorem in the form (6), for inverting transforms. In the next subsection we will use it in the form (5), for computing transforms.

This brief discussion of integro-differential and integral equations is not a prerequisite for subsequent sections.

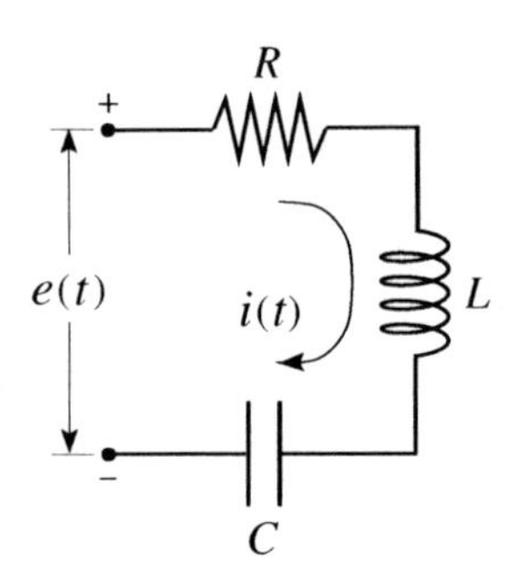

Figure 3. The RLC circuit governed by (17).

5.5.4 Integro-differential equations and integral equations. The electrical current $i(t)$ in the circuit shown in Fig. 3, and the charge $q(t)$ on the condenser, are governed by the differential equation

$$Li' + Ri + \frac{1}{C}q = 0. \tag{17}$$

The current and charge are not independent, but are related (Section 1.3.5) by

$$i = \frac{dq}{dt}. \tag{18}$$

We will show that the problem can be expressed in differential equation form, or in an equivalent "integro-differential equation" form.

Differential equation form. We can eliminate $q(t)$ in favor of $i(t)$ by differentiating (17) and using (18). That step gives

$$Li'' + Ri' + \frac{1}{C}i = 0; \qquad i(0) = i_0,\ i'(0) = 0, \tag{19}$$

in which we've also arbitrarily assigned two initial conditions. Taking the Laplace transform gives

$$I(s) = \frac{Li_0 s + Ri_0}{Ls^2 + Rs + \frac{1}{C}}, \tag{20}$$

inversion of which gives the solution $i(t)$. Such steps have already been covered.

Integro-differential equation form. Above, we eliminated the $q(t)$ in (17), in favor of $i(t)$, by differentiating (17) and then using (18). That step gave a second-order differential equation for $i(t)$. Alternatively, we can obtain $q(t)$ in terms of

$i(t)$ by integrating (18) from 0 to t, which gives $q(t) = \int_0^t i(\tau)\,d\tau + q(0)$. We don't know $q(0)$, but can can infer its value by setting $t = 0$ in (17) and using the initial conditions given in (19). Doing that gives $q(0) = -RCi_0$, so

$$q(t) = \int_0^t i(\tau)\,d\tau - RCi_0. \tag{21}$$

Now putting (21) into (17) gives

$$Li' + Ri + \frac{1}{C}\int_0^t i(\tau)\,d\tau - Ri_0 = 0; \quad i(0) = i_0. \tag{22}$$

The latter is an **integro-differential equation** because it contains derivatives *and* integrals of the unknown function $i(t)$.

We did not include in (22) the second initial condition $i'(0)=0$ that was given in (19) because, as a differential equation, (22) is only of first order. But it doesn't matter if we include it in (22) or not, because (22) automatically satisfies $i'(0) = 0$, as we can see by setting $t=0$ in that equation.

We can solve (22) by the Laplace transform, *because the integral term in it is a convolution integral.* That is, (22) can be written as

$$Li' + Ri + \frac{1}{C}1 * i(t) - Ri_0 = 0; \quad i(0) = i_0. \tag{23}$$

Thus, if we take the Laplace transform of (23), and use the convolution theorem to transform the $1 * i(t)$, we obtain

$$L[sI(s) - i_0] + RI(s) + \frac{1}{C}[\frac{1}{s}I(s)] - \frac{Ri_0}{s} = 0 \tag{24}$$

$L\{1 * i(t)\}$
$= \mathcal{L}\{1\}\mathcal{L}\{i(t)\} = \frac{1}{s}I(s).$

or, solving for $I(s)$,

$$I(s) = \frac{Li_0 s + Ri_0}{Ls^2 + Rs + \frac{1}{C}}, \tag{25}$$

which is the same as (20).

We were able to solve the integro-differential equation (22) by the Laplace transform — because the integral term happened to be of convolution form (which is not always true of integro-differential equations).

EXAMPLE 5. An Integral Equation. We called (22) an integro-differential equation because it contained both derivatives and integrals of the unknown function $i(t)$. As another example, the equation

$$\sinh t = \int_0^t x(\tau)\cos(t-\tau)\,d\tau \tag{26}$$

is an **integral equation** because it contains only integrals of the unknown $x(t)$. We can solve (26) for $x(t)$ by the Laplace transform because the integral is of convolution type. Specifically, (26) can be written as

However, integrals in integral equations are not necessarily of convolution type.

$$\sinh t = x(t) * \cos t. \tag{27}$$

To solve (27), take its Laplace transform. That step gives

$$\frac{1}{s^2-1} = X(s)\frac{s}{s^2+1}. \tag{28}$$

Then

$$X(s) = \frac{s^2+1}{s(s^2-1)} = \frac{A}{s} + \frac{B}{s-1} + \frac{C}{s+1}, \tag{29}$$

and we find that $A = -1$, $B = 1$, and $C = 1$. Thus,

$$x(t) = \mathcal{L}^{-1}\left\{ -\frac{1}{s} + \frac{1}{s-1} + \frac{1}{s+1} \right\} = -1 + e^t + e^{-t}, \text{ or } 2\cosh t - 1. \quad \blacksquare \tag{30}$$

Closure. The Laplace convolution is important because the Laplace transform of the convolution $f*g$ of f and g is simply the product of their individual transforms, $F(s)G(s)$. That result is useful in both directions — in the forms (5) and (6). For instance, in Example 2 we used (6) when we inverted the right-hand side of (10) as the convolution of the inverses of the two factors $1/(ms^2+k)$ and $F(s)$. The inverse of $F(s)$, of course, was simply the original function $f(t)$. And in Example 5 we used (5), observing that the right-hand side of (26) was the convolution of $x(t)$ and $\cos t$. Thus, when we took the transform of the equation, the right-hand side was simply the algebraic product of $X(s)$ and the transform $s/(s^2+1)$ of $\cos t$.

Further, we met integro-differential and integral equations in Section 5.5.4, for the first time in this text, and saw that the convolution theorem can be used to solve them if the integrals in those equations happen to be of convolution form.

EXERCISES 5.5

1. Which of the following are true, and which are false? Explain.

$$\mathcal{L}\{\cos t * t\} = \mathcal{L}\{\cos t\}\mathcal{L}\{t\}, \tag{1.1}$$

$$\mathcal{L}\{te^t\} = \mathcal{L}\{t\}\mathcal{L}\{e^t\}, \tag{1.2}$$

$$\mathcal{L}^{-1}\{\frac{1}{s(s+1)}\} = \mathcal{L}^{-1}\{\frac{1}{s}\} * \mathcal{L}^{-1}\{\frac{1}{s+1}\} \tag{1.3}$$

$$\mathcal{L}^{-1}\{\frac{1}{s(s+1)}\} = \mathcal{L}^{-1}\{\frac{1}{s}\}\mathcal{L}^{-1}\{\frac{1}{s+1}\}. \tag{1.4}$$

2. Given $f(t)$ and $g(t)$, verify that $f*g = g*f$, by working out both of them and showing that they are the same.

(a) $f(t) = 1$, $g(t) = t$ (b) $f(t) = 1$, $g(t) = e^t$
(c) $f(t) = 5$, $g(t) = \sin t$ (d) $f(t) = e^t$, $g(t) = e^{-t}$

3. Find the inverse of the given transform two different ways: using partial fractions, and using the convolution theorem.

(a) $24/[s(s+8)]$ (b) $21/(3s^2+5s-2)$
(c) $1/(s^2+3s+2)$ (d) $5/[(s+1)(3s+2)]$
(e) $1/(s^2+s)$ (f) $2/(2s^2-s-1)$
(g) $12/(s^2+s-2)$ (h) $6/(s^2-9s+20)$
(i) $1/[(s+i)(s-i)]$ (j) $1/[(s+3i)(s-3i)]$

4. We obtained the solution (12) of the IVP (8) by using the Laplace transform, and the convolution theorem.
(a) How could we solve (8) *without* using the Laplace transform? Just outline the method you would use.
(b) Carry out that alternative solution, and show that you do obtain (12).

5. **More Than Two Functions.** The convolution theorem 5.5.1 is for two functions, but it also holds for any finite number of functions (that satisfy the conditions in the theorem).

For instance, for the case of three functions show that

$$\boxed{\mathcal{L}\{f * g * h\} = F(s)G(s)H(s).} \tag{5.1}$$

NOTE: $f * g * h$ may appear to be ambiguous; that is, does it mean $(f * g) * h$ or $f * (g * h)$? According to the associative property (3b), it doesn't matter because these are equal. Hence, parentheses can be omitted without ambiguity.

6. Exercise 5, Continued. In the other direction, (5.1) says that

$$\mathcal{L}^{-1}\{F(s)G(s)H(s)\} = f * g * h. \tag{6.1}$$

In each case break the transform into *three* factors and use (6.1) to invert.

(a) $\dfrac{1}{s^3}$ (b) $\dfrac{1}{s(s-1)(s-2)}$

(c) $\dfrac{1}{s(s^2-9)}$ (d) $\dfrac{1}{(s-2)^2(s+2)}$

7. Use the convolution theorem to evaluate the Laplace transform of each:

(a) $\displaystyle\int_0^t (\sin\tau)e^{t-\tau}\,d\tau$ (b) $\displaystyle\int_0^t (\sin\tau)(t-\tau)^4\,d\tau$

(c) $\displaystyle\int_0^t \tau\sin\tau\,d\tau$ (d) $\displaystyle\int_0^t \tau^3(\tau-t)^4\,d\tau$

(e) $\displaystyle\int_0^t \sin 4(t-\tau)\sinh\tau\,d\tau$ (f) $\displaystyle\int_0^t \cosh(t-\tau)\,d\tau$

8. Conversion of Differential Equations to Integral Equations. Consider the IVP

$$x'' + x = f(t); \quad x(0) = x_0,\ x'(0) = x_0'. \tag{8.1}$$

(a) By integrating the differential equation with respect to t, from 0 to t, obtain the integro-differential equation

$$x'(t) - x_0' + \int_0^t x(\mu)\,d\mu = \int_0^t f(\mu)\,d\mu. \tag{8.2}$$

(b) Now integrate (8.2) from 0 to t and show that

$$x(t) - x_0 - x_0't + \int_0^t\int_0^\tau x(\mu)d\mu\,d\tau = \int_0^t\int_0^\tau f(\mu)d\mu\,d\tau. \tag{8.3}$$

By reversing the order of integration in each of the two iterated integrals, simplify (8.3) to the integral equation

$$x(t) - x_0 - x_0't + \int_0^t (t-\mu)x(\mu)\,d\mu = \int_0^t (t-\mu)f(\mu)\,d\mu. \tag{8.4}$$

(c) Solve the integro-differential equation (8.2), with the initial condition $x(0) = x_0$, for $x(t)$ by the Laplace transform.
(d) Solve the integral equation (8.4) for $x(t)$ by the Laplace transform. [Of course, you should obtain the same result as in part (c).]

9. Solve the given integral equation by the Laplace transform. Use computer software if you wish, to help with any partial fraction expansions or integrations.

(a) $\int_0^t e^{\tau-t}x(\tau)\,d\tau = t + x(t)$

(b) $\int_0^t (4+\tau-t)x(\tau)\,d\tau = 6t^2 - 5x(t)$

(c) $x(t)e^t = \int_0^t e^\tau x(\tau)\,d\tau - 10$

(d) $x(t) = \int_0^t [(t-\tau)x(\tau) + \tau\sin\tau]\,d\tau$

(e) $6x(t) = 6t^2 - \int_0^t (\tau-t)^3 x(\tau)\,d\tau$

10. Show, by integrating twice with respect to t, from 0 to t, that the IVP

$$x'' + p(t)x = 0; \quad x(0) = a,\ x'(0) = 0 \tag{10.1}$$

can be converted to the integral equation

$$x(t) - a + \int_0^t p(\tau)x(\tau)(t-\tau)\,d\tau = 0. \tag{10.2}$$

NOTE: We don't mean to imply that all we need to do is convert a differential equation IVP to an integral equation and then solve it by the Laplace transform and convolution theorem. In this example, (10.1) was difficult because of the nonconstant coefficient $p(t)$, and (10.2) is difficult still, because the integral is the convolution $(p(t)x(t)) * t$, and if we take the Laplace transform of (10.2) we obtain

$$X(s) - \frac{a}{s} + \mathcal{L}\{p(t)x(t)\}\,\mathcal{L}\{t\} = 0. \tag{10.3}$$

The difficulty is that in general we cannot work out the transform $\mathcal{L}\{p(t)x(t)\}$ in terms of $X(s)$ because $p(t)x(t)$ is an ordinary product, not a convolution.

11. Use the definition of convolution to prove the associative property (3b). HINT: The steps will include reversing the order of integration.

12. Integrals of Heaviside Functions. (a) We sometimes encounter integrals of the form $\displaystyle\int_0^t H(\tau - a)f(\tau)\,d\tau$, in which $a > 0$. Show that

$$\boxed{\int_0^t H(\tau-a)f(\tau)\,d\tau = H(t-a)\int_a^t f(\tau)\,d\tau.} \tag{12.2}$$

(b) **Application of (12.2).** For instance, in Example 4 of Section 5.4, when we transformed the IVP $x' + x = p(t)$ with initial condition $x(0) = 3$, we used (10) (in that section) in working out the transform of the piecewise-defined $p(t)$. Then, when we obtained (21) we used (10) again, in the reverse direction, to obtain the inverse. Alternatively, we could have used the convolution theorem and (12.2), as follows: The transform of (15) gives

$$X(s) = \frac{3}{s+1} + \frac{1}{s+1}P(s), \tag{12.3}$$

in which $P(s)$ is the transform of $p(t)$. This time, instead of working out the transform $P(s)$ and then inverting $X(s)$, use the convolution theorem. That is,

$$x(t) = 3e^{-t} + e^{-t} * p(t) = 3e^{-t} + \int_0^t e^{-(t-\tau)}p(\tau)\,d\tau =$$

$$3e^{-t} + e^{-t}\int_0^t e^{\tau}4\tau\,d\tau + e^{-t}\int_0^t e^{\tau}(2-4\tau)H(\tau-2)\,d\tau. \tag{12.4}$$

Then, use (12.2) to evaluate the last integral in (12.4), and show that your result agrees with equation (22) in Section 5.4. CONCLUSION: Generalizing, the idea is that if the forcing function in the differential equation, say "$p(t)$," is defined piecewise, it is efficient *not* to take the function's transform, but to keep it intact as $P(s)$ and to use the convolution theorem to invert, together with the integration formula (12.2), to evaluate the convolution integral that occurs.

5.6 IMPULSIVE FORCING FUNCTIONS; DIRAC DELTA FUNCTION

(a)

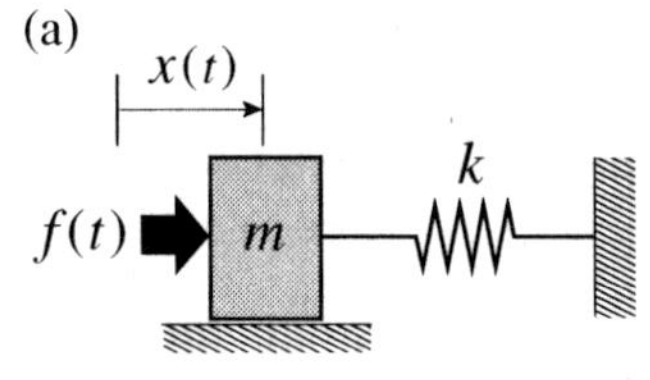

(b)

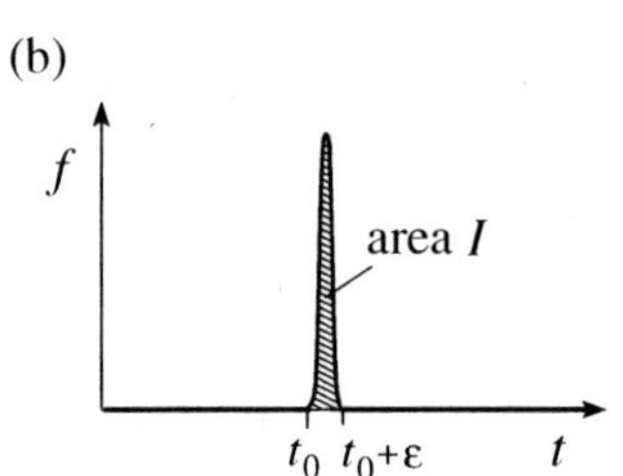

Figure 1. Harmonic oscillator subjected to impulsive force.

5.6.1 Impulsive forces. Consider the mass in Fig. 1a, that sits on a frictionless table and is restrained by a coil spring. Suppose it is initially at rest, and that we strike it with a hammer at time $t = t_0$. The graph of the force $f(t)$ due to the hammer blow will be somewhat as sketched in Fig. 1b, having a large peak value, and acting over a very short interval $t_0 < t < t_0 + \epsilon$. The IVP for the resulting motion $x(t)$ is

$$mx'' + kx = f(t); \quad x(0) = 0,\ x'(0) = 0, \tag{1}$$

in which m and k are the mass and spring stiffness, respectively. Application of the Laplace transform to (1) readily gives

$$X(s) = \frac{F(s)}{ms^2 + k}. \tag{2}$$

To complete the solution, we can evaluate the transform

$$\mathcal{L}\{f(t)\} = F(s) = \int_0^\infty f(t)e^{-st}\,dt, \tag{3}$$

and then invert the right-hand side of (2) to get $x(t)$.

However, for a force such as a hammer blow, we do not know the force history $f(t)$, or in fact even its time duration ϵ, so how can we evaluate the integral in (3), to get $F(s)$? Perhaps we should use the convolution theorem instead, to invert (2), because then we don't need to know $F(s)$, the solution simply being the convolution of the inverse of $1/(ms^2 + k)$ and the inverse $f(t)$ of $F(s)$. In that case, we can obtain the solution without ever evaluating $F(s)$. Trying that approach, first invert the $1/(ms^2+k)$ factor:

$$\mathcal{L}^{-1}\left\{\frac{1}{ms^2+k}\right\} = \frac{1}{m}\mathcal{L}^{-1}\left\{\frac{1}{s^2+(k/m)}\right\} = \frac{1}{m\omega}\mathcal{L}^{-1}\left\{\frac{\omega}{s^2+\omega^2}\right\} = \frac{1}{m\omega}\sin\omega t, \tag{4}$$

in which $\omega = \sqrt{k/m}$. Then it follows from (2) that

$$x(t) = \frac{1}{m\omega}\sin\omega t * f(t) = \frac{1}{m\omega}\int_0^t f(\tau)\sin\omega(t-\tau)\,d\tau. \tag{5}$$

However, in (5) we meet the same difficulty, for how can we carry out the integration without knowing the function $f(t)$ or even its duration ϵ?

Since the solving of (1) is going to involve integrating $f(t)$, one way or another, we must find a way to do that — even without knowing $f(t)$. Of course, we have to know *something* about $f(t)$ to integrate it, and we will assume partial knowledge as follows: first, that the duration of its action, ϵ, is very short; second, that although we don't know the function $f(t)$, we assume that we do know the area of its graph. In fact, in mechanics, the area of the graph of the force, between an initial time t_i and a final time t_f, is important and is called the **impulse** of the force f over that interval,

We assume that f's duration is very short, and that its impulse I is known.

$$I \equiv \int_{t_i}^{t_f} f(t)\,dt. \tag{6}$$

The reason the concept of impulse is important, in mechanics, is that the impulse I equals the resulting change in the momentum of the mass. To see that, just integrate Newton's second law, written in the form $d\,[mv(t)]/dt = f(t)$, in which $v(t)$ is the velocity dx/dt.[1] That step gives

$$mv(t)\Big|_{t_i}^{t_f} = \int_{t_i}^{t_f} f(t)\,dt; \tag{7}$$

that is, the change in the momentum of the mass equals the impulse I, so we can say the impulse "causes" an equal change in the momentum.

Any force $f(t)$, acting over a time interval, has an impulse defined by (6), but only forces that are sharply focused in time, such as the force corresponding to a hammer blow, are called *impulsive forces*, and that is what we will mean here by an impulsive force.

[1]Multiplying the equation of motion, given by Newton's law, by dt and integrating gives an *impulse-momentum* relation; multiplying instead by dx and integrating gives a *work-energy* relation, namely, the work done by f, over a time interval from t_i to t_f, equals the change in the kinetic energy, $(1/2)mv(t)^2|_{t_i}^{t_f}$.

5.6.2 Dirac delta function. The mathematical stumbling block that has arisen, whether in (3) or in (5), is one of integration — how to evaluate an integral of the form

$$\int_0^\infty f(t)h(t)\,dt,$$

In (3), $h(t)$ is e^{-st}, and in (5) it is $\sin\omega(t-\tau)$.

in which $f(t)$ is impulsive and $h(t)$ is any generic continuous function. Although we don't know $f(t)$, or even its duration, say from t_0 to $t_0+\epsilon$, we do know that its duration ϵ is very short, and we assume further that we know its impulse I over that time interval. Then, to evaluate the foregoing integral, proceed as follows:

Understand the steps in (8).

$$\begin{aligned}\int_0^\infty f(t)h(t)\,dt &= \int_{t_0}^{t_0+\epsilon} f(t)h(t)\,dt \approx \int_{t_0}^{t_0+\epsilon} f(t)h(t_0)\,dt \\ &= h(t_0)\int_{t_0}^{t_0+\epsilon} f(t)\,dt = h(t_0)I. \qquad (8)\end{aligned}$$

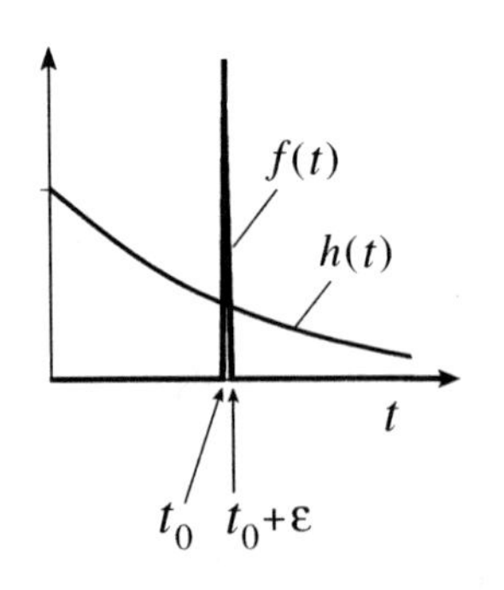

Figure 2. $h(t)$ approaches a constant on $(t_0, t_0+\epsilon)$ as $\epsilon \to 0$.

In the first step, we changed the limits because $f(t) = 0$ for $t < t_0$ and for $t > t_0 + \epsilon$. The second step in (8) is the key. Our reasoning there is that if ϵ really is very short, then the generic continuous founction $h(t)$ is *essentially a constant over the interval*, and can therefore be replaced, approximately, by its value $h(t_0)$ at t_0, as is suggested in Fig. 2.

Since ϵ is very small and its value is not known to us, it seems reasonable, and "cleaner," to adopt an idealization in which $\epsilon \to 0$ while the impulsive function $f(t)$ becomes very large, with its impulse remaining equal to I. In that idealization, we say that $f(t)$ acts "instantaneously" at t_0, at which time it delivers an impulse I. Understand that the detailed shape of $f(t)$ has "washed out" in this idealization. In fact, as $\epsilon \to 0$ (8) becomes an equality — provided that we are willing to idealize our hammer-blow force $f(t)$ as acting over "zero time," at t_0, and yet delivering an impulse I.

If we let $I = 1$, then we call $f(t)$ a unit impulsive force, or unit impulse, and denote it as the **delta function** $\delta(t-t_0)$. To understand the $t-t_0$, think of it this way: A Heaviside function $H(t)$ is a unit step at $t = 0$, and $H(t-t_0)$ is a unit step that acts instead at $t = t_0$; that is, the graph of the step is translated rightward by t_0. Similarly, let $\delta(t)$ be a unit impulse at $t = 0$. Then, $\delta(t-t_0)$ is a unit impulse that acts instead at $t = t_0$.

But, we have backed ourselves into a corner in saying that the graph of the delta function $\delta(t-t_0)$ has zero width and "infinite height," such that its area is unity, because an ordinary function cannot behave like that. Thus, it is *not* an ordinary function, its graph looking more like a spike, and it was rejected by mathematicians when the physicist *P. A. M. Dirac* introduced it in 1929 in connection with his work on quantum mechanics. However, although its definition by Dirac was only heuristic, the delta function *worked*, so mathematicians developed a theory in which its definition and use could be made rigorous.[1]

[1]It was not until the 1940s that the French mathematician *Laurent Schwartz* created a rigorous foundation for Dirac's delta function. Essentially, Schwartz's idea was to focus not on the values of

We will find that the delta function represents not a complication, but a simplification, for we will see how easy it is to use. We will not be interested in the values of a delta function or its graph, the way we are interested in the values and graph of ordinary functions, because we will always end up *integrating* it, as in (3) and (5). Thus, all we will need to know is its integral property that we've already derived in (8):

$$\boxed{\int_0^\infty \delta(t-t_0)h(t)\,dt = h(t_0) \quad (t_0 \geq 0),} \tag{9}$$

Equation (9) is known as the *sifting property* of the delta function, for we can think of $\delta(t-t_0)$ as "sifting" through the values of $h(t)$, and picking out its value at t_0.

for any continuous function $h(t)$. How simple (9) is, for there is no integration to carry out; the answer is simply the inputted function $h(t)$ evaluated at the point of action of the delta function.

Before using (9) to complete our solution of the IVP (1), we have two comments about that formula. First, since f is nonzero only at the point t_0, (9) holds even if we increase the lower integration limit from 0 to any $a < t_0$, and if we decrease the upper limit to any $b > t_0$. More informatively than (9), then, we can write

$$\int_a^b \delta(t-t_0)h(t)\,dt = \begin{cases} h(t_0), & \text{if } a < t_0 < b \\ 0, & \text{if } t_0 < a \text{ or } t_0 > b. \end{cases} \tag{10}$$

Equation (10) is more complete than (9) in that it allows for integration limits other than 0 and ∞.

To understand the zero in (10), just realize that in that case the delta function's spike falls outside the interval of integration, so the integral is zero.

Now let us use the delta function to solve the IVP (1). It seems simpler to *not* use the convolution theorem approach that led to (5), so we simply return to (1) and follow our usual methodology. We will suppose that our hammer blow is modeled as an impulsive force with impulse I. In that case, we must scale the delta function by I, so the forcing function in (1) is $f(t) = I\delta(t-t_0)$. After all, if $\delta(t-t_0)$ has impulse (area) 1, then $I\delta(t-t_0)$ has impulse (area) I. We will need to know the transform of the delta function, so let us obtain that first. It follows immediately from (9) with $h(t) = e^{-st}$:

$$\mathcal{L}\{\delta(t-t_0)\} = \int_0^\infty \delta(t-t_0)e^{-st}\,dt = e^{-st_0}, \tag{11}$$

which we frame for emphasis before continuing:

$$\boxed{\mathcal{L}\{\delta(t-t_0)\} = e^{-st_0}.} \tag{12}$$

The Laplace transform of the delta function.

the delta function, or its graph, as we do for an "ordinary" function, but on its action, namely, what it does to an inputted (continuous) function in the course of integration, as we will see in equation (9). The latter defines the delta function not as a function but as a **functional** or, in Schwartz's terminology, a **distribution**. For a function, the input is a number and the output is a number, but for a functional, the input is a *function* and the output is a number. In the case of the functional defined in (9), the input is the function $h(t)$ and the output is the number $h(t_0)$.

Let us return to (1) now, and solve it. With $f(t) = I\delta(t-t_0)$, the problem is

$$mx'' + kx = I\delta(t-t_0); \quad x(0) = 0, \ x'(0) = 0. \tag{13}$$

The transform of the delta function is given by (12), so transforming (13) gives

$$(ms^2 + k)X(s) = Ie^{-st_0}.$$

Thus,

$$X(s) = \frac{Ie^{-st_0}}{ms^2 + k} = \frac{I}{m}e^{-st_0}\frac{1}{s^2+\omega^2}. \tag{14}$$

To invert $X(s)$, first invert

$$\mathcal{L}^{-1}\left\{\frac{1}{s^2+\omega^2}\right\} = \frac{1}{\omega}\mathcal{L}^{-1}\left\{\frac{\omega}{s^2+\omega^2}\right\} = \frac{1}{\omega}\sin\omega t. \tag{15}$$

Then, use the time-delay inversion formula, item 29, and obtain the solution

$$x(t) = \frac{I}{m\omega}H(t-t_0)\sin\omega(t-t_0), \tag{16}$$

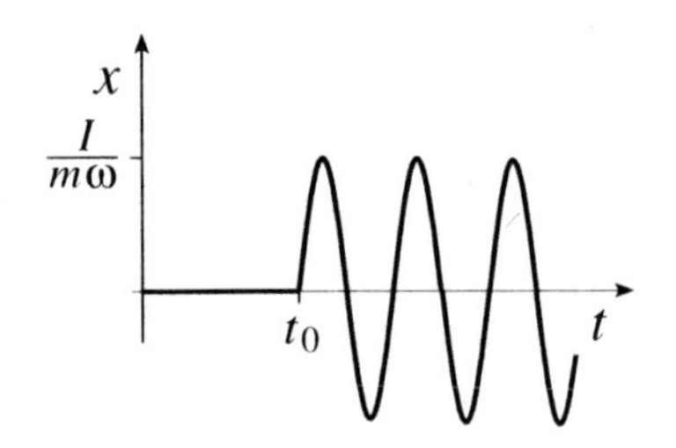

Figure 3. The time-delayed response (16).

or, written out,

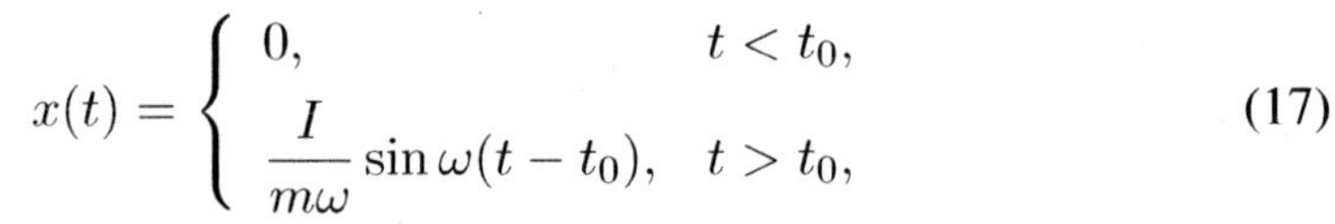

$$x(t) = \begin{cases} 0, & t < t_0, \\ \dfrac{I}{m\omega}\sin\omega(t-t_0), & t > t_0, \end{cases} \tag{17}$$

which is the time-delayed response plotted in Fig. 3.

Observe how simple it was to use the delta function. Its transform was given by (12) and it introduced an e^{-st_0} factor into part of $X(s)$. Inversion of that part of $X(s)$ was handled by the time-delay item 29 in the table. Of course: If we don't give the mass a knock until 2pm, then we don't expect a response to begin until 2pm.

5.6.3 The jump caused by the delta function. That completes the solution of (13), but it will be instructive to think about the discontinuity in the slope $x'(t)$ at t_0 that is observed in the solution (17) and in the figure. That jump in x' is not a surprise, in view of our earlier comments about an impulse causing a change in the momentum, and hence in the slope $x'(t)$, but let us study the matter directly from the differential equation itself. In doing so, we might as well include a damping term cx' as well, and consider the more general equation $mx'' + cx' + kx = I\delta(t-t_0)$.

To focus on the neighborhood of t_0, integrate the differential equation from $t_0-\mu$ to $t_0+\mu$:

Think of μ as small; we will let it tend to zero in a moment.

$$m\int_{t_0-\mu}^{t_0+\mu} x''\,dt + c\int_{t_0-\mu}^{t_0+\mu} x'\,dt + k\int_{t_0-\mu}^{t_0+\mu} x\,dt = I\int_{t_0-\mu}^{t_0+\mu}\delta(t-t_0)\,dt. \tag{18}$$

The integral on the right is 1 because the delta function's unit area is captured within the integration interval, and the integral of x'' is x', so (18) gives

$$mx'(t)\Big|_{t_0-\mu}^{t_0+\mu} + c\int_{t_0-\mu}^{t_0+\mu} x'(t)\,dt + k\int_{t_0-\mu}^{t_0+\mu} x(t)\,dt = I. \tag{19}$$

Now let $\mu \to 0$ in (19). Of the terms on the left, the first gives m times the jump in the velocity, and the third and the second give *zero* because the integrands are the mass' displacement $x(t)$ and velocity $x'(t)$, which are bounded, and the length of the integration intervals, 2μ, is being shrunk to zero. (We state the boundedness of the displacement and velocity without proof.) Thus, we obtain

$$mx'(t)\Big|_{t_0-}^{t_0+} = I, \tag{20}$$

In (20), $x'(t_0+)$ is given by the derivative of the *lower* member of (17), at t_0, and $x'(t_0-)$ is given by the derivative of the *upper* member of (17), at t_0.

in which $x'(t_0+)$ denotes the limit of $x'(t)$ as $t \to t_0$ from the right, and $x'(t_0-)$ denotes the limit of $x'(t)$ as $t \to t_0$ from the left. Thus, the differential equation itself shows that the impulsive force at t_0 will cause an instantaneous jump I/m in the velocity at t_0.[1]

The jump predicted by (20) agrees with what we find in the foregoing solution (17): Differentiating the $t > t_0$ member of (17) gives $x'(t) = (I/m)\cos\omega(t-t_0)$, so $x'(t_0+) = (I/m)\cos 0 = I/m$. Further, differentiating the $t < t_0$ member of (17) gives $x'(t_0-) = 0$, of course, so the jump found in the solution (17) is $x'(t)|_{t_0-}^{t_0+} = I/m$, which agrees with (20).

5.6.4 Caution. In this subsection, we call attention to a specific subtlety that arises if the impulsive force acts at $t_0 = 0$.

EXAMPLE 1. The case in which $t_0 = 0$. Consider the IVP

$$x'' + 4x' + 3x = 3\delta(t); \quad x(0) = 2,\ x'(0) = 1. \tag{21}$$

The transform of (21) gives

$$[s^2X(s) - sx(0) - x'(0)] + 4[sX(s) - x(0)] + 3X(s) = 3e^{0s}. \tag{22}$$

Putting in the initial conditions and solving for $X(s)$ gives

$$X(s) = \frac{2s+12}{s^2+4s+3} = \frac{5}{s+1} - \frac{3}{s+3}, \tag{23}$$

and the solution follows readily from (23) as

$$x(t) = 5e^{-t} - 3e^{-3t}. \tag{24}$$

[1]Could we not have seen that result more readily from our earlier argument, that integrating $mx'' = f(t)$ gives the impulse-momentum relation, namely, that the change in momentum equals the impulse I? No, because our present equation, $mx'' + cx' + kx = I\delta(t-t_0)$ contains the additional terms cx' and kx on the left-hand side, terms that were not present in $mx'' = f(t)$.

In this case, the solution is not of delay type [i,e, there is no Heaviside function in the right-hand side of (24)] because the delta function forcing function in (21) acts at $t = 0$, so there is no delay.

COMMENT. The subtlety that we referred to involves the satisfaction of the initial conditions: (24) gives $x(0) = 2$, which agrees with the given condition, but it also gives $x'(0) = 4$, which does *not* agree with the initial condition $x'(0) = 1$. This is not an error; it is explained by the jump behavior (20). To apply (20), we will need m and I, and from (21) these are $m = 1$ and $I = 3$. Then, (20) gives the jump in $x'(t)$ at $t = 0$ as $x'(0+) - x'(0) = I = 3$. As mentioned above, (24) gives $x'(0+) = 4$, the initial condition is $x'(0) = 1$, and $4 - 1$ does equal 3, so there is no error in the solution (24).

Our point is that if the delta function forcing function acts at $t_0 = 0$, be aware of the jump behavior (20) and the instantaneous jump in x', or we may think that our solution does not satisfy the initial conditions and is therefore incorrect. ▮

5.6.5 Impulse response function. Consider again a forced oscillator, governed by the IVP

$$mx'' + cx' + kx = f(t); \quad x(0) = 0, \; x'(0) = 0. \tag{25}$$

Application of the Laplace transform to (25) gives

$$X(s) = \frac{1}{ms^2 + cs + k} F(s) \equiv Q(s)F(s). \tag{26}$$

With zero initial conditions obtain $X(s) = Q(s)F(s)$, in which $Q(s)$ is the transfer function. The inverse of $Q(s)$ is the system's impulse response function $q(t)$, and the solution is then given by $x(t) = q(t) * f(t)$.

Here, $Q(s) = 1/(ms^2 + cs + k)$ is called the **transfer function** of the system. If $f(t)$ is a delta function $\delta(t)$, then, by (9) with $t_0 = 0$, its transform is $F(s) = 1$. In that case the solution $x(t)$ is simply the inverse $q(t)$ of the transfer function $Q(s)$ and is called the **impulse response function** for the system because it is the response specifically to a unit impulse acting at $t = 0$, for the initial conditions are zero.

The impulse response function $q(t)$ is important because if we know $q(t)$, then the response to *any* forcing function $f(t)$ is seen from (26) to be *the convolution of* $f(t)$ *with* $q(t)$:

$$x(t) = q(t) * f(t) = \int_0^t q(\tau) f(t - \tau)\, d\tau. \tag{27}$$

Similarly, for *any* system governed by a constant-coefficient differential equation with a forcing function, whether of second order or not, the response to the forcing function $f(t)$ is the convolution of the system's impulse response function $q(t)$ with $f(t)$. One reason this result is useful is that in a given application we may not even know the left-hand side of the differential equation! Even if we do know its form, as for instance in equation (2), the physical parameters are not necessarily readily known. For instance, accurate experimental determination of the damping coefficient c is not a trivial matter. Yet, even without knowing c, if we subject the mass to a unit impulse and measure the response $q(t)$ experimentally, then we can use that result, and (27), to predict the system's response to any given forcing function $f(t)$.

Closure. In Section 5.4 we studied discontinuous forcing functions, and in this section we've considered ones that are even more singular — impulsive forces that are spike-like. To solve a problem containing an impulsive forcing function $f(t)$ by the Laplace transform, we must be able to evaluate integrals containing $f(t)$, but how can we do that if that function is not known, as is typical in the case of impulsive forces? Assuming that $f(t)$ is sharply focused in time and that we know its impulse, we can integrate according to the steps in (8). In the idealization in which $\epsilon \to 0$, the result is given by (9) [and in an alternative form by (10)], and if the impulsive force $f(t)$ has unit impulse it is known as the Dirac delta function, $\delta(t-t_0)$. The formulas (9) and (10) are the chief result in this section, and the transform of the delta function follows from it according to (12).

Note that in using the delta function we do not need to know its values, in the way we need to know the values of functions such as e^t, $\sin t$, and so on. We need only its *integral behavior*, given by (9) and (10), and its transform integral that follows from (9) with $h(t) = e^{-st}$ and is given by (12). In fact, the delta function is not an ordinary function, but a "distribution" or "functional." To use it, we do not need to know the theory of distribution or functionals; it suffices to understand the steps in (8) and the resulting equations (9), (10), and (12). Nevertheless, it is difficult to give up the familiar idea of the values and graph of a function, so one can think of the delta function, heuristically, as the limit of the sequence of *ordinary* functions $\delta_\epsilon(t-t_0)$ that is shown in Fig. 4. Note that as ϵ decreases the rectangle gets narrower, and taller, but always has unit area.

Finally, we should mention that delta functions are used not only to represent functions that are focused in time, but also those that are focused in space, such as a point force acting on a beam. In that case the independent variable is not the time t, but a space variable such as x. That case is left for the Additional Exercises.

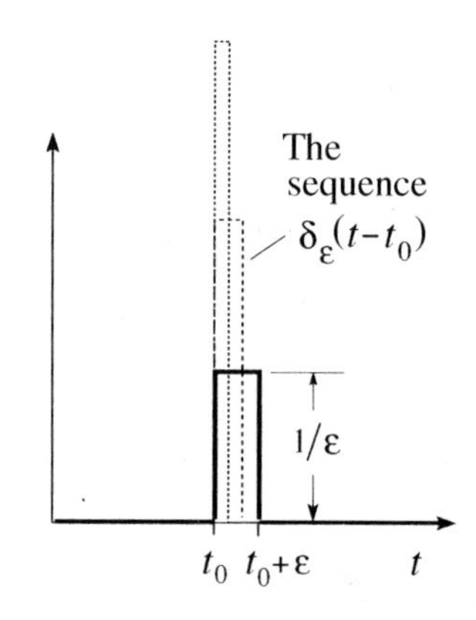

Figure 4. The sequence $\delta_\epsilon(t-t_0)$. We can think of $\delta(t-t_0)$ as the limit of the sequence $\delta_\epsilon(t-t_0)$ as $\epsilon \to 0$. The area of $\delta_\epsilon(t-t_0)$ is $(1/\epsilon)(\epsilon) = 1$ for each ϵ.

EXERCISES 5.6

1. Solve for $x(t)$ on $0 \le t < \infty$.

(a) $x' + x = 500\delta(t-2); \quad x(0) = 0$
(b) $x' + 3x = 10\delta(t-1) - 5\delta(t-4); \quad x(0) = 1$
(c) $x' - x = \delta(t-2) + H(t-3), \quad x(0) = 5$
(d) $x'' - 4x = 6\delta(t-1); \quad x(0) = 0,\ x'(0) = -3$
(e) $x'' - 3x' + 2x = 2 + 4\delta(t-5); \quad x(0) = 0,\ x'(0) = -2$
(f) $x'' + 2x' + x = 10\delta(t-2); \quad x(0) = x'(0) = 0$
(g) $x'' + x' = \delta(t-1) - \delta(t-2); \quad x(0) = x'(0) = 0$
(h) $x'' - x' = 3 + 4\delta(t-2); \quad x(0) = 0,\ x'(0) = 3$
(i) $x''' = 2\delta(t-5); \quad x(0) = 6,\ x'(0) = x''(0) = 0$
(j) $x''' + 3x'' + 2x' = 8\delta(t-5); \quad x(0) = x'(0) = x''(0) = 0$
(k) $x'''' - 4x'' = 8\delta(t-1); \quad x(0) = x'(0) = x''(0) = 0,$ $x'''(0) = 1$

2. Evaluate these integrals.

(a) $\int_0^\infty (1+3t^2)\delta(t-2)\,dt$ (b) $\int_0^\infty H(t-2)\delta(t-5)\,dt$
(c) $\int_0^\infty (\cos t)\,\delta(t-2)\,dt$ (d) $\int_0^\infty (t^2-1)\delta(t-1)\,dt$
(e) $\int_0^\infty (1+3t^2)\delta(t-\pi)\,dt$ (f) $\int_0^\infty \sin(3t-2)\delta(t-2)\,dt$

3. Consider the IVPs
$x_1' = \delta(t-1);\ x_1(0) = 0,$
$x_2'' = \delta(t-1);\ x_2(0) = 0,\ x_2'(0) = 0,$
$x_3''' = \delta(t-1);\ x_3(0) = 0,\ x_3'(0) = 0,\ x_3''(0) = 0,$
$x_4'''' = \delta(t-1);\ x_4(0) = 0,\ x_4'(0) = 0,\ x_4''(0) = 0,\ x_4'''(0) = 0$
Solve them for $x_1(t), \dots, x_4(t)$, sketch their graphs, together, on $0 < t < 3$, and comment about them.

4. Transfer and Impulse Response Functions. Find the transfer function $Q(s)$ and the impulse response function $q(t)$.

(a) $x' + x = f(t)$ (b) $x' - 2x = f(t)$

(c) $x'' = f(t)$ (d) $x'' + 9x = f(t)$
(e) $x'' + x' = f(t)$ (f) $x'' + 2x' + x = f(t)$
(g) $x''' + x' = f(t)$ (h) $x''' + 3x'' + 2x' = f(t)$

5. The Jump in the Next-to-Highest-Order Derivative. For the second-order equation $mx'' + cx' + kx = I\delta(t-t_0)$, equation (18) indicated a jump of magnitude I/m in $x'(t)$ at t_0. More generally, it can be shown that for the nth-order equation

$$x^{(n)} + p_1 x^{(n-1)} + \cdots + p_{n-1}x' + p_n x = I\delta(t-t_0), \quad (5.1)$$

in which I is the impulse of the forcing function (after the coefficient of $x^{(n)}$ has been normalized to unity), there will be a jump in the $(n-1)$st-order derivative, at t_0, given by

$$\boxed{x^{(n-1)}(t)\Big|_{t_0-}^{t_0+} = I.} \quad (5.2)$$

You need not derive (5.2). Just solve the given IVP for $x(t)$ (or use computer software to solve it) and verify the jump formula (5.2).

(a) $6x' - x = 2\delta(t)$; $x(0) = 5$
(b) $x' + 4x = 3\delta(t)$; $x(0) = 6$
(c) $2x' - 4x = \delta(t-3)$; $x(0) = 5$
(d) $6x' - 2x = \delta(t-2)$; $x(0) = 0$
(e) $x'' = \delta(t)$; $x(0) = 3$, $x'(0) = -5$
(f) $x'' + x = 4\delta(t)$; $x(0) = x'(0) = 0$
(g) $4x'' + 2x' - 2x = \delta(t-3)$; $x(0) = 3$, $x'(0) = 0$
(h) $0.2x'' + 2x' + 5x - \delta(t-1)$; $x(0) = x'(0) = 0$
(i) $x''' = \delta(t-2)$; $x(0) = 3$, $x'(0) = 4$, $x''(0) = 10$
(j) $4x''' + x'' = \delta(t-3)$; $x(0) = 0$, $x'(0) = 73$, $x''(0) = 0$

ADDITIONAL EXERCISES

6. An Impulse in Space, not Time; A "Very Spatial" Delta Function. Although the Laplace transform is typically applied to problems in which the independent variable is the time t, that is by no means a restriction for use of the method. In this exercise, we consider an application that is unusual because the independent variable is a spatial variable x, and because the interval is not $0 < t < \infty$, but a finite interval $0 < x < L$, and also because the problem will be of boundary value, rather than initial value, type. The problem will be to determine the deflection of the beam shown in the figure, due to the point load F applied at $x = a$. According to *Euler beam theory*, the downward deflection $y(x)$ of the beam is governed by the differential equation $EIy'''' = -w(x)$ together with boundary conditions at the two ends. Here, E is Young's modulus

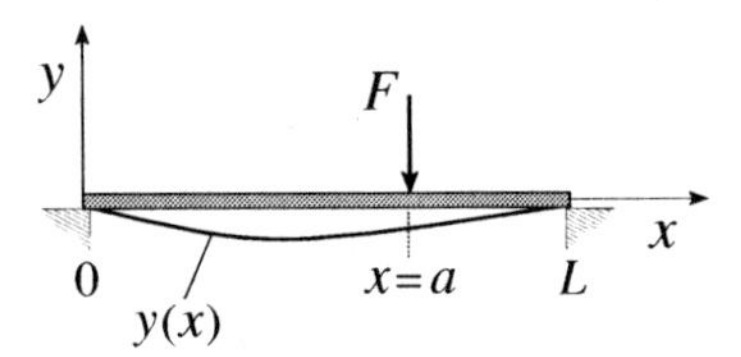

of the beam material and I is the area moment of inertia of the beam cross section about a horizontal axis through its centroid, both of which are known positive constants. (Be careful; here, I denotes inertia, not impulse.) A distributed load on the beam is denoted as $w(x)$, which is downward force per unit length along the beam. Thus, $\int_{x_1}^{x_2} w(x)\,dx$ is the load between any points x_1 and x_2 (with $x_2 > x_1$) and is analogous to the iimpulse $\int_{t_1}^{t_2} f(t)\,dt$ of a force applied over time. Thus, $w(x) = \delta(x-a)$ would model a concentrated force, a point force of unit strength at $x = a$, so the load distribution corresponding to the point force F in the figure corresponds to $w(x) = F\delta(x-a)$. The problem for $y(x)$ is

$$EIy'''' = -F\delta(x-a); \quad y(0) = 0,\ y''(0) = 0,$$
$$y(L) = 0,\ y''(L) = 0, \quad (6.1)$$

in which we state without elaboration that the boundary conditions $y''(0) = 0$ and $y''(L) = 0$ correspond, physically, to there being no moments applied at the two ends.

(a) Solve (6.1) for $y(x)$ by using the Laplace transform, proceeding as follows: Taking the transform, obtain

$$Y(s) = \frac{y'(0)}{s^2} + \frac{y'''(0)}{s^4} - \frac{F}{EI}\frac{e^{-as}}{s^4}. \quad (6.2)$$

The $y'(0)$ and $y'''(0)$ needed in (6.2) are not given, so just call them "A" and "B" and carry them along. Then, show that inversion of (6.2) gives

$$y(x) = Ax + B\frac{x^3}{6} - \frac{F}{6EI}H(x-a)(x-a)^3. \quad (6.3)$$

Now apply the two thus-far-unused boundary conditions $y(L) = 0$ and $y''(L) = 0$ to (6.3), and solve the resulting two algebraic equations for A and B. Thus, show that

$$y(x) = -\frac{F}{6EIL}\big[(a^3 - 3a^2L + 2aL^2)x$$
$$-(L-a)x^3 + LH(x-a)(x-a)^3\big]. \quad (6.4)$$

(b) For definiteness, let $a = 3L/4$. From (6.4), determine the point x at which the maximum deflection occurs. Is it at the point of application of the load F?

7. Point Load on a Guitar String. Now consider not a stiff beam, but a flexible string, stretched by a tension τ, tied at its two ends, and subjected to a point force F at $x = a$, as sketched below. Neglecting the weight of the string (compared to F), and restricting F to be small enough so that the deflection at $x = a$ is small compared to L, it can be shown that the deflection $y(x)$ is governed by the boundary value problem

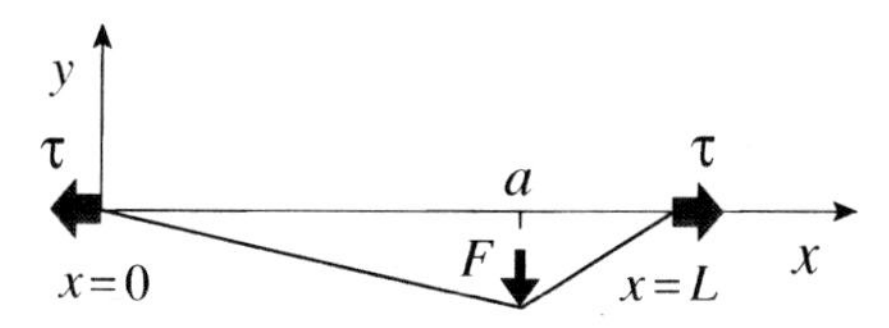

$$\tau y'' = F\delta(x-a); \quad y(0) = 0,\ y(L) = 0. \tag{7.1}$$

You need not derive (7.1), but solve it by the Laplace transform. HINT: The latter is a *boundary* value problem, so can we use the Laplace transform? Yes. [Read part (a) of Exercise 6.] You should obtain

$$y(x) = \begin{cases} -\dfrac{F}{\tau L}(L-a)x, & x < a, \\ -\dfrac{Fa}{\tau L}(L-x), & x > a, \end{cases} \tag{7.2}$$

or, equivalently,

$$y(x) = -\frac{F}{\tau L}\Big[(L-a)x - H(x-a)(x-a)L\Big]. \tag{7.3}$$

NOTE: For a flexible string the differential equation is of *second* order, and the jump given by equation (5.2) in Exercise 5 above, is in the slope $y'(x)$. After all, a flexible string, by definition, has no stiffness, so it can suffer a kink. In contrast, as we've noted, a beam subjected to a point load is governed by a *fourth*-order equation, and the jump is not until the *third* derivative, y'''. The string solution should be no surprise, for we no doubt know from experience that the solution will look like the graph shown in the figure.

8. Approximating an Impulse as a Delta Function; Computer. (a) The delta function approach used in this section hangs on the assumption that the impulsive force is sufficiently focused in time (or space, depending on the application) so that only its impulse matters, and not the details of its graph. To explore this point in a concrete example, consider the IVP

$$x'' + x = f(t); \quad x(0) = 0,\ x'(0) = 0, \tag{8.1}$$

in which $f(t)$ is as shown below. The claim is that if ϵ is sufficiently small, then we can replace the $f(t)$ in (8.1) (which is seen from the figure to have unit impulse) by $\delta(t-3)$.
Here is the problem: Use computer software to solve (8.1), in which $f(t)$ is the function having the graph given in the figure, for the cases where $\epsilon = 1, 0.2,$ and 0.05 and also for the idealized limit in which $f(t) = \delta(t-3)$. Plot the graphs of those four solutions, and comment on your results.

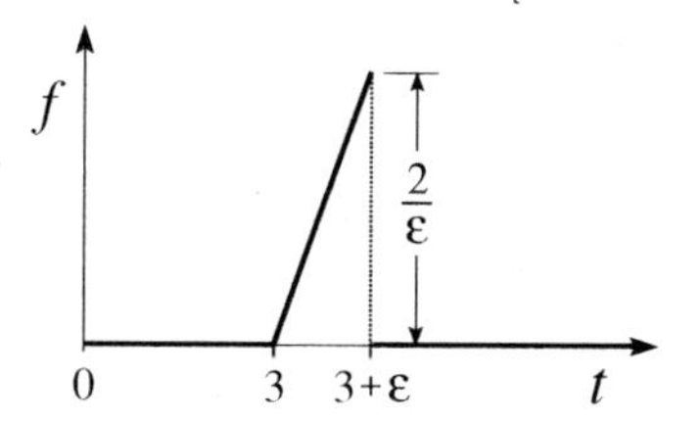

(b) Finally, by "sufficiently focused" we mean that the duration of the impulse, ϵ, is sufficiently small. But "small" is always relative, so we ask: Small compared to what?

9. Pharmacology. Suppose one takes a dose of a certain drug, either orally or intravenously. As the blood circulates, the concentration $c(t)$ of the drug will tend to become uniform throughout the circulatory system. That will probably happen so quickly (particularly if the dose is given intravenously), compared to the time T between doses (such as 24 hours), that we can idealize the situation and model the drug inputs as delta functions. Studies show that following a dose the concentration diminishes with time approximately according to the equation $dc/dt = -kc$, where k is a positive experimentally known constant. Thus, we can model the complete problem as follows:

$$c' + kc = f(t); \quad c(0) = 0, \tag{9.1}$$

where $f(t) = C_1\left[\delta(t) + \delta(t-T) + \delta(t-2T) + \cdots\right]$ and where C_1 is the increase in concentration due to one dose, that is, the dose amount divided by the volume of the circulatory system. Schematically, $f(t)$ is a series of spikes, as shown in the figure.

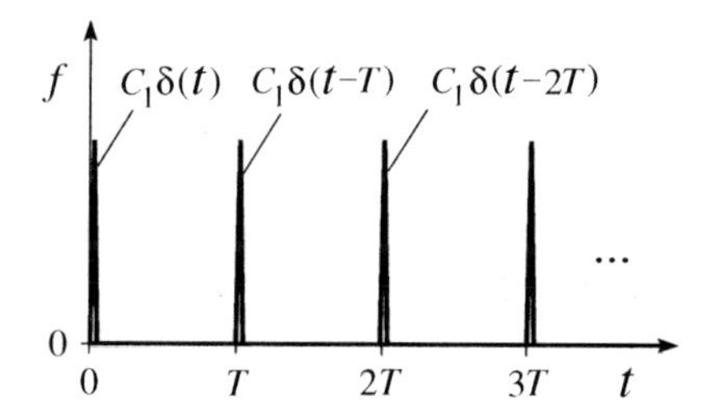

(a) Derive the solution

$$c(t) = C_1\Big[e^{-kt} + e^{-k(t-T)}H(t-T) + e^{-k(t-2T)}H(t-2T) + e^{-k(t-3T)}H(t-3T) + \cdots\Big]. \tag{9.2}$$

Based on (9.2), give a labeled sketch of the graph of $c(t)$. NOTE: This problem is the same as Exercise 33 in Section 1.3, but there we solved for $c(t)$ one time interval at a time.
(b) Generate a computer plot of $c(t)$ from (9.2), on $0 \leq t \leq 6.5$, using $k = T = C_1 = 1$.
(c) Show from (9.2) that $c(t)$ approaches a periodic function, as $t \to \infty$, with period T.
(d) Show, from (9.2), that as $t \to \infty$ the maxima and minima approach these values:

$$c_{\text{max}} = C_1 \frac{1}{1-e^{-kT}}, \quad c_{\text{min}} = C_1 \frac{e^{-kT}}{1-e^{-kT}}. \tag{9.3a,b}$$

10. Harmonic Oscillator Driven by Periodic Hammer Blows. Consider an harmonic oscillator forced periodically by hammer blows with period T. Specifically, consider the IVP

$$x'' + x = \delta(t) + \delta(t-T) + \delta(t-2T) + \cdots, \tag{10.1}$$

with initial conditions $x(0) = 0$, $x'(0) = 0$. Recalling (from Section 3.4) that resonance would occur if the forcing function were harmonic, *at* the natural frequency [which, in (10.1) is 1], it is natural to wonder if the response $x(t)$ in (10.1) will be resonant if $T = 2\pi$, that is, if the hammer blows occur at the natural frequency. Show whether or not that is the case.

CHAPTER 5 REVIEW

This chapter contains many important formulas, so it may be best to use this review to organize them here, by section.

Section 5.2. We defined the Laplace transform of a function $f(t)$ defined on $0 \leq t < \infty$, as

$$\mathcal{L}\{f(t)\} = F(s) = \int_0^\infty f(t) e^{-st}\, dt. \tag{1}$$

and gave conditions for its existence, that is, for the convergence of the integral. Both the transform and its inverse are *linear*:

$$\mathcal{L}\{\alpha f(t) + \beta g(t)\} = \alpha F(s) + \beta G(s), \tag{2}$$

$$\mathcal{L}^{-1}\{\alpha F(s) + \beta G(s)\} = \alpha f(t) + \beta g(t). \tag{3}$$

Section 5.3. The transform of a derivative is

$$\mathcal{L}\{f'(t)\} = sF(s) - f(0). \tag{4}$$

The latter can be used recursively for higher-order derivatives and gives

$$\mathcal{L}\{f''(t)\} = s^2 F(s) - sf(0) - f'(0), \tag{5}$$

and so on.

In addition,

$$\mathcal{L}\{t^n f(t)\} = (-1)^n \frac{d^n F}{ds^n}, \tag{6}$$

$$\lim_{s\to\infty} [sF(s)] = f(0). \quad \text{(initial value formula)} \tag{7}$$

Section 5.4. To deal with the important case of discontinuous functions, we used the Heaviside step function $H(t)$ and its time-shifted version $H(t-a)$ and introduced time-delayed functions $H(t-a)f(t-a)$:

$$\mathcal{L}\{H(t-a)\} = \frac{e^{-as}}{s} \quad \text{(transform of Heaviside function)}, \tag{8}$$

$$\mathcal{L}\{H(t-a)f(t-a)\} = e^{-as}F(s) \quad \text{(transform of time-delayed function)}. \tag{9}$$

In problems with time-delayed forcing functions we used the latter in both directions, first in the form (9) to take the transform of the time-delayed function and then in the reverse direction to invert the resulting transform of the unknown. After all, if the input is time-delayed, then so will the corresponding response.

Section 5.5. The Laplace convolution of $f(t)$ and $g(t)$ is

$$(f * g)(t) = \int_0^{\infty} f(\tau)g(t-\tau)\,d\tau. \tag{10}$$

The convolution theorem:

$$\mathcal{L}\{(f * g)(t)\} = F(s)G(s) \quad \text{or} \quad \mathcal{L}^{-1}\{F(s)G(s)\} = (f * g)(t). \tag{11}$$

In addition,

$$\mathcal{L}^{-1}\left\{\frac{1}{s}F(s)\right\} = 1 * f(t) = \int_0^t f(\tau)\,d\tau \quad \text{(multiplication by } 1/s\text{)}. \tag{12}$$

Section 5.6. The defining property of the Dirac delta function is

$$\int_0^{\infty} \delta(t-t_0)h(t)\,dt = h(t_0) \quad (t_0 \geq 0) \tag{13}$$

for any continuous function $h(t)$. With $h(t) = e^{-st}$, the latter gives

$$\mathcal{L}\{\delta(t-t_0)\} = e^{-st_0} \quad \text{(transform of delta function)}. \tag{14}$$

Chapter 6

SERIES SOLUTIONS

6.1 INTRODUCTION

When we studied linear differential equations of second order and higher in Chapter 2 we were quite successful in terms of the theory. For instance, we found that if the coefficients $p(x)$ and $q(x)$ are continuous on an x-interval I, then the differential equation

$$y'' + p(x)y' + q(x)y = 0 \tag{1}$$

admits two LI (linearly independent) solutions on I; and if $y_1(x)$ and $y_2(x)$ denote any two such solutions then a general solution of (1) is given by an arbitrary linear combination of them,

$$y(x) = C_1 y_1(x) + C_2 y_2(x). \tag{2}$$

However, we were less successful in *finding* solutions, and we developed solution methods only for special cases: constant-coefficient equations and Cauchy–Euler equations. If equation (1) is neither of those types, then we may be forced to turn to numerical solution or to seeking solutions in the form of infinite series. This chapter is about the latter, solutions in series form.

Fortunately, constant-coefficient and Cauchy–Euler equations are of great importance.

Series methods can be used not only for second-order equations but also for equations of first order, and higher order as well. However, second-order equations are, arguably, the most important in applications, and it is that case that we focus on in this chapter.

To introduce the series solution method with a simple example, consider

$$\frac{dy}{dx} + y = 0, \tag{3}$$

a general solution of which is

$$y(x) = Ae^{-x}, \tag{4}$$

with A an arbitrary constant. To illustrate the series solution method, we will put (4) aside and seek a solution to (3) in the form of a power series,

$$y(x) = \sum_{n=0}^{\infty} a_n x^n = a_0 + a_1 x + a_2 x^2 + \cdots, \tag{5}$$

in which the constants $a_0, a_1, \ldots$ are to be determined so that, if possible, (5) satisfies (3). We will proceed rather formally since the purpose of this illustration is only to convey the main ideas.

From (5),

$$\frac{dy}{dx} = \frac{d}{dx}\left(a_0 + a_1 x + a_2 x^2 + \cdots\right) = a_1 + 2a_2 x + 3a_3 x^2 + \cdots \tag{6}$$

and if we put (5) and (6) into (3), we obtain

$$\left(a_1 + 2a_2 x + 3a_3 x^2 + \cdots\right) + \left(a_0 + a_1 x + a_2 x^2 + \cdots\right) = 0, \tag{7}$$

or

$$(a_1 + a_0) + (2a_2 + a_1)\, x + (3a_3 + a_2)\, x^2 + \cdots = 0. \tag{8}$$

For (8) to hold, the coefficient of each power of x must be zero, so $a_1 + a_0 = 0$, $2a_2 + a_1 = 0$, $3a_3 + a_2 = 0$, and so on. Solving these gives

We can solve for a_1, a_2, ... in terms of a_0, with a_0 remaining arbitrary.

$$\begin{aligned} a_1 &= -a_0, \\ a_2 &= -a_1/2 = -(-a_0)/2 = a_0/2, \\ a_3 &= -a_2/3 = -(a_0/2)/3 = -a_0/6, \end{aligned} \tag{9}$$

and so on, in which a_0 remains arbitrary. Finally, putting (9) into (5) gives

$$y(x) = a_0\Big(1 - x + \frac{1}{2}x^2 - \frac{1}{6}x^3 + \cdots\Big) \tag{10}$$

which is indeed the same as (4), though in series form and with "a_0" in place of A.

To develop a systematic approach to obtaining series solutions, particularly for the more difficult case of nonconstant-coefficient equations, we begin with a review of power series and Taylor series.

6.2 POWER SERIES AND TAYLOR SERIES

6.2.1 Power series. A **power series about $x = x_0$** is a series of the form

$$\boxed{\sum_{n=0}^{\infty} a_n (x - x_0)^n = a_0 + a_1 (x - x_0) + a_2 (x - x_0)^2 + \cdots.} \tag{1}$$

in which the **coefficients** $a_1, a_2, \ldots$ and the **center** x_0 are constants. The exponent is the **order** of the term; for instance, $a_3(x-x_0)^3$ is the third-order term. Further, $\sum_{n=0}^{\infty} A_n$ **converges** to s means that $\lim_{N\to\infty}\sum_{n=0}^{N} A_n = s$.

Since the terms in (1) are functions of x rather than constants, they are different in general from one x to another; for instance, the series $\sum_0^\infty x^n$ is $1+1/2+\cdots$ at $x=1/2$, and $1+1/10+\cdots$ at $x=1/10$. Consequently, (1) may converge at some points on the x axis and diverge at others. At the least, it converges at the center x_0 because there it reduces to the single term a_0.

If it converges at other points as well, that convergence will occur in an **interval of convergence** $|x-x_0|<R$ centered at x_0, with the **radius of convergence** R being either finite or infinite. If R is finite (as in Fig. 1), then at each endpoint the series may converge *or* diverge, but convergence or divergence at the endpoints of the interval of convergence will not be relevant in this chapter, so we will not bring up this matter again. To determine R, we can apply one of the standard convergence tests, such as the ratio test or the nth-root test.

The convergence interval $|x-x_0|<R$ is the same as $-R<x-x_0<R$, or $x_0-R<x<x_0+R$.

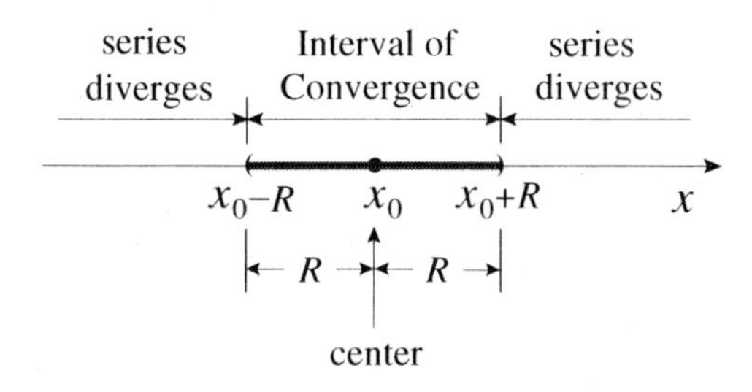

Figure 1. Interval of convergence.

The **ratio test** is as follows. Consider any series of nonzero terms, $\sum_{n=0}^{\infty} A_n$, and suppose $\lim_{n\to\infty}|A_{n+1}/A_n|=\rho$. If $\rho<1$, the series converges; if $\rho>1$ or $\rho=\infty$, the series diverges; and if $\rho=1$, the test is inconclusive.

EXAMPLE 1. Where do these power series converge:

$$\sum_{n=0}^{\infty} n!x^n, \quad \sum_{n=0}^{\infty} n(x+4)^n, \quad \sum_{n=7}^{\infty}\frac{1}{n!}(x-6)^n\ ? \tag{2}$$

Applying the ratio test to the first series in (2), for which A_n is $n!x^n$, gives

$$\lim_{n\to\infty}\left|\frac{(n+1)!x^{n+1}}{n!x^n}\right| = \lim_{n\to\infty}|(n+1)x| = \begin{cases}\infty, & \text{if } x\neq 0\\ 0, & \text{if } x=0,\end{cases} \tag{3}$$

The first series in (2) converges *only* at $x=0$.

which is less than 1 only if $x=0$. Thus, the series converges only at its center, $x=0$. For the second series,

$$\lim_{n\to\infty}\left|\frac{(n+1)(x+4)^{n+1}}{n(x+4)^n}\right| = \lim_{n\to\infty}|x+4| = |x+4|, \tag{4}$$

The second series converges in $|x+4|<1$ and diverges in $|x+4|>1$; $R=1$.

so the series converges in $|x+4|<1$ (hence $R=1$) and diverges in $|x+4|>1$. For the third series,

The third series converges for *all* x; $R = \infty$.

$$\lim_{n\to\infty}\left|\frac{n!(x-6)^{n+1}}{(n+1)!(x-6)^n}\right| = \Big(\lim_{n\to\infty}\left|\frac{1}{n+1}\right|\Big)|x-6| = (0)|x-6| \tag{5}$$

is zero for all x, so the series converges for all x. The lower summation limit being 7 instead of 0, in (2), is immaterial insofar as convergence and divergence, because it is only the limit of the ratio of successive terms as $n\to\infty$ that matters. ▮

6.2.2 Manipulation of power series.

To work with power series, we need to know how they can be manipulated.

Addition and subtraction. If two power series about the same point converge in I, they *can be added or subtracted termwise* (i.e., term by term) within I, and the resulting series will converge within I, so we can write

$$\sum_{n=0}^{\infty} a_n(x-x_0)^n \pm \sum_{n=0}^{\infty} b_n(x-x_0)^n = \sum_{n=0}^{\infty}(a_n \pm b_n)(x-x_0)^n. \tag{6}$$

Multiplication. *Power series can be multiplied termwise.* If $f(x) = \sum_0^\infty a_n(x-x_0)^n$ and $g(x) = \sum_0^\infty b_n(x-x_0)^n$ converge in I, then, with $z = x-x_0$ for brevity,

$$\begin{aligned} f(x)g(x) &= \Big(\sum_{n=0}^{\infty} a_n z^n\Big)\Big(\sum_{n=0}^{\infty} b_n z^n\Big) \\ &= \big(a_0 + a_1 z + +a_2 z^2 + \cdots\big)\big(b_0 + b_1 z + b_2 z^2 + \cdots\big) \\ &= a_0 b_0 + (a_0 b_1 + a_1 b_0)z + (a_0 b_2 + a_1 b_1 + a_2 b_0)z^2 + \cdots \end{aligned} \tag{7}$$

in I.

Differentiation. A power series is *differentiable within its interval of convergence* I. The differentiation can be carried out *termwise*, and the resulting series also converges in I. This can be carried out any number of times. Thus, if

A power series is infinitely differentiable in its interval of convergence, and the differentiations can be carried out termwise.

$$f(x) = a_0 + a_1(x-x_0) + a_2(x-x_0)^2 + \cdots = \sum_{n=0}^{\infty} a_n(x-x_0)^n \tag{8}$$

in I, then

$$\frac{df}{dx}(x) = a_1 + 2a_2(x-x_0) + 3a_3(x-x_0)^2 + \cdots = \sum_{n=1}^{\infty} n a_n (x-x_0)^{n-1}, \tag{9a}$$

$$\frac{d^2 f}{dx^2}(x) = 2a_2 + 6a_3(x-x_0) + \cdots = \sum_{n=2}^{\infty} n(n-1)a_n(x-x_0)^{n-2}, \tag{9b}$$

and so on, in I.

Identity principle. *If two power series converge and are equal within* I*, then their* *corresponding coefficients must be equal.* That is, if

$$\sum_{n=0}^{\infty} a_n(x-x_0)^n = \sum_{n=0}^{\infty} b_n(x-x_0)^n \tag{10}$$

in I, then it must be true that $a_n = b_n$ for each n; this is the **identity principle**. As a special case, if the b_n's are zero in (10), so

$$\sum_{n=0}^{\infty} a_n(x-x_0)^n = 0, \tag{11}$$

then each a_n must be zero.

In summary, convergent power series can be manipulated in essentially the same ways as polynomials; they are extremely "user friendly."

6.2.3 Taylor series.

The **Taylor series** of $f(x)$ about a point x_0, which we will denote here as $\text{TS}f|_{x_0}$, is the series

Every Taylor series is a power series.

$$\begin{aligned} \text{TS}f|_{x_0} &= f(x_0) + \frac{f'(x_0)}{1!}(x-x_0) + \frac{f''(x_0)}{2!}(x-x_0)^2 + \cdots \\ &= \sum_{n=0}^{\infty} \frac{f^{(n)}(x_0)}{n!}(x-x_0)^n. \end{aligned} \tag{12}$$

In this text, $\text{TS}f|_{x_0}$ denotes the Taylor series of f about x_0.

For instance, recall these, with their centers at the origin:

$$e^x = 1 + \frac{1}{1!}x + \frac{1}{2!}x^2 + \frac{1}{3!}x^3 + \cdots = \sum_{n=0}^{\infty} \frac{1}{n!}x^n \quad (|x| < \infty), \tag{13a}$$

$$\frac{1}{1-x} = 1 + x + x^2 + x^3 + \cdots = \sum_{n=0}^{\infty} x^n \quad (|x| < 1). \tag{13b}$$

The series in (13b) is well known as the **geometric series**.

Let $s_N(x)$ denote the Nth-degree polynomial consisting of the terms in (12) up to and including the term of order N. These arc the **partial sums** of the series.

For the series in (13a), for instance, $s_0(x) = 1$, $s_1(x) = 1 + x$, and so on. In Fig. 2 we've plotted the first several partial sums, together with e^x. We see from the figure that $s_0(x)$ gives a good approximation to e^x only if x is very small; $s_1(x)$ stays close to e^x over a broader interval than $s_0(x)$, $s_2(x)$ stays close over a broader interval than $s_1(x)$, and so on. After all, $s_0(x)$ matches only the value of e^x at the point of expansion, $x = 0$; $s_1(x)$ matches not only the function value there but also its first derivative; $s_2(x)$ matches not only the function and its derivative there but also the second derivative (hence the curvature), and so on.

It's true that with N fixed the discrepancy between $s_N(x)$ and e^x becomes infinite as $x \to \infty$, but that circumstance is irrelevant insofar as the convergence of

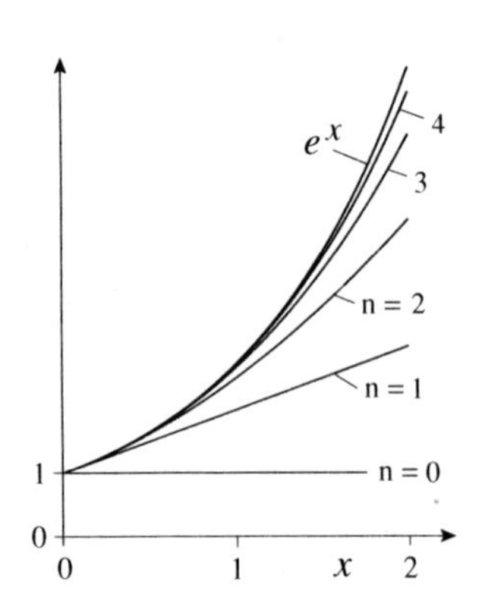

Figure 2. Convergence of the sequence of partial sums $s_n(x)$ of (13a) to e^x.

the series is concerned, because by the convergence of $s_N(x)$ to e^x we mean that $s_N(x)$ tends to e^x, at a *fixed point* x, as $N \to \infty$. To illustrate, choose $x = 1$, and observe how the sequence of partial sums there, $s_0(1) = 1$, $s_1(0) = 2$, $s_2(1) = 2.5$, $s_3(1) = 2.667$, $s_4(1) = 2.708$, $s_5(1) = 2.717$ ..., does seem to be converging to the function value, $e^1 = 2.718$, at that point.

Closely related to Taylor series is the concept of analyticity: f is **analytic** at $x = x_0$ if $\text{TS}f|_{x_0}$ converges to $f(x)$ in $|x - x_0| < R$ for some $R > 0$; if not, it is **singular** there.[1] If $\text{TS}f|_{x_0}$ converges to $f(x)$ in I, then f is analytic in I. For instance, (13a) holds for all x, so e^x is analytic for all x.

We will be particularly interested in rational functions, a **rational function** being the ratio of polynomials, such as $\dfrac{9-4x}{3+x+5x^2}$. For such functions we have the following simple test for analyticity:

THEOREM 6.2.1 *Analyticity of Rational Functions*
Let $f(x) = P(x)/Q(x)$, where P and Q are polynomials, and where any common factors in P and Q have been canceled. Then
(i) f is analytic at x_0 if and only if $Q(x_0) \neq 0$, and
(ii) the radius of convergence of the Taylor series of f about x_0 is the distance from x_0 to the nearest zero of Q in the complex plane.

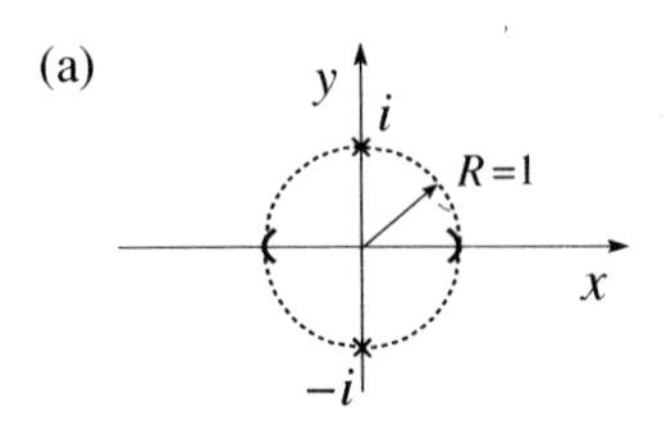

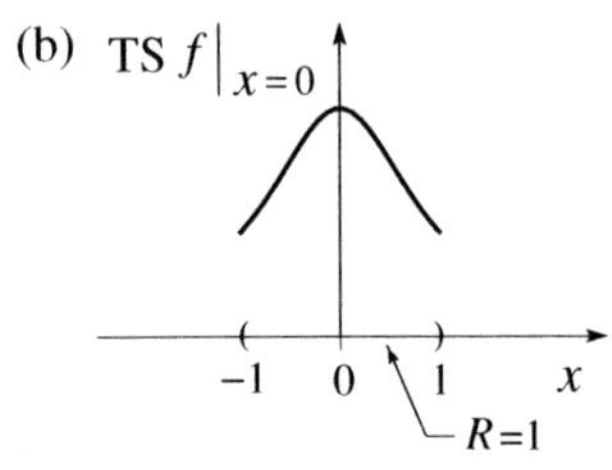

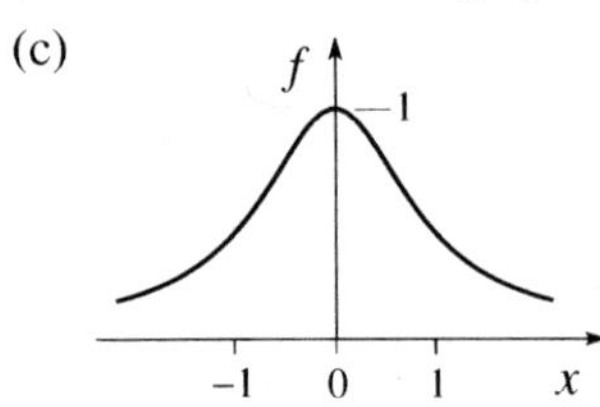

Figure 3. $f(x)$ in Example 2.

As a special simple case, it follows from Theorem 6.2.1 that any polynomial is analytic for all x, because we can take that polynomial to be $P(x)$, and $Q(x) = 1$, the latter being nonzero for all x.

EXAMPLE 2. Application of Theorem 6.2.1. Is the rational function

$$f(x) = \frac{1}{1+x^2} \tag{14}$$

analytic at $x = 0$, for instance? Theorem 6.2.1 tells us it is because the denominator $Q(x) = 1 + x^2$ is not zero at $x = 0$. It's that simple for rational functions.

Next, what does the theorem tell us about the radius of convergence of the Taylor series of $f(x)$ about $x = 0$? To answer, we need to know the zeros of Q in the complex plane, so set $Q(z) = 1 + z^2 = 0$, where $z = x + iy$ is complex.[2] The latter gives $z = \pm i$. Thus, the theorem tells us that the radius of convergence of the Taylor series of $f(x)$, about $x = 0$, is $R = 1$ because that is the distance from the point $x = 0$ (on the real axis) to the nearest zero of Q (namely, the ones at $\pm i$), as depicted in Fig. 3a.

[1]It would be natural to think that to determine if a function $f(x)$ is analytic at a given point $x = a$ it suffices to see if the Taylor series of $f(x)$ about that point converges in some open interval (i.e., rather than just at the point of expansion). However, notice the words "to $f(x)$." Surprising though it may be, it is possible for $\text{TS}f|_{x_0}$ to converge in some open interval, but *not to the function* $f(x)$; such an example is given in the exercises. Indeed, analyticity is a subtle matter, but for our purposes in this chapter Theorem 6.2.1 will suffice. For further discussion, see the Additional Exercises.

[2]It seems better to write $Q(z) = 0$ than $Q(x) = 0$ because we are looking for roots in the complex plane, not just on the real axis. The complex plane is reviewed in Appendix D.

Thus, the Taylor series of f about $x = 0$ converges to f on $|x| < 1$, as indicated in Fig. 3b; it *represents* f on that interval. Outside that interval it diverges and does *not* represent f; that is, we cannot use it to compute the values of $f(x)$.

How about $x = 2$? Is f analytic at $x = 2$? Yes, because $Q(x)$ is nonzero there. And we can see from Fig. 4 that the radius of convergence of its Taylor series about that point is $R = \sqrt{5}$.

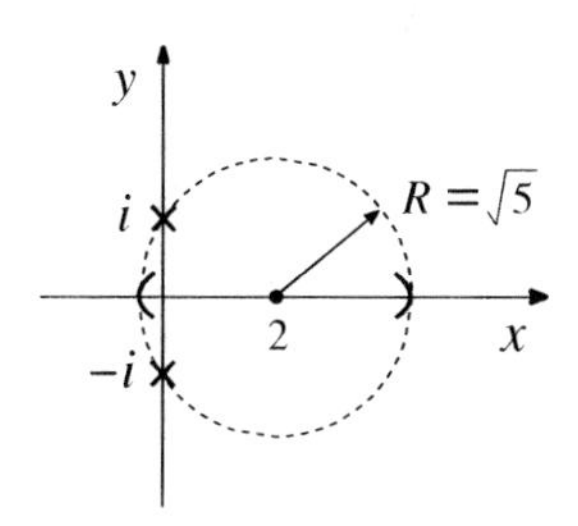

Figure 4. The interval of convergence of the Taylor series of $f(x) = \dfrac{1}{1+x^2}$ about $x = 2$ is $|x - 2| < \sqrt{5}$.

COMMENT. It is striking that the function $f(x) = 1/(1 + x^2)$ is well behaved for *all* x (in fact, it is differentiable an arbitrarily large number of times) and its graph (Fig. 3c) gives no hint of trouble anywhere. Hence, we might have expected its Taylor series about $x = 0$ to converge to $f(x)$ for $-\infty < x < \infty$. That it converges only on $|x| < 1$ is due not to any "bad" behavior of $f(x)$ along the x axis (indeed, there is none) but to the bad behavior of "$f(z) = 1/(1 + z^2)$" *off* of the real axis at $z = \pm i$, where $f(z) = 1/0$ "blows up," is undefined. ▮

EXAMPLE 3. As a final example, consider the rational function

$$f(x) = \frac{x^3 + 3}{(x + 2)(x^2 - 2x + 5)}. \tag{15}$$

In this case $Q(z) = (z + 2)(z^2 - 2z + 5)$ has zeros at $z = 1 \pm 2i$, and at $z = -2$ as well. Thus, $f(x)$ is analytic for all $x \neq -2$; at $x = -2$ it is singular.

To further illustrate the determination of the interval of convergence of Taylor series, consider the Taylor expansions of this f about various points along the x axis, such as $x = 3$ and $x = -1/2$. For the expansion about $x = 3$, show that the radius of convergence is $R = 2\sqrt{2}$ and the interval of convergence is $|x - 3| < 2\sqrt{2}$, because of the zeros of Q at $1 \pm 2i$. (HINT: Use a diagram like the one in Fig. 4, with the zeros of $Q(z)$ marked.) And for the expansion about $x = -1/2$, show that the radius of convergence is $R = 3/2$ and the interval of convergence is $|x - (-1/2)| = |x + 1/2| < 3/2$ because of the zero of Q at -2. ▮

Closure. In this section we reviewed power series and Taylor series and defined analyticity: f is analytic at x_0 if $\mathrm{TS}f\big|_{x_0}$ converges to $f(x)$ in $|x - x_0| < R$ for some (any) $R > 0$; if not, it is singular there. We will be particularly concerned with rational functions, and a simple test for the analyticity of a rational function is given by Theorem 6.2.1.

EXERCISES 6.2

1. Determine the radius of convergence R of the given power series. NOTE: To simplify notation we will often use $\sum_0^\infty$ to mean $\sum_{n=0}^\infty$.

(a) $\displaystyle\sum_0^\infty nx^n$

(b) $\displaystyle\sum_0^\infty (-1)^n n^{1000} x^n$

(c) $\displaystyle\sum_5^\infty e^n x^n$

(d) $\displaystyle\sum_0^{1000} n!\, x^n$

(e) $\displaystyle\sum_0^\infty \left(\frac{x+3}{2}\right)^n$

(f) $\displaystyle\sum_2^\infty (n-1)^3 (x-5)^n$

(g) $\displaystyle\sum_{1}^{\infty} \frac{n^{50}}{n!}(x+7)^n$ (h) $\displaystyle\sum_{0}^{\infty}(n+1)x^{2n}$

(i) $\displaystyle\sum_{3}^{\infty} \frac{(-1)^n}{2^n}(x+2)^{2n}$ (j) $\displaystyle\sum_{100}^{\infty}(-1)^n\frac{n}{2^n}(x-5)^{2n}$

2. Manipulation of Power Series. Fill in the missing "?" information and determine the x-interval on which the result holds.

(a) $\displaystyle\sum_{0}^{\infty}\frac{1}{n+1}x^n + \sum_{0}^{\infty}\left(\frac{x}{e}\right)^n = \sum_{0}^{\infty}(?)$

(b) $\displaystyle\sum_{0}^{\infty}\frac{1}{n+1}x^n - \sum_{0}^{\infty}\frac{1}{n+2}x^n = \sum_{0}^{\infty}(?)$

(c) $\displaystyle\sum_{3}^{\infty}(n+1)(2x)^n + \sum_{3}^{\infty}(n-1)(5x)^n = \sum_{3}^{\infty}(?)$

(d) $\displaystyle\frac{d}{dx}\left(\sum_{1}^{\infty}nx^n - 3\sum_{1}^{\infty}\frac{1}{2^n}x^n\right) = \sum_{1}^{\infty}(?)$

(e) $\displaystyle\frac{d}{dx}\sum_{0}^{\infty}\frac{1}{4^n}x^n = \sum_{?}^{\infty}(?)$

(f) $\displaystyle\frac{d^2}{dx^2}\sum_{1}^{\infty}n^2x^n = \sum_{?}^{\infty}(?)$

3. Work out the product, termwise, through terms of third order.

(a) $\displaystyle\left(\sum_{0}^{\infty}nx^n\right)\left(\sum_{0}^{\infty}x^n\right)$ (b) $\displaystyle\left(\sum_{0}^{\infty}e^nx^n\right)\left(\sum_{0}^{\infty}nx^n\right)$

(c) $\displaystyle\left(\sum_{0}^{\infty}ex^n\right)\left(\sum_{0}^{\infty}\left(\frac{x}{e}\right)^n\right)$ (d) $\displaystyle\left(\sum_{0}^{\infty}\frac{1}{n!}x^n\right)^2$

(e) $\displaystyle\left(\sum_{0}^{\infty}\frac{1}{n!}x^n\right)\left(\sum_{0}^{\infty}\frac{1}{n!}(2x)^n\right)$

(f) $\displaystyle\left(\sum_{0}^{\infty}\frac{1}{n!}x^n\right)\left(\sum_{0}^{\infty}\frac{(-1)^n}{n!}x^n\right)$

4. Rational Functions. Determine the radius of convergence R of the Taylor series of the given rational function $f(x)$ about each of the indicated points; you need not generate the Taylor series — just determine its radius of convergence. Also, for what x's (if any), real or complex, is $f(x)$ singular?

(a) $\displaystyle\frac{6}{x^2}$; $x = -5, 1$

(b) $\displaystyle\frac{1}{x^2+x+1}$; $x = -1, 0, 2$

(c) $\displaystyle\frac{x^4+x^3}{(x-1)(x-2)}$; $x = 0, 50$

(d) $\displaystyle\frac{x^2+2x+3}{(x-1)(x^2+4)}$; $x = -5, 5$

(e) $\displaystyle\frac{x^2+2x+3}{(x-1)^3(x^2+1)}$; $x = -1, 0, 4$

(f) $\displaystyle\frac{6}{1+x+x^2+x^3}$; $x = -3, 0, 3$

(g) $\displaystyle\frac{x^4+1}{(x-1)(x^2+5x+4)}$; $x = -2, 0, 5$

(h) $\displaystyle\frac{x^3}{x^4+x^2-12}$; $x = -4, 1, 8$

(i) $\displaystyle\frac{x^4+1}{x^2+1}$; $x = 0, 2, 10$

5. Verification of Analyticity for $1/(1-x)$ in $|x| < 1$. It is possible for $\mathrm{TS}f|_{x_0}$ to converge on an open interval I, yet not to the given function $f(x)$, and an example of such an unusual function is given in Exercise 6. Thus, although it may be easy to determine the interval of convergence of a Taylor series, it is probably not simple to show, further, that the sum function on that interval is identical to the function $f(x)$ that we started with. However, we *can* do that for the function $f(x) = 1/(1-x)$, as we now outline:

(a) Verify (by multiplying through by $1-x$) that the following is simply an identity, for all $x \neq 1$ and for any integer $k \geq 0$:

$$\frac{1}{1-x} = 1 + x + x^2 + \cdots + x^k + \frac{x^{k+1}}{1-x}. \tag{5.1}$$

(b) By letting $k \to \infty$ in (5.1), show that the Taylor series of $f(x) = 1/(1-x)$ about $x = 0$, which is the geometric series $1 + x + x^2 + \cdots$, not only converges on $-1 < x < 1$, it converges *to the function* $f(x)$. NOTE: See the next exercise as well.

ADDITIONAL EXERCISES

6. A Counterexample. We stated in a footnote that the Taylor series of a given function $f(x)$ can converge on some open interval, but not to $f(x)$. The function that is traditionally used to illustrate that circumstance is

$$f(x) = \begin{cases} e^{-1/x^2}, & \text{if } x \neq 0 \\ 0, & \text{if } x = 0, \end{cases} \tag{6.1}$$

the graph of which is given below. [It appears from the figure that $f(x)$ is identically zero in some open interval near the origin, but it is not; it is extremely small there, but is nonzero for all $x \neq 0$. For instance, $f(0.2) \approx 1.4 \times 10^{-11}$.] It can be shown that *all* derivatives of $f(x)$ are zero at $x = 0$, so the Taylor series of $f(x)$ about that point is

$$\mathrm{TS}f\big|_{x=0} = 0 + 0x + 0x^2 + \cdots. \tag{6.2}$$

Clearly, that Taylor series converges, for all x in fact, but it converges to the zero function, not to $f(x)$ (see the figure)! Thus, since the Taylor series of $f(x)$ about $x = 0$ does not converge to $f(x)$ in any open interval, $f(x)$ is not analytic at $x = 0$; it is singular there.

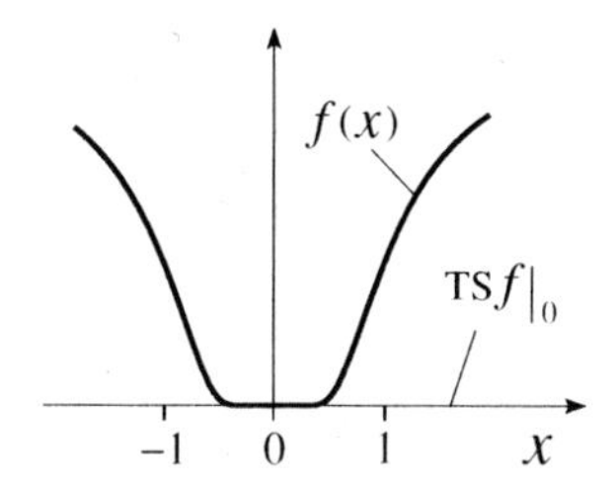

Here is the problem:
(a) Show that $f'(0) = 0$, as we claimed. HINT: Differentiating the function e^{-1/x^2} is of no help because that expression of $f(x)$ holds only for $x \neq 0$. Rather, fall back on the difference quotient definition of the derivative, and then use l'Hôpital's rule to evaluate the limit.
(b) Then show that $f''(0) = 0$.
(c) At this point you should see the pattern whereby all higher derivatives are zero there as well. Explain.

6.3 POWER SERIES SOLUTION ABOUT A REGULAR POINT

6.3.1 Power series solution theorem. We now show how to use the power series solution method to solve the second-order differential equation

$$\boxed{y'' + p(x)y' + q(x)y = 0.} \tag{1}$$

In earlier chapters, existence and uniqueness theorems for the linear differential equation (1) called for the coefficients $p(x)$ and $q(x)$ to be *continuous*; for instance, see Theorem 2.7.1. In this chapter, however, we will have to ask much more of $p(x)$ and $q(x)$ than mere continuity, because we are seeking solutions for $y(x)$ in the form of power series, and power series are infinitely differentiable within their interval of convergence (Section 6.2.2); in fact, they represent analytic functions therein.

Looking ahead to requirements that we will place on $p(x)$ and $q(x)$ in the theorem that follows.

To see a connection between the degree of "niceness" of the solution $y(x)$ of (1), and the niceness that we must require of the coefficients $p(x)$ and $q(x)$, solve (1) for y'' and obtain $y'' = -p(x)y' - q(x)y$. Then, differentiate repeatedly, to get expressions for successively higher order derivatives of $y(x)$: $y''' = -py'' - p'y' - qy' - q'y$, and so on. Those formulas for the higher-order derivatives of y include successively higher-order derivatives of p and q, so that for y to be infinitely differentiable we need p and q to be as well. Going a step further, if we expect $y(x)$ to be analytic, then it seems reasonable that we should ask the same of p and q, and we will do exactly that in the theorem that follows.

But, first, we want to emphasize that requiring p and q to be analytic is to require much more than continuity. After all, a function $f(x)$ can be continuous at x_0 without being even once differentiable there, for its graph can have a "kink" there. In contrast, for $f(x)$ to be analytic at x_0, its derivatives of all orders must exist there, for it to even *have* a Taylor series about x_0.

Theorem 6.3.1, assures us that solutions can be expressed in power series form, so, and this is the point, we will be able to find them by looking for them in that form.

THEOREM 6.3.1 *Power Series Solution*
If p and q are analytic at x_0, then every solution of (1) is too, and can be represented in the power series form

$$y(x) = a_0 + a_1(x - x_0) + a_2(x - x_0)^2 + \cdots = \sum_{n=0}^{\infty} a_n(x - x_0)^n \tag{2}$$

in its interval of convergence $|x - x_0| < R$, where R is *at least* as large as the smaller of the radii of convergence of TS $p|_{x_0}$ and TS $q|_{x_0}$.[1]

Regular points and singular points of (1).

If p and q are indeed analytic at x_0, then x_0 is called a **regular point** of (1); if not, it is a **singular point** of (1). In this section we consider expansions only about regular points. Expansions about singular points are also important, and will be discussed in Sections 6.4 and 6.5.

To illustrate regular points, singular points, and Theorem 6.3.1, consider

$$x(x-1)y'' + 2xy' - y = 0. \tag{3}$$

First, get (3) into the standard form (1) by dividing by $x(x-1)$:

$$y'' + \frac{2}{x-1}y' - \frac{1}{x(x-1)}y = 0. \tag{4}$$

Now we can identify p and q as

$$p(x) = \frac{2}{x-1}, \quad q(x) = -\frac{1}{x(x-1)}. \tag{5}$$

These are rational functions, so, by Theorem 6.2.1, p is analytic for all x except $x = 1$, and q is analytic for all x except $x = 0$ and $x = 1$, where their denominators are zero. Thus, every x is a regular point of (3), except $x = 0$ and $x = 1$. If we avoid those two singular points and choose $x_0 = 5$, for instance, then Theorem 6.3.1 assures us that two LI solutions of (4) can be found in the power series form $\sum_0^\infty a_n(x-5)^n$, with radii of convergence R at least as large as 4, which is the distance from x_0 to the nearer of the two singular points, namely, the one at $x = 1$.

[1]For proof of the theorem, see, for instance, A. L. Rabenstein, *Introduction to Ordinary Differential Equations* (NY: Academic Press, 1966), Section 4.3.

Typically, p and q in (1) will be analytic for all x, or for all x except for a number of x's, such as $x = 0$ and $x = 1$ in the case of equation (4). *Theorem 6.3.1 guides us to a general solution of (1) if we avoid the singular points of the differential equation, if any, when we choose x_0.*

6.3.2 Applications. We now illustrate the use of Theorem 6.3.1 in obtaining power series solutions of linear second-order differential equations.

EXAMPLE 1. A First Example. The purpose of the series solution method is to help us solve *nonconstant*-coefficient differential equations, but we will begin with the familiar constant-coefficient differential equation

$$y'' + y = 0 \tag{6}$$

because we can use its known general solution to check our results.

Since $p(x) = 0$ and $q(x) = 1$ are analytic for all x, every point x is a regular point of (6), so we can choose x_0 to be whatever we like. Let $x_0 = 0$. Then Theorem 6.3.1 assures us that the solutions of (6) can be found in the form

$$y(x) = a_0 + a_1 x + a_2 x^2 + a_3 x^3 + a_4 x^4 + \cdots = \sum_{n=0}^{\infty} a_n x^n, \tag{7}$$

and that these series will have infinite radii of convergence. We can differentiate (7) termwise and obtain

Note the changes in the lower summation limit, from 0 in (7) to 1 in (8), and to 2 in (9).

$$y'(x) = 0 + a_1 + 2a_2 x + 3a_3 x^2 + \cdots = \sum_{n=1}^{\infty} n a_n x^{n-1} \tag{8}$$

and again,

$$y''(x) = \sum_{n=2}^{\infty} n(n-1) a_n x^{n-2}. \tag{9}$$

Then, put (7) and (9) into (6) and obtain

Be sure to understand the steps from (10) to (13) because they come up repeatedly in using the series solution method.

$$\underbrace{\sum_{n=2}^{\infty} n(n-1) a_n x^{n-2}}_{\text{set } n-2=k} + \sum_{n=0}^{\infty} a_n x^n = 0. \tag{10}$$

We want to combine the two sums in (10), but to do that we need both the exponents on x and the summation limits in the two sums to be the same, and they are not. *Work on the exponents first.* We can get the exponents to be the same by making the change of variables $n - 2 = k$ in the first series. That is, let $n = k + 2$, obtaining

If unclear about the dummy index idea, write out the first few terms of the two series, the first series in (11) and the first in (12), and see that they are identical.

$$\sum_{k=0}^{\infty} (k+2)(k+1) a_{k+2} x^k + \sum_{n=0}^{\infty} a_n x^n = 0. \tag{11}$$

The exponents in (11) look different because one is k and one is n, but the difference is superficial because k and n are merely "dummy" indices. We can change the name "k" to "n" in the first, obtaining

$$\sum_{n=0}^{\infty}(n+2)(n+1)a_{n+2}x^n + \sum_{n=0}^{\infty} a_n x^n = 0. \tag{12}$$

Fortuitously, the summation limits are also the same now, so we can combine the two sums, obtaining

$$\sum_{n=0}^{\infty}\left[(n+2)(n+1)a_{n+2} + a_n\right] x^n = 0, \tag{13}$$

from which it follows from the identity principle (in Section 6.2.2) that the bracketed coefficient is zero,

$$(n+2)(n+1)a_{n+2} + a_n = 0, \tag{14}$$

for each $n = 0, 1, 2, \ldots$.

The recursion formula.

Equation (14) is called the **recursion formula** because it gives the a_n's recursively, for if we re-express it as

$$a_{n+2} = -\frac{1}{(n+2)(n+1)}a_n \qquad (n = 0, 1, 2, \ldots) \tag{15}$$

and set $n = 0, 1, 2, \ldots$ in turn, it gives

$$\begin{aligned}
n = 0: &\quad a_2 = -\frac{1}{(2)(1)}a_0,\\
n = 1: &\quad a_3 = -\frac{1}{(3)(2)}a_1,\\
n = 2: &\quad a_4 = -\frac{1}{(4)(3)}a_2 = \frac{1}{(4)(3)(2)(1)}a_0 = \frac{1}{4!}a_0,\\
n = 3: &\quad a_5 = -\frac{1}{(5)(4)}a_3 = \frac{1}{5!}a_1,
\end{aligned} \tag{16}$$

and so on. That is, it evaluates a_2, a_3, a_4, ... , in terms of a_0 and a_1, these two remaining arbitrary. Unfortunately, (15) is not an explicit formula for a_n, but by letting $n = 0, 1, 2, \ldots$ it does enable us to generate as many a_n's as we desire, in terms of a_0 and a_1.

Finally, put (16) into (7) and obtain

$$\begin{aligned}
y(x) &= a_0 + a_1 x - \frac{1}{2!}a_0 x^2 - \frac{1}{3!}a_1 x^3 + \frac{1}{4!}a_0 x^4 + \frac{1}{5!}a_1 x^5 + \cdots\\
&= a_0\left(1 - \frac{1}{2!}x^2 + \frac{1}{4!}x^4 - \cdots\right) + a_1\left(x - \frac{1}{3!}x^3 + \frac{1}{5!}x^5 - \cdots\right)\\
&= a_0 y_1(x) + a_1 y_2(x).
\end{aligned} \tag{17}$$

In this example it is not hard to see the pattern and to anticipate the general term in each of the two series in (17):

$$y_1(x) = \sum_{n=0}^{\infty}(-1)^n \frac{x^{2n}}{(2n)!}, \qquad y_2(x) = \sum_{n=0}^{\infty}(-1)^n \frac{x^{2n+1}}{(2n+1)!}. \tag{18}$$

Each of $y_1(x)$ and $y_2(x)$ is a solution of (6) because a_0 and a_1 are arbitrary — so we could set $a_0 = 1$ and $a_1 = 0$ in (17) and obtain $y(x) = y_1(x)$, or we could set $a_0 = 0$ and $a_1 = 1$ and obtain $y(x) = y_2(x)$. Further, $y_1(x)$ and $y_2(x)$ are LI because they are only two functions, neither one being a constant multiple of the other. Thus, (17) is a general solution of (6), with a_0 and a_1 playing the roles of "C_1" and "C_2."

Alternatively, $y_1(x)$ and $y_2(x)$ are LI because their Wronskian determinant, at $x = 0$, equals 1, which is nonzero.

COMMENT 1. In this introductory example, we can recognize the series in (18) as

$$y_1(x) = \cos x \quad \text{and} \quad y_2(x) = \sin x,$$

but in general such recognition is not possible. Pretending that we do not recognize these series as $\cos x$ and $\sin x$, if we want to use (18) to compute $y_1(x)$ or $y_2(x)$ at a given x we must add enough terms of the series to achieve the desired accuracy. For small values of x (i.e., for x's close to the point of expansion $x_0 = 0$) a few terms may suffice. For example, just the first three terms of $y_1(x)$ give $y_1(0.5) = 0.87760$, and the exact value (to five decimal places) is $y_1(0.5) = \cos 0.5 = 0.87758$. As x increases, more terms are needed for comparable accuracy. The situation is suggested in Fig. 1, in which we've plotted several partial sums of $y_2(x)$, for instance, along with the sum function $\sin x$. The larger we choose n, the broader the x interval on which the approximation $s_n(x)$ stays close to the sum function $\sin x$.

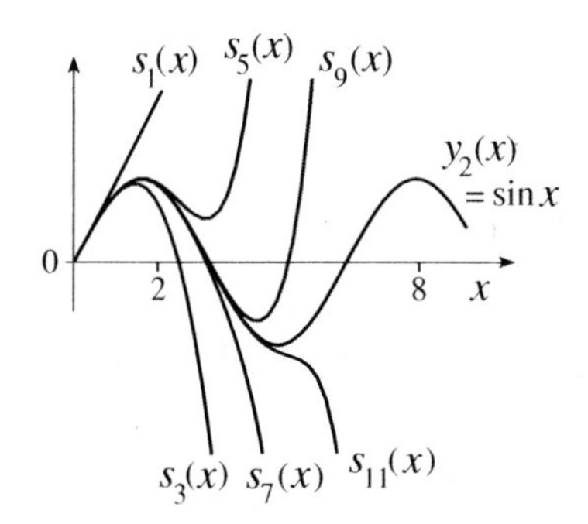

Figure 1. Partial sums of $y_2(x)$, compared with the sum function $y_2(x) = \sin x$. Recall that $s_N(x)$ means the partial sum through the term of order N, that is, through the x^N term.

COMMENT 2. There is no need to determine the radii of convergence of the series solutions in (18) because Theorem 6.3.1 assures us that the power series solutions of (6) will have infinite radii of convergence. Nevertheless, to illustrate, let us use the ratio test to determine the radius of convergence of $y_1(x)$. Since

$$\lim_{n\to\infty}\left|\frac{(-1)^{n+1}x^{2(n+1)}/[2(n+1)]!}{(-1)^n x^{2n}/(2n)!}\right| = \lim_{n\to\infty}\left|\frac{1}{(2n+2)(2n+1)}x^2\right| = 0 \tag{19}$$

for all x, the series converges for all x. Similarly for $y_2(x)$. ∎

EXAMPLE 2. Nonconstant Coefficients. Find a general solution of

$$y'' + (9-x)y = 0. \tag{20}$$

Realize that (20) is neither of constant-coefficient nor Cauchy–Euler types, so it is nonelementary, and none of the solution methods given in the preceding chapters are of help. Thus, we can turn to numerical solution or series solution. Here, we choose the latter.

Both $p(x) = 0$ and $q(x) = 9 - x$ are polynomials, and are therefore analytic for all x, so Theorem 6.3.1 guarantees that we can find two LI power series solutions about any point x_0, with infinite radii of convergence.

Choosing $x_0 = 0$, seek

$$y(x) = \sum_{n=0}^{\infty} a_n x^n. \tag{21}$$

Putting (21) into (20) gives

$$\underbrace{\sum_{n=2}^{\infty} n(n-1)a_n x^{n-2}}_{n-2=k} + \sum_{n=0}^{\infty} 9a_n x^n - \underbrace{\sum_{n=0}^{\infty} a_n x^{n+1}}_{n+1=k} = 0. \tag{22}$$

Remember, to combine the sums we need the same exponents on the x's and the same summation limits. *Dealing with the exponents first,* let $n-2=k$ in the first sum and let $n+1=k$ in the third. Doing so converts (22) to

Deal with the exponents before the summation limits.

$$\sum_{k=0}^{\infty}(k+2)(k+1)a_{k+2}x^k+\sum_{n=0}^{\infty}9a_nx^n-\sum_{k=1}^{\infty}a_{k-1}x^k=0. \tag{23}$$

As in Example 1, change the dummy-index k's to n's, simply to have the same summation index in the three sums. Finally, attend to the summation limits, which are not the same; they are 0, 0, and 1. But, we can change the lower limit to 0 in the third sum if we simply define $a_{-1}\equiv 0$. With these changes, we can now combine the sums and obtain

$$\sum_{n=0}^{\infty}\left[(n+2)(n+1)a_{n+2}+9a_n-a_{n-1}\right]x^n=0. \tag{24}$$

The identity principle requires that the square-bracketed coefficient be zero for each n, so we obtain the recursion formula

$$(n+2)(n+1)a_{n+2}+9a_n-a_{n-1}=0 \qquad (n=0,1,2,\dots), \tag{25}$$

or

$$a_{n+2}=-\frac{9}{(n+2)(n+1)}a_n+\frac{1}{(n+2)(n+1)}a_{n-1} \qquad (n=0,1,2,\dots). \tag{26}$$

Then, setting $n=0,1,2,\dots$ in (26), in turn, and remembering that $a_{-1}=0$, gives

$$\begin{aligned}
n=0:\ a_2&=-\frac{9}{2}a_0+\frac{1}{2}a_{-1}=-\frac{9}{2}a_0+\frac{1}{2}(0)=-\frac{9}{2}a_0,\\
n=1:\ a_3&=-\frac{9}{6}a_1+\frac{1}{6}a_0=\frac{1}{6}a_0-\frac{3}{2}a_1,\\
n=2:\ a_4&=-\frac{9}{12}a_2+\frac{1}{12}a_1=-\frac{3}{4}\left(-\frac{9}{2}a_0\right)+\frac{1}{12}a_1=\frac{27}{8}a_0+\frac{1}{12}a_1,
\end{aligned} \tag{27}$$

and so on. Finally, putting (27) into (21) gives

$$\begin{aligned}
y(x)&=a_0+a_1x-\frac{9}{2}a_0x^2+\left(\frac{1}{6}a_0-\frac{3}{2}a_1\right)x^3+\left(\frac{27}{8}a_0+\frac{1}{12}a_1\right)x^4+\cdots\\
&=a_0\left(1-\frac{9}{2}x^2+\frac{1}{6}x^3+\frac{27}{8}x^4+\cdots\right)+a_1\left(x-\frac{3}{2}x^3+\frac{1}{12}x^4+\cdots\right)\\
&=a_0y_1(x)+a_1y_2(x),
\end{aligned} \tag{28}$$

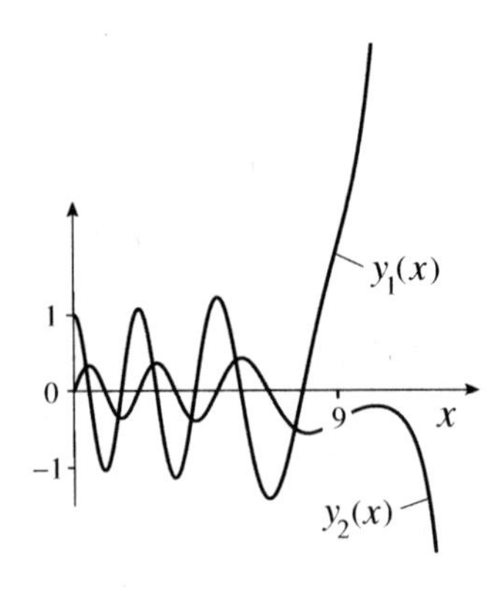

Figure 2. Graph of the LI solutions $y_1(x)$ and $y_2(x)$ of (20), given in (28). These plots are accurate: $y_1(x)$ and $y_2(x)$ really do rise and plummet sharply for $x>9$. See Exercise 5.

in which a_0 and a_1 are arbitrary; (28) is a general solution of (20). The LI solutions $y_1(x)$ and $y_2(x)$ are plotted in Fig. 2 for $0\le x<\infty$.

COMMENT 1. Since $p(x)=0$ and $q(x)=9-x$ are analytic for all x, we are assured by Theorem 6.3.1 that the series solutions $y_1(x)$ and $y_2(x)$ have radii of convergence $R=\infty$; they converge for all x.

COMMENT 2. Observe that if initial conditions are supplied at $x=0$, they are readily applied to (28) because the expansion is about that point. For instance, if the initial conditions are $y(0) = -2$ and $y'(0) = 7$, then applying them to (28) gives $y(0) = -2 = a_0(1) + a_1(0)$ and $y'(0) = 7 = a_0(0) + a_1(1)$, so $a_0 = -2$ and $a_1 = 7$. However, if the initial conditions are at a different point, their application is not as convenient. For instance, suppose the initial conditions are at $x = 4$, with $y(4) = 3$ and $y'(4) = 10$. Applying those conditions to (28) gives

$$y(4) = 3 = a_0 y_1(4) + a_1 y_2(4),$$
$$y'(4) = 10 = a_0 y_1'(4) + a_1 y_2'(4),$$

in which the numbers $y_1(4)$, $y_2(4)$, $y_1'(4)$, and $y_2'(4)$ are given by nonterminating series that must be summed. The moral is that *if initial conditions are specified at a point $x = a$, then expand about that point.* ∎

Closure. The focus of this section is Theorem 6.3.1, which assures us that if p and q are analytic at x_0, then two LI solutions (hence a general solution) of the equation $y'' + p(x)y' + q(x)y = 0$ can be found in the form of power series centered at x_0. If we put the assumed power series form into the differential equation and equate the coefficient of each power of $x - x_0$ to zero, we can solve for the a_n coefficients, recursively, in terms of two that remain arbitrary. Theorem 6.3.1 also assures us that the radii of convergence of those series solutions will be at least as large as the smaller of the radii of convergence of the Taylor series of p and q about x_0. In the important case in which p and q are rational functions, those radii of convergence are simply the distance from x_0 to their nearest singular points of p and q in the complex plane.

If initial conditions are given at a regular point $x = a$, then it is most convenient to choose the point of expansion, x_0, to be that point.

EXERCISES 6.3

1. Constant-Coefficient Equations. Solve for $y(x)$ by seeking power series solutions about $x_0 = 0$. Obtain the recursion formula and the terms through fourth order in each of two LI solutions. NOTE: Here, $p(x)$ and $q(x)$ are constants and hence analytic for all x, so your series solutions will have $R = \infty$; you need not verify that. Nor do you need to verify that your solution is equivalent to the general solution obtained, more simply for these constant-coefficient equations, by seeking solutions in the form $y(x) = e^{rx}$.

(a) $y'' - 9y = 0$ (b) $y'' + y' = 0$
(c) $y'' - 2y' = 0$ (d) $y'' - 3y' + 2y = 0$
(e) $y'' + 3y' - 4y = 0$ (f) $y'' + y' - 2y = 0$
(g) $y'' - 7y' + 12y = 0$ (h) $y'' - 2y' + 2y = 0$

2. Nonconstant-Coefficient Equations. Solve for $y(x)$ by seeking power series solutions about $x_0 = 0$. Obtain the recursion formula and the terms through fourth order in each of two LI solutions. Use Theorem 6.3.1 to determine the minimum radii of convergence of your series. HINT: $z^n = 1$ has n roots that are equally spaced around the unit circle, centered at the origin in the complex plane.

(a) $y'' + (1 + x)y = 0$ (b) $y'' + x^2 y = 0$
(c) $y'' - xy = 0$ (d) $y'' + xy' = 0$
(e) $(2 - x)y'' + y = 0$ (f) $(1 + x^2)y'' - y = 0$
(g) $(1 - x^8)y'' + 9y = 0$ (h) $(x^2 + 2x + 2)y'' + 2y = 0$

3. Least Possible Radius of Convergence. Use Theorem 6.3.1 to determine the least possible value of the radius of convergence of power series solutions about the specified point x_0. You need not find those series.

(a) $y'' - 3y' + 6y = 0;\quad x_0 = 6$
(b) $y'' - e^x y = 0;\quad x_0 = 0$

(c) $xy'' + y' + x^2y = 0; \quad x_0 = 5$
(d) $(x^2 - 1)y'' - (x+1)y' + (x-1)y = 0; \quad x_0 = 0.3$
(e) $(4x^2 + 1)y'' + 3xy' + y = 0; \quad x_0 = 5$
(f) $(x^2 + 2x + 5)y'' + (x^2 + 1)y = 0; \quad x_0 = 1$
(g) $(1-x^2)y'' - 2xy' + 6y = 0; \quad x_0 = 0$
(h) $xy'' + y' + y = 0; \quad x_0 = 4$
(i) $(x^2 + 4x)y'' + y' - y = 0; \quad x_0 = -3$
(j) $(x^3-8)y'' - 2y = 0; \quad x_0 = 5$

4. Least Possible Radius Exceeded. (a) Apply Theorem 6.3.1 to the equation

$$(1-3x+2x^2)y'' + 2xy' - 2y = 0 \tag{4.1}$$

and determine the least possible radius of convergence of power series solutions about the point $x = 0$.
(b) Although (4.1) looks difficult, we designed it to have the simple closed form solution

$$y(x) = C_1\frac{1}{x-1} + C_2\frac{(2x-1)^2}{x-1}. \tag{4.2}$$

By referring to (4.2), explain why *every* power series solution about $x = 0$ has a radius of convergence of at least 1, which is *greater* than the least possible radius given by Theorem 6.3.1.
(c) In fact, show that there are solutions, contained in (4.2) by suitable choice of C_1 and C_2, that even have an *infinite* radius of convergence.

5. Qualitative Reasoning. From a qualitative point of view, explain how the general features of the solutions shown in Fig. 2 could have been anticipated from the form of the differential equation (20).

6. Nonhomogeneous Equations. This section has dealt only with finding two LI solutions of the homogeneous equation $y'' + p(x)y' + q(x)y = 0$. For a nonhomogeneous equation

$$y'' + p(x)y' + q(x)y = f(x) \tag{6.1}$$

we also need a particular solution $y_p(x)$. Recall that the method of undetermined coefficients is specifically for constant-coefficient equations, but we did derive a formula, in Section 2.8, that holds even if the equation is not of constant-coefficient type. It gives $y_p(x)$ in terms of integrals of Wronskians of $f(x)$ and two LI homogeneous solutions $y_1(x)$ and $y_2(x)$. However, working out those integrals is unwieldy if $y_1(x)$ and $y_2(x)$ are in series form, and it is probably simpler to retain the $f(x)$ in (6.1) when implementing the power series solution method, assuming that $f(x)$ is analytic at the chosen expansion point. That will lead to a general solution $y(x) = C_1y_1(x) + C_2y_2(x) + y_p(x)$, not just to the $C_1y_1(x) + C_2y_2(x)$ part.

(a) As a simple example, derive a general solution of

$$y'' + y = 2e^x \tag{6.2}$$

by seeking y in the form

$$y(x) = \sum_0^\infty a_n(x-x_0)^n, \tag{6.3}$$

with $x_0 = 0$. Obtain terms through fifth order. HINT: Besides putting (6.3) into the left-hand side of (6.2) in the usual way, also expand the $2e^x$ about $x = 0$. Finally, equate coefficients of like powers of x on the left- and right-hand sides.
(b) Verify that your results agree (through fifth-order) with the closed-form solution obtained by the methods given in Chapter 2. HINT: Realize that particular solutions are not uniquely determined; if $y_p(x)$ is a particular solution of a given linear differential equation, and $y_1(x)$ and $y_2(x)$ are LI solutions of the homogeneous equation, then $y_p(x) + \alpha y_1(x) + \beta y_2(x)$ is also a particular solution, for any values of the constants α and β.

7 Nonhomogeneous Equations. Use the method described in Exercise 6 to derive a general solution. Expand about $x = 0$ and develop your solution through terms of fifth order.

(a) $y'' - y = 3\sin x$
(b) $y'' + y' = 4xe^x$
(c) $y'' + xy = e^x$
(d) $y'' - xy = x^2e^x$
(e) $y'' + xy' = 100$
(f) $y'' - xy' = 24x$

ADDITIONAL EXERCISES

8. Higher-Order Equations. We've limited our attention in this chapter to second-order equations, but the method can be applied formally to higher-order equations as well.

(a) To solve the third-order equation

$$y''' - 2y'' - y' + 2y = 0, \tag{8.1}$$

seek $y(x) = \sum_0^\infty a_nx^n$. Obtain the recursion formula and, from it,

$$\begin{aligned} y(x) &= a_0\left(1 - \frac{1}{3}x^3 - \frac{1}{6}x^4 - \frac{1}{12}x^5 - \cdots\right) \\ &\quad + a_1\left(x + \frac{1}{6}x^3 + \frac{1}{120}x^5 + \cdots\right) \\ &\quad + a_2\left(x^2 + \frac{2}{3}x^3 + \frac{5}{12}x^4 + \frac{1}{6}x^5 + \cdots\right) \\ &= a_0y_1(x) + a_1y_2(x) + a_2y_3(x), \end{aligned} \tag{8.2}$$

in which a_0, a_1, and a_2 are arbitrary constants.
(b) Alternatively, solve (8.1) by the method of assumed exponential form, as in Chapter 2, and derive the general solution

$$y(x) = C_1 e^{-x} + C_2 e^{x} + C_3 e^{2x}. \tag{8.3}$$

The equivalence of (8.2) and (8.3) is not so obvious because the three series in (8.2) are not the Taylor series of the e^{-x}, e^{x}, and e^{2x} in (8.3). Nevertheless, show that the solution forms (8.2) and (8.3) are indeed equivalent (through fifth-order terms). State your reasoning.
NOTE: Of course, to solve (8.1) it is much simpler to use the method of assumed exponential form, but the purpose of this example is only to illustrate the use of the power series method for higher-order equations.

9. Airy's Equation. The equation

$$\boxed{y'' + xy = 0 \quad (-\infty < x < \infty)} \tag{9.1}$$

is **Airy's equation**, after the British astronomer and mathematician *Sir George Airy* (1801–1892).

(a) By seeking a power series solution about $x = 0$, derive a general solution of (9.1), through terms of tenth order (by which point the pattern of the coefficients should become clear). What are the radii of convergence of your series solutions?

(b) Alternatively, use computer software to obtain graphs of two LI solutions of (9.1). Let the first satisfy the initial conditions $y(0) = 1, y'(0) = 0$, and let the second satisfy the initial conditions $y(0) = 0, y'(0) = 1$. Plot these solutions on $-5 \leq x \leq 20$, $-5 \leq y \leq 5$.

(c) Discuss the qualitative nature of the graphs obtained in part (b) in terms of the form of the differential equation (9.1), and explain why (9.1) is called a **change-of-type** differential equation. HINT: Think of the nature of the solutions of the simpler constant-coefficient equation $y'' + \kappa y = 0$, for negative and positive values of the constant κ.

6.4 LEGENDRE AND BESSEL EQUATIONS

6.4.1 Introduction. In this section we apply the series solution method to two important nonconstant-coefficient differential equations, the Legendre and Bessel differential equations, that arise in a variety of applications — for instance, in modeling the steady-state temperature distribution in spheres and cylinders, respectively.

This section is not a prerequisite for Section 6.5, nor is subsection 6.4.2 a prerequisite for subsection 6.4.3.

Our series solution of the Legendre equation will proceed as in Section 6.3, because we will expand about a regular point. But, for the Bessel equation it is better to expand about a *singular* point of the equation. Thus, for the latter we will proceed into new territory, without Theorem 6.3.1 to guarantee success. In fact, expansions about singular points will be covered in the next section, by the Frobenius theory, so our solution of the Bessel equation, here, will provide a first step in that direction.

6.4.2 Legendre's equation. The **Legendre equation** is

$$\boxed{(1-x^2)y'' - 2xy' + \lambda y = 0 \quad (-1 < x < 1),} \tag{1}$$

The constant λ is a parameter. Assign to it any value that you like.

in which λ is a constant, and is named after the French mathematician *Adrien-Marie Legendre* (1752–1833). Of most interest, in applications of (1), are expansions about $x = 0$. By Theorem 6.2.1, both $p(x) = -2x/(1-x^2)$ and $q(x) = \lambda/(1-x^2)$ are analytic at $x = 0$, and their Taylor series about that point converge in the interval under consideration, $-1 < x < 1$. Hence, Theorem 6.3.1 guarantees that two LI solutions of (1) can be found in the power series form

$$y(x) = \sum_{n=0}^{\infty} a_n x^n. \tag{2}$$

For definiteness, take $\lambda = 6$. Then, putting (2) into (1) gives

$$(1-x^2)\sum_{n=2}^{\infty} n(n-1)a_n x^{n-2} - 2x\sum_{n=1}^{\infty} na_n x^{n-1} + 6\sum_{n=0}^{\infty} a_n x^n = 0 \tag{3}$$

or

$$\sum_{n=2}^{\infty} n(n-1)a_n x^{n-2} - \sum_{n=2}^{\infty} n(n-1)a_n x^n - \sum_{n=1}^{\infty} 2na_n x^n + \sum_{n=0}^{\infty} 6a_n x^n = 0. \tag{4}$$

As usual, work on the exponents first: To get the same exponents, set the exponent $n-2$ in the first sum equal to k. That step gives

Ignore the underlining for a moment.

$$\sum_{k=0}^{\infty}(k+2)(k+1)a_{k+2}x^k - \sum_{n=2}^{\infty} \underline{n(n-1)}a_n x^n - \sum_{n=1}^{\infty} 2\underline{n}a_n x^n + \sum_{n=0}^{\infty} 6a_n x^n = 0. \tag{5}$$

Then change the k's to n's, to have the same dummy summation index in the four sums.

Next, the summation limits: We can change the lower limit from 1 to 0 in the third sum because the underlined n in that sum is zero at $n = 0$ anyhow, and we can change the lower limit from 2 to 0 in the second sum because the underlined $n(n-1)$ is zero at $n = 0$ and $n = 1$. Thus,

$$\sum_{n=0}^{\infty}(n+2)(n+1)a_{n+2}x^n - \sum_{n=0}^{\infty} n(n-1)a_n x^n - \sum_{n=0}^{\infty} 2na_n x^n + \sum_{n=0}^{\infty} 6a_n x^n = 0. \tag{6}$$

Now combine the sums as

$$\sum_{n=0}^{\infty}\left[(n+2)(n+1)a_{n+2} - (n^2+n-6)a_n\right]x^n = 0. \tag{7}$$

Thus, $(n+2)(n+1)a_{n+2} - (n^2+n-6)a_n = 0$ for each $n = 0, 1, 2, \dots$, so we infer from (7) the recursion formula

$$a_{n+2} = \frac{n(n+1)-6}{(n+2)(n+1)} a_n, \qquad n = 0, 1, 2, \dots. \tag{8}$$

The recursion formula.

If we set $n = 0, 1, 2, \dots$ in turn, (8) gives

$$\begin{aligned}
n = 0: &\quad a_2 = \frac{0-6}{2} a_0 = -3a_0, \\
n = 1: &\quad a_3 = \frac{2-6}{6} a_1 = -\frac{2}{3} a_1, \\
n = 2: &\quad a_4 = \frac{6-6}{12} a_2 = 0, \\
n = 3: &\quad a_5 = \frac{12-6}{20} a_3 = \frac{3}{10}\Big(-\frac{2}{3} a_1\Big) = -\frac{1}{5} a_1, \\
n = 4: &\quad a_6 = \frac{20-6}{30} a_4 = 0, \\
n = 5: &\quad a_7 = \frac{30-6}{42} a_5 = \frac{4}{7}\Big(-\frac{1}{5} a_1\Big) = -\frac{4}{35} a_1,
\end{aligned} \tag{9}$$

and so on, with a_0 and a_1 remaining arbitrary. Finally, put the results given in (9) into (2) and obtain

$$\begin{aligned}
y(x) &= a_0 + a_1 x - 3a_0 x^2 - \frac{2}{3} a_1 x^3 - \frac{1}{5} a_1 x^5 - \frac{4}{35} a_1 x^7 + \cdots \\
&= a_0\Big(1 - 3x^2\Big) + a_1\Big(x - \frac{2}{3}x^3 - \frac{1}{5}x^5 - \frac{4}{35}x^7 + \cdots\Big) \\
&= a_0 y_1(x) + a_1 y_2(x),
\end{aligned} \tag{10}$$

which is a general solution of the given differential equation.

Since p and q are singular only at $x = \pm 1$, we are assured by Theorem 6.3.1 that the series solutions converge at *least* in $-1 < x < 1$. Since the series solution $y_1(x)$ terminates (after the x^2 term), it converges for *all* x, but what about $y_2(x)$? If we apply the ratio test to successive terms in that series, we obtain

$$\begin{aligned}
\lim_{n\to\infty} \left| \frac{a_{n+2} x^{n+2}}{a_n x^n} \right| &= \Big(\lim_{n\to\infty} \left| \frac{a_{n+2}}{a_n} \right| \Big) |x|^2 = x^2 \\
&= \lim_{n\to\infty} \left| \frac{n(n+1)-6}{(n+2)(n+1)} \right| x^2 = x^2.
\end{aligned} \tag{11}$$

We used $a_{n+2}x^{n+2}$ in the numerator of (11) instead of $a_{n+1}x^{n+1}$ because only every other term is present in the $y_2(x)$ series. That is, we see from (10) that the *successive terms* in $y_2(x)$ are $a_1 x$, $a_3 x^3$, $a_5 x^5$, and so on.

Since the ratio test requires the latter to be less than 1 for convergence, it follows that the y_2 series converges in $x^2 < 1$, that is, in $-1 < x < 1$, and it diverges in $x^2 > 1$. Observe that even though we did not have a formula for the nth coefficient in the series, *the recursion formula gave us what we needed for the ratio test*, namely, the ratio a_{n+2}/a_n needed in (11).

Striking, about the solutions in (10), is that one of the series *terminates* and is a finite-degree polynomial. Does (1) admit such a solution for any value of the parameter λ, or was there something special about the value $\lambda = 6$? Let us reconsider our series solution for any λ, that is, without specifying λ. Again seek $y(x)$ in the form (2). We can omit the subsequent several steps because we need merely change the 6, in (8), to λ. Then, the recursion formula is this:

$$a_{n+2} = \frac{n(n+1)-\lambda}{(n+2)(n+1)} a_n, \qquad n = 0, 1, 2, \ldots \tag{12}$$

Letting $n = 0, 1, 2, \ldots$ gives $a_2 = -\lambda a_1/2$, $a_3 = (2-\lambda)a_1/6$, and so on, with a_0 and a_1 remaining arbitrary, and the result is the general solution

Or, if you prefer, $y(x) = C_1 y_1(x) + C_2 y_2(x)$. Of course, if $\lambda = 6$ then (13) reduces to (10).

$$\begin{aligned} y(x) &= a_0\left(1 - \frac{\lambda}{2}x^2 - \frac{(6-\lambda)\lambda}{24}x^4 - \frac{(20-\lambda)(6-\lambda)\lambda}{720}x^6 - \cdots\right) \\ &\quad + a_1\left(x + \frac{2-\lambda}{6}x^3 + \frac{(12-\lambda)(2-\lambda)}{120}x^5 + \cdots\right) \\ &= a_0 y_1(x) + a_1 y_2(x). \end{aligned} \tag{13}$$

As above, the ratio test shows that each of the two series in (13) converges in $|x| < 1$ and diverges in $|x| > 1$. At the endpoints $x = \pm 1$ the test gives no information, but it is known that both series diverge there (unless, by special choice of λ, one of them terminates, as we will see) and that their sums are unbounded as $x \to \pm 1$.[1]

We have not presented (1) in the context of a physical application, but *in typical applications unboundedness of the solution on* $-1 < x < 1$ *is unacceptable.* In those applications, λ is not specified in advance; it is a free parameter. The idea, then, is to choose

Bounded solutions of (1) exist only if $\lambda = N(N+1)$, and are the Legendre polynomials.

$$\lambda = N(N+1) \tag{14}$$

for any nonnegative integer N, because then one of the two series in (13) terminates. After all, the right-hand side of (12) becomes zero when n reaches N, so $a_n = 0$ for all $n \geq N$. A terminating series is only a polynomial, so of course it converges and is bounded on $-1 < x < 1$, as desired. Those polynomial solutions of the Legendre equation, with $\lambda = N(N+1)$ and $N = 0, 1, 2, \ldots$, are the **Legendre polynomials**. They are the *only* solutions of the Legendre equation (1) that are bounded on $-1 < x < 1$.

To find the first few Legendre polyonomials, write out (13), choosing λ according to (14) for $N = 0$, 1, 2, and so on.

[1]A function $f(x)$ being **bounded** on an interval I means that there exists a finite constant M such that $|f(x)| < M$ for all x in I.

$\lambda=0$: For $N=0$, (14) gives $\lambda=0$, and putting $\lambda=0$ in (13) gives

$$\begin{aligned} y(x) &= a_0(1) + a_1\left(x + \frac{1}{3}x^3 + \frac{1}{5}x^5 + \cdots\right) \\ &= a_0 y_1(x) + a_1 y_2(x). \end{aligned} \tag{15}$$

$\lambda=2$: For $N=1$, (14) gives $\lambda=2$, and then (13) gives

$$\begin{aligned} y(x) &= a_0\left(1 - x^2 - \frac{1}{3}x^4 - \frac{1}{5}x^6 - \cdots\right) + a_1 x \\ &= a_0 y_1(x) + a_1 y_2(x). \end{aligned} \tag{16}$$

$\lambda=6$: For $N=2$, (14) gives $\lambda=6$, and then (13) gives

$$\begin{aligned} y(x) &= a_0(1-3x^2) + a_1\left(x - \frac{2}{3}x^3 - \frac{1}{5}x^5 - \cdots\right) \\ &= a_0 y_1(x) + a_1 y_2(x), \end{aligned} \tag{17}$$

and so on.

Thus, for each of the values $\lambda = 0, 2, 6, \dots$, the Legendre equation (1) admits one polynomial solution and one infinite series solution that is unbounded on $-1 < x < 1$: for $\lambda=0$ the polynomial solution is the $a_0(1)$ term in (15), for $\lambda=2$ it is the $a_1 x$ term in (16), for $\lambda=6$ it is the $a_0(1-3x^2)$ term in (17), and so on. If we choose the a_0's and a_1's to scale those polynomial solutions to equal 1 at $x=1$, then we obtain the Legendre polynomial solutions $P_N(x)$, the first several of which are

There is no harm in scaling these polynomials to be 1 at $x=1$, since a_0 and a_1 are arbitrary anyhow.

$$P_0(x) = 1 \quad (\text{for } \lambda=0), \tag{18a}$$

$$P_1(x) = x \quad (\text{for } \lambda=2), \tag{18b}$$

$$P_2(x) = \frac{1}{2}(3x^2-1) \quad (\text{for } \lambda=6), \tag{18c}$$

$$P_3(x) = \frac{1}{2}(5x^3-3x) \quad (\text{for } \lambda=12), \tag{18d}$$

$$P_4(x) = \frac{1}{8}(35x^4-30x^2+3) \quad (\text{for } \lambda=20). \tag{18e}$$

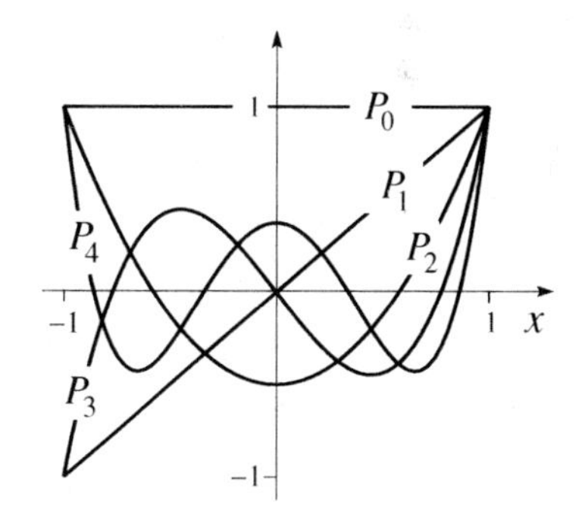

Figure 1. The first several Legendre polynomials.

Their graphs are given in Fig. 1. Realize that the Legendre polynomials are *not* LI solutions of the same equation; each one corresponds to a different value of λ in (1) and hence satisfies a different differential equation.

Useful properties of the Legendre polynomials are given in the exercises.

6.4.3 Bessel's equation.

The equation

$$\boxed{x^2y'' + xy' + (x^2-\nu^2)y = 0 \qquad (0 < x < \infty),} \tag{19}$$

in which ν is a given constant, is the **Bessel equation of order ν**, after the German astronomer *Friedrich Wilhelm Bessel* (1784–1846).[1] Just as the Legendre equation contains a parameter λ, the Bessel equation contains the parameter ν. The equation starts out looking like a Cauchy–Euler equation, due to the x^2y'' and xy' terms, but the x^2y term destroys the Cauchy–Euler pattern. Like the Legendre equation, the Bessel equation is nonelementary, and we will be content to obtain series solutions. It is among the most important linear second-order differential equations with nonconstant coefficients, because of its many applications — for instance, to the steady-state temperature in a cylinder, vibrating circular membranes, water waves in canals of variable cross section, acoustic waves, the buckling of columns, the oscillation of a pendulum of variable length and of a hanging rope or chain, the vibration of a beam of variable cross section, the current density in a wire carrying an alternating current, and the unsteady temperature distribution in a circular disk.

Bessel's equation of order zero. We will consider only the Bessel equation of order zero [i.e., with $\nu = 0$ in (19)],

$$\boxed{xy'' + y' + xy = 0 \qquad (0 < x < \infty).} \tag{20}$$

Divide (20) by x, to put it into the standard form $y'' + p(x)y' + q(x)y = 0$, and identify $p(x) = 1/x$ and $q(x) = 1$. Thus, the only singular point of (20) is $x = 0$ because $p(x) = 1/x$ is singular there. If we avoid $x = 0$ and seek the solution as a power series expansion about a regular point such as $x = 3$, then Theorem 6.3.1 guarantees that we will succeed in finding two LI solutions. However, the distance from $x = 3$ to the singular point at $x = 0$ is 3, so the theorem guarantees the convergence of the series solutions only in $|x - 3| < 3$, that is, in $0 < x < 6$. The solutions thus found are indeed valid only on $0 < x < 6$, which is but a small part of the interval of interest that was stated in (20) to be $0 < x < \infty$. Of course, if we expand about $x = 10$, say, the solution will be valid on $0 < x < 20$, which is broader.

Don't avoid the singular point $x = 0$; expand about it.

However, we can do much better if we expand *about the singular point itself*! That is, let us seek

$$y(x) = \sum_{n=0}^{\infty} a_n x^n, \tag{21}$$

realizing that we are "on our own" now because Theorem 6.3.1 does not cover the case of expansions about singular points.

Putting (21) into (20) gives

$$\underbrace{\sum_{n=2}^{\infty} n(n-1)a_n x^{n-1}}_{n-1=k} + \underbrace{\sum_{n=1}^{\infty} na_n x^{n-1}}_{n-1=k} + \underbrace{\sum_{n=0}^{\infty} a_n x^{n+1}}_{n+1=k} = 0. \tag{22}$$

[1]Although Bessel was the first to study this equation systematically, in connection with the motion of the planets, it also arose in the work of earlier mathematicians — such as John Bernoulli and his brother Daniel, as well as Leonhard Euler.

To obtain the same exponents on the x's, let $n-1=k$ in the first and second sums and let $n+1=k$ in the third. Then (22) becomes

First get the exponents the same, then the summation limits.

$$\sum_{k=1}^{\infty}(k+1)\underline{k}a_{k+1}x^k+\sum_{k=0}^{\infty}(k+1)a_{k+1}x^k+\sum_{k=1}^{\infty}a_{k-1}x^k=0. \tag{23}$$

Next, to obtain the same summation limits for the three sums, we can change the lower limit to 0 in the first because the underlined factor k is 0 at $k=0$ anyhow, and we can change the lower limit to 0 in the third sum if we define $a_{-1}\equiv 0$. Doing so, we can now combine the three sums and obtain

$$\sum_{k=0}^{\infty}\left[(k+1)^2a_{k+1}+a_{k-1}\right]x^k=0, \tag{24}$$

In (24), we've ended up with k's rather than n's. It doesn't matter; change the k's to n's if you like.

from which follows the recursion formula

$$(k+1)^2a_{k+1}+a_{k-1}=0 \qquad (k=0,1,2,\dots). \tag{25}$$

If we write out (25) for $k=0,1,2,\dots$, in turn, remembering that $a_{-1}=0$, we find that $a_1=a_3=a_5=\cdots=0$ and that we can solve for $a_2,a_4,a_6,\dots$ in terms of a_0, with a_0 remaining arbitrary. Putting those results into (21) gives

$$\begin{aligned} y(x)&=a_0\left(1-\frac{1}{2^2}x^2+\frac{1}{4^22^2}x^4-\frac{1}{6^24^22^2}x^6+\cdots\right)\\ &=a_0\sum_{n=0}^{\infty}\frac{(-1)^n}{(n!)^2}\left(\frac{x}{2}\right)^{2n}\equiv a_0y_1(x). \end{aligned}\tag{26}$$

The solution $y_1(x)$ that we have found is called $J_0(x)$, the **Bessel function of the first kind and of order zero.** The latter is plotted in Fig. 2 and looks a bit like a damped cosine function.

Remember that the order of the Bessel equation is the ν in equation (1), which we've taken to be zero.

The good news is that the series in (26) converges with an *infinite* radius of convergence (proof of which is left for the exercises), on $-\infty<x<\infty$, and hence *on the entire interval of interest.*

The bad news is that the expansion (21) about the singular point $x=0$ led to just one solution, the Bessel function

$$\boxed{J_0(x)=1-\frac{1}{2^2}x^2+\frac{1}{4^22^2}x^4-\cdots=\sum_{n=0}^{\infty}\frac{(-1)^n}{(n!)^2}\left(\frac{x}{2}\right)^{2n}.} \tag{27}$$

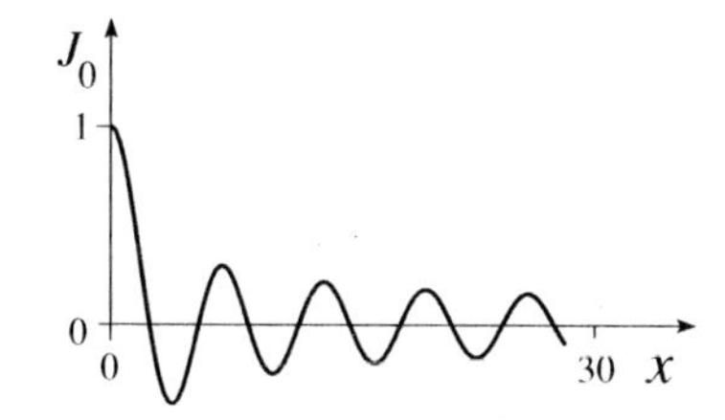

Figure 2. Bessel function of the first kind and order zero.

Evidently, the missing second LI solution to (20), say $y_2(x)$, is not analytic at $x=0$ or else we would have found it when we sought $y(x)$ in the power series form (21). How can we find $y_2(x)$? Recall from Section 2.10 that if $y_1(x)$ is a solution of

$y'' + p(x)y' + q(x)y = 0$ [other than the trivial solution $y(x) = 0$], then a second LI solution is given by the *reduction-of-order formula*

This application illustrates the importance of the reduction-of-order formula.

$$y_2(x) = y_1(x) \int \frac{e^{-\int p(x)\,dx}}{[y_1(x)]^2}\,dx. \tag{28}$$

With $p(x) = 1/x$ and $y_1(x) = J_0(x)$, (28) gives

In (29) we used the steps $e^{-\ln x} = 1/e^{\ln x} = 1/x$.

$$y_2(x) = J_0(x) \int \frac{e^{-\ln x}}{[J_0(x)]^2}\,dx = J_0(x) \int \frac{dx}{x\,[J_0(x)]^2}. \tag{29}$$

The latter integral looks difficult, and it is. But realize that we don't need to evaluate it in closed form; we will accept a series expression for $y_2(x)$ just as (26) is a series expression for $y_1(x)$. To obtain such a series, expand the $1/[J_0(x)]^2$ in the integrand in a power series about $x = 0$. That is, expand

$$\frac{1}{J_0^2(x)} = \sum_{n=0}^{\infty} b_n x^n, \tag{30}$$

in which we must now determine the b_n's. To do that, re-express (30) as

See Exercise 11 for an alternative way to evaluate the b_n's.

$$\begin{aligned}
1 &= \Big(\sum_{n=0}^{\infty} b_n x^n\Big) J_0^2(x) \\
&= (b_0 + b_1 x + b_2 x^2 + \cdots)\left(1 - \frac{1}{2^2}x^2 + \frac{1}{4^2 2^2}x^4 - \cdots\right)^2 \\
&= b_0 + b_1 x + \left(b_2 - \frac{1}{2}b_0\right)x^2 + \left(b_3 - \frac{1}{2}b_1\right)x^3 + \left(b_4 - \frac{1}{2}b_2 + \frac{3}{32}b_0\right)x^4 \\
&\quad + \left(b_5 - \frac{1}{2}b_3 + \frac{3}{32}b_1\right)x^5 + \left(b_6 - \frac{1}{2}b_4 + \frac{3}{32}b_2 - \frac{5}{5184}b_0\right)x^6 + \cdots.
\end{aligned} \tag{31}$$

By the identity principle, equate coefficients of like powers of x on both sides of (31) and obtain $b_0 = 1$, $b_2 = 1/2$, $b_4 = 5/32$, $b_6 = 23/576, \ldots$, and $b_1 = b_3 = b_5 = \cdots = 0$. With these b_n's, put (30) into (29) and, formally integrating termwise, obtain

The terms $x^2/4 - 3x^4/128 + \cdots$, at the end of (32), were obtained by multiplying J_0's series and the series $x^2/4 + 5x^4/128 + \cdots$.

$$\begin{aligned}
y_2(x) &= J_0(x) \int \left(\frac{1}{x} + \frac{1}{2}x + \frac{5}{32}x^3 + \frac{23}{576}x^5 + \cdots\right)dx \\
&= J_0(x) \left[\ln x + \left(\frac{1}{4}x^2 + \frac{5}{128}x^4 + \frac{23}{3456}x^6 + \cdots\right)\right] \\
&= J_0(x)\ln x + \frac{1}{4}x^2 - \frac{3}{128}x^4 + \frac{11}{13824}x^6 + \cdots,
\end{aligned} \tag{32}$$

which is called $\boldsymbol{Y}_0(x)$, the **Neumann function of order zero**. Note that (32) is not a power series, because of the presence of the $\ln x$ term. That is why this solution eluded us when we used the power series form (21).

We have found the two LI solutions $y_1(x) = J_0(x)$ and $y_2(x) = \boldsymbol{Y}_0(x)$, so we can use them to form a general solution of (2). However, it is customary (for reasons to be explained below) to use, in place of $\boldsymbol{Y}_0(x)$, a linear combination of $J_0(x)$ and $\boldsymbol{Y}_0(x)$ for $y_2(x)$, namely,

$$y_2(x) = \left(\frac{2}{\pi}\right) \boldsymbol{Y}_0(x) + \left(\frac{2(\gamma - \ln 2)}{\pi}\right) J_0(x) \equiv Y_0(x), \tag{33}$$

The reason for preferring $Y_0(x)$ over $\boldsymbol{Y}_0(x)$, as a second LI solution, to go with $J_0(x)$, will be explained below.

where

$$\boxed{\begin{aligned} Y_0(x) &= \frac{2}{\pi}\left[\left(\ln\frac{x}{2}+\gamma\right) J_0(x) + \left(\frac{x}{2}\right)^2 - \frac{1+\frac{1}{2}}{(2!)^2}\left(\frac{x}{2}\right)^4 \right. \\ &\qquad\qquad \left. + \frac{1+\frac{1}{2}+\frac{1}{3}}{(3!)^2}\left(\frac{x}{2}\right)^6 - \cdots\right] \\ &= \frac{2}{\pi}\left[\left(\ln\frac{x}{2}+\gamma\right) J_0(x) \right. \\ &\qquad\qquad \left. + \sum_{n=1}^{\infty} \frac{(-1)^{n+1}\left(1+\frac{1}{2}+\cdots+\frac{1}{n}\right)}{(n!)^2}\left(\frac{x}{2}\right)^{2n}\right] \end{aligned}} \tag{34}$$

is the **Bessel function of the second kind, of order zero** and γ is **Euler's constant.**[1] The graphs of $J_0(x)$ and $Y_0(x)$ are given in Fig. 3. It can be shown that the series in (34), like the one in (27), converges for all x.

Important features of $J_0(x)$ and $Y_0(x)$ are that they look a bit like damped cosine and sine functions; for instance, $J_0(0) = 1$ and $J_0'(0) = 0$, like the cosine function, and the zeros of J_0 and Y_0 interlace, as do the zeros of the cosine and sine. However, instead of being zero at the origin like the sine, $Y_0(x) \to -\infty$ as $x \to 0$ because of the $\ln x$ in (34); see Fig. 3.

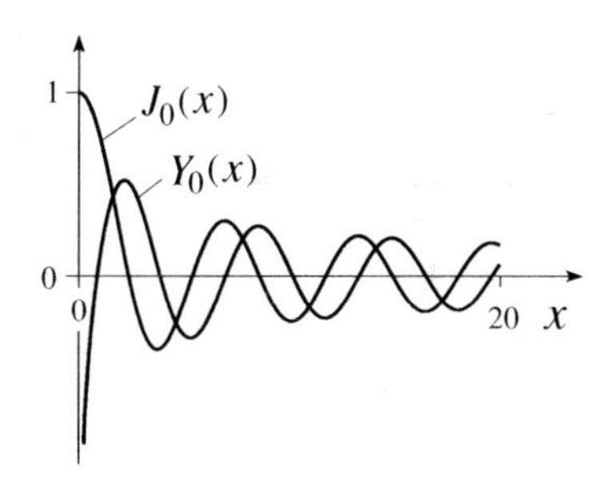

Figure 3. J_0 and Y_0. $Y_0(x)$ tends to $-\infty$ as $x \to 0$.

The asymptotic behavior of $J_0(x)$ and $Y_0(x)$, as $x \to 0$, follows from (27) and (34) as[2]

$$J_0(x) \sim 1, \tag{35a}$$

$$Y_0(x) \sim \frac{2}{\pi} \ln x. \tag{35b}$$

Their asymptotic behavior as $x \to \infty$ cannot be seen from (27) and (34), but it can be shown by other means that

$$J_0(x) \sim \sqrt{\frac{2}{\pi x}} \cos\left(x - \frac{\pi}{4}\right), \tag{36a}$$

$$Y_0(x) \sim \sqrt{\frac{2}{\pi x}} \sin\left(x - \frac{\pi}{4}\right) \tag{36b}$$

[1] To seven decimal places the Euler (or Euler–Mascheroni) constant is $\gamma = 0.5772157$.

[2] We wrote $\ln x$ rather than $\ln(x/2)$ in (35b) because $\ln(x/2) = \ln x - \ln 2 \sim \ln x$ as $x \to 0$.

as $x \to \infty$.

From (36) we finally see why the Bessel function Y_0 ("lightface Y_0") makes a nicer companion for J_0 than Neumann's Bessel function $\boldsymbol{Y}_0$ ("boldface Y_0"), because (33) and (36) imply that

$$\boldsymbol{Y}_0(x) = \frac{\pi}{2} Y_0(x) - (\gamma - \ln 2)\, J_0(x)$$

$$\sim \sqrt{\frac{2}{\pi x}} \left[\frac{\pi}{2} \sin\left(x - \frac{\pi}{4}\right) - (\gamma - \ln 2) \cos\left(x - \frac{\pi}{4}\right) \right] \tag{37}$$

as $x \to \infty$. Surely, (36a) and (36b) make a nicer pair than (36a) and (37)![1]

It might appear, from Fig. 3, that the *zeros* of J_0 and Y_0 [i.e., the roots of $J_0(x) = 0$ and $Y_0(x) = 0$] are equally spaced, as are the zeros of the cosine and sine functions, but they are not. However, (36a) and (36b) show that they *tend* to an equal spacing as $x \to \infty$, because the zeros of $J_0(x)$ tend to the (equally spaced) zeros of $\cos(x - \pi/4)$.

In summary, we have found the general solution

$$\boxed{y(x) = AJ_0(x) + BY_0(x)} \tag{38}$$

of the Bessel equation of order zero, $xy'' + y' + xy = 0$, in which $J_0(x)$ is the Bessel function of order zero of the first kind, defined by (27), and $Y_0(x)$ is the Bessel function of order zero of the second kind, defined by (34).

Closure. The Legendre and Bessel equations are prominent nonconstant-coefficient second-order differential equations. They are nonelementary, and we resorted to obtaining two LI solutions for each, using series solution methods.

The Legendre equation has singular points at $x = \pm 1$. We sought power series solutions about the *regular* point $x = 0$, and hence we obtained series that converge on the interval of interest, $-1 < x < 1$. However, in the applications in which the Legendre equation arises one is only after solutions that are bounded, and such solutions occur only if the parameter λ equals $N(N+1)$ for any nonnegative integer N. For each such λ, one of the two series LI solutions terminates and is bounded because it is only a finite-degree polynomial. Those are well known as the Legendre polynomials.

For the Bessel equation, on $0 < x < \infty$, the equation has only one singular point, at $x = 0$. *If we avoid it, and expand about an* $x_0 > 0$, *then the solutions obtained converge only on* $0 < x < 2x_0$ *rather than on the whole interval of interest,* $0 < x < \infty$. Thus, instead of expanding about such an x_0, we expanded

The first important reason for expanding about the singular point $x = 0$, instead of avoiding it.

[1]If this point is not clear, consider the following simpler example: $y'' + y = 0$ gives solutions $\cos x, \sin x, e^{ix}, e^{-ix}, 6\cos x - 5e^{ix}$, and so on. If we choose $y_1(x) = \cos x$, then we could choose $y_2(x)$ to be *any* of the solutions $\sin x$, e^{ix}, e^{-ix}, $6\cos x - 5e^{ix}$, and so on, but the most *natural* companion for $y_1(x) = \cos x$ is $y_2(x) = \sin x$: $\cos x$ and $\sin x$ form a "handsome couple," but $\cos x$ and $6\cos x - 5e^{ix}$, for instance, do not.

about the *singular* point $x = 0$, even though Theorem 6.3.1 would not be applicable and therefore would not guarantee success. The method did fall short in that it did not lead to two LI solutions, but it did lead to one. A second LI solution, not of power series form, was found using the first solution and the reduction-of-order formula (28). The second solution, $Y_0(x)$, contains a $\ln x$ term and, consequently, is not analytic at $x = 0$; that is why we did not find it when we sought power series solutions about $x = 0$. Note that *only by expanding about the singular point itself did the important singular* $\ln x$ *behavior come into explicit view.*

The second important reason for expanding about the singular point $x = 0$.

For more about Bessel and Legendre functions, see Francis B. Hildebrand's *Advanced Calculus*, 2nd ed. (Englewood Cliffs, NJ: Prentice Hall, 1976), which is more advanced but very readable.

EXERCISES 6.4

LEGENDRE'S EQUATION

1. Rodrigues' Formula. The Legendre polynomials $P_n(x)$ are given explicitly by the formula

$$P_n(x) = \frac{1}{2^n n!}\frac{d^n}{dx^n}\left[(x^2-1)^n\right], \tag{1.1}$$

which is **Rodrigues' formula.** You need not derive (1.1), just use it to generate $P_0(x)$, $P_1(x)$, and $P_2(x)$.

2. Laplace's Integral Form. Besides the derivative expression (1.1), above, there is also the following integral formula for the Legendre polynomials,

$$P_n(x) = \frac{1}{\pi}\int_0^\pi \left(x + \sqrt{x^2-1}\,\cos t\right)^n dt, \tag{2.1}$$

for $n = 0, 1, 2, \ldots$, which is **Laplace's integral form** for $P_n(x)$. You need not derive (2.1). Just verify it by working out the right-hand side for $n = 0$, 1, and 2, and comparing your results with (18a,b,c).

3. Generating Function. If we regard $(1 - 2xr + r^2)^{-1/2}$ as a function of r, with x fixed, and expand it in a Taylor series about $r = 0$ then the coefficients will of course be functions of x. It turns out that those coefficients are the Legendre polynomials $P_n(x)$:

$$\frac{1}{\sqrt{1-2xr+r^2}} = \sum_{n=0}^{\infty} P_n(x)r^n \quad (|x| \le 1,\ |r| < 1). \tag{3.1}$$

Thus, $\left(1 - 2xr + r^2\right)^{-1/2}$ is called the **generating function** for the P_n's; (3.1) is the source of much information about the Legendre polynomials.

(a) Verify (3.1). Specifically, expand the left-hand side in a Taylor series in r, about $r = 0$, through r^2, say, and verify that the coefficients of r^0, r^1, r^2, are indeed $P_0(x)$, $P_1(x)$, $P_2(x)$, respectively.
(b) By changing x to $-x$ and r to $-r$ on both sides of (3.1), show that

$$P_n(-x) = (-1)^n P_n(x), \tag{3.2}$$

so that if the integer n is even, then $P_n(-x) = P_n(x)$ and the graph of $P_n(x)$ is symmetric about $x = 0$, whereas if n is odd then $P_n(-x) = -P_n(x)$ and the graph of $P_n(x)$ is antisymmetric about $x = 0$, as seen in Fig. 1 through $n = 4$.

4. Legendre Polynomial Recursion Formula. (a) Derive the formula

$$(n+1)P_{n+1}(x) = (2n+1)xP_n(x) - nP_{n-1}(x) \tag{4.1}$$

for $n = 1, 2, \ldots$. HINT: Take $\partial/\partial r$ of (3.1), above, multiply both sides by $1 - 2xr + r^2$, then equate coefficients of like powers of r on both sides.
(b) To illustrate the use of (4.1), use it, together with the P_n's listed in (18), to obtain $P_5(x)$.

5. Orthogonality Relation. A useful property of Legendre polynomials is the **orthogonality relation**

$$\boxed{\int_{-1}^{1} P_j(x)P_k(x)\,dx = 0 \quad \text{if } j \neq k.} \tag{5.1}$$

(a) Prove (5.1). HINT: First, note that the Legendre equation

$$(1-x^2)\,y'' - 2xy' + N(N+1)y = 0 \tag{5.2}$$

can be expressed as

$$[(1-x^2)y']' + N(N+1)y = 0. \tag{5.3}$$

Let $P_j(x)$ and $P_k(x)$ be Legendre polynomial solutions of (5.3) corresponding to $N = j$ and $N = k$, respectively, so

$$[(1-x^2)P_j']' + j(j+1)P_j = 0, \tag{5.4a}$$

$$[(1-x^2)P_k']' + k(k+1)P_k = 0. \tag{5.4b}$$

Multiply (5.4a) by $P_k(x)$, (5.4b) by $P_j(x)$, subtract, integrate from -1 to 1, and obtain

$$\int_{-1}^{1} P_k[(1-x^2)P_j{}']'\,dx - \int_{-1}^{1} P_j[(1-x^2)P_k{}']'dx$$
$$= [k(k+1) - j(j+1)] \int_{-1}^{1} P_j P_k\,dx. \tag{5.5}$$

Integrate the terms on the left by parts and show that the left-hand side of (5.5) is zero. Finally, (5.1) will follow because $j \neq k$, by assumption.
(b) Check (5.1) by working out the integral for these representative cases: $j = 1$ and $k = 2$; $j = 1$ and $k = 3$.

6. (a) The orthogonality formula (5.1), above, is for the case where $j \neq k$. For the case where $j = k$, derive the companion formula

$$\boxed{\int_{-1}^{1} [P_n(x)]^2\,dx = \frac{2}{2n+1}} \tag{6.1}$$

for $n = 0, 1, 2, \ldots$. HINT: Square both sides of (3.1), integrate, and obtain

$$\int_{-1}^{1} \frac{dx}{1-2rx+r^2} = \int_{-1}^{1} \sum_{m=0}^{\infty} r^m P_m(x) \sum_{n=0}^{\infty} r^n P_n(x)\,dx. \tag{6.2}$$

Then, integrate the left side, use the orthogonality relation (5.1) to simplify the right side, and obtain

$$\frac{1}{r}\ln\left(\frac{1+r}{1-r}\right) = \sum_{n=0}^{\infty}\left\{\int_{-1}^{1} [P_n(x)]^2\,dx\right\} r^{2n}. \tag{6.3}$$

Finally, expand the left-hand side in a Taylor series in r, and show that (6.1) follows.
(b) Check (6.1), by working out the integral, for the cases $n = 0$, 1, and 2.

7. Electric Field of Two Electric Charges. Consider a positive charge Q and an equal negative charge $-Q$, a distance $2a$ apart, as shown below. The *electric potential* (i.e., the *voltage*) induced by the pair at any point P is

$$V = \kappa\left(\frac{Q}{\rho_+} - \frac{Q}{\rho_-}\right), \tag{7.1}$$

where $\kappa = 1/4\pi\epsilon_0$ and ϵ_0 is the permittivity of free space, a known constant.

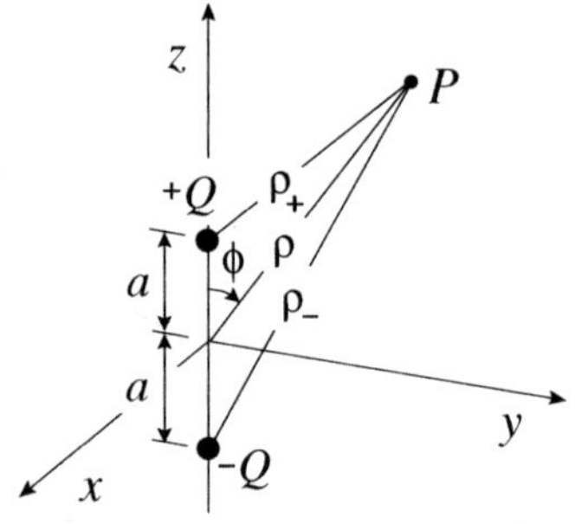

(a) Show from (7.1) that

$$V(\rho,\phi) = \kappa Q\left(\frac{1}{\sqrt{a^2+\rho^2-2a\rho\cos\phi}} - \frac{1}{\sqrt{a^2+\rho^2+2a\rho\cos\phi}}\right). \tag{7.2}$$

(b) Use (3.1), above, to show that

$$V(\rho,\phi) = \frac{2\kappa Q}{\rho}\sum_{n=1,3,\ldots}^{\infty}\left(\frac{a}{\rho}\right)^n P_n(\cos\phi) \tag{7.3}$$

for $\rho > a$.
(c) We can see, from (7.3) or (7.2), that if $a \to 0$ then $V \to 0$; the two charges just cancel each other. However, show that if, as $a \to 0$, we also let $Q \to \infty$, such that Qa is held constant, then in that limit we obtain

$$V(\rho,\phi) = \frac{\mu}{4\pi\epsilon_0}\frac{\cos\phi}{\rho^2}, \tag{7.4}$$

in which $\mu = 2Qa$. The result is called an **electric dipole**, and μ is called the **dipole moment**, which is the "strength" of the dipole.

BESSEL'S EQUATION

8.. Verify, by the ratio test, that the series for $J_0(x)$, given in (27), converges on $-\infty < x < \infty$, as we claimed.

9. From Fig. 2 we see that $J_0(x)$'s oscillation damps out rapidly when x is small, but slowly when x is large. Explain that behavior, qualitatively, from the differential equation (20) itself.

10. Zeros of $J_0(x)$. It was stated, below (37), that the zeros of $J_0(x)$ and $Y_0(x)$ are not equally spaced, as they are for the cosine and sine functions.

(a) Show, from (36), that the spacing between successive zeros, both for $J_0(x)$ and $Y_0(x)$, tends to π as $x \to \infty$.
(b) Using computer software, determine the first five zeros of $J_0(x)$, denote them as z_1 through z_5, compute their differences, and confirm that their differences do appear to be approaching π. NOTE: As a check, you should obtain $z_1 = 2.4048$, $z_2 = 5.5201$, and $z_3 = 8.6537$ for the first three zeros, with differences $z_2 - z_1 = 3.1153$ and $z_3 - z_2 = 3.1336$.

11. Alternative Evaluation of the b_n's. We evaluated the b_n's in (30) by matching coefficients of like powers of x on both sides of (31). Alternatively, since the right-hand side of (30) is the *Taylor series* of $1/J_0^2(x)$ about $x = 0$, we can compute the b_n's from the formula

$$b_n = \frac{1}{n!}\frac{d^n}{dx^n}\left[\frac{1}{J_0^2(x)}\right]\Bigg|_{x=0}. \tag{11.1}$$

Use (11.1), with the help of (27) and chain differentiation, to evaluate the first few b_n's: b_0, b_1, and b_2.

12. Generating Function for $J_0(x)$. If we regard the function $\exp\left[\frac{x}{2}\left(r - \frac{1}{r}\right)\right]$ as a function of r, with x fixed, and expand it in a Taylor series in r, then the coefficients will of course be functions of x. It turns out that the coefficient of the r^0 is $J_0(x)$; hence, $\exp[(x/2)(r-1/r)]$ is called the **generating function** for $J_0(x)$. Verify that claim. HINT: Use the Taylor series to expand the $e^{xr/2}$ in powers of $xr/2$ and the $e^{-x/2r}$ in powers of $x/2r$, multiply those series together, and pick out the r^0 terms.

13. Change of Variables May Give a Bessel Equation. Some equations are not themselves Bessel equations but can be converted to Bessel equations (and hence solved in terms of Bessel functions) by a change of variables. For instance, consider

$$xy'' + y' + \kappa^2 xy = 0 \tag{13.1}$$

which would be the Bessel equation of order 0 if not for the κ^2. [The exponent 2 on κ is included for convenience because we will be taking the square root of the κ^2 in a moment.] Try making a simple change of variables $x = \alpha t$, with the constant α chosen to eliminate the κ^2. If we denote $y(x) = y(\alpha t) \equiv u(t)$, say, show that the new equation is

$$tu''(t) + u'(t) + \kappa^2\alpha^2 tu(t) = 0. \tag{13.2}$$

Now choose α so that (13.2) is a Bessel equation of order zero and thus obtain the general solution

$$y(x) = AJ_0(\kappa x) + BY_0(\kappa x) \tag{13.3}$$

of (13.1).

14. Integral Form for $J_0(x)$. Besides the infinite series representation of $J_0(x)$ given by (27), there is also the following "integral representation"

$$J_0(x) = \frac{1}{\pi}\int_0^\pi \cos(x\sin\theta)\,d\theta. \tag{14.1}$$

Verify (14.1) by showing that it agrees with (27). You need only go as far as the x^8 term. HINT: Use the Taylor series of the cosine function to expand the $\cos(x\sin\theta)$ integrand in powers of its argument $x\sin\theta$ and then integrate term by term. Integrals of the form $\int_0^\pi \sin^m x\,dx$ will arise and you can evaluate them either by computer software or using integral tables.

ADDITIONAL EXERCISES

15. Vibrations of a Circular Drum. In applications, Bessel's equation is more likely to arise indirectly than directly, particularly in the process of solving partial differential equations. To illustrate, consider the vibration of a drumhead such as a musical drum of radius a that is stretched with a uniform tension per unit length τ, and that has a uniform density σ mass per unit area. Let its deflection normal to its undisturbed plane be axisymmetric (i.e., a function of the polar radius r but not of the polar angle θ), and denote it as $w(r,t)$ where t is the time. The motion $w(r,t)$ must satisfy the *partial differential equation*

$$\frac{\tau}{\sigma}\left(\frac{\partial^2 w}{\partial r^2} + \frac{1}{r}\frac{\partial w}{\partial r}\right) = \frac{\partial^2 w}{\partial t^2}$$

or, with $c^2 \equiv \tau/\sigma$ and using subscripts for partial derivatives,

$$c^2\left(w_{rr} + \frac{1}{r}w_r\right) = w_{tt}. \tag{15.1}$$

(a) Let us seek possible vibrational motions of the form

$$w(r,t) = R(r)\sin\omega t, \tag{15.2}$$

in which the "shape factor" or "mode shape" $R(r)$ and the vibrational frequency ω are to be determined. Put (15.2) into (15.1) and show that $R(r)$ must satisfy the ordinary differential equation

$$R'' + \frac{1}{r}R' + \left(\frac{\omega}{c}\right)^2 R = 0. \tag{15.3}$$

(b) The latter is almost but not quite a Bessel equation, and its solution was discussed in Exercise 13. Obtain a general solution for $R(r)$ and apply these two boundary conditions: $R(0)$ must be finite, and $R(a) = 0$ since the drumhead is attached to the drum along its perimeter. The first condition should lead you to set equal to zero the coefficient of the unbounded solution, and the second should give the condition

$$J_0\left(\frac{\omega}{c}a\right) = 0, \tag{15.4}$$

with which to determine the allowable vibrational frequencies ω. With the help of the zeros given in Exercise 10, show that the first few possible vibrational modes are, give or take constant scale factors,

$$w(r,t) = J_0\!\left(2.40\frac{r}{a}\right)\sin\left(\frac{2.40c}{a}t\right),$$
$$J_0\!\left(5.52\frac{r}{a}\right)\sin\left(\frac{5.52c}{a}t\right),\ J_0\!\left(8.65\frac{r}{a}\right)\sin\left(\frac{8.65c}{a}t\right),\ \ldots. \tag{15.5}$$

NOTE: One would adjust the tension τ so that the lowest frequency $2.40c/a$ is the frequency of the desired musical note.
(c) Give a hand sketch or computer plot of the first three mode shapes, $J_0\!\left(2.40\frac{r}{a}\right)$, $J_0\!\left(5.52\frac{r}{a}\right)$, and $J_0\!\left(8.65\frac{r}{a}\right)$.

6.5 THE METHOD OF FROBENIUS

Section 6.4 is not a prerequisite for this section.

6.5.1 Motivation. We know from Section 6.3 that we can generate two LI solutions of a given equation

$$y'' + p(x)y' + q(x)y = 0 \tag{1}$$

by seeking $y(x)$ in the form of a power series about any regular point x_0 of the equation. Nevertheless, such solutions may be less than satisfactory. For instance, consider

$$x^2y'' - (x^2 - 2x + 2)y = 0, \quad (0 < x < \infty) \tag{2}$$

in which $p(x) = 0$ and $q(x) = -(x^2 - 2x + 2)/x^2$ are regular for all x except $x = 0$. Thus, staying away from $x = 0$, we can find two LI solutions as power series about $x_0 = 2$, say. If we do that, and impose initial conditions, for instance $y(2) = 2$ and $y'(2) = 1$, we obtain

$$y(x) = 5 + (x-2) + \frac{1}{2}(x-2)^2 + \frac{1}{12}(x-2)^3 + \cdots. \tag{3}$$

In fact, we've designed (2), and the initial conditions, to have the

$$y(x) = 4\frac{e^{x-2}}{x}, \tag{4}$$

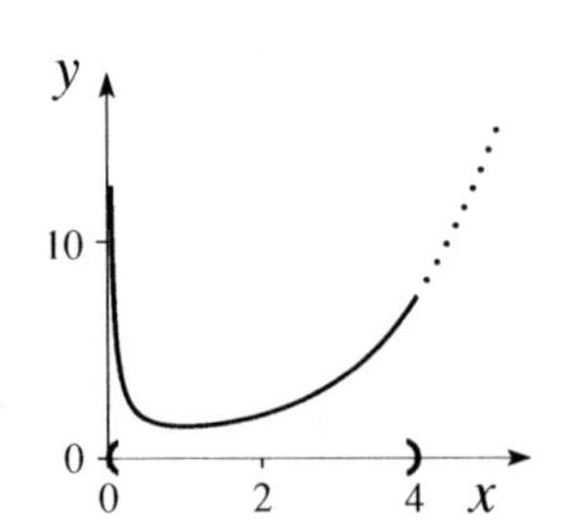

Figure 1. The solution (4) exists on $0 < x < \infty$, but is represented by (3) only on $0 < x < 4$.

plotted in Fig. 1, and (3) is indeed the Taylor series of (4) about $x = 2$. However, (3) has these serious shortcomings: First, it converges only on $0 < x < 4$, because of the singular point of (4) at $x = 0$, so it represents the solution (4) only there, giving *no information* for $x \geq 4$, even though the solution (4) exists on $0 < x < \infty$. Second, whereas a key feature of (4) is its singular behavior $y(x) \sim 4e^{-2}/x$ as $x \to 0$, that behavior cannot be discerned directly from (3). It's true that if we plot the series (3), its computed sum rises sharply as $x \to 0$, but we cannot tell, from the graph, that $y(x) \sim 4e^{-2}/x$ there.

The Frobenius method, due to the German mathematician *F. Georg Frobenius* (1849–1917) overcomes these shortcomings. To present the method, we first distinguish two kinds of singular points of a differential equation (1).

6.5.2 Regular and irregular singular points.

Recall that x_0 is a "singular point of (1)" if $p(x)$ and/or $q(x)$ are singular (i.e., not analytic) at x_0. Now we distinguish singular points as regular or irregular:

DEFINITION 6.5.1 *Regular and Irregular Singular Points of (1)*
Let x_0 be a singular point of (1). It is
(a) a **regular singular point** of (1) if $(x - x_0)p(x)$ and $(x - x_0)^2 q(x)$ are analytic at x_0,
(b) an **irregular singular point** of (1) if it is not a regular singular point.

---

x_0 is a real number, a point on the x axis.

EXAMPLE 1. Classify any singular points of (2). Its only singular point is at the origin, so $x_0 = 0$. According to Definition 6.5.1, examine the products

$$(x - x_0)p(x) = xp(x) = 0, \tag{5a}$$

$$(x - x_0)^2 q(x) = x^2 q(x) = -x^2 + 2x - 2. \tag{5b}$$

Both are analytic at $x = 0$, so the singular point of (2) at $x = 0$ is a regular singular point. The multiplication of $q(x) = -(x^2 + 2x - 2)/x^2$ by x^2 removed q's singularity at $x = 0$. ∎

EXAMPLE 2. Classify any singular points of

$$y'' + \frac{x+2}{x(x-1)}y' + \frac{5}{(x-1)^3}y = 0. \tag{6}$$

The only singular points of (6) are $x = 0$ and $x = 1$. For $x = 0$,

$$(x - x_0)p(x) = (x - 0)p(x) = x\frac{x+2}{x(x-1)} = \frac{x+2}{x-1},$$

$$(x - x_0)^2 q(x) = (x - 0)^2 q(x) = x^2 \frac{5}{(x-1)^3} = \frac{5x^2}{(x-1)^3},$$

both of which are analytic at $x = 0$, so $x_0 = 0$ is a *regular* singular point of (6). (That they are singular elsewhere, at $x = 1$, is irrelevant to the classification of the singular point at

$x = 0$.) Next, consider $x_0 = 1$:

$$(x-x_0)p(x) = (x-1)p(x) = (x-1)\frac{x+2}{x(x-1)} = \frac{x+2}{x},$$
$$(x-x_0)^2q(x) = (x-1)^2q(x) = (x-1)^2\frac{5}{(x-1)^3} = \frac{5}{x-1}.$$

The former is analytic at $x = 1$ but the latter, $5/(x-1)$, is not. Hence, (6) has an *irregular* singular point at $x_0 = 1$. ▮

The term "regular singular point" seems self-contradictory, for how can something singular also be regular? Think of it as meaning singular but not *too* singular. If x_0 is a *regular* singular point of (1), then the Frobenius theorem given below will show how to find two LI solutions; if it is an irregular singular point, the situation is more difficult and is not covered by that theory.

The method of Frobenius is only for expansions about *regular* singular points.

6.5.3 The method of Frobenius. Let $x = 0$ be a *regular singular point* of the equation

$$\boxed{y'' + p(x)y' + q(x)y = 0,} \tag{7}$$

on the interval $0 < x < \infty$. [There is no loss of generality in assuming that the singular point is at the origin, for if it is at $x = x_0 \neq 0$ we can make a change of variable $t = x - x_0$ to move the singular point to the origin of the t axis.]

Understand this motivational point.

We can motivate the Frobenius theorem, given below, as follows. First, multiply (7) by x^2 and group terms by brackets as

$$x^2y'' + x\left[xp(x)\right]y' + \left[x^2q(x)\right]y = 0. \tag{8}$$

Since $x = 0$ is a regular singular point it follows from the definition of a regular singular point that the bracketed functions $xp(x)$ and $x^2q(x)$ are analytic at $x = 0$ and can therefore be expanded about the origin in convergent Taylor series. Thus, we can write (8) in the form

$$x^2y'' + x\left(p_0 + p_1x + \cdots\right)y' + \left(q_0 + q_1x + \cdots\right)y = 0. \tag{9}$$

If, near $x = 0$, we approximate $p_0 + p_1x + \cdots \approx p_0$ and $q_0 + q_1x + \cdots \approx q_0$, then (9) becomes

$$\boxed{x^2y'' + p_0xy' + q_0y = 0.} \tag{10}$$

The approximations $p_0 + p_1x + \cdots \approx p_0$ and $q_0 + q_1x + \cdots \approx q_0$ hold "locally," near $x = 0$, so we have hopes that the solutions of (10) approximate those of (7) locally, as well.

We do *not* plan to simply replace (7) by the approximate equation (10). Rather, we plan to use the *form* of the solutions to (10) to *suggest* appropriate solution forms for (7). Equation (10) is a nonconstant-coefficient equation but an elementary one, a Cauchy–Euler equation. Thus, it has at least one solution of the form x^r for some constant r. If a solution $y(x)$ of (7) behaves as x^r in the neighborhood of $x = 0$, then perhaps we can express it as x^r times a function that is analytic at $x = 0$, that is, in the form x^r times a power series in x:

$$y(x) = x^r \left(a_0 + a_1 x + a_2 x^2 + \cdots\right), \tag{11}$$

which does indeed behave like a constant time x^r as $x \to 0$.

Thus, instead of seeking a power series solution $\sum_0^\infty a_n x^n$, which would work if $x = 0$ were a regular point of (7), it appears that if $x = 0$ is a regular singular point then we should seek solutions of (7) in the more general form

$$\boxed{y(x) = x^r \sum_{n=0}^{\infty} a_n x^n = \sum_{n=0}^{\infty} a_n x^{n+r},} \tag{12}$$

in which we are to determine not only the a_n's, as usual, but also the exponent r. We can assume, without loss of generality, that $a_0 \neq 0$ in (12) so, by definition, $a_0 x^r$ is the first (nonzero) term of the series.

Let a_0 be the first *nonvanishing* coefficient, so $a_0 \neq 0$.

We can determine r as follows. Let us call (10) the **reduced equation** (although that terminology is not standard) because it is a locally reduced version of (7). Since (10) is a Cauchy–Euler equation we can find at least one solution of it in the form x^r. Putting $y(x) = x^r$ into the reduced equation gives $[r(r-1) + p_0 r + q_0]x^r = 0$, so r is found by solving the quadratic equation

$$\boxed{r^2 + (p_0 - 1)r + q_0 = 0,} \tag{13}$$

which is the **indicial equation** for (7). The roots of the indicial equation are the **indicial roots**. With r determined from (13) we can put that value into (12), put (12) into (7), and then proceed with the solution in essentially the same way that we did in Section 6.3.

The reduced equation is a differential equation; the indicial equation is an algebraic equation.

The foregoing reasoning was intended only as heuristic, as a guide. The results, established by Frobenius, are as follows:

THEOREM 6.5.1 *Frobenius' Theorem*
Let $x = 0$ be a regular singular point of

$$y'' + p(x)y' + q(x)y = 0 \qquad (x > 0), \tag{14}$$

with $xp(x) = p_0 + p_1 x + \cdots$ and $x^2 q(x) = q_0 + q_1 x + \cdots$ having radii of convergence R_1, R_2, respectively. Let r_1, r_2 be the roots of the indicial equation

$$r^2 + (p_0 - 1)r + q_0 = 0, \tag{15}$$

numbered so that $r_1 \geq r_2$ if the roots are real; if they are not real, they will be

For definiteness, number the r's so that $r_1 \geq r_2$ if they are real.

complex conjugates and we can take r_1 to mean either of those roots. Seeking $y(x)$ in the form

$$y(x) = x^r \sum_{n=0}^{\infty} a_n x^n = \sum_{n=0}^{\infty} a_n x^{n+r} \quad (a_0 \neq 0) \tag{16}$$

with $r = r_1$ will lead to a solution

$$y_1(x) = x^{r_1} \sum_{n=0}^{\infty} a_n x^n \quad (a_0 \neq 0), \tag{17}$$

in which $a_1, a_2, \ldots$ are known multiples of a_0, with a_0 remaining arbitrary.

The form of a *second LI solution*, $y_2(x)$, depends on r_1 and r_2 as follows:

"An integer" is real, not an integer times i.

(i) r_1 and r_2 *distinct and not differing by an integer.* [Complex conjugate roots belong to this case because their difference will be an imaginary number, not an integer.] Then, with $r = r_2$, (16) yields the form

$$y_2(x) = x^{r_2} \sum_{n=0}^{\infty} b_n x^n \quad (b_0 \neq 0), \tag{18a}$$

in which the b_n's are generated by the same recursion relation as the a_n's, but with $r = r_2$ instead of $r = r_1$; $b_1, b_2, \ldots$ are known multiples of b_0, which is arbitrary.

(ii) *Repeated roots*, $r_1 = r_2 \equiv r$. In this case, $y_2(x)$ can be found in the form

The presence of the $\ln x$ term in (18b) should not be surprising because if $r_1 = r_2 \equiv r$, then the reduced equation (10) has the general solution $(C_1 + C_2 \ln x)x^r$, which contains a $\ln x$ term. NOTE: a_0, b_0, d_0 must be nonzero in (17), (18a), and (18c), as noted in those equations, but c_1 in (18b) *may* be zero.

$$y_2(x) = y_1(x) \ln x + x^r \sum_{n=1}^{\infty} c_n x^n. \tag{18b}$$

(iii) $r_1 - r_2$ *equal to a nonzero integer.* The larger root r_1 gives the solution (17), and $y_2(x)$ can be found in the form

$$y_2(x) = \kappa y_1(x) \ln x + x^{r_2} \sum_{n=0}^{\infty} d_n x^n \quad (d_0 \neq 0), \tag{18c}$$

in which the constant κ *may* turn out to be zero.

The radius of convergence of each of the series in (17) and (18) is at least as large as the smaller of R_1, R_2.

Implementation of Frobenius' theorem is known as the **method of Frobenius**, which we now outline before illustrating the three cases.

The Method of Frobenius

1. Be sure that $x = 0$ is a regular singular point of (14).

2. Evaluate p_0 and q_0. Since these are the leading coefficients in the Taylor expansion of $xp(x)$ and $x^2q(x)$, they can be computed as

$$\boldsymbol{p_0 = \lim_{x\to 0}[xp(x)],} \tag{19a}$$
$$\boldsymbol{q_0 = \lim_{x\to 0}[x^2q(x)].} \tag{19b}$$

If the regular singular point is at $x_0 \neq 0$, then let t, say, equal $x - x_0$ in (14), so the new equation has a regular singular point at $t = 0$.

3. Solve the indicial equation (15) for its roots r_1, r_2.

4a. If r_1 and r_2 are distinct and do not differ by an integer [case (i)] then LI solutions can be found in the forms (17) and (18a).

4b. If $r_1 = r_2 \equiv r$ [case (ii)] then LI solutions can be found in the forms (17) and (18b).

4c. If $r_1 - r_2$ is a nonzero integer [case (iii)] then the larger root r_1 gives one solution in the form (17) and a second LI solution can be found in the form (18c), in which κ *may* turn out to be zero.

EXAMPLE 3. Example of Case (i). Apply the method of Frobenius to

$$6x^2y'' + 7xy' - (1+x^2)y = 0. \qquad (0 < x < \infty) \tag{20}$$

The latter has a singular point at $x = 0$ because $p(x) = 7x/(6x^2) = 7/(6x)$ and $q(x) = -(1+x^2)/(6x^2)$ are singular there, and it is a regular singular point because both $xp(x) = 7/6$ and $x^2q(x) = -(1+x^2)/6$ are analytic there. Next, (19a) and (19b) give $p_0 = 7/6$ and $q_0 = -1/6$, so the indicial equation is $r^2 + (1/6)r - 1/6 = 0$ with roots $r = 1/3$ and $-1/2$. Since r_1 and r_2 are numbered (in Theorem 6.5.1) such that $r_1 \geq r_2$, choose

$$r_1 = \frac{1}{3}, \quad r_2 = -\frac{1}{2}. \tag{21}$$

These are distinct and do not differ by an integer, so this is case (i). Accordingly, we can find one solution using the form (17) with $r_1 = 1/3$, and then another using (18a) with $r_2 = -1/2$. We will outline the steps for r_1, and leave r_2 for the exercises.

First, with $r = r_1 = 1/3$, put

$$y_1(x) = x^{1/3}\sum_{n=0}^{\infty} a_n x^n = \sum_{n=0}^{\infty} a_n x^{n+1/3} \tag{22}$$

into (20), cancel an $x^{1/3}$ from each term, and obtain

$$\sum_{n=0}^{\infty} 6(n+\frac{1}{3})(n-\frac{2}{3})a_n x^n + \sum_{n=0}^{\infty} 7(n+\frac{1}{3})a_n x^n - \sum_{n=0}^{\infty} a_n x^n - \sum_{n=0}^{\infty} a_n x^{n+2} = 0. \tag{23}$$

To get the same exponents, let $n+2=k$ in the last sum, change the k's to n's (to have the same dummy summation index in all the sums), and define $a_{-2}=a_{-1}\equiv 0$ so the last sum in (23) becomes $\sum_0^\infty a_{n-2}x^n$. Combining the sums, set the coefficient of x^n equal to zero and obtain

$$n(6n+5)a_n - a_{n-2} = 0 \tag{24}$$

for $n=0,1,2,\ldots$, and hence the recursion formula

$$a_n = \frac{1}{n(6n+5)}a_{n-2}. \tag{25}$$

Now set $n=0,1,2,\ldots$ to obtain the a_n's. For $n=0$ equation (25) gives the indeterminate result $a_0=0/0$ so, for $n=0$, do not use (25); back up and use (24), which gives $0a_0 - 0 = 0$. Hence, a_0 is an arbitrary number. Then use (25) for $n\geq 1$ and obtain $a_1=0$, $a_2=a_0/34$, $a_3=0$, $a_4=a_0/3944$, $a_5=0$, and so on. Thus, with $a_0=1$, say,

$$y_1(x) = x^{1/3}\left(1+\frac{1}{34}x^2+\frac{1}{3,944}x^4+\frac{1}{970,224}x^6+\cdots\right). \tag{26}$$

The procedure is the same for $r_2=-1/2$. The steps are left for the exercises, and the result is

$$y_2(x) = x^{-1/2}\left(1+\frac{1}{14}x^2+\frac{1}{1,064}x^4+\frac{1}{197,904}x^6+\cdots\right). \tag{27}$$

Then a general solution of (20) is $y(x)=C_1y_1(x)+C_2y_2(x)$ with C_1 and C_2 arbitrary constants. Graphs of $y_1(x)$ and $y_2(x)$ are given in Fig. 2.

What are the radii of convergence R_1 and R_2 of the parenthesized series in (26) and (27)? Determine them from Theorem 6.3.1. The products

$$xp(x)=\frac{7}{6} \quad\text{and}\quad x^2q(x)=-\frac{1}{6}-\frac{1}{6}x^2$$

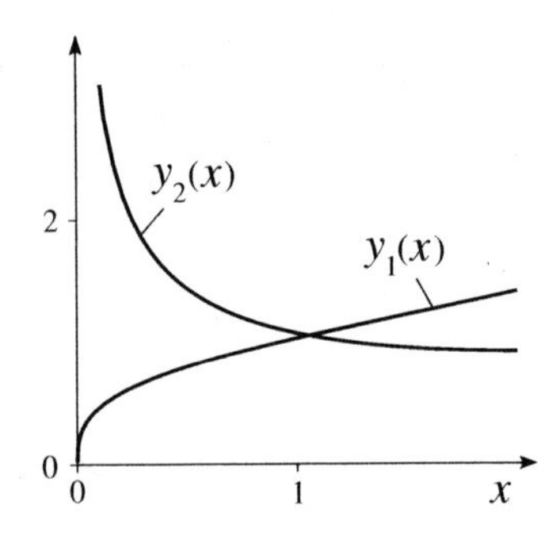

Figure 2. Graphs of the LI solutions $y_1(x)$ and $y_2(x)$ of (20).

happen to already be Taylor series about $x=0$. Since they terminate, their radii of convergence R_1 and R_2, respectively, are infinite. Then, according to the last sentence of Theorem 6.3.1, each of the series in (26) and (27) has an infinite radius of convergence. Thus, (26) and (27) hold on the entire interval $0<x<\infty$ specified in (20).

COMMENT. Observe from Fig. 2 that both $y_1(x)$ and $y_2(x)$ are singular at the origin: $y_1(x)$ has an infinite slope there [which is seen also in (26), because the derivative of the first term $x^{1/3}$ is $x^{-2/3}/3$ which tends to infinity as $x\to 0$] and $y_2(x)$ is not even defined there because $x^{-1/2}\to\infty$ as $x\to 0$. We can think of those singularities in the solutions as "inherited" from the singularities in p and q in the differential equation. This behavior, as $x\to 0$, comes into view in (26) because we expanded about the singular point $x=0$. ▮

EXAMPLE 4. Example of Case (ii). Apply the method of Frobenius to the equation

This equation was studied in Section 6.4, but we will not assume that you've studied that section.

$$xy''+y'+xy=0, \tag{28}$$

which is the **Bessel equation of order zero** that was studied in Section 6.4. The latter has a singular point at $x=0$ because $p(x)=1/x$ is singular there, although $q(x)=1$ is not, and it is a regular singular point because both $xp(x)=1$ and $x^2q(x)=x^2$ are analytic there. Next, (19a) and (19b) give $p_0 = 1$ and $q_0 = 0$ so the indicial equation (15) is simply $r^2 = 0$, with the repeated roots

$$r_1 = r_2 = 0. \tag{29}$$

Since the indicial roots are repeated, this is an example of case (ii). Thus, there can be found one solution, $y_1(x)$, of the form (17) with $r_1 = 0$, and another of the form (18b), with $r = 0$. Finding y_1 proceeds just as in Section 6.3 because, with $r_1 = 0$, $y_1(x) = \sum_0^\infty a_n x^n$ is simply a power series. The result, with $a_0 = 1$, say, is

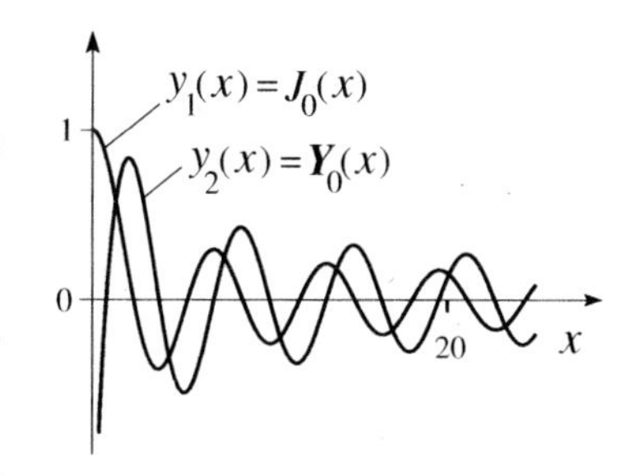

Figure 3. Graphs of the solutions (30) and (31) of the Bessel equation (28).

$$y_1(x) = 1 - \frac{1}{2^2}x^2 + \frac{1}{4^2 2^2}x^4 - \cdots = \sum_{n=0}^{\infty} \frac{(-1)^n}{(n!)^2}\left(\frac{x}{2}\right)^{2n}, \tag{30}$$

which is the solution $J_0(x)$ that was found in Section 6.4.

With y_1 given above, then use (18b) to find y_2. The steps are outlined in the exercises, and the result is

$$y_2(x) = \left(1 - \frac{1}{4}x^2 + \frac{1}{64}x^4 - \frac{1}{2,304}x^6 + \cdots\right)\ln x + \frac{1}{4}x^2 - \frac{3}{128}x^4 + \frac{11}{13,824}x^6 - \cdots, \tag{31}$$

which is the same as the solution $\boldsymbol{Y}_0(x)$ found in Section 6.4 using $J_0(x)$ and the reduction of order formula. These are plotted in Fig. 3.

What are the radii of convergence of the series in (30) and (31)? The products $xp(x) = 1$ and $x^2 q(x) = x^2$ are terminating Taylor series and therefore have infinite radii of convergence: $R_1 = R_2 = \infty$. Hence, according to Theorem 6.5.1, the radii of convergence of the series in (30) and (31) are infinite as well. ▮

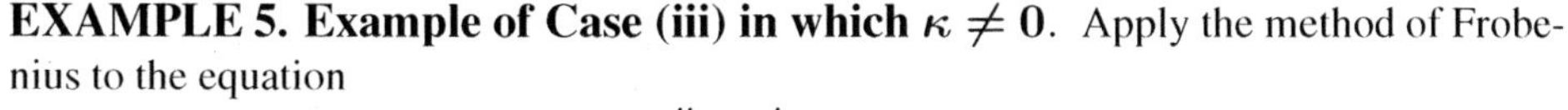

EXAMPLE 5. Example of Case (iii) in which $\kappa \neq 0$. Apply the method of Frobenius to the equation

$$xy'' - y' + y = 0. \tag{32}$$

The latter has a singular point at $x = 0$ because $p(x) = -1/x$ and $q(x) = 1/x$ are singular there, and it is a regular singular point because $xp(x) = -1$ and $x^2 q(x) = x$ are analytic there. Next, (19a) and (19b) give $p_0 = -1$ and $q_0 = 0$, so the indicial equation (15) is $r^2 - 2r = 0$ with roots

$$r_1 = 2, \quad r_2 = 0. \tag{33}$$

Since these differ by a nonzero integer, this example illustrates case (iii).

Accordingly, we can find one solution in the form (17) with $r = r_1 = 2$, and one in the form (18c). The steps are outlined in the exercises, and the results are

In case (iii), r_1 and r_2 need not be integers themselves; their *difference* is an integer.

$$y_1(x) = x^2\left(1 - \frac{1}{3}x + \frac{1}{24}x^2 - \frac{1}{360}x^3 + \cdots\right) = x^2 - \frac{1}{3}x^3 + \frac{1}{24}x^4 - \frac{1}{360}x^5 + \cdots \tag{34}$$

and

$$y_2(x) = \left(x^2 - \frac{1}{3}x^3 + \frac{1}{24}x^4 - \frac{1}{360}x^5 + \cdots\right)\ln x$$
$$-2 - 2x + \frac{4}{9}x^3 - \frac{25}{288}x^4 + \frac{157}{21,600}x^5 - \cdots. \tag{35}$$

By the same reasoning as in Examples 3 and 4, we find the radii of convergence of the series in (33) and (34) to be infinite.

The solutions $y_1(x)$ and $y_2(x)$ of (32) are plotted in Fig. 4. ∎

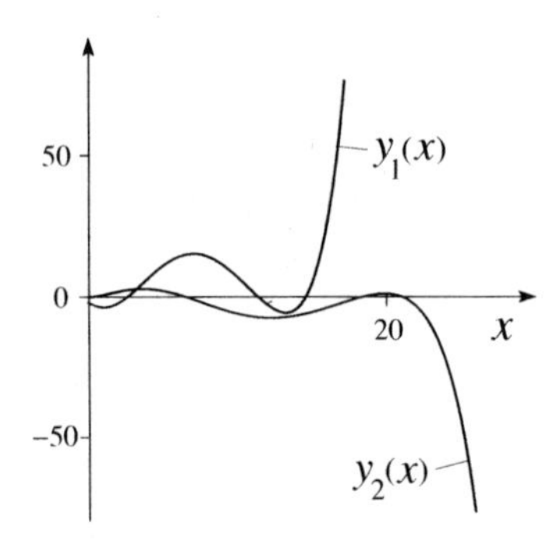

Figure 4. Graphs of the LI solutions (34) and (35) of (32).

Recall from Theorem 6.5.1 that the constant κ in (18c) *may* turn out to be zero, in which case there is no logarithmic term in (18c). That did not occur in the preceding example, so here is one more example of case (iii) — in which κ does turn out to be zero.

EXAMPLE 6. Example of Case (iii) in which $\kappa = 0$. Apply the method of Frobenius to

$$x^2y'' - 3xy + 3y = 0. \tag{36}$$

We don't need the powerful Frobenius theory because (36) is merely a Cauchy–Euler equation, and if we seek $y(x) = x^r$ we readily obtain the general solution

$$y(x) = C_1x^3 + C_2x. \tag{37}$$

But we *could* use the method of Frobenius (for illustration purposes) because (36) has a regular singular point at $x = 0$. In this case, the original equation (36) and its reduced equation (10) are identical, so $p_0 = -3$ and $q_0 = 3$. Thus, the indicial equation is $r^2 - 4r + 3 = 0$ with roots

$$r_1 = 3, \quad r_2 = 1 \tag{38}$$

that differ by a nonzero integer. Hence, (36) is an example of case (iii). Sure enough, the solution $y_1(x) = x^3$ in (37) is of the form (17) [with $r_1 = 3$, $a_0 = 1$ and all subsequent a_n's equal to zero] and $y_2(x) = x$ in (37) is of the form (18c) [with $\kappa = 0$, $r_2 = 1$, $d_0 = 1$, and all subsequent d_n's equal to zero]. If this illustration of case (iii) with $\kappa = 0$ seems too simple, because it is merely a Cauchy–Euler equation, change the equation to $x^2y'' - 3xy' + (3 + x^4)y = 0$, for instance, and you will again obtain case (iii) with $\kappa = 0$, but this time with nonterminating series solutions. ∎

In closing, consider this question: If $x = 0$ is a regular singular point of (1) *must* there be solutions of (1) that are singular at $x = 0$? In more intuitive language, is it possible that the singularity in the coefficients of the differential equation does *not* get passed along in some form to any of its solutions? Yes, Example 6 shows that it can happen, since both $y_1(x) = x^3$ and $y_2(x) = x$ are analytic at $x = 0$ (and indeed for all x) even though the differential equation (36) has a regular singular point at the origin.

Closure. The method of Frobenius enables us to obtain series solutions of a given equation $y'' + p(x)y' + q(x)y = 0$ about a regular singular point of that equation.

Consider the following attempt at a simple explanation of the method: If the *reduced* equation for the given differential equation is

$$x^2y'' + p_0xy' + q_0y = 0,$$

which is a Cauchy–Euler equation, yields distinct roots r_1 and r_2, then its solution is

$$y(x) = Ax^{r_1} + Bx^{r_2},$$

so generalize A and B into power series and seek solutions of the original equation $y'' + p(x)y' + q(x)y = 0$ in the forms

$$y_1(x) = x^{r_1} \sum_{n=0}^{\infty} a_n x^n \quad \text{and} \quad y_2(x) = x^{r_2} \sum_{n=0}^{\infty} b_n x^n. \tag{39}$$

If, instead, the *reduced* equation yields equal roots $r_1 = r_2 = r$, its solution is

$$y(x) = (A + B \ln x) x^r,$$

so generalize A and B into power series and seek solutions of the given differtential equation in the forms

$$y_1(x) = x^r \sum_{n=0}^{\infty} a_n x^n \quad \text{and} \quad y_2(x) = x^r \ln x \sum_{n=0}^{\infty} b_n x^n.$$

That proposed explanation comes close, falling short only in the event that r_1 and r_2 differ by a nonzero integer. In that case we do obtain the $y_1(x)$ solution in (39), but the form for $y_2(x)$ in (39) leads only to the trivial solution. Yet, y_2 can be found from y_1 using the *reduction-of-order formula* from Section 2.5,

$$y_2(x) = y_1(x) \int \frac{e^{-\int p(x)\,dx}}{[y_1(x)]^2}\,dx. \tag{40}$$

Whether or not (40) produces a $\ln x$ term depends on whether or not the coefficient of $1/x$ in the integrand of (40) *happens* to be zero. In this way we obtain the solution that was given by (18c) in the theorem.

To implement the method of Frobenius, follow the steps given below equation (18). Once p_0 and q_0 are determined from (19a,b), and hence r_1 and r_2 from (13), find a solution $y_1(x)$ in the form (17), and you will know from r_1 and r_2 which of (18a,b,c) to use for a second solution $y_2(x)$. From that point, implementation amounts to putting the assumed solution forms into the differential equation and equating coefficients of powers of x in the usual way. The details in Examples 3–6 were kept minimal, and reserved for the exercises.

EXERCISES 6.5

1. Cite and use a theorem in this text to show that the IVP consisting of (2) and the initial conditions $y(2) = 2$, $y'(2) = 1$ has a unique solution on $0 < x < \infty$.

2. Singular Points. Identify all singular points, if any, and classify each as regular or irregular.
NOTE: The only theorem given in Section 6.2 to test for analyticity was for rational functions, which are most important in this chapter. Other functions are encountered in this exercise, so here is additional information that will enable you to work the exercise: $\sin x$, $\cos x$, e^x, and e^{-x} are analytic for all x. Further, if f and g are analytic at x_0, then so is fg, and so is f/g if $g(x_0) \neq 0$; if $g(x_0)$ does equal zero, then f/g is analytic there, nevertheless, if f/g has a finite limit as $x \to x_0$.

(a) $xy'' - (\cos x)y' + 5y = 0$
(b) $(x^2-1)y'' - y = 0$
(c) $x^2(x+1)^3y'' + e^x y = 0$
(d) $(x^4-1)y'' + e^x y' + y = 0$
(d) $y'' + xy' + x^2 y = 0$
(f) $e^x y'' + (\sin^2 x)y = 0$
(g) $x^2 y'' + 3(\sin x)y' - (\cos x)y = 0$
(h) $x^4 y'' + x^3 y' + y = 0$
(i) $(x^2+2)y'' + (\cot x)y = 0$
(j) $(x^3 y')' + 4y = 0$
(k) $(e^x y')' + 2e^{-x} y = 0$

3. Computer. Each differential equation is on $x > 0$ and has a regular singular point at $x = 0$. First, obtain the reduced equation and the indicial roots r_1 and r_2. Second, from those values, state whether the solution is of type (i), (ii), or (iii) in Theorem 6.5.1. Third, use computer software (such as the *Maple* "dsolve" command) to obtain a general solution in series form, about $x = 0$, obtaining the first several terms, and state whether your computer solution is consistent with the type that you predicted. If your solution is of case (iii) type, identify the value of κ from the computer output. Fourth, determine (using Theorem 6.5.1) the minimum possible radii of convergence of the series solutions that you obtained. NOTE: See the note in Exercise 2.

(a) $2x^2y'' + (x^2+5x)y' + y = 0$
(b) $xy'' + e^x y' + y = 0$
(c) $x^3y'' - 3x^2y' + 4(e^x - 1)y = 0$
(d) $xy'' - 2y' + y = 0$
(e) $x^2(1+x)y'' - (2+3x)y = 0$
(f) $xy'' + e^x y = 0$
(g) $x^2y'' + (x^2+x)y' - y = 0$
(h) $4x(x^2+x+1)y'' + y' + y = 0$
(i) $2x^2y'' - 3xy' + (3+x)y = 0$
(j) $(x^2+2x)y'' + y' + y = 0$
(k) $x^3y'' - 5x^2y' + 9(e^x - 1)y = 0$
(l) $x^3y'' + 5(x^2+x^4)y' + 4(\sin x)y = 0$
(m) an equation supplied by your instructor

4. Example 3. With $r_2 = -1/2$, use the form (18a),

$$y_2(x) = x^{-1/2}\sum_0^\infty b_n x^n, \tag{4.1}$$

to derive (27). HINT: The steps will parallel those in the derivation of (26).

5. Example 4. In Example 4 we found that $r_1 = r_2 = 0$, so LI solutions are given by (17) and (18b) as

$$y_1(x) = \sum_0^\infty a_n x^n \quad (a_0 \neq 0) \tag{5.1}$$

and

$$y_2(x) = y_1(x)\ln x + \sum_1^\infty c_n x^n. \tag{5.2}$$

Obtain recursion formulas for the a_n's and c_n's and show that your solutions for y_1 and y_2 agree with (30) and (31). HINT: The $y_1(x)$ part is straightforward, so this hint is for $y_2(x)$. By substitution of (5.2) into the differential equation, show that

$$(xy_1''+y_1'+xy_1)\ln x + 2y_1' - \frac{1}{x}y_1 + \frac{1}{x}y_1$$
$$+\sum_2^\infty n(n-1)c_n x^{n-1} + \sum_1^\infty nc_n x^{n-1} + \sum_1^\infty c_n x^{n+1} = 0. \tag{5.3}$$

Not by accident, the coefficient of $\ln x$ vanishes because y_1 satisfies (28). Recalling (5.1) (in which the a_n's are now known), use

$$y_1' = \sum_1^\infty na_n x^{n-1} = \sum_0^\infty (k+1)a_{k+1}x^k. \tag{5.4}$$

For the three sums in (5.3), adjust the exponents on the x's and the lower summation limits in the usual way, and show that the recursion formula follows as

$$c_{k+1} = -\frac{1}{(k+1)^2}c_{k-1} - \frac{2}{k+1}a_{k+1} \tag{5.5}$$

for $k = 0, 1, 2, \ldots$. Solve the latter for c_1, c_2, $\ldots$ and put them, and $y_1(x)$, into (5.2).

6. Example 5. In Example 5 we found that $r_1 = 2$ and $r_2 = 0$, so linearly independent solutions are given in the form of (17) and (18c) as

$$y_1(x) = x^2\sum_0^\infty a_n x^n \quad (a_0 \neq 0) \tag{6.1}$$

$$y_2(x) = \kappa y_1(x)\ln x + \sum_0^\infty d_n x^n \quad (d_0 \neq 0). \tag{6.2}$$

Obtain recursion formulas for the a_n's and d_n's and show that your solutions for y_1 and y_2 agree with (34) and (35). HINT: The $y_1(x)$ part is straightforward, so this hint is for $y_2(x)$. Substitute (6.2) into the equation $xy'' - y' + y = 0$. You should find that the $\ln x$ term is multiplied by $xy_1'' - y_1' + y_1$, which is zero because y_1 satisfies the differential equation. Adjusting the sums so that all the exponents on the x's and all the summation limits are the same, as usual, obtain the recursion formula

$$2\kappa k a_{k-1} + (k^2-1)d_{k+1} + d_k = 0 \qquad (6.3)$$

for $n = 0, 1, 2, \ldots$. Setting $n = 0, 1, 2, \ldots$, in turn, the equations thus obtained should show that $d_1 = d_0$, $\kappa = -d_0/2$, d_2 = arbitrary, $d_3 = -2d_0/9$, and so on. If you set the arbitrary d_2 equal to zero, and assign a suitable value to d_0 you should obtain the solution $y_2(x)$ given in (35).

7. Example 5, Continued. Consider Example 5 again. Instead of solving for $y_2(x)$ along the lines outlined in Exercise 6, use (34) and the reduction of order formula (40) to derive (35). HINT: Show that

$$\int \frac{e^{-\int p\,dx}}{[y_1(x)]^2}\,dx = \int \left(\frac{1}{x^3} + \frac{2}{3x^2} + \frac{1}{4x} + \cdots\right) dx$$

and, formally integrating the latter termwise, obtain from (40)

$$y_2(x) = \frac{1}{4}y_1(x)\ln x - \frac{1}{2} - \frac{1}{2}x + \frac{29}{144}x^2 + \cdots. \qquad (7.1)$$

Scaling the latter by 4 does indeed give the solution (35), except for the $29x^2/144$ term, and the terms following that one. It is not a mistake. Explain the apparent discrepancy.

ADDITIONAL EXERCISES

8. Complex Indicial Roots. If the indicial roots r_1 and r_2 are complex conjugates, then they are distinct, and their difference is imaginary and therefore not equal to an integer. This situation corresponds to case (i) in Theorem 6.5.1. Let us go through an example together, with you filling in various steps. Consider

$$x^2y'' + xy' + (1+x)y = 0. \qquad (8.1)$$

(a) Show that (8.1) has a regular singular point at $x = 0$ and that its indicial roots are $r = \pm i$.

(b) For each of the indicial roots, derive the recursion formula and thus obtain the LI solutions

$$y_1(x) = x^i\left(1 + \frac{-1+2i}{5}x - \frac{1+3i}{40}x^2 + \frac{9+7i}{1,560}x^3 + \cdots\right), \quad (8.2)$$

$$y_2(x) = x^{-i}\left(1 + \frac{-1-2i}{5}x - \frac{1-3i}{40}x^2 + \frac{9-7i}{1,560}x^3 + \cdots\right). \quad (8.3)$$

(c) However, normally we prefer real-valued solutions, so we would like to obtain from (8.2) and (8.3) two LI real-valued solutions. Preliminary to doing that, show that if $y(x)$ is a solution of $y'' + p(x)y' + q(x)y = 0$, in which $p(x)$ and $q(x)$ are real-valued functions, then the real and imaginary parts of $y(x)$ are solutions as well. That is, if $y(x) = u(x) + iv(x)$ is a solution, in which $u(x)$ and $v(x)$ are real-valued, then each of $u(x)$ and $v(x)$ is a solution as well.

(d) Use the result from part (c) to obtain two LI solutions of (8.1), say $u(x)$ and $v(x)$, as

$$\begin{aligned} u(x) &= \operatorname{Re} y_1(x) \\ &= \cos(\ln x)\left(1 - \frac{1}{5}x - \frac{1}{40}x^2 + \frac{3}{520}x^3 + \cdots\right) \\ &\quad + \sin(\ln x)\left(-\frac{2}{5}x + \frac{3}{40}x^2 - \frac{7}{1,560}x^3 + \cdots\right), \end{aligned} \qquad (8.4)$$

$$\begin{aligned} v(x) &= \operatorname{Im} y_1(x) \\ &= \cos(\ln x)\left(\frac{2}{5}x - \frac{3}{40}x^2 + \frac{7}{1,560}x^3 + \cdots\right) \\ &\quad + \sin(\ln x)\left(1 - \frac{1}{5}x - \frac{1}{40}x^2 + \frac{3}{520}x^3 + \cdots\right). \end{aligned} \qquad (8.5)$$

HINT: Recall (from Section 2.5.3) that

$$x^{\alpha+i\beta} = x^\alpha\left[\cos(\ln(\beta x)) + i\sin(\ln(\beta x))\right] \qquad (8.6)$$

so $x^{\pm i} = \cos(\ln x) \pm i\sin(\ln x)$ in (8.2) and (8.3).

9. Irregular Singular Point. (a) Show that

$$y'' + \sqrt{x}\,y = 0 \quad (0 < x < \infty) \qquad (9.1)$$

has an irregular singular point at $x = 0$, so the Frobenius theory *does not apply*.

(b) However, in this example we happen to be able to convert the irregular singular point to a regular singular point by a change of variables. Specifically, show that if we set $\sqrt{x} = t$, then the equation governing $y(x(t)) \equiv Y(t)$ is

$$tY''(t) - Y'(t) + 4t^4Y(t) = 0, \qquad (9.2)$$

and that the latter has a *regular* singular point at $t = 0$.

(c) Use computer software to generate a Frobenius-type solution of (9.2). Then replace the t's in that solution by $\sqrt{x}$'s and show that you obtain the solution

$$\begin{aligned} y(x) &= C_1\left(1 - \frac{4}{15}x^{5/2} + \frac{1}{75}x^{10/2} - \cdots\right) \\ &\quad + C_2x\left(1 - \frac{4}{35}x^{5/2} + \frac{2}{525}x^{10/2} - \cdots\right). \end{aligned} \qquad (9.3)$$

(d) Are the series solutions in (9.3) of the Frobenius form $y(x) = x^r \sum_0^\infty a_n x^n$? Explain.

CHAPTER 6 REVIEW

In solving nonelementary linear higher-order equations by series solution methods, we considered only the second-order equation

$$y'' + p(x)y' + q(x)y = 0. \tag{1}$$

After reviewing power series in Section 6.2, we began in Section 6.3 with the simplest case — in which the point of expansion x_0 is a regular point of (1), a point at which both p and q in (1) are analytic. In earlier chapters, theorems routinely called only for the continuity of the coefficients. The assumption of analyticity here is much more demanding, but if p and q are indeed analytic at x_0, then (Theorem 6.3.1) all solutions $y(x)$ of (1) will be analytic there too and, consequently, that *two LI solutions can be found in power series form.*

Since power series can be differentiated termwise any number of times and added termwise, within their intervals of convergence, we can put the assumed series into (1) and equate the coefficient of each power of x on the left-hand side to zero by the identity principle. Furthermore, minimum radii of convergence of the solutions thus obtained are given by the theorem in terms of the location of the singular points (if any) of p and q in the complex plane.

In Section 6.4, we considered two equations that are prominent in applications, the Legendre equation

$$\left(1-x^2\right)y'' - 2xy' + \lambda y = 0 \tag{2}$$

in which λ is a constant, and the Bessel equation of order ν,

$$x^2y'' + xy' + \left(x^2 - \nu^2\right)y = 0, \tag{3}$$

in which ν is a constant.

The Legendre equation has singular points only at $x = \pm 1$, so when we expanded about the origin $x = 0$ we were able to find two LI solutions as power series expansions, both of which converge in the interval of interest, $-1 < x < 1$. If λ happens to be of the form $N(N+1)$, where N is any nonnegative integer, then one of the two series terminates. These finite-degree polynomial solutions, corresponding to $N = 0, 1, 2, \ldots$, are the well known Legendre polynomials $P_0(x)$, $P_1(x), \ldots$. They are the only solutions of (2) that are bounded on $-1 \le x \le 1$.

For the Bessel equation we studied only the equation of order $\nu = 0$. Rather than avoid the singular point at the origin, we sought power series solutions about that point. In that case Theorem 6.3.1 did not apply, but we proceeded nonetheless and were partially successful, finding just one such solution. Then we used the reduction-of-order formula from Section 2.10 to find a second LI solution. The latter was not of power series form; it was singular at the origin because it contained a $\ln x$ term. Besides finding two LI solutions of the Bessel equation, we observed that if we expand about a *singular point* of (1), then modified solution forms are needed to find two LI solutions.

That was our agenda in Section 6.5, to determine what those modified solution forms might be. We began by distinguishing singular points as regular or irregular. Roughly put, a regular singular point is singular but "not too singular." The classical Frobenius theorem, Theorem 6.5.1, covered the case of expansions about regular singular points just as completely as Theorem 6.3.1 covered the case of expansions about regular points. It assures us that even if $x = 0$ is a singular point, we can successfully find two LI series solutions about that point if it is a *regular* singular point. The key difference between the power series solutions about regular points (Theorem 6.3.1) and the Frobenius series solutions about regular singular points (Theorem 6.5.1) is that the Frobenius series contain the functions x^r, and/or $\ln x$. Both of those functions are singular at the origin (if r is other than a nonnegative integer) and their presence in solutions is the result of the singularity in the differential equation. Frobenius solutions occur in one of three different forms, depending on the indicial roots associated with the local Cauchy–Euler approximation of (1), its "reduced equation."

Chapter 7

SYSTEMS OF NONLINEAR DIFFERENTIAL EQUATIONS

7.1 INTRODUCTION

We used the *phase line* for the qualitative study of nonlinear first-order equations that are *autonomous*, $dx/dt = f(x)$; that is, the right-hand side is $f(x)$, not $f(x,t)$.

Now we return to that idea and extend it, in Sections 7.2–7.5, to systems of *two* autonomous first-order equations, in which case we obtain a phase *plane* instead of a phase line. Phase plane analysis will be a powerful qualitative tool for the study of nonlinear differential equations.

Historically, interest in nonlinear differential equations is as old as the subject of differential equations, which dates back to Newton, but little progress was made until the late 1880's when the great mathematician and astronomer *Henri Poincaré* (1854 – 1912) took up a systematic study of the subject in connection with celestial mechanics. Realizing that nonlinear equations are rarely solvable analytically, and not having the benefit of computers to generate solutions numerically, he sidestepped the search for solutions and, instead, sought to answer questions about the qualitative and topological nature of solutions of nonlinear differential equations without actually finding them.

Poincaré was motivated primarily by problems of celestial mechanics, but the subject attracted broader attention during and following World War II, especially in connection with nonlinear control theory. In the postwar years, interest was stimulated further by the publication in English of N. Minorsky's *Nonlinear Mechanics* (Ann Arbor, MI: J. W. Edwards) in 1947. Also important were the translation of A. Andronov and C. Chaikin's 1937 book *Theory of Oscillations* into English by Solomon Lefschetz (Princeton: Princeton University Press, 1949), and J. J. Stoker's *Nonlinear Vibrations* (New York: Interscience, 1950). With these available as texts, the subject appeared in university curricula by the end of the 1950's. Assisted greatly by the advent of digital computers around that time, the subject of nonlinear dynamics expanded into what is now known as *dynamical systems*, with applications in such diverse subjects as biology, the social sciences, economics, and chemistry.

In the final section, we complement our discussion of the phase plane — which is a blend of analysis, computation, and graphics — with a strictly numerical approach, the numerical solution of systems of differential equations by Euler's method. The latter is a generalization of the method used in Section 1.9 to solve a single differential equation, and is applicable whether the system is linear or nonlinear.

7.2 THE PHASE PLANE

7.2.1 Phase plane method. The phase plane method can be applied to an autonomous system

The system (1) is **autonomous** because P and Q are functions only of x and y, not t.

$$\frac{dx}{dt} = P(x,y), \qquad \frac{dy}{dt} = Q(x,y), \tag{1a,b}$$

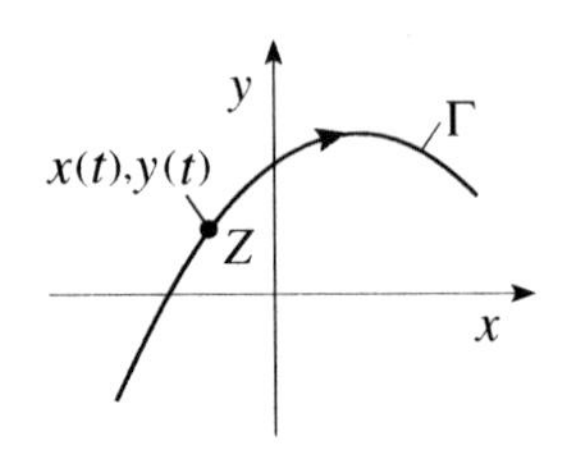

Figure 1. A portion of a trajectory Γ in the x, y phase plane.

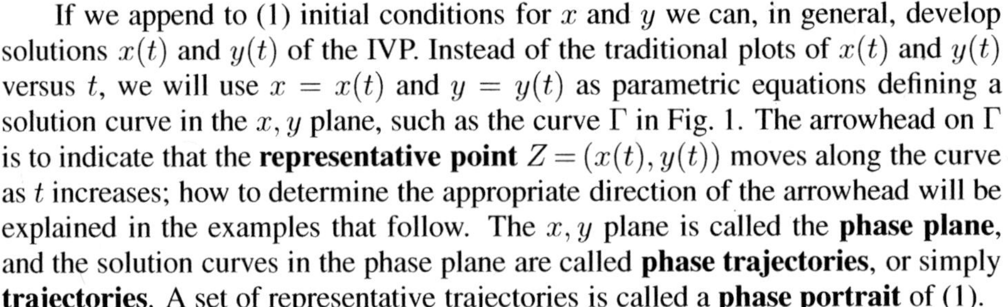

in which the independent variable t, typically, is the time, and in which the physical significance of the dependent variables x and y differs from one application to another.

If we append to (1) initial conditions for x and y we can, in general, develop solutions $x(t)$ and $y(t)$ of the IVP. Instead of the traditional plots of $x(t)$ and $y(t)$ versus t, we will use $x = x(t)$ and $y = y(t)$ as parametric equations defining a solution curve in the x, y plane, such as the curve Γ in Fig. 1. The arrowhead on Γ is to indicate that the **representative point** $Z = (x(t), y(t))$ moves along the curve as t increases; how to determine the appropriate direction of the arrowhead will be explained in the examples that follow. The x, y plane is called the **phase plane**, and the solution curves in the phase plane are called **phase trajectories**, or simply **trajectories**. A set of representative trajectories is called a **phase portrait** of (1).

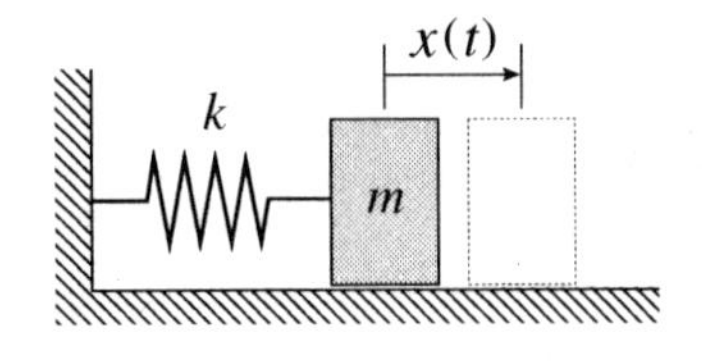

Figure 2. Frictionless mechanical harmonic oscillator.

EXAMPLE 1. Harmonic Oscillator. As a first example, consider the frictionless mass–spring system shown in Fig. 2, modeled by the by-now-familiar equation

$$mx'' + kx = 0. \tag{2}$$

Of course we can readily obtain the general solution $x(t) = C_1 \cos \omega_n t + C_2 \sin \omega_n t$, in which $\omega_n = \sqrt{k/m}$. We can apply initial conditions, evaluate C_1 and C_2, and plot the graph of x versus t. However, we will not do that. Rather, we will re-express (2) in the form (1), and generate its phase portrait.

To convert (2) to the form (1), define $x' = y$ (so, in this example, the auxiliary variable y happens to have physical significance as the velocity of the mass). Then the x'' in (2) is

y', so the single second-order equation (2) gives the pair of first-order equations

$$\frac{dx}{dt} = y, \tag{3a}$$

$$\frac{dy}{dt} = -\frac{k}{m}x, \tag{3b}$$

which is equivalent to (2). The first serves to define y and the second then follows from (3a) and (2). The system (3) is indeed of the autonomous form (1), with $P(x, y) = y$ and $Q(x, y) = -kx/m$.

Next, divide (3b) by (3a), obtaining

$$\frac{dy}{dx} = -\frac{k}{m}\frac{x}{y} \qquad \text{or} \qquad my\,dy + kx\,dx = 0, \tag{4}$$

integration of which gives

$$\frac{1}{2}my^2 + \frac{1}{2}kx^2 = C. \tag{5}$$

Since y is $x'(t)$, (5) is actually a first-order differential equation for $x(t)$, so it is called a **first integral** of the original second-order equation (2). If we were to solve (5) by algebra for y, as a function of x, and then replace y by x', we would have a first-order differential equation for $x(t)$. By solving that differential equation for $x(t)$ we would eventually arrive at the general solution $x(t) = C_1 \cos \omega_n t + C_2 \sin \omega_n t$ of (2) that was cited above [and that is found from (2) much more readily by seeking $x(t)$ in the exponential form e^{rt}]. However, instead we will take the first integral (5) as our final result, and will plot the elliptical phase trajectories defined by (5) in the x, y phase plane for several values of the integration constant C (Fig. 3).

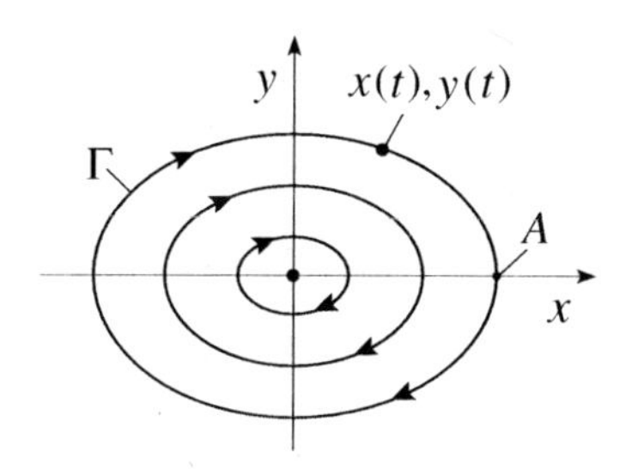

Figure 3. Phase portrait of (3).

In this example, C has physical significance as the total energy [the kinetic energy $mx'(t)^2/2$ of the mass plus the potential energy $kx^2(t)/2$ of the spring], which is "conserved;" it does not vary with t. If we set $t = 0$ in (5) we can evaluate C in terms of the initial conditions as $C = mx'(0)^2/2 + kx(0)^2/2$. In particular, if the mass is initially at rest, in equilibrium, with $x(0) = x'(0) = 0$, then $C = 0$. In that case, (5) corresponds to the "point ellipse" $x = y = 0$, which is denoted by the heavy dot in Fig. 3. As C is increased we obtain larger ellipses, corresponding to motions of larger amplitude.

Notice that any reference to the time t was eliminated when we divided (3b) by (3a) and obtained (4). We reiterate that in the phase plane t enters only as a parameter, through the parametric representation $x = x(t)$, $y = y(t)$, and we can visualize the representative point $(x(t), y(t))$ as a dot that moves in time along a given trajectory, as suggested by the arrows in Fig. 3. The direction of the arrows is implied by (3a): $dx/dt = y$, so if $y > 0$ then $dx/dt > 0$ and $x(t)$ is increasing; and if $y < 0$, then $x(t)$ is decreasing. Thus, *in this example*, in the upper half-plane the arrows are to the right, and in the lower half-plane they are to the left, as in Fig. 3.

Direction of the arrows.

We can see from (5), by suitable choice of C, that there is an elliptical trajectory through *every* point of the phase plane in Fig. 3 (including the origin if we call that one a "point trajectory," consisting only of a single point), and if we showed them all, we would have an all-black picture. Rather, plot just enough representative trajectories to show the key features of the phase portrait.

COMMENT. Observe that we never did obtain the functions $x(t)$ and $y(t)$. We were content to proceed only as far as (5), which gave the phase trajectories, and to add arrows showing the flow direction along those trajectories. ∎

Connection between phase line and phase plane. Recall from Section 1.3 that for a first-order autonomous differential equation we can construct a phase line, and that for a second-order autonomous equation we have instead a phase plane.

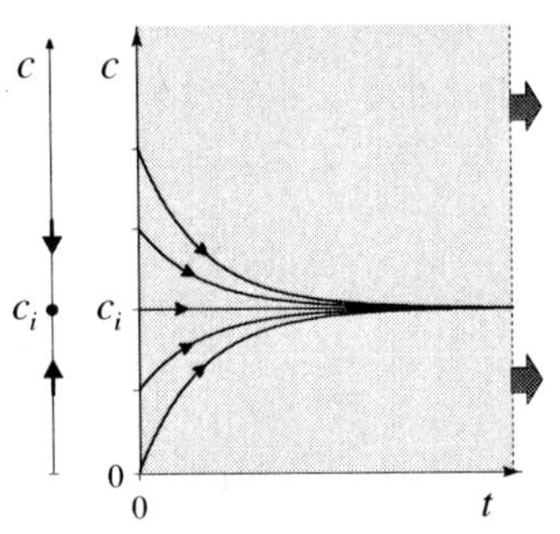

Figure 4. Plots of $c(t)$ versus t, and the c phase line, from an example in Section 1.3; same as Fig. 7 in Section 1.3.

To clarify the connection, recall the phase line idea. We've reproduced here (Fig. 4) a figure from an example in Section 1.3. The latter is a plot of the concentration $c(t)$ of a certain solute in a mixing tank versus the time t; $c(t)$ was modeled by a first-order autonomous differential equation. As noted in Section 1.3, we can imagine clicking on the right-hand edge of the c, t plot and dragging it to the right (as indicated by the large arrows) or left. Dragging it to the left compresses the time until it has disappeared altogether and we are left with the phase line, arranged vertically at the left, and a one-dimensional flow along that line. The simplicity of the resulting phase line is at the expense of losing the time dimension in the display.

The same idea applies for the phase plane. For the system (3), imagine plotting solutions in Cartesian x, y, t space, with the t axis perpendicular to the x, y plane. If we compress the time we end up with the two-dimensional flow in the x, y phase plane, shown in Fig. 3. Again, the simplicity gained by compacting the solution curves into the phase plane is at the expense of having lost the time dimension in the display.

Solving for the trajectories. In Example 1, dividing (3b) by (3a) eliminated the time and gave the first-order differential equation (4), integration of which gave the equation of the trajectories. More generally, divide (1b) by (1a) and obtain the differential equation

$$\boxed{\frac{dy}{dx} = \frac{Q(x,y)}{P(x,y)}.} \tag{6}$$

Of course, (6) may be too difficult to solve. However, it is always possible to solve it if (1) is linear, and it may be possible to solve it even if (1) is nonlinear.

EXAMPLE 2. Nonlinear System. For instance, consider the nonlinear system

$$x' = (1-y)x, \tag{7a}$$
$$y' = (-1+x)y. \tag{7b}$$

Dividing (7b) by (7a) gives

$$\frac{dy}{dx} = \frac{(-1+x)y}{(1-y)x}, \tag{8}$$

and even though (7) and (8) are nonlinear, we can solve (8) because it is separable. That step gives the equation for the phase trajectories as

$$xye^{-(x+y)} = C. \tag{9}$$

This example will be studied in detail in Section 7.4.2, in connection with *predator–prey population dynamics*. Here, our purpose is only to illustrate that we *may* be able to solve (6) even if it is nonlinear, by methods developed in Chapter 1.

By the way, notice that the system (1) can arise in either of two ways: directly, as in this example, or it can arise when we are given a single second-order equation and use an auxiliary dependent variable to re-express that equation as an equivalent system of two first-order equations, as in Example 1. ▮

In the predator–prey model (7), $x(t)$ is the population of the prey and $y(t)$ is the population of the predator.

The "flow" in the phase plane. Fig. 3 suggests that we think of (1) as defining a **flow** in the x, y phase plane. From (1), the x and y velocity components of the flow are $x' = P(x, y)$ and $y' = Q(x, y)$, respectively, and in general these vary from point to point in the plane. As depicted in Fig. 4, the speed of the representative point along the trajectory is ds/dt, where s is arc length along the trajectory, and is the **phase velocity**. By the Pythagorean theorem, $s' = \sqrt{x'^2 + y'^2}$, so at any given x, y point it is

$$s' = \sqrt{P^2(x, y) + Q^2(x, y)}. \tag{10}$$

For instance, for the flow in Fig. 3, (10) and (3) give $s' = \sqrt{y^2 + (k/m)^2 x^2}$.

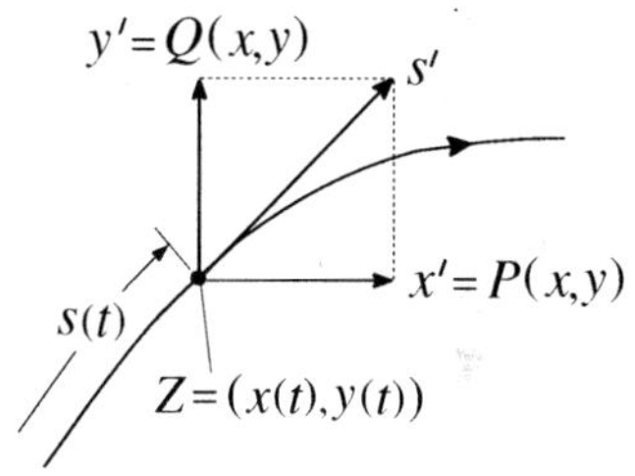

Figure 5. The phase velocity components x' and y', and the phase velocity s' of the representative point.

Equilibrium points. A point at which the phase velocity is zero is an **equilibrium point**; thus, (x_0, y_0) is an equilibrium point of (1) if both x' and y' are zero there, that is, if

$$\begin{aligned} P(x_0, y_0) &= 0, \\ Q(x_0, y_0) &= 0. \end{aligned} \tag{11a,b}$$

Equations to find equilibrium points of (1).

Thus, equilibrium points of (1), if any, are found as the solutions of equations (11a,b). For instance, in Example 1 those equations were $P(x, y) = y = 0$ and $Q(x, y) = -kx/m = 0$, the only solution of which was the point at the origin.

The phase trajectory through an equilibrium point (x_0, y_0) is a **point trajectory** because if the representative point starts there, it remains there; $x(t) = x_0$ and $y(t) = y_0$ for all $t \geq 0$. Thus, the trajectory is the single point (x_0, y_0).

7.2.2 Application to nonlinear pendulum.

To illustrate the value of phase plane analysis, consider a classical nonlinear system in some detail, the free oscillation of a pendulum, neglecting any friction and air resistance (Fig. 6).

One learns in physics that for a rigid body undergoing pure rotation about an axis, the inertia of the body about that axis times the body's angular acceleration equals the moment (or "torque") applied about that axis.[1] For the pendulum, the

[1]The moment of a force about a point P is defined as the force times the perpendicular distance

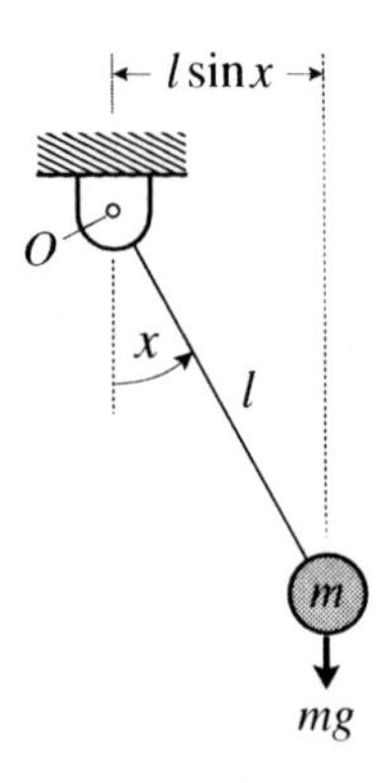

Figure 6. The pendulum.

pivot axis is normal to the paper at the pin O, the inertia of the mass m about O is ml^2, the angular acceleration is $x''(t)$, and the moment about O is the force mg times its moment arm $l \sin x$ (Fig. 6). Thus, the equation of motion is $ml^2 x'' = -mgl \sin x$, with the minus sign included because the moment is in the negative x direction, so

$$x'' + \frac{g}{l} \sin x = 0, \tag{12}$$

which is nonlinear because of the $\sin x$ term. For simplicity, let $g/l = 1$ so (12) becomes

$$\boxed{x'' + \sin x = 0.} \tag{13}$$

Following the program outlined above, re-express (13) as the equivalent system

$$x' = y \qquad \equiv P(x, y), \tag{14a}$$
$$y' = -\sin x \equiv Q(x, y). \tag{14b}$$

Whereas (12) was a challenging second-order nonlinear equation, (15) is a simple separable first-order equation!

Next, divide (14b) by (14a) and obtain

$$\frac{dy}{dx} = -\frac{\sin x}{y}. \tag{15}$$

The latter is nonlinear but separable, and its solution is

(16) gives *all* the phase trajectories, each one corresponding to a particular value of C.

$$\frac{1}{2} y^2 = \cos x + C \tag{16}$$

for the phase plane trajectories.

Finally, plotting the trajectories (16) for various values of C gives the phase portrait in Fig. 7.

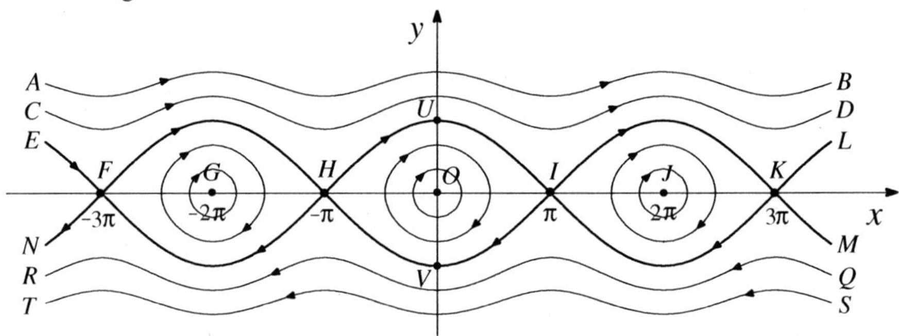

Figure 7. Phase portrait for the pendulum equation $x'' + \sin x = 0$.

How did we decide in which direction to draw the flow arrows in Fig. 7? In this example (as in Example 1) we see from (14a) that if $y > 0$ then $x' > 0$ (because

(the "moment arm") from P to the line of action of the force.

$x' = y$), and if $y < 0$ then $x' < 0$. Thus, the flow is rightward in the upper half-plane and leftward in the lower half-plane.

To find any equilibrium points, set

$$P(x,y) = y = 0 \quad \text{and} \quad Q(x,y) = -\sin x = 0. \tag{17a,b}$$

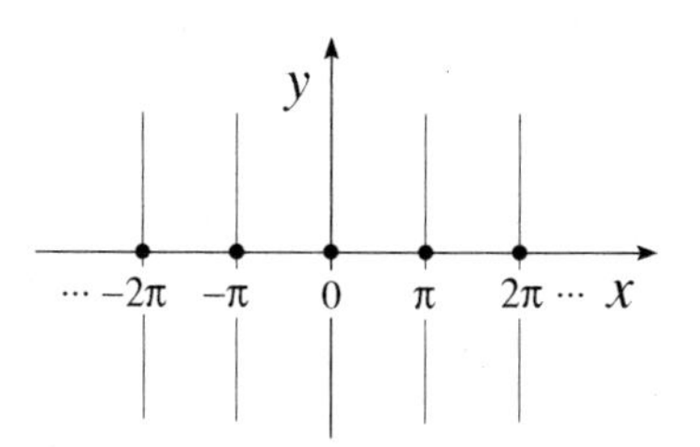

Figure 8. The equilibrium points of (14) are the intersections of the horizontal line $y = 0$, from (17a), and the vertical lines $x = n\pi$, from (17b).

Now, $y = 0$ is the x axis, and $\sin x$ equals zero on the vertical lines $x = 0, \pm\pi, \pm 2\pi, \ldots$, respectively, so the solution set of (17) is the intersection of those lines, the dotted points on the x axis in Fig. 8. Thus, the points $O, I, H, J, G, K, F, \ldots$ in Fig. 7 are the equilibrium points of (14).

Of particular interest in Fig. 7 is the **separatrix**, which is the collection of trajectories linking E and N with L and M, and which is depicted with slightly thicker lines for emphasis. It is called a separatrix because it separates qualitatively different flows, an interior flow consisting of closed loops and an exterior flow consisting of open paths on $-\infty < x < \infty$. (We're not in a position to give a precise definition of the term separatrix, but the idea that it is a trajectory or collection of trajectories that separates qualitatively different flows should suffice for present purposes.)

What value of C in (16) gives the separatrix? At the point I, for instance, which lies on the separatrix, $x = \pi$ and $y = 0$. With those values, (16) gives $0 = -1 + C$, so $C = 1$. Thus, the upper and lower branches of the separatrix are given by (16) as

$$\frac{1}{2}y^2 = \cos x + 1, \quad \text{that is, by} \quad y = \pm\sqrt{2\cos x + 2}. \tag{18}$$

For instance, at $x = 0$ equation (18) gives $y = +2$ at U and $y = -2$ at V.

Now we can explain how we selected C values in (16) to generate Fig. 7. We arbitrarily chose to have five trajectories cross the positive y axis and to have five cross the negative y axis, including the upper and lower branches of the separatrix. Since $y = +2$ at U and -2 at V, we sought the trajectories through $(0, \pm 2/3)$, $(0, \pm 4/3)$, $(0, \pm 2)$, $(0, \pm 8/3)$, and $(0, \pm 10/3)$. Putting $(0, \pm 10/3)$, for instance, into (16) gives $C = 41/9$ so the implicit form

$$\frac{1}{2}y^2 = \cos x + \frac{41}{9} \tag{19}$$

or, equivalently, the explicit form

$$y = \pm\sqrt{2\cos x + \frac{82}{9}}, \tag{20}$$

gives the trajectories AB and ST. Similarly, $(0, \pm 8/3)$ gives $C = 23/9$, $(0, \pm 2)$ gives $C = 1$, $(0, \pm 4/3)$ gives $C = -1/9$, and $(0, \pm 2/3)$ gives $C = -7/9$. These C values yield all the remaining trajectories shown in Fig. 7 [including the four almost circular trajectories near G and J, since $(\pm 2\pi, \pm 4/3)$ once again gives $C = -1/9$ and $(\pm 2\pi, \pm 2/3)$ once again gives $C = -7/9$].

We're denoting the point $(x(t), y(t))$ that we're tracking as Z, here and in Fig. 9.

To understand Fig. 7 in physical terms, imagine an experiment: If the pendulum hangs straight down ($x = 0$) and is at rest ($x' = 0$), it will remain at rest. That "motion" corresponds to a point solution, the equilibrium point O. If, instead, with $x(0) = 0$ we impart an initial velocity $x'(0) = y(0) > 0$ to the mass by striking it with a hammer, then, beginning at W (Fig. 9) at $t = 0$, the representative point Z will traverse the closed loop trajectory Γ_1 repeatedly so that its projection x (i.e., onto the x axis) undergoes a periodic motion of amplitude A. Physically, the pendulum swings back and forth with angular amplitude A.[1] If we strike the mass harder, so as to start the motion closer to U, there results a periodic motion Γ_2 of larger amplitude, and so on. As we impart more initial energy, the resulting

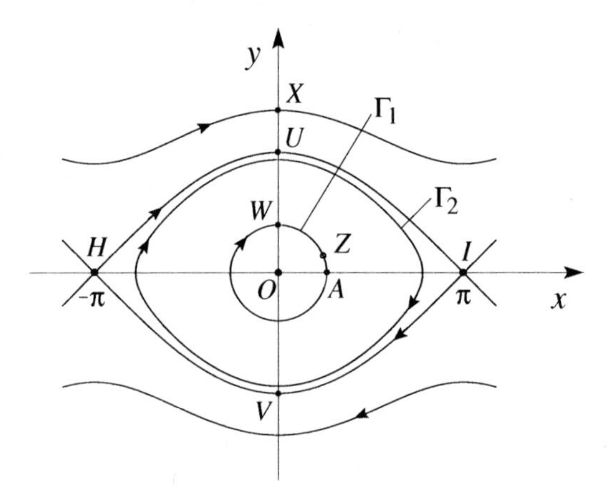

Figure 9. The periodic motions inside $HUIVH$.

"orbit" not only becomes larger, it also becomes distorted so as to conform to the pointed "football" shape of the separatrix $HUIVH$. By the **period** of an orbit Γ we mean the time that it takes, beginning at any point on Γ, to traverse Γ one time and return to the starting point. If we strike the mass hard enough so as to start at the point U *on* the separatrix, between H and I, then the point Z, on HUI, will approach I as $t \to \infty$ but not reach it in finite time. We leave the details of that claim to the exercises, and only suggest its truth by noting that the phase velocity

$$s' = \sqrt{x'^2 + y'^2} = \sqrt{y^2 + \sin^2 x} \tag{21}$$

is not only zero *at* the equilibrium point I, it tends to zero as the representative point Z *approaches* I. Thus, $HUIVH$ is not itself a periodic orbit, for if we begin at U, for instance, we never get past I because we approach I only as $t \to \infty$.

If we impart still more energy so as to start at X, say, then the pendulum does slow down as the pendulum approaches the inverted position $x = \pi$ (i.e., we see from the figure that the velocity $x' = y$ decreases as $x \to \pi$), but has enough energy to go "up and over." See CD in Fig. 7.

[1]Of course, to generate the motion Γ_1 we don't need to start at W, we can impose initial conditions to start at any point on Γ_1.

Do you see how we can initiate motions within the other "footballs?" To obtain an orbit around J, say, in Fig. 7, rotate the pendulum by hand counterclockwise through one revolution, then let it hang motionless. That step puts the representative point at J. If at time $t = 0$ we then strike the mass so as to impart a nonzero value of $y(0)$ [with $-2 < y(0) < 2$ so the representative point remains within the separatrix], we will generate an orbit within the separatrix between I and K.

7.2.3 Singular points and their stability. At an equilibrium point a unique slope is not defined because

$$\frac{dy}{dx} = \frac{Q(x,y)}{P(x,y)} \tag{22}$$

gives $\dfrac{dy}{dx} = \dfrac{0}{0}$ there. After all, an equilibrium point is a point solution, and a single point does not have a slope. Thus, we say that an equilibrium point of (1) is also a **singular point** of (1) in the sense that dy/dx is undefined there.

The concept of an equilibrium point is closely related to the concept of stability. For instance, a marble is in equilibrium in a valley or on a hill, but the former is stable and the latter is unstable. Similarly, we classify equilibrium points (i.e., singular points) in the phase plane as stable or unstable as follows:

A singular point S of the autonomous system (1) is **stable** *if motions (i.e., trajectories) that start out close to S remain close to S.* To make that intuitively stated definition precise, let us understand the distance between any two points $Z_1 = (x_1, y_1)$ and $Z_2 = (x_2, y_2)$ in the phase plane to be the usual Euclidean distance $\sqrt{(x_1 - x_2)^2 + (y_1 - y_2)^2}$, and let $Z(t) = (x(t), y(t))$ denote the representative point, as usual. Then, a singular point S of (1) is stable if, for any $\epsilon > 0$ (i.e., as small as we wish) there corresponds a $\delta > 0$ such that $Z(t)$ will remain closer to S than ϵ for all $t > 0$ if it starts out (i.e., at $t = 0$) closer to S than δ (Fig. 10a).[1] If S is not stable, it is **unstable**.

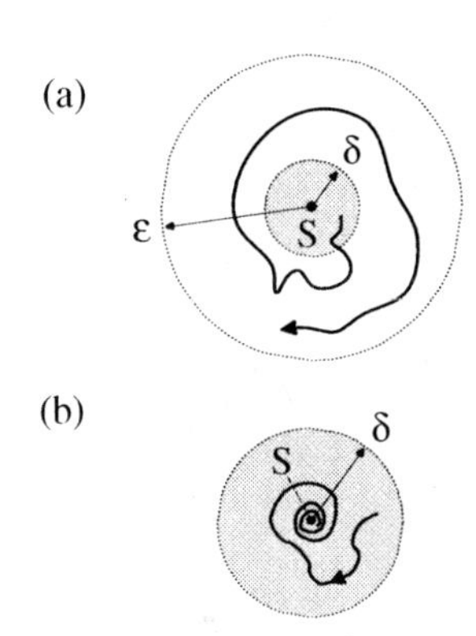

Figure 10. The singular point at S: (a) its stability; (b) its asymptotic stability.

Further, S is not only stable but is **asymptotically stable** if motions that start out sufficiently close to S not only stay close to S but *approach* S as $t \to \infty$. That is, S is asymptotically stable if there is a $\delta > 0$ such that the distance between $Z(t)$ and S *tends to zero as $t \to \infty$* if $Z(t)$ starts closer to S than δ (Fig. 10b).

For instance, in Fig. 7 the singular point G is stable because the trajectories that start out close to G remain close to G, but it is not asymptotically stable because those trajectories do not approach G as $t \to \infty$. Similarly for O and J (and all the singular points on the x axis at $x = 0, \pm 2\pi, \pm 4\pi, \ldots$). If we were to include an x' damping term in (13), then the flow would change and there would be *asymptotically stable* equilibrium points at those points instead; we leave that case for the exercises in Section 7.4.

However, the singular point F is unstable. It is true that initial points on EF and the lower branch of HF tend to F as $t \to \infty$ but, no matter how small we

[1] In general, we can expect that the smaller we choose ϵ to be, the smaller δ will have to be; $\delta(\epsilon)$ will be a function of ϵ.

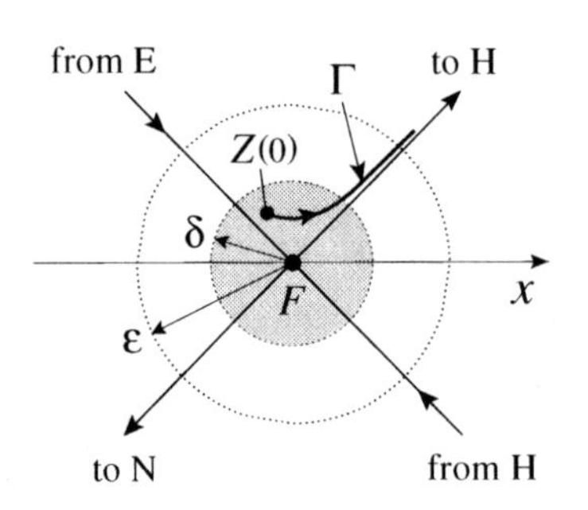

Figure 11. The instability of the singular point F in Fig. 7.

make δ, we cannot keep *all* trajectories originating within the δ-circle (such as Γ in Fig. 11) within a circle of radius $\epsilon = 0.1$, say, for all $t > 0$. Similarly for H, I, and K.

Closure. To apply the phase plane method to an autonomous system

$$\frac{dx}{dt} = P(x, y), \quad \frac{dy}{dt} = Q(x, y), \tag{23}$$

divide the second of these by the first to obtain

$$\frac{dy}{dx} = \frac{Q(x, y)}{P(x, y)}, \tag{24}$$

in which the time t does not appear. Rather than try to solve the coupled equations (23) for $x(t)$ and $y(t)$, which may be too difficult if (23) is nonlinear, try to solve the single equation (24). We can definitely solve (24) if (23) is linear, and possibly even if (23) is nonlinear. We will see, in subsequent sections, that even if we are not able to solve (23) we can still develop its phase portrait using a blend of singular point analysis and computer software.

To illustrate the phase plane method, we studied the nonlinear pendulum equation $x'' + \sin x = 0$ or, equivalently, the system

$$x' = y, \quad y' = -\sin x, \tag{25}$$

and found that its phase portrait (Fig. 7) showed, in a single plot, the variety of motions that are possible — periodic motions corresponding to the closed loop trajectories inside the separatrix, more energetic "up and over" motions corresponding to the trajectories above and below the separatrix, and the borderline motions on the separatrix itself. This compact overview of the system's behavior is a striking feature of phase plane analysis. Furthermore, the phase portrait was obtained by solving only the simple separable first-order equation $dy/dx = -\sin x/y$; we never solved the much more challenging second-order equation $x'' + \sin x = 0$.

The equilibrium points played an important role, so we focused attention on them and their stability in Section 7.2.3. Since at an equilibrium point both $P(x, y)$ and $Q(x, y)$ are zero, we see from (24) that the slope dy/dx is not defined there. In that sense an equilibrium point is a singular point of the differential equation (24) and we will subsequently call such points *singular points*. The term *critical point* is used by some authors.

CAUTION regarding existence and uniqueness.

We close this review with a word of caution. Regarding questions of existence and uniqueness, we must be clear whether we are thinking of the system (1) for $x(t)$ and $y(t)$, or the equation $dy/dx = Q(x, y)/P(x, y)$ for $y(x)$. For the latter we have Theorem 1.5.1, and for the former a suitable theorem is given in Section 7.6.2.

EXERCISES 7.2

1. Determine the phase trajectories and sketch representative trajectories. Use arrows to indicate the direction of movement along those trajectories. Determine any singular points and identify them in your phase portrait by heavy dots (or heavy lines if appropriate).

(a) $x' = y, \quad y' = -x$ (b) $x' = -y, \quad y' = x$
(c) $x' = xy, \quad y' = -x^2$ (d) $x' = y^2, \quad y' = -9xy$
(e) $x' = y, \quad y' = -y$ (f) $x' = x, \quad y' = 2x$

2. Is it possible for every point in the plane to be a singular point of (1)? Explain.

3. Computer. Generate Fig. 7 using computer software. HINT: Since we were able to integrate (15) you can simply plot (16) for the values of C that were cited in the text discussion of Fig. 7.

4. Graphical Interpretation of Phase Velocity. Consider the system (1) for the special case where $P(x, y) = y$. [This case is important because it occurs whenever we reduce a second-order equation $x'' = f(x, x')$ to a system of two first-order equations by setting $x' \equiv y$, as we did in (3).]

(a) From the accompanying sketch, show that in that case the phase velocity s' can be interpreted graphically as

$$\boxed{s' = a,} \tag{4.1}$$

where a is the distance, perpendicular to s', from $(x(t), y(t))$ to the x axis. [Remember that (4.1) holds only if $P(x, y) = y$.]

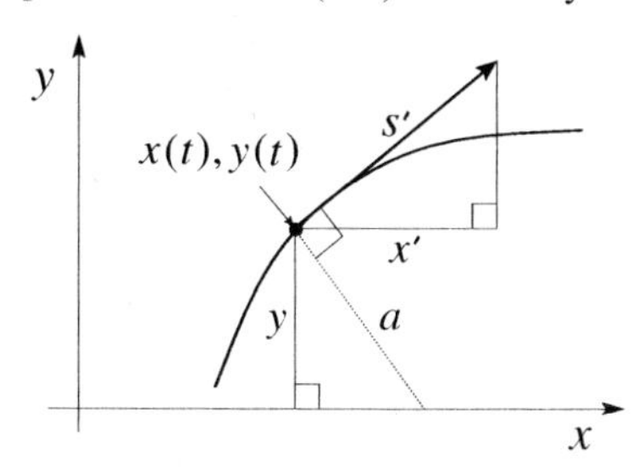

(b) To illustrate the use of (4.1), consider a phase trajectory that is a rectangle with vertices at the (x, y) points $(-A, B)$, (A, B), $(A, -B)$, and $(-A, -B)$, where $A > 0$ and $B > 0$. Denote those points as P_1, P_2, P_3, P_4, respectively, and the point $(0, B)$ as P_5. Show that the period of the periodic motion around that trajectory is $4A/B$, and give a labeled sketch of the graph of x versus t if the initial point is P_5. Show at least two periods and mark the points, on your plot of x versus t, that correspond to the P_j points in the phase plane.

5. Continuation From Exercise 4. Suppose $P(x, y) = y$ in (1) so that (4.1) holds, and consider a trajectory Γ that crosses the x axis at an angle other than $90°$, as sketched below at the left. Since $x' = y$, it follows that the flow directions must be as indicated by the arrows, and that the point $(x_0, 0)$ is necessarily an equilibrium point. We want to know if points on Γ approach the equilibrium point $(x_0, 0)$ only as $t \to \infty$, or if they reach it in finite time. Let it suffice to consider the simpler case where Γ is straight, as shown in the right-hand figure. *The problem is this*: Show that if the representative point Z starts on either the upper or lower part of Γ, it does *not* reach $(x_0, 0)$ in finite time. HINT: Write the equation of Γ as $y = -m(x - x_0)$, where $m > 0$, and remember that y is x'.

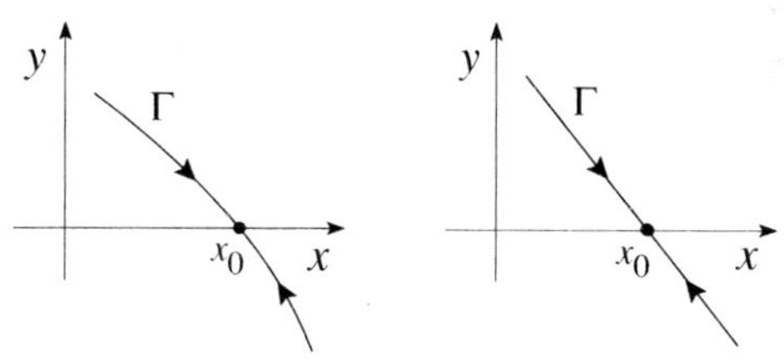

6. Nonlinear Pendulum; Computer. Consider the orbit Γ_1 in Fig. 9. We are interested in the period of that orbit as a function of its amplitude A, for $0 < A < \pi$. At one extreme, as $A \to 0$, we claim that the period corresponding to the full equation $x'' + \sin x = 0$ approaches the period 2π corresponding to the linearized version of the equation, $x'' + x = 0$. And at the other extreme, as $A \to \pi$, we claim that the period tends to infinity. Essentially, the idea is that not only is the flow velocity zero *at* the equilibrium point I, it is very small in the neghborhood of I. Thus, the closer the orbit Γ_2 comes to $HUIVH$ the longer it takes the representative point Z to pass through the neighborhood of I.

Here, we ask you to explore those two claims with calculations. Specifically, use computer software to solve (14) subject to the initial conditions $x(0) = A$, $y(0) = 0$, for $A = 0.1, 0.5, 1, 2, 2.5, 3, 3.1$, and to plot $x(t)$ versus t for these cases. From those plots measure the period for each case, and then plot (a hand plot will suffice) the period versus A, for those A values. Do your results appear to support the foregoing claims for the limiting cases $A \to 0$ and $A \to \pi$?

7. Predator–Prey System (7). Consider the predator–prey population dynamic system (7). Since x and y are populations, they cannot be negative, so we are interested only in the

first quadrant of the phase plane: $0 \leq x < \infty, 0 \leq y < \infty$.
(a) Determine, from the differential equations, any equilibrium points.
(b) Use computer software to obtain the direction field on the rectangle $0 \leq x \leq 3$ and $0 \leq y \leq 3$.
(c) Directly on your direction field sketch, by hand, the trajectories through the points $(2, 0), (0, 2), (1, 0.25), (1, 0.5)$, and $(1, 0.8)$.
(d) Determine, by any means, the periods of the orbits through $(1, 0.25), (1, 0.5)$, and $(1, 0.8)$.

8. Computer. (a) Just as we developed the phase portrait for the nonlinear pendulum equation

$$x'' + \sin x = 0 \tag{8.1}$$

in Fig. 2, we ask you to do the same for the equation

$$x'' + x - x^3 = 0. \tag{8.2}$$

HINT: This time you will find only three equilibrium points; be sure to identify them. There is a separatrix; be sure to include it and to label it in your phase portrait.
(b) Give the equation of the separatrix.

9. Linear Equations. Consider the system

$$x' = ax + by, \tag{9.1a}$$

$$y' = cx + dy, \tag{9.1b}$$

in which a, b, c, d are given constants. Show, in each case, that (9.1) has just one singular point, namely, at $x = y = 0$. Then, determine whether that singular point is stable or unstable. If it is stable, determine whether or not it is asymptotically stable. HINT: Solve (9.1) for $x(t)$ and $y(t)$ by any method studied in Chapter 4, and examine those solutions.
(a) $a = 0, b = 1, c = -9, d = 0$
(b) $a = 0, b = 1, c = 4, d = 0$
(c) $a = 0, b = 1, c = -5, d = -2$
(d) $a = 0, b = 1, c = -5, d = 2$
(e) $a = 1, b = -5, c = 2, d = -1$
(f) $a = 1, b = 1, c = -1, d = 1$
(g) $a = 1, b = 1, c = 1, d = 2$
(h) $a = 1, b = 2, c = 2, d = 1$

10. Conservation of Energy.
(a) Show that (16) is a statement of conservation of energy (i.e., the kinetic energy plus the potential energy is a constant).
(b) More generally, suppose that Newton's second law gives

$$mx'' = F(x), \tag{10.1}$$

that is, in which the force F is an explicit function of x but not of x' or t. Show that if we multiply (10.1) by dx and integrate, we obtain

$$\frac{1}{2}mx'^2 + V(x) = \text{constant}, \tag{10.2}$$

where $V(x) = -\int F(x)\, dx$. The latter is a statement of conservation of energy and $V(x)$ is called the *potential energy* of the system. The system (10.1) is therefore said to be *conservative*.

ADDITIONAL EXERCISES

11. Another Nonlinear Oscillator; Duffing's Equation. We assumed in the harmonic oscillator equation $mx'' + cx' + kx = 0$ that the spring is linear, namely, that the spring force F_s is given by Hooke's law as $F_s = kx$. For small amounts of stretch or compression, this formula can be quite accurate. However, for larger displacements, many springs behave somewhat as sketched in the figure below, that is, the slope increases as x increases in magnitude. For instance, for a coil spring one finds that the spring becomes stiffer

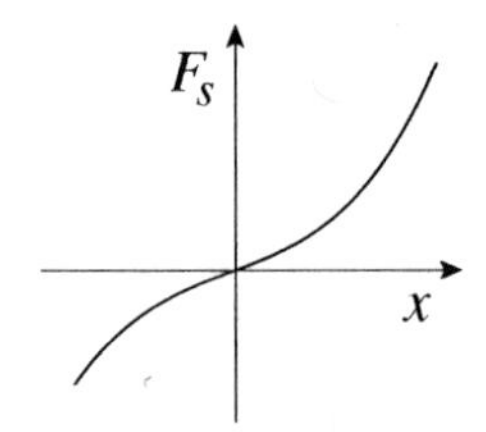

as the extension or compression is increased. To include that effect, replace the linear function $F_s(x) = kx$ by the nonlinear function

$$F_s(x) = \alpha x + \beta x^3, \tag{11.1}$$

in which α and β are empirically known positive constants. If we use (11.1), then in place of the linear harmonic oscillator equation $mx'' + cx' + kx = 0$ we have the *nonlinear* equation

$$mx'' + cx' + \alpha x + \beta x^3 = 0 \tag{11.2}$$

known as **Duffing's equation** after *G. Duffing*, who studied it around 1918. For definiteness, let $m = \alpha = \beta = 1$ and $c = 0$, so

$$x'' + x + x^3 = 0. \tag{11.3}$$

(a) Setting $x' = y$, it follows from (11.3) that $y' = -x - x^3$. Show that the equation of the phase trajectories is

$$\frac{1}{2}y^2 + \frac{1}{2}x^2 + \frac{1}{4}x^4 = C. \tag{11.4}$$

NOTE: Notice how readily the equation of the trajectories, (11.4), is found even though (11.3) is nonlinear.
(b) Using computer software, obtain the trajectories through the initial points $(0.25, 0), (1, 0), (2, 0), (3, 0)$. Your results should agree with the solid curves in the figure given at the right.
(c) One way to see the effect of the nonlinear term on the solutions, is to draw a circle centered at the origin through each of the four initial points given above, as indicated in the figure by the dotted curves, because that's what the trajectories *would* be if the x^3 term in (11.3) were omitted. Briefly discuss the deviation between the solid trajectories and those circles.
(d) Show that seeking a solution of (11.3) in the form $x(t) = e^{rt}$ does *not* work.

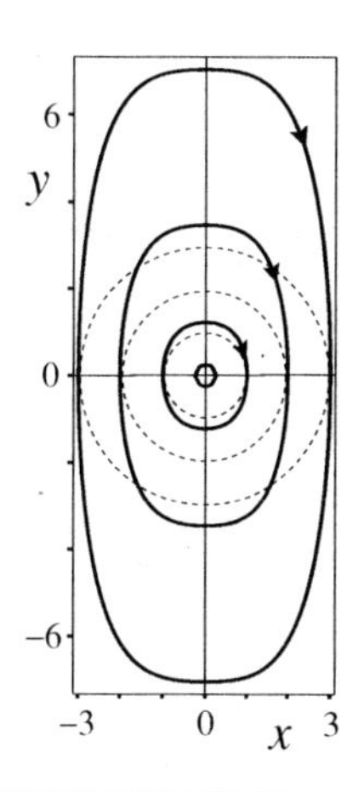

7.3 LINEAR SYSTEMS

7.3.1 Introduction. With the basic phase plane definitions and concepts established, we can begin a systematic study — of linear systems in this section, and nonlinear systems in the next section. Consider the linear system

PREREQUISITE: Section 4.6 on the eigenvalue problem.

$$\frac{dx}{dt} = ax + by, \qquad \frac{dy}{dt} = cx + dy, \tag{1a,b}$$

with real constant coefficients a, b, c, d.

Recall that singular points of (1) are found by setting $dx/dt = P(x, y) = 0$ and $dy/dt = Q(x, y) = 0$:

$$P(x, y) = ax + by = 0, \tag{2a}$$
$$Q(x, y) = cx + dy = 0, \tag{2b}$$

which has the unique solution $x = y = 0$ if

$$\begin{vmatrix} a & b \\ c & d \end{vmatrix} = ad - bc \neq 0. \tag{3}$$

In this section we assume (3) holds, so (1) has only the one singular point at the origin. For an application in which (3) is not satisfied, see Exercise 15.

If $ad - bc$ *is* zero, then the singular points of (1) [namely, the solutions of (2)] constitute either a line through the origin, or even the entire x, y plane, the latter occurring in the uninteresting case in which $a = b = c = d = 0$. Here, we assume that (3) is satisfied, so (1) has only the one singular point at the origin.

Divide (1b) by (1a), obtaining

Then solve (4) to get the phase trajectories.

$$\frac{dy}{dx} = \frac{cx + dy}{ax + by}. \tag{4}$$

The latter is nonlinear (unless $b = 0$), but it is homogeneous and can therefore be solved (Section 1.8.2). Its solution can be quite messy, but it does give the phase trajectories.

Alternatively, we could solve (1) for $x(t)$ and $y(t)$ using either of the methods given in Chapter 4 — elimination, or the method of assumed exponential form. We will use the latter, and thus seek a solution of (1) in the form

$$x(t) = q_1 e^{rt}, \quad y(t) = q_2 e^{rt}. \tag{5}$$

Put (5) into (1), cancel the e^{rt}'s, and obtain

$$\begin{bmatrix} a & b \\ c & d \end{bmatrix} \begin{bmatrix} q_1 \\ q_2 \end{bmatrix} = r \begin{bmatrix} q_1 \\ q_2 \end{bmatrix} \tag{6}$$

or

$$\mathbf{A}\mathbf{q} = r\mathbf{q}.$$

The exceptional case, in which $\lambda_1 = \lambda_2$, will be considered last.

By Theorem 4.7.1, if $\mathbf{A}$ has distinct eigenvalues λ_1 and λ_2 with corresponding eigenvectors

$$\mathbf{e}_1 = \alpha \begin{bmatrix} e_{11} \\ e_{12} \end{bmatrix} \quad \text{and} \quad \mathbf{e}_2 = \beta \begin{bmatrix} e_{21} \\ e_{22} \end{bmatrix}, \tag{7}$$

then (1) has the general solution

Whether we use the α and β from (7) for the arbitrary constants, or change them to C_1 and C_2 to use the notation in Theorem 4.7.1, surely doesn't matter.

$$\begin{bmatrix} x(t) \\ y(t) \end{bmatrix} = \mathbf{e}_1 e^{\lambda_1 t} + \mathbf{e}_2 e^{\lambda_2 t} = C_1 \begin{bmatrix} e_{11} \\ e_{12} \end{bmatrix} e^{\lambda_1 t} + C_2 \begin{bmatrix} e_{21} \\ e_{22} \end{bmatrix} e^{\lambda_2 t} \tag{8}$$

or, in scalar form,

$$x(t) = C_1 e_{11} e^{\lambda_1 t} + C_2 e_{21} e^{\lambda_2 t}, \tag{9a}$$
$$y(t) = C_1 e_{12} e^{\lambda_1 t} + C_2 e_{22} e^{\lambda_2 t}, \tag{9b}$$

in which C_1 and C_2 are arbitrary constants.

The eigenvalues of $\mathbf{A}$ are the roots of the characteristic equation

$$\det(\mathbf{A} - \lambda\mathbf{I}) = \begin{vmatrix} a-\lambda & b \\ c & d-\lambda \end{vmatrix} = \lambda^2 - (a+d)\lambda + (ad - bc) = 0, \tag{10}$$

namely, $\lambda = [a + d \pm \sqrt{(a+d)^2 - 4(ad - bc)}\,]/2$, or

$$\lambda = \frac{a + d \pm \sqrt{(a-d)^2 + 4bc}}{2}. \tag{11}$$

Neither of these λ's can be zero because zero is a root of (10) only if $ad - bc = 0$, and we've assumed that $ad - bc \neq 0$.

The qualitative nature of the flow in the phase plane depends entirely upon the eigenvalues given by (11), and there are four possibilities:

1. purely imaginary complex conjugates, $\lambda = \pm i\omega$
2. complex conjugates with nonzero real part, $\lambda = \mu \pm i\omega$
3. real, of the same sign
4. real, of opposite sign

Consider these cases, in turn.

7.3.2 Purely imaginary eigenvalues (CENTER). To get purely imaginary eigenvalues, we see from (11) that we need $a + d = 0$ and $(a-d)^2 + 4bc < 0$. With $d = -a$, the latter inequality becomes $4(a^2 + bc) < 0$. Thus, if $a + d = 0$ and $a^2 + bc < 0$ then (11) gives purely imaginary eigenvalues

$$\lambda = \pm i\sqrt{|a^2 + bc|} \equiv \pm i\omega, \tag{12}$$

in which we call the square root ω for brevity: take $\lambda_1 = +i\omega$ and $\lambda_2 = -i\omega$. Since the $e^{\lambda_1 t}$ and $e^{\lambda_2 t}$ terms in (8) are $e^{i\omega t}$ and $e^{-i\omega t}$, both the $x(t)$ and $y(t)$ motions will be linear combinations of $\cos \omega t$ and $\sin \omega t$ and will therefore be oscillatory. We state without proof that *the trajectories in the x, y phase plane will be a family of ellipses. Their principal axes will be either the x, y axes, or axes rotated relative to the x, y axes.* The singular point at the origin is called a **center**.

EXAMPLE 1. Center. To illustrate, consider

$$x' = 4y, \tag{13a}$$
$$y' = -x. \tag{13b}$$

Then $a = 0$, $b = 4$, $c = -1$, and $d = 0$ so (11) gives $\lambda = \pm 2i$. Since the λ's are purely imaginary, the singularity is a center. As we mentioned, integrating (4) can lead to a messy relation on x and y for the phase trajectories, but in this case it is simple, because if we divide (13b) by (13a) and integrate, we obtain

$$x^2 + 4y^2 = C, \tag{14}$$

which is the family of ellipses shown in Fig. 1. We've included the direction field in this figure, but in subsequent figures in this section we will omit it.

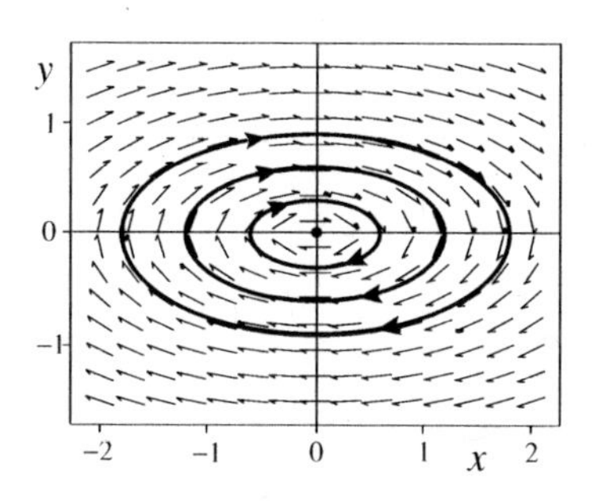

Figure 1. Example 1; a center. In this case the ellipses are not rotated relative to the Cartesian axes.

COMMENT 1. For completeness, let us also obtain the solution for $x(t)$ and $y(t)$ in the form (8). If we solve the eigenvalue problem for $\mathbf{A} = \begin{bmatrix} 0 & 4 \\ -1 & 0 \end{bmatrix}$, we obtain

$$\lambda_1 = 2i,\ \mathbf{e}_1 = \alpha \begin{bmatrix} 2 \\ i \end{bmatrix}; \qquad \lambda_2 = -2i,\ \mathbf{e}_2 = \beta \begin{bmatrix} 2 \\ -i \end{bmatrix}, \tag{15}$$

so (8) gives a general solution of (13) as

$$\begin{bmatrix} x(t) \\ y(t) \end{bmatrix} = C_1 \begin{bmatrix} 2 \\ i \end{bmatrix} e^{i2t} + C_2 \begin{bmatrix} 2 \\ -i \end{bmatrix} e^{-i2t}. \tag{16}$$

COMMENT 2. You may wonder how we could plot the solutions given in (16) since they appear to be complex. However, if you apply real initial conditions $x(t_0) = x_0$ and $y(t_0) = y_0$ to (16), solve for C_1 and C_2, and use Euler's formulas for the e^{i2t_0} and e^{-i2t_0}, you will find that C_1 and C_2 are complex conjugates, and that $x(t)$ and $y(t)$ are real. Put differently, have confidence that (16) *must* be consistent with the ellipses found in (14). ▮

EXAMPLE 2. Typically, However, the Principal Axes are Rotated. In Example 1 the ellipses were aligned with the x, y axes. More typically, if (1) is a center, the ellipses are rotated relative to those axes. For instance, consider

$$x' = x - 3y, \tag{17a}$$

$$y' = 2x - y. \tag{17b}$$

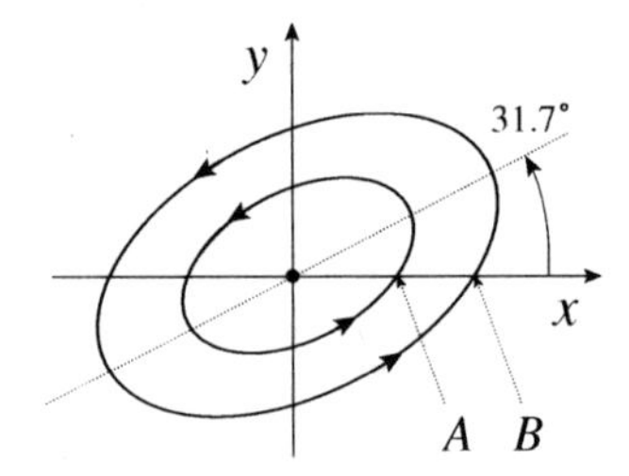

Figure 2. Example 2; ellipses with rotated principal axes.

Then $a = 1, b = -3, c = 2$, and $d = -1$ so (11) gives $\lambda = \pm\sqrt{5}i$. Hence the singularity at the origin is indeed a center. If we divide (17b) by (17a) and solve (Exercise 4), we obtain the trajectories

$$2x^2 - 2xy + 3y^2 = C \tag{18}$$

which, like (14), is a family of ellipses, but this time with their axes rotated counterclockwise by $31.7°$, as indicated in Fig. 2 (Exercises 4 and 5).

COMMENT. In Example 1 the flow was clockwise because (13a) showed that $x' > 0$ in the upper half plane and that $x' < 0$ in the lower half plane. In the present example it suffices to look on the x axis, by setting $y = 0$ in (17a). Then $x' = x$, so the x velocity component is positive at A and B (Fig. 2). Hence, the flow is counterclockwise. ▮

7.3.3 Complex conjugate eigenvalues (SPIRAL).

Next, suppose the eigenvalues are complex, but not purely imaginary. That is, $a + d$ in (11) is nonzero, and $(a-d)^2 + 4bc < 0$. Then we can express the eigenvalues in the form $\lambda = \mu \pm i\omega$, and we see from (8) that $x(t)$ and $y(t)$ are linear combinations of terms of the form

$$e^{\mu t} e^{\pm i\omega t},$$

so each is $e^{\mu t}$ times a linear combination of $\cos \omega t$ and $\sin \omega t$. Thus, the motion is an oscillation that grows exponentially in magnitude if $\mu > 0$, or decays exponentially if $\mu < 0$. The trajectories spiral away from the origin or toward the origin, so the singularity is called a **spiral**. If $a + d < 0$ the spiral is *asymptotically stable*; if $a + d > 0$, it is *unstable*.

EXAMPLE 3. Spiral. Consider

$$x' = 2x - 4y, \tag{19a}$$

$$y' = 3x - y. \tag{19b}$$

Then $a = 2$, $b = -4$, $c = 3$, and $d = -1$ so (11) gives $\lambda = \frac{1}{2} \pm i\frac{\sqrt{39}}{2}$. Since the λ's are complex conjugates with a positive real part, the singularity is an unstable spiral. We could divide (19b) by (19a) and solve the differential equation, but the steps would be tedious. Instead, we've used computer software to obtain the representative trajectories shown in Fig. 3a.

COMMENT. To illustrate a *stable* spiral, we can simply reverse all signs on the right-hand side of (19), obtaining

$$x' = -2x + 4y, \tag{20a}$$

$$y' = -3x + y. \tag{20b}$$

Then the ratio $dy/dx = (-3x + y)/(-2x + 4y) = (3x - y)/(2x - 4y)$ is the same as for (19). Hence, the trajectories of (20) are the same as for (19), but the flow direction is reversed because the signs of x' and y' are reversed (Fig. 3b). ▮

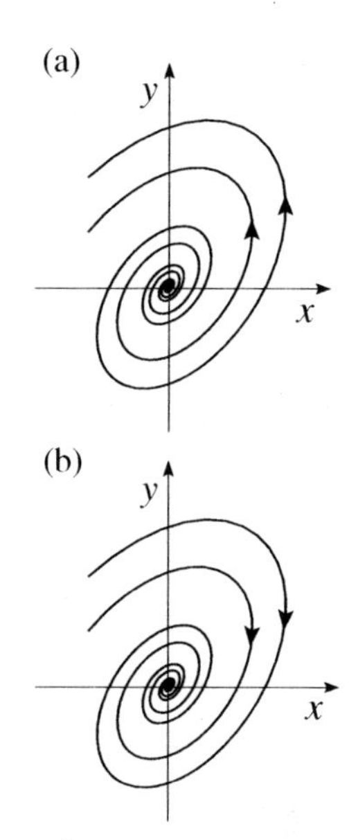

Figure 3. (a) The system (19) has an unstable spiral. (b) The system (20) has a stable spiral.

7.3.4 Real eigenvalues of the same sign (NODE).

Recall from (8) that

$$\begin{bmatrix} x(t) \\ y(t) \end{bmatrix} = \mathbf{e}_1 e^{\lambda_1 t} + \mathbf{e}_2 e^{\lambda_2 t} = C_1 \begin{bmatrix} e_{11} \\ e_{12} \end{bmatrix} e^{\lambda_1 t} + C_2 \begin{bmatrix} e_{21} \\ e_{22} \end{bmatrix} e^{\lambda_2 t}. \tag{21}$$

If the eigenvalues are real and *both negative* or *both positive*, the singularity is a **node**.

Since both λ_1 and λ_2 are real, the exponentials $e^{\lambda_1 t}$ and $e^{\lambda_2 t}$ are monotone functions of t, and the motion is nonoscillatory. From (21), if the λ's are negative the solutions (for any values of C_1 and C_2) tend to zero as $t \to \infty$, so the origin is an asymptotically stable singular point, a **stable node**. Conversely, if both λ's are positive then the solutions grow unboundedly as $t \to \infty$, and the origin is an **unstable node**. Here are two examples:

For nodes, both λ's are real, with both negative or both positive.

EXAMPLE 4. Stable Node. Consider

$$x' = -4x + 3y, \tag{22a}$$

$$y' = x - 2y. \tag{22b}$$

Here $a = -4$, $b = 3$, $c = 1$, and $d = -2$, so (11) gives $\lambda_1 = -1$ and $\lambda_2 = -5$. Since both λ's are real and negative, the singularity is a stable node. The corresponding eigenvectors are as follows:

$$\lambda_1 = -1,\ \mathbf{e}_1 = \alpha \begin{bmatrix} 1 \\ 1 \end{bmatrix}; \qquad \lambda_2 = -5,\ \mathbf{e}_2 = \beta \begin{bmatrix} 3 \\ -1 \end{bmatrix}, \tag{23}$$

so a general solution of (22) is

$$\begin{bmatrix} x(t) \\ y(t) \end{bmatrix} = \mathbf{e}_1 e^{\lambda_1 t} + \mathbf{e}_2 e^{\lambda_2 t} = C_1 \begin{bmatrix} 1 \\ 1 \end{bmatrix} e^{-t} + C_2 \begin{bmatrix} 3 \\ -1 \end{bmatrix} e^{-5t}, \tag{24}$$

and the phase portrait is shown in Fig. 4.

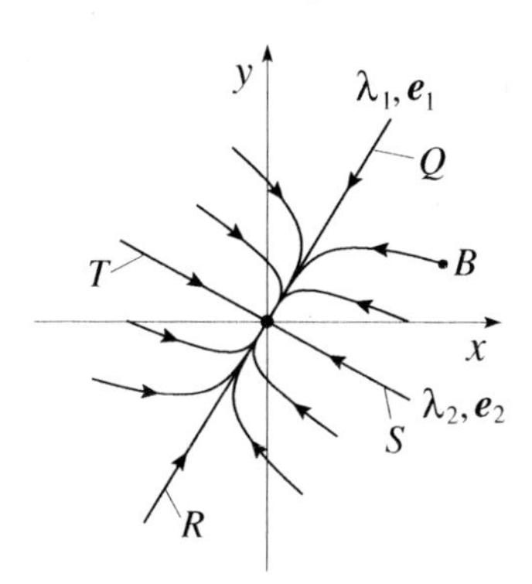

Figure 4. Phase portrait of (22). Both λ's negative; a stable node.

The phase portrait includes *straight line trajectories*, labeled Q, R, S, and T, corresponding to the eigenvectors $\mathbf{e}_1$ and $\mathbf{e}_2$. If the initial conditions give $C_2 = 0$ in (24), then the trajectory is along the $\mathbf{e}_1$ line: Q if $C_1 > 0$, and R if $C_1 < 0$. Similarly, if the initial conditions result in $C_1 = 0$ then the trajectory is along the $\mathbf{e}_2$ line: S if $C_2 > 0$ and T if $C_2 < 0$. The approach to the origin along S or T is much faster than the approach along Q or R because the $e^{\lambda_2 t} = e^{-5t}$ factor tends to zero exponentially faster than the $e^{\lambda_1 t} = e^{-t}$ factor. Thus, the $\mathbf{e}_2$ line (containing S and T) is called the **fast manifold** and the $\mathbf{e}_1$ line (containing the trajectories Q and R) is the **slow manifold**.

By $\mathbf{x}(t)$, we mean $\begin{bmatrix} x(t) \\ y(t) \end{bmatrix}$.

As typical of the other trajectories, consider the one through the initial point B in Fig. 4. From (24), $\mathbf{x}(t) \sim \mathbf{e}_1 e^{\lambda_1 t} = \mathbf{e}_1 e^{-t}$ as $t \to +\infty$ and $\mathbf{x}(t) \sim \mathbf{e}_2 e^{\lambda_2 t} = \mathbf{e}_2 e^{-5t}$ as $t \to -\infty$. That is, if we start at B and go *back* in time, the trajectory becomes parallel to $\mathbf{e}_2$ as $t \to -\infty$, and if we go *forward* in time, the trajectory turns and aligns with $\mathbf{e}_1$ as $t \to +\infty$.

COMMENT 1. To create the phase portrait in Fig. 4, first determine the straight line trajectories (defined by $\mathbf{e}_1$ and $\mathbf{e}_2$) and include an initial point on each one of the four, and then choose a number of additional initial points to fill out the flow field.

COMMENT 2. It looks as though there are two straight-line trajectories, but the $\mathbf{e}_1$ and $\mathbf{e}_2$ lines in Fig. 4 are not single trajectories; each is comprised of *three*: For $\mathbf{e}_1$ there are the half-line trajectories Q and R, and the point trajectory at the origin, and similarly for $\mathbf{e}_2$. ∎

EXAMPLE 5. Unstable Node. To illustrate an unstable node, consider

$$x' = 2x + y \tag{25a}$$

$$y' = x + 2y. \tag{25b}$$

The eigenvalues and eigenvectors of $\mathbf{A} = \begin{bmatrix} 2 & 1 \\ 1 & 2 \end{bmatrix}$ are

$$\lambda_1 = 3,\ \mathbf{e}_1 = \alpha \begin{bmatrix} 1 \\ 1 \end{bmatrix}; \qquad \lambda_2 = 1,\ \mathbf{e}_2 = \beta \begin{bmatrix} 1 \\ -1 \end{bmatrix}, \tag{26}$$

so a general solution of (25) is given by (8) as

$$\begin{bmatrix} x(t) \\ y(t) \end{bmatrix} = C_1 \begin{bmatrix} 1 \\ 1 \end{bmatrix} e^{3t} + C_2 \begin{bmatrix} 1 \\ -1 \end{bmatrix} e^{t}. \tag{27}$$

This time both λ's are positive, so the singularity is an unstable node. Representative trajectories are shown in Fig. 5. ∎

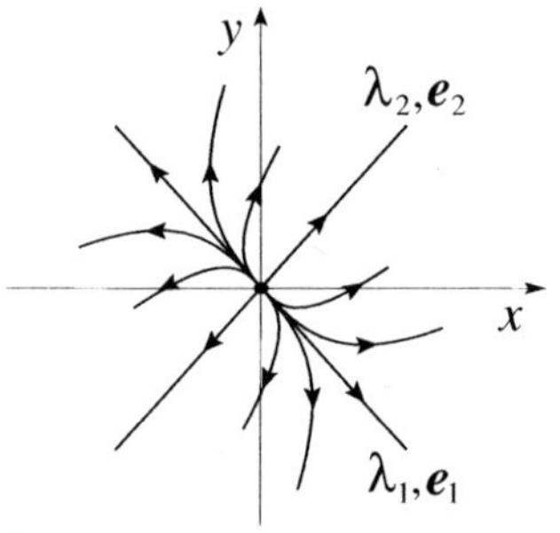

Figure 5. Both λ's positive, so (25) has an unstable node. The slow and fast manifolds are perpendicular because the eigenvectors $[1, 1]^{\mathrm{T}}$ and $[1, -1]^{\mathrm{T}}$ are perpendicular to each other.

In the foregoing discussion of nodes we have assumed that the two eigenvalues are distinct, in which case we classify the node as a stable or unstable **ordinary node**. If they are equal there are two cases, depending on whether the corresponding eigenspace is one- or two-dimensional (i.e., depending on whether there are one or two LI eigenvectors).

Suppose first that

$$\mathbf{A} = \begin{bmatrix} a & 0 \\ 0 & a \end{bmatrix} \qquad (a \neq 0); \tag{28}$$

that is, $b = c = 0$ and $d = a$. Then we have the repeated eigenvalue $\lambda_1 = \lambda_2 = a$ and the *two*-dimensional eigenspace

$$\mathbf{e} = \begin{bmatrix} \alpha \\ \beta \end{bmatrix} = \alpha \begin{bmatrix} 1 \\ 0 \end{bmatrix} + \beta \begin{bmatrix} 0 \\ 1 \end{bmatrix}, \tag{29}$$

two-dimensional because the $[1,0]^{\mathrm{T}}$ and $[0,1]^{\mathrm{T}}$ in (29) are LI. In this case (1) is of the form

$$x' = ax, \quad y' = ay, \tag{30}$$

which is readily solved because it is not coupled. Its general solution is

$$x(t) = C_1 e^{at}, \tag{31a}$$
$$y(t) = C_2 e^{at}, \tag{31b}$$

in which C_1 and C_2 are arbitrary constants. Then x and y are proportional to one another, so the trajectories are rays, as in Fig. 6a for $a < 0$ and as in Fig. 6b for $a > 0$. In this case the node is called a **star**, stable if $a < 0$ and unstable if $a > 0$. It is not difficult to show that *the special form (28) is the only form of* $\mathbf{A}$ *that gives a repeated eigenvalue with two LI eigenvectors*, but we leave that point for the exercises (Exercise 9).

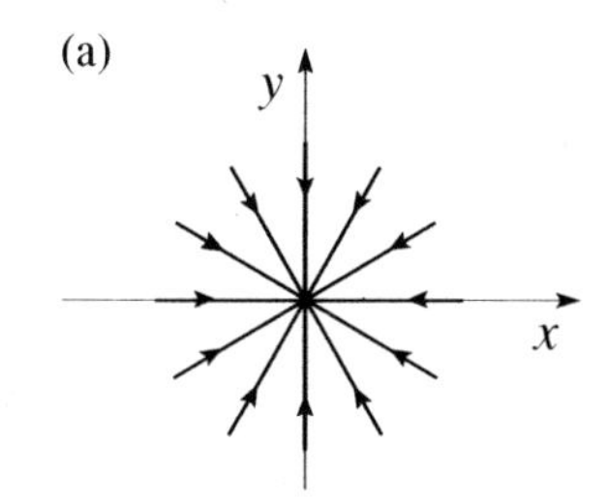

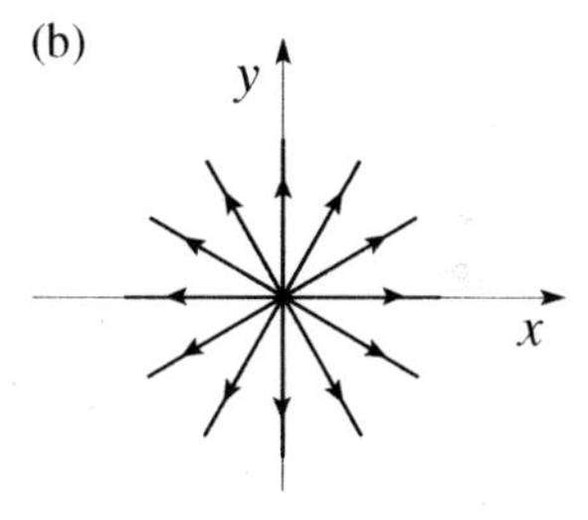

Figure 6. Repeated eigenvalue with two-dimensional eigenspace: (a) stable star for $a < 0$ in (28), (b) unstable star for $a > 0$ in (28).

Now consider the other case, where $\lambda_1 = \lambda_2$ is a repeated eigenvalue having only a one-dimensional eigenspace. In that case we come up short, and instead of obtaining a general solution of the form (21) we obtain only

$$\begin{bmatrix} x(t) \\ y(t) \end{bmatrix} = \mathbf{e}_1 e^{\lambda_1 t} + ?, \tag{32}$$

with the question mark reminding us that we're missing a second LI solution. To find the missing solution, we can seek it in the form

$$\mathbf{x}(t) = (\mathbf{q} + \mathbf{p}t)e^{\lambda_1 t}. \tag{33}$$

The modified form (33) was discussed in Section 4.7.4, but that section is not a prerequisite for this one.

To find the vectors $\mathbf{q}$ and $\mathbf{p}$, put (33) into $\mathbf{x}' = \mathbf{A}\mathbf{x}$ and cancel the $e^{\lambda_1 t}$'s, obtaining

$$\mathbf{p} + \lambda_1(\mathbf{q} + \mathbf{p}t) = \mathbf{A}\mathbf{q} + t\mathbf{A}\mathbf{p}$$

or

$$(\mathbf{p} + \lambda_1 \mathbf{q})1 + (\lambda_1 \mathbf{p})t = (\mathbf{A}\mathbf{q})1 + (\mathbf{A}\mathbf{p})t. \tag{34}$$

The 1's and t's in (34) are LI, so we can match coefficients and write

$$1: \quad \mathbf{A}\mathbf{q} = \lambda_1 \mathbf{q} + \mathbf{p}, \tag{35a}$$
$$t: \quad \mathbf{A}\mathbf{p} = \lambda_1 \mathbf{p}. \tag{35b}$$

It follows from (35b) that $\mathbf{p}$ is the eigenvector $\mathbf{e}_1$ (which, along with λ_1, was already found when we solved the eigenvalue problem corresponding to the matrix $\mathbf{A}$). If we put $\mathbf{p} = \mathbf{e}_1$ into (35a) and solve the latter by Gauss elimination for $\mathbf{q}$, we will have $\mathbf{q}$ and $\mathbf{p}$ in (33) and hence the second LI solution needed in (32). Thus, we can write a general solution of (1) in the form

The underbraced term in (36) is the solution that was missing in (32).

$$\mathbf{x}(t) = \mathbf{e}_1 e^{\lambda_1 t} + \underbrace{(\mathbf{q} + \mathbf{e}_1 t)e^{\lambda_1 t}}. \tag{36}$$

As $t \to \infty$, the $\mathbf{e}_1 t e^{\lambda_1 t}$ is dominant in (36) because of the extra t, so

$$\mathbf{x}(t) \sim \mathbf{e}_1 t e^{\lambda_1 t} \tag{37}$$

as $t \to +\infty$. The upshot is that *all trajectories align with* $\mathbf{e}_1$ *as* $t \to \infty$. The following example illustrates this case.

EXAMPLE 6. Repeated Eigenvalue and One-Dimensional Eigenspace. For the system

$$x' = -x + y, \tag{38a}$$
$$y' = -x - 3y, \tag{38b}$$

we find that $\lambda_1 = \lambda_2 = -2$ with the one-dimensional eigenspace $\mathbf{e}_1 = \alpha[1, -1]^{\mathrm{T}}$. Using the steps described above we can determine $\mathbf{q}$ and $\mathbf{p}$ and hence the general solution. Namely, with $\mathbf{p} = \alpha[1, -1]^{\mathrm{T}}$ and $\lambda_1 = -2$, (35a) becomes

$$\begin{bmatrix} -1 & 1 \\ -1 & -3 \end{bmatrix} \begin{bmatrix} q_1 \\ q_2 \end{bmatrix} = -2 \begin{bmatrix} q_1 \\ q_2 \end{bmatrix} + \alpha \begin{bmatrix} 1 \\ -1 \end{bmatrix},$$

which gives $q_1 + q_2 = \alpha$ and $-q_1 - q_2 = -\alpha$, with solution $q_2 = C$ (arbitrary constant) and $q_1 = \alpha - C$. Then the general solution of (38) is given by (36) as

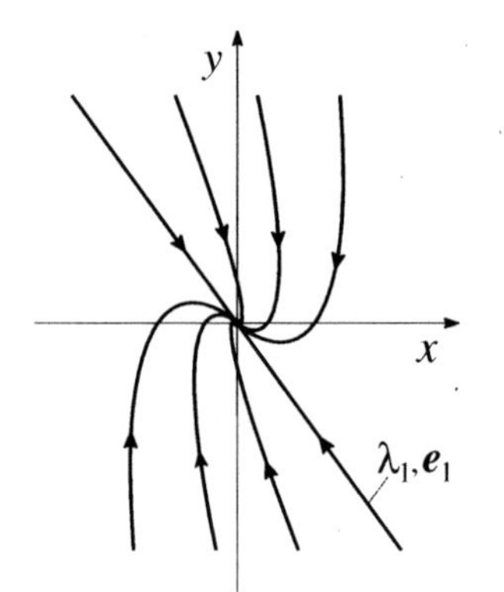

Figure 7. Phaseportrait of (38), a degenerate node.

$$\begin{aligned} \begin{bmatrix} x(t) \\ y(t) \end{bmatrix} &= \alpha \begin{bmatrix} 1 \\ -1 \end{bmatrix} e^{-2t} + \left(\begin{bmatrix} \alpha - C \\ C \end{bmatrix} + \alpha \begin{bmatrix} 1 \\ -1 \end{bmatrix} t \right) e^{-2t} \\ &= C \begin{bmatrix} -1 \\ 1 \end{bmatrix} e^{-2t} + \alpha \left(\begin{bmatrix} 1 \\ -1 \end{bmatrix} + \begin{bmatrix} 1 \\ 0 \end{bmatrix} + \begin{bmatrix} 1 \\ -1 \end{bmatrix} t \right) e^{-2t} \\ &= C_1 \begin{bmatrix} 1 \\ -1 \end{bmatrix} e^{-2t} + C_2 \left(\begin{bmatrix} 2 \\ -1 \end{bmatrix} + \begin{bmatrix} 1 \\ -1 \end{bmatrix} t \right) e^{-2t}. \end{aligned} \tag{39}$$

In the last step we renamed C as $-C_1$ and α as C_2, just to be consistent with our usual notation. The phase portrait for (38) is shown in Fig. 7, and we see the approach of *all* trajectories to the $\mathbf{e}_1$ line as $t \to \infty$, as promised by (37). ▮

This case, a repeated eigenvalue with only a one-dimensional eigenspace, is called a **degenerate node** because the two "eigenmanifolds" (e.g., the $\mathbf{e}_1$ and $\mathbf{e}_2$ lines in Fig. 4) coalesce, or degenerate, into one, as in Fig. 7. The degenerate node in Example 6 was stable because $\lambda_1 = -2 < 0$. If λ_1 were positive, there would again be a single eigenmanifold, but the flow would be away from the origin and the singularity would be an unstable degenerate node.

Summary for nodes: If both λ's are real and of the same sign, then (1) has a node — stable if they are negative, unstable if they are positive. If the λ's are distinct, the node is an ordinary node; but if they are equal, there are two cases: If their eigenspace is two-dimensional, we have a star; and if it is one-dimensional, we have a degenerate node.

7.3.5 Real eigenvalues of opposite sign (SADDLE). Finally, suppose the eigenvalues are real and of *opposite* sign, with $\lambda_1 < 0$ and $\lambda_2 > 0$ for definiteness. In that case the singularity at the origin is a **saddle**.

The general solution is given by (8) as

For saddles, both λ's are real, with one positive and one negative.

$$\begin{bmatrix} x(t) \\ y(t) \end{bmatrix} = \mathbf{e}_1 e^{\lambda_1 t} + \mathbf{e}_2 e^{\lambda_2 t} = C_1 \begin{bmatrix} e_{11} \\ e_{12} \end{bmatrix} e^{\lambda_1 t} + C_2 \begin{bmatrix} e_{21} \\ e_{22} \end{bmatrix} e^{\lambda_2 t}, \tag{40}$$

in which $\lambda_1 < 0$ and $\lambda_2 > 0$. If the initial conditions give $C_2 = 0$, then the solution $C_1[e_{11}, e_{12}]^{\mathrm{T}} e^{\lambda_1 t}$ gives a straight-line trajectory along the $\mathbf{e}_1$ line; and if instead the initial conditions give $C_1 = 0$, then the solution $C_2[e_{21}, e_{22}]^{\mathrm{T}} e^{\lambda_2 t}$ gives a straight-line trajectory along the $\mathbf{e}_2$ line. All other trajectories (i.e., for which both C_1 and C_2 are nonzero) approach these lines asymptotically. Specifically, since $\lambda_1 < 0$ and $\lambda_2 > 0$, it follows from (40) that

$$\begin{bmatrix} x(t) \\ y(t) \end{bmatrix} \sim \mathbf{e}_2 e^{\lambda_2 t} \tag{41a}$$

as $t \to +\infty$, and

$$\begin{bmatrix} x(t) \\ y(t) \end{bmatrix} \sim \mathbf{e}_1 e^{\lambda_1 t} \tag{41b}$$

as $t \to -\infty$.

Let us illustrate.

EXAMPLE 7. Saddle. Consider

$$x' = 5x + 3y, \tag{42a}$$

$$y' = -3x - 5y. \tag{42b}$$

We obtain

$$\lambda_1 = -4,\ \mathbf{e}_1 = \alpha \begin{bmatrix} 1 \\ -3 \end{bmatrix}; \quad \lambda_2 = +4,\ \mathbf{e}_2 = \beta \begin{bmatrix} 3 \\ -1 \end{bmatrix}.$$

Since the λ's are real and of opposite sign, the singularity is a saddle. Representative trajectories are shown in Fig. 8. Note the asymptotic approach to the $\mathbf{e}_2$ line as $t \to +\infty$, and to the $\mathbf{e}_1$ line as $t \to -\infty$, in accordance with (41a) and (41b), respectively. ∎

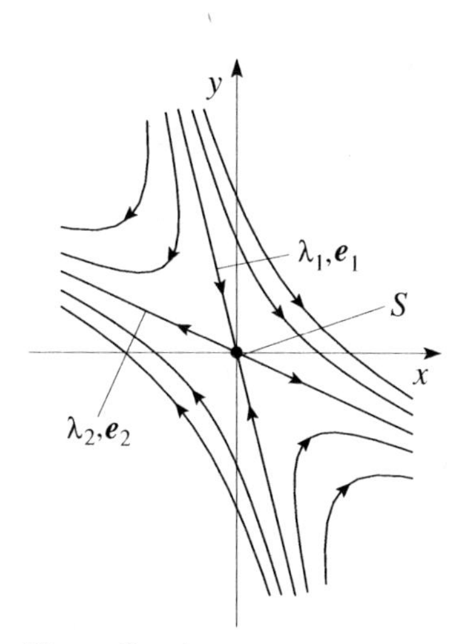

Figure 8. The system (42) has a saddle.

The $\mathbf{e}_1$ and $\mathbf{e}_2$ lines are called the stable and unstable manifolds, respectively. By the **stable manifold** of a saddle point (S in Fig. 8) we mean the set of all initial points (x_0, y_0) such that $(x(t), y(t)) \to S$ as $t \to +\infty$ if $(x(0), y(0)) = (x_0, y_0)$, and by the **unstable manifold** we mean the set of all initial points (x_0, y_0) such that $(x(t), y(t)) \to S$ as $t \to -\infty$ if $(x(0), y(0)) = (x_0, y_0)$.

Every saddle is unstable because there are points that are initially close to the origin that do not remain close to it. (In fact, *every* point that not on the stable manifold gets "sent to infinity" as $t \to +\infty$.)

This completes our discussion of the four types of (isolated) singularity of the linear system (1) — centers, spirals, nodes, and saddles. The type of singularity depends entirely on the eigenvalues of the **A** matrix, and the main results are summarized in the table.

Table 6.3.1 Classifying the type of singularity of $\mathbf{x}' = \mathbf{A}\mathbf{x}$ at the origin.

Eigenvalues of A	Singularity	Stability
Purely imaginary $a + d = 0$	CENTER	Stable
Complex conjugates	SPIRAL	Stable if $a + d < 0$ Unstable if $a + d > 0$
Real of same sign Distinct: fast and slow manifolds Repeated: star or degenerate node	NODE	Stable if λ's < 0 Unstable if λ's > 0
Real of opposite sign	SADDLE	Unstable

Recall that $a+d$ is the real part of λ.

Closure. We've considered the *linear* system (1) and have assumed that $\begin{vmatrix} a & b \\ c & d \end{vmatrix} = ad - bc \neq 0$ so that the only singular point of (1) is at the origin. There are four types of flow (in the x, y phase plane), depending on the eigenvalues of $\mathbf{A} = \begin{bmatrix} a & b \\ c & d \end{bmatrix}$, and these are listed in Table 6.3.1. These four types (center, spiral, node, and saddle) will serve as "building blocks" when we study *nonlinear* systems in the next section.

For centers and spirals there are no straight line trajectories, but a saddle has two (really *four* half-lines, of which two are paths of attraction and two are paths of repulsion) and a node typically has two, but one in the case of a degenerate node and an infinite number of them in the case of a star.

Remember that we've considered only the case where $ad - bc \neq 0$, so the origin $x = y = 0$ is an isolated singular point of (1). Nonisolated singular points *can* occur in applications, as illustrated in the exercises.

Looking ahead, in this section we considered the linear system (1) with an isolated singular point at the origin, but in Section 7.4 we will encounter linear systems with singular points not at the origin. However, we will be able to reduce them to the standard form (1) by a simple translational change of variables.

EXERCISES 7.3

1. For the given values of a, b, c, and d, classify the singularity of (1) at $x = y = 0$. If it is a spiral or a node, is it stable or unstable? If a node, is it a star or a degenerate node?

(a) $a = 2, b = -5, c = 4, d = -2$
(b) $a = -1, b = 4, c = -1, d = -1$
(c) $a = 1, b = -2, c = 2, d = 1$
(d) $a = 4, b = 1, c = 3, d = 1$
(e) $a = -2, b = 0, c = 1, d = -4$
(f) $a = 2, b = 3, c = 2, d = -3$
(g) $a = 1, b = 2, c = 3, d = 4$
(h) $a = 1, b = 2, c = -2, d = -2$
(i) $a = 4, b = -4, c = 1, d = 8$
(j) $a = 2, b = 3, c = -3, d = -4$
(k) $a = -2, b = -1, c = 1, d = -4$
(l) $a = 1, b = 4, c = 1, d = -1$
(m) $a = 2, b = 4, c = -4, d = -2$

2. (a)-(m) **Computer.** Use computer software to generate the phase portrait of the system defined in the corresponding part of Exercise 1. Include enough trajectories (especially any straight line trajectories) to clarify the flow pattern, and use arrows to show the flow direction. NOTE: It may take several tries before you find a set of initial points that gives a nice-looking phase portrait.

3. Damped Mechanical Oscillator. We can express the oscillator equation $mx'' + px' + kx = 0$ as the system

$$x' = y, \;\; y' = -(k/m)x - (p/m)y,$$

in which we can identify the coefficients a, b, c, d in (1): $a = 0$, $b = 1$, $c = -k/m$, and $d = -p/m$. Here, $m > 0$, $p \geq 0$, and $k > 0$.

(a) If we begin with $p = 0$ and then increase it, the singularity at the origin of the phase plane changes from one type to another. Determine the values of p at which those changes occur, and classify the corresponding singularities that are encountered.
(b) Although the singularity changes type abruptly, as p passes through those critical values, the phase portrait does *not* change abruptly. Explain that claim.

4. Example 2. Derive (18) from (17). HINT: Dividing (17b) by (17a) gives a first-order equation. Recall from Chapter 1 that solution methods include separation of variables, the first-order linear equation, exact equations, and homogeneous equations. See if your "favorite method" applies. If so, use it. If not, see if your next-favorite method applies, and so on.

5. Example 2. In Example 2 we claimed that the trajectories

$$2x^2 - 2xy + 3y^2 = C \tag{5.1}$$

are ellipses with their axes rotated counterclockwise by 31.7°, as indicated in Fig.2. Verify that claim by introducing a rotated coordinate system as follows:

(a) If $\bar{x}, \bar{y}$ is a Cartesian coordinate system that is rotated counterclockwise through an angle α, as shown below, show that the x, y and $\bar{x}, \bar{y}$ coordinates of any given point are related, through the trigonometry, according to

$$x = \bar{x}\cos\alpha - \bar{y}\sin\alpha, \tag{5.2a}$$

$$y = \bar{x}\sin\alpha + \bar{y}\cos\alpha. \tag{5.2b}$$

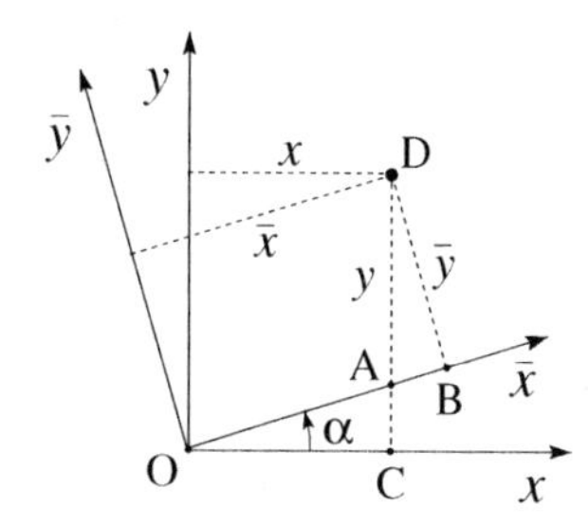

(b) Put (5.2) into (5.1) and choose α so as to eliminate the "mixed" term, the $\bar{x}\bar{y}$ term, so that the result will be a family of ellipses having the $\bar{x}$ and $\bar{y}$ axes as their principal axes. Show that those steps give $\alpha = 31.7°$ and another value as well. Explain the significance of the other value.

6. Figure 2. We established that the flow in Fig. 2 is counterclockwise by checking at the specific points A and B: we noted from (17a) that the x velocity component is positive at A and B [or from (17b) that the y velocity component $y' = 2x$ is positive at A and B]. But if the arrow is counterclockwise at A and B, does it follow that it is counterclockwise *everywhere* on those ellipses? Explain why it does follow, that is, why the flow cannot be counterclockwise on some parts of a trajectory and clockwise on other parts of it.

7. Changing to Polar Coordinates. It is sometimes helpful to re-express a system $x' = P(x, y),\ y' = Q(x, y)$ in terms of polar coordinates r, θ, where

$$x = r\cos\theta, \quad y = r\sin\theta. \tag{7.1a,b}$$

(a) To put (7.1) into $x' = P(x, y),\ y' = Q(x, y)$, differentiate $x(t) = r(t)\cos\theta(t)$ with respect to t and obtain $x' = r'\cos\theta - r\theta'\sin\theta$; similarly, differentiate $y(t) = r(t)\sin\theta(t)$. Solve those equations for r' and θ', and show that

$$\boxed{r' = P\cos\theta + Q\sin\theta, \quad \theta' = \frac{Q\cos\theta - P\sin\theta}{r}.} \tag{7.2a,b}$$

P and Q denote $P(r\cos\theta, r\sin\theta)$ and $Q(r\cos\theta, r\sin\theta)$.
(b) With P and Q given by the right-hand sides of (17a,b), respectively, examine (7.2b) and show that $\theta' > 0$, so the flow in Fig. 2 is counterclockwise.
(c) With P and Q given by the right-hand sides of (19a,b), respectively, examine (7.2b) and show that $\theta' > 0$, so the flow in Fig. 3a is counterclockwise.

8. Example 3. In Example 3 we obtained the phase portrait by using computer software. In this exercise we ask you to proceed analytically instead and to show that the phase trajectories are given by

$$\ln(4y^2 - 3xy + 3x^2) - \frac{2}{\sqrt{39}}\arctan\left(\frac{8y - 3x}{\sqrt{39}x}\right) = C. \tag{8.1}$$

HINT: Solve $\dfrac{dy}{dx} = \dfrac{3x - y}{2x - 4y}$ by the method given in Section 1.8.2. You may use an integral table or computer software to evaluate any integrals that occur.

9. Stars. We showed that if $\mathbf{A}$ is of the form (28) (with $a \neq 0$, because we assumed in this section that $ad - bc \neq 0$) then it has a repeated eigenvalue with a two-dimensional eigenspace, so the phase portrait is a star (Fig. 6). Show that $\mathbf{A}$ *must* be of the form (28) to obtain a repeated eigenvalue with a two-dimensional eigenspace.

10. Degenerate Nodes. Show, for the given system, that there is a repeated eigenvalue with a one-dimensional eigenspace, so the singularity is a degenerate node. Use the steps outlined in (32)–(37) to derive the solution for $x(t)$ and $y(t)$.

(a) $x' = 8x + y,\ y' = -4x + 4y$
(b) $x' = -4x - 3y,\ y' = 3x + 2y$
(c) $x' = 4x - y,\ y' = x + 2y$
(d) $x' = -2x - 3y,\ y' = 3x + 4y$

11. Straight-Line Trajectories. We've seen that the phase portrait of the linear system (1) can include lines through the origin. Show that there can be one, two, or an infinite number of such lines, but not a finite number of them greater than two. HINT: Put $y = mx$ in (1).

12. Parametric Study. In each case β is a real constant, a parameter, with $-\infty < \beta < \infty$. There exist one or more transition points on the β axis such that the system has one type of singularity (at the origin of the x, y plane) for β's to the left of that transition point, and a different type of singularity for β's to the right of it. Those points break the β axis into two or more subintervals. Give inequalities for those subintervals and state the singularity type on each subinterval, including its stability or instability (if it is a spiral or node).

(a) $x' = x + y,\ y' = x + \beta y$
(b) $x' = x + y,\ y' = -x + \beta y$
(c) $x' = \beta y,\ y' = x + y$
(d) $x' = -x - y,\ y' = \beta x$
(e) $x' = 2x + \beta y,\ y' = x - 2y$
(f) $x' = 3x + \beta y,\ y' = -x + 2y$
(g) $x' = x + \beta y,\ y' = -x + y$

13. Mixing Tanks. Each of the two tanks shown contains V gal of initially clear water. Brine enters one at the constant rate Q gal/min, with a salt concentration c_i lb/gal. The tanks are kept stirred so their salt concentrations c_1 and c_2 are uniform, and are functions only of the time t. There is an exchange of q gal/min, so the governing equations for $c_1(t)$ and $c_2(t)$ are

$$Vc_1' = Qc_i - qc_1 + qc_2 - Qc_1, \tag{13.1a}$$

$$Vc_2' = qc_1 - qc_2. \tag{13.1b}$$

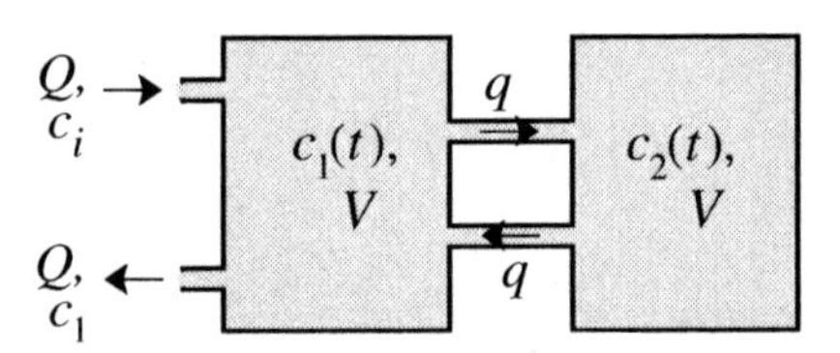

(a) Show that in steady state $c_1 = c_2$, and denote it as c_s. Solve for c_s in terms of V, Q, and c_i. Thus, in the $c_1,\ c_2$ phase plane there is one equilibrium point (singular point) at (c_s, c_s).
(b) It is convenient, though not essential, to move the singular point to the origin, which is readily accomplished by the translational change of variables

$$x = c_1 - c_s, \quad y = c_2 - c_s. \tag{13.2}$$

Make that change of variables and show that (13.1) becomes

$$x' = -\alpha x + \beta y, \quad y' = \beta x - \beta y, \tag{13.3a,b}$$

where $\alpha = (q+Q)/V$ and $\beta = q/V$.
(c) Show that the singularity is necessarily a stable node.

14. Nonisolated Singular Point. If $ad - bc = 0$ in (1), then the singular point at x, y is not isolated; either there is a *line* of singular points, or every point in the *plane* is a singular point. In each case identify the singular points (with a heavy line) and give a labeled sketch of the phase portrait.

(a) $x' = x + y,\ y' = 0$
(b) $x' = x + y,\ y' = 3x + 3y$
(c) $x' = 0,\ y' = x - y$
(d) $x' = x + y,\ y' = -3x - 3y$
(e) $x' = -8x + 4y,\ y' = -2x + y$
(f) $x' = 8x - 4y,\ y' = -2x + y$

15. Two-Compartment Diffusion; Example of a Nonisolated Singular Point. Consider the two-compartment biological system shown schematically in the figure. The compartments have volumes V_1 and V_2 and contain liquid in which there is a solute of interest, with concentrations c_1 and c_2 that are assumed to be spatially uniform in each compartment and are functions only of the time t. They are separated by a permeable membrane through which the solute diffuses, from the side of higher concentration to the side of lower concentration. The rate of diffusion will be proportional to the area A of the membrane and the concentration difference across the membrane, and inversely proportional to the membrane thickness δ. Hence, the rate of increase of solute in compartment 1, $V_1 c_1'(t)$, will be proportional to $A(c_2 - c_1)/\delta$, and similarly for compartment 2, so the governing differential equations are

$$c_1' = -\frac{DA}{V_1\delta}(c_1 - c_2), \quad c_2' = \frac{DA}{V_2\delta}(c_1 - c_2), \qquad \text{(15.1a,b)}$$

in which the proportionality constant D is considered as known. Let $V_1 = 4$, $V_2 = 1$, and $DA/\delta = 0.1$.

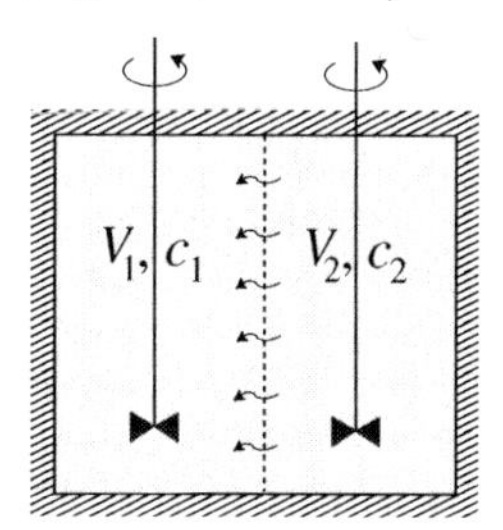

(a) Determine the singular point(s), and obtain the phase portrait in the c_1, c_2 phase plane.
(b) Suppose $c_1(0) = 0.2$ and $c_2(0) = 0.7$. Show the phase trajectory (for $0 \le t < \infty$) corresponding to those initial conditions, and determine the limits of $c_1(t)$ and $c_2(t)$ as $t \to \infty$.

7.4 NONLINEAR SYSTEMS

We studied linear autonomous systems in the phase plane in Section 7.3 to prepare for the case of primary interest, the *nonlinear* autonomous system

$$\frac{dx}{dt} = P(x,y), \qquad \frac{dy}{dt} = Q(x,y). \qquad \text{(1a,b)}$$

In general, the differential equation

$$\frac{dy}{dx} = \frac{Q(x,y)}{P(x,y)} \qquad (2)$$

for the phase trajectories is too difficult to solve. How then can we proceed?

To motivate the idea in an intuitive way, consider the triangle shown in Fig. 1a. The latter is not a phase portrait, but it *is* a curve that has singularities — the three

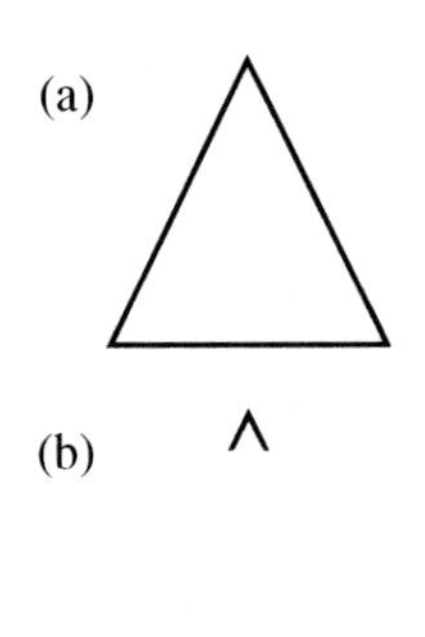

Figure 1. In (b), the singular features (the corners) suggest the whole (the triangle).

corners at which the tangent is discontinuous. Observe from Fig. 1b how the local information, in the neighborhood of the three "singular points," captures the essential structure of the triangle.

In the same spirit, the key features of a phase portrait are the singular points (if any) and their types. For instance, consider the phase portrait shown in Fig. 7 of Section 7.2, corresponding to the motion of a pendulum. The singular points are on the x axis and are alternately centers and saddles. Observe from that figure how the entire phase portrait seems established by those singularities.

Thus, our attack on a nonlinear problem (1) will begin with finding the singular point(s) of (1), and determining their types. That information will enable us to determine the structure of the phase portrait.

7.4.1 Local linearization. Let (x_s, y_s) be a singular point of (1); that is, $P(x_s, y_s) = 0$ and $Q(x_s, y_s) = 0$. We assume that it is an **isolated singular point**, namely, that there exists some circle, centered at (x_s, y_s), inside of which there are no other singular points.

To determine the phase portrait in the neighborhood of (x_s, y_s) we will use linear approximations of the functions $P(x, y)$ and $Q(x, y)$, in (1), in the neighborhood of that point. That idea will not be new to you, for recall from the calculus that to approximate a differentiable function of a single variable $f(x)$ in the neighborhood of a given point ξ, a simple approximation is provided by the **tangent line** at that point, which we denote here by $f_{\mathrm{T}}(x)$:

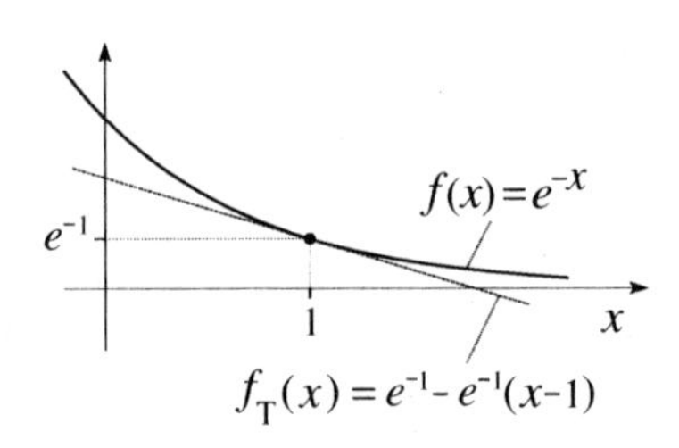

Figure 2. Tangent-line approximation of e^{-x} at $x = 1$.

$$f(x) \approx f_{\mathrm{T}}(x) \equiv f(\xi) + f'(\xi)(x - \xi), \tag{3}$$

as illustrated in Fig. 2 for the function $f(x) = e^{-x}$, with $\xi = 1$. If we expand $f(x) = e^{-x}$ in a Taylor series about $x = 1$, and cut off the series after the first-order term, the result is the tangent-line approximation shown in the figure.

Analogously for a function of two variables, if $f(x, y)$ is differentiable at (ξ, η) then it admits a **tangent-plane** approximation[1]

$$f(x, y) \approx f_{\mathrm{T}}(x, y) \equiv f(\xi, \eta) + f_x(\xi, \eta)(x - \xi) + f_y(\xi, \eta)(y - \eta) \tag{4}$$

The subscripted x and y in (4) denote partial derivatives with respect to the first and second arguments of f, respectively.

there. The tangent-line and tangent-plane approximations are called *linear* approximations because $f_{\mathrm{T}}(x)$ in (3) is a linear function of x, and $f_{\mathrm{T}}(x, y)$ in (4) is a linear function of x and y.

To study (1) in the neighborhood of a singular point (x_s, y_s), approximate $P(x, y)$ and $Q(x, y)$ by their tangent-plane approximations at (x_s, y_s), $P_{\mathrm{T}}(x, y)$ and $Q_{\mathrm{T}}(x, y)$, respectively:

$$P(x, y) \approx P_{\mathrm{T}}(x, y) = P(x_s, y_s) + P_x(x_s, y_s)(x - x_s) + P_y(x_s, y_s)(y - y_s), \tag{5a}$$
$$Q(x, y) \approx Q_{\mathrm{T}}(x, y) = Q(x_s, y_s) + Q_x(x_s, y_s)(x - x_s) + Q_y(x_s, y_s)(y - y_s). \tag{5b}$$

[1] A function $f(x, y)$ of *two* variables is **differentiable** at (a, b) if the partial derivatives f_x and f_y are continuous in some small disk centered at (a, b). In case it is not clear that the right-hand equality in (4) is the equation of a plane, note that the general equation of a plane in x, y, z space, $\alpha(x - a) + \beta(y - b) + \gamma(z - c) = 0$, can (if $\gamma \neq 0$) be expressed as $z = c - \frac{\alpha}{\gamma}(x - a) - \frac{\beta}{\gamma}(y - b)$, which is of the same form as (4), with f_{T} playing the role of "z."

If we put (5) into (1) and remember that $P(x_s, y_s) = 0$ and $Q(x_s, y_s) = 0$ [because (x_s, y_s) is a singular point of (1)], we obtain the approximate equations

$$\boxed{\begin{aligned} \frac{dx}{dt} &= P_x(x_s, y_s)(x-x_s) + P_y(x_s, y_s)(y-y_s), \\ \frac{dy}{dt} &= Q_x(x_s, y_s)(x-x_s) + Q_y(x_s, y_s)(y-y_s). \end{aligned}} \quad \text{(6a,b)}$$

Whereas (1) was nonlinear, the approximate system (6) is linear. Realize that the four coefficients in (6), $P_x(x_s, y_s)$ through $Q_y(x_s, y_s)$, are numbers, not functions.

We've used equal signs in (6) instead of approximately equal signs, but will keep in mind that those equations are approximate.

Here is the point: *The approximate system (6) is linear and hence readily solved (Section 7.3). Our hope is that its solution gives a good approximation to the solution of the original nonlinear system (1) in the neighborhood of the singular point (x_s, y_s) in essentially the same way that the tangent-plane approximations on the right-hand sides of (5) give good approximations to the functions $P(x, y)$ and $Q(x, y)$ near (x_s, y_s).*

For that hope to be realized we need $P(x, y)$ and $Q(x, y)$ in (1) to be sufficiently well behaved. It will suffice to assume, as we will in this section, that the partial derivatives P_x, P_y, Q_x, Q_y are continuous in some neighborhood of (x_s, y_s).

Since our interest is in the local flow, *near* (x_s, y_s), let us move the origin to that point by the change of variables

$$X = x - x_s, \quad Y = y - y_s, \quad (7)$$

which translates the axes so that the origin of the X, Y system is at the singular point. If, in addition, we denote $P_x(x_s, y_s)$, $P_y(x_s, y_s)$, $Q_x(x_s, y_s)$, $Q_y(x_s, y_s)$ in (6) as a, b, c, d, respectively, then (6) takes the compact form

$$\boxed{\begin{aligned} \frac{dX}{dt} &= aX + bY, \\ \frac{dY}{dt} &= cX + dY, \end{aligned}} \quad \text{(8a,b)}$$

which is the linear system studied fully in Section 7.3.

Remember that we are considering only singular points that are isolated, so we continue to assume that $ad - bc \neq 0$.

Keep in mind our hope that the linearization (8) gives a good approximation of the flow locally, in the neighborhood of the singular point. But, it is not obvious that if we accurately approximate the terms $P(x, y)$ and $Q(x, y)$ in the differential equations [by the tangent-plane approximations (5)], then we will obtain accurate approximations of the *solutions* of those differential equations. The details regarding this point are well beyond our scope, but we can say *yes, the local flow, near (x_s, y_s), is faithfully captured by the linearized differential equations (8) in the sense that if (8) has a spiral, an ordinary node (i.e., distinct eigenvalues), or a saddle, then the original nonlinear system (1) does too.* That is, in those cases, (1) exhibits flows that are qualitatively the same as (8), so we say that (1) has a spiral, ordinary node, or saddle, respectively.

However, in the following borderline cases the linear and nonlinear flows *may* differ:

Table 7.4.1 Borderline cases where the linearized and nonlinear flows may differ.

Eigenvalues of A	Linearized System (8)	Nonlinear System (1)
1. Purely imaginary $a+d=0$	Center	Center *or* Stable or Unstable Spiral
2. Repeated, real $\lambda_1 = \lambda_2$	Star or Degenerate Node	Star or Degenerate Node *or* Spiral *or* Ordinary Node

We say these are borderline cases because $a+d=0$ is borderline between $a+d<0$ and $a+d>0$, and $\lambda_1 = \lambda_2$ is borderline between $\lambda_1 < \lambda_2$ and $\lambda_1 > \lambda_2$.

To appreciate the source of the possible discrepancies listed in the table between the behaviors of (1) and (8), consider, for instance, case 1, in which the linearized analysis gives $a+d$ *exactly* equal to zero. If $a+d$ were ever-so-slightly positive, the singularity would be an unstable spiral, and if it were ever-so-slightly negative, the singularity would be a stable spiral. Heuristically, think of the error incurred by the linearization as capable of pushing the effective value of $a+d$ slightly one way or the other, or not at all.

We will illustrate singularity point analysis by considering two classical applications to population dynamics.

7.4.2 Predator – prey population dynamics. Consider an ecological system containing two species, one a predator such as hawks, and the other its prey such as field mice. Denote their populations by $y(t)$ and $x(t)$, respectively. According to the *Lotka–Volterra* model of predator–prey dynamics, the populations are governed by the system of differential equations

Predator population $y(t)$, prey population $x(t)$.

$$\boxed{\begin{aligned}\frac{dx}{dt} &= (\alpha - \beta y)x,\\ \frac{dy}{dt} &= (-\gamma + \delta x)y,\end{aligned}} \tag{9a,b}$$

in which $\alpha, \beta, \gamma, \delta$ are empirically known positive constants.

Think of the field mouse equation (9a) as being of the form $x' = \kappa_1 x$ in which the net birth rate $\kappa_1 = \alpha - \beta y$ diminishes linearly with the predator population y, as seems a reasonable approximation. Similarly, think of the hawk equation

Net birth rate means birth rate minus death rate.

(9b) as being of the form $y' = \kappa_2 y$ in which the net birth rate $\kappa_2 = -\gamma + \delta x$ increases linearly with the supply of mice. Naturally, if $x = 0$, then $\kappa_2 = -\gamma$ is negative — because the hawks have no food.

For definiteness, let $\alpha = \beta = \gamma = \delta = 1$, so (9) becomes

$$x' = (1-y)x, \tag{10a}$$

$$y' = (-1+x)y. \tag{10b}$$

If we take the population units to be in thousands, say, then $x = 1$ and $y = 1$ means one thousand mice and one thousand hawks, not one mouse and one hawk.

Setting $P(x,y) = (1-y)x = 0$ and $Q(x,y) = (-1+x)y = 0$ gives two singular points: $(0,0)$ and $(1,1)$. Recall that singular points are *equilibrium* points; for instance, if $x(0) = y(0) = 1$, then $x(t) = y(t) = 1$ for all $t \geq 0$.

$(\mathbf{0},\mathbf{0})$: Consider first the singular point $(x,y) = (0,0)$. If we use the linear approximation (8) [with $P(x,y) = (1-y)x$, $Q(x) = (-1+x)y$, and $x_s = y_s = 0$], then the coefficients in (8) are $a = P_x(0,0) = 1$, $b = P_y(0,0) = 0$, $c = Q_x(0,0) = 0$, and $d = Q_y(0,0) = -1$. Then let $X = x - 0 = x$ and $Y = y - 0 = y$ and obtain, as the linearized equations,[1]

$$X' = 1X + 0Y \tag{11a}$$

$$Y' = 0X - 1Y. \tag{11b}$$

To solve (11), proceed as in Section 7.3: seek $X(t) = q_1 e^{rt}$, $Y(t) = q_2 e^{rt}$, and obtain an eigenvalue problem for r and $\mathbf{q}$.

The eigenvalues and eigenvectors of the coefficient matrix $\begin{bmatrix} 1 & 0 \\ 0 & -1 \end{bmatrix}$ in (11) are $\lambda_1 = -1$, $\mathbf{e}_1 = \alpha[0,1]^{\mathrm{T}}$ and $\lambda_2 = 1$, $\mathbf{e}_2 = \beta[1,0]^{\mathrm{T}}$. Take $\alpha = 1$ and $\beta = 1$, say. Since the λ's are of opposite sign, the singularity of (11) is a saddle, and the singularity of the original system (10) is a saddle as well.

We know from Section 7.3 that for a saddle there are straight-line trajectories through the singular point, defined by $\mathbf{e}_1$ and $\mathbf{e}_2$: $\mathbf{e}_1 = [0,1]^{\mathrm{T}}$ gives the Y axis, and the flow along it is toward the origin because $\lambda_1 = -1 < 0$, and $\mathbf{e}_2 = [1,0]^{\mathrm{T}}$ gives the X axis, and the flow along it is away from the origin because $\lambda_2 = 1 > 0$.

By solving $dY/dX = -Y/X$ we find the full set of trajectories to be the family of hyperbolas $XY =$ constant, as suggested in Fig. 3.

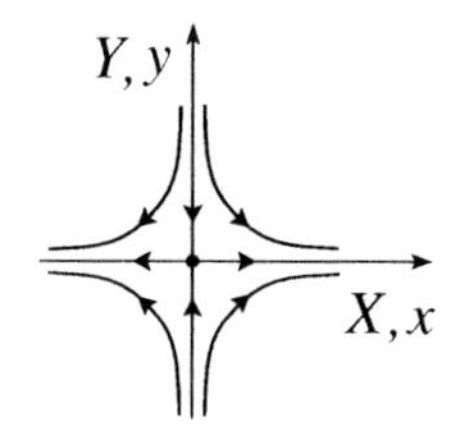

Figure 3. Saddle at $x = y = 0$.

$(\mathbf{1},\mathbf{1})$: Next, consider the singular point $(x_s, y_s) = (1,1)$. The linear approximation (6) followed by the change of variables $X = x - 1$, $Y = y - 1$ gives

$$X' = 0X - 1Y, \tag{12a}$$

$$Y' = 1X - 0Y. \tag{12b}$$

We find from (12) that $\lambda = \pm i$, so this singularity is a center, and if we solve $dY/dX = -X/Y$ we obtain for the trajectories the family of circles $X^2 + Y^2 =$ constant sketched in Fig. 4.

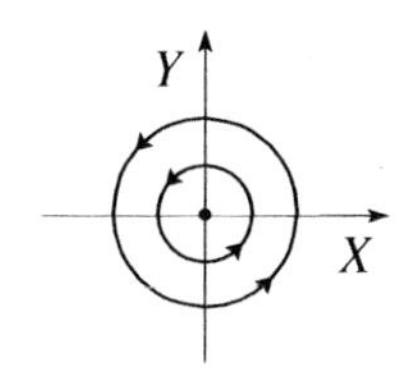

Figure 4. Center at $x = y = 1$.

[1] Of course we don't really need to change letters from x, y to X, Y if $X = x$ and $Y = y$, but we do so because the X, Y notation reminds us that we are looking at the locally linearized equations rather than at the original nonlinear equations.

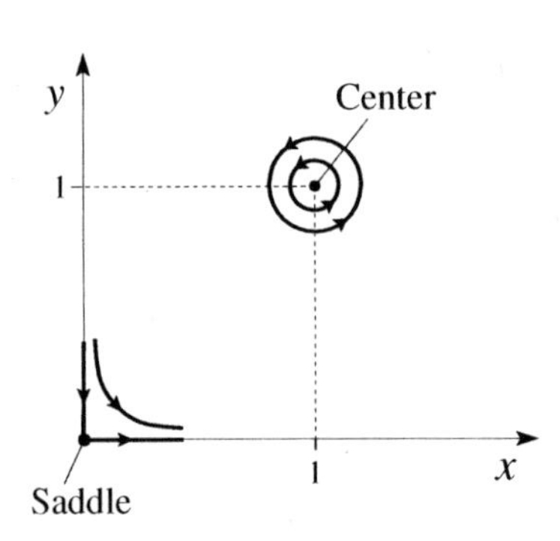

Figure 5. The singular point flows.

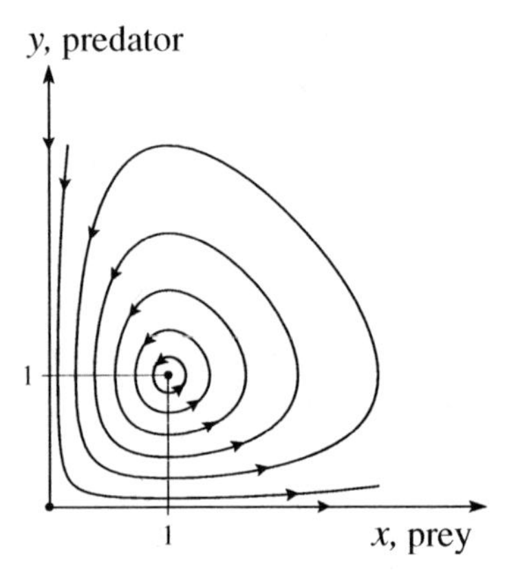

Figure 6. Phase portrait for (10).

Thus far, we have the two singular point flows sketched in Fig. 5 (in which we consider only the first quadrant because x and y are populations and cannot be negative). Next, we will use computer software to plot the trajectories through some chosen set of initial points, to complete the phase portrait. We can use Fig. 5 to help us select the initial points. For instance, we might choose a number of points on the line connecting the two singular points. To generate the phase portrait shown in Fig.6 we used six points on that line, namely, $(0.19, 0.19)$, $(0.375, 0.375)$, $(0.53, 0.53)$, $(0.67, 0.67)$, $(0.8, 0.8)$, and $(0.9, 0.9)$, because they give a "nice" phase portrait, as well as a point on each of the positive x and positive y axes.

Recall, from Table 7.4.1, that whereas the linearized system (12) has a center at $(1, 1)$, the original nonlinear system (10) can have a center *or* a spiral there. The computer plot in Fig. 6 indicates that the singularity there is a center, not a spiral. Except for the trajectories consisting of the positive x axis, the positive y axis, and the "point trajectories" at $(0, 0)$ and $(1, 1)$, each trajectory is a closed orbit around the singular point at $(1, 1)$ and corresponds to a periodic motion.

To complement the phase portrait given in Fig. 6, we've also plotted x and y versus t in Fig. 7, for two different initial conditions: $x(0) = 1$, $y(0) = 0.75$, and $x(0) = 1$, $y(0) = 0.25$. These solutions correspond to an almost-circular orbit close to the point $(1, 1)$ in Fig. 6 and to a larger more distorted orbit, respectively.

What can we say about the period of these motions? In the limit as the size of the orbit about $(1, 1)$ tends to zero the motion approaches the linearized approximation given by (12), and if we apply elimination to (12) we obtain the harmonic oscillator equations $X'' + X = 0$ and $Y'' + Y = 0$ for $X(t)$ and $Y(t)$, the solutions of which have the period $T = 2\pi \approx 6.28$. In fact, if we measure the period of the smaller orbit (Fig. 7a), for $x(t)$ or $y(t)$, we obtain $T \approx 6.5$.

For the larger orbit (Fig. 7b) we measure the period to be $T \approx 7.2$. In fact, for larger orbits the period T increases without bound. To understand why, let the initial point be $x(0) = 1$ and $y(0) = \epsilon$, for instance. Then the period $T(\epsilon) \to \infty$ as $\epsilon \to 0$ because as ϵ decreases, the trajectory passes closer and closer to the origin (Fig. 8), which is an equilibrium point, at which the phase velocity is zero, and in the neighborhood of which the flow is very slow. Thus, the representative point (x, y) passes through the neighborhood of the origin very slowly, as though "swimming through molasses," which increases the period.

Besides being an interesting application of the phase plane concept, predator–prey models are of considerable interest in the field of *mathematical ecology*. Courses on that subject are generally offered in Biology departments, but the subject is also quite mathematical and is studied in Mathematics and Applied Mathematics departments as well. In fact, Volterra (1860–1940) was a mathematician.

COMMENT. Even though (10) is nonlinear, the equation

$$\frac{dy}{dx} = \frac{(-1 + x)y}{(1 - y)x} \tag{13}$$

for the trajectories is separable and readily solved. Its solution is

$$xye^{-(x+y)} = K, \tag{14}$$

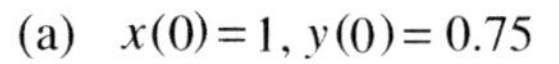

The smaller orbit

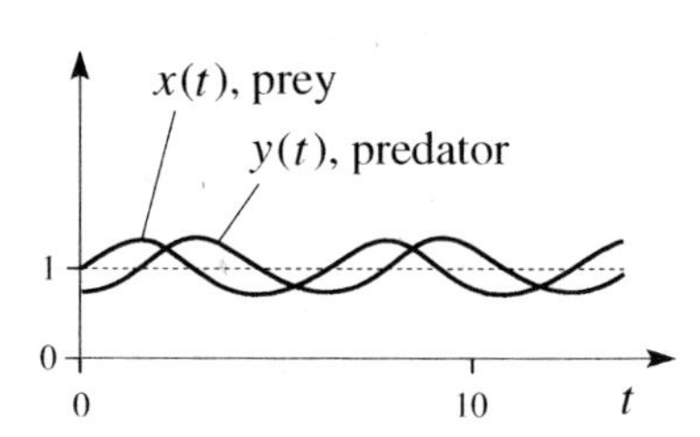

(b) x(0) = 1, y(0) = 0.25

The larger orbit

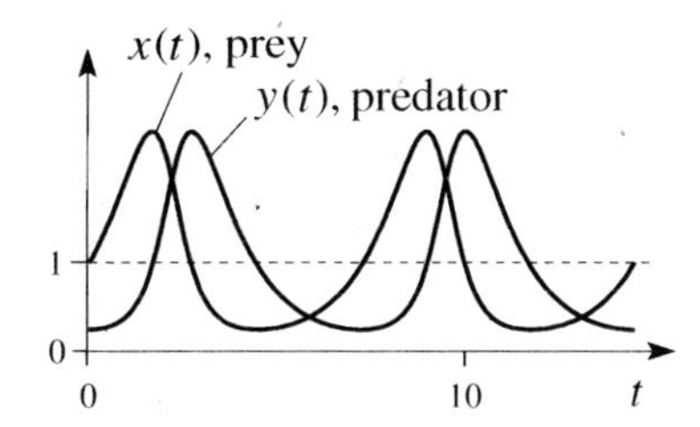

Figure 7. The solutions x and y versus t, of (10), for two initial conditions.

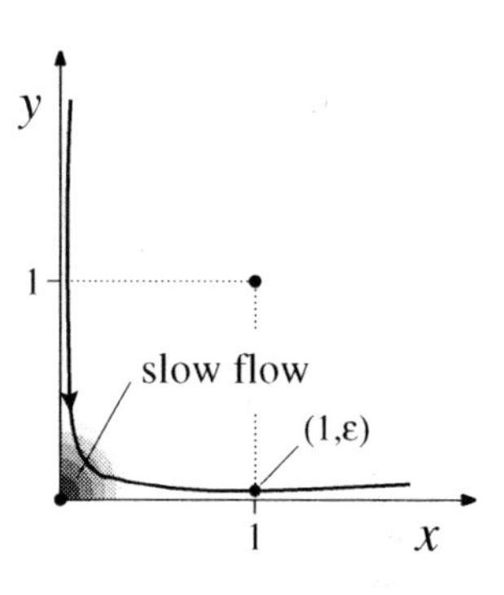

Figure 8. Moving slowly near the origin.

where K is an integration constant. Isn't it striking that (14) gives the closed orbits shown in Fig. 6? After all, orbits close to the singular point are almost circular, yet (14) doesn't look at all like the familiar equation of a circle. This, and other points, are left for the exercises.

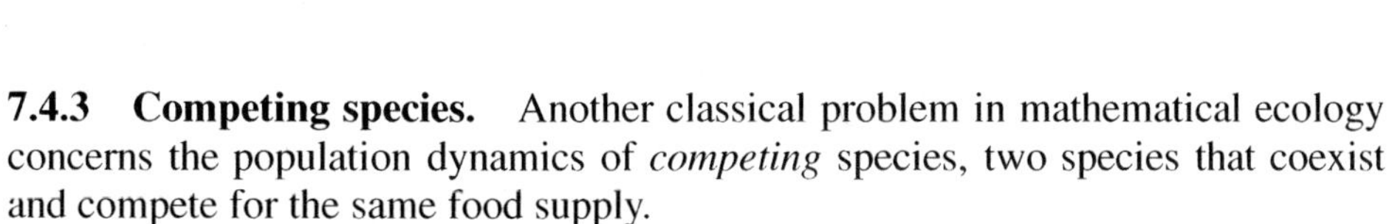

7.4.3 Competing species. Another classical problem in mathematical ecology concerns the population dynamics of *competing* species, two species that coexist and compete for the same food supply.

Let their populations be $x(t)$ and $y(t)$, and suppose that each species existing *alone* is modeled by a logistic equation:

The logistic equation was studied in Section 1.6.1.

$$x' = (\alpha - \beta x)x, \tag{15a}$$
$$y' = (\gamma - \delta y)y, \tag{15b}$$

in which $\alpha, \beta, \gamma, \delta$ are positive constants. To account for the effect of species y on species x, it seems reasonable to expect that the net birth rate $\alpha - \beta x$ of species x in (15a) will be reduced in proportion to y, to $\alpha - \beta x - \epsilon y$ because of the competition from y. Similarly for the effect of x on y, so modify the governing equations (15) as

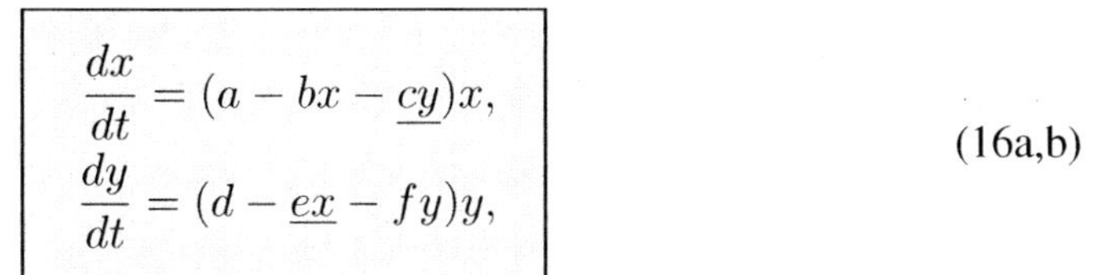

$$\frac{dx}{dt} = (a - bx - \underline{cy})x, \qquad \frac{dy}{dt} = (d - \underline{ex} - fy)y, \tag{16a,b}$$

The cy in (16a) is negative because the net birth rate of x will be *reduced* due to competition from y. Similarly for the ex in (16b).

in which $a, \ldots, f$ are positive constants. The underlined terms express the competition between the two species. For definiteness, let $a = 2$, $d = e = 3$, and $b = c = f = 1$, say, so

$$x' = (2 - x - y)x, \tag{17a}$$
$$y' = (3 - 3x - y)y. \tag{17b}$$

To find the singular points (x_s, y_s) of (17), set

$$P(x,y) = (2 - x - y)x = 0, \tag{18a}$$
$$Q(x,y) = (3 - 3x - y)y = 0. \tag{18b}$$

These give $(x_s, y_s) = (0,0)$, $(0.5, 1.5)$, $(2,0)$, and $(0,3)$. Consider these singular points, in turn.

$(\mathbf{0}, \mathbf{0})$: With $P(x,y) = 2x - x^2 - xy$ and $Q(x,y) = 3y - 3xy - y^2$, the coefficients in (6) are given by $P_x(0,0) = 2$, $P_y(0,0) = 0$, $Q_x(0,0) = 0$, and $Q_y(0,0) = 3$, so the linearized equations (8), at $(0,0)$, are

$$X' = 2X + 0Y, \tag{19a}$$
$$Y' = 0X + 3Y. \tag{19b}$$

The eigenvalues and eigenvectors of the coefficient matrix $\mathbf{A} = \begin{bmatrix} 2 & 0 \\ 0 & 3 \end{bmatrix}$ are

$$\lambda_1 = 2, \; \mathbf{e}_1 = \begin{bmatrix} 1 \\ 0 \end{bmatrix}; \quad \lambda_2 = 3, \; \mathbf{e}_2 = \begin{bmatrix} 0 \\ 1 \end{bmatrix}. \tag{20}$$

Since both λ's are positive, the singular point of (19) is an unstable node. The eigenvectors $\mathbf{e}_1$ and $\mathbf{e}_2$ correspond to the slow and fast manifolds, respectively, because $\lambda_1 < \lambda_2$.

$(\mathbf{0.5}, \mathbf{1.5})$: $P_x(0.5, 1.5) = -0.5$, $P_y(0.5, 1.5) = -0.5$, $Q_x(0.5, 1.5) = -4.5$, and $Q_y(0.5, 1.5) = -1.5$, so the linearized equations at $(0.5, 1.5)$ are

$$X' = -0.5X - 0.5Y, \tag{21a}$$
$$Y' = -4.5X - 1.5Y, \tag{21b}$$

which give

$$\lambda_1 = 0.5811, \; \mathbf{e}_1 = \begin{bmatrix} 1 \\ -2.1623 \end{bmatrix}; \quad \lambda_2 = -2.5811, \; \mathbf{e}_2 = \begin{bmatrix} 1 \\ 4.1623 \end{bmatrix}. \tag{22}$$

These λ's are real and of opposite sign so the singularity is a saddle, with $\mathbf{e}_1$ and $\mathbf{e}_2$ corresponding to the straight-line trajectories of repulsion (because $\lambda_1 > 0$) and attraction (because $\lambda_2 < 0$), respectively.

$(\mathbf{2}, \mathbf{0})$: $P_x(2,0) = -2$, $P_y(2,0) = -2$, $Q_x(2,0) = 0$, and $Q_y(2,0) = -3$, so the linearized system at $(2,0)$ is

$$X' = -2X - 2Y, \tag{23a}$$
$$Y' = 0X - 3Y. \tag{23b}$$

The latter gives

$$\lambda_1 = -2,\ \mathbf{e}_1 = \begin{bmatrix} 1 \\ 0 \end{bmatrix};\quad \lambda_2 = -3,\ \mathbf{e}_2 = \begin{bmatrix} 2 \\ 1 \end{bmatrix}, \tag{24}$$

so $(2, 0)$ is a stable node with fast manifold along $\mathbf{e}_2$ and slow manifold along $\mathbf{e}_1$.

$(\mathbf{0}, \mathbf{3})$: $P_x(0, 3) = -1$, $P_y(0, 3) = 0$, $Q_x(0, 3) = -9$, $Q_y(0, 3) = -3$, so the linearized system at $(0, 3)$ is

$$X' = -1X + 0Y, \tag{25a}$$

$$Y' = -9X - 3Y. \tag{25b}$$

The latter gives

$$\lambda_1 = -1,\ \mathbf{e}_1 = \begin{bmatrix} 1 \\ -4.5 \end{bmatrix};\quad \lambda_2 = -3,\ \mathbf{e}_2 = \begin{bmatrix} 0 \\ 1 \end{bmatrix}, \tag{26}$$

so $(0, 3)$ is a stable node with fast manifold along $\mathbf{e}_2$ and slow manifold along $\mathbf{e}_1$.

The foregoing results establish the local flows near the four singular points, and these are sketched in Fig. 9. Turning to computer software to develop the phase

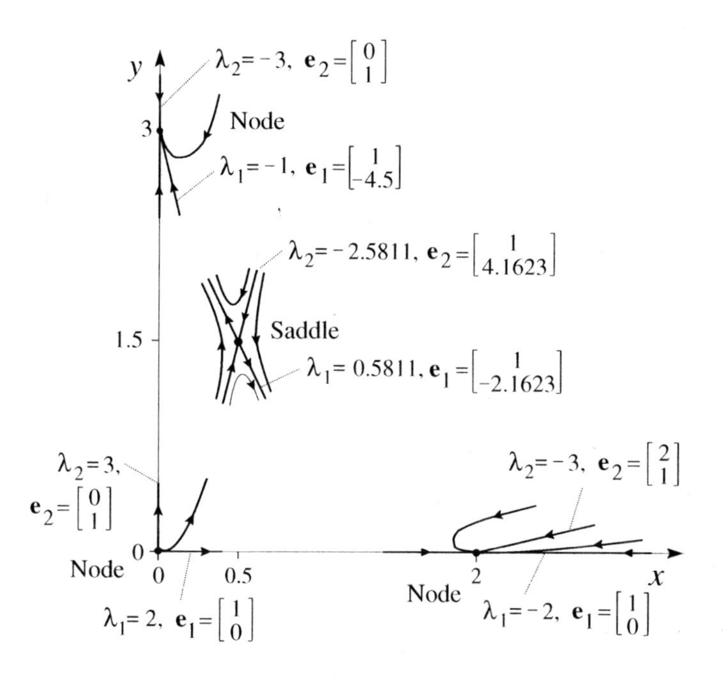

Figure 9. The local flows near the four singular points.

portrait, we can obtain trajectories through any specified set of initial points, and we can use Fig. 9 to help us choose those points. It will be important to obtain the two trajectories that depart from the saddle along the eigenvector $\mathbf{e}_1$, and the two that approach it along the eigenvector $\mathbf{e}_2$, namely, the $\mathbf{e}_1$ and $\mathbf{e}_2$ given in (22) and in Fig. 9. Initial points for these trajectories can be taken as

$$(x, y) = (0.5, 1.5) \pm s(1, -2.1623) \tag{27}$$

and

$$(x, y) = (0.5, 1.5) \pm s(1, 4.1623), \tag{28}$$

respectively, where s is any nonzero real number. But realize that the saddle flow shown in Fig. 9 is accurate only locally, and we can expect the straight fast and slow manifolds to bend (curve) as they leave the neighborhood of the saddle point. Thus, we should make s very small, so our initial points are very close to $(0.5, 1.5)$. With $s = 0.001$, say, (27) and (28) give the initial points $P_1 = (0.501, 1.498)$, $P_2 = (0.499, 1.502)$, $P_3 = (0.501, 1.504)$, and $P_4 = (0.499, 1.496)$.

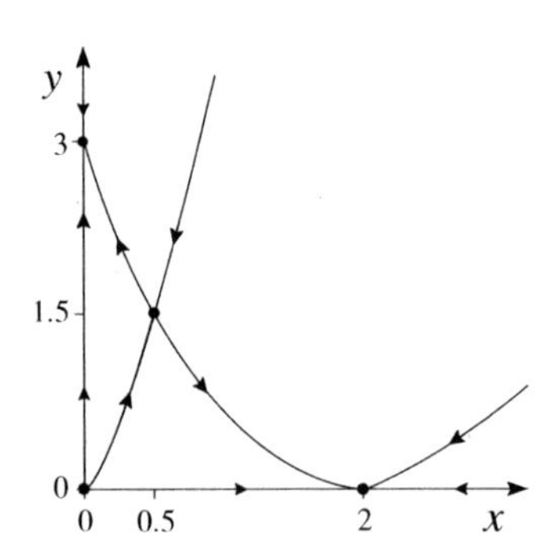

Figure 10. The key trajectories for (17); a skeleton of the phase portrait.

For the node at $(0, 3)$, we want the trajectory that approaches $(0, 3)$ along $\mathbf{e}_1$, but that trajectory *appears* to be the same as the one that leaves the saddle in the eigenvector direction $[-1, 2.1623]^{\mathrm{T}}$ and that will be obtained by the initial point P_2 "northwest" of the saddle. There is no need to choose initial points above and below $(0, 3)$ on the y axis because we can see by inspection of (17) that those trajectories will simply be along the y axis, since $x(t) = 0$ satisfies equation (17a). Similarly, for the node at $(0, 0)$ there is no need to obtain the trajectories leaving in the $\mathbf{e}_1$ and $\mathbf{e}_2$ directions because we can see from (17) that these will simply be along the x and y axes, respectively.

Finally, for the node at $(2, 0)$ we can obtain the trajectory approaching $(2, 0)$ along $\mathbf{e}_2$ by choosing an initial point $P_5 = (2, 0) + 0.001(2, 1) = (2.002, 0.001)$.

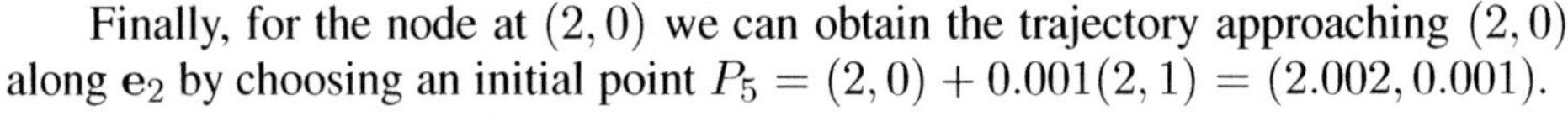

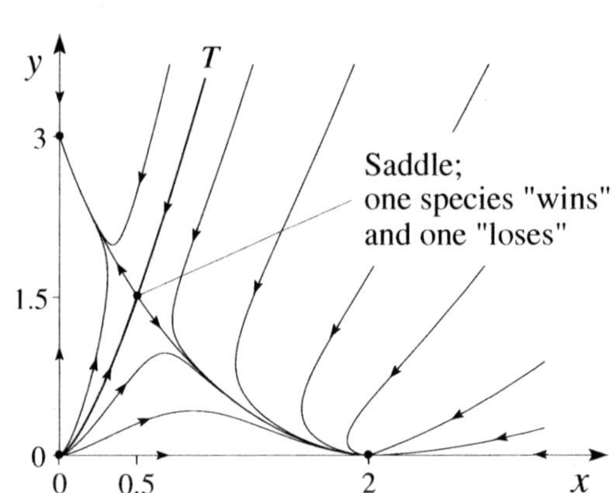

Figure 11. Phase portrait for (17). Stable coexistence not possible due to the strong competition. At (0,3) y has won and x has lost; at (2,0) x has won and y has lost.

Using these five initial points, computer software gives the trajectories shown in Fig. 10, and with this "skeleton" in hand it is now easy to choose additional initial points to fill out the phase portrait. Our final result is shown in Fig. 11.

COMMENT 1. No doubt our initial points $P_1, \ldots, P_5$ did not lie exactly on the trajectories that we were after because we located those points by moving in straight lines in the eigenvector directions whereas we can see from Fig. 10 that the desired trajectories bend as they depart from the singular points. However, these discrepancies were too small to have observable consequences in Figs. 10 and 11.

COMMENT 2. In Fig. 11, T denotes the stable manifold of the saddle. The latter is of special importance because all initial points to its left end up (as $t \to \infty$) at the equilibrium point $(0, 3)$, and all those to its right end up (as $t \to \infty$) at the equilibrium point $(2, 0)$. In the former case species x dies out as $t \to \infty$, and only species y is left, its final number being the *environmental carrying capacity* $y = 3$ implied by the logistic equation $y' = (3 - y)y$ that is obtained by setting $x = 0$ in (17b). And in the latter case, species x dominates and species y dies out, the final x being its environmental carrying capacity $x = 2$ implied by the logistic equation $x' = (2 - x)x$ obtained by setting $y = 0$ in (17a). We call the region to the left of T the **basin of attraction** of the node at $(0, 3)$, and the region to the right of T the basin of attraction of the node at $(2, 0)$.[1]

COMMENT 3. Observe from Fig. 11 that for the system (17) the equilibrium point $(0.5, 1.5)$ represents a state of coexistence of x and y. However, the latter is unstable, and if the initial point is to the right of the separatrix T then x "wins the competition:" $x(t) \to 2$ and $y(t) \to 0$ as $t \to \infty$. If instead the initial point is to the left of the separatrix, then y wins: $x(t) \to 0$ and $y(t) \to 3$ as $t \to \infty$.

But this is only one example, for one set of values of the constants $a, \dots, f$ in (16). Is it possible, for different values of those constants, to obtain a *stable* state of coexistence? Perhaps we can if we reduce the competition, by reducing c and e in (16). In fact, if we reduce c from 1 to 0.5 and e from 3 to 1, for instance, then the resulting phase portrait in Fig. 12 does contain a stable equilibrium point (i.e., a stable state of coexistence), namely, the stable node at the point $(1, 2)$. The basin of attraction for that equilibrium point is the entire quadrant $(x > 0, y > 0)$. (It does make sense that at that equilibrium point x and y are less than their environmental carrying capacities — due to the competition.)

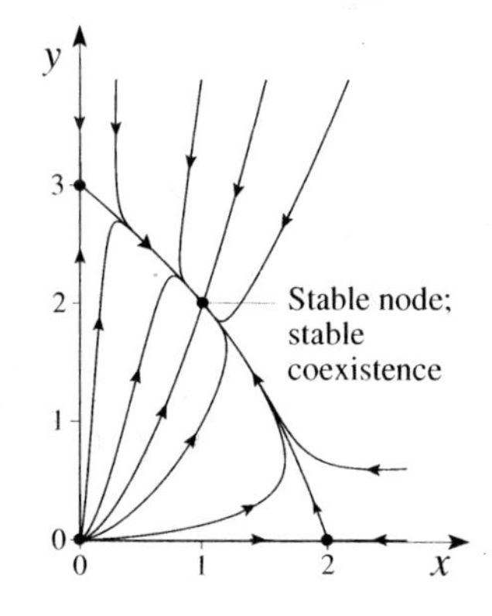

Figure 12. Weak competition results in stable coexistence; $x' = (2 - x - 0.5y)x$, $y' = (3 - x - y)y$.

In fact, it can be shown that if the "competition coefficients" c and e in (16) are small enough, namely, if

$$ce < bf, \tag{29}$$

then stable coexistence will occur, in the form of a stable node, as in Fig. 12. For the case shown in Fig. 12, $ce = (0.5)(1) = 0.5$ and $bf = (1)(1) = 1$, so (29) is indeed satisfied. For the case shown in Fig. 11, however, $ce = 3$ and $bf = 1$, so (29) was not satisfied and stable coexistence was not obtained.

COMMENT 4. Finally, recall that the saddle at $(0.5, 1.5)$ in Fig. 11 and the node at $(1, 2)$ in Fig. 12 correspond to the singular point of (16) that exists at the intersection

[1]A basin of attraction of an asymptotically stable equilibrium point P is the set of points (a, b) such that if $(x(0), y(0)) = (a, b)$, then $(x(t), y(t)) \to P$ as $t \to \infty$; that is, it is the set of points that are attracted to P.

of the lines $a - bx - cy = 0$ and $d - ex - fy = 0$, namely, the point

$$x_s = \frac{af - cd}{bf - ce}, \quad y_s = \frac{bd - ae}{bf - ce}. \tag{30}$$

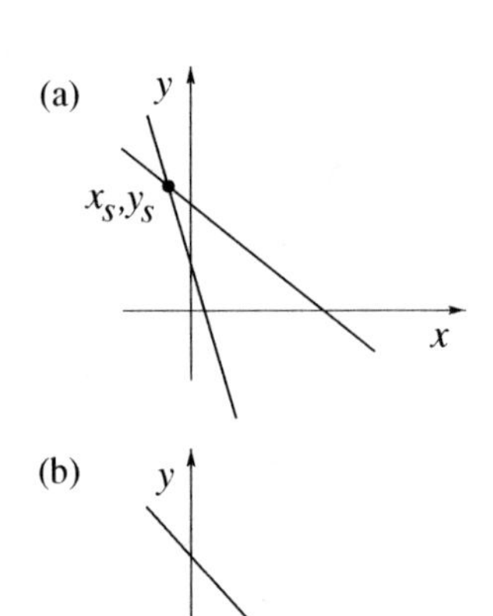

Figure 13. The singular point at the intersection of $a-bx-cy=0$ and $d-ex-fy=0$ can fall outside the first quadrant, namely, in the second or fourth quadrants.

Besides falling in the first quadrant (as in Figs. 11 and 12), that point could, depending on the coefficients $a, \ldots, f$, fall outside of the first quadrant — in the second or fourth quadrants, as in Fig. 13. (It cannot fall in the third because it follows from the positiveness of $a, \ldots, f$ that each of the lines $a - bx - cy = 0$ and $d - ex - fy = 0$ has positive intercepts on the x and y axes.) To obtain conditions on $a, \ldots, f$ that correspond to (x_s, y_s) falling in the second or fourth quadrant, observe from Fig. 13 that in each of those cases x_s and y_s are of opposite sign. Now look at (30). Since the denominators are identical, it follows that (x_s, y_s) will fall outside of the first quadrant only if the numerators are of opposite sign, namely, if

$$bd - ae > 0 \quad \text{and} \quad af - cd < 0 \tag{31a}$$

or if

$$bd - ae < 0 \quad \text{and} \quad af - cd > 0. \tag{31b}$$

What is the outcome? If (31a) holds, then y wins; that is, $x(t) \to 0$ and $y(t)$ tends to its saturation value d/f as $t \to \infty$. And if (31b) holds, then x wins; that is, $y(t) \to 0$ and $x(t)$ tends to its saturation value a/b as $t \to \infty$. Proof is left for the exercises. These two cases are illustrated in Figs. 14a and 14b, respectively.

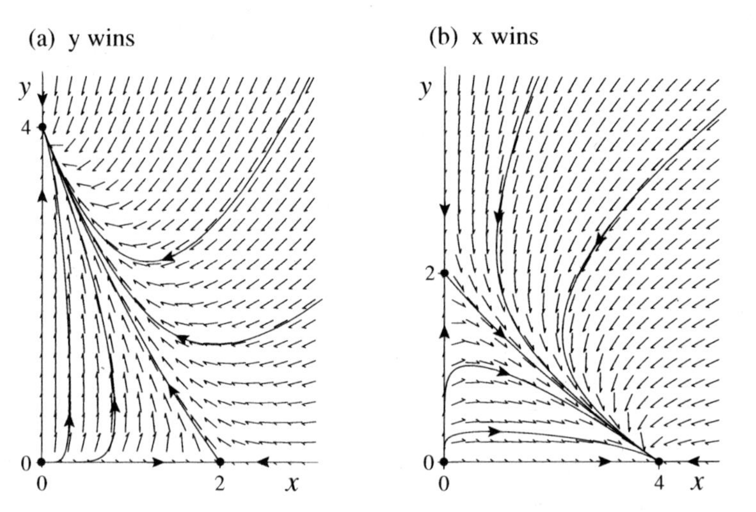

Figure 14. (a) $x' = (2-x-2y)x$, $y' = (12-4x-3y)y$; (31b) satisfied, y wins.
(b) $x' = (12-3x-4y)x$, $y' = (2-2x-y)y$; (31a) satisfied, x wins.

Closure. In this section we see the power of the phase plane in studying nonlinear autonomous systems

$$x' = P(x, y),$$
$$y' = Q(x, y).$$

Note, for instance, how clearly Fig. 11 reveals the key features of the population dynamics of the competing species x and y modeled by equations (17): stable coexistence is not possible; for every initial state to the left of the separatrix T, $x(t) \to 0$ and $y(t) \to 3$ as $t \to \infty$; and for every initial state to the right of T, $x(t) \to 2$ and $y(t) \to 0$ as $t \to \infty$. Imagine how difficult it would be to obtain that information if we embarked on a program of solving (17) numerically for $x(t)$ and $y(t)$ for a great many different initial conditions. Can we conclude, then, that a phase portrait is better than the usual plots of x and y versus t? Not at all. The two approaches are complementary, each providing information and a viewpoint that the other does not.

Here is a summary of the phase plane method that we've used:

1. Be sure the system is of the autonomous form (1).

2. Find any singular points by solving the equations $P(x,y) = 0$ and $Q(x,y) = 0$. Be sure they are isolated.

3. Starting at any one of the singular points, evaluate the coefficients P_x, P_y, Q_x, Q_y in (6), and thereby obtain the linearized version (8).

4. Evaluate the eigenvalues and eigenvectors of the $\mathbf{A}$ matrix in (8).

5. Carry out steps 3 and 4 for each singular point, in turn.

6. Knowing all the singular points and the eigenvalues and eigenvectors associated with each, develop a sketch analogous to the ones in Figs. 5 and 9.

7. With that figure in hand, choose a set of initial points to use in obtaining a computer-generated phase portrait, being sure to obtain any trajectories that approach the singular points as t tends to plus or minus infinity, and any separatrices that may be present.

8. Use computer software to generate the phase portrait.

EXERCISES 7.4

1. Locate and classify all singular points, and use a combination of singular point analysis and computer software to generate the phase portrait — on a rectangle that is large enough to contain the essential features of the flow.

(a) $x' = y - x, \; y' = y - x^2$
(b) $x' = y - x^2, \; y' = y + x^2 - 8$
(c) $x' = y + x - 1, \; y' = y - x^2 + 1$
(d) $x' = y + 2\sin x, \; y' = x$
(e) $x' = y + y^2, \; y' = -x + y^2$
(f) $x' = y, \; y' = -x + y^2$
(g) $x' = y - 3, \; y' = -2x + x^2 - y$
(h) $x' = (2 + x - 2y)x, \; y' = (12 + 4x - 4y)y$
(i) $x' = y, \; y' = -x + x^3/6$
(j) $x' = -y - x(x^2 + y^2), \; y' = x - y(x^2 + y^2)$

2. A system $x' = P(x,y)$, $y' = Q(x,y)$ does not necessarily have any singular points. Make up an example of such a system.

3. Isolated Singular Points? Suppose (1) has singular points at $x = 1/n$ and $y = 0$, for each $n = 1, 2, \ldots$, and also at $x = y = 0$. Which, if any of them, is not isolated? Explain.

4. Damped Nonlinear Pendulum. The undamped nonlinear pendulum was studied in Section 7.2.2, which discussion you may want to quickly review. Continuing to let $g/l = 1$ for definiteness, now consider the *damped* nonlinear pendulum, governed by the equation

$$x'' + \epsilon x' + \sin x = 0, \tag{4.1}$$

in which the positive constant ϵ is a damping coefficient. As in Section 7.2.2, let $x' = y$, and thus re-express (4.1) as the equivalent system

$$x' = y, \quad y' = -\epsilon y - \sin x. \tag{4.2}$$

(a) Locate and classify all singular points of (4.2).
(b) For $\epsilon = 0.2$, develop a phase portrait, on the rectangle $-5 \le x \le 12$, $-3 \le y \le 3$. SUGGESTION: Try sketching it by hand, before turning to the computer.
(c) Same as part (b), but for $\epsilon = 2.1$.

CHANGING TO POLAR COORDINATES

5. It is sometimes helpful (as in the next exercise) to re-express the system $x' = P(x,y)$, $y' = Q(x,y)$ in terms of polar coordinates r, θ, where

$$x = r\cos\theta, \quad y = r\sin\theta. \tag{5.1a,b}$$

(a) To put (5.1) into $x' = P(x,y)$, $y' = Q(x,y)$, differentiate $x(t) = r(t)\cos\theta(t)$ with respect to t and obtain $x' = r'\cos\theta - r\theta'\sin\theta$; similarly, differentiate $y(t) = r(t)\sin\theta(t)$. Solve those equations for r' and θ', and show that

$$r' = P\cos\theta + Q\sin\theta, \quad \theta' = \frac{Q\cos\theta - P\sin\theta}{r}. \tag{5.2a,b}$$

in which P and Q denote $P(r\cos\theta, r\sin\theta)$ and $Q(r\cos\theta, r\sin\theta)$.
(b) To illustrate, show that the system $x' = y + y^2$, $y' = -x + y$ becomes

$$r' = r\sin^2\theta\,(r\cos\theta + 1), \tag{5.3a}$$

$$\theta' = \sin\theta\cos\theta - r\sin^3\theta - 1. \tag{5.3b}$$

6. A Borderline Case: Center or Spiral? (a) Show that the system

$$x' = y + \alpha x(x^2 + y^2), \tag{6.1a}$$

$$y' = -x + \alpha y(x^2 + y^2), \tag{6.1b}$$

in which α is a parameter, has one singular point — at the origin.
(b) Show that the linearized system has a center, so that the nonlinear equations (6.1) can have a center *or* a stable or unstable spiral.
(c) Which is it? Perhaps if we change to polar coordinates we can tell. Use equations (5.2), above, to show that (6.1) becomes

$$r' = \alpha r^3, \tag{6.2a}$$

$$\theta' = -1, \tag{6.2b}$$

and show from (6.2) that whereas the linearized system has a center, the nonlinear system has a stable spiral if $\alpha < 0$, a center if $\alpha = 0$, and an unstable spiral if $\alpha > 0$. NOTE: This example was designed so that the resulting system (6.2), in polar coordinates, would be simple. Unfortunately, that won't always be the case; for instance, the system (5.3) above does not appear to be any simpler than the original system $x' = y + y^2$, $y' = -x + y$.

7. Another Borderline Case: Node or Spiral? Consider the system

$$x' = \alpha x + y(x^2 + y^2), \tag{7.1a}$$

$$y' = \alpha y - x(x^2 + y^2), \tag{7.1b}$$

in which α is a constant.

(a) Show that (7.1) has a singularity only at the origin, which can be a star, a degenerate node, a spiral, or an ordinary node.
(b) To determine which it is, try changing to polar coordinates (Exercise 5), and thus obtain

$$r(t) = r_0 e^{\alpha t}, \tag{7.2a}$$

$$\theta(t) = \theta_0 + \frac{r_0^2}{2\alpha}(1 - e^{2\alpha t}) \tag{7.2b}$$

if $\alpha \neq 0$, and

$$r(t) = r_0, \quad \theta(t) = \theta_0 - r_0^2 t \tag{7.3a,b}$$

if $\alpha = 0$, where r_0 is $r(0)$ and θ_0 is $\theta(0)$.
(c) From (7.2) and (7.3), you should now be able to determine the singularity type. Do that, for the cases $\alpha < 0$, $\alpha = 0$, and $\alpha > 0$.

PREDATOR–PREY

8. Computer Plots. (a) Use computer software to generate the phase portrait shown in Fig. 6. As usual, label the figure and add flow arrows by hand.
(b) Also obtain plots of $x(t)$ and $y(t)$ versus t, for the initial conditions $x(0) = 1$, $y(0) = 0.02$.

9. We stated that a general solution of (13) is given by (14).
(a) Derive that solution.
(b) Further, verify by direct substitution that it satisfies (13).

10. Computer Calculation of the Period. Let the initial conditions for (10) be $x(0) = 1$ and $y(0) = \epsilon$. We stated in the

text that the period $T(\epsilon)$ tends to 2π as $\epsilon \to 1$ and to ∞ as $\epsilon \to 0$.
(a) As a heuristic test of these claims, use computer software to plot $x(t)$ versus t for the cases where $\epsilon = 0.9$ and 0.99, and measure the period $T(\epsilon)$ from those graphs. Does $T(\epsilon)$ appear to be tending to 2π as $\epsilon \to 1$?
(b) Then, use computer software to plot $x(t)$ versus t for $\epsilon = 0.1, 0.001$, and 0.00001, and measure T from those graphs. Do your results appear to support the claim that $T(\epsilon) \to \infty$ as $\epsilon \to 0$?

11. Circles Near $(1, 1)$. We stated that the orbits around $(1, 1)$ in Fig. 6 tend to circles centered at $(1, 1)$ as they are made smaller and smaller. Yet, the exact solution (14) hardly resembles the equation of such a family of circles,

$$(x-1)^2 + (y-1)^2 = \text{constant}. \tag{11.1}$$

Nevertheless, expand the function $xye^{-(x+y)}$ in Taylor series about $x = 1$ and $y = 1$ and, keeping terms only through second order, show that (14) does give the form (11.1) near $(1, 1)$. HINT: The function is actually the product of xe^{-x} and ye^{-y}. Expand them each, and multiply.

COMPETING SPECIES

12. (a) Use computer software to obtain the phase portrait in Fig. 11. NOTE: Of course, the trajectories that you choose to fill out the phase portrait need not be the same as the ones we used to develop Fig. 11.
(b) Use computer software to obtain the phase portrait shown in Fig. 12.

13. Regarding the Conditions (31a,b). We noted from (30) that the singular point (x_s, y_s) of (16), at the intersection of the lines $a - bx - cy = 0$ and $d - ex - fy = 0$, falls outside the first quadrant if

$$bd - ae > 0 \quad \text{and} \quad af - cd < 0, \tag{13.1a}$$

or if

$$bd - ae < 0 \quad \text{and} \quad af - cd > 0. \tag{13.1b}$$

Then we claimed that these outcomes follow: if (13.1a) holds then y wins, and if (13.1b) holds then x wins. Prove that claim. HINT: Consider the singular points on the x and y axes.

NONELEMENTARY SINGULARITIES

14. We don't want to leave you with the incorrect impression that a singularity is necessarily elementary, that is, admitting a local linearization $X' = aX + bY,\ Y' = cX + dY$ with $ad - bc \neq 0$ and hence corresponding to a center, spiral, saddle, or node. In these four problems the singularities are not elementary. Use whatever analysis and/or computer software you need to develop the phase portrait.

(a) $x' = y^2,\ y' = x^2$ (b) $x' = 2y,\ y' = 3x^2$
(c) $x' = xy,\ y' = x^2$ (d) $x' = y,\ y' = -4x^2 + y^2$

ADDITIONAL EXERCISES

15. The Chemostat and its Stability. To conduct experiments on a particular strain of bacteria it is convenient to have a stock supply of it on hand. The chemostat is a laboratory device used to maintain a continuous culture of that bacteria, from which batches may be withdrawn, as needed. A schematic diagram of a chemostat is given in the figure. Nutrient is pumped into

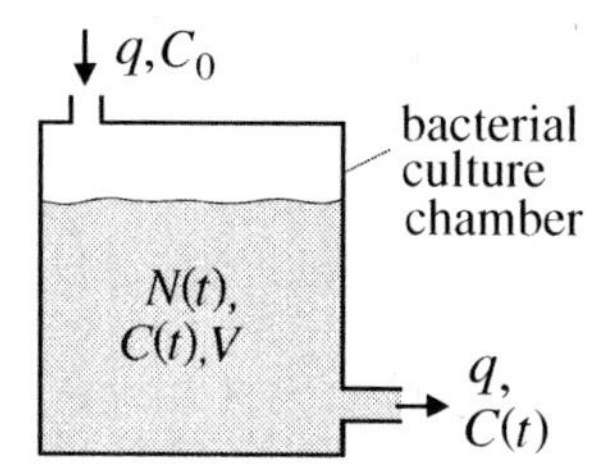

the chamber at a constant rate q (volume per unit time) and at a constant concentration C_0 (nutrient mass per unit volume). The contents are kept stirred so that the distribution of nutrient and bacteria is spatially uniform. The dependent variables are the bacteria population $N(t)$ and the nutrient concentration in the chamber, $C(t)$, and the independent variable is the time t. The outflow rate q equals the inflow rate, so the volume V of the culture is a constant. The governing differential equations, that are derived in Leah Edelstein-Keshet's *Mathematical Models in Biology* (New York: Random House, 1988), are these:

$$N' = \kappa(C)N - \frac{q}{V}N, \tag{15.1a}$$

$$C' = -\alpha\kappa(C)N - \frac{q}{V}C + \frac{q}{V}C_0, \tag{15.1b}$$

in which the net birth-minus-death rate $\kappa(C)$ in (15.1a) will surely depend on the nutrient concentration C. The last term in (15.1a) accounts for the loss of bacteria due to the outflow. The last two terms in (15.1b) account for the loss of nutrient due to the outflow, and for the gain of nutrient due to the inflow, respectively. Following Edelstein-Keshet, let us model $\kappa(C)$ according to *Michaelis–Menten kinetics*, as

$$\kappa(C) = \frac{\beta C}{\gamma + C}, \tag{15.2}$$

in which the asymptotic vaiue β, attained as $C \to \infty$, is called the *saturation value*, and γ is the value of C at which κ attains half of its saturation value, as indicated in the figure given below. Essentially, γ is a measure of the *rate* at which $\kappa(C)$ approaches its saturation level. We assume that the constants

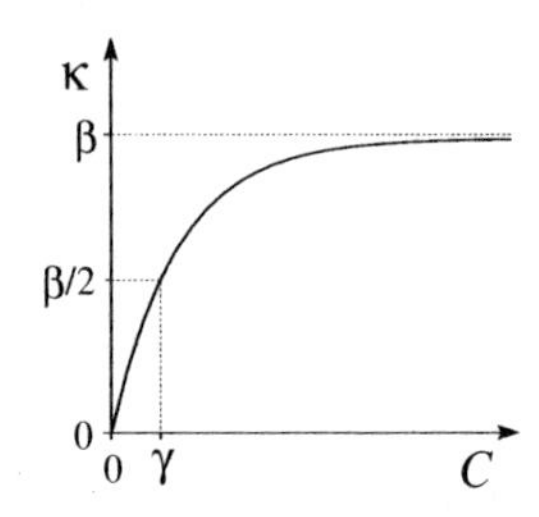

β and γ are known empirically. The existence of a saturation phenomenon is reasonable in that we expect there to be a limit to the rate at which nutrient can be consumed, no matter how much is available. Finally, how can we understand the $-\alpha\kappa(C)N$ term in (15.1b)? We can think of the consumption of nutrient as comprised of two parts, a part due to the birth-by-cell-division process and a part due to subsequent feeding, or "maintenance," of the bacteria. The former will be proportional to the $\kappa(C)N$ rate-of-generation term in (15.1a), the constant of proportionality being "α," and a minus sign being needed because the process consumes rather than generates nutrient. The latter would be proportional to N and is not found in (15.1b) so, evidently, Edelstein-Keshet is assuming that the maintenance part is relatively small. Thus, we are considering the system

$$N' = \frac{\beta C}{\gamma + C} N - \frac{q}{V} N, \tag{15.3a}$$

$$C' = -\alpha \frac{\beta C}{\gamma + C} N - \frac{q}{V} C + \frac{q}{V} C_0. \tag{15.3b}$$

Studying (15.3) will be tedious because of the presence of five parameters: $\beta, \gamma, q/V, \alpha$, and C_0. To reduce this number as much as possible, let us begin by scaling the independent and dependent variables.

(a) Specifically, let us scale N, C, and t by defining

$$\bar{N} = \frac{\alpha\beta V}{\gamma q} N, \tag{15.4a}$$

$$\bar{C} = \frac{1}{\gamma} C, \tag{15.4b}$$

$$\bar{t} = \frac{q}{V} t. \tag{15.4c}$$

Show that these changes of variables reduce (15.3) to the form

$$\bar{N}' = \delta \frac{\bar{C}}{1 + \bar{C}} \bar{N} - \bar{N}, \tag{15.5a}$$

$$\bar{C}' = -\frac{\bar{C}}{1 + \bar{C}} \bar{N} - \bar{C} + \epsilon, \tag{15.5b}$$

where primes now denote $d/d\bar{t}$ and

$$\delta = \beta V/q, \quad \epsilon = C_0/\gamma. \tag{15.6}$$

To simplify notation, let us drop all overhead bars in (15.5) and remember that "N, C, t" henceforth mean $\bar{N}, \bar{C}, \bar{t}$. Thus,

$$N' = \delta \frac{C}{1 + C} N - N, \tag{15.7a}$$

$$C' = -\frac{C}{1 + C} N - C + \epsilon. \tag{15.7b}$$

NOTE: The purpose of the change of variables (15.4) is that the new equations (15.7) contain fewer parameters — the *two* parameters δ and ϵ in place of the original *five*! The new variables $\bar{C}$, $\bar{N}$, $\bar{t}$ are nondimensional (dimensionless). To see that, note from the denominator in (15.2) that γ has the same dimensions as C, so $\bar{C} = C/\gamma$ is nondimensional. Also, V/q has units of time, so $\bar{t} = t/(V/q)$ is dimensionless. Similarly for $\bar{N}$. Nondimensionalizing the dependent and independent variables is a standard step in research, to reduce the number of parameters.

(b) Show that (15.7) has two equilibrium points (singular points):

$$(N_1, C_1) = (0, \epsilon), \tag{15.8a}$$

$$(N_2, C_2) = \left(\frac{\delta}{\delta - 1}(\epsilon\delta - \epsilon - 1), \frac{1}{\delta - 1} \right). \tag{15.8b}$$

(c) Now the problem is coming into focus. This application is unusual in that we are not so much interested in the dynamics of the system and the details of the phase portrait. Rather, our interest is in the *stability* of the two equilibrium points: We want to design the system (specifically, we want to choose values for q, V, and C_0) so that the singular point $(N_1, C_1) = (0, \epsilon)$ is *unstable* and hence not an attractor (because it causes annihilation, $N = 0$) and, also, so that the singular point (N_2, C_2) falls in the first quadrant of the phase plane (i.e., $N > 0$ and $C > 0$) and is *stable*. Show that for (N_2, C_2) to be in the first quadrant δ and ϵ must satisfy the inequalities

$$\delta > 1 \quad \text{and} \quad \epsilon > \frac{1}{\delta - 1}. \tag{15.9a,b}$$

In terms of the original parameters, realize that we have no control over the values of the physical constants β and γ, but we can indeed vary V, q, C_0 so that δ and ϵ [defined by (15.6)] satisfy the inequalities (15.9a,b) for stability.

(d) Next, we must also determine whatever constraints on the parameters are needed to ensure, if possible, that the singular points (N_1, C_1) and (N_2, C_2) are unstable and stable, respectively. Show that they are an *unstable* node and a *stable* node, respectively, for any values of δ and ϵ satisfying (15.9a,b). Hence, the chemostat will function properly for any values of the five (positive) quantities $\beta, \delta, q/V, \alpha$, and C_0 such that δ and ϵ satisfy those two inequalities. We suggest that you use computer software to help you through the analysis.
NOTE: The upshot is that the equilibrium points (N_1, C_1) and (N_2, C_2) will *always* be unstable and stable, respectively, if δ and ϵ satisfy (15.9), that is, if they fall in the shaded part of the figure given below.

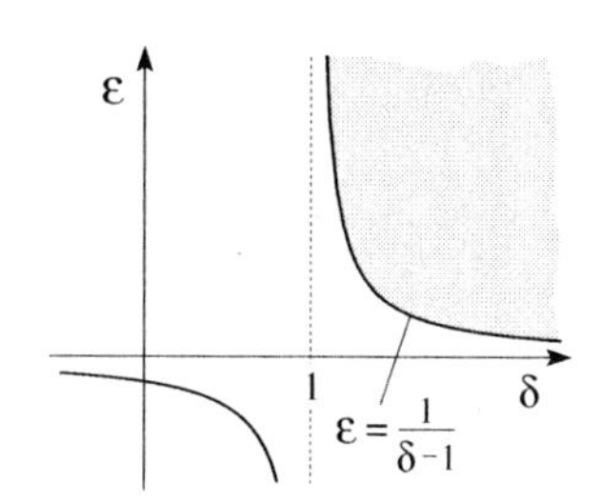

(e) It is interesting that the system (15.7) happens to have a straight-line trajectory, that is, of the form $N = aC + b$, whether or not δ and ϵ satisfy the stability conditions (15.9). Determine a and b.
NOTE: You may wish to obtain a computer-generated direction field, for the first quadrant of the N, C plane, once for a δ, ϵ pair that satisfies (15.9) and once for a δ, ϵ pair that does not satisfy (15.9). In each case you should be able to discern the straight-line trajectory.

7.5 LIMIT CYCLES

The differential equation

$$\boxed{x'' - \mu(1 - x^2)x' + x = 0,} \qquad (\mu > 0), \tag{1}$$

which is nonlinear because of the x^2x' term, was studied by *Balthasar van der Pol* (1889–1959), in connection with electric current oscillations in a certain vacuum tube circuit and then in connection with the modeling of the beating of the human heart.[1] The **van der Pol equation** is among the most prominent nonlinear differential equations in applied mathematics.

Consider the parameter μ to be positive. To study (1) in the phase plane, first re-express it, equivalently, as the system

$$x' = y, \tag{2a}$$
$$y' = -x + \mu(1 - x^2)y, \tag{2b}$$

[1] B. van der Pol, "On Relaxation Oscillations," *Philosophical Magazine*, Vol. 2, 1926, pp. 978–992, and B. van der Pol and J. van der Mark, "The Heartbeat As a Relaxation Oscillation, and An Electrical Model of the Heart," *Philosophical Magazine*, Vol. 6, 1928, pp. 763–775.

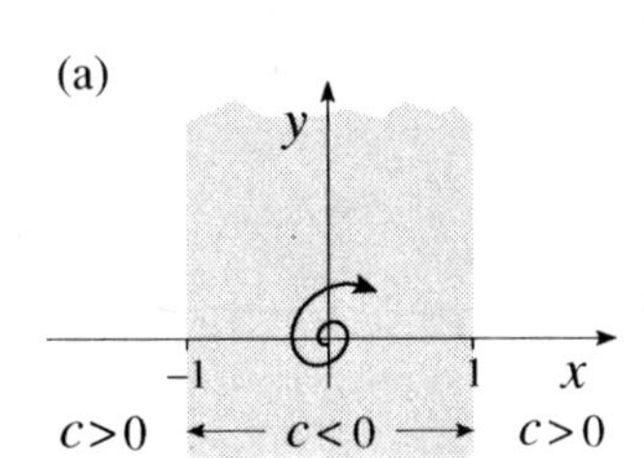

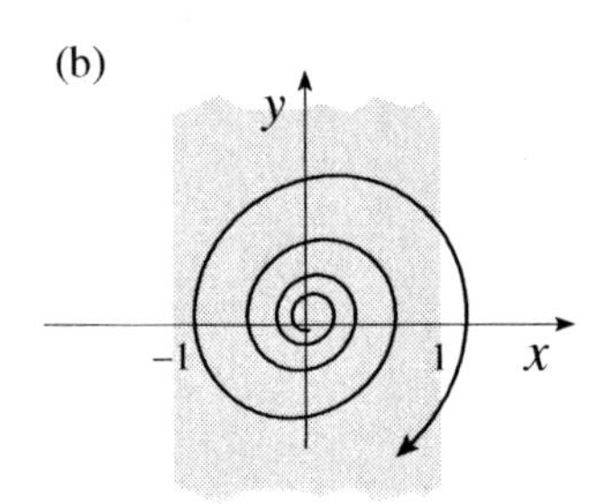

Figure 1. Van der Pol equation: unstable spiral at $(0,0)$ if $\mu < 2$.

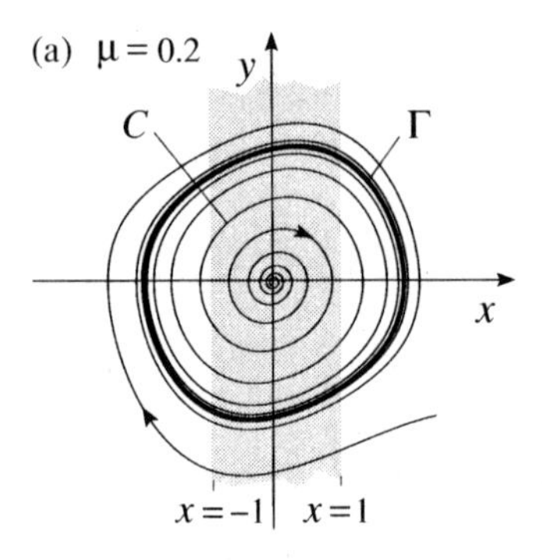

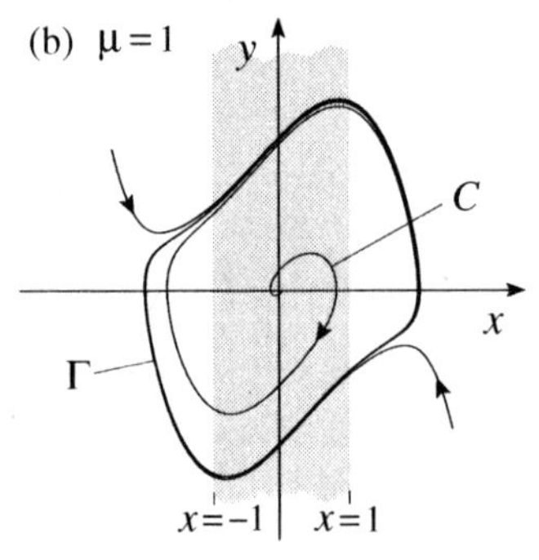

Figure 2. The van der Pol limit cycle, for $\mu = 0.2$ and 1.

which has one singular point, at $(0, 0)$. To determine its type, linearize (2) about $(0, 0)$, and obtain

$$x' = y, \tag{3a}$$
$$y' = -x + \mu y. \tag{3b}$$

[We've not bothered to change x, y in (3) to X, Y, as we did in Section 7.4.] Thus, $\mathbf{A} = \begin{bmatrix} 0 & 1 \\ -1 & \mu \end{bmatrix}$, the eigenvalues of which are

$$\lambda = \frac{\mu \pm \sqrt{\mu^2 - 4}}{2}. \tag{4}$$

We see from (4) that the singularity of the linearized system (3) at the origin is an *unstable spiral* if $\mu < 2$ and an *unstable node* if $\mu > 2$. This result makes sense because (3) is equivalent to the ODE $x'' - \mu x' + x = 0$, which (from Chapter 3) is the familiar equation of a damped harmonic oscillator, but with *negative damping*.

Near the origin in the x, y phase plane the flow is accurately described by the linearized system (3) and is shown (for a representative $\mu < 2$) in Fig. 1a. As the point $(x(t), y(t))$ spirals out of the neighborhood of the origin, the approximate linearized system (3) grows less accurate, and we wonder how the trajectory shown in Fig. 1a continues to develop as t increases. We can generate trajectories using computer software, but first think about what to expect. Similar to (1) is the familiar linear equation

$$mx'' + cx' + kx = 0 \tag{5}$$

that models a mechanical mass/spring/damping system. If we compare (5) with (1), we have $m = 1$ and $k = 1$, and we can think of the $-\mu(1-x^2)$ in (1) as the damping coefficient c in (5), but only heuristically because it is not a constant, but a function of x. Since $c = -\mu(1-x^2)$ in (1) is negative in the vertical strip $|x| < 1$ (shaded in Fig. 1a) we expect the spiral in Fig. 1a to continue to grow. Eventually it will break out of the $|x| < 1$ strip (Fig. 1b). As the point $(x(t), y(t))$ spends more and more time outside the strip, where $c = -\mu(1-x^2) > 0$, the effect of the positive damping in $|x| > 1$ will increase, relative to the effect of the negative damping within $|x| < 1$, and it is natural to wonder if the trajectory might therefore approach some limiting closed orbit as $t \to \infty$, over which the effects of the positive and negative damping are exactly in balance.

Computer plots are given in Fig. 2 for the values $\mu = 0.2$ and 1. In both cases we see that there does indeed exist such a closed orbit around the origin, labeled Γ, that is approached asymptotically as $t \to \infty$, both by trajectories initiated inside of Γ and by trajectories initiated outside of Γ.[1] Such a closed orbit is called a **limit cycle**. It can be shown that as $\mu \to 0$ the van der Pol limit cycle tends to a circle of radius 2 centered at the origin. From part (a) of Fig. 2 we see that even for $\mu = 0.2$ the limit cycle Γ is indeed approximately a circle of radius 2, and from part (b) we see that as μ is increased Γ becomes distorted.

[1]Of course, if the initial point is *on* Γ then the solution $x(t)$ doesn't *approach* a periodic solution, it will *be* a periodic solution.

The van der Pol limit cycle is *stable* because trajectories *approach* it, both from the inside and from the outside. *The stable limit cycle* Γ *is the most important feature of the van der Pol equation* because it represents the steady-state oscillation, the place where all trajectories (except the point trajectory at the origin) end up as $t \to \infty$. Besides the phase portraits in Fig. 2 we've also plotted $x(t)$ versus t in Fig. 3 corresponding to the trajectories labeled C in Fig. 2, each of which starts close to, but not at, the origin. Observe in Fig. 3 the relatively slow approach to the steady-state limit cycle oscillation for $\mu = 0.2$ and the rapid approach for $\mu = 1$, which can be seen in Fig. 2b as well.

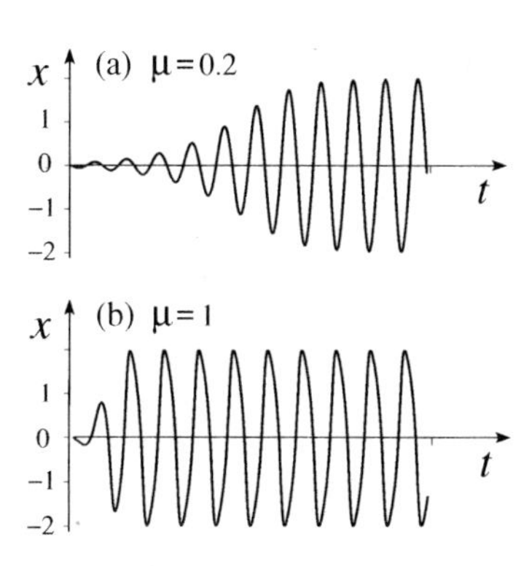

Figure 3. Plots of $x(t)$ versus t for the van der Pol equation (1), corresponding to the trajectories labeled C in Fig. 2.

Consider the limiting cases $\mu \to 0$ and $\mu \to \infty$. As we've mentioned, as $\mu \to 0$ the limit cycle tends to a circle of radius 2, which corresponds to a harmonic motion with amplitude 2.[1] But as μ is increased, the limit cycle develops kinks near $(x, y) = (2, 0)$ and $(-2, 0)$, and the solution $x(t)$ even becomes discontinuous as $\mu \to \infty$. For large μ the solution is called a **relaxation oscillation**, and is comprised of relatively long t intervals of gradual change separated by relatively short t intervals of rapid change, as can be seen in Fig. 4 for $\mu = 10$.

The periodic van der Pol limit cycle oscillation has a form and a period (for a given μ) that are entirely independent of the initial conditions. This property is crucial in a variety of biological phenomena such as the heart beat, respiration, and the firing of a neuron, and it provides a sort of **biological clock** for that process. The oscillation is *self-sustained* because there is no forcing function in (1). It is *not* the response to a periodic forcing function; there *is no* forcing function.

The van der Pol limit cycle is but one example of a limit cycle. More generally, a limit cycle is an isolated periodic orbit, isolated in the sense that neighboring trajectories are not closed; they wind toward or away from the limit cycle as $t \to \infty$. For instance, the periodic orbits in Fig.6 of Section 7.4 are not limit cycles.

If the limit cycle attracts neighboring solutions from both sides (i.e., from inside and outside) it is **stable**, an **attractor**; if it repels neighboring solutions then it is **unstable**, a **repellor**, and if it attracts on one side and repels on the other side, then it is **semistable**.

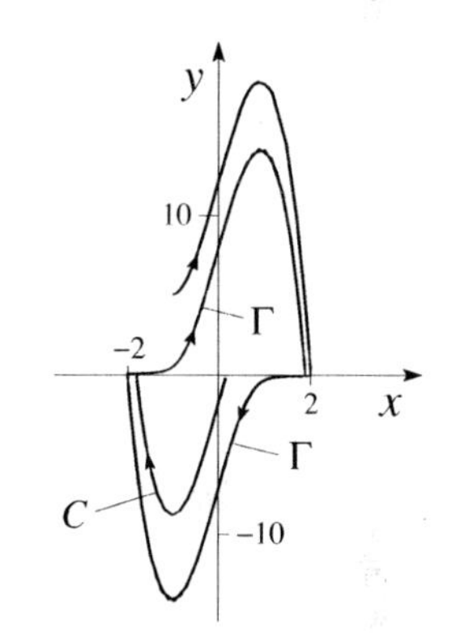

Limit cycles of an autonomous system $x' = P(x, y), y' = Q(x, y)$ *are possible only for nonlinear systems.* The simplest way to see that is to observe that the general *linear* autonomous system[2]

$$x' = ax + by, \tag{6a}$$

$$y' = cx + dy \tag{6b}$$

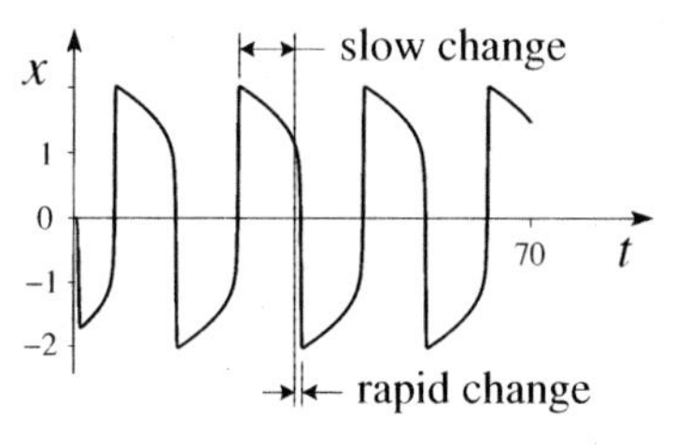

Figure 4. The large-μ behavior of the van der Pol limit cycle; $\mu = 10$.

can be solved, and its solutions examined. We did that in Section 7.3 (for the case where $ad - bc \neq 0$) and found only centers, spirals, nodes, and saddles — no limit cycles. It is not difficult to show that even if $ad - bc = 0$, limit cycles are not possible.

[1]By harmonic we mean of the form $A \sin(\omega t + \phi)$.

[2]Actually, the most general linear autonomous system includes additive constants on the right-hand sides of (6), but such additional terms merely amount to a translation of the x, y coordinate system.

Closure. Limit cycles correspond to periodic phenomena and arise, for instance, in modeling the human heart beat and the repetitive firing of the neuron (nerve cell). The van der Pol equation (1) is particularly well known and has a stable limit cycle, which is the most important feature of that equation. For that equation two limiting cases are of particular interest: In the small -μ limit the limit cycle tends to a circle with corresponding periodic solution for $x(t)$ that is purely harmonic. And in the large-μ limit the limit cycle becomes kinky and the corresponding periodic solution is comprised of successive t-intervals of slow and rapid change, as observed in Fig. 4.

EXERCISES 7.5

1. Some Simple Limit Cycles, in Polar Coordinates. To make up systems that have limit cycles, let us use polar coordinates $r(t), \theta(t)$ instead of $x(t), y(t)$. In each case sketch (or obtain a computer plot of) the x, y phase portrait, identify any limit cycles, and state whether they are stable, unstable, or semistable.

(a) $r' = (1-r)r,\ \theta' = 1$
(b) $r' = (3-r)r,\ \theta' = -2$
(c) $r' = 1-r^2,\ \theta' = 4$
(d) $r' = (1-r)^2,\ \theta' = 3$
(e) $r' = r(r-1),\ \theta' = 1$

NOTE: We do *not* mean to imply that if we re-express a system $x' = P(x,y),\ y' = Q(x,y)$ in polar coordinates then any limit cycles will be observable merely by inspection of the r, θ differential equations, as they are in this exercise.

2. The Van Der Pol Limit Cycle for Small μ. To examine the van der Pol limit cycle in the limit as $\mu \to 0$, let us begin by simply setting $\mu = 0$ in (1). Then (1) becomes the undamped harmonic oscillator equation

$$x'' + x = 0 \tag{2.1}$$

or, in the x, y phase plane,

$$x' = y,\ y' = -x. \tag{2.2a,b}$$

Integration of $dy/dx = --x/y$ gives the trajectories as the family of circles

$$x^2 + y^2 = C \tag{2.3}$$

— rather than the single van der Pol limit cycle trajectory that we were seeking. Nevertheless, since (2.3) shows that each of these orbits is circular and has period 2π we do expect that in the limit as $\mu \to 0$ the van der Pol limit cycle should tend to a circle, and that the period T of that cycle should tend to 2π. But there is no clue, from the preceding, as to the radius of that limit cycle.

(a) Recall that we argued [below (5)] that in $|x| > 1$ the term $-\mu(1-x^2)x'$ in (1) amounts to positive damping and absorption of energy, that in the strip $|x| < 1$ it amounts to negative damping, and that on the limit cycle trajectory these effects are in perfect balance. Thus, to find the radius of the small-μ limit cycle, let us proceed formally by requiring that the work done by the force term $-\mu(1-x^2)x'$ in (1) over one cycle be zero:

$$\text{Work} = -\mu \int (1-x^2)x'\,dx = 0. \tag{2.4}$$

The initial point on the limit cycle doesn't matter, so let us take it to be the point where the limit cycle crosses the positive x-axis, at $(A, 0)$. (We don't yet know the value of A, which is the radius of the limit cycle, which we're trying to determine.) With those initial conditions [$x(0) = A$ and $y(0) = 0$] the solution of (2.1) for x is $x(t) = A\cos t$. Put that expression for $x(t)$ into (2.4), evaluate the integral, and show that you obtain $A = 2$. Although the foregoing derivation has been only formal, the result is correct, namely, that in the small μ limit the radius of the van der Pol limit cycle is 2.
(b) Choose a very small value of μ, and any convenient initial conditions, and use computer software to generate the x, y phase portrait and the x versus t plot, as we did in Figs.2 and 3. Do your results seem consistent with the claim that as $\mu \to 0$ the van der Pol limit cycle tends to a circle of radius 2, with period 2π?

3. The Van Der Pol Limit Cycle for Large μ. We stated in the text that as $\mu \to \infty$ the van der Pol limit cycle develops kinks and the $x(t)$ response becomes a discontinuous "relaxation oscillation." Figure 4, which is for the moderately large value $\mu = 10$, appears to be consistent with that claim.

(a) Increase μ further, to 20, and use computer software to obtain both the phase portrait and the graph of x versus t. We're interested here only in the limit cycle, so a single initial condition (of your choice, and not necessarily *on* the limit cycle) will suffice.
(b) Measure the period T from your graph of x versus t and compare it with the known asymptotic result

$$T \sim (3-2\ln 2)\mu \quad \text{as} \quad \mu \to \infty. \tag{3.1}$$

(c) Observe from your x versus t graph that the graph is almost discontinuous, and that there are four points worth labeling for any one given period: at the beginning and at the end of each rapid rise, and at the beginning and one at the end of each rapid fall. Label them as A,B,C,D. Then label the corresponding points on the phase portrait as A,B,C,D.

4. Negative μ in the Van Der Pol Equation. Perhaps if μ in (1) is negative, instead, then (1) still yields a limit cycle, this time unstable? Use computer software to explore that possibility. If you do find an unstable limit cycle, does its shape appear to be the same, for a given negative μ, as for the corresponding positive value of μ? Discuss.

5. From the form of the differential equation itself, see if you can predict any limit cycles in the x, y phase plane (with $x' = y$), and indicate their stability. (Of course, you could check your hypothesis by using computer software to generate some phase trajectories.)

(a) $x'' + (x^2+x'^2-1)x' + x = 0$
(b) $x'' + (x^2+x'^2-1)x' + 4x = 0$
(c) $x'' + (9-x^2-x'^2)x' + 3x = 0$
(d) $x'' + (1-x^2-x'^2)(4-x^2-x'^2)x' + x = 0$

6. A Theorem for the Existence of a Limit Cycle. We state the following theorem without proof. Use it to show that the equations given in (a)–(c), below, have limit cycles.

THEOREM 7.5.1. Let $f(x)$ be an even function [i.e., $f(-x) = f(x)$ for all x], and continuous for all x. Let $g(x)$ be an odd function [i.e., $g(-x) = -g(x)$ for all x] with $g(x) > 0$ for all $x > 0$, and let $g'(x)$ be continuous for all x. With

$$\int_0^x f(\xi)\,d\xi \equiv F(x) \quad \text{and} \quad \int_0^x g(\xi)\,d\xi \equiv G(x), \tag{6.1}$$

suppose that: (i) $G(x) \to \infty$ as $x \to \infty$, and (ii) there is an $x_0 > 0$ such that $F(x) < 0$ for $0 < x < x_0$, $F(x) > 0$ for $x > x_0$, and $F(x)$ is monotonically increasing for $x > x_0$, with $F(x) \to \infty$ as $x \to \infty$. Then the **generalized Liénard equation**

$$\boxed{x'' + f(x)x' + g(x) = 0} \tag{6.2}$$

has a single periodic solution, the trajectory of which is a closed curve encircling the origin in the x, y phase plane (with $x' = y$). All other trajectories (except the trajectory consisting of the single point at the origin) spiral toward the closed trajectory (limit cycle) as $t \to \infty$.

(a) $x'' - \mu(1-x^2)x' + x = 0 \quad (\mu > 0)$
(b) $x'' - (1-x^2)x' + x^3 = 0$
(c) $x'' + x^2(5x^2-3)x' + x = 0$

7. Hopf Bifurcation. (a) Show that if we change to polar coordinates r, θ (see Exercise 4 in Section 7.4), then the system

$$x' = \epsilon x + y - x(x^2+y^2), \tag{7.1a}$$
$$y' = -x + \epsilon y - y(x^2+y^2) \tag{7.1b}$$

simplifies to

$$r' = r(\epsilon - r^2), \tag{7.2a}$$
$$\theta' = -1 \tag{7.2b}$$

(b) Show that (7.1) has only one singular point — at the origin, and that it is a stable spiral if $\epsilon < 0$ and an unstable spiral if $\epsilon > 0$.
(c) Show, from (7.2), that $r(t) = \sqrt{\epsilon}$ is a trajectory (if $\epsilon > 0$) — in fact, a stable limit cycle.
NOTE: We say that zero is a **bifurcation value** of ϵ: Consider an ϵ axis (shown below). For negative ϵ there is only a stable

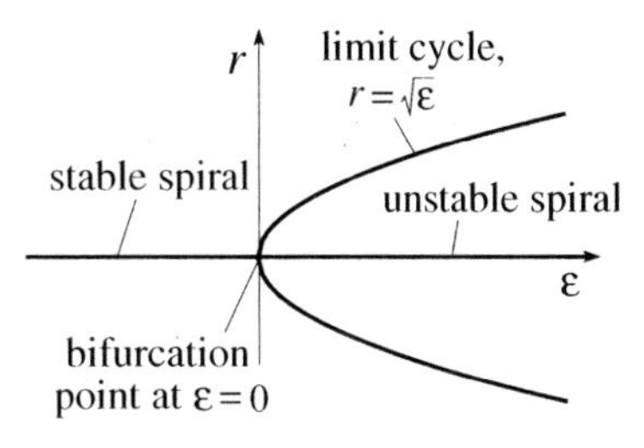

spiral at the origin ($r = 0$). As ϵ increases, the stable spiral loses its stability and becomes an unstable spiral when ϵ becomes positive, and a limit cycle is born (i.e., it "bifurcates" somewhat as the branch of a tree does), the radius of which increases with ϵ according to the parabola $r = \sqrt{\epsilon}$. This phenomenon is an example of a **Hopf bifurcation**, and $\epsilon = 0$ is the **bifurcation value** of ϵ.

(d) How can we modify (7.1) so that it gives an *unstable* limit cycle at $r = \sqrt{\epsilon}$ instead?

7.6 NUMERICAL SOLUTION OF SYSTEMS BY EULER'S METHOD

PREREQUISITE: Section 1.9.

In Section 1.9 we introduced the numerical solution of differential equations by showing how to solve the IVP

$$y' = f(x, y); \quad y(a) = b \tag{1}$$

by Euler's method. The discussion was only introductory since there are far more accurate methods than Euler's, such as the fourth-order Runge–Kutta method. In this section we again limit discussion to the simple Euler's method as we extend the method to systems of first-order differential equations and to higher-order equations. In addition, we consider not only initial value problems (IVPs), but also boundary value problems (BVPs). The latter were not relevant in Section 1.9 because a BVP must have at least two boundary conditions, whereas for a single first-order equation only one initial condition is appropriate.

7.6.1 Initial value problems. Recall that Euler's method for the discrete approximate solution of the single first-order IVP

$$x'(t) = f(t, x); \quad x(a) = x_0 \tag{2}$$

is the algorithm

$$\boxed{x_{n+1} = x_n + f(t_n, x_n)\, h \quad (n = 0, 1, 2, \ldots),} \tag{3}$$

where h is the step size, $t_n = a + nh$, and the x_n's are the approximate Euler-generated values of the exact solution $x(t)$ at the discrete points t_n.

To apply Euler's method to a *system* of first-order IVPs such as

$$x'(t) = f(t, x, y); \quad x(a) = x_0, \tag{4a}$$
$$y'(t) = g(t, x, y); \quad y(a) = y_0, \tag{4b}$$

To apply Euler's method to the system (4), we apply (3) to *each* of (4a) and (4b).

we simply apply it to each of the problems (4a) and (4b), as follows:

$$\boxed{\begin{aligned} x_{n+1} &= x_n + f(t_n, x_n, y_n)\, h, \\ y_{n+1} &= y_n + g(t_n, x_n, y_n)\, h \end{aligned}} \tag{5a,b}$$

for $n = 0, 1, 2, \ldots$, bringing the x_n's and y_n's along together. That is, we do *not* use (5a) to evaluate $x_1, x_2, \ldots$ and then (5b) to evaluate $y_1, y_2, \ldots$. Rather, we use (5a) and (5b) to evaluate x_1 and y_1, then x_2 and y_2, and so on.

EXAMPLE 1. Illustration of Euler's Method. To illustrate (5), consider

$$x' = t + y; \qquad x(0) = 0, \tag{6a}$$
$$y' = xy^2; \qquad y(0) = 1. \tag{6b}$$

The system (6) is nonlinear because of the y^2 in (6b), but the algorithm (5) is insensitive to whether (4) is linear or not. For the system (6), the Euler algorithm (5) is this:

$$x_{n+1} = x_n + (t_n + y_n)\, h; \qquad x_0 = 0, \tag{7a}$$
$$y_{n+1} = y_n + \left(x_n y_n^2\right) h; \qquad y_0 = 1. \tag{7b}$$

With $h = 0.1$, say, let us carry out (7) through the first two steps.

$n = 0:$

$$x_1 = x_0 + (t_0 + y_0)h = 0 + (0 + 1)(0.1) = \underline{0.1},$$
$$y_1 = y_0 + (x_0 y_0^2)h = 1 + (0)(1)^2(0.1) = \underline{1}.$$

$n = 1:$

$$x_2 = x_1 + (t_1 + y_1)h = 0.1 + (0.1 + 1)(0.1) = \underline{0.21},$$
$$y_2 = y_1 + (x_1 y_1^2)h = 1 + (0.1)(1)^2(0.1) = \underline{1.01}.$$

Carrying out two more steps, the results are as follows:

n	t_n	x_n	y_n
0	0	0	1
1	0.1	0.1	1
2	0.2	0.21	1.01
3	0.3	0.331	1.0314
4	0.4	0.4641	1.0666

where $n = 0$ corresponds to the initial conditions. The calculations for $n = 1$ and 2 are given above; those for $n = 3$ and 4 are left for the exercises. ▮

If the given system contains any higher-order equations, we can first use **auxiliary dependent variables** to obtain an equivalent system of first-order equations, to which we can then apply Euler's method.

EXAMPLE 2. Using Auxiliary Variables. Consider

$$x'' + tx' - y^3 = \sin t; \qquad x(0) = 2,\ x'(0) = -1, \tag{8a}$$
$$y' + x^2 - y = 0; \qquad y(0) = 1. \tag{8b}$$

By introducing an auxiliary variable $x' = u$ we obtain the equivalent system of *first*-order equations

$u(0) = -1$ because $u(t)$ is $x'(t)$, and (8a) specified that $x'(0) = -1$.

$$x' = u; \qquad x(0) = 2, \tag{9a}$$
$$u' = \sin t - tu + y^3; \qquad u(0) = -1, \tag{9b}$$
$$y' = -x^2 + y; \qquad y(0) = 1. \tag{9c}$$

The latter can now be solved by the Euler algorithm

$$x_{n+1} = x_n + u_n h; \qquad x_0 = 2, \tag{10a}$$
$$u_{n+1} = u_n + \left(\sin t_n - t_n u_n + y_n^3\right) h; \qquad u_0 = -1, \tag{10b}$$
$$y_{n+1} = y_n + \left(-x_n^2 + y_n\right) h; \qquad y_0 = 1. \tag{10c}$$

If $h = 0.1$, say, then $t_n = 0.1n$ and (10) gives

$$x_1 = 2 + (-1)(0.1) = \underline{1.9},$$
$$u_1 = -1 + (0 - 0 + 1)(0.1) = \underline{-0.9},$$
$$y_1 = 1 + (-4 + 1)(0.1) = \underline{0.7},$$

and so on.

COMMENT. Of course, we may not be interested in the u_n values because u is only an auxiliary variable. Thus, if we present the results in a table or graph we may wish to leave it out, but we cannot leave it out of the calculations because (10) forces us to generate all three: the x_n's, u_n's, and y_n's. ▮

EXAMPLE 3. A Single Higher-Order Equation. To solve

Equation (11) is nonlinear because of the xx'' and x'^2 terms, but the nonlinearity in no way hinders Euler's method.

$$x''' - 2txx'' - x'^2 = 0; \quad x(1) = 3, \; x'(1) = -1, \; x''(1) = 2 \tag{11}$$

by Euler's method, introduce *two* artificial variables, $x' = u$ and $u' = v$ (i.e., $x'' = v$), and express (11) in the first-order form

$$\begin{aligned} x' &= u; & x(1) &= 3, \\ u' &= v; & u(1) &= -1, \\ v' &= 2txv + u^2; & v(1) &= 2. \end{aligned} \tag{12}$$

Then Euler's method becomes

$$\begin{aligned} x_{n+1} &= x_n + u_n h; & x_0 &= 3, \\ u_{n+1} &= u_n + v_n h; & u_0 &= -1, \\ v_{n+1} &= v_n + \left(2t_n x_n v_n + u_n^2\right) h; & v_0 &= 2, \end{aligned} \tag{13}$$

with $t_n = 1 + nh$. ▮

7.6.2 Existence and uniqueness for nonlinear systems. The numerical methods discussed above are equally applicable to linear and nonlinear systems, but Theorem 4.5.1, for existence and uniqueness, was only for linear systems. To fill this gap, we state the following existence and uniqueness theorem.

THEOREM 7.6.1 *Existence and Uniqueness for IVPs; Nonlinear Case*
Consider the IVP

$$\begin{aligned} x_1' &= f_1(t, x_1, \dots, x_k); \qquad x_1(a) = b_1 \\ &\vdots \\ x_k' &= f_k(t, x_1, \dots, x_k); \qquad x_k(a) = b_k \end{aligned} \tag{14}$$

or, in more compact notation,

$$\boxed{\mathbf{x}' = \mathbf{f}(t, \mathbf{x}); \qquad \mathbf{x}(a) = \mathbf{b}.} \tag{15}$$

The x in $\mathbf{f}(t, \mathbf{x})$ in (15) is standard shorthand for $x_1, \dots, x_k$; that is, $f(t, \mathbf{x})$ means $f(t, x_1, \dots, x_k)$.

Let the functions $f_1, \dots, f_k$ and their first-order partial derivatives with respect to $x_1, \dots, x_k$ be continuous in a region R defined by $c < t < d$, $\xi_1 < x_1 < \eta_1$, ..., $\xi_k < x_k < \eta_k$, and let the initial point $(a, b_1, \dots, b_k)$ be in R. Then the IVP (15) has a solution defined on some open t interval containing $t = a$, and that solution is unique.

On *some* open t interval; like Theorem 1.5.1, Theorem 7.6.1 does not tell us how broad that t interval will be.

The latter is a k-dimensional generalization of Theorem 1.5.1. As in the case of Theorem 1.5.1, the conditions given in Theorem 7.6.1 are sufficient, not necessary and sufficient.

EXAMPLE 4. Application of Theorem 7.6.1. Apply Theorem 7.6.1 to the problem

$$x'' + cx' + x + x^3 = F_0 \cos \Omega t; \qquad x(0) = x_0, \quad x'(0) = x_0'. \tag{16}$$

Equation (16) is a *forced Duffing equation.* It is nonlinear because of the x^3. It was the subject of Exercise 18 of Section 3.4.

First, put (16) into the form (14) by using an auxiliary variable u:

$$\begin{aligned} x' &= u; & x(0) &= x_0, & \text{(17a)} \\ u' &= x - x^3 - cu + F_0 \cos \Omega t; & u(0) &= x_0'. & \text{(17b)} \end{aligned}$$

In the theorem k is 2, x_1 is x, and x_2 is u. From (17), $f_1(t, x, u) = u$, $f_2(t, x, u) = x - x^3 - cu + F_0 \cos \Omega t$, $\partial f_1/\partial x = 0$, $\partial f_1/\partial u = 1$, $\partial f_2/\partial x = 1 - 3x^2$, and $\partial f_2/\partial u = -c$, all of which are continuous on $-\infty < t < \infty$, $-\infty < x_1 < \infty$, and $-\infty < x_2 < \infty$. Thus, (16), and hence (17), does admit a unique solution on some open t interval containing the initial point $t = 0$. ▮

7.6.3 Linear boundary value problems. Boundary value problems are more difficult than IVPs, as we will see, but a *linear* BVP can be solved by solving a set of IVPs.

EXAMPLE 5. A Linear BVP. Consider the problem

The interval I is $0 \leq x \leq 2$.

$$y''' - x^2 y = -x^4; \qquad y(0) = 0,\ y'(0) = 0,\ y(2) = 4, \tag{18}$$

which is of boundary value type because at least one condition is given at each end of the interval, at $x = 0$ and at $x = 2$. To solve numerically, first recast (18) as the system of first-order equations

$$\begin{aligned} y' &= u; & y(0) &= 0,\ y(2) = 4, & \text{(19a)}\\ u' &= v; & u(0) &= 0 & \text{(19b)}\\ v' &= x^2 y - x^4. & & & \text{(19c)} \end{aligned}$$

The problem is that we are missing $v(0)$ in (19c), so we cannot get the numerical solution method started.

Trouble: We cannot apply the numerical solution techniques discussed above because (19c) does not have an initial condition, so we cannot get the solution started. Whereas (19c) is missing an initial condition on v, (19a) has an extra condition, the right end condition $y(2) = 4$, but that extra condition is of no help in a numerical scheme that develops a solution, step by step, beginning at the left endpoint $x = 0$.

However, the linearity of (18) saves the day and permits us to work with an initial value version instead, as we now explain. Suppose we solve (numerically) the four IVPs

In place of the one BVP (18), we have the four IVPs (20).

$$\begin{aligned} LY_1 &= 0, & Y_1(0) &= 1, & Y_1'(0) &= 0, & Y_1''(0) &= 0,\\ LY_2 &= 0, & Y_2(0) &= 0, & Y_2'(0) &= 1, & Y_2''(0) &= 0,\\ LY_3 &= 0, & Y_3(0) &= 0, & Y_3'(0) &= 0, & Y_3''(0) &= 1,\\ LY_p &= -x^4, & Y_p(0) &= 0, & Y_p'(0) &= 0, & Y_p''(0) &= 0, \end{aligned} \tag{20}$$

where $L = d^3/dx^3 - x^2$ is the differential operator in (18). We chose the nine initial conditions in the first three lines of (20) so that their determinant would be nonzero; it is 1. Consequently, and this is the point, $Y_1(x), Y_2(x), Y_3(x)$ comprise a fundamental set of solutions of the homogeneous equation $LY = 0$ (Exercise 4). We chose the three initial conditions for the particular solution $Y_p(x)$ as zero merely for simplicity; any values will do since any particular solution will do.

Suppose the four IVPs in (20) have now been solved by the numerical methods discussed above. Then $Y_1(x), Y_2(x), Y_3(x), Y_p(x)$ are known on the x interval of interest $[0, 2]$, and from them we can construct the general solution

$$y(x) = C_1 Y_1(x) + C_2 Y_2(x) + C_3 Y_3(x) + Y_p(x) \tag{21}$$

of the differential equation $Ly = -x^4$ in (18). Finally, we can evaluate the constants C_1, C_2, C_3 by imposing on (21) the boundary conditions given in (18):

On the right-hand side of (22) we've used the initial conditions defined in (20).

$$\begin{aligned} y(0) &= 0 = C_1 + 0 + 0 + 0,\\ y'(0) &= 0 = 0 + C_2 + 0 + 0,\\ y(2) &= 4 = C_1 Y_1(2) + C_2 Y_2(2) + C_3 Y_3(2) + Y_p(2). \end{aligned} \tag{22}$$

Regarding the solution of (22) for C_1, C_2, C_3, two cases present themselves. First, suppose the computed value of $Y_3(2)$ is nonzero. Then (22) gives the unique solution $C_1 = C_2 = 0$ and $C_3 = [4 - Y_p(2)]/Y_3(2)$, and hence the unique solution of (18) as

$$y(x) = \frac{4 - Y_p(2)}{Y_3(2)} Y_3(x) + Y_p(x). \tag{23}$$

The question of the existence and uniqueness of a solution $y(x)$ of (18) comes down to the question of the existence and uniqueness of a solution C_1, C_2, C_3 of the linear algebraic equations (22).

In fact, since $C_1 = C_2 = 0$ the functions $Y_1(x)$ and $Y_2(x)$ are not present in (23), and we don't need to calculate them. All we need are $Y_3(x)$ and $Y_p(x)$, and these are found by the numerical solution of their IVPs,

$$\begin{aligned} Y_3' &= U_3; & Y_3(0) &= 0, \\ U_3' &= V_3; & U_3(0) &= 0, \\ V_3' &= x^2 Y_3; & V_3(0) &= 1, \end{aligned} \tag{24}$$

for $Y_3(x)$, and

$$\begin{aligned} Y_p' &= U_p; & Y_p(0) &= 0, \\ U_p' &= V_p; & U_p(0) &= 0, \\ V_p' &= x^2 Y_p - x^4; & V_p(0) &= 0, \end{aligned} \tag{25}$$

for $Y_p(x)$, which correspond to the third and fourth lines of (20), respectively. The problems (24) and (25) can be solved by the methods discussed above, and this is left for the exercises.

As the second case, suppose that $Y_3(2)$ in (22) does turn out to be zero. Then the last of the equations in (22) is $4 = 0C_2 + Y_p(2)$. Thus, if the computed value of $Y_p(2)$ happens not to be 4, then the latter equation cannot be satisfied by any choice of C_2, and there is *no solution* of (18). But, if the computed value of $Y_p(2)$ does happen to be 4, then the latter equation is satisfied for *any* value of C_3 and we have the *nonunique solution* $y(x) = C_3 Y_3(x) + Y_p(x)$ of (18), in which C_3 is arbitrary.

If these results seem strange, remember (from Section 3.5.1) that BVPs can have a unique solution, no solution, or a nonunique solution. ▮

Closure. Euler's method for a system of first- or higher-order IVPs is a straightforward extension of the Euler algorithm given in Section 1.9 for a single first-order equation IVP. If the equation, or system of equations, contains any derivatives of order higher than one, then the first step is to use one or more auxiliary dependent variables to obtain a system of first-order equations. Then, apply the algorithm to all equations in the set, from one step to the next.

If the set is expressed in vector form as

$$\mathbf{x}'(t) = \mathbf{f}(t, \mathbf{x}); \quad \mathbf{x}(a) = \mathbf{b}, \tag{26}$$

then the Euler algorithm for (26) can be expressed as

$$\mathbf{x}_{n+1} = \mathbf{x}_n + \mathbf{f}(t_n, \mathbf{x}_n)\, h; \quad \mathbf{x}_0 = \mathbf{b},\ t_n = a + nh. \tag{27}$$

Boundary value problems are more difficult than IVPs since, when expressed as a system of first-order equations, at least one initial condition is missing, so we cannot get the numerical calculation started. We considered only *linear* BVPs because in that case we can solve the BVP by solving a set of IVPs, as we illustrated in Example 5. Keep in mind that BVPs can have a unique solution, or none, or a nonunique solution.

EXERCISES 7.6

INITIAL VALUE PROBLEMS

1. Example 1. In Example 1 we used Euler's method to calculate x_1, y_1, x_2, y_2. Continuing, calculate x_3, y_3, x_4, y_4, and verify that your results agree with those in the table.

2. Euler Hand Calculation. Reduce the given equation or system to a system of first-order equations, if necessary. Then apply two Euler steps (by hand) using $h = 0.2$. In part (b), for instance, that amounts to computing $x(t)$ and $y(t)$ at $t = 5.2$ and 5.4.

(a) $x' = x + y^2; \quad x(0) = 1$
$\quad y' = tx; \quad y(0) = 0$

(b) $x' = x^2 + y^2; \quad x(5) = 1$
$\quad y' = x^2 + 2t; \quad y(5) = 0.4$

(c) $x'' = x + y^2; \quad x(0) = 2,\ x'(0) = 0$
$\quad y' = 50t + xy; \quad y(0) = 1$

(d) $x' = x + y; \quad x(1) = 2$
$\quad y'' = x + 4; \quad y(1) = 0,\ y'(1) = -1$

(e) $x' = x + y; \quad x(0) = 5$
$\quad y' = y + z; \quad y(0) = 0$
$\quad z' = y + z; \quad z(0) = 3$

(f) $x' = xz; \quad x(0) = 3$
$\quad y' = xy; \quad y(0) = 0$
$\quad z' = -x; \quad z(0) = 2$

(g) $x'' + 4(t+1)x = 0; \quad x(0) = 2,\ x'(0) = 0$

(h) $x''' - x = 6; \quad x(1) = 2,\ x'(1) = 1,\ x''(1) = 1$

(i) $x''' - xx' = 10t; \quad x(2) = 3,\ x'(2) = -1,\ x''(2) = 2$

3. Computer Euler Calculation; Reducing the Step Size. Use computer software to evaluate $x(t)$ and $y(t)$ by Euler's method, through $t = 2$, and print the results at $t = 1$ and $t = 2$. Do so for $h = 0.1$, 0.01, and 0.001, and compare your results, in tabular form, with values computed from the exact solution (given in brackets) at those two points. (We constructed these examples to have simple explicit solutions, for comparison with the numerical solutions.) NOTE: This exercise is to promote understanding of the convergence of the Euler algorithm, not to encourage you to seek accurate solutions by using Euler with a tiny step size. For accurate results, a higher-order algorithm, such as fourth-order Runge–Kutta, would be better.

(a) $x' = y; \quad x(0) = 1 \quad [x(t) = \cos t]$
$\quad y' = -x; \quad y(0) = 0 \quad [y(t) = -\sin t]$

(b) $x' = -y^2/x; \quad x(0) = 1 \quad [x(t) = e^{-t}]$
$\quad y' = -x; \quad y(0) = 1 \quad [y(t) = e^{-t}]$

(c) $x' = x^2 - y - 2t + 1; \quad x(0) = 1 \quad [x(t) = t + 1]$
$\quad y' = 2x - 2; \quad y(0) = 1 \quad [y(t) = t^2 + 1]$

(d) $x' = txy - t^4 + 1; \quad x(0) = 0 \quad [x(t) = t]$
$\quad y' = y - x^2 + 2t; \quad y(0) = 0 \quad [y(t) = t^2]$

(e) $x' = y - x - 4; \quad x(0) = 1 \quad [x(t) = e^{-t}]$
$\quad y' = x^2 - e^{-2t}; \quad y(0) = 4 \quad [y(t) = 4]$

(f) $x' = x^{3/2} - y + 2t; \quad x(0) = 0 \quad [x(t) = t^2]$
$\quad y' = x + 2t^2; \quad y(0) = 0 \quad [y(t) = t^3]$

BOUNDARY VALUE PROBLEMS

4. Example 5. Below (20), we said that the three solutions $Y_1(x), Y_2(x), Y_3(x)$ of the homogeneous equation $y''' - x^2 y = 0$ will be LI because of our choice of the initial conditions in the first three lines of (20). Explain why that claim is true, citing any relevant theorem(s).

5. Instructions. Study Example 5, then write out step-by-step instructions for the numerical solution of a BVP of the form

$$y'' + p(x)y' + q(x)y = f(x); \quad y(a) = y_1,\ y(b) = y_2 \quad (7.1)$$

on $a \leq x \leq b$, where $p(x)$, $q(x)$, and $f(x)$ are prescribed functions that are continuous on the interval. Assume that the user knows how to solve a system of first-order IVPs numerically, so you need not explain that part.

6. Computer. Use the method explained in Example 5 to reduce the given BVP to a system of IVPs. Use computer software to solve those IVPs using Euler's method with $h = 0.05$. If you believe there is no solution or a nonunique solution, state your reasoning. Carry four or five significant figures. NOTE: In part (a), for instance, you will need to solve the IVPs

$$Y_1'' - 2xY_1' + Y_1 = 0; \qquad Y_1(0) = 1,\ Y_1'(0) = 0,$$
$$Y_2'' - 2xY_2' + Y_2 = 0; \qquad Y_2(0) = 0,\ Y_2'(0) = 1,$$
$$Y_p'' - 2xY_p' + Y_p = 3\sin x; \quad Y_p(0) = 0,\ Y_3'(0) = 0.$$

for $Y_1(x)$, $Y_2(x)$, and $Y_p(x)$. If you were going to solve those three IVPs by hand, or to program them for solution by Euler's method, you would first break each of them down into a system of two first-order ODEs, analogous to the systems (24) and (25) in Example 5. However, for purposes of this exercise you need *not* do that. Simply solve them directly by computer software.

(a) $y'' - 2xy' + y = 3\sin x; \quad y(0) = 1, \quad y(2) = 3.$
Determine $y(x)$ at $x = 0.5,\ 1,\ 1.5$.

(b) $y'' + (\cos x)y = 0; \quad y(0) = 3, \quad y(10) = 2.$
Determine $y(x)$ at $x = 2$.

(c) $y'' - [\ln(x+1)]\,y' - y = 2\sin 3x + 1:$
$y(0) = 3, \quad y(2) = -1.$
Determine $y(x)$ at $x = 0.5,\ 1,\ 1.5$.

(d) $y'' + y' - xy = x^3; \quad y(0) = 0, \quad y(5) = 20.$
Determine $y(x)$ at $x = 1,\ 2,\ 3,\ 4$.

(e) $y''' + xy = 2x^3; \quad y(1) = y'(1) = 0, \quad y''(2) = -3.$
Determine $y(x)$ at $x = 1.5,\ 2$.

(f) $y''' - xy = 0; \quad y'(0) = 4, \quad y(2) = y'(2) = 0.$
Determine $y(x)$ at $x = 0,\ 0.5,\ 1,\ 1.5$.

ADDITIONAL EXERCISES

7. Completion of Example 5. (a) Solve (24) and (25) *by hand*, using Euler's method with $h = 0.5$, and then use those results and (23) to compute $y(x)$. You should obtain these results:

x	$Y_3(x)$	$Y_p(x)$	Euler $y(x)$ $h = 0.5$	Exact $y(x)$ $y(x) = x^2$
0	0	0	0	0
0.5	0	0	0	0.25
1	0.25	0	0.6680	1
1.5	0.75	0	2.0039	2.25
2	1.5	-0.0078	4.0	4.0

In fact, we made up (18) so as to have a simple exact solution, $y(x) = x^2$, which is included as the final column of the table. Of course, the boundary conditions force the Euler values to agree with the exact values at $x = 0$ and at $x = 2$, but at the three internal points the Euler results corresponding to $h = 0.5$ give only a rough approximation.
(b) If h is reduced, for greater accuracy, hand calculation becomes impractical, but we can program the Euler solution to be implemented on a computer or calculator. To illustrate, reduce h to 0.1, say, program the calculation and carry it out on a computer. Thus compute $Y_3(x)$ and $Y_p(x)$ at $x = 0.1, 0.2, \ldots, 2.0$ and use (23) to calculate $y(x)$ at the representative points $x = 0.5,\ 1,\ 1.5$. You should obtain the results given in the third column of the following table.

x	Euler $h = 0.5$	Euler $h = 0.1$	Exact $y(x) = x^2$
0	0	0	0
0.5	0	0.21134	0.25
1	0.6680	0.95079	1
1.5	2.0039	2.21596	2.25
2	4	4	4

The $h = 0.1$ results are still not particularly accurate. To obtain agreement with the exact values, to several significant figures, we would need to reduce h to around 0.001; you need not do that.

8. Temperature In an Annular Plate. The steady state temperature distribution $u(r)$ in the annular plate shown in the figure is governed by the BVP

$$k\frac{1}{r}\frac{d}{dr}\left(r\frac{du}{dr}\right) - h(u - u_\infty) = 0, \quad (a < r < b) \qquad (8.1)$$

with boundary conditions $u(a) = u_1$, $u(b) = u_2$; the $h(u - u_\infty)$ term corresponds to heat lost, due to Newton cooling, from the uninsulated annular flat faces to the environment, which is at constant temperature u_∞, h is the corresponding

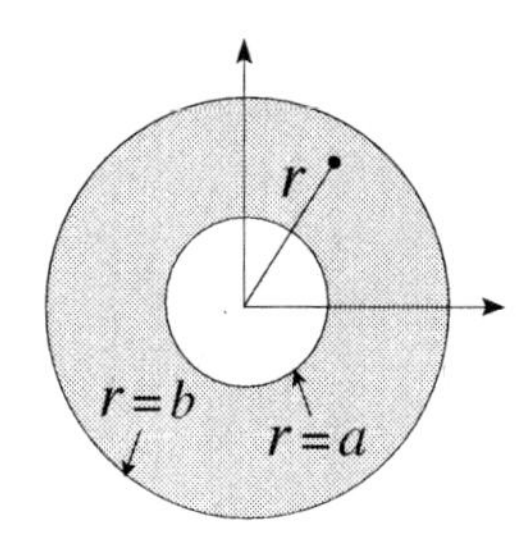

heat transfer coefficient, and k is the thermal conductivity of the plate material. Let $k = h = 1$, $u_\infty = 0$, $a = 1$, $b = 2$, $u_1 = 0$, and $u_2 = 100$. (Problems of this sort were covered in Section 3.5 and in the Exercises therein.) Solve this linear BVP using the method explained in Section 7.6.3. Use computer software to solve the IVPs that arise, by Euler's method, with stepsize $h = 0.05$, and obtain those solutions at $r = 1, 1.2, 1.4, 1.6, 1.8, 2$. From those results, compute $u(r)$ at those points.

CHAPTER 7 REVIEW

This chapter on the phase plane method, for autonomous systems of the form

$$\frac{dx}{dt} = P(x, y), \tag{1a}$$

$$\frac{dy}{dt} = Q(x, y), \tag{1b}$$

complements most of the material presented in the earlier chapters in that the approach is highly geometric, emphasizing the qualitative aspects of the flow in the x, y phase plane rather aiming at the traditional plots of the solutions $x(t)$ and $y(t)$.

By dividing (1b) by (1a) we obtain the differential equation

$$\frac{dy}{dx} = \frac{Q(x, y)}{P(x, y)} \tag{2}$$

in which t is no longer present. Consequently, the flow pattern in the x, y phase plane does not vary with the time t. The elimination of any time dependence, which results from the dividing of (1b) by (1a), has both positive and negative aspects. On the positive side, (2) is only a single first-order equation whereas (1a,b) were two coupled first-order equations. Even if (2) is nonlinear, we may be able to solve it because it is only of first order. On the negative side, if we do obtain from (2) the phase portrait, we do not know how t varies along a given trajectory because any explicit information about t has been lost; the t dimension, normal to the x, y plane, has essentially been "squashed" — compressed onto the x, y plane.

A classic example of the phase plane method is afforded by the nonlinear pendulum equation $x'' + \sin x = 0$ (if we set $g/l = 1$). To study it in the phase plane we reduced it to an equivalent system of first-order equations by defining an auxiliary variable y according to $x' = y$, obtaining

$$x' = y, \tag{3a}$$

$$y' = -\sin x, \tag{3b}$$

and hence the first-order equation $dy/dx = -\sin x/y$, which, though it is nonlinear, is separable and readily integrated to give the equation $y^2/2 = \cos x + C$ for the trajectories. The phase portrait (given as Fig. 6 in Section 7.2) clearly and compactly revealed the qualitatively different types of motions: periodic motions, motions along the separatrix that approach the straight-down or straight-up position only as $t \to \infty$, and "up-and-over" motions. Even though in this example we were able to obtain the equation of the trajectories, it was important to identify the singular points, at which both $x' = P(x,y) = 0$ and $y' = Q(x,y) = 0$: $x = \pm\pi, \pm 2\pi, \ldots$ on the x axis. The reason is that the trajectories connecting those singular points at $\pm\pi, \pm 3\pi, \ldots$ defined a separatrix, the boundary between the regions of periodic and nonperiodic motions.

In general, however, we may very well not be successful in integrating (2). In such cases, the phase plane technique that we followed is this: First solve the algebraic equations $P(x,y) = 0$ and $Q(x,y) = 0$ for the singular points. Then, study each singular point, in turn. Linearize the system (1) about the singular point; classify it as a center, spiral, node, or saddle (if the condition $ad - bc \neq 0$ is met at that singular point so the singular point is indeed of one of those four types); and determine any straight-line trajectories through it. That information should suffice to give a skeleton of the phase portrait, such as the ones in Figs. 5 and 9 in Section 7.4. Finally, we are in a position to choose a set of initial points through which to generate solutions using computer software, being careful not to miss any separatrices.

We also studied limit cycles, in Section 7.5, which occur for some nonlinear systems, and stressed their importance in biological applications. The classical example of the limit cycle phenomenon is given by the van der Pol equation

$$x'' - \mu(1 - x^2)x' + x = 0, \tag{4}$$

which was first studied in connection with the modeling of the human heartbeat.

Finally, in **Section 7.6** we extended the simple numerical solution method of Euler, given in Section 1.9 for a single first-order equation, to systems of first-order equations. We also showed how to solve boundary value problems numerically, but only for linear systems.

APPENDIX A

REVIEW OF PARTIAL FRACTION EXPANSIONS

One uses partial fraction expansions in the integral calculus, to express a difficult integral as a linear combination of simpler ones. For example,

$$\begin{aligned}\int \frac{dx}{x^2+4x+3} &= \int \left(\frac{1}{2}\frac{1}{x+1} - \frac{1}{2}\frac{1}{x+3} \right) dx \\ &= \frac{1}{2}\int \frac{dx}{x+1} - \frac{1}{2}\int \frac{dx}{x+3} \\ &= \frac{1}{2}\ln|x+1| - \frac{1}{2}\ln|x+3| + \text{constant}.\end{aligned}$$

In this text we will use partial fractions occasionally in Chapter 1, to carry out integrations in the solution of separable equations (e.g., the logistic equation in Section 1.6). More significant will be its use in Chapter 5 in the "inversion of Laplace transforms" such as $F(s) = (5s-2)/(s^2+3s+8)$. There, the variable will be s rather than x, but in this review we will continue to use x.

Let $p(x)$ and $q(x)$ be *polynomials in* x, of degree P and Q, respectively. Then

$$f(x) = \frac{p(x)}{q(x)} \tag{1}$$

is a **rational function** of x. To apply the method of partial fractions we need $P < Q$. That will be the case in the applications encountered in this text, so we will assume that $P < Q$ and proceed.

Distinct roots. First, factor the denominator $q(x)$. That is, solve $q(x) = 0$ and obtain its roots $x_1, x_2, \ldots, x_Q$. Recall that even if the coefficients of the polynomial $q(x)$ are real, there may be complex roots; if there are, they will necessarily occur in complex conjugate pairs.

Suppose, first, that the Q roots are distinct. Then $f(x)$ admits the **partial**

fraction expansion

$$f(x) = \frac{p(x)}{q(x)} = \frac{a_1}{x - x_1} + \frac{a_2}{x - x_2} + \cdots + \frac{a_Q}{x - x_Q}, \tag{2}$$

in which the a_j's are constants. To determine the a_j's, recombine the terms on the right-hand side over a common denominator [namely, $q(x)$] and require its numerator to be identical to $p(x)$.

Factor the denominator, then expand.

EXAMPLE 1. Expand $f(x) = \dfrac{2x^2 + 3}{x(x^2 + x - 12)}$ in partial fractions. The denominator has three distinct roots, 0, 3, and -4, so we can expand $f(x)$ according to (2) as

$$f(x) = \frac{2x^2 + 3}{x(x-3)(x+4)} = \frac{a_1}{x} + \frac{a_2}{x-3} + \frac{a_3}{x+4}. \tag{3}$$

To determine the a_1, a_2, and a_3, recombine the fractions on the right-hand side of (3), over a common denominator, as

$$\begin{aligned} \frac{2x^2 + 3}{x(x-3)(x+4)} &= \frac{a_1(x-3)(x+4) + a_2 x(x+4) + a_3 x(x-3)}{x(x-3)(x+4)} \\ &= \frac{(a_1 + a_2 + a_3)x^2 + (a_1 + 4a_2 - 3a_3)x - 12a_1}{x(x-3)(x+4)} \end{aligned} \tag{4}$$

For the numerators on the left- and right-hand sides to be identical, match coefficients of like powers of x:

$$\begin{aligned} x^2 &: \quad 2 = a_1 + a_2 + a_3, \\ x^1 &: \quad 0 = a_1 + 4a_2 - 3a_3, \\ x^0 &: \quad 3 = -12a_1. \end{aligned}$$

These equations give $a_1 = -1/4$, $a_2 = 1$, and $a_3 = 5/4$, so the desired partial fraction expansion is given by (3) as

$$\frac{2x^2 + 3}{x(x^2 + x - 12)} = -\frac{1}{4}\frac{1}{x} + \frac{1}{x-3} + \frac{5}{4}\frac{1}{x+4}. \quad \blacksquare$$

EXAMPLE 2. Expand $f(x) = \dfrac{1}{(x^2 + 9)(x^2 + 1)}$ in partial fractions. In this case the roots are complex, but the same steps apply. Factor the denominator and write

$$\frac{1}{(x+3i)(x-3i)(x+i)(x-i)} = \frac{a_1}{x+3i} + \frac{a_2}{x-3i} + \frac{a_3}{x+i} + \frac{a_4}{x-i}. \tag{5}$$

To solve for the a_j's we can proceed as above, but let us indicate a simple method (for cases with nonrepeated roots) that you may prefer: To determine a_1, multiply both sides of (5) by $x + 3i$, cancel quantities where possible, and then set $x = -3i$. Those steps give $a_1 = \dfrac{1}{(-6i)(-2i)(-4i)} = -\dfrac{1}{48i^3} = \dfrac{1}{48i} = -\dfrac{i}{48}$. This method is sometimes called the "coverup method" because it can be accomplished (even without writing, after a bit of practice) by covering up the $x + 3i$ in the left-hand side, and everything on the right except the a_1. For a_2, multiply both sides of (5) by $x - 3i$, cancel quantitites where possible, set $x = 3i$, and obtain $a_2 = \dfrac{1}{(6i)(4i)(2i)} = \dfrac{i}{48}$, and so on. Similarly for the other a_j's. The result is

Is it legitimate to set $x = -3i$? Yes, (5) is to be an identity, true for all x's. Thus, it must be true for any value(s) of x that we choose.

$$\frac{1}{(x+3i)(x-3i)(x+i)(x-i)} = -\frac{i}{48}\frac{1}{x+3i} + \frac{i}{48}\frac{1}{x-3i} + \frac{i}{16}\frac{1}{x+i} - \frac{i}{16}\frac{1}{x-i}. \tag{6}$$

COMMENT. We reiterate that our purpose, in expanding in partial fractions, as in (6), is to obtain a representation of the given rational fraction that is simpler to work with. In Chapter 5 this same example will arise (in Example 2 of Section 5.3), and a representation that is simplest to work with will be a "partial" expansion rather than the "full" expansion given by (6). To explain, suppose we re-combine the first two terms on the right-hand side of (6), and likewise the second two terms. The result will be of the form

$$\frac{1}{(x^2+9)(x^2+1)} = \frac{a+bx}{x^2+9} + \frac{c+dx}{x^2+1}. \tag{7}$$

Example of a "partial expansion." (It seems simpler to use the letters a, b, c, d, rather than subscripted ones.)

It will turn out that (7) will be more convenient (within the context of Section 5.3) than (6). To obtain a, b, c, d in (7) we could, of course, combine the first two terms on the right-hand side of (6), and the last two terms, as we mentioned. But, if you prefer to seek (7) directly, without first generating (6), combine the two terms on the right-hand side of (7) and obtain

$$\frac{1}{(x^2+9)(x^2+1)} = \frac{(b+d)x^3 + (a+c)x^2 + (b+9d)x + (a+9c)}{(x^2+9)(x^2+1)}. \tag{8}$$

Then, matching coefficients of powers of x in the numerators gives

$$\begin{aligned} x^3 &: \quad 0 = b + d, \\ x^2 &: \quad 0 = a + c, \\ x^1 &: \quad 0 = b + 9d, \\ x^0 &: \quad 1 = a + 9c. \end{aligned}$$

Solving gives $a = -1/8$, $c = 1/8$, and $b = d = 0$, so the desired "partial expansion" (7) is

$$\frac{1}{(x^2+9)(x^2+1)} = -\frac{1}{8}\frac{1}{x^2+9} + \frac{1}{8}\frac{1}{x^2+1}. \quad \blacksquare$$

Repeated roots. If any root x_j of $q(x)$ is of multiplicity k [i.e., $(x-x_j)^k$ is a factor of $q(x)$ but $(x-x_j)^{k+1}$ is not], with $k > 1$, then the corresponding term on the right-hand side of (2), the jth term, must be changed to the group of terms

$$\boxed{\frac{a_{j1}}{x-x_j} + \frac{a_{j2}}{(x-x_j)^2} + \cdots + \frac{a_{jk}}{(x-x_j)^k}.} \tag{9}$$

The expression (9) does not replace the right-hand side of (2); it replaces only the jth term.

Let us illustrate.

EXAMPLE 3. To expand $f(x) = \dfrac{3x^2+x+6}{(x-1)^3(x+1)}$ in partial fractions, let us use the form (9) for the terms coming from the repeated root, and write

$$\frac{3x^2+x+6}{(x-1)^3(x+1)} = \underbrace{\frac{a}{x-1} + \frac{b}{(x-1)^2} + \frac{c}{(x-1)^3}}_{\text{according to (9)}} + \frac{d}{x+1}, \tag{10}$$

where the a, b, c, d notation will be simpler than the subscripted a_{jk}'s used in (9). Now proceed as usual: re-combine the fractions on the right-hand side of (10) over a common denominator, and then match coefficients of like powers of x in the numerators of the left- and right-hand sides. Thus,

$$\begin{aligned}\frac{3x^2+x+6}{(x-1)^3(x+1)} &= \frac{a(x-1)^2(x+1)+b(x-1)(x+1)+c(x+1)+d(x-1)^3}{(x-1)^3(x+1)} \\ &= \big[(a+d)x^3 + (-a+b-3d)x^2 + (-a+c+3d)x \\ &\quad +(a+b-3d)\big]/[(x-1)^3(x+1)],\end{aligned} \tag{11}$$

and matching coefficients of powers of x gives $0 = a+d,\ 3 = -a+b-3d,\ 1 = -a+c+3d,\ 6 = a = b-3d$. Solving these gives $a=1,\ b=1,\ c=5,\ d=-1$, so

$$\frac{3x^2+x+6}{(x-1)^3(x+1)} = \frac{1}{x-1} + \frac{1}{(x-1)^2} + \frac{5}{(x-1)^3} - \frac{1}{x+1}.$$

To illustrate what we mean by *solvable*, the equations $x+2y=3$, $2x+5y=$ *are* solvable, and the equations $x+2y=3$, $2x+4y=5$ are *not*.

Getting the form right in (10) was the key. With hindsight, we can now be certain that the form (10) was correct because the four equations for a, b, c, d were *solvable*. Thus, we were indeed able to make (10) an identity. ▮

EXERCISES

Derive the partial expansion of

(a) $\dfrac{x+5}{x^2+3x+2}$ (b) $\dfrac{x^2+12x+9}{x(x^2+9)}$ (c) $\dfrac{x^2-2}{x(x^2+2x+1)}$ (d) $\dfrac{x-1}{x^3(x+1)}$

ANSWERS

(a) $-\dfrac{3}{x+2} + \dfrac{4}{x+1}$ (b) $-\dfrac{1}{x} + \dfrac{1-2i}{x-3i} + \dfrac{1+2i}{x+3i}$

(c) $-\dfrac{2}{x} + \dfrac{3}{x+1} + \dfrac{1}{(x+1)^2}$ (d) $-\dfrac{2}{x} + \dfrac{2}{x^2} - \dfrac{1}{x^3} + \dfrac{2}{x+1}$

APPENDIX B

REVIEW OF DETERMINANTS

Determinants. Every $n \times n$ matrix $\mathbf{A} = \{a_{ij}\}$ has associated with it a number called the **determinant** of **A** and denoted as

If the term "matrix" is unfamiliar to you see Section 5.3.

$$\det \mathbf{A} = \begin{vmatrix} a_{11} & a_{12} & \cdots & a_{1n} \\ a_{21} & a_{22} & \cdots & a_{2n} \\ \vdots & \vdots & & \vdots \\ a_{n1} & a_{n2} & \cdots & a_{nn} \end{vmatrix}. \tag{1}$$

The latter is an ***n*th-order** determinant because it is $n \times n$, having n rows and n columns. The determinant of **A** is defined by an expansion about any of **A**'s rows or columns. If we expand about the first row, for instance, then det**A** is the following linear combination of the elements of the first row:

The sum of each a_{ij} in the first row, times its minor determinant M_{ij}, with a plus sign if $i + j$ is even and a minus sign if it is odd.

$$\begin{aligned} \det \mathbf{A} &= (-1)^{1+1} a_{11} M_{11} + (-1)^{1+2} a_{12} M_{12} + \cdots + (-1)^{1+n} a_{1n} M_{1n} \\ &= +a_{11} M_{11} - a_{12} M_{12} + \cdots + (-1)^{1+n} M_{1n}, \end{aligned} \tag{2}$$

where M_{jk} is the **minor** associated with the a_{jk} element, namely, the $(n-1)\times(n-1)$ determinant that is left if we delete (or "cover up") the row and column through that element. The power of -1 multiplying each $a_{jk} M_{jk}$ is simply $+1$ or -1 depending on whether $j + k$ is even or odd, respectively.

For instance, if $n = 5$, then (2) gives the 5×5 determinant of **A** as a linear combination of five 4×4 determinants. Then we can use (2) again to expand each of the five 4×4 determinants as a linear combination of four 3×3 determinants, and so on.

However, we have not really defined the determinant; all we've done is show how to express an $n \times n$ determinant as a linear combination of $(n-1) \times (n-1)$ determinants. To complete the definition we define the determinant of a 2×2 matrix as

For a 2×2 determinant it seems overkill to use the a_{jk} notation; thus, we've just used a, b, c, d for the elements.

$$\boxed{\begin{vmatrix} a & b \\ c & d \end{vmatrix} \equiv ad - bc.} \tag{3}$$

(We won't be interested, here, in 1×1 determinants.) Let us illustrate the calculation for the 3×3 matrix

$$\mathbf{A} = \begin{bmatrix} 2 & 0 & -4 \\ 4 & 3 & 5 \\ 6 & 2 & -1 \end{bmatrix}. \tag{4}$$

Using (2) to expand about the first row gives

$$\begin{aligned} \det\mathbf{A} &= (-1)^{1+1}a_{11}M_{11} + (-1)^{1+2}a_{12}M_{12} + (-1)^{1+3}a_{13}M_{13} \\ &= +a_{11}M_{11} - a_{12}M_{12} + a_{13}M_{13} \\ &= +(2)\begin{vmatrix} 3 & 5 \\ 2 & -1 \end{vmatrix} - (0)\begin{vmatrix} 4 & 5 \\ 6 & -1 \end{vmatrix} + (-4)\begin{vmatrix} 4 & 3 \\ 6 & 2 \end{vmatrix} \\ &= +(2)(-3-10) - (0)(-4-30) + (-4)(8-18) \\ &= 14, \end{aligned} \tag{5}$$

where we used (3) to evaluate each of the 2×2 determinants. (Of course, we didn't need to evaluate the second one because it is multiplied by 0.)

What if we had expanded about a different row, or about one of the columns? The same value is obtained, for $\det\mathbf{A}$, no matter which row or column we expand about. Let us illustrate that claim for the matrix given above by also expanding about the second row and about the third column.

Expansion about second row:

a_{21}, a_{22}, a_{23} are the elements of the second row.

$$\begin{aligned} \det\mathbf{A} &= (-1)^{2+1}a_{21}M_{21} + (-1)^{2+2}a_{22}M_{22} + (-1)^{2+3}a_{23}M_{23} \\ &= -a_{21}M_{21} + a_{22}M_{22} - a_{23}M_{23} \\ &= -(4)\begin{vmatrix} 0 & -4 \\ 2 & -1 \end{vmatrix} + (3)\begin{vmatrix} 2 & -4 \\ 6 & -1 \end{vmatrix} - (5)\begin{vmatrix} 2 & 0 \\ 6 & 2 \end{vmatrix} \\ &= -(4)(0+8) + (3)(-2+24) - (5)(4-0) \\ &= 14, \end{aligned} \tag{6}$$

Expansion about third column:

a_{13}, a_{23}, a_{33} are the elements of the third column.

$$\begin{aligned} \det\mathbf{A} &= (-1)^{1+3}a_{13}M_{13} + (-1)^{2+3}a_{23}M_{23} + (-1)^{3+3}a_{33}M_{33} \\ &= +a_{13}M_{13} - a_{23}M_{23} + a_{33}M_{33} \\ &= +(-4)\begin{vmatrix} 4 & 3 \\ 6 & 2 \end{vmatrix} - (5)\begin{vmatrix} 2 & 0 \\ 6 & 2 \end{vmatrix} + (-1)\begin{vmatrix} 2 & 0 \\ 4 & 3 \end{vmatrix} \\ &= +(-4)(8-18) - (5)(4-0) + (-1)(6-0) \\ &= 14, \end{aligned} \tag{7}$$

For the problems encountered in this text, this expansion process should suffice. But, for large n's the expansion process is exceedingly laborious. For $n = 10$, say, the expansion gives a linear combination of ten 9×9 determinants. In turn, each of these ten is a linear combination of nine 8×8 determinants, and so on! It is interesting to see how serious this predicament is, and how spectacular is the remedy. For estimating purposes, let us count each multiplication, addition, and subtraction as one "calculation." It can be shown that the number of calculations $N(n)$ required in the evaluation (by the expansion method) of an $n \times n$ determinant grows as

$$N(n) \sim e\,n! \tag{8}$$

as $n \to \infty$, where $e \approx 2.718$ is the base of the natural logarithm, and $n!$ is n factorial. If each calculation takes approximately one microsecond, then some rough time estimates are as follows. Faster computers offer no hope. For instance, a com-

n	Computing Time
5	0.0003 sec
10	10 sec
15	4×10^6 sec $\approx$ 40 days
20	7×10^{12} sec $\approx$ 210,000 years
25	4×10^{19} sec $\approx 10^{12}$ years

puter that is a million times as fast would still take around 10^6 years to evaluate a 25×25 determinant, and scientific calculations can easily involve determinants that are 250×250 or much larger.

The problem is not with the determinant concept but with the laborious expansion algorithm described above. To design a better algorithm we need to know some of the *properties* of determinants.

First, we define "triangular matrices." A square matrix $\mathbf{A} = \{a_{ij}\}$ is **upper triangular** if $a_{ij} = 0$ for all $j < i$ and **lower triangular** if $a_{ij} = 0$ for all $j > i$. That is,

$$\begin{bmatrix} a_{11} & a_{12} & \cdots & a_{1n} \\ 0 & a_{22} & \cdots & a_{2n} \\ \vdots & \vdots & \ddots & \vdots \\ 0 & 0 & \cdots & a_{nn} \end{bmatrix} \quad \text{and} \quad \begin{bmatrix} a_{11} & 0 & \cdots & 0 \\ a_{21} & a_{22} & \cdots & 0 \\ \vdots & & \ddots & \vdots \\ a_{n1} & a_{n2} & \cdots & a_{nn} \end{bmatrix} \tag{9}$$

are upper triangular and lower triangular, respectively. If a matrix is upper triangular or lower triangular it is **triangular**.

Here are the three properties that we will need.

PROPERTIES OF DETERMINANTS

D1. If any row (or column) of $\mathbf{A}$ is modified by adding α times the corresponding elements of any other row (or column) to it, yielding a new matrix $\mathbf{B}$, then $\det\mathbf{B} = \det\mathbf{A}$.

D2. If any two rows (or columns) of $\mathbf{A}$ are interchanged, yielding a new matrix $\mathbf{B}$, then $\det\mathbf{B} = -\det\mathbf{A}$.

D3. If $\mathbf{A}$ is triangular, then $\det\mathbf{A}$ is simply the product of its diagonal elements, $\det\mathbf{A} = a_{11}a_{22}\cdots a_{nn}$.

Let us illustrate the use of properties D1–D3 instead of the expansion process described above.

A better way, using D1–D3. Consider the $\mathbf{A}$ matrix given by (4) again. Using properties D1–D3, we obtain

$$\det\mathbf{A} = \begin{vmatrix} 2 & 0 & -4 \\ 4 & 3 & 5 \\ 6 & 2 & -1 \end{vmatrix} = \begin{vmatrix} 2 & 0 & -4 \\ 0 & 3 & 13 \\ 6 & 2 & -1 \end{vmatrix} = \begin{vmatrix} 2 & 0 & -4 \\ 0 & 3 & 13 \\ 0 & 2 & 11 \end{vmatrix}$$

$$= \begin{vmatrix} 2 & 0 & -4 \\ 0 & 3 & 13 \\ 0 & 0 & \frac{7}{3} \end{vmatrix} = (2)(3)(\frac{7}{3}) = 14, \tag{10}$$

as obtained above. Following the second equality we added -2 times the first row to the second row, and following the third equality we added -3 times the first row to the third row, which steps left the determinant unchanged (D1). Following the fourth equality we added $-\frac{2}{3}$ times the second row to the third row, which step left the determinant unchanged (D1). Those steps produced a triangular matrix and, finally, we used D3.

The point, then, is to use some combination of D1 and D2 steps to reduce the determinant to triangular form (upper or lower), so it can then be evaluated readily by D3. We will call this method **triangularization**. In the foregoing example we did not need D2, but see Exercise 3.

It is hard to tell, from the modest 3×3 calculation given above, whether the method is really more efficient than the cofactor expansion. However, it can be shown that using triangularization the number of calculations $N(n)$ is

$$N(n) \sim \frac{2n^3}{3} \tag{11}$$

as $n \to \infty$. Again assuming one microsecond per calculation, (11) gives a computing time of around 0.005 second for $n = 20$ and 0.01 second for $n = 25$, compared with 210,000 years and 10^{12} years, respectively!!

The upshot is that except for small hand calculations we should avoid the cofactor expansion, and use triangularization instead.

Although properties D1 – D3 suffice for the efficient calculation of determinants, other properties are useful and are listed below.

ADDITIONAL PROPERTIES OF DETERMINANTS

D4. If all the elements of any row or column are zero, then $\det\mathbf{A} = 0$.

D5. If any two rows or columns are proportional to each other, then $\det\mathbf{A} = 0$.

D6. If all the elements of any row or column are scaled by α, yielding a new matrix $\mathbf{B}$, then $\det\mathbf{B} = \alpha \det\mathbf{A}$.

D7. A determinant of the form

$$\begin{vmatrix} 1 & 1 & \cdots & 1 \\ \lambda_1 & \lambda_2 & \cdots & \lambda_n \\ \vdots & \vdots & & \vdots \\ \lambda_1^{n-1} & \lambda_2^{n-1} & \cdots & \lambda_n^{n-1} \end{vmatrix}, \tag{12}$$

is a **Vandermonde determinant**. The key property of a Vandermonde determinant is that *it is nonzero if and only if the* λ_j*'s are distinct.*

D8. $\det(\mathbf{AB}) = (\det\mathbf{A})(\det\mathbf{B})$.

D9. If the elements of $\mathbf{A}$ are functions of t, say, then

$$\frac{d}{dt}\begin{vmatrix} a_{11}(t) & \cdots & a_{1n}(t) \\ a_{21}(t) & \cdots & a_{2n}(t) \\ \vdots & & \vdots \\ a_{n1}(t) & \cdots & a_{nn}(t) \end{vmatrix}$$

$$= \begin{vmatrix} a'_{11} & \cdots & a'_{1n} \\ a_{21} & \cdots & a_{2n} \\ \vdots & & \vdots \\ a_{n1} & \cdots & a_{nn} \end{vmatrix} + \begin{vmatrix} a_{11} & \cdots & a_{1n} \\ a'_{21} & \cdots & a'_{2n} \\ \vdots & & \vdots \\ a_{n1} & \cdots & a_{nn} \end{vmatrix} + \cdots + \begin{vmatrix} a_{11} & \cdots & a_{1n} \\ a_{21} & \cdots & a_{2n} \\ \vdots & & \vdots \\ a'_{n1} & \cdots & a'_{nn} \end{vmatrix}.$$

Primes denote d/dt.

D10. If any jth row $[a_{j1}, \dots, a_{1n}]$ is decomposed as $[b_{j1}+c_{j1}, \dots, b_{jn}+c_{jn}]$, then $\det\mathbf{A}$ is the sum of two determinants, one having $[b_{j1}, \dots, b_{jn}]$ as its jth row and the other having $[c_{j1}, \dots, c_{jn}]$ as its jth row.

Cramer's rule. If, for a set of n linear algebraic equations in n unknowns, the determinant of the coefficient matrix is nonzero, then there is a unique solution. One way to find that solution utilizes determinants, and is known as Cramer's rule. It is straightforward, so a single example should suffice to explain the method. Consider the system

$$\begin{aligned} 2x + 0y - 4z &= 4, \\ 4x + 3y + 5z &= 5, \\ 6x + 2y - z &= 10 \end{aligned} \tag{13}$$

for x, y, z. Cramer's rule gives the first unknown x as the ratio of two determinants; the one in the denominator is the determinant of the coefficient matrix, and the one in the numerator is the same, but with the first (because x is the first unknown) column deleted and replaced by the column of numbers on the right-hand side of (13):

$$x = \frac{\begin{vmatrix} \mathbf{4} & 0 & -4 \\ \mathbf{5} & 3 & 5 \\ \mathbf{10} & 2 & -1 \end{vmatrix}}{\begin{vmatrix} 2 & 0 & -4 \\ 4 & 3 & 5 \\ 6 & 2 & -1 \end{vmatrix}} = \frac{28}{14} = 2. \tag{14}$$

For y, replace the *second* column in the numerator by the numbers 4,5,10 on the right side of (13), and similarly for z:

$$y = \frac{\begin{vmatrix} 2 & \mathbf{4} & -4 \\ 4 & \mathbf{5} & 5 \\ 6 & \mathbf{10} & -1 \end{vmatrix}}{\begin{vmatrix} 2 & 0 & -4 \\ 4 & 3 & 5 \\ 6 & 2 & -1 \end{vmatrix}} = \frac{-14}{14} = -1, \quad z = \frac{\begin{vmatrix} 2 & 0 & \mathbf{4} \\ 4 & 3 & \mathbf{5} \\ 6 & 2 & \mathbf{10} \end{vmatrix}}{\begin{vmatrix} 2 & 0 & -4 \\ 4 & 3 & 5 \\ 6 & 2 & -1 \end{vmatrix}} = \frac{0}{14} = 0. \tag{15}$$

We've shown the "replacement" columns in bold so you can see what we've done.

If the determinant of the coefficient matrix in the denominator is zero, then Cramer's rule does not apply.

EXERCISES

1. We evaluated the determinant of the matrix given in (4) by expanding about the first row, second row, and third column, and by triangularization. Show that you obtain the same result, 14, by expanding about the third row, first column, and second column.

2. In Section 3.2 we gave (39) for any 3×3 determinant. Verify that formula by expanding the determinant about any row or column.

3. Use triangularization to show that $\begin{vmatrix} 0 & 1 & 2 \\ 3 & 4 & -1 \\ 1 & 1 & 5 \end{vmatrix} = -18$. HINT: First, interchange the first row with the third, say, and put a minus sign out in front of the determinant to compensate for the change in sign caused by the interchange (D2). Next, add -3 times the first row to the second. Next, add -1 times the second row to the third. Then use D3.

4. Show that $\begin{vmatrix} 2 & 1 & 1 & 1 \\ 6 & 2 & -1 & 1 \\ 0 & 3 & 2 & 0 \\ 4 & 0 & -2 & -3 \end{vmatrix} = -68$ two ways: First, expand about any row or column. (The third row will be the most convenient because of the two zeros in it.) Second, use triangularization. HINT: Add -3 times the first row to the second and -2 times the first row to the fourth; then add 3 times the second row to the third and -2 times the second row to the fourth; then add $2/5$ times the third row to the fourth; then use D3 and obtain $(2)(-1)(-10)(-17/5) = -68$.

5. For the A matrix given, use property D9 to show that $\frac{d}{dt}\mathbf{A} = 9t^2 \sin t + 3t^3 \cos t - 20$, then check your result by working out det$A$ and taking its derivative.

$$\mathbf{A} = \begin{bmatrix} t^2 & t & 2 \\ 0 & 3t & 1 \\ 4 & 0 & \sin t \end{bmatrix}.$$

6. To illustrate property D10, express

$$\begin{vmatrix} 2 & 3 & 1 \\ 5 & 6 & 3 \\ 2 & -1 & 0 \end{vmatrix} = \begin{vmatrix} 2 & 3 & 1 \\ 3+2 & 1+5 & 0+3 \\ 2 & -1 & 0 \end{vmatrix} = \begin{vmatrix} 2 & 3 & 1 \\ 3 & 1 & 0 \\ 2 & -1 & 0 \end{vmatrix} + \begin{vmatrix} 2 & 3 & 1 \\ 2 & 5 & 3 \\ 2 & -1 & 0 \end{vmatrix},$$

the second equality following from D10. Work out the determinant on the left and the two on the right, and thus verify the foregoing equality.

7. Cramer's Rule. Use Cramer's rule to show that the solutions of

$$3x - 2y = 8$$
$$x + 4y = -2$$

and

$$x + y + 0z = 3$$
$$2x - y + z = 2$$
$$x + y + 3z = -9$$

are $x = 2,\ y = -1$ and $x = 3,\ y = 0,\ z = -4$, respectively.

APPENDIX C

REVIEW OF GAUSS ELIMINATION

Consider n simultaneous linear algebraic equations

$$\begin{aligned} a_{11}x_1 + a_{12}x_2 + \cdots + a_{1n}x_n &= c_1, \\ a_{21}x_1 + a_{22}x_2 + \cdots + a_{2n}x_n &= c_2, \\ &\vdots \\ a_{n1}x_1 + a_{n2}x_2 + \cdots + a_{nn}x_n &= c_n, \end{aligned} \tag{1}$$

in which the a_{ij} coefficients and the c_j's are known, and $x_1, x_2, \dots, x_n$ are the unknowns. Such systems occur in this text in connection with higher-order linear differential equations and systems of linear differential equations, and we will need to know how how to solve them.

Let us denote the determinant of the a_{ij} coefficients as det**A**. Then:

1. If det**A** $\neq 0$, there exists a *unique solution* of (1), which can be found by Cramer's rule (Appendix B), whereby each x_j is given by the ratio of two $n \times n$ determinants, the one in the denominator being det**A**.

2. If det**A** $= 0$, either there is *no solution, or a nonunique solution* (indeed, an infinite number of solutions).

Thus, depending on whether det**A** is nonzero or zero, the number of solutions of (1) is 1, 0, or ∞.

If there exist one or more solutions, then the system (1) is said to be **consistent**, and if there are none, then it is **inconsistent**.

When we meet systems of the type (1) in connection with the "eigenvalue problem" (in Chapters 5 and 7), det**A** will *always* be zero, so Cramer's rule will not apply. Alternatively, a solution method of successive elimination, known as **Gauss elimination**, applies *whether or not* det**A** is nonzero. It is vastly more efficient if n is large, and it will be our solution method of choice in most of this text. This appendix is a review of that method. We will rely on examples.

EXAMPLE 1. Unique Solution. Consider the system

$$x_1 + x_2 - 2x_3 = 1, \tag{2a}$$
$$3x_1 + x_2 - 2x_3 = 6, \tag{2b}$$
$$x_1 - x_2 + 4x_3 = 8. \tag{2c}$$

Equations (3b) and (3c) can be obtained by solving (2a) for x_1, in terms of x_2 and x_3, and substituting that into (2b) and (2c) for x_1. But, follow the steps of the Gauss algorithm instead.

To solve by Gauss elimination, proceed as follows: Keep the first equation intact, add -3 times the first equation to the second (as a replacement for the second), and then add -1 times the first equation to the third (as a replacement for the third). Those steps yield the "indented" system

$$x_1 + x_2 - 2x_3 = 1, \tag{3a}$$
$$-2x_2 + 4x_3 = 3, \tag{3b}$$
$$-2x_2 + 6x_3 = 7. \tag{3c}$$

Next, keep the first two of these equations intact and add -1 times the second equation to the third (as a replacement for the third), thus obtaining another indentation:

$$x_1 + x_2 - 2x_3 = 1, \tag{4a}$$
$$-2x_2 + 4x_3 = 3, \tag{4b}$$
$$2x_3 = 4. \tag{4c}$$

Finally, multiplying the second of these by $-1/2$ and the third by 1/2, to normalize the leading coefficients, gives

$$x_1 + x_2 - 2x_3 = 1, \tag{5a}$$
$$x_2 - 2x_3 = -3/2, \tag{5b}$$
$$x_3 = 2. \tag{5c}$$

Think of the original system (2) as a tangle of string that we wish to unravel. From experience, we know that the first step is to find a loose end and that is, in effect, what the foregoing process of successive indentations has done for us; (5c) is the "loose end." With that in hand we can unravel (5) just as we would unravel a tangle: Putting $x_3 = 2$ into (5b) gives $x_2 = 5/2$, then putting $x_3 = 2$ and $x_2 = 5/2$ into (5a) gives $x_1 = 5/2$. Thus, we obtain

$$x_3 = 2, \quad x_2 = 5/2, \quad x_1 = 5/2. \tag{6}$$

COMMENT 1. In this example, calculation shows that det**A** $= -8$. Since the latter is nonzero, (2) must have a unique solution; indeed, that is what we found, the unique solution given by (6).

COMMENT 2. The system (2) was a "tangle" because the equations were **coupled**; that is, equations contained more than one unknown. Actually, the final system (5) is coupled too, since (5a) contains all three unknowns and (5b) contains two of them. But the coupling in (5) is not troublesome because (5) is in "triangular form," so we were able to solve (5c) for x_3. We put that value into (5b) and solved for x_2, then we put those values into (5a) and solved for x_1. More generally, converting the original system to indented triangular form is the purpose of Gauss elimination.

COMMENT 3. Regarding the existence and uniqueness of solutions, it is helpful to think of (2) in geometrical terms. Each of (2a,b,c) can be regarded as the equation of a plane in a Cartesian x_1, x_2, x_3 space. Recall from the calculus that the normal vector to a plane $ax + by + cz = d$ is $[a, b, c]$. Since the normal vectors to the planes (2a) and (2b), $[1, 1, -2]$ and $[3, 1, -2]$, are not collinear, it follows that those planes are not parallel and hence intersect along a line. Evidently, that line of intersection pierces the third plane (2c) at the unique solution point defined by (6). ▮

In terms of the familiar unit vectors $\mathbf{i}, \mathbf{j}, \mathbf{k}$, $[a, b, c]$ means $a\mathbf{i} + b\mathbf{j} + c\mathbf{k}$.

EXAMPLE 2. No Solution. Consider the system

$$\begin{aligned} 2x_1 + 3x_2 - 2x_3 &= 4, &\text{(7a)}\\ x_1 - 2x_2 + x_3 &= 3, &\text{(7b)}\\ 7x_1 - x_3 &= 2. &\text{(7c)} \end{aligned}$$

Example 1 illustrated the case of a unique solution. The other possibilities, no solution and a nonunique solution, are illustrated in in Examples 2 and 3, respectively.

Keep the first equation intact, add $-\frac{1}{2}$ times the first equation to the second, and add $-\frac{7}{2}$ times the first to the third:

$$\begin{aligned} 2x_1 + 3x_2 - 2x_3 &= 4, &\text{(8a)}\\ -\tfrac{7}{2}x_2 + 2x_3 &= 1, &\text{(8b)}\\ -\tfrac{21}{2}x_2 + 6x_3 &= -12. &\text{(8c)} \end{aligned}$$

Keep the first two equations intact, and add -3 times the second equation to the third:

$$\begin{aligned} 2x_1 + 3x_2 - 2x_3 &= 4, &\text{(9a)}\\ -\tfrac{7}{2}x_2 + 2x_3 &= 1, &\text{(9b)}\\ 0 &= -15. &\text{(9c)} \end{aligned}$$

Any solution x_1, x_2, x_3 of (9) must satisfy *each* of those three equations, but there are *no* values of x_1, x_2, x_3 that can satisfy (9c) (which is really $0x_1 + 0x_2 + 0x_3 = -15$). Thus, (9) is inconsistent, it has no solution, so (7) is inconsistent as well.

COMMENT. In this case, planes (7a) and (7b) have different normals, so they intersect along a line. Evidently, this time that line does not pierce the third plane, (7c); it is parallel to it, and displaced from it. Hence, there are *no* points common to all ▮

It is wasteful to keep writing the x_j's and the equal signs when carrying out Gauss elimination, because it is only the coefficients on the left-hand sides and the numbers on the right-hand sides that participate in the calculation. It is more efficient to work with the numerical array

$$\begin{bmatrix} a_{11} & a_{12} & \cdots & a_{1n} & c_1 \\ a_{21} & a_{22} & \cdots & a_{2n} & c_2 \\ \vdots & \vdots & & \vdots & \vdots \\ a_{n1} & a_{n2} & \cdots & a_{nn} & c_n \end{bmatrix}, \qquad (10)$$

In (10), the first row, for instance, is shorthand for the equation $a_{11}x_1 + a_{12}x_2 + \cdots + a_{1n}x_n = c_1$ in (1).

namely, the coefficient array augmented by the column of c_j's. The bracketed array in (10) is called the **augmented array**, or **augmented matrix**, that is, of coefficients a_{ij} augmented by the column of c_j's.

Applying the Gauss algorithm to a given system (1) amounts to carrying out a finite sequence of **elementary row operations** on the augmented array (10). These operations are as follows:

1. Addition of a multiple of one row to another
   Symbolically: jth row $\to \alpha\,(k$th row$) + j$th row

2. Multiplication of a row by a nonzero constant
   Symbolically: jth row $\to \alpha\,(j$th row$)$

3. Interchange of two rows
   Symbolically: jth row $\leftrightarrow k$th row

These elementary *row* operations on (10) correspond to elementary *equation* operations on (1). For instance, the third corresponds to interchanging the jth and kth equations. These operations enable us to simplify the system into triangular form, *while not affecting its solution set.* For instance, surely scaling one of the equations by 6, or interchanging two of the equations, does not change the solution set. Thus, when we solved (5) we were really solving (2); of course you can verify (6) by substituting those values into (2).

We will use this augmented array format in the next example.

EXAMPLE 3. Nonunique Solution. Consider

$$x_1 - 2x_2 + x_3 = 1, \tag{11a}$$

$$x_1 - x_2 + 2x_3 = 3, \tag{11b}$$

$$x_1 + x_2 + 4x_3 = 7. \tag{11c}$$

If an x_j is missing, that means its coefficient is 0. For example, if (11a) were $x_1 - 2x_2 = 1$ instead, then the first row in (12) would be $1, -2, 0, 1$.

The augmented array is

$$\begin{bmatrix} 1 & -2 & 1 & 1 \\ 1 & -1 & 2 & 3 \\ 1 & 1 & 4 & 7 \end{bmatrix}. \tag{12}$$

Replace the second row by -1 times the first row plus the second row, then replace the third row by -1 times the first row plus the third row:

$$\begin{bmatrix} 1 & -2 & 1 & 1 \\ 0 & 1 & 1 & 2 \\ 0 & 3 & 3 & 6 \end{bmatrix}. \tag{13}$$

Then add -3 times the second row to the third:

$$\begin{bmatrix} 1 & -2 & 1 & 1 \\ 0 & 1 & 1 & 2 \\ 0 & 0 & 0 & 0 \end{bmatrix}. \tag{14}$$

In the second row (which represents the equation $x_2 + x_3 = 2$) we can let x_3 be arbitrary, say α, and solve for x_2, obtaining $x_2 = 2 - \alpha$. Then, putting $x_3 = \alpha$ and $x_2 = 2 - \alpha$ into the first row gives $x_1 = 1 - \alpha + 2(2 - \alpha) = 5 - 3\alpha$. Thus,

$$x_1 = 5 - 3\alpha, \qquad x_2 = 2 - \alpha, \qquad x_3 = \alpha, \tag{15}$$

in which α is an arbitrary constant. Since α is arbitrary, (13) is a nonunique solution, an infinity of solutions.

COMMENT 1. In geometrical terms, the planes (11a) and (11b) intersect along a line. Evidently, from (15), that line lies *in* the third plane (11c), so all points along it are solutions of (11); (15) gives parametric equations for the line, the parameter being α. As α varies between $-\infty$ and $+\infty$, (15) generates the line of solution points.

COMMENT 2. Look at the pattern in (14) again. Instead of letting $x_3 = \alpha$ and solving for x_2, we could have let x_2 be β (which, like α, is an arbitrary constant). In that case, $x_2 = \beta$, $x_3 = 2 - x_2 = 2 - \beta$, and $x_1 = 1 - x_3 + 2x_2 = -1 + 3\beta$, which looks different from the solution given in (15), but is equivalent to it. [To see the equivalence, set $5 - 3\alpha = -1 + 3\beta$, $2 - \alpha = \beta$, and $\alpha = 2 - \beta$, all three of which are satisfied if the arbitrary constants α and β are related by $\beta = 2 - \alpha$.] ▮

The foregoing examples have been for systems of three equations, but the Gauss method applies for any n. In fact, it applies even if there are m equations in n unknowns with $m < n$ or $m > n$, that is, with $m \neq n$. We will not meet that case in this text and, therefore, have not covered it in this review.

EXERCISES

Solve each by Gauss elimination. Answers are given in brackets; α and β are arbitrary constants.

1. $x_1 + x_2 = 5,\ 2x_1 + x_2 = 7$ $[x_1 = 2, x_2 = 3]$

2. $x_1 + 3x_2 = 5,\ 2x_1 + 6x_2 = 7$ [no solution]

3. $x_1 + 3x_2 = 5,\ 2x_1 + 6x_2 = 10$ $[x_2 = \alpha, x_1 = 5 - 3\alpha]$

4. $x_1 + x_2 - x_3 = 3,\ 2x_1 + x_2 - x_3 = 2,$ $3x_1 - x_2 - x_3 = 3$ $[x_1 = -1, x_2 = -1, x_3 = -5]$

5. $x_1 + x_2 + 2x_3 = 2,\ 2x_1 + x_2 - x_3 = -1,$ $4x_1 + 3x_2 + 3x_3 = 3$ $[x_3 = \alpha, x_2 = 5 - 5\alpha, x_1 = -3 + 3\alpha]$

6. $x_1 + x_2 + 2x_3 = 2,\ 2x_1 + x_2 - x_3 = -1,$ $4x_1 + 3x_2 + 3x_3 = 4$ [no solution]

7. $x_2 + x_3 = 2,\ x_1 - x_2 + 3x_3 = 1,\ 2x_1 + x_2 + x_3 = 8$ $[x_1 = 3, x_2 = 2, x_3 = 0]$ HINT: As your first elementary row operation, interchange the first and third rows.

8. $x_1 + x_2 + x_3 = 2,\ 2x_1 + 2x_2 + 2x_3 = 4,\ 3x_1 + 3x_2 + 3x_3 = 6$ $[x_3 = \alpha, x_2 = \beta, x_1 = 2 - \alpha - \beta]$ HINT: You should obtain $x_1 + x_2 + x_3 = 2,\ 0 = 0,\ 0 = 0$. Discard the latter two equations ($0 = 0$). In the first, you can let x_3 be arbitrary (say α) and also let x_2 be arbitrary (say β). Then solve for x_1, obtaining $x_1 = 2 - x_3 - x_2 = 2 - \alpha - \beta$.

9. $x_1 + x_2 = 0,\ x_1 + x_2 + x_3 = 0,\ x_2 + x_3 + x_4 = 0,\ x_3 + x_4 = 1$ $[x_1 = 1, x_2 = -1, x_3 = 0, x_4 = 1]$

10. $x_1 + 2x_2 + 2x_3 + x_4 = 0,\ 2x_1 + x_2 + x_3 + 2x_4 = 0,\ -x_1 + x_2 = 0,\ 2x_1 + 4x_2 + 3x_3 + 3x_4 = 0$ $[x_4 = \alpha, x_3 = \alpha, x_2 = -\alpha, x_1 = -\alpha]$

11. $2x_1 + x_2 = 0,\ x_1 + 2x_2 + x_3 = 0,\ x_2 + 2x_3 + x_4 = 0,\ x_3 + 2x_4 = 5$ $[x_4 = 4, x_3 = -3, x_2 = 2, x_1 = -1]$ HINT: Use these steps: Interchange 1st and 2nd rows (just to avoid the entry of fractions); add -2 times 1st to 2nd; interchange 2nd and 3rd; add 3 times 2nd to 3rd; interchange 3rd and 4th; add -4 times 3rd to 4th; then solve from bottom up.

APPENDIX D

REVIEW OF COMPLEX NUMBERS AND THE COMPLEX PLANE

Complex numbers were created several hundred years ago through the study of algebraic equations. If one allowed only real numbers, then equations such as $x^2 + 1 = 0$ and $x^2 + 2x + 4 = 0$ had no solution. In a step that was slow to gain acceptance, a broader number system was devised so that the equations given above, and indeed every polynomial equation, possess solutions within that number system.[1] Those numbers, eventually called **complex numbers**, were of the form $a + ib$ where a and b are real and where i satisfies the equation $i^2 = -1$, or $i = \sqrt{-1}$.

The real numbers a and b are the **real part** and **imaginary part** of $a + ib$, respectively, and are denoted by Re and Im:

$$\text{Re}(a+ib) \equiv a, \qquad \text{Im}(a+ib) \equiv b \quad (b, \text{ not } ib). \tag{1}$$

For instance, $\text{Re}(2 + 3i) = 2$ and $\text{Im}(2 - 3i) = -3$. Complex numbers are *equal* if their real and imaginary parts, respectively, are equal; that is,

$$a_1 + ib_1 = a_2 + ib_2 \tag{2}$$

if and only if $a_1 = a_2$ and $b_1 = b_2$.

To use complex numbers we must define arithmetic rules for them. The addition, subtraction, and multiplication of complex numbers are defined by

$$\boxed{(a_1 + ib_1) + (a_2 + ib_2) \equiv (a_1 + a_2) + i(b_1 + b_2),} \tag{3}$$

$$\boxed{(a_1 + ib_1) - (a_2 + ib_2) \equiv (a_1 - a_2) + i(b_1 - b_2),} \tag{4}$$

[1]The reluctance to accept complex numbers may be better appreciated if we mention that, before complex numbers, there had even been reluctance to accept *negative* numbers. After all, how could one have a negative amount of something? That was before the invention of the credit card.

and

$$(a_1+ib_1)(a_2+ib_2) \equiv (a_1a_2-b_1b_2)+i(a_1b_2+b_1a_2). \tag{5}$$

For instance, $(3-2i)+(5+i)=8-i$, $(3-2i)-(5+i)=-2-3i$, and $(3-2i)(5+i)=17-7i$. The definitions (3), (4), and (5) are designed to mimic the corresponding operations for real numbers. For instance, consider (5): just as $(a+b)(c+d)=ac+bd+ad+bc$ for real numbers, if we write out the left-hand side of (4) in the same manner we get $a_1a_2+i^2b_1b_2+a_1ib_2+ib_1a_2$ or, using $i^2=-1$ and grouping real and imaginary terms, $(a_1a_2-b_1b_2)+i(a_1b_2+b_1a_2)$, which is the same as the right-hand side of (4).

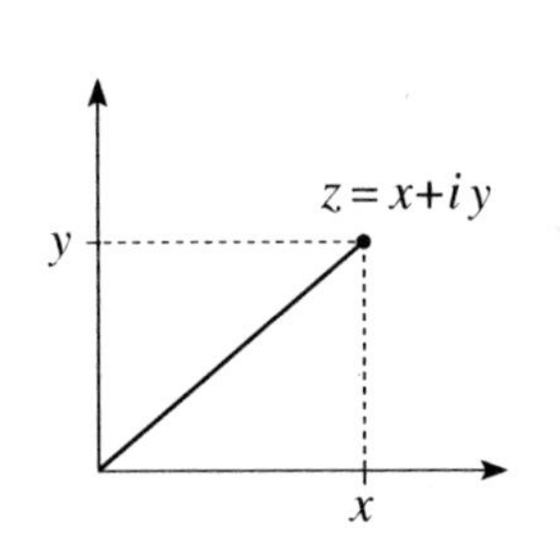

Figure 1. Complex z plane.

We often denote a complex number by the single letter z. Thus, $z=a+ib$. With the definitions given above, the familiar rules of algebra hold for complex numbers. For instance, if we denote any three complex numbers as z_1, z_2, and z_3, then $z_1+z_2=z_2+z_1$, $z_1z_2=z_2z_1$, and so on.

Just as we represent a real number, graphically, as a point on a real axis, we represent a complex number $z=x+iy$ as a point in a Cartesian plane, the **complex z plane**, as indicated in Fig. 1. The horizontal axis is the **real axis**, and the vertical axis is the **imaginary axis**.

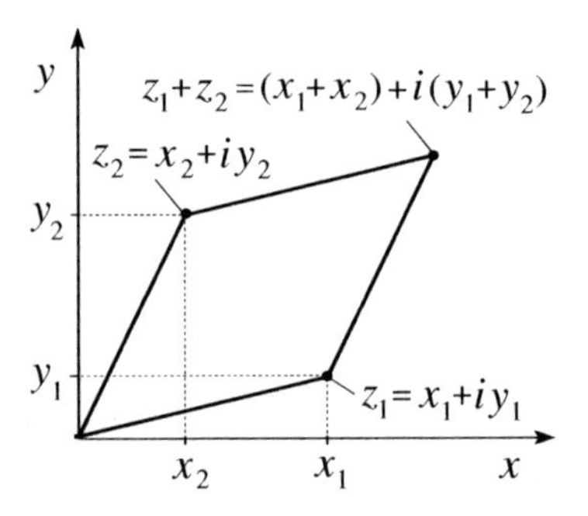

Figure 2. Addition in the z plane.

Observe from Fig. 2 that the addition of complex numbers defined by (4) satisfies the parallelogram law for the addition of vectors so, in terms of addition, we can think of complex numbers as vectors. That is, a "z vector" is the vector from the origin to the point z. With that viewpoint in mind, we can revise Fig. 2 as the vector diagram shown in Fig. 3.

The distance from the origin to the point z (i.e., the "length of the z vector") is called the **modulus** of z and, by analogy with the absolute value of a real number, is denoted as $|z|$, or as mod(z). Then, from the Pythagorean theorem (Fig. 1),

$$|z|=\sqrt{x^2+y^2}. \tag{6}$$

For example, $|2-i|=\sqrt{2^2+(-1)^2}=\sqrt{5}$. From (5) and (6) it can be shown that

$$|z_1z_2|=|z_1|\,|z_2|; \tag{7}$$

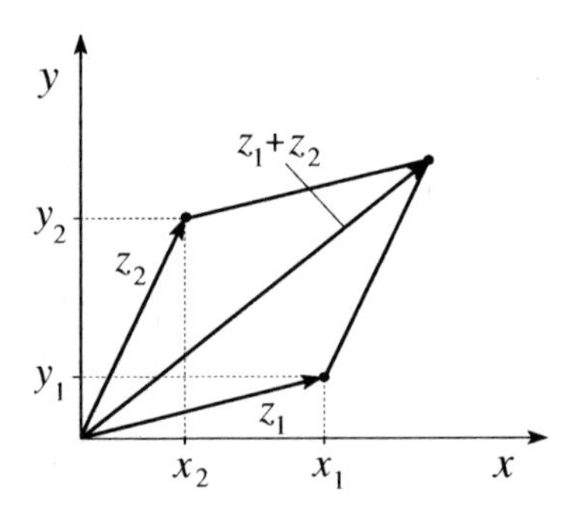

Figure 3. Addition in the z plane, from a vector point of view.

that is, the modulus of a product equals the product of the moduli of the factors.

Note that the complex numbers are not *ordered* as real numbers are. For instance, whereas the inequalities $-6<2$ and $10>7$ are meaningful, analogous statements such as $z<0$, $z>3$, and $4+3i>1+2i$ are not. However, statements such as Re $z<6$, Im $z<6$, $|z|>4$, and $|1-i|<|3-i|$ *are* meaningful because Re z, Im z, and $|z|$ are real numbers.

Besides $z=x+iy$, it is useful to define the **complex conjugate** of z as

$$\overline{z} \equiv x-iy. \tag{8}$$

Thus, if $z_1 = 8 + 3i$ and $z_2 = -4i$, then $\overline{z}_1 = 8 - 3i$ and $\overline{z}_2 = 4i$. It is shown from (3), (5), and (8) that

$$\overline{z_1 + z_2} = \overline{z}_1 + \overline{z}_2, \tag{9a}$$

$$\overline{z_1 z_2} = \overline{z}_1 \overline{z}_2, \tag{9b}$$

and

$$|z| = \sqrt{z\overline{z}}. \tag{10}$$

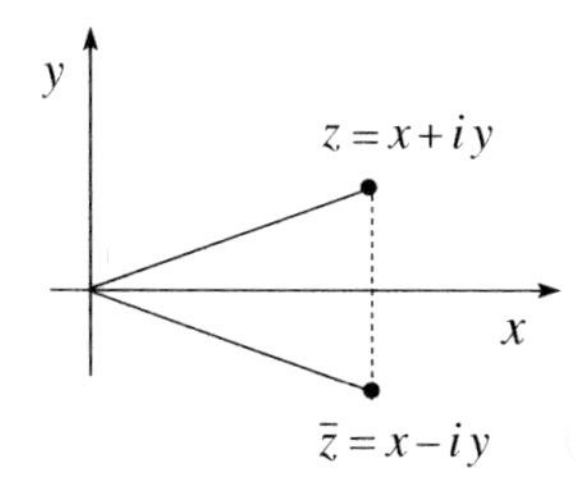

Figure 4. Complex conjugate $\overline{z}$ of z.

Graphically, $\overline{z}$ is simply the reflection of z about the real axis (Fig. 4).

The complex conjugate is useful in evaluating the quotient of two complex numbers, z_1/z_2. By "evaluating" any complex quantity we mean getting it into the standard Cartesian form $a + i\,b$. To evaluate z_1/z_2, multiply numerator and denominator by the conjugate of the denominator, that is, by $\overline{z}_2$:

$$\frac{z_1}{z_2} = \frac{z_1}{z_2}\frac{\overline{z}_2}{\overline{z}_2}, \tag{11}$$

because whereas the denominator on the left is complex, the denominator on the right, $z_2\overline{z}_2 = |z|^2$, is real. Writing it out,

$$\begin{aligned}\frac{z_1}{z_2} = \frac{z_1\overline{z}_2}{z_2\overline{z}_2} &= \frac{x_1 + iy_1}{x_2 + iy_2}\frac{x_2 - iy_2}{x_2 - iy_2} = \frac{(x_1x_2 + y_1y_2) + i(x_2y_1 - x_1y_2)}{x_2^2 + y_2^2} \\ &= \left(\frac{x_1x_2 + y_1y_2}{x_2^2 + y_2^2}\right) + i\left(\frac{x_2y_1 - x_1y_2}{x_2^2 + y_2^2}\right).\end{aligned} \tag{12}$$

Actually, there are two standard forms, the Cartesian form and also *polar form*. In this book Cartesian form will suffice.

EXAMPLE 1. For instance,

$$\frac{2+i}{3-4i} = \frac{2+i}{3-4i}\frac{3+4i}{3+4i} = \frac{(6-4)+(8+3)i}{3^2-(4i)^2} = \frac{2+11i}{9+16} = \frac{2}{25} + \frac{11}{25}i,$$

which result can be checked by showing that $3 - 4i$ times $\frac{2}{25} + \frac{11}{25}i$ equals $2 + i$. ∎

EXERCISES

1. Let $z = x + iy$, $z_1 = x_1 + iy_1$, and $z_2 = x_2 + iy_2$ be any complex numbers. Verify (7), (9a), (9b), and (9c) by working out both sides of each and showing that they are equal.

2. Show that

(a) $|z_1z_2z_3| = |z_1|\,|z_2|\,|z_3|$

(b) $|z_1z_2z_3z_4| = |z_1|\,|z_2|\,|z_3|\,|z_4|$

3. Show that

(a) $|z^3| = |z|^3$

(b) $\overline{\overline{z}} = z$

(c) z is real if and only if $z = \overline{z}$
(d) $\overline{z^3} = \overline{z}^3$

4. Evaluate the following.

(a) i^3 (b) $2 + 5i - \dfrac{1}{i^3}$ (c) $3i^2 - 4i^5$

(d) $(2 - i)^3$ (e) $\dfrac{1}{1 - 2i}$ (f) $\dfrac{i}{2 + 5i}$

(g) $\dfrac{1+i}{1-i}$ (h) $\left(\dfrac{1+i}{2-i}\right)^3$ (i) $\dfrac{1}{(1+i)^3}$

(j) $\operatorname{Re} \dfrac{2+3i}{4+5i}$ (k) $\operatorname{Im}(1+i)^3$ (l) $\left(\operatorname{Re} \dfrac{1}{1+i}\right)^3$

(m) $\operatorname{Im}(i^6 + 5i^8)$ (n) $\operatorname{Re} \dfrac{1}{1-i}$ (o) $\operatorname{Im} \dfrac{1}{(2+i)^3}$

(p) $\left|\dfrac{2-2i}{1+i}\right|$ (q) $\left|\dfrac{(2-i)^3}{(1+3i)^2}\right|$ (r) $\left|\left(\dfrac{1+i}{1+2i}\right)^2\right|$

ANSWERS

2. (a) Use (7) twice: $|z_1 z_2 z_3| = |z_1 z_2||z_3| = |z_1||z_2|z_3|$.
(b) First, $|z_1 z_2 z_3 z_4| = |z_1 z_2 z_3||z_4|$ by (7), then use the result from part (a), and obtain $|z_1 z_2 z_3 z_4| = |z_1||z_2||z_3||z_4|$.

4. (a) $-i$ (b) $2 + 4i$ (c) $-3 - 4i$ (d) $2 - 11i$ (e) $\dfrac{1}{5} + \dfrac{2}{5}i$ (f) $\dfrac{5}{29} + \dfrac{2}{29}i$
(g) i (h) $-\dfrac{26}{125} - \dfrac{18}{125}i$ (i) $-\dfrac{1}{4} - \dfrac{1}{4}i$ (j) $\dfrac{23}{41}$ (k) 2 (l) $\dfrac{1}{8}$ (m) 0 (n) $\dfrac{1}{2}$
(o) $-\dfrac{11}{125}$ (p) 2 (q) $\dfrac{\sqrt{5}}{2}$ (r) $\dfrac{2}{5}$

ANSWERS TO EXERCISES

NOTE: The following are answers to exercises with underlined exercise numbers.

CHAPTER 1

Section 1.1

1. (a) First. Yes, yes, no (c) First. No, yes, no (e) First. No, yes (g) Second. Yes, yes (i) Third. Yes, no, yes (k) First. No, yes
2. (a) $A = 2$ (c) $A = -e^{24}$
3. (a) $A=0$, $B = 2$ (c) $A=B=0$ (e) $A=1/2$, $B=-1/2$
4. (a) Linear (d) Nonlinear (g) Nonlinear (j) Nonlinear (m) Linear
5. (a) -3 (c) $1, 2$ (e) $-1, 3$ (g) $0, 1, -1$ (i) ± 1, $\pm\sqrt{5}$
6. (a) -1 (c) 0 (e) ± 3 (g) $-1 \pm \sqrt{3}$
8. (a) One, $y = x - 1$ (c) An infinite number, $y = \pm 3x + b$, where b is arbitrary (e) None (g) One, $y = 2x + 3$ (i) One, $y = -2x$ (k) None

Section 1.2

2. (a) $y(x) = \exp(2x^3 - 54)$, $-\infty < x < \infty$
(c) $y(x) = e^{(\sin x - \sin 3)}$, $-\infty < x < \infty$
(e) $y(x) = 27/x^3$, $0 < x < \infty$
(g) $y(x) = \sin x / \sin 3$, $-\infty < x < \infty$
(i) $y(x) = e^{1/3} e^{-1/x}$, $0 < x < \infty$
(k) $y(x) = (2/3)x/(5-x)$, $-\infty < x < 5$
(m) $y(x) = e^{-1} e^{5/(x+2)}$, $-2 < x < \infty$
(o) $y(x) = 256/(x+1)^4$, $-1 < x < \infty$
3. (a) $y(x) = (3x-6)e^x$, $-\infty < x < \infty$
(c) $y(x) = 2/x - 8/x^3$, $0 < x < \infty$
(e) $y(x) = 2x^3 - 64/x^2$, $0 < x < \infty$
(g) $y(x) = x - 1 - e^{-(x-2)}$, $-\infty < x < \infty$
(i) $y(x) = (\cos 2 - \cos x + 2\sin x - 2\sin 2)/x$, $0 < x < \infty$
(k) $y(x) = (\cos 2 - \cos x)/x$, $0 < x < \infty$
4. (a) $y(x) = x/2$, $-\infty < x < \infty$
(c) $y(x) = (x + \sqrt{x^2+4})/2$, $-\infty < x < \infty$
7. (7.1) is 6(b), (7.2) is 6(c), (7.3) is 6(a), (7.4) is 6(d)
10. (a) $y = 2x + 5$ (c) $y = 1 - 3x$ (e) None (g) $y = mx + 3m - 1$ for any (finite) m
12. (a) $(1-x)y' + y = 1$ (c) $(\sinh x)y' - (\cosh x)y = -1$ (e) None
15. (a) $n = 0$: $y(x) = \sigma(x)^{-1}\left(\int \sigma(x) q(x)\, dx + C\right)$, where $\sigma(x) = e^{\int p\, dx}$ and C is arbitrary. $n = 1$: $y(x) = Ae^{-\int (p-q)\, dx}$, where A is arbitrary.
16. (a) $y(x) = e^{-x}/(3x+1)$, $-1/3 < x < \infty$
(c) $y(x) = (x+5)^2/(x+1)^2$, $-1 < x < \infty$
17. (a) $-\infty < x < \infty$ (c) $-\infty < x < \infty$ (e) $0 < x < \infty$ (g) $0 < x < \pi$ (i) $0 < x < \infty$ (k) $0 < x < 5$ (m) $-2 < x < \infty$ (o) $-1 < x < \infty$

Section 1.3

1. $\kappa = 0.065$, $N(0) = 3,250$
3. 53.3 yr to double, 84.5 yr to triple
6. $N(20) = 2,457,384$
9. (a) 847 yr (b) 81 yr
11. 217 days
13. $T = (t_2 - t_1)\ln 2 / \ln(m_1/m_2)$
15. (a) For $t < T$, $c(t) = c_i\left(1 - e^{-Qt/v}\right)$. For $t > T$, $c(t) = c_i\left(1 - e^{-QT/v}\right)e^{-Q(t-T)/v}$
(b) We asked for a hand sketch, but will give a computer-generated graph:

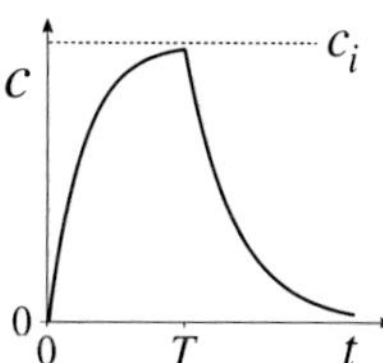

17. $c(60\text{ min}) = 0.0022$ lb/ft^3
18. (a) $c' + [5/(1{,}000-7t)]c = 10/(1{,}000-7t)$; $c(0) = 0$. $c(t) = 2 - 2(1-0.007t)^{5/7}$ on $0 \le t < 1{,}000/7$
(b) $c(t) \to 2$
20. For $t < T$, $c(t) = c_i - [c_i - c(0)]e^{-Qt/v}$ and, for $t > T$, $c(t) = c_i - [c_i - c(0)]e^{-QT/v}$, which is constant. The latter holds only for $T < t < T + v/Q$ because for $t > T + v/Q$ the tank is empty.
22. (a) $c(0) = 0,\ c(5) = 0.04264,\ c(10) = 0.04576$

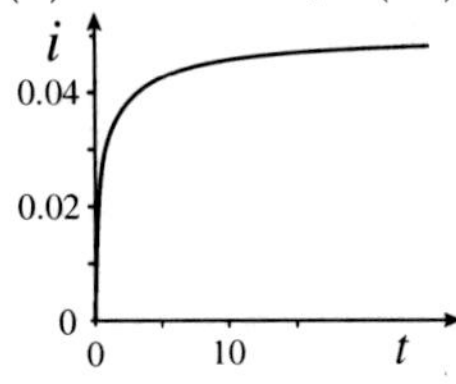

23. (a)

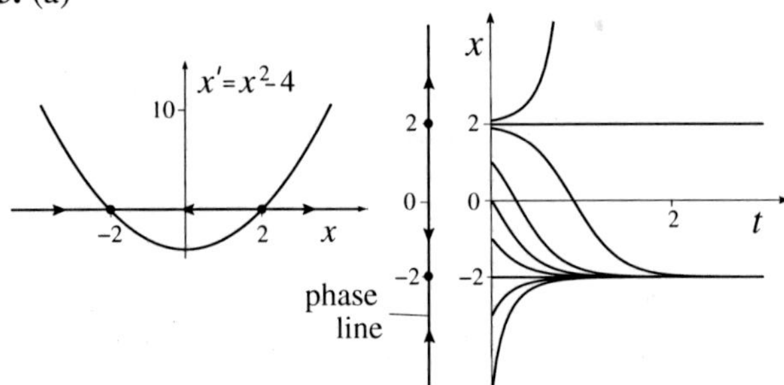

Equilibrium points: $x = -2$ (stable), 2 (unstable)
(d)

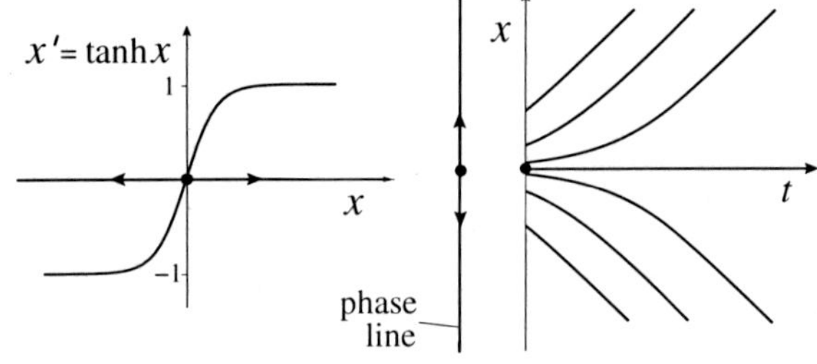

Equilibrium point at $x = 0$ (unstable)
24. (a) 0.230 in
26. (b) $S(1)/S_0 = 1.05$ (yearly), 1.05116 (monthly), 1.05125 (weekly), 1.051267 (daily), 1.051271 (continuously)
27. (b) $\mu = 0.00104$
29. (a) $i(t) = (5\sin t - 25\cos t + 25e^{-t/5})/52$ for $t < 6\pi$
$i(t) = -0.47e^{-(t-6\pi)/5}$ for $t > 6\pi$
30. (b) 87.8 min, 76.4° (d) 2.21 hr

32. (d) Time between doses $= (1/k)\ln(C_{\max}/C_{\min})$; dosage $= (C_{\max} - C_{\min})v$

Section 1.4

1. (a) $y(x) = \ln(x^3 + 1)$ on $-1 < x < \infty$

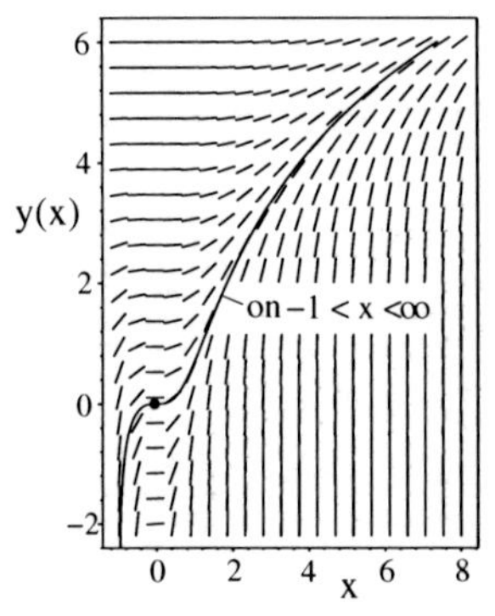

(d) $y(x) = 1/\sqrt{9 - 8x}$ on $-\infty < x < 9/8$

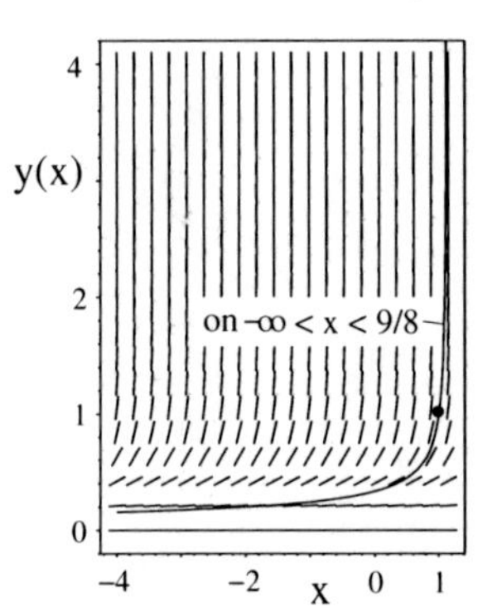

(g) $y(0) = -3:\ y(x) = -(2x + 3)/(2x + 1)$ on $-1/2 < x < \infty$
$y(0) = -1:\ y(x) = -1$ on $-\infty < x < \infty$
$y(0) = 3:\ y(x) = -(4x + 3)/(4x - 1)$ on $-\infty < x < 1/4$

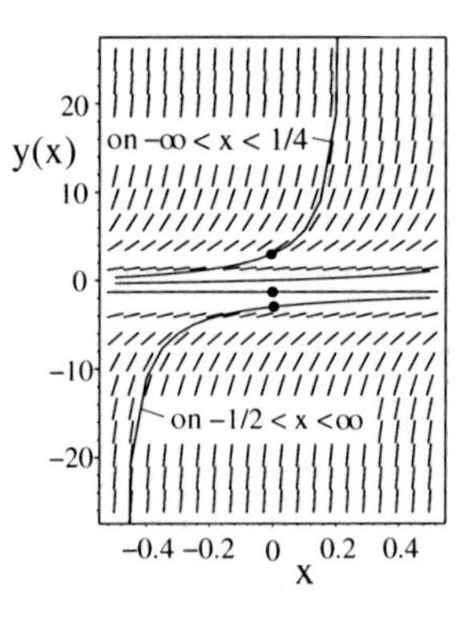

(j) $y(x) = x$ on $-\infty < x < \infty$

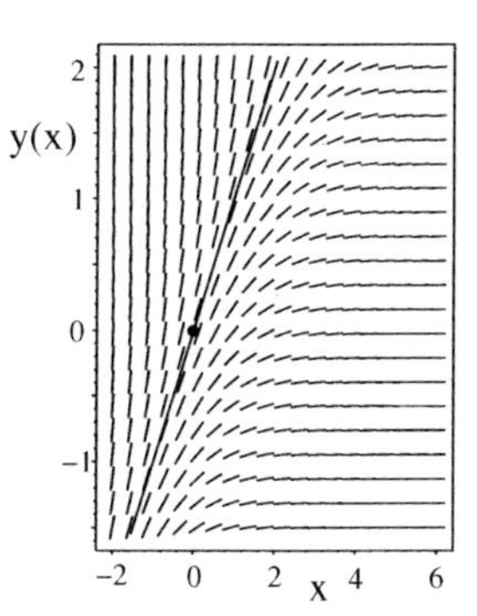

(m) $y(x) = 4 + \arctan(x-3)$ on $-\infty < x < \infty$

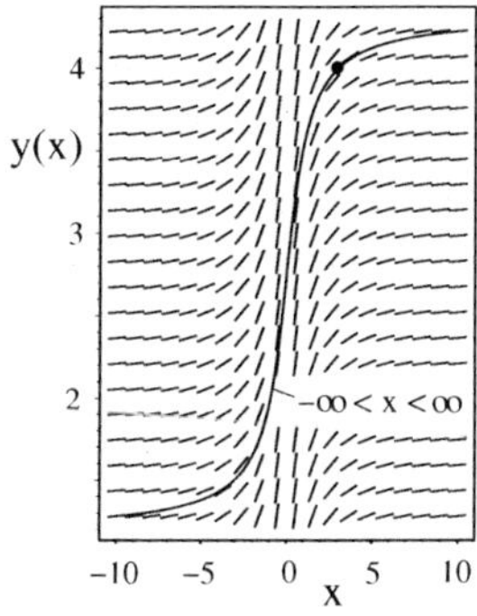

2. (a) For each of the two initial conditions the integration constant happens to be the same, and the solution is $\sin y + e^x = 1$ on $-\infty < x < \ln 2$. In this case we can solve for y as $y(x) = \arcsin(1-e^x)$. The latter is multi-valued and the solution corresponding to $y(0) = 0$ is the portion $-\pi/2 < y < \pi/2$, and to $y(0) = 2\pi$ it is the portion $3\pi/2 < y < 5\pi/2$. (If this is not clear, examine a computer plot of $\sin y + e^x = 1$.) In each case the interval of existence is $-\infty < x < \ln 2$.
(c) Implicit solution is $xy = 2e^{2-y}$. From a computer plot of its graph it appears that the interval of existence is $0 < x < \infty$, but things are clearer if we solve explicitly for x as $x = 2e^{2-y}/y$, from which we see that as $y \to \infty$, $x \to 0$, and as $y \to 0$, $x \to \infty$. The interval of existence is indeed $0 < x < \infty$.
(e) $e^x = y - e^{-y} + e^2 + 1$. It follows that x and y tend to infinity together, with $y \sim e^x$ as $x \to +\infty$. And as $x \to -\infty$, y tends to the (single) root of $0 = y - e^{-y} + e^2 + 1$, namely, $y = -1.874$. Interval of existence is $-\infty < x < \infty$.
(g) $x + 5y - 3y^5 = 0$, which we can at least solve for x: $x = 3y^5 - 5y$. A plot of the graph shows two vertical tangents, which can be found by noting that the DE gives $y' = \infty$ at $y = \pm 1/3^{1/4}$. At those y's the solution gives $x = \pm 4/3^{1/4} = \pm 3.039$, so interval of existence is $-3.039 < x < 3.039$.
(i) $y^2 + 0.01y^6 + 4x^2 = 1.01$, which is a slightly distorted ellipse (distorted by the y^6 term) with x and y axes as minor and major axes, respectively. Its lower half passes through the initial point $y(0) = -1$. Interval of existence is $-\sqrt{1.01/4} < x < \sqrt{1.01/4}$, or, $-0.5025 < x < 0.5025$.
5. The solution corresponding to $y(5) = 0.4$ is given implicitly by $xy = e^{1.093}e^{-y}$ on $0 < x < \infty$, and the solution corresponding to $y(5) = -0.2$ is given implicitly by $xy = -e^{-0.2}e^{-y}$ on $2.225 < x < \infty$.

Section 1.5

1. (a) $y(x) = 2\exp(x^2)$. Theorem 1.2.1: $p(x) = -2x$ and $q(x) = 0$ are continuous for all x, so the solution exists and is unique on $-\infty < x < \infty$.
(d) Each of the two solutions $y(x) = \pm x$ exists on $-\infty < x < \infty$; nonunique solution. $f = x/y$ and $f_y = -x/y^2$ are not continuous in any neighborhood of $(0,0)$, so Theorem 1.5.1 gives no information.
(g) No solution. Theorem 1.2.1 gives no information because the continuity condition is not satisfied by $p(x) = 1/x$.
(j) Nonunique solution $y(x) = Ax^2$ where A is arbitrary. In fact, any left half-parabola can be fitted to any right half-parabola to form a solution. For example, if we call the origin O, then AOB, AOC, and AOD are all solutions.

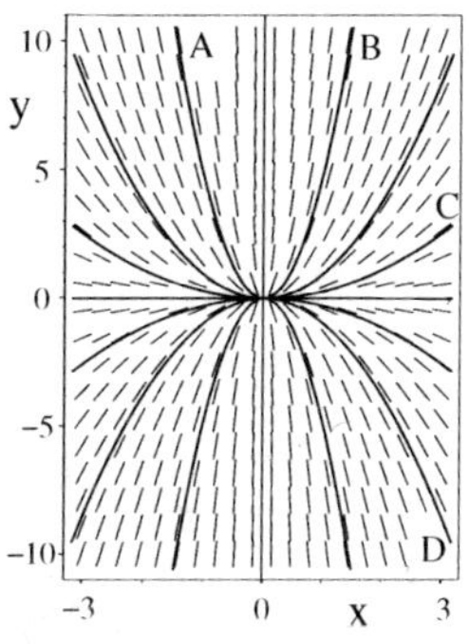

Theorem 1.2.1 gives no information because $p(x) = -2/x$ is discontinuous at the initial point $x = 0$.
(m) The solution, in implicit form, is $\ln|\sin y| = x + \ln|\sin(-3)| = x - 1.958$, the graph of which is the bold curve, below. Unique solution on $-\infty < x < 1.958$.

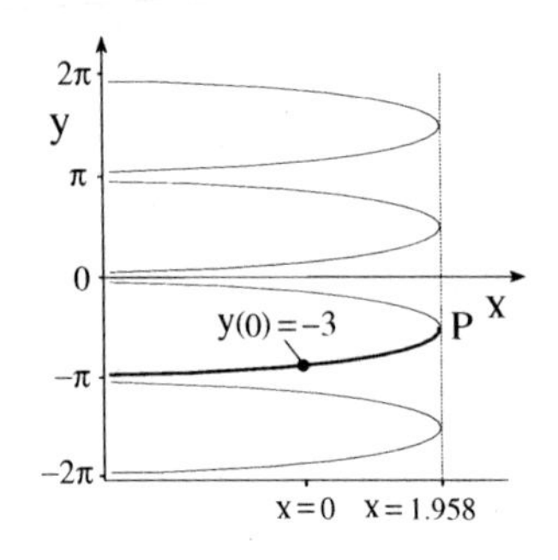

3. (a) There is one envelope, $x = y^2/4$.

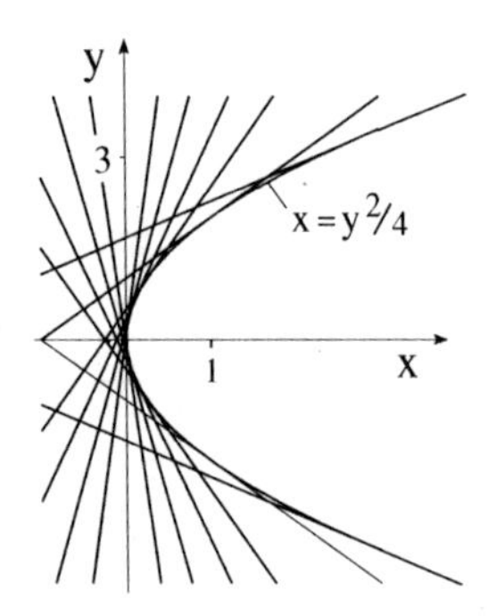

(c) One envelope, $y = 0$.

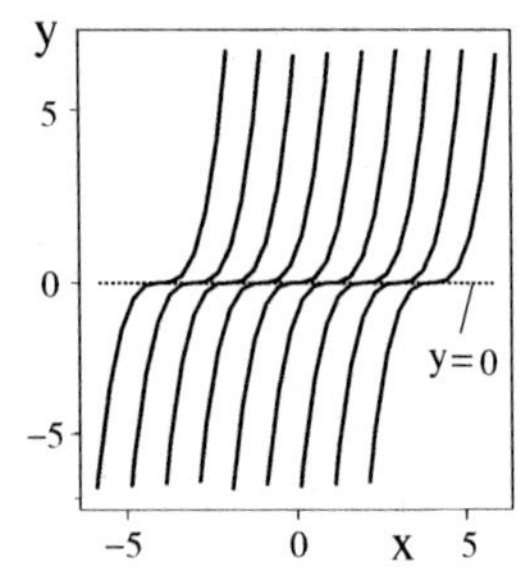

7. (b) $y_1(x) = 1 + x^2$, $y_2(x) = 1 + x^2 + x^4 + x^6/3$,
$y_3(x) = 1 + x^2 + x^4 + x^6 + 2x^8/3 + x^{10}/3 + x^{12}/9 + x^{14}/63$
(c)

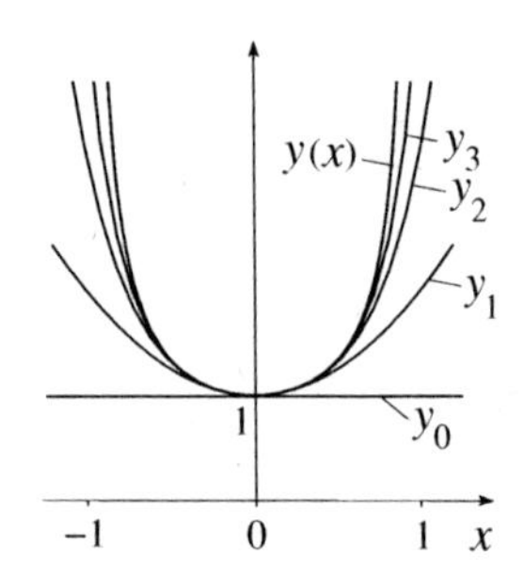

Section 1.6

1. $h_{cr} = a^2/4b$; need $h < a^2/4b$ to prevent driving the fish to extinction.

4. (a) Unstable equilibrium point at $x = \ln 10$

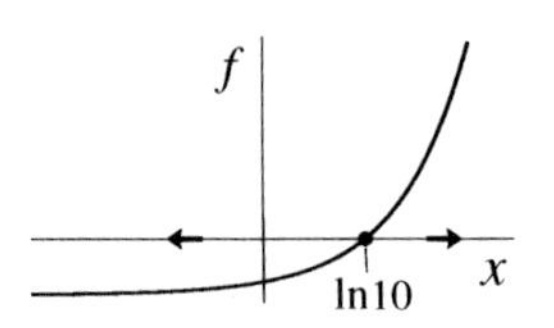

(c) Stable equilibrium points at $x = -2, 1$; unstable ones at $x = -1, 2$

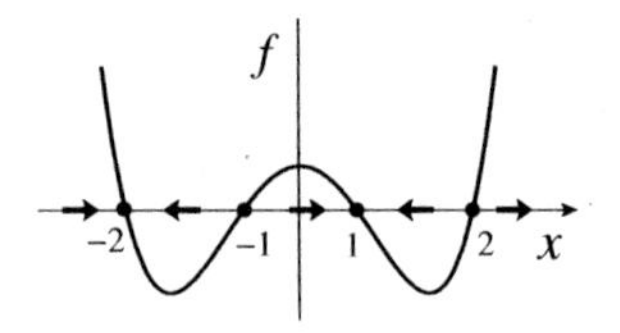

(e) Stable equilibrium point at $x = -1$, unstable at $x = 1$

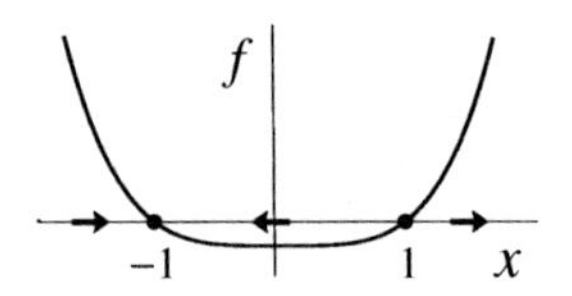

5. (a) $x_{eq} = \ln 10$: $f'(\ln 10) = 10 > 0$, so unstable according to (14b) (c) $x_{eq} = \pm 1, \pm 2$: $f'(-1) = 6 > 0$, so -1 is unstable. $f'(+1) = -6 < 0$, so $+1$ is stable. $f'(-2) = -12 < 0$, so -2 is stable. $f'(+2) = 12 > 0$, so $+2$ is unstable. (e) $x_{eq} = \pm 1$: -1 is stable, $+1$ is unstable.

7. (a) $c_{eq} = c_0$; stable.

9. (a) $y = Ae^{-4x}$ and $x = 2y^2 + B$

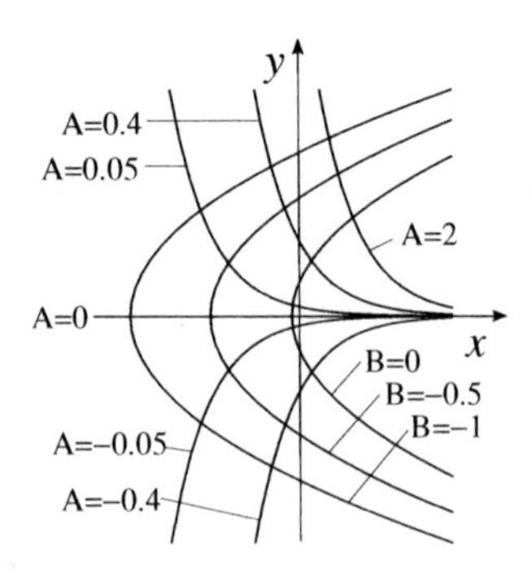

(c) $xy = A$ and $y^2 = x^2 + B$

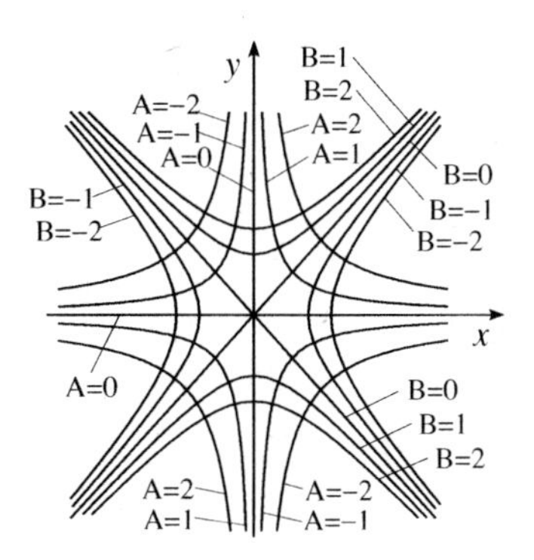

10. (a) Orthogonal families: $y = Ce^{2x}$ and $y^2 + x = A$
(c) Orthogonal families: $y = Cx^3$ and $x^2 + 3y^2 = A$

Section 1.7

1. (a) $y = 3x + C$
(c) $10 \sin 2x + e^{-5y} = C$
(e) $e^x + xy + \cos y = C$
(g) $x \sin y + y \sin x = C$; $x \sin y + y \sin x = 2\pi$
(i) $x^2 y + ye^y = C$
(k) $x^2 y \ln y + y^2 = C$

3. (a) The equation is exact if and only if $b = A$

4. (a) $\sigma(x) = e^{3x}$, $y(x) = Ce^{-3x}$
(c) $\sigma(y) = 1/y$, $x \ln y + y = C$
(e) $\sigma(x) = 1/x$, $y = -\ln x + C$
(g) No integrating factors $\sigma(x)$ or $\sigma(y)$
(i) $\sigma(x) = x$, $y = x - C/x^2$
(k) $\sigma(y) = 1/y$, $x \ln y + x^2 y = C$

6. (a) $\sigma(x, y) = x^{-2} y^{-2}$

9. (a) $2xy - x^2 + y^2 = C$
(c) $t^2 + te^{-v} = C$

Section 1.8

1. (a) $p = -2/x$, $q = x^2$, $n = 2$, $y = 5x^2/(C - x^5)$
(c) $p = q = 2/(1+x)$, $n = 1/2$, $y = (x+C)^2/(x+1)^2$
(e) $p = 1/3$, $q = x/3$, $n = -1/2$, $y = (x - 2 + Ce^{-x/2})^{2/3}$
(g) $p = 0$, $q = x$, $n = 1/2$, $y = (C + x^2/4)^2$

3. (a) For $y(2) = 0$: $y = 2x - \sqrt{3x^2 + 4}$ on $-\infty < x < \infty$
For $y(2) = 4 - 2\sqrt{3}$: $y = (2 - \sqrt{3})x$ on $-\infty < x < \infty$
For $y(2) = 1$: $y = 2x - \sqrt{3x^2 - 3}$ on $1 < x < \infty$
For $y(2) = 7$: $y = 2x + \sqrt{3x^2 - 3}$ on $1 < x < \infty$
For $y(2) = 4 + 2\sqrt{3}$: $y = (2 + \sqrt{3})x$ on $-\infty < x < \infty$
For $y(2) = 8$: $y = 2x + \sqrt{3x^2 + 4}$ on $-\infty < x < \infty$

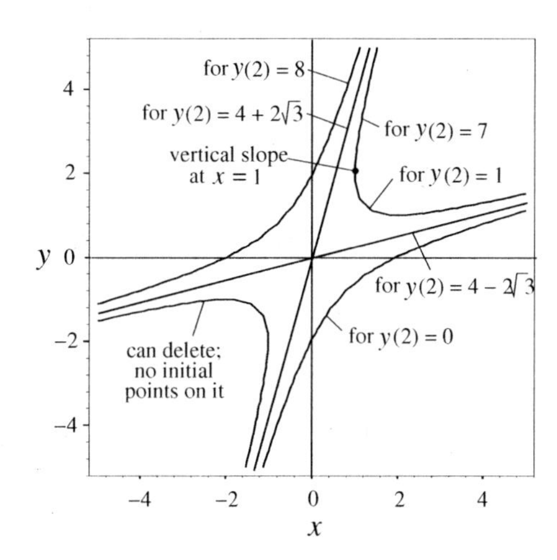

(c) $y(x) = 2x/(1 - 4 \ln x)$ on $0 < x < e^{1/4}$
(e) $y(x) = x \arcsin(x/2)$ on $-2 < x < 2$

4. (b) $h = 1$, $k = -1$, $y - x + 2 = C(y+1)^2$ with $-\infty < C < \infty$, plus the additional solution $y = -1$
(d) Implicit solution is $(y+1)^2 - 2(x+1)(y+1) - (x+1)^2 = C$ with $-\infty < C < \infty$

5. $y(x) = -x \ln(C - x^3/3)$ with $-\infty < C < \infty$

6. (b) $y(x) = (2x^2 - 2Cx + 1)/(C - x)$ with $-\infty < C < \infty$

8. (a) $y(x) = 4/(Ce^{-4x} - 1)$ with $-\infty < C < \infty$
(c) $y(x) = 2/(Ce^x + e^{-x})$ with $-\infty < C < \infty$

9. (a) $y' - (1/x)y = (1/\sqrt{x})y^{1/2}$ is a Bernoulli equation with $n = 1/2$, so let $v = y^{1/2}$
(d) $y' = xy^2 - 2x^2 y + x^3 + 1$ is a Riccati equation (Exercise 7) with $\eta(x) = x$, so let $y = x + 1/v$
(g) $y' = [(y/x) + 1]^2/(y/x)^2$ is homogeneous, so let $v = y/x$
(j) Let $y - 5x = v$, as in Exercise 6.

Section 1.9

2. (a) $y_1 = 0.8$, $y_2 = 0.64$, $y_3 = 0.512$ (c) $y_1 = 0.6$, $y_2 = 0.6848$, $y_3 = 0.77225$ (e) $y_1 = 1.8$, $y_2 = 3.2$, $y_3 = 3.464$
(g) $y_1 = 5.30718$, $y_2 = 9.69287$, $y_3 = 18.1404$

3. Exact values given in brackets.
(a) $y_{100} = 0.367695\ [0.367879]$, $y_{200} = 0.135200\ [0.135335]$, $y_{300} = 0.049712\ [0.049787]$
(c) $y_{100} = 0.999368\ [1]$, $y_{200} = 1.999135\ [2]$, $y_{300} = 4.999050\ [5]$
(e) $y_{100} = 4.020695\ [4.020726]$, $y_{200} = 4.160057\ [4.160168]$, $y_{300} = 4.497749\ [4.497941]$
(g) $y_{100} = -0.249986\ [-1/4]$, $y_{200} = -0.333277\ [-1/3]$, $y_{300} = -0.499771\ [-1/2]$

4. We will give the values at $x = 3$ using $h = 0.1, 0.01, 0.001, 0.0001$, followed by the exact value of $y(3)$ in brackets.
(a) 0.042391, 0.049041, 0.049712, 0.049780 $[0.049787]$
(c) 4.904239, 4.990490, 4.999050, 4.999905 $[5]$
(e) 4.478675, 4.496016, 4.497749, 4.497922 $[4.497941]$
(g) -0.479368, -0.497735, -0.499771, -0.499977 $[-1/2]$

CHAPTER 2

Section 2.1

8. (a) Not in general. For instance, $y_1(x) = e^x$ is a solution of $y'' - 3y' + 2y = 0$, and so is $y_1^2(x) = e^{2x}$. But $y_1(x) = e^x$ is a solution of $y'' - y = 0$ and $y_1^2(x) = e^{2x}$ is not.

9. (a) There is no violation of the theorem because the differential equation is not homogeneous.
(b) There is no violation of the theorem because the differential equation is not linear.

Section 2.2

1. (a) $y(x) = C_1 + C_2e^{-5x}$
(c) $y(x) = (C_1 + C_2x)e^{-x}$
(e) $y(x) = C_1e^x + C_2e^{2x}$
(g) $y(x) = (C_1 + C_2x)e^{-4x}$
(i) $y(x) = C_1e^{-2x} + C_2e^{-4x}$
(k) $y(x) = C_1e^{-4x} + C_2e^{-3x}$
(m) $y(x) = C_1e^{3x} + C_2e^{-x/2}$
(o) $y(x) = C_1e^{(-3+\sqrt{5})x/2} + C_2e^{(-3-\sqrt{5})x/2}$
(q) $y(x) = e^{-5x/2}(C_1e^{\sqrt{21}x/2} + C_2e^{-\sqrt{21}x/2})$

2. (a) $y(x) = \frac{1}{5}(14 + e^{-5x})$
(c) $y(x) = (3 + 2x)e^{-x}$
(e) $y(x) = 7e^x - 4e^{2x}$
(g) $y(x) = (3 + 11x)e^{-4x}$
(i) $y(x) = \frac{1}{2}(11e^{-2x} - 5e^{-4x})$
(k) $y(x) = 11e^{-3x} - 8e^{-4x}$
(m) $y(x) = \frac{1}{7}(e^{3x} + 20e^{-x/2})$
(o) $y(x)=\frac{1}{10}e^{-3x/2}[(15+7\sqrt{5})e^{\sqrt{5}x/2}+(15-7\sqrt{5})e^{-\sqrt{5}x/2}]$
(q) $y(x) = \frac{1}{42}e^{-5x/2}[(63 + 13\sqrt{21})e^{\sqrt{21}x/2} + (63-13\sqrt{21})e^{-\sqrt{21}x/2}]$

3. (a) $y(x) = 4(1-e^{-5(x-5)})/5$
(c) $y(x) = (4x-20)e^{-(x-5)}$
(e) $y(x) = 4e^{2(x-5)} - 4e^{x-5}$
(g) $y(x) = (4x-20)e^{-4(x-5)}$
(i) $y(x) = 2e^{-2(x-5)} - 2e^{-4(x-5)}$
(k) $y(x) = 4e^{-3(x-5)} - 4e^{-4(x-5)}$
(m) $y(x) = 8(e^{3(x-5)} - e^{-(x-5)/2})/7$
(o) $y(x) = 4e^{-3(x-5)/2}[e^{\sqrt{5}(x-5)/2} - e^{-\sqrt{5}(x-5)/2}]/\sqrt{5}$
(q) $y(x) = 4e^{-5(x-5)/2}[e^{\sqrt{21}(x-5)/2} - e^{-\sqrt{21}(x-5)/2}]/\sqrt{21}$

4. (a) $y'' - 8y' + 12y = 0$ (c) $y'' - 5y' = 0$
(e) $25y'' + 5y' - 2y = 0$ (g) $y'' + 4y' + 4y = 0$
(i) $y'' - 40y' + 400y = 0$ (k) $y'' - 0.3y' + 0.02y = 0$

5. (a) $y'' + y' = 0$ (d) $y'' = 0$ (g) $y'' - y = 0$

8. Following is only a hint: If C_3 and C_4 were $C_3 = C_1 + C_2$ and $C_4 = 3C_1 + 3C_2$ instead, then $C_4 = 3C_3$, so it would *not* be true that C_3 and C_4 are arbitrary constants

9. (a) $y'' + y' - 6y = (D + 3)(D - 2)y = 0$, so set $(D - 2)y = u$ and $(D + 3)u = 0$. These give $u(x) = Ae^{-3x}$, so $y' - 2y = Ae^{-3x}$. The latter gives $y(x) = e^{-x}(\int e^{-2x}Ae^{-3x}\,dx + B) = \text{"}C_1\text{"}e^{-3x} + \text{"}C_2\text{"}e^{2x}$.
(d) $y(x) = C_1e^{-4x} + C_2e^{-x}$

Section 2.3

1. (a) $-e^2$, that is, $-e^2 + 0i$ (c) $(1/\sqrt{2}) + (1/\sqrt{2})i$
(e) $\cos 2 + i \sin 2$ (g) $\cos(2\pi/5) - i \sin(2\pi/5)$
(i) $(e^{-2}\sin 2) + i(e^{-2}\cos 2)$ (k) $(e^5\cos 3) - i(e^5\sin 3)$

2. (a) e^3 (c) 1 (e) 3

3. (a) No roots (d) No roots

8. (a) $y(x) = e^{4x}(C_1\cos 3x + C_2\sin 3x)$
(d) $y(x) = e^{x/2}[C_1\cos(\sqrt{3}x/2) + C_2\sin(\sqrt{3}x/2)]$
(g) $y(x) = e^{-x/4}[C_1\cos(\sqrt{7}x/4) + C_2\sin(\sqrt{7}x/4)]$
(j) $y(x) = C_1\cos(x/2) + C_2\sin(x/2)$

9. (a) $y(x) = e^{4x}(3\cos 3x - 2\sin 3x)$
(d) $y(x) = 2e^{x/2}\cos(\sqrt{3}x/2)$
(g) $y(x) = (10/7)e^{-x/4}[7\cos(\sqrt{7}x/4) + \sqrt{7}\sin(\sqrt{7}x/4)]$
(j) $y(x) = 4\cos(x/2)$

10. (a) $p_1^2 - 4p_2 < 0$

11. (a) $y(x) = e^{3x}(C_1\cos 4x + C_2\sin 4x)$, $y'' - 6y' + 25y = 0$
(c) $y(x) = e^{3x}(C_1\cos 5x + C_2\sin 5x)$, $y'' - 6y' + 34y = 0$
(e) $y(x) = e^{-2x}(C_1\cos x + C_2\sin x)$, $y'' + 4y' + 5y = 0$

12. (a) $y'' - 6y' + 13y = 0$ (c) $y'' + 4y' + 29y = 0$
(e) $y'' - 8y' + 25y = 0$ (g) $y'' + 10y' + 26y = 0$
(i) No such equation exists (k) No such equation exists

Section 2.4

1. Call the functions in the set $u_1(x), u_2(x)$, and so on.
(a) LD: $0u_1(x) + 7u_2(x) = 0$ (c) LD: $-5u_1(x) + 1u_2(x)$

$= 0$ (e) LI (g) LD: $u_1(x) + 3u_2(x) - 6u_3(x) = 0$
(i) LD: $3u_1(x) - u_2(x) + 0u_3(x) + 0u_4(x) = 0$ (k) LI
(m) LD: $u_1(x) + u_2(x) + 0u_3(x) + 0u_4(x) = 0$
(o) LD: $u_1(x) - 9u_2(x) + 3u_3(x) = 0$

8. (a) $A = 4,\ B = 12,\ C = 9$ (d) No (g) $A = 2,\ B = -1/2,\ C = 1$ (j) $A = -1,\ B = 2,\ C = -1$

9. (a) Yes, because $W[y_1, y_2](a) = 1 \neq 0$ shows, by Theorem 2.4.4(c), that $\{y_1, y_2\}$ is LI on I
(c) No, because $W[y_1, y_2](a) = 0$ shows, by Theorem 2.4.4(c), that $\{y_1, y_2\}$ is LD on I (e) No

10. (a) To apply the theorem we must verify that the coefficients $p_1(x)$ and $p_2(x)$ are continuous on I, that the two functions do satisfy the differential equation, and that they are LI on I. The coefficients are constants and hence continuous on I, both functions are indeed solutions, and $W[y_1, y_2](x) = 3e^{2x} \neq 0$, so they are also LI. Thus, yes, y_1, y_2 are a basis for the given differential equation, on I. (c) No, they are both solutions, but they are not LI on I (e) No, y_2 is not a solution (g) Yes (i) Yes (k) Yes

12. (a) Left side is -2 and right side is C, so (18) is satisfied with $C = -2$ (d) Left side is 0 and right side is Ce^x, so (18) is satisfied with $C = 0$ (g) Left side is x^5 and right side is Cx^5, so (18) is satisfied with $C = 1$

Section 2.5

2. (a) $y_2(x) = xe^{3x}$ (c) $y_2(x) = e^x$ (e) $y_2(x) = xe^{-\sqrt{2}x}$ (g) $y_2(x) = xe^{\sqrt{5}x}$ (i) $y_2(x) = (\ln x)/x$ (k) $y_2(x) = x^2 \ln x$

4. (a) $y_2(x) = x$ (c) $y_2(x) = \cos(x^2)$
(e) $y_2(x) = \dfrac{x}{2} \ln\left(\dfrac{1+x}{1-x}\right) - 1$

Section 2.6

4. (a) $y(x) = C_1 x^{-2} + C_2 x^3,\ y(x) = x^3$ on $-\infty < x < \infty$ (d) $y(x) = (C_1 + C_2 \ln x)x^5,\ y(x) = [(1 + 5\ln 2) - 5\ln x]x^5$ on $0 < x < \infty$ (g) $y(x) = (C_1 + C_2 \ln x)/x,\ y(x) = 18/x$ on $0 < x < \infty$ (j) $y(x) = C_1 + C_2 \ln x,\ y(x) = 3 - 2\ln x$ on $0 < x < \infty$ (m) $y(x) = C_1 \cos(\ln x) + C_2 \sin(\ln x)$, $y(x) = 3\cos(\ln x) + 2\sin(\ln x)$ on $0 < x < \infty$ (o) $y(x) = \sqrt{x}\,[C_1 \cos(\ln x) + C_2 \sin(\ln x)],\ y(x) = \sqrt{x}\,[6\cos(\ln x) + \sin(\ln x)]$ on $0 < x < \infty$

6. (a) $y(x) = C_1 x^3 + C_2 x^{-2}$ (c) $y(x) = (C_1 + C_2 \ln x)x^2$ (e) $y(x) = C_1 x + C_2 x^2$ (g) $y(x) = (C_1 + C_2 \ln x)/x$ (i) $y(x) = C_1 \cos(\ln x) + C_2 \sin(\ln x)$ (k) $y(x) = \sqrt{x}\,[C_1 \cos(\ln x) + C_2 \sin(\ln x)]$

8. (a) $y(x) = -5\ln(-x)$ on $-\infty < x < 0$, $y(-10) = -11.51$ (c) $y(x) = x + 6\sqrt{-x}$ on $-\infty < x < 0$, $y(-10) = 8.974$

10. (b) Get $xu' - 5u = 0$ and $xy' + 2y = u$, so $u(x) = Ax^5$ and $y(x) = C_1 x^5 + C_2 x^{-2}$

Section 2.7

1. (a) $y(x) = C_1 + C_2 e^{\sqrt{2}x} + C_3 e^{-\sqrt{2}x}$ or, equivalently, $y(x) = C_1 + C_2 \cosh \sqrt{2}x + C_3 \sinh \sqrt{2}x$ (c) $y(x) = C_1 e^x + (C_2 + C_3 x)e^{2x}$ (e) $y(x) = C_1 e^x + C_2 \cos 3x + C_3 \sin 3x$ (g) $y(x) = (C_1 + C_2 x)e^x + C_3 e^{5x}$ (i) $y(x) = C_1 e^x + C_2 e^{-x} + C_3 e^{2x} + C_4 e^{-2x}$ (k) $y(x) = C_1 e^x + C_2 e^{-x} + C_3 \cos x + C_4 \sin x$

2. (a) $y(x) = C_1 e^x + C_2 e^{2x} + C_3 e^{3x}$
(c) $y(x) = (C_1 + C_2 x + C_3 x^2)e^x$

3. (a) $y''' - 2y'' - y' + 2y = 0$ (d) $y''' - 5y'' + 6y' = 0$ (g) $y'''' - 17y'' + 16y = 0$ (j) $y'''' - 2y''' + 11y'' - 18y' + 18y = 0$

4. (a) The characteristic roots are $r = 0, -1, 2$ so the characteristic equation is $(r-0)(r+1)(r-2) = r^3 - r^2 - 2r = 0$, so $y''' - y'' - 2y' = 0$ (c) $y''' - y' = 0$ (e) $y''' - 3y'' + 3y' - y = 0$ (g) $y''' - y'' + 9y' - 9y = 0$ (i) $y''' - 3y'' + 4y' - 2y = 0$

5. (a) $y(x) = (C_1 + C_2 x + C_3 x^2)e^x$
(d) $y(x) = C_1 + C_2 x + C_3 x^2 + C_4 e^{5x}$
(g) $y(x) = (C_1 + C_2 x)\cos x + (C_3 + C_4 x)\sin x$
(j) $y(x) = e^{2x}[(C_1 + C_2 x)\cos 3x + (C_3 + C_4 x)\sin 3x]$

9. (a) $y(x) = C_1 + C_2 \ln x + C_3 x^4$ (c) $y(x) = C_1 x + C_2 x^{-1} + C_3 x^2$ (e) $y(x) = [C_1 + C_2 \ln x + C_3 (\ln x)^2]x^5$ (g) $y(x) = C_1 x + [C_2 \cos(3\ln x) + C_3 \sin(3\ln x)]x^2$

10. (a) $x^3 y''' - 3x^2 y'' + 6xy' - 6y = 0$
(d) $x^3 y''' + x^2 y'' + 9xy' = 0$

11. (a) $x^3 y''' - 4x^2 y'' + 8xy' - 8y = 0$ (c) $x^3 y''' - 2x^2 y'' + 4xy' - 4y = 0$ (e) $x^3 y''' + xy' - y = 0$ (g) $x^3 y''' + 3x^2 y'' + 2xy' = 0$ (i) $x^3 y''' - x^2 y'' + 11xy' - 20y = 0$

13. (a) Left side is -8 and right side is C, so (11.1) is satisfied with $C = -8$. (d) Left side is $-16x$ and right side is Cx, so (11.1) is satisfied with $C = -16$.

Section 2.8

1. (a) $y(x) = x^2 + x + 2 + C_1 e^x$ (d) $y(x) = 2x^2 + C_1 + C_2 e^{-x}$ (g) $y(x) = 3\cos x + (C_1 + C_2 x)e^x$ (j) $y(x) = xe^{-x} + C_1 e^{-x} + C_2 e^{-3x}$ (m) $y(x) = -x^2 + C_1 e^x + C_2 e^{-x} + C_3 \cos x + C_4 \sin x$ (p) $y(x) = 4x^2 + C_1 x^3 + C_2 x^{-2}$

2. We will put $y_p(x)$ in brackets; the rest is $y_h(x)$.
(a) $y(x) = [2e^{2x} - 6x^2] + C_1 + C_2x$ (c) $y(x) = [4x^3] + C_1 + C_2x + C_3x^2$ (e) $y(x) = [2x^5/5] + C_1 + C_2x + C_3x^2 + C_4x^3$ (g) $y(x) = [12x^2] + C_1x$

3. (a) $y_p(x) = 9/2$ (d) $y_p(x) = 20$ (g) $y_p(x) = 7e^x/4$

4. (a) $y_p(x) = 4x^2 - 8$ (d) $y_p(x) = (8\cos x - 4\sin x)/5$ (g) $y_p(x) = 5\sin x$ (j) $y_p(x) = -\sin x - 5$

6. (a) $y_p(x) = 25 - 4e^x - 2e^{3x}$ (d) $y_p(x) = 4x^2e^x + 2e^{-x}$ (g) $y_p(x) = 4x^4 - 8x^3 - 18x^2$

Section 2.9

1. (a) It gives $A = 0$ and $A = -3/4$, which is impossible; thus, the form doesn't work. (c) Get $A = 0$, $B = -3/4$, $C = 0$, $D = -3/8$, and hence the same result as in Example 1. Including the extra Ax^3 term doesn't hurt because we obtain $A = 0$.

2. (a) $y_p(x) = -9x^3 - 9x^2 - 6x - 2$ (c) $y_p(x) = 2x^3e^{2x}$ (e) $y_p(x) = 4x^2 + 14x + 17$ (g) $y_p(x) = (18x^2 - 6x)e^{3x} - 2$ (i) $y_p(x) = 50x^2e^x$ (k) $y_p(x) = 8x^3$ (m) $y_p(x) = 2x^4 - 8x^3 + 24x^2$ (o) $y_p(x) = -2x^5 - 5x^4 - 10x^3$ (q) $y_p(x) = 2 - 2\cos 2x + 16\sin 2x$ (s) $y_p(x) = 6 + 2\cos 2x + 3\sin 2x$

3. (a) $y_p(x) = 2(\cos x + \sin x)$ (c) $y_p(x) = (3\cos x + \sin x)/10 + 3\cos 2x + 2\sin 2x$ (e) $y_p(x) = -50\sin x$ (g) $y_p(x) = -4\cos 3x - 2\sin 3x$ (i) $y_p(x) = (\cos 2x - 12\sin 2x)/29$

5. (a) $y_p(x) = 24x^3/7$ (c) $y_p(x) = -2x^3$ (e) $y_p(x) = (4/7)x^5\ln x$

7. The chest hits bottom at $t = 10.375$ sec; $x(10.375) = 82.91$ ft.

8. (a) $y(x) = C_1e^x + C_2e^{2x} + 2e^{5x}$ (c) $y(x) = C_1e^x + C_2e^{-x} - 2\cos x$

Section 2.10

1. NOTE: Abbreviate undetermined coefficients as UC here. (a) $y_p(x) = e^{2x}$. UC would work. (c) $y_p(x) = 3x^3$. Nonconstant coefficients, so UC would not work. (However, see Exercise 5 in Section 2.9.) (e) $y_p(x) = 2\sin x$. Nonconstant coefficients, and $\sec x$ does not have a finite number of LI derivatives, so UC would not work. (g) $y_p(x) = (\cos x)\ln(\cos x) + x\sin x$. $\sec x$ does not have a finite number of LI derivatives, so UC would not work. (i) $y_p(x) = (20x - 5)e^x - 80$. UC would work. (k) $y_p(x) = 3 + 4x(1 - \ln x)e^x$. e^x/x does not have a finite number of LI derivatives, so UC would not work, (m) $y_p(x) = 2e^{2x}$. UC would work. (o) $y_p(x) = -8$. Nonconstant coefficients, so UC would not work. (However, see Exercise 5 in Section 2.9.) (q) $y_p(x) = 3x\ln x - 2x$. Nonconstant coefficients, so UC would not work. (However, see Exercise 5 in Section 2.9.)

2. (a) x (c) x^2 (e) $-x$

5. (a) $y_p(x) = 3e^{3x}$ (c) $y_p(x) = 2x\ln x + 3x$

CHAPTER 3

Section 3.2

1. (a) $x(t) = (2 - 3i/2)e^{it} + (2 + 3i/2)e^{-it}$, $4\cos t + 3\sin t$, $5\sin(t + 0.927)$
(c) $x(t) = (-2 - i/2)e^{i2t} + (-2 + i/2)e^{-i2t}$, $-4\cos 2t + \sin 2t$, $\sqrt{17}\sin(2t - 1.326)$
(e) $x(t) = (-2 - 5\sqrt{2}i/4)e^{i\sqrt{2}t} + (-2 + 5\sqrt{2}i/4)e^{-i\sqrt{2}t}$, $-4\cos\sqrt{2}t + (5/\sqrt{2})\sin\sqrt{2}t$, $\sqrt{57/2}\sin(\sqrt{2}t - 0.847)$
(g) $x(t) = (-4 + i/6)e^{i3t} + (-4 - i/6)e^{-i3t}$, $-8\cos 3t - (1/3)\sin 3t$, $\sqrt{577/9}\sin(3t + 4.67)$

4. (a) 3.260 ft (b) If L is too short by around 0.001 ft the clock runs slow by 1 min/wk

5. (b) $\omega_n = \sqrt{3g/2L}/2\pi$ cycle/sec. With $g \approx 32$ ft/sec^2, $L \approx 3$ ft, and a stride of around 4 ft/cycle for an average size adult, the walking speed is, roughly, 2.5 ft/sec or 1.7 mi/hr.

Section 3.3

2. (b) mg/k
(c) Underdamped:

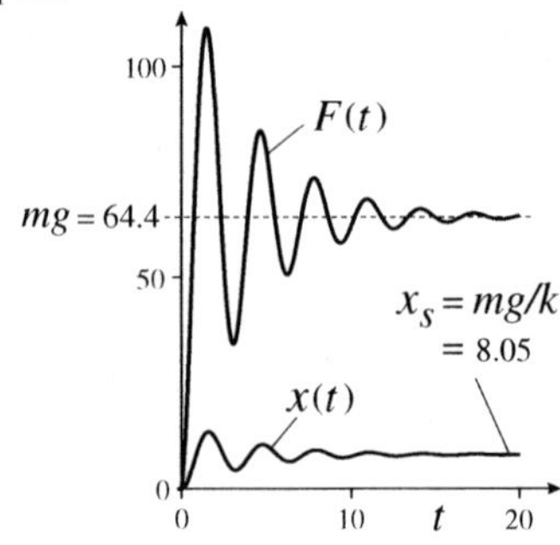

Section 3.4

8. $x(t) = (F_0/k)(1 - \cos\omega_n t)$ for $0 < t < T$;
$x(t) = (F_0/k)[(\cos\omega_n T - 1)\cos\omega_n t + (\sin\omega_n T)\sin\omega_n t]$ for $T < t < \infty$

14. $C = 1$

Section 3.5

1. (a) For all values of b_1, b_2 there is a unique solution, $y(x) = b_1 + (b_2 - b_1)x/L$.
(c) No solution if $b_1 \neq b_2$; if $b_1 = b_2 \equiv b$ there is the nonunique solution $y(x) = bx + C$, with C an arbitrary constant.

2. (a) $y(x) = A\cos x + B\sin x$. The boundary conditions give $50 = A$ and $100 = -A$, which are inconsistent. No solution.
(c) Unique solution $y(x) = 0$.
(e) Unique solution $y(x) = 50\cos x$.
(g) Nonunique solution $y(x) = A\cos x + 25\sin x$, with A an arbitary constant.

3. (a) The nontrivial solutions are $y(x) = B\sin(n\pi x/L)$, corresponding to $\lambda = (n\pi/L)^2$, for $n = 1, 2, \ldots$, with B an arbitrary constant.
(c) The nontrivial solutions are $y(x) = B\sin(n\pi x/2L)$, corresponding to $\lambda = (n\pi/2L)^2$, for $n = 1, 3, 5, \ldots$, with B an arbitrary constant.

4. (a) $u(x) = (f_0/2k)x(L-x)$, unique
(c) $u(x) = 20 + 30x/L + (f_0/2k)x(L-x)$, unique
(e) If $f_0 \neq 0$ then no solution, if $f_0 = 0$ then nonunique solution $u(x) = C + 10x$ with C arbitrary

6. (a) $u_L(x) = f_0 Lx/8k$ and
$u_R(x) = (f_0/8k)(-L^2 + 5Lx - 4x^2)$

10. (a) $u(r) = \left(\frac{u_1\ln b - u_2\ln a}{\ln b - \ln a}\right) + \left(\frac{u_2 - u_1}{\ln b - \ln a}\right)\ln r$, unique
(c) Doesn't exist
(e) $u(r) = 50\left(\frac{2b\ln a - b\ln b - 2}{b\ln a - b\ln b - 2}\right) - 50\left(\frac{b}{b\ln a - b\ln b - 2}\right)\ln r$, unique

12. (a) $u(\rho) = \left(\frac{au_1 - bu_2}{a-b}\right) + \frac{ab(u_2 - u_1)}{a-b}\bigg)\frac{1}{\rho}$, unique
(c) Doesn't exist
(e) $u(r) = 50\left(\frac{2a + 2b^2 - ba}{b^2 - ba + 2a}\right) - 50\left(\frac{ab^2}{b^2 - ba + 2a}\right)\frac{1}{\rho}$, unique

Section 3.6

1. (c) $\lambda_0 = 0$ and $y_0(x) = 1$, $\lambda_n = n^2$ and $y_n(x) = \cos nx$ for $n = 1, 2, \ldots$
(e) $\lambda_n = 1 + n^2$ and $y_n(x) = e^{-x}\sin nx$ for $n = 1, 2, \ldots$

CHAPTER 4

Section 4.1

1. (a) $x(t) = C_1e^{3t} + C_2e^{-3t} + 4 - \cos t$, $y(t) = -3C_1e^{3t} + 3C_2e^{-3t} + 9\sin t$; $C_1 = -7/3$, $C_2 = -2/3$
(c) $x(t) = C_1e^{4t} + C_2e^{-4t} + 2t$,
$y(t) = C_1e^{4t}/3 - C_2e^{-4t} + 2t - 1$; $C_1 = 9/2$, $C_2 = -9/2$
(e) $x(t) = C_1 + C_2t$, $y(t) = C_2 - C_1 - C_2t - 50$;
$C_1 = 0$, $C_2 = 55$
(g) $x(t) = C_1e^{2t} + C_2e^{-2t} + t - 2$,
$y(t) = 3C_1e^{2t} - C_2e^{-2t} + t + 7$; $C_1 = 0$, $C_2 = 2$
(i) $x(t) = C_1e^t + C_2e^{-t} - 3t$,
$y(t) = C_1e^t - C_2e^{-t} - 3$; $C_1 = 9e^{-2}$, $C_2 = e^2$
(k) $x(t) = C_1e^t + C_2e^{3t} - 9$, $y(t) = -C_1e^t + C_2e^{3t} + 18$;
$C_1 = 13e^{-2}$, $C_2 = 0$
(m) $x(t) = C_1e^{-t} + C_2e^{3t} - e^t$, $y(t) = -C_1e^{-t} + C_2e^{3t}$;
$C_1 = (e^4 - e^2)/2$, $C_2 = (e^2 + 9)/(2e^6)$

2. (c) $x(t) = A + Be^t$, $y(t) = Ce^{2t} - Be^t$,
$z(t) = Ce^{2t} - Be^t - A$

3. (a) $x(t) = A + (B + Ct)e^{-2t} + 4e^t$,
$y(t) = -2e^t - A + (B + C + Ct)e^{-2t}$
(c) $x(t) = A + Bt + Ce^{2t}$, $y(t) = B - A - 4 - Bt + Ce^{2t}$
(e) $x(t) = 2e^t + At^3 + Bt^2 + (6A + C)t + (2B + D)$,
$y(t) = e^t + At^3 + Bt^2 + Ct + D$

4. (a) $V_1c_1' + 12c_1 - 2c_2 = 10c_{1i}$, $V_2c_2' + 17c_2 - 12c_1 = 5c_{2i}$

11. (a) $x(t) = Ae^{2t} + Be^{-2t} + t - 2$,
$y(t) = 3Ae^{2t} - Be^{-2t} + t + 7$
(c) $x(t) = A + Be^{3t} + e^t + 6te^{3t}$,
$y(t) = 2Be^{3t} - A + (12t + 6)e^{3t} - 2e^t$
(e) $x(t) = At^3 + Bt^2 + Ct + D$, $y(t) = At^3 + Bt^2 + Ct + D - 6At - 2B$

12. $x(t) = (-B - C) + Ae^{3t}$, $y(t) = C + Ae^{3t}$, $z(t) = B + Ae^{3t}$. Your solution may look different, but be equivalent to the latter.

Section 4.2

1. (a) $x_1(t) = C_1\cos t + C_2\sin t + C_3\cos 2t + C_4\sin 2t$,
$x_2(t) = C_1\cos t + C_2\sin t - C_3\cos 2t - C_4\sin 2t$
(b) $x_1(t) = G\sin(t + \phi) + H\sin(2t + \psi)$,
$x_2(t) = G\sin(t + \phi) - H\sin(2t + \psi)$
(c) $x_1(0) = a$, $x_2(0) = a$, $x_1'(0) = b$, $x_2'(0) = b$ for any values of a and b, not both zero

2. (a) $x_1(t) = C_1\cos t + C_2\sin t + C_3\cos 2t + C_4\sin 2t$,
$x_2(t) = (3/2)(C_1\cos t + C_2\sin t - C_3\cos 2t - C_4\sin 2t)$
(b) $x_1(t) = G\sin(t + \phi) + H\sin(2t + \psi)$,
$x_2(t) = (3/2)[G\sin(t + \phi) - H\sin(2t + \psi)]$
(c) $x_1(0) = a$, $x_2(0) = 3a/2$, $x_1'(0) = b$, $x_2'(0) = 3b/2$ for any values of a and b, not both zero

14. (b) $1/\sqrt{LC}$ and $\sqrt{3/LC}$ radians per unit time

Section 4.3

1. (a) $[11, -2, 9, 4]^{\mathrm{T}}$ (d) $[7, -4, 15, 2]^{\mathrm{T}}$
(g) $[17, 25, -2, 6]^{\mathrm{T}}$ (j) $[-14, 0, -6, -6]^{\mathrm{T}}$

2. (a) $\mathbf{u}+\mathbf{v} = [u_1+v_1, \ldots, u_n+v_n] = [v_1+u_1, \ldots, v_n+u_n] = \mathbf{v}+\mathbf{u}$. The key is the second equality, which follows from the commutativity of *scalar* addition. The first and third equalities are by the definition of the addition of vectors.
(c) $\alpha(\beta\mathbf{u}) = \alpha[\beta u_1, \ldots, \beta u_n] = [\alpha(\beta u_1), \ldots, \alpha(\beta u_n)] = [(\alpha\beta)u_1, \ldots, (\alpha\beta)u_n] = (\alpha\beta)[u_1, \ldots, u_n] = (\alpha\beta)\mathbf{u}$. The key is the third equality, which follows from the associative rule of *scalar* multiplication. The first two equalities are by the definition of scalar multiplication (4b), the fourth is by (4b) again, and the fifth is simply by the definition of $\mathbf{u}$.

4. (a) $[-1/3, 2, 3, 2/3]^{\mathrm{T}}$ (c) $[-11, 2, 3, -2]^{\mathrm{T}}$
(e) $[-2, 4, 6, 1]^{\mathrm{T}}$ (g) $[-1/8, 3/8, -3/2, 1/4]^{\mathrm{T}}$

5. (a) $[5, 2, 4]^{\mathrm{T}}$ (d) $[6, 3, 3]^{\mathrm{T}}$ (g) $[5, 2, -2]^{\mathrm{T}}$ (j) $[9, -3, 3]^{\mathrm{T}}$

6. (a) $\mathbf{A} = \begin{bmatrix} 1 & 4 & -1 \\ 6 & 2 & 1 \\ 3 & -1 & -1 \end{bmatrix}$ (d) $\mathbf{A} = \begin{bmatrix} 0 & 0 & 1 \\ 0 & 1 & 0 \\ 1 & 0 & 0 \end{bmatrix}$
(g) $\mathbf{A} = \begin{bmatrix} 1 & 0 & 1 \\ 0 & 0 & 0 \\ 1 & 0 & 1 \end{bmatrix}$ (j) $\mathbf{A} = \begin{bmatrix} 0 & 0 & 5 \\ 0 & 0 & 5 \\ 0 & 0 & 5 \end{bmatrix}$

9. (a) $\mathbf{x}(t)$ on $-\infty < t < \infty$, $\mathbf{A}(t)$ on $0 < t < \infty$
(d) $\mathbf{x}(t)$ on $0 < t < \infty$, $\mathbf{A}(t)$ on $-\infty < t < \infty$
(g) $\mathbf{x}(t)$ on $-2 < t < 3$, $\mathbf{A}(t)$ on $0 < t < 5$

10. (a) $[1, 4t^3, 7t^6]^{\mathrm{T}}$ (d) $[-1, -32\sin 4t]^{\mathrm{T}}$ (g) $[e^t - e^{-t}, 0]^{\mathrm{T}}$

11. (a) $[2t^2-1, t]^{\mathrm{T}}$, $(-\infty, \infty)$
(c) $[2\ln(t+3) - 2\ln 3 + 3, t+4]^{\mathrm{T}}$, $(-3, \infty)$
(e) $[t^2+2, 21 - \ln(\cos t)]^{\mathrm{T}}$, $(-\pi/2, \pi/2)$
(g) $[2t, -2\cos 3t + 2]^{\mathrm{T}}$, $(-\infty, \infty)$

Section 4.4

1. (a) No, if it is not one then it is the other.

3. (a) $3\mathbf{u}_1 + \mathbf{u}_2 = \mathbf{0}$ (d) $7\mathbf{u}_1 + 5\mathbf{u}_2 - 8\mathbf{u}_3 = \mathbf{0}$
(g) $\mathbf{u}_1 - \mathbf{u}_2 + 0\mathbf{u}_3 = \mathbf{0}$ (j) $2\mathbf{u}_1 + 2\mathbf{u}_2 - \mathbf{u}_3 = \mathbf{0}$

4. (a) LD. E.g., $-\mathbf{u}_1 + \mathbf{u}_2 + \mathbf{u}_3 + 0\mathbf{u}_4 = \mathbf{0}$. (c) LI
(e) LD. E.g., $-3\mathbf{u}_1 + 2\mathbf{u}_2 + 0\mathbf{u}_3 + 0\mathbf{u}_4 = \mathbf{0}$. (g) LI
(i) LD. E.g., $0\mathbf{u}_1 + 0\mathbf{u}_2 + 5\mathbf{u}_3 = \mathbf{0}$.
(k) LD. E.g., $6\mathbf{u}_1 + 3\mathbf{u}_2 = \mathbf{0}$. (m) LI (o) LI

9. (a) LI (c) LD (e) LD (g) LI (i) LI (k) LI
(m) LI (o) LD

Section 4.5

1. (a) $\mathbf{A}(t) = \begin{bmatrix} t & 2 \\ 1 & e^t \end{bmatrix}$, $\mathbf{f}(t) = \begin{bmatrix} 0 \\ \sin t \end{bmatrix}$. Both continuous on $-\infty < t < \infty$.
(c) $\mathbf{A}(t) = \begin{bmatrix} \tan t & 1 \\ 1 & 5 \end{bmatrix}$, $\mathbf{f}(t) = \begin{bmatrix} 3 \\ 6t-1 \end{bmatrix}$. $\mathbf{A}(t)$ continuous on $-\pi/2 < t < \pi/2$ and $\mathbf{f}(t)$ on $-\infty < t < \infty$.
(e) $\mathbf{A}(t) = \begin{bmatrix} t\sec t & \sec t \\ \tan 2t & -\tan 2t \end{bmatrix}$, $\mathbf{f}(t) = \begin{bmatrix} 10\tan t \\ 20\tan 2t \end{bmatrix}$.
Both continuous on $-\pi/4 < t < \pi/4$.

3. (a) $c_1 = -6$, $c_2 = 11$ (c) $c_1 = 4$, $c_2 = -4$
(e) $c_1 = -2.207$, $c_2 = 0.9473$

6. (a) $x_1(t) = C_1 + C_2e^{2t}$, $x_2(t) = -C_1 + C_2e^{2t}$; $\mathbf{x}_1(t) = [1, -1]^{\mathrm{T}}$, $\mathbf{x}_2(t) = [e^{2t}, e^{2t}]^{\mathrm{T}}$; $\mathbf{x}_p(t)$ not needed since system is homogeneous; $\mathbf{x}(t) = [4 + e^{2t}, -4 + e^{2t}]^{\mathrm{T}}$
(c) $x_1(t) = C_1e^{4t} + C_2e^{-3t} - 1 + 12t$, $x_2(t) = C_1e^{4t} - (3C_2/4)e^{-3t} - 33 - 36t$; $\mathbf{x}_1(t) = [e^{4t}, e^{4t}]^{\mathrm{T}}$, $\mathbf{x}_2(t) = [e^{-3t}, -3e^{-3t}/4]^{\mathrm{T}}$; $\mathbf{x}_p(t) = [-1 + 12t, -33 - 36t]^{\mathrm{T}}$; $\mathbf{x}(t) = [13e^{4(t+1)} - 1 + 12t, 13e^{4(t+1)} - 33 - 36t]^{\mathrm{T}}$
(e) $x_1(t) = C_1e^t + C_2e^{3t} + 2t$, $x_2(t) = -C_1e^t + C_2e^{3t} - t$; $\mathbf{x}_1(t) = [e^t, -e^t]^{\mathrm{T}}$, $\mathbf{x}_2(t) = [e^{3t}, e^{3t}]^{\mathrm{T}}$; $\mathbf{x}_p(t) = [2t, -t]^{\mathrm{T}}$; $\mathbf{x}(t) = [2e^{3t} + 2t, 2e^{3t} - t]^{\mathrm{T}}$
(g) $x_1(t) = C_1e^t + C_2e^{7t} - 3e^{2t}$, $x_2(t) = -C_1e^t + 2C_2e^{7t} + 4e^{2t} - 1$; $\mathbf{x}_1(t) = [e^t, -e^t]^{\mathrm{T}}$, $\mathbf{x}_2(t) = [e^{7t}, 2e^{7t}]^{\mathrm{T}}$; $\mathbf{x}_p(t) = [-3e^{2t}, 4e^{2t} - 1]^{\mathrm{T}}$; $\mathbf{x}(t) = [-3e^t - 3e^{2t}, 3e^t + 4e^{2t} - 1]^{\mathrm{T}}$

7. (a) We'll omit the straightforward verification that the given functions are solutions. To show they're LI, obtain $\det[\mathbf{x}_1, \mathbf{x}_2, \mathbf{x}_3] = -9e^{3t} \neq 0$, so the set is LI on the interval. General solution $\mathbf{x}(t) = C_1[2, -1, -1]^{\mathrm{T}} + C_2[1, -2, 1]^{\mathrm{T}} + C_3[e^{3t}, e^{3t}, e^{3t}]^{\mathrm{T}}$, and the initial conditions give $C_1 = 9$, $C_2 = -11$, $C_3 = 8$
(c) We'll omit the straightforward verification that the given functions are solutions. To show they're LI, obtain $\det[\mathbf{x}_1, \mathbf{x}_2, \mathbf{x}_3] = -30e^{7t} \neq 0$, so the set is LI on the interval. General solution $\mathbf{x}(t) = C_1[13, 1, 0]^{\mathrm{T}} + C_2[3e^{5t}, e^{5t}, 5e^{5t}]^{\mathrm{T}} + C_3[3e^{2t}, e^{2t}, 2e^{2t}]^{\mathrm{T}} + [-30e^{-t}, 2e^{-t}, 7e^{-t}]^{\mathrm{T}}$, and the initial conditions give $C_1 = 39/10$, $C_2 = 8/5$, $C_3 = -15/2$

11. (c) $\mathbf{x}(t) = 12\mathbf{u}(t) - 6\mathbf{v}(t)$

Section 4.6

1. (a) $\lambda_1 = 0$, $\mathbf{e}_1 = \alpha[3, 1]^{\mathrm{T}}$; $\lambda_2 = 1$, $\mathbf{e}_2 = \beta[1, 0]^{\mathrm{T}}$. Both eigenspaces one-dimensional.
(d) $\lambda_1 = 2$ (multiplicity 2), $\mathbf{e}_1 = \alpha[1, 1]^{\mathrm{T}}$. One-dimensional eigenspace.
(g) $\lambda_1 = -1$, $\mathbf{e}_1 = \alpha[1, 1]^{\mathrm{T}}$; $\lambda_2 = -5$, $\mathbf{e}_2 = \beta[-3, 1]^{\mathrm{T}}$. Both eigenspaces one-dimensional.

(j) $\lambda_1 = -3$, $\mathbf{e}_1 = \alpha[-2,1]^{\mathrm{T}}$; $\lambda_2 = 4$, $\mathbf{e}_2 = \beta[1,3]^{\mathrm{T}}$. Both eigenspaces one-dimensional.
(m) $\lambda_1 = 3$, $\mathbf{e}_1 = \alpha[1,0,0]^{\mathrm{T}}$; $\lambda_2 = 4$ (multiplicity 2), $\mathbf{e}_2 = \beta[6,1,0]^{\mathrm{T}}$. Both eigenspaces one-dimensional.
(p) $\lambda_1 = 0$, $\mathbf{e}_1 = \alpha[-1,1,0]^{\mathrm{T}}$; $\lambda_2 = -1$, $\mathbf{e}_2 = \beta[-1,-1,1]^{\mathrm{T}}$; $\lambda_3 = 2$, $\mathbf{e}_3 = \gamma[1,1,2]^{\mathrm{T}}$. Each eigenspace one-dimensional.
(s) $\lambda_1 = 2$ (multiplicity 3), $\mathbf{e}_1 = \alpha[0,1,0]^{\mathrm{T}}$. One-dimensional eigenspace.
(v) $\lambda_1 = 0$ (multiplicity 4), $\mathbf{e}_1 = [\gamma,\beta,\alpha,0]^{\mathrm{T}} = \alpha[0,0,1,0]^{\mathrm{T}} + \beta[0,1,0,0]^{\mathrm{T}} + \gamma[1,0,0,0]^{\mathrm{T}}$. Three-dimensional eigenspace.

4. (a) No (b) Yes, $\lambda = -6$ (c) No

5. (a) $\mathbf{e} = \alpha[-1,-1,1,1]^{\mathrm{T}}$, one-dimensional (c) $\mathbf{e} = \alpha[1,0,0,-1]^{\mathrm{T}} + \beta[1,-1,0,0]^{\mathrm{T}} + \gamma[1,0,-1,0]^{\mathrm{T}}$, three-dimensional (e) $\mathbf{e} = \alpha[0,0,1,5]^{\mathrm{T}}$, one-dimensional

7. No. Let $\mathbf{Ae} = \lambda_1\mathbf{e}$ and $\mathbf{Ae} = \lambda_2\mathbf{e}$. Subtraction gives $(\lambda_1 - \lambda_2)\mathbf{e} = \mathbf{0}$. If the λ's are different, then it follows that $\mathbf{e} = \mathbf{0}$, which is impossible since an eigenvector must be nonzero.

10. (a) $\lambda_1 = 2i$, $\mathbf{e}_1 = \alpha[-2i,1]^{\mathrm{T}}$; $\lambda_2 = -2i$, $\mathbf{e}_2 = \beta[2i,1]^{\mathrm{T}}$
(c) $\lambda_1 = 3+2i$, $\mathbf{e}_1 = \alpha[-i,1]^{\mathrm{T}}$; $\lambda_2 = 3-2i$, $\mathbf{e}_2 = \beta[i,1]^{\mathrm{T}}$
Scaling the $\mathbf{e}$'s, you may prefer $\mathbf{e}_1 = \alpha[1,i]^{\mathrm{T}}$ and $\mathbf{e}_2 = \beta[1,-i]^{\mathrm{T}}$. (e) $\lambda_1 = 0$, $\mathbf{e}_1 = \alpha[5,1,0]^{\mathrm{T}}$; $\lambda_2 = i$, $\mathbf{e}_2 = \beta[1,0,i]^{\mathrm{T}}$; $\lambda_3 = -i$, $\mathbf{e}_3 = \gamma[1,0,-i]^{\mathrm{T}}$

Section 4.7

8. (a) $\mathbf{x}(t) = C_1\begin{bmatrix}-2\\1\end{bmatrix} + C_2\begin{bmatrix}1\\3\end{bmatrix}e^{7t}$

(c) $\mathbf{x}(t) = C_1\begin{bmatrix}1\\-4\end{bmatrix} + C_2\begin{bmatrix}3\\-5\end{bmatrix}e^{7t}$

(e) $\mathbf{x}(t) = C_1\begin{bmatrix}1\\-4\\5\end{bmatrix} + C_2\begin{bmatrix}1\\-3\\2\end{bmatrix}e^{t} + C_3\begin{bmatrix}1\\-2\\1\end{bmatrix}e^{2t}$

(g) $\mathbf{x}(t) = C_1\begin{bmatrix}0\\1\\0\end{bmatrix}e^{t} + C_2\begin{bmatrix}-2\\0\\1\end{bmatrix}e^{t} + C_3\begin{bmatrix}1\\1\\1\end{bmatrix}e^{4t}$

(i) $\mathbf{x}(t) = C_1\begin{bmatrix}-1\\0\\1\end{bmatrix}e^{3t} + C_2\begin{bmatrix}-1\\1\\0\end{bmatrix}e^{3t} + C_3\begin{bmatrix}1\\1\\1\end{bmatrix}e^{6t}$

(k) $\mathbf{x}(t) = C_1\begin{bmatrix}-2\\1\\0\end{bmatrix} + C_2\begin{bmatrix}-3\\0\\1\end{bmatrix} + C_3\begin{bmatrix}1\\1\\3\end{bmatrix}e^{12t}$

9. (a) $\mathbf{x}(t) = \beta\begin{bmatrix}2\\1\end{bmatrix}e^{4t} + \alpha\begin{bmatrix}-1+2t\\t\end{bmatrix}e^{4t}$

(c) $\mathbf{x}(t) = \beta\begin{bmatrix}-1\\2\end{bmatrix}e^{t} + \alpha\begin{bmatrix}1+2t\\-4t\end{bmatrix}e^{t}$

(e) $\mathbf{x}(t) = \beta\begin{bmatrix}1\\3\end{bmatrix}e^{-2t} + \alpha\begin{bmatrix}1+3t\\9t\end{bmatrix}e^{-2t}$

(g) $\mathbf{x}(t) = C_1\begin{bmatrix}1\\-1\\0\end{bmatrix}e^{t} + C_2\begin{bmatrix}1+2t\\-t\\t\end{bmatrix}e^{-6t} + C_3\begin{bmatrix}2\\-1\\1\end{bmatrix}e^{-6t}$

(i) $\mathbf{x}(t) = C_1\begin{bmatrix}1\\3\\1\end{bmatrix}e^{-t} + C_2\begin{bmatrix}t-1\\1\\4t\end{bmatrix}e^{2t} + C_3\begin{bmatrix}1\\0\\4\end{bmatrix}e^{2t}$

(k) $\mathbf{x}(t) = C_1\begin{bmatrix}6-4t+t^2\\-4+2t\\t^2\end{bmatrix}e^{t} + C_2\begin{bmatrix}t-2\\1\\t\end{bmatrix}e^{t} + C_3\begin{bmatrix}1\\0\\1\end{bmatrix}e^{t}$

Section 4.8

1. (a) $\mathbf{u}\cdot\mathbf{v} = 10$, $\|\mathbf{u}\| = 5$, $\|\mathbf{v}\| = \sqrt{5}$, $\theta = 0.464\,\mathrm{rad} = 26.6^\circ$
(c) $\mathbf{u}\cdot\mathbf{v} = -16$, $\|\mathbf{u}\| = \sqrt{14}$, $\|\mathbf{v}\| = \sqrt{29}$, $\theta = 2.49\,\mathrm{rad} = 143^\circ$
(h) $\mathbf{u}\cdot\mathbf{v} = -30$, $\|\mathbf{u}\| = 4$, $\|\mathbf{v}\| = \sqrt{77}$, $\theta = 2.60\,\mathrm{rad} = 149^\circ$
(m) $\mathbf{u}\cdot\mathbf{v} = 58$, $\|\mathbf{u}\| = \sqrt{88}$, $\|\mathbf{v}\| = \sqrt{75}$, $\theta = 0.776\,\mathrm{rad} = 44.4^\circ$

2. (a) orthogonal (d) not orthogonal (g) orthogonal (j) orthogonal

9. (a) $(\mathbf{A}+\mathbf{B})^2 = (\mathbf{A}+\mathbf{B})(\mathbf{A}+\mathbf{B}) = \mathbf{A}(\mathbf{A}+\mathbf{B}) + \mathbf{B}(\mathbf{A}+\mathbf{B}) = \mathbf{A}^2 + \mathbf{AB} + \mathbf{BA} + \mathbf{B}^2 = \mathbf{A}^2 + 2\mathbf{AB} + \mathbf{B}^2$, only if $\mathbf{AB} = \mathbf{BA}$. [The second equality followed from (22c) and the third from (22d).]

10. (a) Don't use guesswork, be systematic. Choose a simple $\mathbf{A}$, such as $\begin{bmatrix}0&0\\1&0\end{bmatrix}$, and let $\mathbf{B} = \begin{bmatrix}a&b\\c&d\end{bmatrix}$ and $\mathbf{C} = \begin{bmatrix}e&f\\g&h\end{bmatrix}$. Now try to arrange $\mathbf{B}$ and $\mathbf{C}$ so that $\mathbf{AB} = \mathbf{AC}$ without $\mathbf{B}$ being the same as $\mathbf{C}$. Working out $\mathbf{AB}$ and $\mathbf{AC}$, we find that we need $a = e$ and $b = f$, with c, d, g, h remaining free. Thus, for the $\mathbf{A}$ given above, if $\mathbf{B}$ and $\mathbf{C}$ are of the forms $\mathbf{B} = \begin{bmatrix}a&b\\c&d\end{bmatrix}$ and $\mathbf{C} = \begin{bmatrix}a&b\\g&h\end{bmatrix}$, respectively, then $\mathbf{AB} = \mathbf{AC}$ even if $\mathbf{B} \neq \mathbf{C}$. For instance, we could take $\mathbf{B} = \begin{bmatrix}1&2\\3&4\end{bmatrix}$ and $\mathbf{C} = \begin{bmatrix}1&2\\8&5\end{bmatrix}$.

12. (a) The most general $\mathbf{B}$ is $\begin{bmatrix}-3c/2&-3d/2\\c&d\end{bmatrix}$, with c and d arbitrary.

14. (a) $\mathbf{A}^{100} = \begin{bmatrix}2^{99}&2^{99}\\2^{99}&2^{99}\end{bmatrix}$

(d) $\mathbf{D}^2 = \begin{bmatrix} 0 & 0 & 4 & 17 \\ 0 & 0 & 0 & 24 \\ 0 & 0 & 0 & 0 \\ 0 & 0 & 0 & 0 \end{bmatrix}$, $\mathbf{D}^3 = \begin{bmatrix} 0 & 0 & 0 & 24 \\ 0 & 0 & 0 & 0 \\ 0 & 0 & 0 & 0 \\ 0 & 0 & 0 & 0 \end{bmatrix}$,
and $\mathbf{D}^n$ is the zero matrix if $n \geq 4$.

16. (a) $\mathbf{A} = \begin{bmatrix} 1 & 0 & 0 & -3 \\ 0 & 2 & 1 & 0 \end{bmatrix}$

(c) $\mathbf{A} = \begin{bmatrix} 1 & 1 & 0 & 0 \\ 0 & 1 & 1 & 0 \\ 0 & 0 & 1 & 1 \end{bmatrix}$ (e) $\mathbf{A} = \begin{bmatrix} 0 & 0 & 0 & 1 \\ 0 & 0 & 1 & 0 \\ 0 & 1 & 0 & 0 \\ 1 & 0 & 0 & 0 \end{bmatrix}$

17. (a) $[25, 305, 201, 52]^{\mathrm{T}}$
(c) $\begin{bmatrix} -7836 & 1215 & -4716 & -532 \\ 16520 & -10805 & 12940 & 3120 \\ 15482 & -8171 & 11416 & 2450 \\ 519 & -1317 & 762 & 335 \end{bmatrix}$
(e) $\begin{bmatrix} -1435 & -1560 \\ 3120 & 1685 \end{bmatrix}$

18. (a) $\lambda_1 = 4$, $\mathbf{e}_1 = [1, 2]^{\mathrm{T}}$, $\lambda_2 = -1$, $\mathbf{e}_2 = [-2, 1]^{\mathrm{T}}$;
$\mathbf{e}_1 \cdot \mathbf{e}_2 = -2 + 2 = 0$
(c) $\lambda_1 = 3$, $\mathbf{e}_1 = [1, 2]^{\mathrm{T}}$, $\lambda_2 = -2$, $\mathbf{e}_2 = [-2, 1]^{\mathrm{T}}$;
$\mathbf{e}_1 \cdot \mathbf{e}_2 = -2 + 2 = 0$
(e) $\lambda_1 = 0$, $\mathbf{e}_1 = \alpha[-1, 0, 1]^{\mathrm{T}} + \beta[-1, 1, 0]^{\mathrm{T}}$, $\lambda_2 = 3$,
$\mathbf{e}_2 = [1, 1, 1]^{\mathrm{T}}$; $\mathbf{e}_1 \cdot \mathbf{e}_2 = \alpha(-1 + 1) + \beta(-1 + 1) = 0$
Geometrically, this is in 3-space; the eigenspace $\mathbf{e}_2$ is a plane through the origin, and the eigenspace $\mathbf{e}_1$ is the line through the origin that pierces the plane and is perpendicular to it.
(g) $\lambda_1 = -1$, $\mathbf{e}_1 = [-1, -1, 1]^{\mathrm{T}}$, $\lambda_2 = 0$,
$\mathbf{e}_2 = [-1, 1, 0]^{\mathrm{T}}$, $\lambda_3 = 2$, $\mathbf{e}_3 = [1, 1, 2]^{\mathrm{T}}$;
$\mathbf{e}_1 \cdot \mathbf{e}_2 = 0$, $\mathbf{e}_1 \cdot \mathbf{e}_3 = 0$, $\mathbf{e}_2 \cdot \mathbf{e}_3 = 0$

20. (a) $\begin{bmatrix} -1/5 & 2/5 \\ 3/10 & -1/10 \end{bmatrix}$ (c) $\begin{bmatrix} 1/4 & 1/4 \\ 1/2 & -1/2 \end{bmatrix}$

(e) $\begin{bmatrix} -1/2 & 1/4 & 1/4 \\ 1/2 & -3/4 & 1/4 \\ 1/2 & 5/4 & -3/4 \end{bmatrix}$ (g) $\begin{bmatrix} 1 & -1 & 0 \\ 0 & 1 & -1 \\ 0 & 0 & 1 \end{bmatrix}$

(i) $\mathbf{A}$ is singular (k) $\begin{bmatrix} 5/2 & 0 & -1 \\ 0 & 1/3 & 0 \\ -2 & 0 & 1 \end{bmatrix}$

(m) $\begin{bmatrix} 0 & 0 & 0 & 1/4 \\ 0 & 0 & 1/3 & -1/4 \\ 0 & 1/2 & -1/3 & 0 \\ 1 & -1/2 & 0 & 0 \end{bmatrix}$

24. (a) It is given that $\mathbf{Ae} = \lambda\mathbf{e}$. Multiply through by α, getting $\alpha\mathbf{Ae} = \alpha\lambda\mathbf{e}$ or, $(\alpha\mathbf{A})\mathbf{e} = (\alpha\lambda)\mathbf{e}$, from which it follows that the matrix $\alpha\mathbf{A}$ has an eigenvalue $\alpha\lambda$ and eigenvector $\mathbf{e}$.

26. See the answers to the corresponding parts of Exercise 20.

Section 4.9

2. (a) $\mathbf{x}(t) = [12 - 24t + 48t^2, 96t, 24]^{\mathrm{T}}$

5. (a) $e^{\mathbf{A}t} = \begin{bmatrix} 1 & 3t & 2t + 6t^2 \\ 0 & 1 & 4t \\ 0 & 0 & 1 \end{bmatrix}$, so a fundamental set is
$[1, 0, 0]^{\mathrm{T}}$, $[3t, 1, 0]^{\mathrm{T}}$, $[2t + 6t^2, 4t, 1]^{\mathrm{T}}$

7. (a) $e^{\mathbf{A}t} = \dfrac{1}{2}\begin{bmatrix} e^{2t} + e^{4t} & -e^{2t} + e^{4t} \\ -e^{2t} + e^{4t} & e^{2t} + e^{4t} \end{bmatrix}$,
$\mathbf{x}(t) = (1/2)[e^{2t} + e^{4t}, -e^{2t} + e^{4t}]^{\mathrm{T}}$
(c) $e^{\mathbf{A}t} = \dfrac{1}{2}\begin{bmatrix} 1 + e^{2t} & -1 + e^{2t} \\ -1 + e^{2t} & 1 + e^{2t} \end{bmatrix}$,
$\mathbf{x}(t) = (3/2)[1 - e^{2t}, -1 - e^{2t}]^{\mathrm{T}}$
(e) $e^{\mathbf{A}t} =$
$\dfrac{1}{6}\begin{bmatrix} e^{4t} + 3 + 2e^{-2t} & e^{4t} - 3 + 2e^{-2t} & 2e^{4t} - 2e^{-2t} \\ e^{4t} - 3 + 2e^{-2t} & e^{4t} + 3 + 2e^{-2t} & 2e^{4t} - 2e^{-2t} \\ 2e^{4t} - 2e^{-2t} & 2e^{4t} - 2e^{-2t} & 4e^{4t} + 2e^{-2t} \end{bmatrix}$,
$\mathbf{x}(t) = \dfrac{4}{3}[e^{4t} + 3 + 2e^{-2t}, e^{4t} - 3 + 2e^{-2t}, 2e^{4t} - 2e^{-2t}]^{\mathrm{T}}$
(g) $e^{\mathbf{A}t} = \dfrac{1}{6}\begin{bmatrix} e^{6t} + 5 & e^{6t} - 1 & e^{6t} - 1 \\ 2e^{6t} - 2 & 2e^{6t} + 4 & 2e^{6t} - 2 \\ 3e^{6t} - 3 & 3e^{6t} - 3 & 3e^{6t} + 3 \end{bmatrix}$,
$\mathbf{x}(t) = (5/6)[1 - e^{6t}, 2 - 2e^{6t}, -3 - 3e^{6t}]^{\mathrm{T}}$

Section 4.10

2. (a) Do it both ways; the results are the same:
$(\mathbf{AB})' = \begin{bmatrix} 3t^2 \sin t + 2e^t + t^3 \cos t & -4t^3 \\ e^t(\sin t + 4 + \cos t) & -(t + 1)e^t \end{bmatrix}$
(c) $(\mathbf{Ax})' = [2, 4]^{\mathrm{T}}$

7. (a) $\mathbf{x}(t) = \begin{bmatrix} 2 & e^{7t} \\ -1 & 3e^{7t} \end{bmatrix}\mathbf{c}_0 + \begin{bmatrix} 5e^t - 2t - 2/7 \\ -3e^t + t - 6/7 \end{bmatrix}$

(c) $\mathbf{x}(t) = \begin{bmatrix} e^t & 2e^{8t} \\ -3e^t & e^{8t} \end{bmatrix}\mathbf{c}_0 + \begin{bmatrix} -18 - 16t \\ 55 + 56t \end{bmatrix}$

(e) $\mathbf{x}(t) = \begin{bmatrix} e^{4t} & e^{6t} \\ -e^{4t} & e^{6t} \end{bmatrix}\mathbf{c}_0 + \begin{bmatrix} -4e^t \\ e^t \end{bmatrix}$

(g) $\mathbf{x}(t) = \begin{bmatrix} 2e^{5t} & e^{-t} \\ e^{5t} & -e^{-t} \end{bmatrix}\mathbf{c}_0 + \begin{bmatrix} -2e^t + 4e^{2t} \\ e^t - e^{2t} \end{bmatrix}$

(i) $\mathbf{x}(t) = \begin{bmatrix} 1 & t^3 \\ 0 & 3t^2 \end{bmatrix}\mathbf{c}_0 + \begin{bmatrix} 2t^3 \\ 0 \end{bmatrix}$

(k) $\mathbf{x}(t) = \begin{bmatrix} 1 & e^{-t} & e^{2t} \\ -1 & e^{-t} & e^{2t} \\ 0 & -e^{-t} & 2e^{2t} \end{bmatrix}\mathbf{c}_0 + \begin{bmatrix} -2e^t \\ -2e^t \\ -2e^t \end{bmatrix}$

CHAPTER 5

Section 5.2

3. (a) $1/s^2$ for $s>0$ (c) $e^{-3}/(s-2)$ for $s>2$ (e) $[(\cos 2)s+\sin 2]/(s^2+1)$ for $s>0$ (g) $\frac{1}{2}\left(\frac{e}{s-1}-\frac{e^{-1}}{s+1}\right)$ or $\frac{(\sinh 1)s+\cosh 1}{s^2-1}$ for $s>1$ (i) $\frac{s+1}{(s-1)(s+3)}$ or $\frac{s+1}{(s+1)^2-4}$ for $s>1$

4. (a) $50(e^{-3s}-e^{-5s})/s$; yes (c) $100e^{-5s}/s$; yes (e) $[1-e^{-2(s-1)}]/(s-1)$; yes (g) $e^{-3s}/(s+1)$; yes

7. (a) $12/(s^2+4)$ (d) $1/s+6/(s+1)^4$ (g) $2s/(s^2+1)^2+2s/(s^2-1)^2$ (j) $s/(s^2+25)-(4s^2-100)/(s^2+25)^2$ (m) $1/(s-1)-1/(s-2)-4/s^2$ (p) $2a^2/[s(s^2+4a^2)]$

8. (a) Yes; any $K\geq 5$, any $c\geq 4$, any $T>0$ (c) Yes; $\sinh 2t=(e^{2t}-e^{-2t})/2<e^{2t}/2$ for all $t>0$, so can take $K=0.5$ and $c=2$ (e) $\frac{\sinh t^2}{e^{ct}}\sim\frac{1}{2}e^{t^2-ct}\to\infty$ no matter how large we make c. Thus, $\sinh t^2$ is not of exponential order (g) Yes; $|\cos t^3|\leq 1e^{0t}$, so can take any $K\geq 1$, any $c\geq 0$, any $T>0$ (i) Yes; the function tends to 1 from below as $t\to\infty$ and $|f(t)|\leq 1e^{0t}$ for all $t>0$, so can take any $K\geq 1$, any $c\geq 0$, and any $T>0$ (k) Yes, can take any $K\geq 9$, any $c\geq 2$, and any $T>0$

9. (a) Using items 1 and 7, or just 7, get $t^2/2-5t$ (c) Using item 7, get $5-t^3$ (e) Using item 17, get $(5/39366)t^3e^{t/9}$ (g) Using item 17, get $t^4e^{-t}/24$ (i) Using items 5 and 6, get $(3\sinh 2t-7\cosh 2t)/2$ (k) Using items 5 and 7, get $t^2+4\sqrt{3}\sinh\sqrt{3}t$ (m) Using item 17, get $(51t^3/2+30t^2+6t)e^{5t}$

10. (a) $\frac{1}{4}\frac{1}{s-3}-\frac{1}{4}\frac{1}{s+1}$ gives $\frac{1}{4}e^{3t}-\frac{1}{4}e^{-t}$ or $\frac{1}{2}e^t\sinh 2t$ (c) $\frac{4}{s-2}-\frac{4}{s-1}$ gives $4e^{2t}-4e^t$ (e) $\frac{1}{s}-\frac{1}{s+6}$ gives $1-e^{-6t}$ (g) $3e^t-e^{-2t}$

Section 5.3

2. (a) $x(t)=3-\cos t$ (c) $x(t)=5e^{6t}-2e^t$ (e) $x(t)=2-t+t^3$ (g) $x(t)=-1+2t+\cos t-2\sin t$ (i) $x(t)=(1+4t)e^{-t}-e^{-3t}$ (k) $x(t)=2t^3e^{2t}$ (m) $x(t)=\sinh t-(t^2+t)e^{-t}$ (o) $x(t)=2-6t-t^3+6\sinh t$

4. (a) $x(t)=12\cosh(t-5)$ (c) $x(t)=e^{3t}+3e^{-4-t}-4e^{-3}$

5. (a) $x(t)=7+t-7e^{-2t}$, $y(t)=-2+t+7e^{-2t}$ (c) $x(t)=10-2e^{3t}-2e^{-3t}-\cos t$, $y(t)=9\sin t+12\sinh 3t$ (e) $x(t)=2+2\cosh\sqrt{2}t$, $y(t)=-2+2\cosh\sqrt{2}t$

9. (a) Let $U/\alpha\equiv u$ and $V/\alpha\equiv v$. With the help of partial fractions, get $X(s)=\frac{v}{s}+\frac{u\alpha}{s^2+\alpha^2}-\frac{vs}{s^2+\alpha^2}$, $Y(s)=-\frac{u}{s}+\frac{v\alpha}{s^2+\alpha^2}+\frac{us}{s^2+\alpha^2}$, and $Z(s)=0$. These give $x(t)=v+u\sin\alpha t-v\cos\alpha t$, $y(t)=-u+v\sin\alpha t+u\cos\alpha t$, $z(t)=0$

11. (a) $y(x)=10-e^x$ (c) $y(x)=2x$

14. (d) $x(t)=3t^2+3t^{-2}$

Section 5.4

1. (a) $f(t)=50t+H(t-20)(1000-50t)$ (c) $f(t)=10t^2+H(t-2)(25-10t^2)+H(t-5)5$

2. (a) 0 on $0<t<\pi/2$, then $\sin t$ on $\pi/2<t<\infty$ (c) Graph is identical to $\cosh t$ (e) Same as Fig. 6c, with $a=\pi$ (m) 1 for $0<t<1$ and $2<t<\infty$, 0 for $1<t<2$

3. (a) $x'(1.3)=2.6$, $x'(4)=8-2e^{-8}$, $x''(4)=2+4e^{-8}$ (c) $x'(1.3)=0$, $x'(4)=6\cos 4$, $x''(4)=-6\sin 4$

4. (a) $e^{-s}/(s^2-1)$ (c) $4e^{6-2s}/(s-3)$ (e) $\frac{e^{-2s}}{s^2+1}[(\cos 2)s-\sin 2]$ (g) $e^{-s}\left(\frac{1}{s}+\frac{2}{s^2}+\frac{2}{s^3}\right)$ (i) $e^{-2s}\left(\frac{5}{s^2}+\frac{17}{s}\right)$

5. (a) $H(t-4)(t-4)^2/2$ (c) $H(t-5)\left(e^{3(t-5)}-1\right)/3$ (e) $H(t-3)(7-t)$ (g) $H(t-1)(t-1)-2H(t-2)(t-2)^2$

10. (a) $x(t)=x_0e^{-2t}+25\left[H(t-1)(1-e^{-2t})-H(t-2)(1-e^{-2(t-2)})\right]$ (c) $x(t)=x_0e^t-40+40e^t+20H(t-5)(1-e^{t-5})+20H(t-10)(1-e^{t-10})$ (e) $x(t)=\frac{1}{17}\left(e^{4t}-\cos t-4\sin t+H(t-\pi)\left[e^{4(t-\pi)}-\cos(t-\pi)-4\sin(t-\pi)\right]\right)$ (g) $x(t)=\frac{1}{3}\left[H(t-1)(e^{3(t-1)}-1)+H(t-2)(e^{3(t-2)}-1)+H(t-3)(e^{3(t-3)}-1)\right]$

12. (a) $x(t)=50H(t-4)\left[\cosh(t-4)-1\right]-50H(t-6)\left[\cosh(t-6)-1\right]$ (c) $x(t)=-11/2-5t+(11/2)e^{2t}+H(t-4)\left[-5/2-5(t-4)+5e^{2(t-4)}/2\right]$ (e) $x(t)=H(t-1)[3+(t-1)-3e^{t-1}+2(t-1)e^{t-1}]$ (g) $x(t)=-5t^2+10H(t-3)[1+(t-3)^2/2-\cosh(t-3)]$

14. (a) Obtain $X(s) = \dfrac{1}{(s+1)(s^2+1)}\big[1 + 2e^{-\pi s} + 2e^{-2\pi s} + 2e^{-3\pi s} + \cdots\big]$. Let $g(t) = (e^{-t} + \sin t - \cos t)/2$. Then, $x(t) = g(t) + 2H(t-\pi)g(t-\pi) + 2H(t-2\pi)g(t-2\pi)$, as far as 3π, because the next terms have $H(t-3\pi)$, $H(t-4\pi)$, $\ldots$, which are zero for $t > 3\pi$:

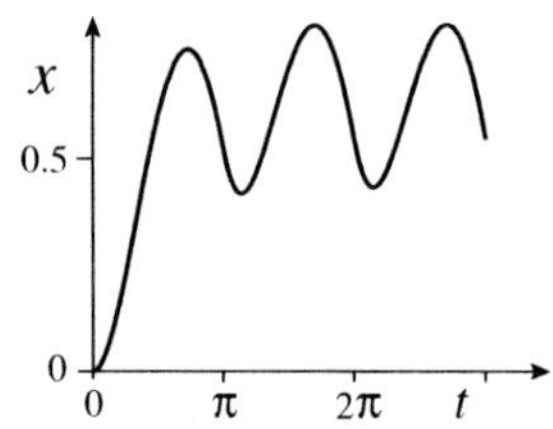

(c) With $g(t) \equiv t - 1 + e^{-t}$, get $x(t) = [g(t) + H(t-2)g(t-2) + H(t-4)g(t-4)] - [H(t-1)(t-1) + H(t-3)(t-3) + H(t-5)(t-5)]$, on the plotting interval $0 < t < 6$:

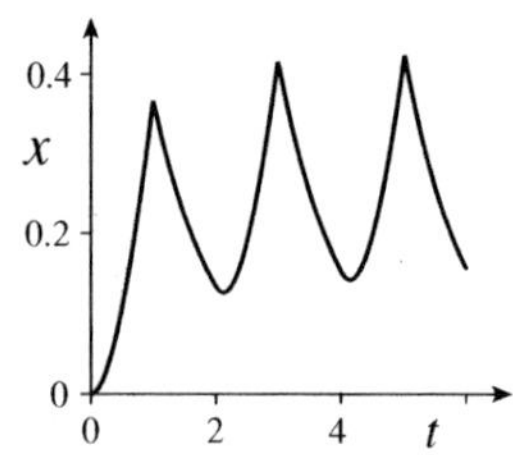

(e) With $g(t) \equiv 50(1 - e^{-t})$, get $x(t) = [H(t-9)g(t-9) + H(t-19)g(t-19) + H(t-29)g(t-29)] - [H(t-10)g(t-10) + H(t-20)g(t-20)]$, on the plotting interval:

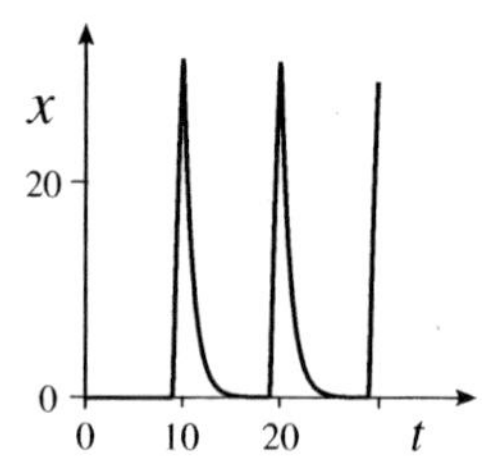

(g) With $g(t) = 1 - e^{-t}$, $x(t) = e^{-t} + t - 1 - H(t-1)g(t-1) - H(t-2)g(t-2)$ through $t = 3$. The graph shows an increasing function of t, though not monotonely increasing, with "peaks" at $t = 1, 2 \ldots$.

15. (b) $x(t) - x(t+3) = e^{-(t+3)} - e^{-(t+1)}$ is not zero [so $x(t)$ is not periodic with period 3], but it does tend to zero as $t \to \infty$, so it *tends* to a periodic function with period 3.

Section 5.5

1. We'll do (1.1) and (1.2) and leave (1.3) and (1.4) for you. (1.1) must be true because $\mathcal{L}\{f(t) * g(t)\} = F(s)G(s)$, but you could check it if you wish: $\cos t * t = 1 - \cos t$ and $\mathcal{L}\{f * g\} = 1/s - s/(s^2+1) = 1/[s(s^2+1)]$. Next, $F(s)G(s) = [s/(s^2+1)][1/s^2]$, which is indeed the same. Now (1.2): We don't expect (1.2) to be true because $\mathcal{L}\{f(t)g(t)\} \neq \mathcal{L}\{f(t)\}\mathcal{L}\{g(t)\}$ in general. But it *could* be true in a given case, so we need to work it out and see. We find that $\mathcal{L}\{te^t\} = 1/(s-1)^2$, which is not the same as $\mathcal{L}\{t\}\mathcal{L}\{e^t\} = (1/s^2)[1/(s-1)]$. Thus, (1.2) is not true.

3. (a) $3 - 3e^{-8t}$ (c) $e^{-t} - e^{-2t}$ (e) $1 - e^{-t}$ (g) $4e^t - 4e^{-2t}$ (i) $12\sin t$

6. (a) $\mathcal{L}^{-1}\{\frac{1}{s}\frac{1}{s}\frac{1}{s}\} = 1 * 1 * 1 = (1 * 1) * 1 = \left(\int_0^t (1)(1)d\tau\right) * 1 = t * 1 = \int_0^t \tau(1)d\tau = t^2/2$ (c) $(\cosh 3t - 1)/9$

7. (a) The key is to identify the integral as a convolution integral (if indeed it is one). $f(t) = \sin t * e^t$, so $F(s) = \mathcal{L}\{\sin t\}\mathcal{L}\{e^t\} = \dfrac{1}{(s^2+1)(s-1)}$. (c) $F(s) = \dfrac{2}{(s^2+1)^2}$ (e) $F(s) = 4/[(s^2+16)(s^2-1)]$

9. (a) $x(t) = -t - t^2/2$ (c) $x(t) = -10$ (e) $x(t) = \cosh t - \cos t$

Section 5.6

1. (a) $x(t) = 500H(t-2)e^{-(t-2)}$
(c) $x(t) = 5e^t + H(t-2)e^{t-2} - H(t-3)(1 - e^{t-3})$
(e) $x(t) = 1 - e^{2t} + 4H(t-5)\left[e^{2(t-5)} - e^{t-5}\right]$
(g) $x(t) = H(t-1)[1 - e^{-(t-1)}] - H(t-2)[1 - e^{-(t-2)}]$
(i) $x(t) = 6 + H(t-5)(t-5)^2$

2. (a) 13 (c) $\cos 2$ (e) $1 + 3\pi^2$

4. (a) $Q(s) = 1/(s+1)$, $q(t) = e^{-t}$ (c) $Q(s) = 1/s^2$, $q(t) = t$ (e) $Q(s) = 1/[s(s+1)]$, $q(t) = 1 - e^{-t}$ (g) $Q(s) = 1/[s(s^2+1)]$, $q(t) = 1 - \cos t$

5. (a) Identify $n = 1$, $p_0 = 6/2 = 3$, $t_0 = 0$, so (20) gives the jump in $x(t)$ at $t = 0$ as 1/3. Next, solve the IVP by Laplace transform and get $x(t) = (16/3)e^{t/6}$. From that, the jump in $x(t)$ at $t = 0$ as $16/3 - 5 = 1/3$, which does agree with (20).
(c) Find that $x(t) = 5e^{2t} + (1/2)H(t-3)e^{2(t-3)}$, which gives the jump in $x(t)$ at $t = 3$ as 1/2. Also, $p_0 = 2$, so the jump is $1/p_0 = 1/2$.
(e) Find that $x(t) = 3 - 4t$ for $t > 0$, which gives the jump in $x'(t)$ at $t = 0$ as $-4 - (-5) = 1$. Also, $p_0 = 1$, so the jump is $1/p_0 = 1$.
(g) Find that $x(t) = 2e^{t/2} + e^{-t} + (1/6)H(t-3)[e^{(t-3)/2} - e^{-(t-3)}]$, which gives the jump in $x'(t)$ at $t = 3$ as $1/4 - 0 = 1/4$. Also, $p_0 = 4$, so the jump is $1/p_0 = 1/4$.
(i) Find that $x(t) = 5t^2 + 4t + 3 + H(t-2)(t-2)^2/2$, which

gives the jump in $x''(t)$ at $t = 2$ as $1-0 = 1$. Also, $p_0 = 1$, so the jump is $1/p_0 = 1$.

CHAPTER 6

Section 6.2

1. (a) $R = 1$ (d) $R = \infty$ (g) $R = \infty$ (j) $R = \sqrt{2}$

2. (a) Holds in $|x| < 1$. "?" is $\left(\frac{1}{n+1} + \frac{1}{e^n}\right) x^n$

(d) Holds in $|x| < 1$. "?" is $\left(n^2 - \frac{3n}{2^n}\right) x^{n-1}$

3. (a) $x + 3x^2 + 6x^3 + \cdots$ (c) $1 + (1+e^2)\left(\frac{x}{e}\right) + (1 + e^2 + e^4)\left(\frac{x}{e}\right)^2 + (1 + e^2 + e^4 + e^6)\left(\frac{x}{e}\right)^3 + \cdots$

4. (a) Singular at $x = 0$. For $x = -5$, $R = 5$; for $x = 1$, $R = 1$. (c) Singular at $x = 1, 2$. For $x = 0$, $R = 1$; for $x = 50$, $R = 48$. (e) Singular at $x = 1, \pm i$. For $x = -1$, $R = \sqrt{2}$; for $x = 0$, $R = 1$; for $x = 4$, $R = 3$. (g) Singular at $x = -4, -1, 1$. For $x = -2$, $R = 1$; for $x = 0$, $R = 1$; for $x = 5$, $R = 4$. (i) Singular at $\pm i$. For $x = 0$, $R = 1$; for $x = 2$, $R = \sqrt{5}$; for $x = 10$, $R = \sqrt{101}$.

Section 6.3

1. (a) Recursion formula: $a_{n+2} = 9a_n/[(n+2)(n+1)]$, $n = 0, 1, \ldots$; $y_1(x) = 1 + 9x^2/2! + 9^2x^4/4! + \cdots$, $y_2(x) = x + 9x^3/3! + \cdots$. Though not asked for, note that the foregoing is consistent with closed-form solution because $y_1(x)$ is $\cosh 3x$ and $y_2(x)$ is $(1/3)\sinh 3x$. Thus, $y(x) = a_0 \cosh 3x + (a_1/3)\sinh 3x = C_1 \cosh 3x + C_2 \sinh 3x$.
(c) Recursion formula: $a_{n+2} = 2a_{n+1}/(n+2)$, $n = 0, 1, \ldots$; $y_1(x) = 1$, $y_2(x) = x + x^2 + 2^2x^3/3! + 2^3x^4/4! + \cdots$. Not asked for, but note that the foregoing is consistent with closed-form solution because $y(x) = a_0(1) + (a_1/2)[(2x) + (2x)^2/2! + (2x)^3/3! + \cdots] = a_0 + (a_1/2)[e^{2x} - 1] = (a_0 - a_1/2) + (a_1/2)e^{2x} = C_1(1) + C_2e^{2x}$.
(e) Recursion formula: $a_{n+2} = -3a_{n+1}/(n+2) + 4a_n/[(n+2)(n+1)]$, $n = 0, 1, \ldots$; $y_1(x) = 1 + 2x^2 - 2x^3 + (13/6)x^4 + \cdots$, $y_2(x) = x - (3/2)x^2 + (13/6)x^3 - (17/8)x^4 + \cdots$.
(g) Recursion formula: $a_{n+2} = 7a_{n+1}/(n+2) - 12a_n/[(n+2)(n+1)]$, $n = 0, 1, \ldots$; $y_1(x) = 1 - 6x^2 - 14x^3 - (37/2)x^4 + \cdots$, $y_2(x) = x + (7/2)x^2 + (37/6)x^3 + (175/24)x^4 + \cdots$.

2. (a) Recursion formula: $a_{n+2} = -(a_n + a_{n-1})/[(n+2)(n+1)]$, $n = 0, 1, \ldots$; $y_1(x) = 1 - x^2/2 - x^3/6 + x^4/24 + \cdots$, $y_2(x) = x - x^3/6 - x^4/12 + \cdots$. $R = \infty$.
(c) Recursion formula: $a_{n+2} = a_{n-1}/[(n+2)(n+1)]$, $n = 0, 1, \ldots$ with $a_{-1} \equiv 0$. $y_1(x) = 1 + x^3/6 + \cdots$, $y_2(x) = x + x^4/12 + \cdots$. $R = \infty$.
(e) Recursion formula: $a_{n+2} = na_{n+1}/[2(n+2)] - a_n/[2(n+2)(n+1)]$, $n = 0, 1, \ldots$; $y_1(x) = 1 - x^2/4 - x^3/24 + \cdots$, $y_2(x) = x - x^3/12 - x^4/48 + \cdots$. R is at least 2.
(g) Recursion formula: $a_{n+2} = (n-6)(n-7)a_{n-6}/[(n+2)(n+1)] - 9a_n/[(n+2)(n+1)]$, $n = 0, 1, \ldots$ with $a_{-1} = a_{-2} = \cdots = a_{-6} \equiv 0$; $y_1(x) = 1 - 9x^2/2! + 81x^4/4! - \cdots$, $y_2(x) = x - 9x^3/3! + \cdots$. R is at least 1.

3. (a) ∞ (d) 0.7 (g) 1

7. (a) $y(x) = a_0(1 + x^2/2 + x^4/24 + \cdots) + a_1(x + x^3/6 + x^5/120 + \cdots) + (x^3/2 + \cdots)$
(c) $y(x) = a_0(1 - x^3/6 + \cdots) + a_1(x - x^2/2 - x^4/12 + \cdots) + (x^2/2 + x^3/6 + x^4/24 - x^5/60 + \cdots)$
(e) $y(x) = a_0(1 + \cdots) + a_1(x - x^3/6 + x^5/40 + \cdots) + (50x^2 - 25x^4/3 + \cdots)$ The first series is only the 1 since $y(x) =$ constant is an obvious homogeneous solution.

Section 6.5

2. (a) Regular singular point at 0 (d) Regular singular points at ± 1 (g) Regular singular point at 0 (j) Irregular singular point at 0

3. (a) $r = -1/2, -1$, case (i). Computer gives (up to but not including terms of order 5, not counting the x^r out in front of the series): $y(x) = C_1 x^{-1}[1 + x + O(x^5)] + C_2 x^{-1/2}\left[1 + \frac{1}{6}x - \frac{1}{120}x^2 + \frac{1}{1680}x^3 - \frac{1}{24192}x^4 + O(x^5)\right]$. Consistent with (17) and (18a). $R = \infty$ for both series.
(d) $r = 0, 3$, case (iii). Computer gives $y(x) = C_1x^3\left[1 - \frac{1}{4}x + \frac{1}{40}x^2 - \frac{1}{720}x^3 + \frac{1}{20160}x^4 + O(x^5)\right] + C_2\Big[\ln x\big[-x^3 + \frac{1}{4}x^4 + O(x^5)\big] + \big[12 + 6x + 3x^2 - \frac{5}{16}x^4 + O(x^5)\big]\Big]$. Consistent with (17) and (18c), with $\kappa \neq 0$. $R = \infty$ for both series.
(g) $r = \pm 1$, case (iii). Computer gives $y(x) = C_1x\big[1 - \frac{1}{3}x + \frac{1}{12}x^2 - \frac{1}{60}x^3 + \frac{1}{360}x^4 + O(x^5)\big] + C_2x^{-1}\big[-2 + 2x - x^2 + \frac{1}{3}x^3 - \frac{1}{12}x^4 + O(x^5)\big]$. Consistent with (17) and (18c), with $\kappa = 0$. $R = \infty$ for both series.
(j) $r = 0, 1/2$, case (i). Computer gives $y(x) = C_1x^{1/2}\big[1 - \frac{1}{4}x + \frac{7}{160}x^2 - \frac{19}{1920}x^3 + \frac{247}{92160}x^4 + O(x^5)\big] + C_2\big[1 - x + \frac{1}{6}x^2 - \frac{1}{30}x^3 + \frac{1}{120}x^4 + O(x^5)\big]$, which is consistent with (17) and (18a). The series converge with $R = 2$ at least.

CHAPTER 7

Section 7.2

1. (a) Trajectories: $x^2 + y^2 = C$

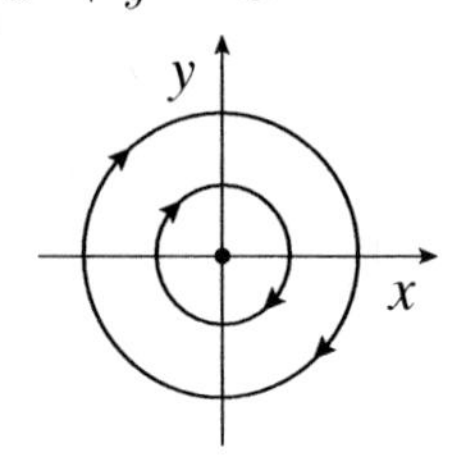

(c) Trajectories: $x^2 + y^2 = C$

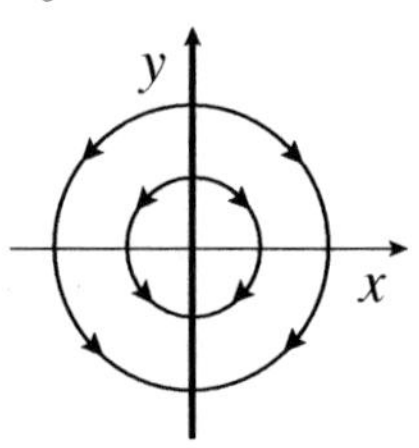

(e) Trajectories: $x + y = C$

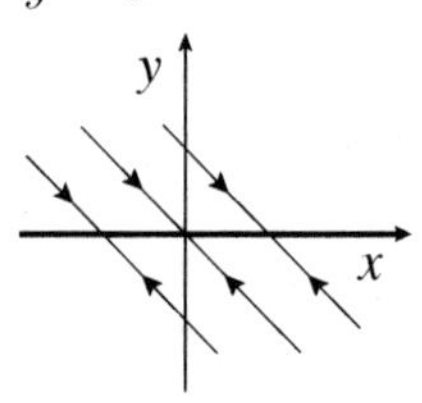

5. On Γ, $y = -m(x - x_0)$, so $x' + mx = mx_0$. Solve, with an initial condition not at $(x_0, 0)$, and show that $x(t) \to x_0$ as $t \to \infty$, not in finite time.

6. Here are a few amplitude-period pairs, to check your results: (1,6.6), (2,8.3), and (2.5,10.3).

7. (a) (0,0) and (1,1) (d) Partial answer: the period of the orbit through (1,0.25) is around 7.

8. (a)

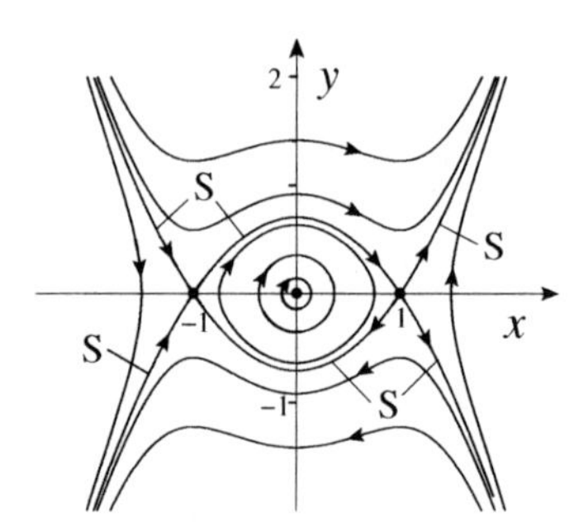

(b) $y = \pm(x^2 - 1)/\sqrt{2}$

9. (a) Stable but not asymptotically stable (c) Asymptotically stable (e) Stable but not asymptotically stable (g) Unstable

11. (c) Our point here is simply that for the small motions (i.e., with $|x| << 1$) the trajectories $x^2 + x^4/2 + y^2 = C$ approach the circles $x^2 + y^2 = C$ because $x^2 + x^4/2 \sim x^2$ as $x \to 0$.

Section 7.3

1. (a) Center (c) Unstable spiral (e) Stable node (g) Saddle (i) Unstable degenerate node

2. (a)

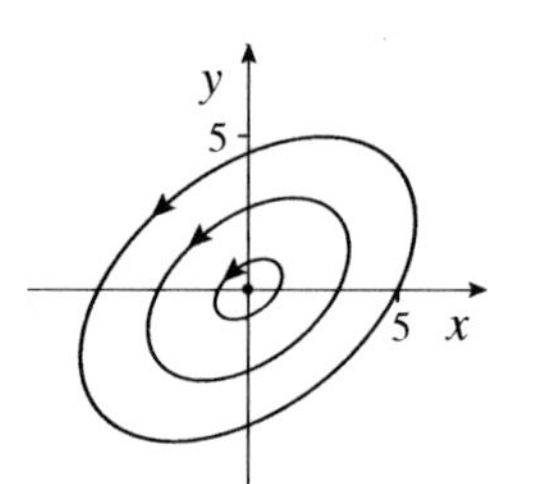

(c)

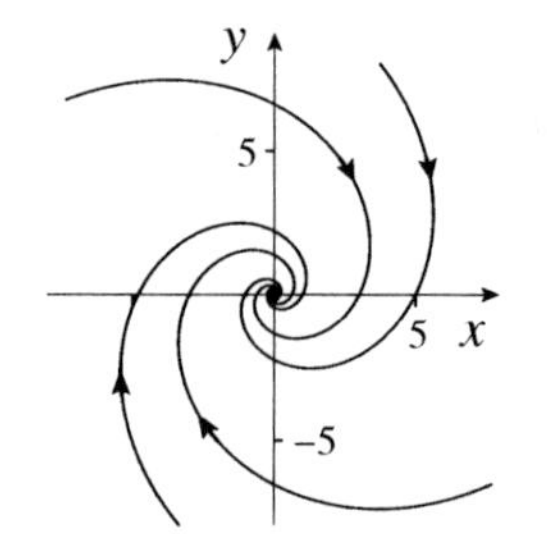

(e)

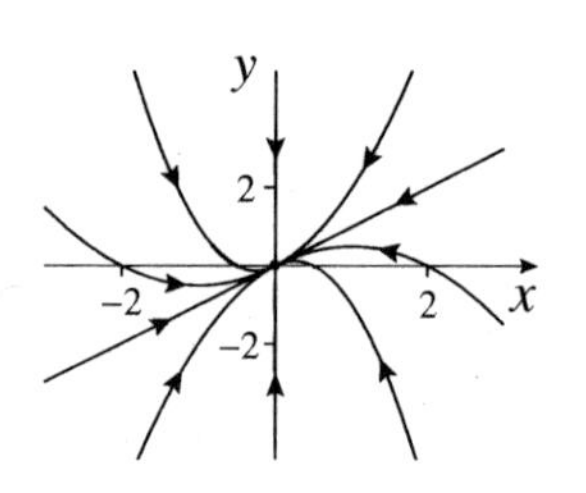

(g)

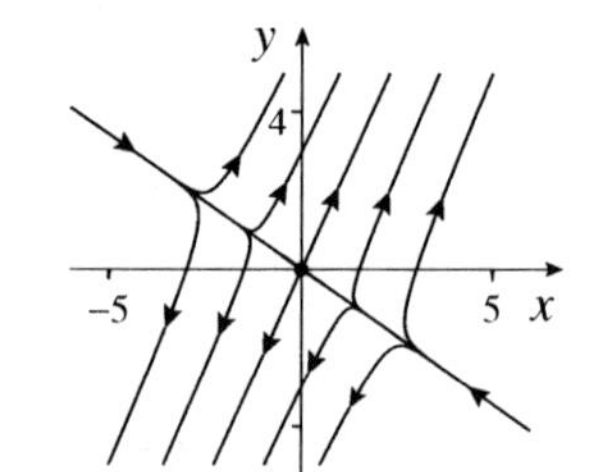

(i)

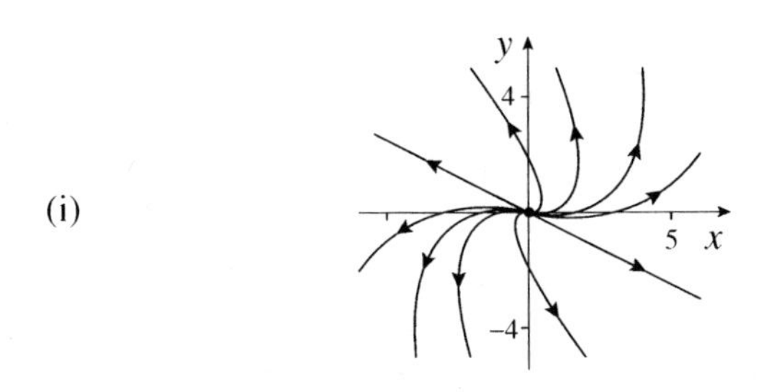

10. (a) $\lambda_1 = \lambda_2 = 6$, $\mathbf{e}_1 = \alpha[1,-2]^{\mathrm{T}}$, $\mathbf{x}(t) = (\mathbf{q} + \mathbf{p}t)e^{6t} = \left(\begin{bmatrix} \beta \\ \alpha - 2\beta \end{bmatrix} + \alpha \begin{bmatrix} 1 \\ -2 \end{bmatrix} t\right) e^{6t}$ or, in scalar form (with "C" notation), $x(t) = (C_1 + C_2 t)e^{6t}$, $y(t) = [(C_2 - 2C_1) - 2C_2 t]e^{6t}$
(c) $x(t) = [(C_1 + C_2) + C_2 t]e^{3t}$, $y(t) = (C_1 + C_2 t)e^{3t}$

12. (a) $\beta < 1$ gives saddle, $\beta > 1$ gives unstable node
(c) $\beta < -1/4$ gives unstable spiral, $-1/4 < \beta < 0$ gives unstable node, $\beta > 0$ gives saddle
(e) $\beta < -4$ gives center, $\beta > -4$ gives saddle

14. (a)

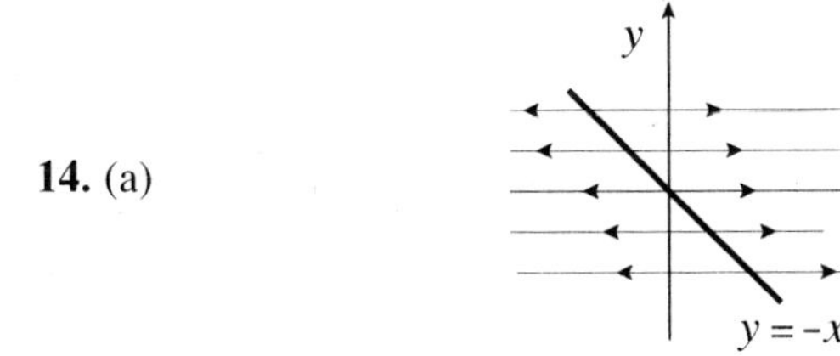

(c)

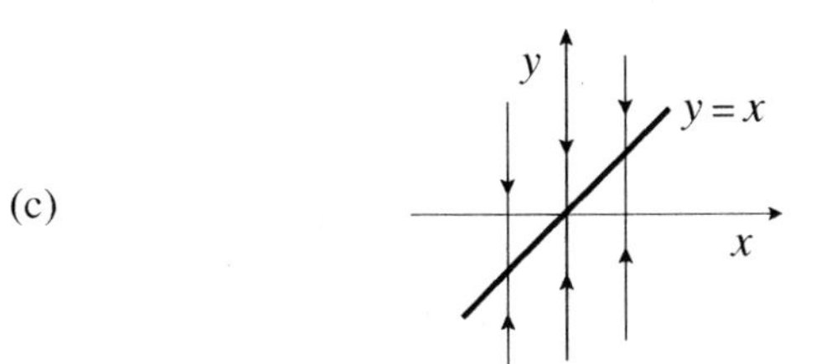

(e)

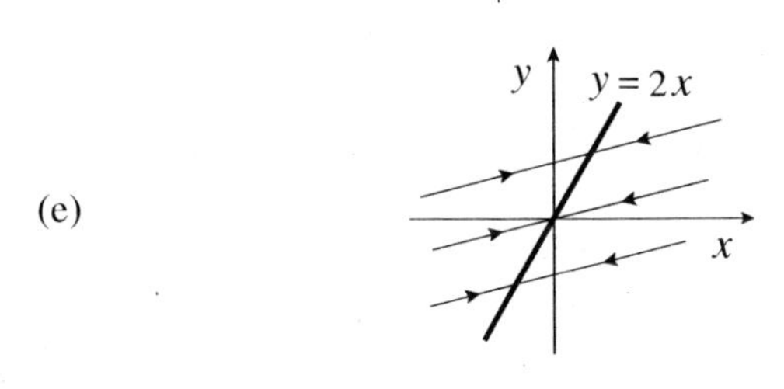

Section 7.4

1. (a) Saddle at (0,0) and center at (1,1). Use the eigenvectors at the saddle to get initial points on the separatrix.

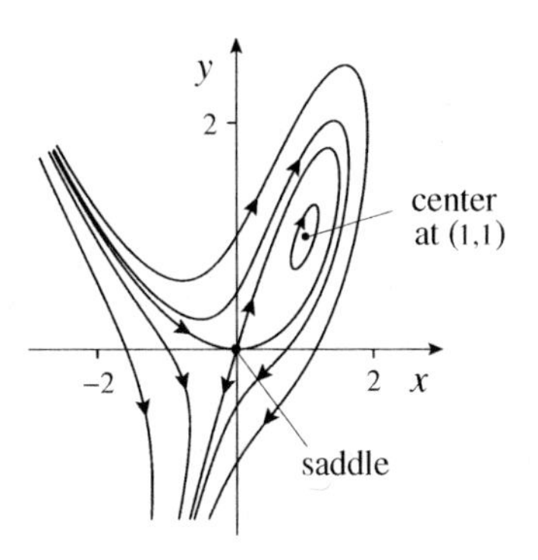

(c) Unstable spiral at (1,0) and saddle at $(-2, 3)$.

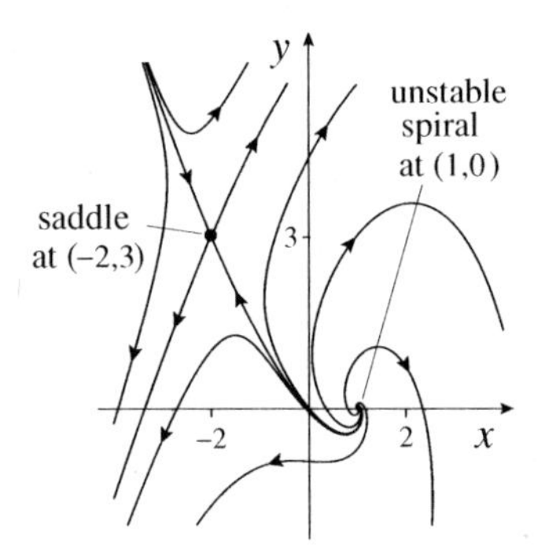

(e) Stable spiral at (0,0) and saddle at $(1, -1)$.

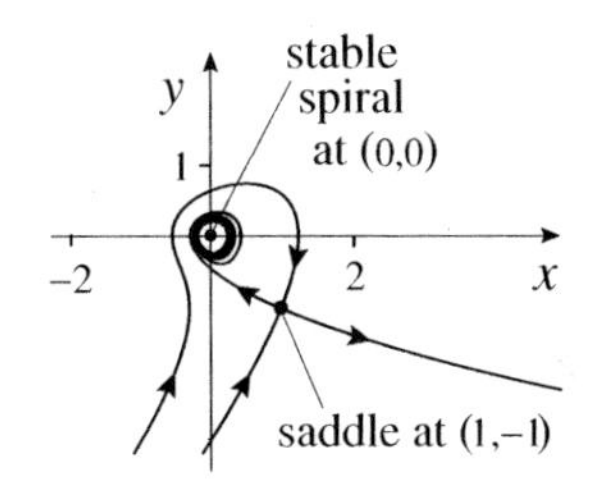

(g) Stable spiral at $(-1, 3)$ and saddle at (3,3).

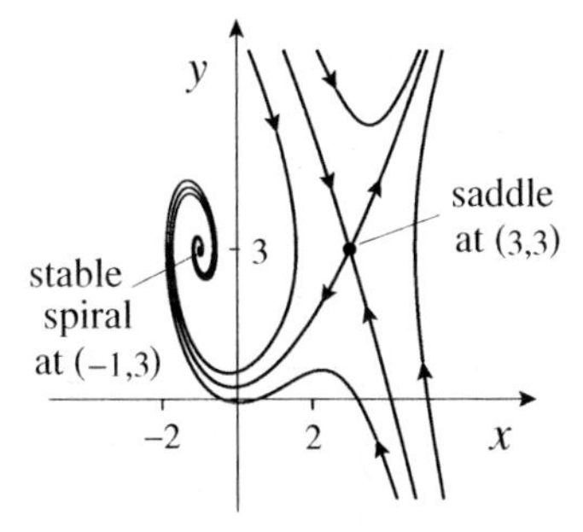

4. (a) Singular points at $x = n\pi, y = 0$, for $n = 0, \pm 1, \pm 2, \ldots$. For even n: stable spiral if $\epsilon < 2$, stable node if $\epsilon > 2$. For odd n: saddles.

10. (a) Partial answer: the period corresponding to $y(0) = 0.9$ is around 6.31.
(b) Partial answer: the period corresponding to $y(0) = 0.001$ is around 13.2.

14. (a)

(c)

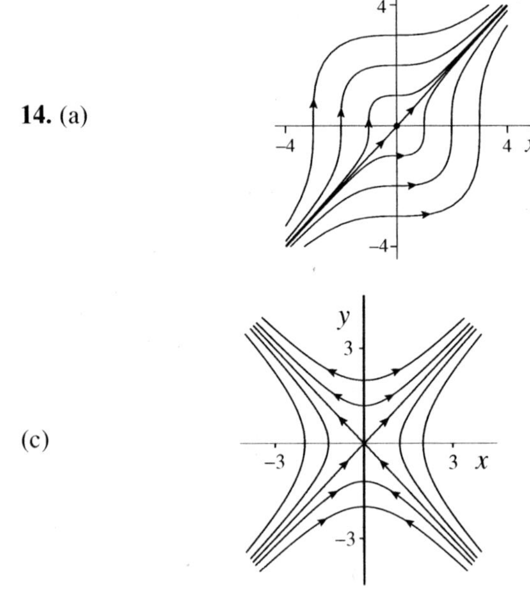

Do you see why this is *not* a saddle? Note that the entire y axis is made up of singular points.

Section 7.5

1. (a) Stable limit cycle on $r = 1$

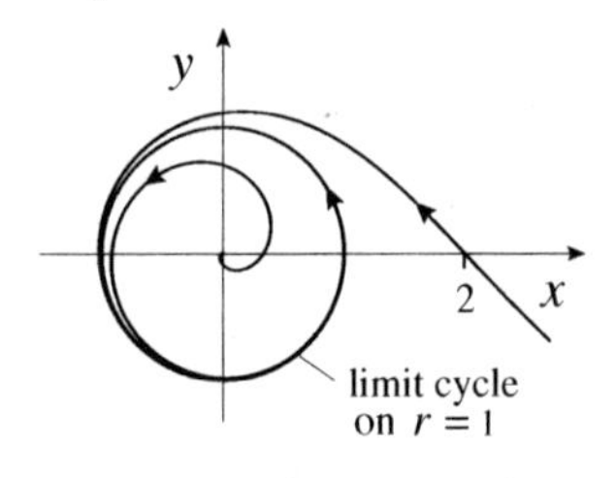

5. (a) Expect a stable limit cycle on $r = 1$. On that curve the damping term is zero. Not only is the damping zero on that circle, $r = 1$ is indeed a trajectory of the reduced equation $x'' + x = 0$ that results from dropping the damping term.

6. (a) Of course, this is the van der Pol equation. Identify $f(x) = -\mu(1 - x^2)$ and $g(x) = x$. Then obtain $F(x) = -\mu(x - x^3/3)$ and $G(x) = x^2/2$. $f(x)$ is even and continuous for all x; $g(x)$ is odd with $g(x) > 0$ for all $x > 0$, and $g'(x) = 1$ is continuous for all x. Further, $G(x) \to \infty$ as $x \to \infty$, $F(x) < 0$ for $0 < x < \sqrt{3}$, $F(x) > 0$ for $x > \sqrt{3}$, $F(x)$ is monotonically increasing for $x > \sqrt{3}$, and $F(x) \to \infty$ as $x \to \infty$. Thus, the conditions of Theorem 7.5.1 are satisfied, so there is a single limit cycle, stable, that encircles the origin.

Section 7.6

2. (a) $x_1 = 1.2,\ y_1 = 0,\ \ x_2 = 1.44,\ y_2 = 0.048$ (c) $x_1 = 2,\ u_1 = 0.6,\ y_1 = 1.4,\ \ x_2 = 2.12,\ u_2 = 1.392,\ y_2 = 3.96$ (e) $x_1 = 6,\ y_1 = 0.6,\ z_1 = 3.6,\ \ x_2 = 7.32,\ y_2 = 1.44,\ z_2 = 4.44$ (g) $x_1 = 2,\ u_1 = -1.6,\ \ x_2 = 1.68$, u_2 not needed in finding x_1, x_2

3. We will give the computed values corresponding to $h = 0.1,\ 0.01,\ 0.001$, and exact, in that order:
(a) $x(1) = 0.57079,\ 0.54304,\ 0.54057,\ 0.54030$;
$y(1) = -0.88251,\ -0.84567,\ -0.84189,\ -0.84147$;
$x(2) = -0.45302,\ -0.42027,\ -0.41656,\ -0.41615$;
$y(2) = -1.0075,\ -0.91846,\ -0.91021,\ -0.90930$
(c) $x(1) = 2.1136,\ 2.0171,\ 2.0018,\ 2$;
$y(1) = 1.9419,\ 1.9976,\ 1.9998,\ 2$;
$x(2) = 17.744,\ 5.5231,\ 3.2055,\ 3$;
$y(2) = 7.9848,\ 5.7697,\ 5.0750,\ 5$
(e) $x(1) = 0.34404,\ 0.36549,\ 0.36764,\ 0.36788$;
$y(1) = 3.9844,\ 3.9984,\ 3.9998,\ 4$;
$x(2) = 0.10494,\ 0.13223,\ 0.13502,\ 0.13534$;
$y(2) = 3.9721,\ 3.9971,\ 3.9997,\ 4$

4. Let I be any open interval containing the interval of interest $0 \le x \le 2$. Since the Wronskian of $Y_1(x), Y_2(x), Y_3(x)$, at $x = 0$ is $\begin{vmatrix} 1 & 0 & 0 \\ 0 & 1 & 0 \\ 0 & 0 & 1 \end{vmatrix} = 1$, which is nonzero, it follows from Theorem 4.5.4 that $\{Y_1, Y_2, Y_3\}$ is LI on I.

6. (a) First, solve

$$LY_1 = 0; \quad Y_1(0) = 1,\ Y_1'(0) = 0$$

$$LY_2 = 0; \quad Y_2(0) = 0,\ Y_2'(0) = 1$$

$$LY_p = 3\sin x; \quad Y_p(0) = 0,\ Y_p'(0) = 0,$$

using Euler's method, with $h = 0.05$, and print their solutions at $x = 0,\ 0.5,\ 1,\ 1.5,\ 2$:

x	0	0.5	1	1.5	2
Y_1	1	0.8819	0.3964	−1.046	−6.470
Y_1	0	0.5157	1.179	2.417	6.243
Y_p	0	0.0466	0.5132	2.506	11.00

A general solution of $Ly = 3\sin x$ is $y(x) = C_1Y_1(x) + C_2Y_2(x) + Y_p(x)$. Application of the boundary conditions, using values in the tabulation above, gives $C_1 = 1$ and $C_2 = -0.2451$. Thus, obtain

x	0	0.5	1	1.5	2
y	1	0.802	0.621	0.868	3

(c) $y(0.5) = 1.333$, $y(1) = 0.496$, $y(1.5) = -0.034$
(e) $y(1.5) = -1.448$, $y(2) = -5.649$

Index

U

V

W

Z

LAPLACE TRANSFORMS

NOTE: As in Chapter 7, s is regarded as real. Also, the constants a and b are real except in item 2, where a is permitted to be complex. *Gamma*, H, *delta* are the gamma, Heaviside, and delta functions, respectively. J_0 is the Bessel function of the first kind and order zero.

$f(t)$	$F(s) = \int_0^\infty f(t)e^{-st}\,dt$
1. 1	$\dfrac{1}{s}$
2. e^{at} (a permitted to be complex)	$\dfrac{1}{s-a}$
3. $\sin at$	$\dfrac{a}{s^2+a^2}$
4. $\cos at$	$\dfrac{s}{s^2+a^2}$
5. $\sinh at$	$\dfrac{a}{s^2-a^2}$
6. $\cosh at$	$\dfrac{s}{s^2-a^2}$
7. t^n ($n =$ positive integer)	$\dfrac{n!}{s^{n+1}}$
8. t^p $(p > -1)$	$\dfrac{\Gamma(p+1)}{s^{p+1}}$
9. $e^{at}\sin bt$	$\dfrac{b}{(s-a)^2+b^2}$
10. $e^{at}\cos bt$	$\dfrac{s-a}{(s-a)^2+b^2}$
11. $e^{at}\sinh bt$	$\dfrac{b}{(s-a)^2-b^2}$
12. $e^{at}\cosh bt$	$\dfrac{s-a}{(s-a)^2-b^2}$
13. $t\sin at$	$\dfrac{2as}{(s^2+a^2)^2}$
14. $t\cos at$	$\dfrac{s^2-a^2}{(s^2+a^2)^2}$
15. $t\sinh at$	$\dfrac{2as}{(s^2-a^2)^2}$
16. $t\cosh at$	$\dfrac{s^2+a^2}{(s^2-a^2)^2}$
17. $t^n e^{at}$ ($n =$ positive integer)	$\dfrac{n!}{(s-a)^{n+1}}$
18. $t^p e^{at}$ $(p > -1)$	$\dfrac{\Gamma(p+1)}{(s-a)^{p+1}}$
19. $H(t-a)$ $(a \geq 0)$	$\dfrac{e^{-as}}{s}$
20. $\delta(t-a)$ $(a \geq 0)$	e^{-as}
21. $J_0(t)$	$\dfrac{1}{\sqrt{s^2+1}}$